1 MONTH OF
FREE
READING

at
www.ForgottenBooks.com

By purchasing this book you are eligible for one month membership to ForgottenBooks.com, giving you unlimited access to our entire collection of over 1,000,000 titles via our web site and mobile apps.

To claim your free month visit: www.forgottenbooks.com/free1303203

ISBN 978-0-428-68163-0
PIBN 11303203

This book is a reproduction of an important historical work. Forgotten Books uses
state-of-the-art technology to digitally reconstruct the work, preserving the original format
whilst repairing imperfections present in the aged copy. In rare cases, an imperfection in
the original, such as a blemish or missing page, may be replicated in our edition. We do,
however, repair the vast majority of imperfections successfully; any imperfections that
remain are intentionally left to preserve the state of such historical works.

PREFACE.

THE following pages contain the results of several years'
study of one of the most interesting and difficult branches of
American Conchology. My MS. was completed in 1865,
and I find, upon freshly taking up the subject, that I am
inclined to question many of the conclusions at which I had
then arrived. A more enlarged acquaintance with fresh-water
shells convinces me that a much greater reduction of the
number of species than I have attempted must eventually be
made; but until the prolific waters of the southern states
have been systematically explored and a great collection of
specimens obtained, which shall represent every portion of
those streams and include as many transitional forms as can
be procured, a definitive monograph of our Melanians cannot
be written. I am indebted to several kind friends for assistance in preparing this work; first of all, to Dr. Isaac Lea,
who not only gave me constant access to his noble collection,
but on many occasions aided me by comparing specimens and
elucidating knotty questions in synonymy. Mr. John G.
Anthony, Prof. S. S. Haldeman and the late Dr. Aug. A.
Gould, with great liberality, sent to me their types; and in
these collections and that of the Academy of Natural Sciences
of Philadelphia, I also found types of many of the species
described by Say and Conrad. Most of my synonymy is
derived from the direct comparison of these typical shells, and
to this extent I believe my work will prove to be reliable.

G. W. T., Jr.

November, 1873.

ADVERTISEMENT.

THE Smithsonian Institution, realizing the lack of knowledge in reference to the land and fresh-water shells of North America, issued a circular, several years ago, to its correspondents and the friends of science generally, asking contributions of specimens from as many localities as possible, with a view of publishing a report on the subject. In the course of a few years a gratifying response was made to this appeal from all parts of the continent, in the form of extensive collections of specimens, embracing not only the several species, but those illustrating geographical distribution.

The specimens thus obtained were placed by the Institution in the hands of specialists, for the preparation of a series of monographs to bear the general title of " Land and Fresh-water Shells of North America." This was subdivided into: I, *Pulmonata Geophila*, terrestrial univalve shells, breathing free air; II, *Pulmonata Limnophila* and *Thalassophila*, free air breathing univalves, but usually living in or near fresh waters (*Limnophila*) or the sea (*Thalassophila*); III, all the operculated land and fresh-water mollusks (excepting the *Strepomatidæ* or American Melanians) and embracing the *Ampullariidæ*, *Valvatidæ*, *Viviparidæ*, *Rissoidæ*, *Cyclophoridæ*, *Truncatellidæ*, *Neritidæ* and *Helicinidæ*; IV, the *Strepomatidæ*; V, the *Corbiculadæ*; and VI, the *Unionidæ*.

Of these monographs, Parts II and III, by Mr. W. G. Binney, were published in September, 1865. Part I, by Mr. Binney and Mr. T. Bland, in February, 1869; and Part V, by Mr. Temple Prime, December, 1865. An elaborate monograph of the *Hydrobiinæ*, a subfamily of *Rissoidæ*, treated in less detail by Mr. Binney in Part

III, from the pen of Dr. Wm. Stimpson, was published in August, 1865.

Of the two remaining monographs, Part IV is given in the following pages, as prepared by Mr. G. W. Tryon, Jr., and will, it is hoped, tend to facilitate the study of a very intricate group, little understood. No special arrangement has been made by the Institution in reference to a monograph of the *Unionidæ* (which would form a Part VI) since the many illustrated papers and synopses of the group, published by Mr. Isaac Lea in the Memoirs of the Academy of Natural Sciences, and of the American Philosophical Society, as well as printed privately, render this less necessary. The present work by Mr. Tryon, therefore, completes the series of works on "Land and Fresh-water Shells of North America," as originally contemplated by the Institution.

JOSEPH HENRY,

Secretary Smithsonian Institution.

SMITHSONIAN INSTITUTION,

Washington, December, 1873.

LAND AND FRESH-WATER SHELLS

OF

NORTH AMERICA.

IV.

PRELIMINARY OBSERVATIONS ON THE

Family STREPOMATIDÆ.*

1. *Classification.*—Swainson, who may be considered the originator of the modern system of classification of the families and genera of Mollusca (as he was the first general conchologist who, breaking through the trammels of Lamarckian nomenclature, inaugurated the work since so boldly and successfully continued by Dr. Gray and Messrs. H & A. Adams), had, unfortunately, very little knowledge of the affinities with the other Mollusca, of the so-called Melanians inhabiting both America and the Old World, since he has confounded them with marine shells under his family *Turbidæ;* but, notwithstanding this error in the disposition of the whole group, he had the sagacity to separate into numerous, and generally well-characterized, genera, the incongruous material which Lamarck had allowed to remain under one generic name,—*Melania.*

Messrs. H. & A. Adams† approach more closely to the present ideas of conchologists relating to this subject, by separating from, but placing in close neighborhood to, the *Cerithiadæ,* their family *Melaniidæ,* of which they admit two subfamilies, *Melani-*

* Reprinted from the American Journal of Conchology, Vol. i, No. 2, 1865.
† Genera of Recent Mollusca, i, 293.

inæ including those shells with "aperture simple in
front, without a distinct notch," = various genera
of Melanians; and a second subfamily, characterized
by a notched aperture to the shell, including *Melanopsis*, Lam. Dr. Gray, the only other recent sys-
tematist who has investigated the subject,[*] adopts a
family *Melaniadæ*, including the subfamilies *Risso-
aina*, *Melaniaina*, *Triphorina*, *Scalarina*, and *Liti-
opina*, with a heterogeneous assemblage of marine
and fluviatile genera; the *Melaniaina* comprising
all the genera of American and exotic Melanians,
the Cerithians, and the shells which I recently separated under the family name of *Amnicolidæ*.

It is strange that neither European nor American
conchologists who have studied this family have
availed themselves until quite recently of the obvious differences, both in shell and animal, between
the American and Oriental forms, for their complete
separation, notwithstanding the fact that Prof. Haldeman showed our Melanians to have a plain or
entire mantle-margin, whilst the Oriental species
have the mantle-margin fringed, thus allying the
latter more closely with the Cerithians than with
the so-called American Melanians.[†]

Dr. Brot, a gentleman who has devoted much
attention to the Melanians, remarks[‡] that the generally adopted classification of the family is very
confused and uncertain, but does not attempt to
propose a new one.

Mr. Lovell Reeve, who has published an elaborate
monograph of the family,[§] in his preface assigns to
the animals of *all* the species a fringed mantle-margin.

Prof. S. S. Haldeman was the first naturalist who
detected the difference between our own and the

[*] List of the Genera of Recent Mollusca. — Proceed. Zool. Soc.,
London, 1847.

[†] The American species are oviparous, the oriental species ovovivip-
arous; a more important distinction first pointed out by Dr. Wm.
Stimpson in Am. Jour. Sci., xxxviii, July, 1864.

[‡] Cat. Syst. des Espèces qui composent la Famille des Melaniens.

[§] Conchologica Iconica,—*Melania, Anculotus, Io, Melatoma.*

Oriental Melanians;[*] but he did not at that time apply the results of his examinations to their obvious separation into two families.

Mr. Isaac Lea in 1862 proposed a new genus of Melanians, *Goniobasis,*[†] which, with other genera previously admitted, and *including Melania,* Lam., he still continued to regard as belonging to the family *Melaniidæ,* although in a foot-note he writes, "I very much doubt if we have a single species in the United States which properly belongs to this genus."

Mr. Theodore Gill, in a recent paper on the classification of our fluviatile Mollusca,[‡] assigns the following characters to the family *Melaniidæ:*—

"Teeth of lingual membrane, 3·1·3; gills concealed; rostrum moderately produced and entire or simply notched; foot not produced beyond the head; branchiæ uniserial; lateral jaws present.

"Aperture of shell acuminate behind; generally channelled at front; size moderate.

"The family of *Melaniidæ* is here restricted to exclude *Faunus,* Montford (= *Pyrena,* Lam.,) *Melanatria,* Bowditch, *Melatoma,* Sw. (= *Clionella ?* Gray), *Melanopsis,* Lam., *Vibex,* Oken, and *Hemisinus,* Sw. These appear to belong to a distinct family, equally distinguished by the projecting foot of the animal and the notch of the aperture of its shell.

"The family may be named *Melanopidæ.*

"The other genera or subgenera that have been proposed scarcely appear to exist in nature. * *

"The American *Melaniidæ* form a peculiar subfamily—*Ceriphasinæ.*"

Subsequently, in a foot-note,[§] Mr. Gill mentions the reason which caused him to make the above

[*] Amer Jour. Science, xli, 1, 21. Icon. Encyc. (Am Ed.), ii, Mollusca. p. 84.

[†] Proceed. Acad. Nat. Sciences, May, 1862.

[‡] Systematic Arrangement of the Mollusks of the Family Viviparidæ, and others, inhabiting the United States.—Proc. Acad. Nat. Sci., p. 33, Feb., 1863. [§] *ibid,* p. 35.

subfamily. "The American *Melaniidæ*, so far as I know, have not a fringed mantle, and, consequently, belong to a different group." We readily admit the propriety of separating the *Melanopidæ* from *Melaniidæ*, as a distinct family, and only wonder that Mr. Gill did not make a *family* of *Ceriphasinæ*, as the distinctive characters of the animal, so far as known to us, and of the shell undoubtedly, are quite as important as those which he assigns to his *Melanopidæ*. When we come to consider the geographical distribution of the two groups, the reasons for this separation are still more obvious. We find the *Melanopidæ* distributed over both hemispheres, while the *Ceriphasinæ* are entirely restricted to North America, to the exclusion almost entirely of the *Melanopidæ*, and totally of the fringe-mantled *Melaniidæ*. We find them inhabiting this faunal province in immense numbers of species, exuberantly varied in form, size, weight and color, presenting a number of genera—in fact, exhibiting all that redundancy of character and isolation of position which are the sure indications of a primordial separate existence.*

---

*It has become fashionable lately to disparage the value of the *mere* shells as a means of distinguishing generic and family groups, and to rely wholly on such differences as may be found in the animals. Without denying the great importance which should properly be accorded to the latter, we would insist that, in general, the *expression* of these differences may be observed in the shell, and that at least very few generic distinctions have been made from the study of the animals which have not been also indicated plainly enough by the shells. The study of Malacology is yet in its infancy, and those who figure in it are very apt to give undue importance to the characters on which they rely for building up their systems. To investigate how many characters of form or function have successively been called forth as the most important to stand godfathers at the baptisms of new genera, would be curious, but lamentable.

One thing is certain, that genera founded on the shells alone are always found to be corroborated by the study of the animals, while many genera founded on differences in the animal have remained unverified, and will continue so, owing to the undue importance given to the difference of form relied on for the generic distinction.

We do not regard the differences, so far as discovered, in the animals of our so-called Melanians from the Oriental *Melaniidæ*, as alone of sufficient importance to justify their separation; we are contented to separate them upon considerations connected with the shell

The publication of Mr. Gill's paper redirected Prof. Haldeman's attention to the subject, which he had left unfinished in his investigations at an earlier period; and the result is the publication of a short but important paper in the Proceedings of the Academy of Natural Sciences, September, 1863, entitled, "On Strepomatidæ as a Name for a Family of Fluviatile Mollusca usually confounded with Melania," wherein he finally separates our species as a distinct family, remarking that the Oriental Melanians are not so nearly allied to ours as they are to the *Cerithiadæ* — with which conclusion we cordially agree.

We have, therefore, adopted the name *Strepomatidæ* as indicating a distinct family, in preference to the prior name of *Ceriphasinæ*, the adoption of which would still leave our species in connection, as a subfamily, with shells to which they are not at all closely related.

In endeavoring to eliminate, from the rather confused synonymy, generic and subgeneric groups of *Strepomatidæ*, some difficulty is encountered at the threshold, on account of the various opinions held by the different naturalists who have studied them, regarding the relative importance which should be assigned to various characters of the shell, in constituting these divisions.

The genus *Hemisinus*, Swainson (*Basistoma*, Lea), belongs to Mr. Gill's family *Melanopidæ*. The little *Paludomus brevis*, D'Orb., of the West Indies, is apparently the American representative of an exotic genus; the large tuberculate Melanians of Central America, and the smooth Pachycheili of that country and of Mexico, probably do not belong to our family *Strepomatidæ*.

Thus the range of the species of the family may

also, and with geographical distribution, believing, however, that other and more important distinctive characters will reward the industry and skill of some future malacologist.

be considered as restricted within the borders of the United States.*

Swainson formed the following curious generic system for the shells under consideration :†

Family TURBIDÆ.

(Subfamilies *Ampullarinœ, Melanianœ, Turbinœ, Janthinœ.*)

Subfamily MELANIANÆ.

Genus PALUDOMUS, Swainson.

Subgenus ANCULOSA, Say.

Genus MELANIA, Lam.

Subgenus HEMISINUS, Swainson.

Genus MELANOPSIS, Lam.

Subgenus MELAFUSUS, Swainson.

Subfusiform, the base contracted, and the aperture and spire nearly equal. 1 species. America. (= Io.)

Fig. 1.

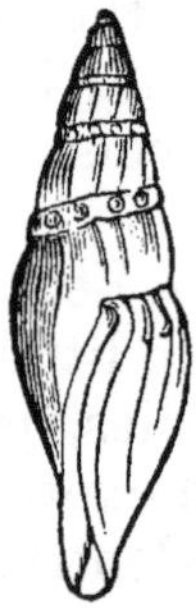

Subgenus MELATOMA, Swainson.

Fusiform, longitudinally ribbed; a deep sinus at the top of the outer lip; base contracted, channel wide. *M. costata.* (This species, mistaken by some for our genus Schizostoma, is actually an exotic marine shell = genus *Clionella.* A copy of Swainson's figure is subjoined (fig. 1).

Genus CERITHIDEA, Swainson.

Clavate, cerithiform; aperture subemarginate.

Subgenus CERITHIDEA, Swainson.

Shell light, decollated; outer lip semicircular, dilated by a flattened border; aperture emarginate. *C. lineolata,* Griff. Cuv., t. 14, f. 4. *C. fragilis,* Ibid , t. 32, f. 12. (= POTA-MIDES.)

---

* Three or four are extra-limital, inhabiting Cuba and Mexico; but these do not constitute one per cent. of the whole number of species.
† Manual of Malacology, 1840.

Subgenus Ceriphasia, Swainson.

Cerithiform; outer lip thin, dilated at the base; aperture small, slightly emarginate, without any internal groove; inner lip thin. *C. sulcata*, Sw., fig. 38 (figs. 2 and 3 of this work). Founded on certain Ohio shells resembling *Cerithidea?*

Fig. 2.   Fig. 3.

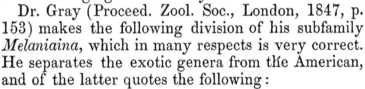

It will be noticed that in the above classification *Melafusus* is a subgenus of *Melanopsis,* which belongs to the family *Melanopidæ,* while *Ceriphasia* is a subgenus of *Cerithidea,* which includes shells belonging to the family *Cerithiidæ!*

Dr. Gray (Proceed. Zool. Soc., London, 1847, p. 153) makes the following division of his subfamily *Melaniaina,* which in many respects is very correct. He separates the exotic genera from the American, and of the latter quotes the following:

Anculotus, Say, 1825.

Anculosa, Swains., 1840 — *A. præmorsa*, Say.
Melanopsis, sp., Moricand — *M. crenocarina.**
Anculosa, sp., Anthony — *Anc. rubiginosa.*
Melania, sp., Say — *Melan. obovata.*

Melatoma, Anthony, 184–? not Swains., 1840.

Melat. altilis, Anthony.

Io, Lea, 1832.

Fusus, sp., Say, 1825.   } *Fusus fluviatilis*, Say.
Melafusus, Swains., 1840. }
Melania, sp., Say—*Mel. armigera*, Say.

Ceriphasia, Swains., 1840.

Gray, Syn., 1844.

Melania, sp., Say—*Ceriphasia sulcata*, Swains.
? *Telescopella.*
Melania, sp., Say—*Mel. undulata*, Say.

Glotella, Gray.

Melania armigera, Say.

* = *Verena*, H. & A. Adams; certainly not an *Anculosa.*—T.

Messrs. H. & A. Adams (Genera of Recent Mollusca) propose the following classification: *—

" CERIPHASIA, Swainson (i, p. 297.)

Shell subfusiform, whorls transversely sulcate, the last angulated; spire acuminated; aperture small, produced in front, with a small groove-like canal at the fore part; outer lip thin, posteriorly sinuated.

Syn. *Telescopella*, Gray.

Ex. *C. canaliculata*, Say, t. 31, f. 6.

The shell of *Ceriphasia* is covered with a dark-green epidermis, and is more like that of *Io* than any other of this family; it may, however, be distinguished from *Io* by the beak being shorter, and by the whorls being sulcated and not spiny."

acuta, Lea.	*luteosa*, Gould.
Alexandrensis, Lea.	*Ordiana*, Lea.
annulifera, Conr.	*regularis*, Lea.
canaliculata, Say.	*spurca*, Lea.
elongata, Lea.	*subularis*, Lea.
exarata, Lea.	*sulcosa*, Lea.
Haleiana, Lea.	*symmetrica*, Hald.
Kirtlandiana, Lea.	*Vainafa*, Gould.
lugubris, Lea.	*Virginica*, Gmel.†

" Genus PACHYCHEILUS, Lea (i, 298.)

Operculum suborbicular, of several whorls. Shell subfusiformly conical, smooth, solid; aperture ovate, entire anteriorly; columellar lip thickened posteriorly; outer lip thick.

The chief peculiarity of this genus is the thickened outer lip; it differs from *Melanopsis* in having no sinus at the fore part of the aperture, and from *Melania* in having a callous columella.

* We quote the full lists of species given by Messrs. Adams, in order that the insufficiency of their genera may become more apparent from the incongruous assemblage of shells of which they have composed them. Prof. Haldeman writes (Proceed. Acad. Nat. Sciences, p. 274, Sept. 1863): "The groups of Messrs. H. & A. Adams often indicate merely sections; and sectional names given as generic are scientifically erroneous, because they erect certain species into genera and subgenera only when they belong to extensive groups, requiring numerous specific names, whilst the same amount of character goes for nothing in groups which have but few species."

† The species here assembled are principally *Goniobases*, but are included in *Ceriphasia* evidently because they are "transversely sulcate." *M. Virginica* and its synonyme *multilineata* are again introduced in *Juga*, a subgenus of *Vibex*, Oken ! *M. canaliculata*, Say, is introduced, but *undulata*, Say, does not appear, while *filum*, Lea, a very closely allied species, is placed in *Elimia*, a subgenus of *Io*.

The operculum has the nucleus subcentral, and is composed of two or three spiral revolutions.

dubiosus, Say. *ferrugineus*, Lea. *simplex*, Say.*

"Subgenus POTADOMA, Swainson (i, 299.)

Shell ovate, solid; spire short, whorls smooth; inner lip somewhat thickened; aperture produced in front; outer lip acute, simple.

depygis, Say.	*ovoideus*, Lea.
gracilis, Lea.	*rufescens*, Lea.
inornatus, Anth.	*sordidus*, Lea.
lævigatus, Lea.	*subcylindraceus*, Lea.
Niagarensis, Lea.	*subsolidus*, Lea.
Ocoeensis, Lea.	*Warderianus*, Lea.†

"Genus Io, Lea (i, p. 299.)

Shell subfusiform, whorls spinose; aperture large, ovate, dilated anteriorly, produced in front into a grooved beak; outer lip simple, acute.

Syn. *Melafusus*, Swains., *Glotella*, Gray.

Ex. *I. fluviatilis*, Say, t. 31, f. 8. Operculum, f. 8, *a*, *b*.

The species of *Io* inhabit the rivers of North America; the shells, like those of most of the *Melaniidæ*, are covered with a brown, black or olivaceous epidermis, and are remarkable for the peculiar elongation of the axis anteriorly, and for the spinose nature of the last whorl.

armigera, Lea.	*pernodosa*, Lea.
Duttoniana, Lea.	*plicata*, Lea.
Florentiana, Lea.	*robulina*, Anthony.
fluviatilis, Say.	*spinigera*, Lea.
fusiformis, Say.	*spinosa*, Lea.
nobilis, Lea.	*tenebrosa*, Lea.
pagodula, Gld.	*tuberculata*, Lea. ‡

"Subgen. ELIMIA, H. & A. Adams (i, p. 300.)

Shell fusiformly ovate; whorls reticulate or nodulose, carinate in the middle; aperture greatly produced anteriorly; outer lip thin, simple, acute.

* The genus *Pachycheilus* was instituted by Mr. Lea to comprise a certain form of shells attaining their greatest numerical development in Central America. There are no shells inhabiting the United States which are congeneric with these; and Messrs. Adams have entirely mistaken the scope of the genus in including such species as *simplex*.

† = *simplex*, Say, which Messrs. Adams place in the genus *Pachycheilus* as typical!

‡ Among the species here enumerated are *Angitremæ*, *Anculosæ*, *Lithasiæ*, *Strephobases*, *Goniobases*, and *Pleuroceræ*. *I. pagodula* is an exotic species, and does not belong to the genus.

acuticarinata, Lea.	*catenoides*, Lea.
apis, Lea.	*elevata*, Lea.
bella, Conrad.	*filum*, Lea.
Boykiniana, Lea.	*Holstonia*, Lea.
caliginosa, Lea.	*nodulosa*, Lea.
cancellata, Say.	*Potosiensis*, Lea.
carinocostata, Lea.	*spinalis*, Lea.
catenaria, Say.	*torta*, Lea.

" MELANIA, Lamarck.

Subgen. MELASMA, H. & A. ADAMS (i. p. 300.*)

Shell solid; spire elevated, whorls smooth, longitudinally plicate; aperture produced anteriorly; inner lip simple, thin; outer lip acute, simple.

blanda, Lea.	*Deshayesiana*, Lea.
brevispira, Anthony.	*Edgariana*, Lea.
clavæformis, Lea.	*laqueata*, Say.
Comma, Conr.	*Lecontiana*, Lea.
concinna, Lea.	*nitens*, Lea.
costulata, Lea.	*plicatula*, Lea.
crebricostata, Lea.	*plicifera*, Lea.
Curreyana, Lea.	

" Genus HEMISINUS, Swainson (i, 302.)

Shell subulate; whorls smooth, simple, numerous; aperture ovate, anteriorly contracted, canaliculate and emarginate in front; outer lip thin, crenulated at the edge.

Syn. *Tania*, Gray, *Basistoma*, Lea.

Ex. *H. lineolatus*, Wood, t. 32, f. 2, *a*, *b*.

This genus comprises many fine species of fresh-water shells, principally from South America, though a few have been regarded as inhabitants of other countries.

bulbosus, Gould.   *symmetricus*, Conr.   *lineolatus*, Wood.†

" Genus VIBEX, Oken (i, 303.)

Shell turreted; whorls tuberculated, spirally ridged or muricate; aperture subcircular, produced, and broadly channelled in front; outer lip thin, simple.

Syn. *Claviger*, Hald., *Melania*, Swains., not Lamarck.

" Subgenus JUGA, H. & A. Adams (i, 304.)

Shell thin; whorls rounded, transversely lirate or furnished with elevated transverse lines; aperture produced anteriorly; outer lip simple, acute.

---

*This genus = the plicate species of *Goniobasis*. *M. brevispira*, however, is never plicate, although included with the species.

† The first two enumerated do not belong to this genus, nor have they the slightest affinity with any of its species.—G. W. T., JR.

Buddii, Say.*
circincta, Lea.
exilis, Hald.
multilineata, Say.
obruta, Lea.
occata, Hinds.
proteus, Lea.

proxima, Say.
Schiedeana, Phil.
silicula, Gld.
striata, Lea.
Troostiana, Lea.
Virginica, Say.

" Genus GYROTOMA, Shuttleworth (i, 305.)

Shell ovate, turreted ; whorls transversely sulcate ; aperture oblong ; inner lip thickened, with a posterior callosity ; outer lip thin, with a deep, narrow, posterior fissure.

Syn. *Schizostoma*, Lea, not Bronn, *Melatoma*, Anthony, not Swainson, *Schizocheilus*, Lea.

Ex. *G. ovoidea*, Shuttleworth, t. 32, f. 4, *a*, *b*.

The fissure in the outer lip is wanting or obsolete in the subgenus *Megara*, the species of which in other respects closely resemble those of *Gyrotoma* proper. Both groups are American in their geographical distribution.

altilis, Anthony.
Babylonica, Lea.
Buddii, Lea.
conica, Say.
constricta, Lea.
curta, Migh.?
curvata, Say.
cylindracea, Migh.?

excisa, Lea.
Foremani, Lea.
funiculata, Lea.
incisa, Lea.
laciniata, Lea.
ovoidea, Shuttl.
pagoda, Lea.
pyramidata, Shuttl.†

" Subgenus MEGARA, H. & A. Adams (i, p. 306.)

Shell ovate, solid ; whorls transversely sulcate ; aperture ovate-oblong, subcanaliculated anteriorly ; outer lip thin, simple, acute.

alveare, Conr.
arctata, Lea.
auriculæformis, Lea.
basalis, Lea.
brevis, Lea.
crebristriata, Lea.
harpa, Lea.
Haysiana, Lea.

Hoeydei, Lea.
impressa, Lea.
lateralis, Lea.
lima, Conr.
oliva, Lea.
olivula, Conr.
ovalis, Lea.
pumila, Lea.

---

*Should read *Buddii*, Lea. *M. exilis*, Hald., and *proxima*, Say, certainly do not belong here. I have already remarked upon *M. Virginica* and *multilineata*.

†Mr. Anthony never described *Gyrotoma altilis*, ranked among these species. *G. conica*, Say, is the young of *Pleurocera canaliculata*. There are, besides, frequent mistakes in all these lists, in misquoting authorities. — T.

solida, Lea.   *undulata*, Say.
torquata, Lea.   *Vanuxemiana*, Lea.*

" Genus Leptoxis, Rafinesque (i, 307.)

Shell ovate or globose, solid, subperforate ; spire very short? aperture oval ; inner lip with a posterior callosity, often ante- riorly callous and produced ; outer lip thin, sinuous with a posterior, ascending canal.

Syn. *Anculotus*, Say, *Anculosa*, Swains., *Ancylotus*, Herm.
Ex. *L. prærosa*, Say, t. 32, fig. 6, *a*, *b*.

The species of this genus are peculiar to the North American rivers ; the spire of the shell has a truncated, eroded apex, and, in the typical species, the shell is solid and subglobose, with the aperture simple in front.

abrupta, Lea.   *pilula*, Lea.
angulata, Conr.   *pisum*, Hald.
crassa, Hald.   *plicata*, Conr.
flammata, Lea.   *prærosa*, Say.
fuliginosa, Lea.   *pumilis*, Conr.
fusca, Hald.   *rubiginosa*, Lea.
fusiformis, Lea.   *squalida*, Lea.
gibbosa, Lea.   *subglobosa*, Say.
globula, Lea.   *tæniata*, Say.
Griffithsiana, Lea.   *tintinnabulum*, Lea.
Hildrethiana, Lea.   *trivittatus*, DeKay.
integra, Say.   *Troostiana*, Lea.
melanoides, Conr.   *turgida*, Hald.
Nickliniana, Lea.   *variabilis*, Lea.
nigrescens, Conr.   *virgata*, Lea.
obtusa, Lea.   *viridis*, Lea.†
picta, Conr.

" Subgenus Nitocris, H. & A. Adams (i, 308.)

Shell thin, subglobose ; whorls angulated, often carinate ; inner lip subtruncate, or ending in a tubercle.

carinata, Lea.   *dilatata*, Conr.
costata, Lea.   *dissimilis*, Say.
dentata, Couth.   *ebena*, Lea.

---

*Here we find shells belonging to several groups, as *pumila*, Lea, *alveare*, Conr., and *torquata*, Lea, to *Strephobasis; lima*, Conr., and *solida*, Lea, to *Lithasia; undulata*, Say, to *Pleurocera*. *Hoeydei*, Lea, was never described. Can it be intended for *Hydei*, Conr. ? The spe- cies are generally, however, the ponderous *Goniobases* of Northern Alabama.

†In the species of this genus there are several errors, some quite elongated forms being included ; also, a species of *Lithasia*.

Kirtlandiana, Anth. *Rogersii,* Conr.
monodontoides, Gld. *subcarinata,* Hald.

"Subgenus LITHASIA, Lea (i, 308).

Shell thick, solid, ovate; whorls gibbose or tuberculated at the hind part; aperture subcanaliculated and produced in front; inner lip with a callus posteriorly, subtruncate anteriorly.
genicula, Hald. *salebrosa,* Conr.
neritiformis, Desh.† *semigranulosa,* Desh."
obovata, Say.

Chenu (Manuel de Conchyliologie) principally follows the arrangement of Messrs. Adams.

Lovell Reeve monographs separately *Io, Hemisinus, Anculotus* and *Melatoma,* and treats all the species not included in those genera as *Melaniæ.* He says, "Advantage might have been taken of the labors of systematists to have distributed them into further genera; but more materials are needed for their elucidation than we at present possess.‡

R. J. Shuttleworth (Mittheil. der Nat.-forsch. Gesellsch. in Bern., No. 50, p. 88) proposed, July 22, 1845, a new American genus of fluviatile shells, which he characterized as follows:—

* = in some respects *Mudalia,* Hald., and *Somatogyrus,* Gill.

† *Neritiformis,* Desh., is an *Anculosa,* and is a syn. of *A. prærosa,* Say.

‡ It is very much to be regretted that Mr. Reeve did not make some kind of a division, however arbitrary, of the immense material entering into his magnificent monograph of *Melania,* as he has published it. Species from all countries, without regard to external resemblances, are, in many cases, grouped on its plates indiscriminately, rendering the identification of shells by its aid exceedingly difficult. Even several of the species are duplicated in description and illustration in the monographs of *Melania, Io* and *Anculotus.*

While on the subject of Mr. Reeve's monograph we cannot refrain from condemning the substitution of new descriptions of the species for those originally given. The descriptions of Mr. Reeve in numerous cases entirely neglect the most important specific characters. The plates frequently do not represent the species for which they are intended; but in this Mr. Reeve has been undoubtedly deceived by wrongly-named specimens.

It is a strange fact that, notwithstanding the length of time which has elapsed since very many of our *Melanians* and *Unios* have been described, and the large number which have been sent to Europe in scientific exchanges, European conchologists are still to a great extent ignorant of the most prominent and important specific characters.

"GYROTOMA.—Shell turreted; columella incurved, above callously thickened; aperture oval, subeffuse at the base; lip simple, acute, narrowly profoundly fissured above.

"Animal.—Operculum corneous, spiral."

This forms one of the most distinct of the genera of *Strepomatidæ*. Mr. Lea, however, anticipated Mr. Shuttleworth's discovery.

Dr. Brot, in his admirable "Systematic Catalogue of the Melanians," proposes, instead of the genera of H. & A. Adams, a series of sections, which are generally excellent, for the arrangement of the species. The following is his plan:—

1. *Operculum concentric.*

Genus PALUDOMUS, Swainson.

2. *Operculum spiral or subspiral.*

* *Aperture entire.*

Genus LEPTOXIS, Raf.
(*Anculotus*, Say; *Anculosa*, Conr.)

Genus MELANIA, Lam.

Group *a*, type *canaliculata*, Say.
" *b*, " *curvilabris*, Anth.
" *c*, " *Haysiana*, Lea.

	a, type	*Virginica*, Say.
	b, "	*costulata*, Lea.
" *d*,	*c*, "	*perangulata*, Conr.
	d, "	*simplex*, Say.
	e, "	*Warderiana*, Lea.

" *e*, " *nupera*, Say.
f, (European.)

	a, "	*lævissima*, Sowb.
" *g*,	*b*, "	*glaphyra*, Morelet.
	c, "	*nigritina*, Morelet.

(All the other groups of this section, thirteen in number, are exotic.)

** *Aperture produced in front.*
Genus Io, Lea.

*** *Aperture truncate in front.*
(MELANOPSIS, HEMISINUS.)

**** *Aperture posteriorly sinuate.*
Genus GYROTOMA, Shuttlw.

***** *Aperture sinuate in front and posteriorly.*
(PIRENA, Lam.)

Passing to American authors, we find Mr. Say was the first to eliminate a native genus from the genus *Melania.* In his description of *Melania prærosa,* he says, " This shell does not seem to correspond with the genus to which I have for the present referred it; and, owing to the configuration of the base of the columella, if it is not a *Melanopsis,* it is probable its station will be between the genera *Melania* and *Agathina.* I propose for it the generic name of *Anculosa.*

He also remarks, in his subsequent description of *M. subglobosa,* " It is a second species of my proposed genus *Anculotus.*"

Mr. Say never *described* his genus; but the above citation and description of two species, both of which are well known, and whose identity with his descriptions has never been questioned, entitle his generic name to be received as authority.

Rafinesque published the following genera, which have been referred to *Strepomatidæ :—*

" *Pleurocera,* Raf. (Jour. de Phys. Bruxelles, vol. lxxxviii, p. 423, 1819). Shell spiral, oval, or pyramidal, of numerous convex volutions. Aperture obliquely oblong, the base prolonged and twisted, sharp above. Outer lip thin, the inner lip appressed, twisted, without umbilicus. Animal with a membranaceous operculum.

" Head proboscidiform, inserted on the back ; tentacles two, lateral, subulate, sharp, with eyes at their exterior bases.

" Family of *Neritacea.* Species numerous, of which I have already twelve, all fluviatile, from rivers and creeks, as well as the following genera."*

* Rafinesque previously described *Pleurocera* in a short paper published in the American Monthly Magazine and Critical Review, iii, p. 354, 1818 (Binney & Tryon's edit. of Rafinesque, p. 22), as follows :—
" Shell variable oboval or conical, mouth diagonal crooked, rhomboidal, obtuse and nearly reflexed at the base, acute above the connection, lip and columella flexuose entire. Animal with an operculum membranaceous, head separated from the mantle inserted above it, elongated, one tentaculum on each side at its base, subulate acute, eyes lateral exterior at the base of the tentacula."

This description was doubtless intended for all the elongate species of Melanians from the Ohio River then known to him, but he afterwards amended it as above.

In his " Enumeration and Account" (Binney & Tryon, p. 67), Rafinesque describes several species of *Pleurocera,* and remarks, " My G.

By some strange mistake, this genus is referred by Messrs. H. & A. Adams to *Vivipara*.

Rafinesque published several species; one of which, *P. verrucosa*, is identical with *Lithasia nupera*, Say, and therefore belongs to an entirely different group. Others, however, are evidently closely related to *M. canaliculata*, Say, and *M. elevata*, Say. The genus is certainly well characterized, and clearly includes those shells which Mr. Swainson has subsequently distinguished as *Ceriphasia*, and Mr. Lea as *Trypanostoma*.

In the same Journal (p. 26), Rafinesque described a genus "*Leptoxis*" as follows: "*Leptoxis*. Differs from *Lymnula* by an oval shell, inflated, the spire of two or three whorls; aperture oval, almost as large as the whole shell. Eyes exterior. About four species, fluviatile, lacustrine and palustrine."

There can be no doubt that this description was *intended* for *Anculosa*, Say, as is proved by a manuscript work by Rafinesque ("Conchologia Ohioensis") in the possession of the Smithsonian Institution, in which there is a rude pen-and-ink drawing of the animal and shell of a *Leptoxis*. The name has been adopted by Prof. Haldeman and others. But as the *published* description refers equally well to species of *Amnicolidæ* or *Viviparidæ*, and as manuscript authority is not recognized in questions of priority, we are compelled to throw aside this name and adopt that given by Say.

In the manuscript quoted above, occurs the description of a new genus called *Strepoma*, together with the figure of a species; which appears to represent a section of *Pleurocera*. It is unnecessary to quote the description, as it was never published:

Pleurocera, 1819, is perhaps a S. G. of *Melania*, but the animal is different, with lateral feelers; the shell is always conical oblong, with the opening oblong oblique acute at both ends, columella flexuose twisted;" and, further, "I leave the name of *Melania* to the shells with the opening obtuse at the end; or they may form the S. G. *Ambloxis*."

it is only mentioned here because Prof. Haldeman adopts it as a generic name in a late paper on the classification of these shells.*

For the same reason we do not adopt the genus *Ambloxis* described in the American Monthly Magazine, p. 355, 1818:—

"Univalve.—Shell thick oboval, mouth oval, rounded at the base, obtuse above, with a thick appendage of the lip, columella flexuose, a small rugose umbilic."

This, the only description, would apply equally well to a *Paludina, Anculosa* or a *Goniobasis* of Lea, and in 1831 (Enum. and Account), although he renders it plain that he intended the latter, still he does not adopt the name for his species there described, and seems disposed to doubt the value of his former division.

The three following genera were published in Journal de Physique, Bruxelles, tome 88, p. 423 *et seq.* :—

"*Ellipstoma*, Raf.—Shell thick, oval, obtuse. Mouth oblique, narrow, elliptic, lips thickened, united and obtusely decurrent posteriorly. A narrow, oblong umbilicus, half covered by the interior lip. Animal unknown. Fluviatile genus of 4 species, *E. gibbosa, E. vittata, E. zonalis* and *E. marginula.*

"From the Ohio, Mississippi, etc."

"*Oxytrema*, Raf.—Differs from *Pleurocera* by an oval oblong or ventricose shell, less number of whorls, the last forming nearly the whole ; mouth sharp on both sides, and anteriorly prolonged into a long, sharp point. 3 fluviatile species."

"*Campeloma*, Raf.—Shell oval ; mouth oval, base truncated, lip reflected, united in a posterior point. No umbilicus. Animal unknown. I have only one species, found in the Ohio, —*C. crassula.* Four whorls of the spire reversed, apex acute, shell thick, mouth more than half the total length."

Messrs. H & A. Adams, with *very* doubtful propriety, refer this genus to *Melanopsis*. Prof. S. S. Haldeman, in an article on Mollusca, contributed by him to the American edition of Heck's Iconographic Encyclopædia, ii, p. 84, remarks that :—

* Proceed. Acad. Nat. Sciences, p. 274, September, 1863.

"Say's *Melania armigera* (and also Lea's *M. Duttoniana* and *M. catenoides*) belongs to Rafinesque's genus *Pleurocera*, in which there is a short, straight canal anteriorly, and when this canal is lengthened, as in *Fusus*, the genus *Io*, of Lea, is the result.

"*Strepoma* of Rafinesque (or *Ceriphasia* of Swainson) are slightly different forms, in which the aperture and the vertical plate formed by the anterior portion of the whorls, bear some resemblance to the same parts in *Cerithium telescopium*."

In October, 1840, Prof. Haldeman published a supplement to his "Monograph of the Limniades," containing, among other matter, the following proposed

"Subgenera of Anculosa.

"*Anculosa*, Say.— Substance of the shell thick and heavy, labium much thickened.

"*Lithasia*, Hald.— Shell heavy, having protuberances; aperture with a notch in the nacre above and below.

"*Paludomus*, Swains.— Shell smooth, margin of the outer lip crenated, labium very thick and enamelled.

"*Hemimitra*, Swains.—Like *Paludomus*, but with coronated whorls.

"*Mudalia*, Hald.— Shell smooth, thin in texture, labium without enamel."

In his description of a species of *Anculosa* published upon the same occasion, Prof. Haldeman refers to "*Paludina (Mudalia) dissimilis*, Say," so that there can be no doubt as to the section of *Anculosa* indicated by the subgenus *Mudalia*. On the cover of No. 2 of the monograph (January, 1841) is the description of "subgenus *Angitrema*. Shell spinous, aperture subrhomboidal, with an anterior sinus. Ex. *Melania armigera*, Say."

I adopt *Angitrema* as a *genus*, with *Lithasia* as a subgenus of it. *Mudalia* cannot stand in the system, because its characters are not constant, *Anc. dissimilis* having frequently a heavy deposit of nacre on the columella.

Mr. Lea has described several new genera of shells eliminated from the American *Melaniæ*. He early recognized in Mr. Say's genus *Anculosa* a good natural genus, and adopted it in his descriptions. In Philos. Trans., VIII, p. 163, he proposed to separate the species of *Melania* according to certain obvious, external (by no means generic) characters, for facility in their determination. He described a large number of species under the following divisions :—

"1. Smooth.   4. Sulcate.   7. Granulate.
2. Plicate.   5. Striate.   8. Cancellate.
3. Carinate.   6. Tuberculate.   9. Spinose."

Perhaps this division of the species suggested to Messrs. Adams the genera which they have adopted in their classification.

In Philos. Trans., IV, p. 122, Mr. Lea proposed to institute a new genus, *Io*, for the *Fusus fluvialis* of Say. His description is, "Io.—Shell fusiform ; base canaliculate ; spire elevated ; columella smooth and concave."

In his description of *Melania excisa*, and *Anculosa incisa*, published in Philos. Proc., II, p. 242, Dec., 1842, Mr. Lea suggested the name *Schizostoma* for those species having a pleurotomose sutural slit in the outer lip. The genus thus proposed, and which bears the same relation to *Goniobasis* as *Schazicheila* does to *Helicina*, was sometime afterwards characterized by Mr. Shuttleworth, from independent observation, under the name of *Gyrotoma*.

In Philos. Proc., Aug., 1845, and in the Transactions, X, p. 67, 1853, Mr. Lea published the following description of his genus :—

"SCHIZOSTOMA, LEA. Shell conical or fusiform. Lip fissured above. Aperture ovate, columella smooth, incurved. Operculum.—

"No operculum has come under my notice ; but I can scarcely doubt that it will be found to be horny, and to resemble, in other respects, that of *Melania*."

Subsequently (vol. X, p. 295), Mr. Lea says, "When I proposed the name of *Schizostoma* for a genus of *Melaniana* with a cut at the superior portion of the aperture, I was not aware that M. Bronn had already used that name for a fossil genus. I now propose to substitute *Schizochilus.*"

In the Proceedings of the Academy of Natural Sciences of Philadelphia, 1860, p. 53, Mr. John G. Anthony makes some lengthy remarks on this genus, as follows :—

" *Gyrotoma.* As some confusion exists regarding the name of this genus, the following notes are given : —

" 'The genus *Melatoma* was established by Swainson, and first given to the world in 1840, in his 'Treatise on Shells and Shell Fishes,' published in London, founded, as he says (p. 202), 'upon a remarkable Ohio shell sent him many years before by his old friend, Prof. Rafinesque.' 'It has,' he remarks, 'the general form of a *Pleurotoma* and of a *Melafusus*, with a well-defined sinus or cleft near the top of the outer lip, while the inner, though thin, is somewhat thickened above.' The other characters named by him are such as are generally considered rather specific than generic, and the pleurotomose cut in the outer lip, as applied to a fluviatile univalve, is altogether insufficient to indicate a new genus. The specimen alluded to by Swainson, and from which his generic description was drawn, was an imperfect one ; and the species has not since been identified by American naturalists. This is less to be wondered at when we consider how very local the genus has always been, and how few specimens have found their way into our collections. The waters of Alabama have, as yet, monopolized this interesting genus ; and it is probable that even there it is confined almost if, not quite, exclusively to the Coosa and its tributaries.

" On p. 342 Swainson gives the following generic description, adding a figure : —

" ' Fusiform, longitudinally ribbed ; a deep sinus at the top of the outer lip ; base contracted ; channel wide.'

" Mr. Swainson's figure is quite unsatisfactory. His genus *Melatoma* is referred doubtfully to *Clionella* by H. & A. Adams, and has not prevailed for this genus in America or Europe. I have, therefore, decided not to make use of it in this case.

" Subsequently this genus has been noticed by various authors, and other names have been applied to it. In 1841 or 1842, Dr. J. W. Mighels sent me specimens of one species, under the name of *Apella scissura ;* but his generic name was never published, and his species, if not identical with any which Mr. Lea afterwards described, seems to have been overlooked and forgotten.

"On the 14th of December, 1842, Mr. Lea read a paper before the American Philosophical Society, in which he describes *Melania excisa* and *Anculosa incisa*. In his remarks upon these species he alludes to the pleurotomose cut in the superior part of the upper lip, and at the time suggests the necessity, in consequence of this character, to construct a new genus, which he proposed to call '*Schizostoma*.' Mr. Lea, finding his name '*Schizostoma*,' preoccupied in palæontology, changed it to '*Schizochilus*' (March 5, 1851, Obs., v, p. 51). In a paper read May 2, 1845, Mr. Lea, in a foot-note to p. 93, first indicates the generic characters of *Schizostoma*, as follows ; 'Testa vel conica vel fusiformis ; labrum superne fissura ; aperture ovata ; columella lævis, incurva,'—and describes six additional species.

"In the above concise definition of the genus, it will at once be noted that the fissure at the upper part of the outer lip is, after all, the *essential* character ; and Mr. Lea himself seems to be aware of this, since, of the six species then described, he states the aperture to be elliptical in five cases and rhomboidal in the other, although his generic character is 'aperture ovate.' Indeed, in the species described by him, but a single one has the aperture ovate, and that one is described as an *Anculosa*.

"It may be doubted whether Mr. Lea's first name will not eventually prevail, since, before he published *Schizostoma*, Bronn's genus of the same name (Lethea Geogn., I, 95, 1835–37) had been called a synonyme of *Bifrontia* (*Omalaxis*) of Deshayes. (Vide Desh. in Lam., IX, p. 104.) Indeed, H. & A. Adams (Gen. Rec. Moll., I, 305) do not appear correct in giving preference to *Gyrotoma* over *Schizostoma*, Lea, on account of *Schizostoma*, Bronn, since (on p. 244) the latter is placed in the synonymy of *Omalaxis*.

"Another generic name *Schizostoma* is quoted in Hermannson's Index. I have not obtained access to the work containing this description ; but its date is said to be anterior to Mr. Lea's description.

"Mr. Lea's second name, *Schizochilus*, had been previously used in Coleoptera, but withdrawn after Mr. Lea's description was published.

"Mr. Shuttleworth, in July, 1845 (Mittheilungen der Naturforschenden Gesellschaft in Bern, p. 88), gives another description of the genus under the name of *Gyrotoma*, founded on two species from the Coosa River, descriptions of which are also given.

"The generic name of Mr. Shuttleworth has been adopted in H. & A. Adams' Genera of Recent Mollusca (I, p. 305, Feb., 1854).

"Dr. Gray also (Guide to Mollusca, I, p. 103, 1857) adopts Shuttleworth's name.

Such being the confused state of the synonymy of the genus, we have decided to adopt, at least temporarily, the earliest name concerning which no doubt exists."

To the above, Mr. Lea made the following reply, upon occasion of describing some new species belonging to the genus, in Proc. Acad. Nat. Sciences, Philada., May, 1860 : —

"Genus SCHIZOSTOMA.

"It will be observed that I have here adopted my first name (*Schizostoma*) for the division of those *Melanidœ* which have a cut or fissure in the upper portion of the last whorl. This name I proposed in December, 1842. Subsequently, finding that it was used by Bronn in 1835, I abandoned it, and proposed the name of *Schizochilus* as a substitute (Obs. on the Genus Unio, v, 5, p. 51, 1852, and Trans. Am. Phil. Soc., 1852). I am now satisfied that Bronn's name was applied to the same genus — *Euomphalus* — which Sowerby established in 1814 (Min. Conch., tab. 45). This evidently liberates my original name, and Hermannsen, in the appendix to his "Generum Malacozorum," very properly restores it. It was supposed that this was the *Melatoma* of Swainson, and Mr. Anthony adopted this name. But it is evident that Mr. Swainson's *Melatoma* is not my *Schizostoma*. By reference to his figure (Malacology, p. 342, f. 104) it will be observed at once that there has never been observed in the United States any of the group of which that figure is the type, while it is known that they exist in the islands of the Indian Ocean. Mr. Swainson says (p. 202), that his *Melatoma* was 'founded upon a remarkable Ohio shell' sent by Rafinesque. Now, as no member of the family *Melanidœ* with a cut in the lip has ever been found in the Ohio, where such hosts of active collectors have since pursued their investigations, it is perhaps beyond the bounds of possibility that the specimen sent by Rafinesque, so eminently careless and reckless as he always was, should ever have been found there. Indeed, if the specimen figured was sent by Mr. Rafinesque to Mr. Swainson, then the question would arise whether it had not been obtained by Mr. R. from some dealer or collector, who may have obtained it from Asia. I have no doubt of the *Melatoma costata*, which Mr. Swainson has figured, being exotic, and belonging to a group probably from the Philippine Islands. Mr. Anthony says, page 64, Proc. A. N. S., 1860, that 'it may be doubted whether Mr. Lea's first name will not eventually prevail, since, before he published *Schizostoma*, Bronn's genus of the same name had been called a synonyme of *Bifrontia*, Desh.' And that 'H. & A. Adams (Gen. Rec. Moll., I, 105) do not appear correct in giving preference to *Gyrotoma* over *Schizostoma*, Lea,' &c. Notwithstanding this, Mr. Anthony in this paper, where he describes nine supposed new species of this genus, adopts the generic name of *Gyrotoma*. It may be added here, that Dr. Gray, in his *Genera of Recent Mollusca*, gives *Melatoma* to Mr. Anthony, not to Swainson, while he does not notice the name of *Schizostoma*. Mr. A. does not pretend to claim it, of course, but adopts *Gyrotoma*, Mr. Shuttleworth's name, proposed in 1845, which, being three years later, cannot have precedence.

"The genus *Schizostoma* seems to be capable of being divided into two natural groups in the form of the *fissura*, the cut in the lip. In one group this fissura is deep and direct, that is, parallel with the suture or upper edge of the whorl; in the other it is not deep and is oblique to the suture."

In the same Journal (April, 1862), was published a new genus, with the following name, description and remarks: —

"Genus TRYPANOSTOMA, Lea.

"Shell conical; aperture rhomboidal, subcanaliculate below. Lip expanded. Columella smooth, twisted below. Operculum corneous, commencing spiral.

"The enormous number of species in the genus *Melania* has made it very desirable to eliminate as many as possible, by founding new genera, where well characterized groups can be established. With this view I proposed, in the Proceedings of the Academy, in April last, the genus *Strephobasis*. The genus now proposed under the name of *Trypanostoma*, will include all the well known *Melania* with an *auger-shaped aperture*, the type of which may be considered to be Mr. Say's *Melania canaliculata*, a very common and well known species from the basin of the Ohio River. It will include a number of large species; indeed, nearly all of the large and ponderous species of the United States. Many new ones will be found in this paper. Objections may be raised against now increasing the number of genera without the aid of the examination of the soft parts. But there is no validity in this objection, from the fact that, in the present condition of the science of Malacology, we are becoming acquainted with a vast number of new and interesting forms, without the hope at present of seeing the organic portion of the animals. These may at some future time, and no doubt will, be examined and carefully described by zoologists who may dwell near the waters where these numerous and highly-developed species reside. Until this takes place, we can only group them upon the characters which are presented by their outward hard portions which are accessible to us now.

"In proposing this new genus, I am aware that European Zoologists have made many genera and subgenera in this Family, but none have made groups of our numerous species by which they can be properly divided. They have mixed them up, with all the time and care they have bestowed upon them, in a manner so as to make great confusion.

"Mr. Swainson, in his ' Treatise on Malacology,' proposed a subgenus of *Melania* under the name of *Ceriphasia*, and gives a figure, page 204 (*C. sulcata*), stating that it came from Ohio. It is evident, on looking at this figure, that it does not represent any Ohio species, neither in the aperture nor in the revolving ribs. Dr. Gray and Messrs. Adams adopt the genus, and the latter give a figure (pl. 31, fig. 6) of *canaliculata*, Say, as the type, which I do not think answers to the description or figure of Mr. Swainson. Dr. Gray, in his excellent 'List of the Genera of Recent Mollusca,' in the Proc. Zool. Soc., expressed a doubt whether his *Telescopella* may not be the same with *Ceriphasia*."

In April, 1861, Mr. Lea proposed another genus, as follows: —

"*Strephobasis*, Lea. — Shell cylindrical; aperture subquadrate; columella thickened and retro-canaliculate below.

"Operculum commencing spiral, corneous.

"The mollusk, for which I propose this genus, was sent to me by Wm. Spillman, M. D., of Columbus, Miss., and I have before me over a dozen specimens from a third to nearly an inch in length. The very great number of species of the genus *Melania* makes it desirable to eliminate any group, with characters sufficiently distinct to permanently recognize it. The very remarkable retrorse callus at the base of the column, causing a lateral sinus, is characteristic of this genus."

Next, we have the genus *Goniobasis*, intended to include most of the vast residue of species not previously eliminated. This genus, proposed in Proc. Acad. Nat. Sciences, May, 1862, is described as follows: —

"*Goniobasis*, Lea. — Shell conical or fusiform. Aperture subrhomboidal, subangulate below. Columella thickened somewhat above. Operculum commencing spiral, corneous.

"In my paper on the genus *Trypanostoma*, proposed by me (Proc. Acad. Nat. Sci., 1863, p. 169), I mentioned the importance of eliminating as many species as possible from the genus *Melania*, which is so enormously extended as almost to prevent the possibility of finding suitable names for the species. In the Proceedings of the Academy, Dec., 1861, I stated that

Prof. Haldeman's genus *Lithasia* formed a very excellent group. In working up a very large number of the family *Melanidæ*, obtained from the Southern and Western States, I have, notwithstanding the divisions which had been made, found myself embarrassed with that form of aperture which is quite different from the auger-mouthed (*Trypanostoma*) species and the *Lithasia*, to which latter they are most nearly allied. I mean those which usually, though not always, have a slight thickening of the upper part of the columella and no callus below, and which are also without the notch of *Lithasia*, although subangular at base. In this subangular character they differ from *Melania* proper, which are round or loop-like at the base. For this group I propose the name of *Goniobasis*,* which will give us for our American *Melanidæ* the following genera, all of them having opercula : —

"*Melania*, † Lam., *Anculosa*, Say, *Io*, Lea, *Lithasia*, Hald., *Schizostoma*, Lea, *Strephobasis*, Lea, *Trypanostoma*, Lea, *Goniobasis*, Lea, and *Amnicola*, Gould and Hald.

"They may be known by,

"*Melania* having a regular loop-form aperture.

"*Anculosa* having a rounded aperture and a callous columella.

"*Io* having a greater or lesser elongate channel or spout at the base.

"*Lithasia* having a callus on the columella above and below, and a notch at the base.

"*Schizostoma* having a cut in the upper part of the outer lip.

"*Strephobasis* having a retrorse callus at base, and usually a squarish aperture.

"*Trypanostoma* having an expanded outer lip and an auger-shaped aperture.

"*Goniobasis* having usually a subrhomboidal aperture, subangular at base and without a channel.

"*Amnicola*‡ having a round mouth and no callus."

In Proc. Academy of Nat. Sciences, January, 1864, Mr. Lea proposed the following : —

"*Meseschiza*.— Shell fusiform, imperforate. Aperture rhomboidal, below canaliculate. Lip expanded, slit in the middle. Columella smooth, incurved. Operculum corneous, spiral.

* Adams' *Elimia* takes in part of this genus.

† Cuvier describes *Melania* as having long tentacula, the eyes being on the exterior side about the third of the length. The eyes of *Melania Virginica*, Say, are at the base of short tentacula. I very much doubt if we have a single species in the United States properly belonging to this genus, of which Cuvier considered *amarula* as the type, and Lamarck, *asperata* as the type.

‡ *Amnicola*, although much like *Paludina*, is more nearly allied to the *Melanidæ*. The operculum is spiral, and, therefore, very different in this character from *Paludina*.

"The little shell which I now propose as a new genus, has so distinct a character in the incision of the middle of the outer lip, as to mark perfectly its place in the *Melanidæ* of the United States. It differs entirely in the character of the cut from that in *Schizostoma*, which has, in all the many species I have seen, a more or less deep incision immediately under the suture. The living soft parts have not yet been observed. They may, when examined, prove to have some characteristics quite different from *Schizostoma.*"

Eurycœlon.— In remarks on *Goniobasis umbonata* (Proc. Acad., p. 3, Jan. 1864), " This is the fourth species of a natural group which I have described and which have a large ear-shaped aperture. If they be not entitled to a generic place, they may at least be considered a subgenus, for which I propose the name of *Eurycœlon*, the aperture being larger than in the *Melanidæ* generally. All the species of *Eurycœlon* have a callus on the collumella above, but not below, as in *Lithasia*, and the base is more or less angular, which is not the case with *Anculosa.* Those which we have considered as varieties of *Anculosa prærosa*, Say, which have an angular base, properly belong, I think, to *Eurycœlon*, as well also *Anthonyi*, Redfield, *turbinata*, and *tintinnabulum* (nobis), and some others. When the soft parts shall be examined, they will, I think, be found to differ from *Goniobasis*, *Trypanostoma* and *Lithasia*, to which genera they seem nearest allied. The operculum of the only one I have seen is the same as *Goniobasis*, and the *Melanidæ* generally." *

Dr. James Lewis (Proc. Acad. Nat. Sciences, Dec., 1862, pp. 588–90) describes the soft parts of *Melania subularis* and *Melania exilis*, and remarks in conclusion, that " the following features of the two species above considered may suffice for placing them apart in subgenera : —

" 1. The presence of a sinus or fold in the sides of the foot and neck of *M. subularis*, and its absence in *M. exilis*.

" 2. The extension of the anastomosing black lines from the margin of the lateral portions of the foot upwards along the side of the neck in *M. subularis*, and the restriction of these lines to a narrow zone along the lateral portions of the foot of *M. exilis*.

" 3. A well-defined dark band around the tentacle in *M. exilis*, not observable or at most only faintly indicated in *M. subularis.*"

* Mr. Lea probably did not intend to include his *tintinnabulum* in *Eurycœlon*, but did so inadvertently. I would add to the description as given above — shell generally obovate, longitudinally humped or angled; columella truncate below. The genus may be placed between the *Lithasiæ* and *Goniobases.*

Dr. Lewis endeavors, by these differences, to indicate respectively the genera *Trypanostoma* and *Goniobasis* of Mr. Lea; but, unfortunately, the only *important* character of distinction mentioned by him, is only a sexual difference.[*]

And now, having cited all that has been done in the classification of these animals by American and foreign naturalists, we will first ascertain the *sequence* of the genera, and then give their names and limitation as we propose to adopt them.

Swainson commenced with the species having an entire aperture, then he described genera possessing a truncated aperture (*Hemisinus, Melanopsis*), and, finally, those with a more or less developed channel at the base.

Dr. Gray's arrangement does not differ essentially; he *adds*, however, *Glotella*, an intermediate form between the *Trypanostomoid* and *Goniobasic* groups.

Messrs. Adams *commence* with the canaliculate species, but not with the highest developed type of that form, *Io*. They give the preference to *Ceriphasia*, Swainson, and next give *Pachycheilus*, which is certainly more of a *Goniobasic* form, and *then* give *Io*.

Dr. Brot's "Groups" represent nearly the following value and sequence in genera: *Leptoxis, Trypanostoma, Goniobasis, Lithasia, Pachycheilus, Io, Melanopsis, Gyrotoma, Pirena*.

Mr. Lea, in remarks on his description of *Goniobasis*, gives the list of genera (which we have quoted), but apparently in the order of their publication.

The sequence of genera in the foregoing examples, can certainly be much improved; *Io* may be considered as the highest development of the canaliculate shell, and is also the largest in size; we find, moreover, as Mr. Lea has justly remarked, the most ponderous species among the *Trypanostomæ* (*Pleuroceræ*). I would then commence with *Io*, and proceed thus: *Io, Pleurocera, Angitrema, Lithasia,*

[*] See Stimpson "On the Structural Characters of the so-called Melanians of North America," Am. Jour. Sci., xxxviii, July, 1864.

Strephobasis, Eurycælon, Goniobasis, Schizostoma, Meseschiza, Anculosa.

We thus proceed from a long canaliculate aperture to one in which the aperture is entire, we also commence with the largest and close with the smallest species. *Pachycheilus* is not included in the above, because it represents an extra-limital group, and will probably be found to belong to another family or subfamily. The same may be said of *Hemisinus* and *Paludomus*.

With regard to nomenclature, we will examine —

1. *Io*, Lea.—We find this genus universally recognized European authors, however, do not seem to understand its true limits, and include species of *Lithasia*.

2. *Pleurocera*, Raf.—Nothwithstanding Mr. Lea's assertion that Swainson's figure of *Ceriphasia sulcata* does not represent a species of this genus, nor his description correspond to it, I believe that *Ceriphasia* was certainly intended for that group of *Trypanostomoid* shells represented by *canaliculata*, Say, and that the figure represents some such shell as *T. moriforme*, Lea. Gray, also, in 1847, proposed *Telescopella* for *Melania undulata*, Say, which belongs to the same group.

Thus, Mr. Lea's *Trypanostoma* is unquestionaoly a synonyme.

Pleurocera, Rafinesque, is the same shell, and having priority over all the other names, I adopt it without hesitation.

Strepoma, Raf., manuscript, applies to the same genus, and *Oxytrema*, Raf. (Jour. de Physique) may be intended for some immature form of *canaliculata*, or its allies, which possesses the sharp-pointed aperture described,—as *Io variabilis*, Lea, for instance.

Messrs. Adams adopt *Ceriphasia*, but they separate certain species, reticulate, or nodulosely carinate in the middle, to form their genus *Elimia*. Their *Megara*, also, consists of species of this genus.

Of course these names are not founded on generic characters, and, at best, can only be used to designate groups.

3. *Lithasia,* Haldeman.—This genus is recognized by Messrs. Adams, but Mr. Reeve and Dr. Brot confound its species with *Io.* Prof. Haldeman first proposed it as a subgenus of *Anculosa.* "Shell heavy, having protuberances." This character applies only to certain species; but the genus is now recognized by American naturalists to include all the species with the columella thickened above and below.

Prof. Haldeman's subgenus *Angitrema* is synonymous with, and has priority over, *Glotella,* Gray, both adopting *Melania armigera,* Say, for their type. As this subgenus really exhibits the highest development of the species, I have concluded to adopt it as a genus, using *Lithasia* as a subgenus for the smaller, smooth forms.

4. *Strephobasis,* Lea.

5. *Eurycœlon,* Lea.

6. *Goniobasis,* Lea, May, 1862.—This genus will retain Mr. Lea's name. *Potadoma,* Swainson, as understood by Messrs. H. & A. Adams, embraces certain species only. These gentlemen take some species of this, *Strephobasis* and *Pleurocera,* to make their *Megara,* a subgenus of *Gyrotoma* (*Schizostoma!*)

They make of the plicate group, *Melasma,* and of the striate species they form *Juga.* These names may be retained as sections of the genus, possessing no really generic characters.

7. *Schizostoma,* Lea, Dec., 1842.—Messrs. Adams, Brot and Anthony, adopt *Gyrotoma,* Shuttleworth, July 22, 1845, because *Schizostoma* was preoccupied.

Mr. Lea was himself of the same opinion, and changed the name to *Schizochilus* (also preoccupied). He subsequently reclaimed the original name, and I give him the genus as first published, having

two and one-half years' priority over Shuttleworth.
I entirely agree with Mr. Lea, that *Melatoma*, Swainson, represents an exotic, and not an American,
group. Mr. Anthony is ignorant how *his* name
came to be used in connection with *Melatoma*. It
was first so used by Dr. Gray* (perhaps through
inadvertence), and afterwards by Mr. Reeve.

8. *Meseschiza*, Lea.

9. *Anculosa*, Say.—*Leptoxis*, Rafinesque, as already mentioned, is not described definitely enough
to justify its substitution for Say's name. Prof.
Haldeman, with the aid of Rafinesque's manuscript
work, identified the genus and used the name. He
has been followed by Messrs. Adams, Brot, and Binney, while Messrs. Lea, Conrad, Anthony, and Reeve,
have adhered to the old name. I think that *Ellipstoma*, Raf. (Jour. de Phys.) really applies to this
genus much better than *Leptoxis*, and might be
readily taken to represent such a form of it as *crassa*,
Hald.

Prof. Haldeman proposed a subgenus *Mudalia* for
certain thin species without enamel on the labium,
and probably *intended* to include such globose forms
as *altilis*, Lea, &c., but the only species which he
cites under the name, are *dissimilis*, Say, and *turgida*,
Hald., both *carinate* shells. I am convinced, from
studying numerous examples, that the characters
of *Mudalia* are not persistent. The *globose* form
of so-called *Anculosæ*, represented by *altilis*, does
not belong to the family. Mr. Gill has proposed
for it the generic name *Somatogyrus*, and it is now
included in *Amnicolidæ*.

These same Virginia and Ohio thin species,
together with the dentate forms, compose the subgenus *Nitocris*, H. & A. Adams, a synonyme, anyhow,
and otherwise of no value. Mr. Anthony proposes
to me to call such shells as *Anculosa monodontoides*,

---

* Mr. Anthony never described such a shell as *Melatoma altilis*,
Anth., referred to by Dr. Gray.

"*Spirodon*," but the toothed columella is not even a constant specific character.

The characters assigned to *Io, Pleurocera, Angitrema, Lithasia, Strephobasis, Eurycælon, Goniobasis, Schizostoma, Meseschiza* and *Anculosa*, are by no means of equal value. I regard the first five as members of the *Trypanostomoid* section of the family, of which *Io* is a genus, with *Pleurocera* for a subgenus. *Lithasia* should, perhaps, be considered a subgenus only of *Angitrema*, which is the highest development of this form, having the thickened columella.

Strephobasis occupies a position between *Lithasia* and *Goniobasis*, but I think that it, also, might be considered a subgenus of *Angitrema*.

Goniobasis, Schizostoma and *Anculosa*, are certainly distinct genera; the first two approximate, forming the *Goniobasic* group or section;* and the last forms a section by itself, characterized by an entire aperture.

Yet this arrangement is liable to exception, as *all* the species of a genus do not fulfil the ideas here conveyed. Some species, on the contrary, remind one of genera which do not immediately succeed or precede them. Moreover, anatomical researches will enable us probably to separate the *natural* genera of this family much more sharply than we are now doing, and may enable us to seize on *corroborative characters* of the shell, which are now overlooked, or whose importance, in this connection, has been thus far under-estimated.

---

* *Eurycælon* will be retained as a genus in this work although I suspect now that the species should merge into *Goniobasis* and *Anculosa*. *Meseschiza*, as I am convinced, represents an abnormal condition of growth in very young shells from a single locality. Unlike *Schizostoma*, there is in *Meseschiza* every evidence that injury to the shell causes the slit in the body whorl. In this case also I retain the genus, simply because otherwise I should not know where to place its single species.— *May*, 1872.

SYNOPSIS OF GENERA OF STREPOMATIDÆ.

I. Aperture produced into a more or less obvious canal in front.

Trypanostomoid Section.

  1. Shell fusiform inflated on the periphery.
Spire and canal produced; columella without deposit
of nacre.—[Fig. 4.]   Io, Lea.

  2. Shell conical, or oval, canal not so much produced.—
[Figs. 5, 6.]   Subgenus Pleurocera, Raf.

Shell oval, or turbiniform, or fusiform, with a revolving
row of nodules on the periphery, canal short. Columella callously thickened above and below.
[Fig. 7.]   Angitrema, Hald.

Shell oval or oblong, smaller, either smooth or adorned
with nodules around the upper portion of the body
whorl.—[Fig. 8.]   Subgenus Lithasia, Hald.

   Canal retrorse.—[Fig. 9.] Subgenus Strephobasis, Lea.

II. Aperture merely angulated in front, with no canal, and the columella
not twisted, frequently callously thickened above.

Goniobasic Section.

  3. Shell obovate, heavy, nodulosely angled, aperture
ear-shaped; columella oval, truncate.—[Figs. 10, 11.]
                    Eurycælon, Lea.

Fig. 4.   Fig. 5.   Fig. 6.

Fig. 7.   Fig. 8.   Fig. 9.   Fig. 10.   Fig. 11.

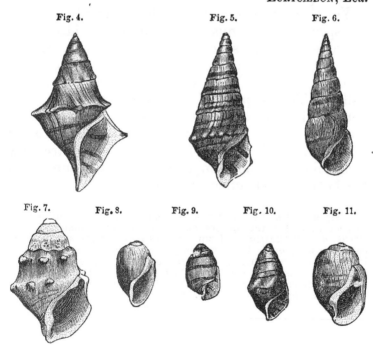

4. Shell heavy, oval, truncate, oblong, or turreted; aperture entire above.—[Figs. 12, 13.] Goniobasis, Lea.

5. Aperture with a sutural, pleurotomose slit above.—[Fig. 14.] Mesesch iza,Lea.
6. Lip slit in the middle.—[Fig. 15.] Schizostoma, Lea.

III. Aperture entire and rounded in front.

7. Shell oval, heavy; columella callously thickened above.—[Fig. 16.] Anculosa, Say.

Fig. 12. Fig. 13. 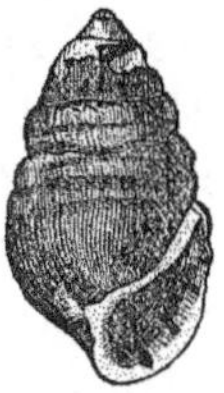 Fig. 14. 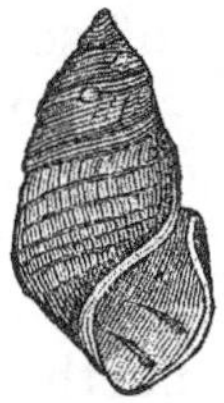 Fig. 15.  Fig. 16.

2. *Geographical Distribution.*—We have, in North America, nearly five hundred recognized species of the shells belonging to the various genera of *Strepomatidœ*. So considerable a moiety of these are found to be inhabitants of the upper Tennessee River and its branches in East Tennessee and North Alabama, and of the Coosa River in the latter State, that we quite agree with Mr. Lea in regarding that region as the great centre of this kind of animal life. We have ascertained that, leaving out the species inhabiting the Pacific States and those which in the descriptions have their habitats designated by States only,* of the remainder, fully two-thirds belong to the above two streams; including three entire genera, nearly all the species in several others, and a *majority* of the species of every genus except one (*Meseschiza*) of a single species.

The *Strepomatidœ* do not appear to flourish in the neighborhood of the sea, and nowhere have the

* As the localities of nearly all of these are "Tennessee" or "Alabama," the most of *them* also were probably obtained from the Tennessee and Coosa Rivers.

species been found numerous within a hundred miles of our coasts; nor do they approach the more northern latitudes of the Middle and Western States, very few species being found so far north as the Ohio River.

The Mississippi River also, seems to have formed, from the junction of the Ohio until its mouth, an insurmountable barrier to the geographical dispersion of these shells.

Thus, we find the district of our country, which they inhabit in such profuse numbers of species and individuals, to be really of somewhat limited extent, and may give its boundaries as follows:—*North,* the Tennessee River and tributaries. The Cumberland Mountains prevent the dispersion of the species of this river to the northward until its course is directed into Alabama. Here the character of its species (which we shall again allude to further on) changes, and they become gradually less numerous and of greater geographical dispersion, as the river runs towards the west. *East,* the mountain range of the Blue Ridge, running southwestwardly into the interior of Northern Georgia. Thence, the Chattahoochee River and tributaries, to within about a hundred miles of the Gulf. *South,* the species are restrained from spreading by the influence of the Gulf of Mexico. *West,* the Alabama, Cahawba and Black Warrior Rivers and their tributaries, those of the latter reaching almost to Florence, on the Tennessee River, which may represent the northwestern point of our boundary.

These limits are necessarily imperfect, but nevertheless include at least three-fourths of our species within an area of three hundred miles extent, either north and south, or east and west.

Of course, where the rivers alone form the boundaries, many of their species have spread into the adjacent streams; but in East Tennessee, southwestern Virginia, western North Carolina and northwestern Georgia, where several parallel mountain ranges completely enclose the valleys of the rivers, almost all the species inhabiting them appear to be confined within their limits. And here, a space of one hundred and fifty miles in length, by fifty in breadth, will include the waters occupied by probably more than a hundred and fifty species of *Strepomatidæ*.

The following table, representing the arrangement of the *Strepomatidæ* followed in my "Synonymy" of the species, published in the Proceedings of the Academy of Natural Sciences, 1863–4, will show both the *total* number of species, and the absolute and relative strength of the genera. A few species since published have not all been included, as we are not sufficiently well acquainted with them : —

NUMBER OF SPECIES OF STREPOMATIDÆ.

1. *Trypanostomoid Section.*

IO 5
smooth 2
spinose 3

PLEUROCERA . . . 84
tuberculate 7
sulcate 8
striate, angulate . . . 12
carinate 8
plicate 2
smooth, angulate . . . 15
smooth, not angulate . . 32

ANGITREMA . . . 12
with a coronal of tubercles 4
with two rows of tubercles 1
with a central row of tu-
bercles 7

LITHASIA 17
large, oval, inflated . . 5
small, compact 7
obliquely flattened . . . 2
subcylindrical 3

STREPHOBASIS . . 8
ovate conical 3
cylindrical 5

2. *Goniobasic Section.*

EURYCÆLON . . . 6

GONIOBASIS . . . 274
spirally ridged 1
tuberculate 18
plicate 85
angulate 16
bi-multi-angulate . . . 11
carinate 4
smooth, short 26
smooth, elevated . . . 43
striate, elevated . . . 8
compact, ponderous . . 62

SCHIZOSTOMA . . 26
fissure narrow 14
fissure wide 12

MESESCHIZA . . .

Third Section.

ANCULOSA . . . 31
nodulous 1
sulcate 2
striate 3
angulate 4
subglobose, or } 21
campanulate . . }

Total in 1st section . . . . 126 species,
"   2d   "   307   "   } 464 species in all.
"   3d   "   31   "   }

We find that, while some groups of species extend over a very wide territorial space, other groups are extremely restricted, and yet are frequently characterized by as great variation in form, size, ornamentation, etc., as the former. The *Goniobasic Group* occupy the entire extent of our country, represented by the sole species of our Northern Atlantic States, the very few forms of the great Northern Lakes and the species of the Pacific States, while they also occupy the entire southern country, with one or two species in Mexico and Cuba.

The *Trypanostomoid Section*, on the contrary, is very much more restricted, being confined principally to the streams tributary to the Mississippi and the Gulf of Mexico. The Mississippi appears to form their western boundary.

While the *Trypanostomoid* forms attain their maximum development in size and number in the Tennessee River, they are, to a very great extent, replaced by the *Goniobasic* forms in the Coosa River, which is undoubtedly the metropolis of the latter. The most striking genus of each of these groups is absolutely confined to the respective streams in which the groups had their origin. Thus, *Io* and *Schizostoma* are inhabitants, the first of the Tennessee and branches, the second of the Coosa, and neither of them is elsewhere found.

Assuming the Ohio River as a dividing line, we find that ninety-five per cent. of all the species originate south of it. Even a smaller proportion inhabit the rivers east of the Alleghany, and west of the Rocky Mountains. In the west, no species of *Strepomatidæ* have been discovered in higher latitudes than the northern boundary of the United States, while in the east, the St. Lawrence River and tributaries appear to be the northern limit of the family.

We thus find the *Strepomatidæ* to be distributed almost exclusively within the limits of the United

States, a distribution coextensive with our *Viviparidæ* and other families of Mollusca; clearly indicating that our country constitutes a distinct faunal province. For, as the *Viviparidæ* are replaced in Mexico by *Ampullaria,* so, for the *Strepomatidæ,* are substituted the more ponderous *Pachychili.* Between the former and the latter extend the broad plains of Texas, with rivers devoid of species, forming a barrier to the intermingling of the two groups. Besides this, the Mississippi River, from the junction of the Ohio to its mouth, appears to have formed a barrier to the westward progression of the *Strepomatidæ,* which but very few species have been able to surmount. We believe that one species only,—the *Goniobasis sordida,* of Lea,—is common to *both* sides of that great stream, while several forms, all of *Goniobasis,* are found inhabiting the western tributary streams exclusively.

Of course, our great river does not interpose such a formidable barrier in the northwest, where its volume is much less, and we here find the species of the great lakes not only inhabiting its waters in abundance, but extending into its western branches.

The species of the great lakes, though few in number and small in size, are very numerous in individuals, yet they fade out as completely on approaching the Ohio River as do the southern species; we are, therefore, compelled to admit in this case the plausibility of the theory of a separate creation of a small group of species, adapted to withstand the rigors of a climate which effectually forbids the introduction of the meridional species.

We may discover in the paucity of species, their small size and scant ornamentation, but multiplicity of individuals, and in their very extended distribution, a striking parallelism with the distribution of boreal marine Mollusca. Like the *Unionidæ,* the *Viviparidæ,* the *Amnicolidæ* and the *Limnæidæ,* of the same latitudes, the intercommunication afforded by our waters has induced the plentiful distribution of the same species from Iowa and Wisconsin to Western New York, and even into Lake Champlain.

We have already alluded to the total separation of the species of our West Coast States. The barrier of the Rocky Mountains has, of course, proved with them even a greater obstacle than with our *Helices*. We find, accordingly, that the few species (all *Goniobases*) mostly partake of two common type characters, being either plicately ribbed* or spirally striated. The *Strepomatidæ* are entirely absent from the waters of the New England States, the exclusion being due probably not only to the severe climate, for they inhabit streams in even higher latitudes, but probably also their proximity to the sea. There is no *natural* method by which the species of the lakes could extend into the head waters of the New England rivers, and none of the species have as yet been transported by accident across the intervening land.

That the proximity of the sea exercises a great disturbing influence on the very few species which are exposed to and able to endure it, is proved by the great mutations of form which characterize *Gon. Virginica* and *Anc. dissimilis* in the Atlantic, and *Gon. plicifera* in the Pacific States.

The very great influence which our two great chains of mountains has exercised, in restricting the distribution of our species, may be inferred from what has already been said, and requires no further allusion.

The following observations on the geographical distribution of the various genera and smaller groups, will exhibit some very curious facts.

---

* Which strangely enough, equally characterizes a group of *Goniobases* of East Tennessee. Our West Coast Helices are all of different species and generally of quite distinct groups; *Vivipara* is excluded, and the *Amnicolidæ* belong to different genera from those of the Atlantic States, yet the same species of *Physa*, *Lymnæa* and *Planorbis*, abound equally in either section!

10.

Of this genus, the type of the *Trypanostomoid* form, there are five species, two of which are smooth and three spinose; they are of extremely localized distribution, being confined to the head waters and tributaries of the Tennessee River, and principally to the Holston, in Southern West Virginia and East Tennessee. They are very numerous in individuals, as Mr. Anthony, during a visit made to this region several years ago, selected and brought home several thousand specimens. Prof. Haldeman also was very successful in collecting them.

PLEUROCERA.

Of the eighty-four species, only thirteen are found so far northward as the Ohio River, and only five of them originate in that stream or its northern tributaries. The Tennessee River and branches claim thirty-three species, of which twenty-one appear to be confined to its waters. The Cumberland River contains four species identical with those of the Tennessee, and about a dozen that are not found in the latter stream. The Alabama River contains fourteen species, three of which seem to be peculiar to it. These species are generally confined, however, to those portions of the Coosa and branches that approach to East Tennessee. A few species also inhabit the Tombigbee, of Mississippi.

About a dozen species have the simple habitat "Tennessee" stated; nine have "Alabama," and two "South Carolina." I doubt very much whether the latter is correct.

There is very good reason to believe that *all* the large tuberculate, sulcate and angulate species inhabit the Tennessee River, the most ponderous ones extending from the Coosa, through Middle and West Tennessee, to the Ohio River. Among the angulate forms two, *trivittatum* and *tortum*, are reported only from the Tombigbee and Chattahoochee Rivers respectively. None of the carinate group — inhabitants of Tennessee River — extend northward to the Ohio; but, strangely enough, the North-western States furnish two peculiar species,— *P. subulare* of Niagara River, and *P. Lewisii* of Illinois River.

But two plicate *Pleuroceræ* have yet been discovered, although this form is so very common to the *Goniobases* inhabiting the same region. These shells are found in the Clinch and Cumberland Rivers.

Of the smooth species, several extend to the Ohio River.

ANGITREMA.

The four species of the first group are inhabitants of the Tennessee River. *A. salebrosa* has been gathered in the Holston, in East Tennessee, and in the Tennessee at Florence, Alabama.

A. Jayana inhabits Caney Fork, Tennessee.

The five species of the third group are, with the exception of *A. rota*, very closely allied.

A. armigera has an extensive distribution. It was described from the Ohio River, and has since been found in the Wabash, Indiana, along with several other nodulous and plicate species, whose range is otherwise confined to more southern rivers. Kentucky and Tennessee are also given as habitats for this species; and in the latter State it doubtless originated. *A. Duttoniana* and *Stygia* are both reported from Cumberland River, and the former inhabits the Tennessee. The fourth group contains two species not easily distinguished, but differing very much in their range of habitat; for, while *A. lima* is confined to the *lower* waters of the Tennessee, *A. verrucosa* has a range coextensive with that of *armigera*. It occurs in the Holston River and the whole extent of the Tennessee, the

Cumberland, the lower parts of the Ohio, and is very plentiful in the Wabash.

LITHASIA.

While the typical *Angitremæ* are essentially a Tennessee group, the subgenus *Lithasia* extends further southwards. Its large inflated species, five in number, all occur in the Tennessee River at Florence, Alabama, and vicinity, while the more numerous, compact, heavy species, approaching in form to the typical *Goniobases*, are almost confined to the Coosa and Cahawba Rivers. The exceptions are a small group of three species, of which *obovata* is the type, which inhabit the Ohio River and its Kentucky and Indiana tributaries, and one singular subcylindrical species reported from the Cumberland.

Mr. Anthony assigns Tennessee as the habitat of his *nucleola;* but I think he is mistaken, as I have specimens from the Coosa.

STREPHOBASIS.

Several of the species are reported only from East Tennessee, while two of them occur in the branches of the Alabama River. One of these is found in both rivers. Prof. Haldeman is in error in assigning Ohio River as the habitat of his *St. curta.* It has never been found there, but is one of the most plentiful shells of the Tennessee River, and as such, is in all our cabinets.

Goniobasic Section.

These shells constitute three-fifths of the species of *Strepomatidæ*. They are naturally divided into two type forms: the first, heavy, compact, with large subcylindrical body and short spire is eminently characteristic of the Coosa River; while the second, containing narrow, elongated species, with high spires of many whorls, although more extensively distributed, is still very characteristic of the waters of the Tennessee River and branches.

To the first of these forms undoubtedly belongs *Eurycælon*, a new genus, which probably includes more species than have yet been assigned to it;— and *Schizostoma*. Of the six species of the former, one is from the Holston, another from the Cumberland, and the balance from the tributaries of the Alabama River.

SCHIZOSTOMA.

This genus, embracing twenty-six species, divided into two distinct groups of nearly equal respective numbers, inhabits the Coosa River only, and in this limited space exhibits all the range of variation in form, size and ornamentation, belonging to genera which possess a more extended geographical distribution.

MESESCHIZA

Contains at present only the type species. It is a very small, fragile shell, inhabiting the Wabash River, and does not appear to be of mature growth.[*]

GONIOBASIS.

This very large and widely-extended genus embraces over two hundred and fifty species — more than half of all the *Strepomatidæ* — and includes the only representatives of the family west of the Rocky Mountains, or south of the United States.

One species, beautifully ridged with sharp, revolving ribs — the G. *proscissa*, of Anthony — is reported simply from northern Alabama. There are eighteen tuberculate species; the heavy, compact ones being principally from the branches of Alabama River, while the elongated ones are found in the Tennessee.

In the latter is included a very distinct group, typified by *Postellii*, of Lea, belonging to the tributaries of the Tennessee, in Northwest Georgia. Two or three allied species are found in Florida.

---

[*] The validity of the genus is doubtful. No specimens have been collected since the type series, and they all appear to have been injured.

Among the tuberculate species, I have included
G. *occata*, Hinds, — a California shell, of very
doubtful generic character.

The plicate species number eighty-five, of which
about half inhabit the Tennessee River. A few of
these extend into the Cumberland, and one or two
to the Green River, of Kentucky.

On the other side, a very few (five only) of the
plicate species are found also in the Coosa and
Black Warrior Rivers. Five species occur in Oregon
and California. One species is reported from South
Carolina, and two from Florida. The Ohio and
Illinois Rivers each possess a species; and several
occur in the Flint and Savannah Rivers, of Georgia.

G. *suturales*, Haldeman, reported from Ohio, is
more likely a Georgia species, identical with one
recently described by Mr. Lea.

Twenty-seven angulate species are about equally
distributed in the Coosa and Tennessee Rivers.
One of them, *sordida*, Lea, occurs both in the
Cumberland and in Saline River, Arkansas.

G. *Potosiensis*, Lea, is found in St. Francis River,
Missouri.

G. *proxima*, Say, occurs in the Holston and Santee
Rivers.

G. *bicincta*, Anth., inhabits the Cahawba, Chattahoochee, Savannah, Roanoke, and is also reported
from North Carolina and Arkansas!

Mr. Anthony's habitat, "Ohio," for his G. *tecta* is
an error; the shell is known to come from the
Coosa River.

It is also very doubtful whether the specimens
of Mr. Lea's G. *Spartanburgensis*, from the Ohio
River and from South Carolina, really belong to
the same species. In such cases the authority for
the alleged habitats should be rigorously investigated.

Of the twenty-six short, clavate, smooth species,
a small group, with dark-colored, inflated shells, is
quite characteristic of East Tennessee and southern

West Virginia. Five species are found in the Ohio
River and the Lakes, and two, both of which will
probably be found to be sometimes plicate, occur in
the rivers of the Pacific States.

There are forty-three smooth, elevated *Goniobases*, of which about one-fourth inhabit the Tennessee, and the same number the Alabama River.
Seven or eight occur in the Ohio River and Great
Lakes, and two are found in California.

Three species inhabit Louisiana, and are the only
Strepomatidæ reported from that State. Neither of
them occurs east of the Mississippi.

G. semicarinata, one of the species of this division,
extends from Tennessee and Kentucky, throughout
all the Western States and the Lakes, and rejoices in
twelve synonymes!

There are eight striate species, of which one,
G. *Virginica*, Say, is the only *Goniobasis* inhabiting
the rivers of New York, Pennsylvania and Maryland.
Through the Erie canal it is extending to the
Western Lakes.*

Very close relatives to this shell are *latitans*,
Anth., and *sulcosa*, Lea, the former from Green
River, Kentucky, and the latter from Tennessee.

There are over sixty species in the group which I
have designated as "compact, ponderous," for want
of a better name. They are essentially a distinct
group from the other *Goniobases*, and *all the species,
except three, are peculiar to the branches of the
Alabama River.*

ANCULOSA.

Thirteen species inhabit the Coosa River, three of
which are common to the Tennessee, and one of
them, *A. prærosa*, extends northward to the Ohio.
Two others are peculiar to the Tennessee. Three
species are found in the Dan, Roanoke and Tar
Rivers.

* *Vide* Dr. James Lewis, Proc. Acad. Nat. Sci.

A peculiar group of shells, possessing an inflated form and much lighter texture, is found in the Potomac and Susquehanna Rivers, the Kanawha and the upper Ohio. They are—*A. dissimilis, dilatata, costata* and *trilineata*.

Concluding Observations.

In studying the species of *Strepomatidæ*, especial care must be taken not to consider young shells to be adult species. All of our conchologists who have described species of this family have fallen into this error. The aspects assumed by young or half-grown shells are frequently so very different from their appearance when mature, as to be liable to mislead experienced naturalists.

All quite young shells are characterized by a thin texture, very light color, and very sharp acuminated spire, and in most cases by the base of the aperture being acuminate also.

Nearly every species, even when smooth in its adult state, presents the first few whorls either sharply carinate, or plicate, or striate. Occasionally they are either one or the other *in the same species*. Hence, in describing shells as carinate, or plicate or angulate, the appearance presented by the adult only should be thus described.

In some of the species, however, these lines, plicæ or carinæ, are persistent in the old shell, under favorable circumstances, but in most specimens are not seen. This is *one* difficulty which has caused the multiplication of synonymic names, generally unavoidably, on account of the scarcity of specimens, known to be from the same locality, for comparison.

When a specimen exhibits a perfect spire in the adult state (rare among the *Strepomatidæ*) and the initial whorls are plicate or carinate, they cannot be regarded as affording reliable data for specific discrimination. And it is only when these marks extend quite, or more than half-way, to the body-whorl, that the species should be regarded as plicate or carinate. Whether species not usually *plicate* do not in some localities *become so*, from the absence of disturbing influences of the waters, is a question that we cannot as yet definitely decide; its decision in favor of such occasional development of plicæ would affect the validity of many species which are now regarded as established.

The development of carinæ or tubercles on the body-whorl of the adult shells is not nearly so constant a character as would, at first sight, appear to be the case, and several species are in doubt on this account. *Generally*, however, these may be regarded as more permanent characters when developed on the body than on the spire, as an *adult* shell is not subject to the same mutations of form as a juvenile individual.

Of course, the relations of size and texture are applicable to adults only; and *then* the former is subject to much variation from external influences. Texture is an important, because a tolerably permanent, discriminative guide.

Color, external or internal, generally should not be much relied on, nor the presence or absence of bands, or maculations; but in exceptional cases it is *very* characteristic, as in *P. viridulum*, Anth., for instance. Perhaps color in the *interior* is a more reliable feature than epidermal or *external* hues.

In *some* species, however, the presence or absence of bands forms a prominent distinctive feature.

Form, though subject to variation, may be relied on as one of the best characteristics; the length, number, and the convexity of the whorls, relative size of the aperture to that of the entire shell, shape of the outer lip and of the columella, are all *generally* reliable.

To repeat; in distinguishing a species of *Strepo-matidæ*, of course the first step is to ascertain whether it is *adult*. The signs of juvenility are — sharp extremities, thin texture, *particularly* the outer lip, which is frequently, on this account, broken, the very light color in the quite young and the absence of callosity upon the columella.

A comparison of shape, angle of divergence of the whorls, etc., with specimens of adult shells, or with figures and descriptions, will generally suffice to detect half-grown shells.

Many of the ponderous Alabama *Goniobases* are *bulbous* in the half-grown state; the spire at first narrowly acuminate, then suddenly and very convexly expanding, resembling the growth of certain West India *Cylindrellæ*. As with these terrestrials, the subulate portion invariably disappears in the adult, leaving a somewhat pupæform shell.

We thus find that no one character (with very few exceptions) can be relied on in specific discrimination; but rather a *combination* of characters, with a general idea of the necessary allowance for variation pervading other species of the same general type, or contiguous locality.

NOTE ON THE LINGUAL DENTITION OF THE STREPOMATIDÆ.*

As lingual dentition has been adopted as a very important character (somewhat hastily, I think) in the classification of the Mollusca, it may be well to ascertain how far it may be corroborative with other differences in the genera of North American *Strepomatidæ*. Troschel, in his magnificent work "Das Gebiss der Schnecken," divides the *Melanians* into several groups, of which the following contain American species:

Ancyloti. The peculiarity of the dentition of the forms belonging to this group is that the Rhachidian tooth is broader than long, rounded behind, and swollen out before (*ausgebuchtet*). The laterals have a rhombic form with the outer posterior angle somewhat drawn out, and the inner Uncini always possess a smaller quantity of denticulations than the outer ones. The jaw exhibits numerous small scales which appear of a polygonal, mostly hexagonal form.

In this group are included *Ancylotus, Melania depygis* (*Goniobasis*), *Gyrotoma* and *Io.*
We copy the figure given by Troschel:—

Fig. 17. *Ancylotus prærosus.* Fig. 20. *Melania depygis.*
"   18.   "   *costatus.* "   21. *Gyrotoma ovoidea.*
"   19.   "   *dissimilis.* "   22. *Io spinosa.*

It will be noticed, by an inspection of these figures, that the differences in the form of the dentition are so slight as to be of no value for the purpose of separating the genera. Indeed Troschel acknowledges that he can find no difference of suffi-

*From American Journal of Conchology, ii, 134, 1866.

cient importance for the separation of *Melania depygis*, or of *Gyrotoma** from *Ancylotus*.

Pachychili. There is in this group also a marked distinctness of form. As we have excluded this genus from the family *Strepomatidæ* on considerations entirely conchological, it is very interesting to find in the dentition differences quite as marked as those existing in the shell. To show the very peculiar form of the Rhachidian tooth, we copy from Troschel the following for comparison : —

Fig. 23.   *Pachychilus lævissimus.*
Fig. 24.   "   *Schiedeanus.*

It is curious, however, and shows how little dependence can be placed on any one character in the grouping of Mollusca, to find *Pirena* and *Melanopsis* placed by this author together with *Pachychilus*, on account of their almost identical dentition, when they differ so much in conchological characters and in geographical distribution.

Dr. William Stimpson, nearly two years since, published a paper in the "American Journal of Science and Arts, "On the Structural Characters of the so-called Melanians of North America," containing the results of observations of the animals of several of our species, including an *Io, Anculosa,* and *Goniobasis.* The individuals of these three very distinct genera were not found to differ one from another in any structural character, although readily distinguished from Oriental species. We will state the differences in their relative importance, as they appear to us. 1st. By being oviparous, while the latter are ovo-viviparous. 2d. By the mantle-margin being plain in the American, and fringed in the exotic family. 3d. By difference in dentition. To these may be added a sufficient conchological difference to justify the separation into two families, even if the soft parts were undistinguishable.

---

* He curiously regrets that the nearly-allied genus *Schizostoma*, Lea, is unknown to him!

Fig. 17.

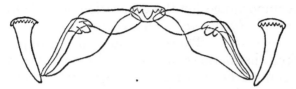

Ancylotus prærosus.

Fig. 18.

Ancylotus costatus.

Fig. 19.

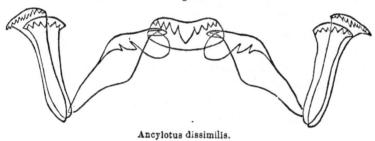

Ancylotus dissimilis.

Fig. 20.

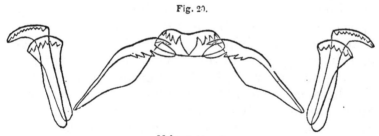

Melania depygis.

Fig. 21.

Gyrotoma ovoidea.

Fig. 22.

Iospinosa.

Fig . 23.

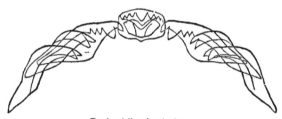

Pachychilus lævissimus.

Fig. 24.

Pachychilus Schiedeanus.

MONOGRAPH OF STREPOMATIDÆ.

FAMILY STREPOMATIDÆ, HALDEMAN.

Strepomatidæ, HALD., Proc. Acad. Nat. Sci., Sept., 1863.

Melaniana, LAM., Extr. d'un Cours., 1812. Hist. Anim. sans. Vert., vi, p. 163, 1822; edit. 2, viii, p. 425, 1838. DESHAYES, Encyc. Meth., iii, pp. 431 and 553, 1832. REEVE, Zool. Proc., p. 76, 1841. Conch. Syst., ii, p. 119, 1842. SOWERBY, Conch. Man., ed. 2, p. 187, 1842. CATLOW, Conch. Nomenc., p. 185, 1845.

Melanidæ (part), LATREILLE, Fam. Nat., 1825. LEA, Proc. Philos. Soc., iii, p. 164, 1843.

Melanianæ (part), SWAINSON, Malacol., pp. 198, 340, 1840.

Melaniadæ (part), GRAY, Syn. Brit. Mus., 1840. Zool. Proc., part 15, p. 152, 1847. TURTON's Manual, ed. 2, p. 79, 85.

Melaniidæ (part), ADAMS, Genera, p. 293, 1854.

Ceriphasinæ, GILL, Proc. Acad. Nat. Sci., pp. 34, 35, Feb., 1863.

IO, LEA.

Io, LEA, Trans. Phil. Soc., iv, p. 122, 1831.* SOWERBY, Conch. Man. 2d edit., p. 167, 1842. DEKAY, Moll., New York, p. 103, 1843. HERMANNSON, Indicis Generum Malacozoorum, p. 562, 1846.

Io (sp.), Lea, GRAY, Proc. Zool. Soc., pt. 15, p. 153, 1847. JAY, Catalogue, 4th edit., p. 277, 1852. H. and A. ADAMS Genera, i, p. 299. CHENU, Man. de Conchyl., i, p. 290, 1859. ANTHONY, Proc. Acad. Nat. Sci., p. 69, 1860. REEVE, Monog. Io, April, 1860. BINNEY, Check List, June, 1860. BROT, Cat. Syst. des Mélaniens, p. 29, 1862.

Melafusus, SWAINSON, Malacol., pp. 201, 341, 1840. WOODWARD, Manual, p. 131, 1851.

Fusus (sp.), SAY, Jour. Acad. Nat. Sci., 1st series, v, pt. 1, p. 129, Nov., 1825.

Melania (sp.), CATLOW and REEVE, Conch. Nomenc., 1845.

---

* Date of title page of the volume, 1834, but the part containing Mr. Lea's Memoir was printed and distributed in 1831.

Description.—Shell fusiform; base canaliculate; spire elevated; columella smooth and concave.—*Lea.*

Fig. 25.

Geographical Distribution.—The few species comprising this genus appear to inhabit exclusively the waters of Middle and East Tennessee and southwestern Virginia.

Observations.—Mr. Lea has recently described eight species which he proposes to consider a distinct group of *Io*, but I cannot distinguish them from *Pleurocera.* The longer fuse, sharp lip and fragile texture of most of these species, show them to be immature shells, and in several instances I had no difficulty in proving them identical with mature shells described by Mr. Lea as *Trypanostoma* ($=$ *Pleurocera*), by means of series of specimens of different ages.

Excluding these, twelve species have been described; of which we propose to retain five, regarding the others as synonymes. Many naturalists consider the genus to be restricted to one valid species, and cite the nearly uniform size of the shells, their similar ornamentation and restricted habitat as proofs of the correctness of their opinion; there appears to me to be a well-founded division of the species into two groups, the one containing shells which are smooth or obscurely tuberculate, and the second those developing distinct spines. Endeavors have been made to connect *Io fluvialis* and *spinosa*, the respective types of the two groups, by series of specimens, but no *fluvialis* has been found with better developed protuberances than the shell described by Mr. Reeve as *verrucosa*, which is still a long way from the *spinosa.* In the young shells the differences are very much better shown than in mature individuals, and no one would think of connecting the quite young of the two.

Species.—There are very many groups in the other genera of Strepomatidæ in which the species resemble one another quite as closely as in *Io;* we may instance the close resemblance of *Angitrema armigera* and *Duttoniana;* of *verrucosa* and *lima;* of *geniculata, salebrosa* and *subglobosa;* of *Anculosa prærosa* and *tæniata;* of the species of *Schizostoma;* of the heavy cylindrical *Goniobases* of North Alabama; and many like instances will occur to those who have studied the family. —*Am. Jour. Conch.*, i, p. 41, 1865.

In a figure included in the introductory portion of this work will be found the lingual dentition of a species of this genus, *Io spinosa*, Lea (fig. 22).

SYNOPSIS OF SPECIES.

A. Shell smooth or somewhat tuberculated.

    1. Io FLUVIALIS, Say.

        Io tenebrosa, Lea.
        Io verrucosa, Reeve.

    2. Io INERMIS, Anthony.

        Io lurida, Anthony.

B. Shell spinose.

    3. Io SPINOSA, Lea.

        Var. *Io crassa*, Anthony.
            (Monstrosity) *Io gibbosa*, Anthony.
        Var. *Io recta*, Anthony.
        Var. *Io rhombica*, Anthony.

    4. Io BREVIS, Anthony.

        Io spirostoma, Anthony.

    5. Io TURRITA, Anthony.

SPECIES.

A.   *Shell smooth or only slightly tuberculate.*

1. I. fluvialis, SAY.

Fusus fluvialis, SAY, Jour. Acad. Nat. Sci., v, p. 129, Nov., 1825. CONRAD, New
    Fresh-Water Shells, p. 12, 1834.
Io fluvialis, Say, BINNEY, Check List, p. 12, June, 1860.
Io fluviatilis, Say, WOODWARD, Manual, t. 8, f. 27. HANLEY, Conchological Misc.
    Melania, t. 6, f. 50. REEVE, Monog. Io, t. 1, f. 5. H. & A. ADAMS. Genera, i, 299.
    BROT, Cat. des Mélaniens, p. 29. BROT, Malacol. Blatt, ii, 114, 1860.
Pleurocera fluvialis, Say, HALDEMAN, Iconog. Encyc., ii, p. 84.
Io fusiformis, LEA, Philos. Trans., iv, p. 122, t. 15, f. 37a, b; Observations, i, p. 132,
    t. 15, f. 37a, b. RAVENEL, Catalogue, p. 11. TROOST, Catalogue Shells of Ten-
    nessee. CHENU, Man. Conchyl., i, f. 1977. DEKAY, Mollusca New York, p. 103.
    WHEATLEY, Cat. Shells U. S., p. 28. SAY, Catalogue, 4th edit., p. 277. REEVE,
    Monog. Io, t. 1, f. 6.
Io tenebrosa, LEA, Philos. Proceedings, ii, p. 34, April, 1841; Philos. Trans., ix, p.
    17; Observations, iv, p. 17. WHEATLEY, Cat. Shells U. S., p. 29. BINNEY,
    Check List, No. 404. H. & A. ADAMS, Genera, i, 299.
Io verrucosa, REEVE, Monograph Io, t. 1, f. 2, April, 1860. BROT, Cat. des Méla-
    niens, p. 29.

Description.—Shell fusiform, olive-green or brownish; spire much
elevated, gradually tapering; volutions nearly six, wrinkled across,
and with a series of elevated un-
dulations on the middle; suture
consisting only of an impressed
line; aperture somewhat fusiform,
within whitish, more or less with
dull reddish, and with several lines
of that color sometimes confluent;
labrum on the inner margin immaculate, edge undulated; canal
rounded at tip; columella very
concave.

Fig. 26.   Fig. 27.

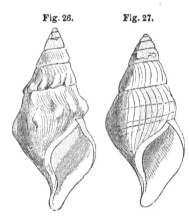

Length, 1 8-10 inches; aperture,
19-20 of an inch; greatest breadth,
19-20 of an inch.

Observations.—Professor Vanuxem found this curious and highly interesting shell (Fig. 27) on the north fork of the Holston River, near
the confluence of a brook of salt water. From the name of the genus
it might reasonably be supposed to be a marine shell, but it has never
been discovered on the coast, and seems to be limited to a very small
district of the Holston River, in company with *Unio cariosus, sub-
tentus*, nobis, *Melania subglobosa*, nobis, and no doubt other fluviatile

shells. When the inhabitant becomes known it may authorize the formation of a new genus, but there appears no character in the conformation of the shell that would readily distinguish it from *Fusus.*— *Say.*

Mr. Lea, upon instituting the genus *Io*, renamed *fluvialis* as *fusiformis*, Lea, in accordance with a custom very usual among naturalists, but very reprehensible. He has recently done Mr. Say and himself the justice of restoring the original name — an example worthy to be followed.

A young, very dark colored specimen of this species, Mr. Lea named *Io tenebrosa.* He now agrees with me in considering it to be a synonyme of *fluvialis.*

The following is the description, together with a figure from the type specimen, of

Io tenebrosa.— Shell fusiform, rather thin, nearly black, smooth; spire conical; sutures scarcely impressed; whorls six, flattened; aperture irregularly pear-shaped; within purple.

Fig. 28.

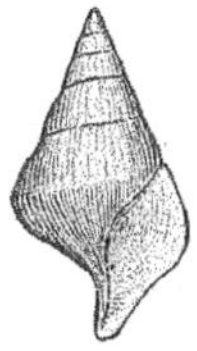

Habitat.— Tennessee.

Diameter, ·48; length, ·75 of an inch.

Observations.— A single specimen only was brought by Mr. Edgar from Tennessee. It is a small specimen, and may be immature. After a good deal of hesitation I have determined to give it a place among the species. It seems to me to be very distinct in color. The channel is more curved to the left and backward than in Mr. Say's species. It has no trace of spines or tubercles, and is dark all over. I do not know if it ever occurs banded.— *Lea.*

The two accompanying figures represent respectively smaller and larger specimens than Mr. Lea's type. The full grown shell is very frequently entirely smooth, though it sometimes develops a few nodules upon the periphery, but these do not attain to the size of the "spines" which characterize *Io spinosa*, and I have not found, among numerous specimens, any that would connect the two species. The color of *fluvialis* varies from yellow through various shades of light and dark green and brown to black. Some specimens are

Fig. 29.

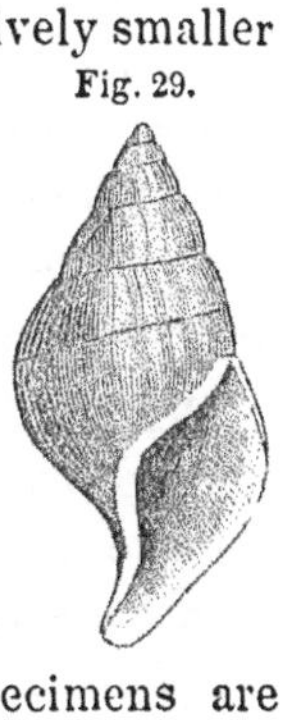

Fig. 30.

beautifully banded.

Fig. 31.

The following description by Mr. Reeve is founded on a shell more than usually noduled; the figure is a copy from his plate.

Io verrucosa.—Shell fusiform, greenish-olive, purple tinged and banded; whorls six, sloping, the first plicately crenulated, the rest tumidly noduled at the periphery; columella attenuately elongated.

Habitat.— Tennessee.

Observations.— In this species, which is of a greenish hue, the periphery of the whorls is furnished with a row of swollen, wart-like nodules, the early whorls of the shell being rippled with small concentric folds.— *Reeve.*

2. I. inermis, ANTHONY.

Io inermis, ANTHONY, Proc. Acad. Nat. Sci., Feb. 1860, p. 70. BINNEY, Check List, No. 401. REEVE, Monog. Io, t. 3, f. 21.
Io lurida, Anthony, REEVE, Monog. Io, t. 3, f. 20.

Description.—Shell conical, smooth, thick; moderately elevated; composed of 7–8 flattened whorls; suture very distinct; upper whorls slightly coronated by an obscure row of low spines, nearly concealed by the preceding whorl; shell otherwise perfectly smooth, or only occasionally or obscurely nodulous on the body-whorl; lines of growth very strong and much curved; aperture pyriform, curved to the left, banded within; columella twisted, callous, thickened above; sinus long and curved.

Fig. 32.

Length of shell, 2 1-16 inches; breadth of shell, 1 inch; length of aperture, 1 inch. Breadth of aperture, ½ inch.— *Anthony.*

Remarkable mainly for its plain, unadorned exterior, and smooth epidermis; its color also is lighter than "*spinosa*" or "*fluviatilis.*" No spines are visible on the body-whorl of this species generally, but I have a few specimens which may perhaps belong to it, and which have a few obscure spines near the aperture; these are, however, little more than knobs. Some hundreds of this species have come under my notice. *Io lurida* was first described by Mr. Reeve. It is only a dark variety of *inermis*. Indeed, Mr. Anthony himself writes to me to that effect.

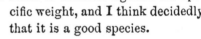

Fig. 33.

The following is the description and figure from the type specimen of

Io lurida. — Shell straightly fusiform; lurid-purple within and without; whorls smooth, unarmed, concavely impressed round the upper part, tumidly gibbous round the middle; columella scarcely twisted.

Habitat.— Southern United States.

Observations.— A smooth, straightly fusiform shell, of a dull, lurid-purple color throughout.—*Reeve.*

This species is considered by many conchologists to be a variety of *fluvialis:* it may be so, but the material before me does not enable me to make a decision against its specific weight, and I think decidedly that it is a good species.

Fig. 34.

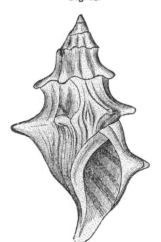

3. I. spinosa, LEA.

Io spinosa, LEA, Philos. Trans., v, p. 112, t. 19, f. 79. Obs., i, p. 224. TROOST, Cat. WHEATLEY, Cat. Shells U. S., p. 29. JAY, Cat., 4th edit., p. 277. BINNEY, Check List, No. 402. REEVE, Monog. Io, t. 1, f. 7. HANLEY, Conch. Misc., t. 6, f. 51.
Io gibbosa, Anthony, REEVE, Monog. Io, t. 3, f. 17.
Io recta, Anthony, REEVE, Monog. Io, t.3,f.21.
Io rhombica, Anthony, REEVE, Monog. Io, t. 3, r. 16.

Description. — Shell obtusely turreted, wide, horn-color, under the epidermis banded, furnished with large spines; whorls seven; mouth elongate, one-half the length of the shell.

Habitat.— Holston River, Washington county, Virginia.

Observations.— This species resembles very much the *Io fusiformis* (nobis), *Fusus fluviatilis*, Say, but may be distinguished by its large, transversely compressed spines, the *fusiformis* having some longitudinal tubercles. I am not acquainted with any fluviatile shell which has such large spines (there being about seven on each whorl), nor any which has such a general resemblance to a marine shell.

Prof. Troost informs me that they are rare in the river, that they had been observed in the graves of the aborigines; and as it was generally believed that these were "conch shells," consequently coming from the sea, it was urged that the inhabitants who possessed them must have come over the sea. It does not appear that they had been observed in their native element, though living at the very doors of the person who had remarked them in the tumuli.— *Lea.*

The accompanying figure is from a half-grown specimen in the Smithsonian Collection. In the shells described by Mr. Reeve, quoted in the above synonymy, I cannot recognize specific characters, although *Io recta* may possibly rank as a variety.

Fig. 35.

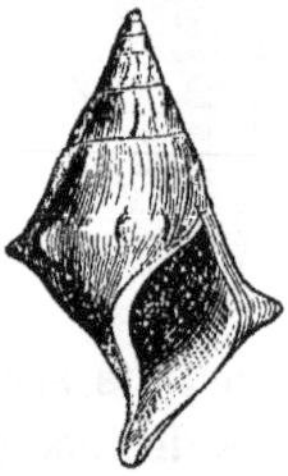

The descriptions of the various synonymes are appended, with figures from the type specimens.

Io gibbosa.— Shell stoutly fusiform, fulvous; whorls rudely obliquely plicated, obtusely tubercled in the middle, last whorl spirally plicately ribbed around the lower part, rib swollen, gibbous; columella arcuately twisted, canal broadly effused.

Habitat.— Southern United States.

Observations.— The gibbous ridge which encircles the lower portion of the body-whorl of this species "is not," writes Mr. Anthony, "a mere accidental aberration; I have seen others like it."— *Reeve.*

Fig. 36.

The extensive suite of *spinosa* that I have examined proves that the gibbous ridge *is* "a mere accidental aberration," being found in all stages of development on specimens which are otherwise distorted in growth, as Mr. Anthony's type, figured above, undoubtedly is.

Io recta.— Shell somewhat elongately fusiform, straight, rather solid, fulvous-olive, whorls concavely sloping around the upper part, conspicuously tubercled at the angle; tubercles rather small; columella arcuately twisted; canal broadly appressed; aperture oblong; interior banded and stained with reddish - purple. — *Reeve.*

Fig. 37.

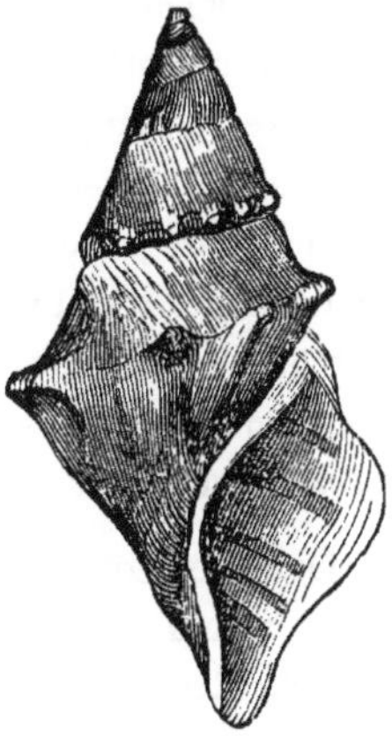

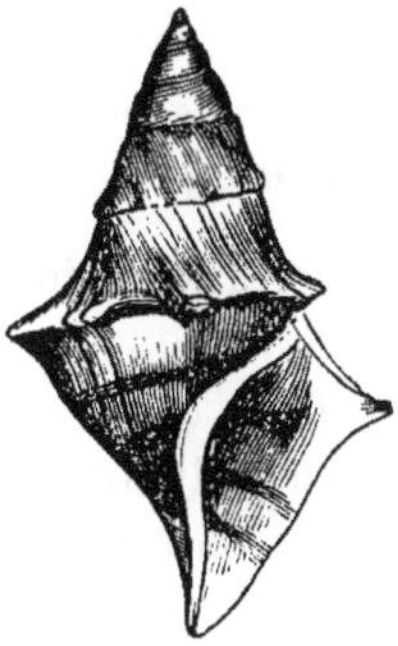

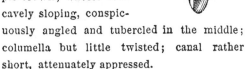

Habitat.— Tennessee.

Io rhombica. — Shell striately fusiform, fulvous-olive, encircled with four bands of purple-brown; whorls concavely sloping, conspicuously angled and tubercled in the middle; columella but little twisted; canal rather short, attenuately appressed.

Fig. 38.

Habitat.— Southern United States.

Observations.— The specimen which Mr. Anthony has here named *I. rhombica*, is of more regular growth than *I. spinosa*, with less twist in the columella, and the whorls are more concavely sloping.—*Reeve.*

4. I. brevis, ANTHONY.

Io brevis, ANTHONY, Proc. Acad. Nat. Sci., Feb., 1860, p. 69. BINNEY, Check List, No. 399. REEVE, Monog. Io, t. 1, f. 4.
Io spirostoma, ANTHONY, Proc. Acad. Nat. Sci., Feb., 1860, p. 70. BINNEY, Check List, No. 403. REEVE, Monog, Io, t. 1, f. 1.

Fig. 39.

Description. — Shell conic, ovate, horn-colored, spinous; spines short, thick, five on each whorl; whorls about seven; aperture elliptical or pyriform, one-half the length of the shell; columella rounded and sinuous near the base, forming with the outer lip a broad, well defined canal at the base.

Length of shell, 2 in.; breadth of shell, 1¼ in. Length of aperture, 1 in.; breadth of aperture, ¾ inch.

Habitat.— Tennessee.

Observations.— Another of the short, heavy forms in this genus, so unlike the normal type of *Io spinosa;* we think no one need confound it with any other species; its short, heavy, flattened spines jutting out like so many miniature spear-heads, and its peculiarly twisted columella will readily characterize it. The columella is also covered with a dense callous deposit, increased in thickness at its upper part and often blotched with dark red at that point; irregular, ill-defined, but broad bands are seen in the interior, often faintly visible on. the epidermis. Appears to be a rather common species in some localities, of which I possess some hundreds of specimens.—*Anthony.*

Dr. Brot considers this, and all the other species of *Io* identical with *I. fluvialis.*

Mr. Reeve suspects the specific identity of *Io brevis* and *spirostoma*, and I am convinced that the latter is only an aberration of growth like *I. gibbosa;* it is, however, a very graceful and beautiful shell.

The following is the description, together with a figure from the type specimen, of

Io spirostoma.— Shell conical, broadly ovate, horn-colored, spinous; spines short, thick, seven to eight on each whorl; whorls about nine;

Fig. 40.

aperture ovate, about half the length of the shell; columella and outer lip much and regularly twisted, and forming a well-defined sinus at base.

Length of shell, 1¾ inches; breadth of shell, 1¼ in. Length of aperture, 15–16 of an inch; breadth of aperture, ½ inch.

Habitat.— Tennessee.

Observations.— This is truly a most remarkable species of this highly interesting genus of mollusks; its difference from the ordinary type of *Io spinosa* is too marked to admit of its being confounded with that, or indeed with any other species; its stout, ovate form, short, heavy spines, and, above all, the peculiar and graceful curvature of its outer lip, are prominent characteristics and readily distinguish it.

Among several thousand specimens of *Io* in my possession, but three adult individuals of this species have been noticed, although I have a dozen or more which seem to be immature forms of it; it may therefore be considered as not only one of the most aberrant and beautiful forms of *Io*, but also one of the rarest.— *Anthony.*

5. I. turrita, ANTHONY.

Io turrita, ANTHONY, Proc. Acad. Nat. Sci., Feb., 1860, p. 69. BINNEY, Check List, No. 405. REEVE, Monog. Io, t. 3, f. 19*a*.

Fig. 41.

Description.— Shell conic, elevated, horn-colored, spinous; spines rather short and heavy, about seven on each whorl; whorls nine; aperture pyriform, about one-third the length of the shell, and irregularly banded within; columella rounded, slightly twisted and forming a short, narrow canal at base.

Habitat.— Tennessee.

Length of shell, $2\frac{1}{2}$ inches; breadth of shell, $\frac{3}{4}$ inch. Length of aperture, $\frac{7}{8}$ inch; breadth of aperture, 7–16 of an inch.

Observations.— This is the most slender and elongate species of this genus which has come under my notice, and although a single specimen only has yet been discovered, its claims to rank as a species will hardly be questioned; its long, slender form, stout, closely-set spines, and small aperture will at once distinguish it from its congeners; two faint bands traverse each whorl, one of which lies precisely in the plane of the spines; lines of growth very distinct, nearly varicose.

This species is farther removed from *Io fluvialis* than any of the others, and appears to be very distinct. Mr. Reeve's figure 19*b*, of which I have seen the original specimen, I would refer to *spinosa* rather than *turrita*. Numerous specimens occur in the collection of Mr. Lea, who is well assured, also, of its specific weight. The illustration is from the type specimen.

SPURIOUS SPECIES.

Io nodosa, Lea. ⎫
Io robusta, Lea. ⎪
Io variabilis, Lea. ⎪
Io Spillmanii, Lea. ⎪
Io modesta, Lea.   ⎬   PLEUROCERA.
Io viridula, Lea. ⎪
Io gracilis, Lea. ⎪
Io nobilis, Lea.   ⎭

Mr. Lea proposes to consider the above a distinct group of
Io, but I cannot distinguish them from *Pleurocera*. The
longer fuse, together with the sharp lip and fragile texture of
most of the shells, shows them to be immature, and indeed, as
already stated, I have had no difficulty in several instances in
identifying them with species of *Pleurocera*, by the comparison of specimens in various stages of growth.

Besides the above, numerous species of *Angitrema*, etc.,
have been referred to *Io* by European authors.

Genus ANGITREMA, HALDEMAN.

Angitrema, HALDEMAN, Cover of No. 2, Monog. Limniades, Jan., 1841.
Potadoma (sp.), Swainson, H. & A. ADAMS, Genera, i, p. 299, 1854.
Glotella, GRAY, Zool. Proc., pt. 15, p. 154, 1847.
Io (sp.), Lea, H. & A. ADAMS, Genera, i, p. 299, 1854. CHENU, Man.
      Conchyl., i, p. 290, 1859. REEVE, Monog. Io, April, 1860. BROT,
      Syst. Cat. Mel., p. 29, 1862.
Lithasia (sp.), Haldeman, H. & A. ADAMS, Genera of Recent Mollusca,
      i, p. 308, 1854.
Anculotus (sp.), Say, JAY, Cat. Shells, 4th edit., p. 276, 1850.
Melania (sp.), AUTHORS.
Juga (sp.), CHENU, Man. de Conchyl.

Description.—Shell spinous; aperture subrhomboidal, with
an anterior sinus; columella with a callous deposit anteriorly
and posteriorly.—*Hald.*

Geographical Distribution.—With two exceptions, the typical
species of this genus are confined in their geographical range

to Tennessee and Northern Alabama. These exceptions are *A. verrucosa* and *armigera*, both of which extend northward into Indiana, inhabiting the Wabash River.*

Unlike the species of *Pleurocera*, those of this genus are with one or two exceptions well defined and easily distinguishable one from another.

SYNOPSIS OF THE SPECIES OF ANGITREMA.

A. Body-whorl with a coronal of tubercles, with frequently an inferior row revolving parallel with it.

  1. *A. geniculata*, HALD.   3. *A. subglobosa*, LEA.

  2. *A. salebrosa*, CONR.   4. *A. Tuomeyi*, LEA.

B. Body-whorl encircled above the aperture by two rows of tubercles, of which the inferior one is the more prominent.

  5. *A. Jayana*, LEA.

C. Body-whorl with a central row of tubercles.

  6. *A. rota*, REEVE.   9. *A. Wheatleyi*, TRYON.

  7. *A. armigera*, SAY.   10. *A. stygia*, SAY.

  8. *A. Duttoniana*, LEA.

D. Body-whorl with numerous tubercles, in parallel rows.

  11. *A. lima*, CONR.   12. *A. verrucosa*, RAF.

A. *Body-whorl with a coronal of tubercles.*

1. A. geniculata, HALDEMAN.

Lithasia geniculata, HALDEMAN, Suppl. to No. 1, Monog. of Limniades, Oct., 1840. BINNEY, Check List, No. 299.
Anculotus geniculatus, Haldeman, JAY, Cat. Shells, 4th edit., p. 276. HANLEY, Conch. Misc., t. 5, f. 41. REEVE, Monog. Anculotus, t. 1, f. 7.
Leptoxis geniculata, Haldeman, BROT, List, p. 24.
Lithasia genicula, Lea, WHEATLEY, Cat. Shells U. S., p. 28. ADAMS, Genera, i, 308.

---

* It is a curious fact that many of the tuberculate and plicate species of *Strepomatidæ* inhabit the Wabash, so far north of their geographical centre. Mr. Lea informs me that the same curious distribution prevails with certain southern species of *Unionidæ.*

Description. — Shell short and ponderous; body-whorl crowned

Fig. 43.   Fig. 42.   Fig. 44.

with a row of conical tubercles; labium with a callus above and below; aperture elliptic, with a sinus at each extremity.

Length, ¾ inch.

Habitat.— East Tennessee.

Observations.— Differs from *Melania salebrosa,* Conrad, in having but a single row of tubercles, and a more abrupt shoulder.— *Haldeman.*

Generally but one row of tubercles is developed on this species, but occasionally a second and less prominent row is visible. The whorls are more shouldered, and the tubercles larger and less numerous than in *L. salebrosa,* Conrad. In general form it approaches *L. Tuomeyi,* Lea. It is the largest and most ponderous species of the genus.

Mr. Lea considers *geniculata* to be the same as *salebrosa.*

2. A. salebrosa, CONRAD.

Melania salebrosa, CONRAD, New Fresh-Water Shells, p. 51, t. 4, f. 5, 1834. CHENU, Reprint, p. 24, t. 4, f. 13. DEKAY, Moll. N. Y., p. 100. WHEATLEY, Cat. Shells U. S., p. 25. JAY, Cat., 4th edit., p. 274.

Anculotus salebrosus, Conrad, REEVE, Monog. Anc., t. 1, f. 6 (bad figure).

Leptoxis salebrosa, Conrad, BROT, List, p. 25.

Lithasia salebrosa, Conrad, BINNEY, Check List, No. 303. ADAMS, Genera, i, 308.

Description.— Shell short suboval; thick, ventricose, with a series

Fig. 45.   Fig. 46.

of very elevated nodes on the shoulder of the body-whorl, and generally two series of smaller nodes beneath; spire very short; apex much eroded; aperture about half the length of the shell, contracted; within purplish; columella with a callus above, and another near the base.

Observations.— This singular shell approaches the genus *Anculotus* in form, but the aperture is that of a *Melania*. I found it adhering to logs in the Tennessee River, at Florence, where it is abundant. My friend, Wm. Hodgson, Jr., found it also in the Holston River, in Tennessee.— *Conrad.*

This species is allied to No 1, but may be distinguished by its smaller size and much smaller shoulder, by its crowded tubercles, and by the constant presence of one or more inferior rows. On the other hand it is closely allied with *L. subglobosa*, Lea. Like the former, it is a very abundant species. I think the locality in East Tennessee, quoted by Mr. Conrad, an error.

3. A. subglobosa, Lea.

Lithasia subglobosa, Lea, Proc. Acad. Nat. Sci., p. 55, Feb., 1861. Jour. Acad. Nat Sci., v, pt. 3, p. 261, t. 35, f. 70. Obs., ix, p. 83.

Description.— Shell tuberculate, subglobose, thick, yellowish horn-color, double-banded; spire scarcely exserted; sutures impressed; whorls five, the last very large, towards the shoulder tuberculate; aperture large, rhomboidal, within white and double-banded, channelled at the base; columella very much thickened above and below; outer lip expanded, acute at the margin.

Fig. 47.   Fig. 48.

Operculum rather small, very dark brown, subovate, with the polar point within the lower left edge.

Habitat.— Tennessee; Prof. G. Troost.

Diameter, ·48; length, ·60 inch.

Observations.— Two specimens of this remarkably globose species have been in my possession for a long time. I had doubts of their being only the young of *Melania (Lithasia) salebrosa*, Conr., but they are so different from any young of that species which I have seen that I cannot now doubt their being entirely distinct. I know of no species which has so obtuse a spire. In this it resembles *Anculosa*, but the well characterized columella forbids its being at all confounded with any species of that genus. The callus above and below is unusually strong; below it amounts almost to a fold. One of the specimens is full grown, and has five turbercles on the shoulder of the outer half of the last whorl, and near the edge there are three above those five. The smaller one is little more than half grown, and has not as yet formed any tubercles. The two broad bands are below the row of tubercles. The last whorl is so large that it nearly covers all the others, leaving merely a point to mark the vertex. The two bands are well pronounced interiorly as well as exteriorly.— *Lea.*

Over fifty specimens of this species are before me. They are closely allied to *salebrosa*, but uniformly much smaller, and generally wider. Besides, the spire is shorter, and but very few of them exhibit a slight tendency towards tuberculation below the upper row. The whorls are not shouldered except in very old individuals. A very constant character of the species consists in the two broad, revolving bands of brown; a few specimens, however, have instead four narrow bands approximating in pairs, and two or three are of uniform color, without bands. The young differ much from the adult shells in appearance.

4. A. Tuomeyi, Lea.

Lithasia Tuomeyi, Lea, Proc. Acad. Nat. Sci., p. 55, Feb., 1861. Jour. Acad. Nat.
 Sci., v, pt. 3, t. 35, f. 68. Obs., ix, p. 8'.
Anculotus Florentianus, Lea, Reeve, Monog. Anc., t. 1, f. 4.

Description.— Shell tuberculate, much inflated, rather thick, dark horn-color, spire obtusely conoidal; sutures impressed; whorls five,

Fig. 49.

the last large, below the sutures obliquely tuberculate; aperture large, rhomboidal, whitish within, obscurely banded, channelled at the base; columella very much incurved, thickened above and below; outer lip expanded, acute at the margin.

Habitat.— North Alabama; Prof. Tuomey.

Diameter, ·64; length, 1·04 inches.

Observations.— A single specimen only was sent to me by Prof. Tuomey. It was with *L. imperialis*, herein described. Being 1·04 inches in length and ·64 in diameter, it will be seen that the proportions differ very much from that species. It cannot be confounded with *Lithasia semigranulosa*, for that species is always more raised in the spire and studded with numerous rather small tubercles. It is more closely allied to *Lithasia salebrosa*, Conr.,* but that species has a lower spire, has larger and usually more tubercles, and these,

---

*Mr. Lea considers *L. salebrosa* and *L. geniculata* dentical. It is with the latter species that the comparison is intended to be made.

if not vertical, incline to the left, while those on *Tuomeyi* are irregular and incline very much to the right, the number on the specimen before me being five on half of the last whorl. It is closely allied to *Lithasia Florentiana*, nobis, but differs much in the tubercles, in being a heavier shell, less acuminate, in being thicker on the columella and open in the channel. The *Tuomeyi* is much thicker above and below on the columella, has the obscure band within, and the outer lip is thickened and white inside the edge.

This species and *imperialis* were accompanied by many specimens of *semigranulosa* and *Florentiana*. The exact habitat was not mentioned. I have peculiar pleasure in dedicating this species to my friend, the late Professor Tuomey, whose able report on the geology of South Carolina and Alabama has justly gained him so much reputation.— *Lea.*

B. *Body-whorl encircled above the aperture by two rows of tubercles, of which the inferior one is most prominent.*

5. A. Jayana, Lea.

Melania Jayana, Lea, Philos. Proc., ii, p. 83. Philos. Trans., ix, p. 20. Obs., iv,
    p. 20. Wheatley, Cat. Shells U. S., p. 25. Jay, Cat. Shells, 4th edit., p. 274,
    Binney, Check List, No. 154.
Io Jayana, Lea, Brot, List, p. 29. Mal. Blatt., v, 115, 1860.
Melania robulina, Anthony, Bost. Proc., iii, p. 263, Dec. 1850. Binney, Check List,
    No. 230.
Io robulina, Anthony, Reeve, Monog. Io, sp. 15. Chenu, Man. Conchyl., i, f. 1976.

Description.— Shell tuberculate, subfusiform, thick, pale horn-color; spire exserted; sutures linear and curved, whorls rather convex; impressed in the middle, surrounded by a double series of tubercles; columella incurved, thickened above; aperture trapezoidal, whitish within.

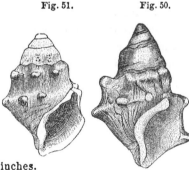

Fig. 51.   Fig. 50.

Habitat.— Caney Fork, De-Kalb county, Tennessee.

Diameter, ·78; length, 1·20 inches.

Observations.— Dr. Jay had two specimens of this species, and I owe to his kindness the possession of one of them. It very closely resembles the *M. armigera* (Say), in most of its characters, but may at once be distinguished by the double row of tubercles, the *armigera* never possessing distinctly more than one row; below the sutures, however, there are sometimes imperfect tubercles, which are caused by the protrusion of the tubercles of the superior whorl. This protrusion also takes place in the *Jayana*, but causes in it only a constant curvature in the linear suture.

The apex of the specimen is much eroded, and consequently I am not sure of the number of the whorls, probably eight or nine. The aperture may be rather more than one-third the length of the shell, and is acutely angular at the base, with rather a deep sinus. The callus above causes a considerable sinus there.

The operculum is dark brown, the radii converging at the lower interior edge.—*Lea.*

This shell and Mr. Anthony's *M. robulina* are entirely identical in every respect, the species being a very constant one in all its characters, as I am unable to select from a considerable number of specimens any which exhibit variations from the type form. It is an exceedingly abundant species, and very remarkable for its peculiar armature and the narrowed canal, suggestive of the genus *Io.*

The following is the description of

Melania robulina. — Shell solid, ovately rhomboidal, corneous, encircled with brown bands; whorls six, bearing a double series of nodules, the upper one immersed in the suture; aperture rhomboidal produced into a rostrum, callous behind.

Habitat.— Cumberland River, Tennessee.

Long. 1; lat. 5-8 poll.

Observations.—Of the same size as *M. armigera*, Say, but differs in coloration; the rostrum is much longer, and the posterior series of tubercles much more developed.— *Anthony.*

C. *Body-whorl with a central row of tubercles.*

6. A. rota, REEVE.

Io rota, REEVE, Monog. Io, sp. 13, April, 1860. BROT, List, p. 29.

Description.— Shell globosely turreted, thick, ponderous, yellowish, encircled at the base by a brown band, olive; whorls few, rudely concavely sloping, faintly striated, encircled round the periphery with large, obliquely compressed tubercles; columella short, but little twisted.

Fig. 52.

Habitat.— United States.

Observations.— A solid, globosely turreted shell, prominently armed with tubercles, which are compressed obliquely into fans, like the fans of a water-wheel.— *Reeve.*

The figure is copied from Reeve. I have never seen this species, the type of which was in the collection of the late Hugh Cuming, Esq., London; it may be only a remarkable specimen of *A. Jayana*, Lea.

7. A. armigera, SAY.

Melania armigera, SAY, Jour. Acad. Nat. Sci , 1st ser., ii. p. 178, Jan., 1821 BINNEY'S Reprint, p. 71. BINNEY, Check List, No. 21. DEKAY, Moll. N. Y., p. 93. JAY, Cat , 4th edit., p. 272. TROOST, Cat. WHEATLEY, Cat. Shells U. S., p. 24. CATLOW. Conch. Nomenc., p. 185. HANLEY, Conch. Misc. Melania, t. 7, f. 60.
Io armigera, Say, REEVE, Monog. Io, f. 11. ADAMS, Genera, i, 299.

Description.— Shell tapering, brownish horn-color; volutions about six, slightly wrinkled; spire near the apex eroded, whitish; body-whorl with a revolving series of about five or six distant, prominent tubercles, which become obsolete on the spire, and are concealed by the revolution of the succeeding whorls, in consequence of which arrangement there is the appearance of a second, smaller, and more obtuse subsutural series of tubercles on the body-whorl; two or three obso-

Fig. 53a.   Fig. 53.

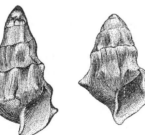

lete revolving reddish-brown lines; aperture bluish-white within; a distinct sinus at the base of the columella.

Habitat.—Ohio River.

Length about one inch.

Distinguished from other North American species, by the armature of tubercles.—*Say.*

This beautiful and extensively distributed species is allied only to *L. Duttoniana*, Lea (for distinctive characters see description of that species); from all others it is very distinct. Besides the original locality, Jay and Troost give Tennessee, and Mr. Wheatley, Kentucky, as its habitat. I have before me a series of the young shells presented to the Philad. Acad. Nat. Sciences, by Mrs. Say, which were collected in the Wabash River, Ind.

This shell Prof. Haldeman has made the type of his subgenus *Angitrema*. He has also (Icon. Encyc., ii, p. 84) referred it to Rafinesque's genus *Pleurocera*.

8. A. Duttoniana, Lea.

Melania Duttoniana, LEA, Philos. Proc., ii, p. 15. Philos. Trans., viii, p. 189, t. 6, f. 54. Obs., iii, p. 26. CATLOW, Conch. Nomenc., p. 186. BINNEY, Check List, No. 92. JAY, Cat. 4th edit., p. 273.

Io Duttoniana, Lea, REEVE, Monog. Io, f. 9. BROT, List, p. 29. CHENU, Man. Conchyl., i, f. 1974.

Io fasciolata, REEVE, Monog. Io, f. 14.

Description.— Shell tuberculate, fusiform, rather thick, yellowish,

Fig. 54.   Fig. 54*a*.

banded; spire elevated, pointed at the apex; sutures irregularly lined; whorls seven, depressed above; aperture elongated, angular and channelled at the base, within whitish.

Habitat.— Waters of Tenn. Duck River, Maury Co., Tenn.

Diameter, ·57; length, 1·09 inches.

Observations.— This is a beautiful species. The most perfect specimens are remarkable for their fusiform shape and their long aperture, which presents a curved columella and extended sinus somewhat like the genus *Io.* The bands in some individuals are numerous and distinct, the largest being nearest the base. The tubercles form a row round the middle of the whorls of most specimens, but in some, though rarely, this part is carinate or rounded. Some are slightly tuberculated below the suture. Among the young specimens some are costate near the apex, others entirely smooth and without bands. I owe the fine specimen figured to Mr· Dutton, after whom I name it.— *Lea.*

This species is smaller and more fragile than *L. armigera.* It is also elegantly banded, which is more rarely the case with *armi-gera;* and it differs also in having smaller, frequently obsolete tubercles, and in the aperture being much less channelled.

Fig. 55.

I do not hesitate in agreeing with M. Brot in considering *fasciolata*, Reeve, as a synonyme.

The original description and copy of Reeve's figure are given below.

Io fasciolata.— Shell shortly fusiform, yellowish-green, encircled with narrow bands of olive, whorls 5 to 6, convexly sloping, the first smooth, the last gibbously angled, tubercled at the periphery, tubercles distant; aperture diamond-shaped, scarcely channelled.

Habitat.— United States.

Observations.— Closely allied to *L. Duttoniana*, but less channelled, and more widely apertured, owing to the more gibbously angled circumference of the last whorl.— *Reeve.*

9. A. Wheatleyi, TRYON.

Angitrema Wheatleyi, TRYON, Am. Journal of Conchol., vol. ii, p. 4, t. 2, f. 1, 1866.

Description.— Shell conoidal, inflated, rather thin; spire conical, sharp pointed, suture not much impressed; whorls about six, those of the spire flattened, the body-whorl large, rather flattened above the somewhat angled periphery, con-vex below, and somewhat attenuate at the base; the periphery is ornamented with a single prominent row of slightly compressed tubercles, and above is rugosely wrinkled, with a tendency towards tuberculation; aperture large, subrhomboidal, half the length of the shell, somewhat attenuate below, columella nearly perpendicular, a little twisted. Bright horn-color, with four broad, equidistant brown bands.

Fig. 56.

Habitat.— Elk River, at Winchester, Tenn.

Diameter, 16 mill.; length, 25 mill.

Observations.—This species is much more inflated, and has more numerous tubercles than *A. Duttoniana*, Lea; it is in appearance more like an obese variety of *A. verrucosa*, Raf., but that species is heavier in texture, and has several rows of tubercles. The well-developed tubercles and inferiorly contracted aperture will readily distinguish this species from *Lithasia fuliginosa*, Lea.— *Tryon.*

10. A. stygia, SAY.

Melania stygia, SAY, New Harmony Dissem., p. 261. Aug. 28. 1829; reprint, p. 17.
   BINNEY'S Reprint, p. 142. BINNEY, Check List, No. 251. WHEATLEY, Cat.
   Shells U. S., p. 27. JAY, Cat., 4th edit., p. 275. DEKAY, Moll. N. Y., p. 93.
   REEVE, Monog. Mel., sp. 400. BROT, List, p. 40.
Melania tuberculata, LEA, Philos. Trans., iv, p. 101, t. 15, f. 31. Obs., i, p. 111.
   DEKAY, Moll. N. Y., p. 93. WHEATLEY, Cat. Shells U. S., p. 27. BINNEY, Check
   List, No. 277. JAY, Cat., 4th edit., p. 275. CATLOW, Conch. Nomenc., p. 189.
Juga tuberculata, Lea, CHENU, Man., Conchyl., i, f. 2017.
Melania Spixiana, LEA, Philos. Trans., vi, p. 93. Obs., v, p. 93.
Melania nodata, REEVE, Monog. Mel., fig. 422.
Io tuberculata, ADAMS, Genera, i, 299.

Description.— Shell robust, ovate conic, black; spire rather larger than the aperture, eroded at tip; volutions five, hardly convex; wrinkles obsolete, excepting a few larger ones; suture not profoundly indented; aperture narrowed at base into a slight sinus and suban-

Fig. 57.

gulated; much wider in the middle; labrum much arcuated in the middle.

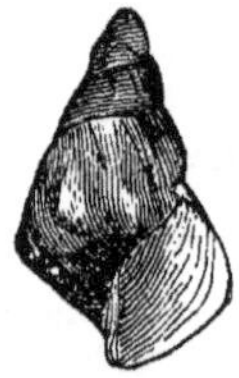

Greatest breadth, less than half an inch, length, three-fourths.

Observations.— A specimen of this shell was given to me by Mr. Lesueur; several were found in Cumberland River by Dr. Troost. In form it resembles *armigera*, nob., more than any other species, but that shell is armed with tubercles and ornamented by colored lines, its suture also is only a simple line.— *Say.*

The following is Mr. Lea's description of

Io tuberculata.—Shell obtusely turreted, wide, very dark brown or black; apex obtuse; whorls, five; middle of the last whorl furnished with tubercles; outer lip irregularly curved; base angulated; aperture purple and one-half the length of the shell.

Habitat.— Tennessee River; Prof. Vanuxem.

Diameter, ·5; length, ·9 of an inch.

Observations.— This species is somewhat allied to the *M. armigera*

(Say), but is smaller and much less ponderous. The tubercles are more numerous and less elevated.

In the *tuberculata* the impressed band, which exists in the *armigera* above the armature, is wanting. In color it differs altogether.—*Lea*'

In Phil. Trans., vi, p. 82, Mr. Lea changed the name of his species, as the original name was preoccupied by Spix. He therefore proposed, instead of *tuberculata*, the name *Spixiana*. Mr. Reeve, finding *tuberculata* preoccupied by Spix, and not having seen Mr. Lea's change of name, proposed *nodata*. These names must all yield, however, to Say's *stygia*, which is the first published description of the species. Mr. Say himself (cover of Conchology, No. 6) decided Mr. Lea's species to be a synonyme—an opinion in which he has been sustained by several of our conchologists.

Through the kindness of Mr. Lea I have been permitted to examine a number of specimens in his cabinet. They exhibit every gradation, from a smooth to a tuberculate surface.

D. *Body-whorl with numerous tubercles, in parallel rows.*

11. A. lima, CONRAD.

Melania lima, CONRAD, New Fresh-Water Shells, p. 54, t. 8, f. 8, 1834. CHENU, Reprint. DEKAY, Moll. N. Y., p. 97. WHEATLEY, Cat. Shells U. S., p. 26. JAY, Cat., 4th edit., p. 274. CATLOW, Conch. Nomenc., p. 187. BROT, List, p. 33. MULLER, Synopsis, p. 46.
Anculotus lima, Conrad, REEVE, Monog. Anc., t. 1, f. 1.
Lithasia lima, Conrad, BINNEY, Check List, No. 300.
Megara lima, Conrad, ADAMS, Genera, i, 306.

Description.— Shell conic, or subfusiform; with approximate nodulous, spiral lines of unequal size; body-whorl angulated; angle with a series of prominent tubercles; base with two lines, the superior one nodulous; aperture nearly half the length of the shell, contracted, and acutely angular above, and obtusely pointed at base; labrum very thin; color olive; within with purple bands.

Fig. 58.

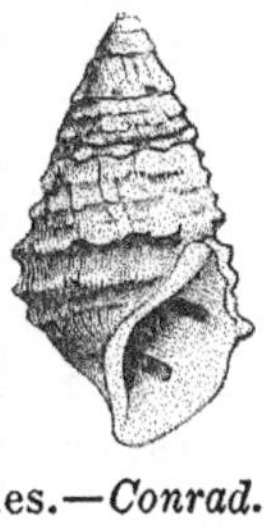

Observations.— A fine species, easily recognized by its numerous tubercles, and ventricose form. Inhabits Elk River, Alabama, adhering to stones, and is a common species.—*Conrad.*

Distinguished from *L. verrucosa*, Raf. (*nupera*, Say), by its angulated body-whorl, conical spire, acute apex, and by the irregularity in the size of its tubercles.

Mr. Reeve originally described this species as *nupera*, and *vice versa*, but subsequently corrected the error. It occurs also in Tennessee River.

12. A. verrucosa, Rafinesque.

Pleurocera verrucosa, RAFINESQUE, Annals of Nature, p. 11, 1820.
Melania nupera, SAY, New Harmony Dissem., p. 260. Amer. Conch., pt. 1, t. 8, f. 1, 2.
   BINNEY'S Repiint, p. 157, t. 8.   CHENU'S Reprint, p. 16, t. 2, f. 3.   DEKAY,
   Moll. N. Y., p. 97.   WHEATLEY, Cat. Shells U. S., p. 26.   BROT, List, p. 40.   JAY,
   Cat. Shells, 4th edit., p. 274.

Description.— Ellipsoidal, top very obtuse, base of the opening obtuse, inside lip thickly plaited; four spires, the last two flattened,

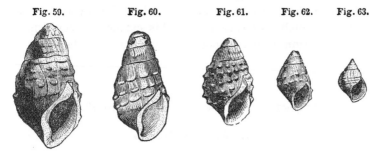

| Fig. 59. | Fig. 60. | Fig. 61. | Fig. 62. | Fig. 63. |

the other large, with several rows of warts; back of the opening wrinkled; color olivaceous-brown, opening whitish.

Habitat.— The lower parts of the Ohio.

Length, about two-thirds of an inch, not quite double the breadth.— *Rafinesque.*

With no disposition to give place to the description of Mr. Rafinesque, at the expense of naturalists of honesty and reputation, I am still constrained, in this instance, to quote his name for the shell that is so well known amongst us as Mr. Say's *nupera*. Indeed, I cannot find any description of a species of shell, by Rafinesque, which indicates so unmistakably the shell intended by him, as does the one here quoted. It may be mentioned, not as proof in itself, but merely as collateral evidence of the correctness of my views of this species, that in a manuscript by Rafinesque, entitled "Conchologia Ohioensis," belonging to the Smithsonian Institution, a rough pen sketch of *Pleurocera verrucosa* is given, which is a very good representation of Mr. Say's *nupera*.

The description of the latter species is as follows:—

Melania nupera. — Shell oblong suboval; volutions five, slightly rounded; body-whorl with about three revolving series of subequal, equidistant granules or tubercles, not higher than wide, occupying the superior portion of the surface; second volution with but two series; remaining volutions with slightly elevated, longitudinal lines instead of tubercles, often obsolete; spire decorticated towards the tip; suture not deeply impressed; aperture longer or as long as the spire; sinus of the superior angle profound; labium concave, with a callus near the superior angle; columella with a slight, obtuse, hardly prominent angle above the incipient sinus, which is obvious; labrum not abbreviated above, nor much produced near the base.

Observations. — This species is common in the Wabash River; the spire is almost invariably so much decorticated that no trace of the longitudinal lines remains; in the young only are the lines distinct, and even in these they are sometimes obsolete or altogether wanting. It varies in the number of its series of tubercles, some specimens having but one, and others, though these are rare, as many as five or six.— *Say.*

Melania Holstonia.—Shell grained, conical, somewhat thick, black; spire somewhat elevated; sutures impressed; whorls flat- Fig. 64. tened above; aperture ovate, purple.

Habitat. — Holston River, Tennessee.

Diameter, ·38; length, ·79 of an inch.

Observations. — A very distinct species with four series of small, rather sharp elevations round the whorls, the two inferior ones rather indistinct. Only two specimens have come under my notice, and both have the apex decollated.—*Lea.*

The figure of *Holstonia* is copied from Mr. Lea's plate. The locality of "Holston River, Tenn.," may well be doubted.

The species is a very common one in North Alabama, and exhibits considerable variation in size and proportions. A specimen in Coll. Haldeman is labelled "Nashville."

As for Deshayes' *Melanopsis semigranulosa*, its identity is proved by his quotation of Mr. Say's species as a synonyme, in his description. Say published in 1829, Deshayes in 1830. It therefore appears that the great French naturalist, upon removing the species to the genus *Melanopsis*, seized the occasion to deprive Mr. Say of his species, a meanness that has unfortunately found many advocates amongst naturalists (?) whose sole ambition appears to be, to write "nobis" as frequently as possible. But, like M. Deshayes, these gentlemen, although sometimes successful for a period, will all eventually find themselves quoted where they have placed the authors they have endeavored to despoil, — *among the synonymes.*

Subgenus LITHASIA, HALDEMAN.

Lithasia, HALDEMAN, Supplement to Monog. Limniades, No. 1, Oct. 1840. BINNEY, Check List of Fluviatile Univalve Shells, June, 1860. LEA, Proc. Acad. Nat. Sci., p. 54, Feb. 1861. Jour. Acad. Nat. Sci., v, pp. 258 and 354, March, 1863. Observations, ix, pp. 80 and 176, March, 1863.
Lithasia, Haldeman (part.), H. & A. ADAMS, Genera, i, p. 308, Feb., 1864.
Lithasia, Lea, 1845, CHENU, Man. Conchyl., i, p. 296, 1859.
Megara (part.), ADAMS, Genera, i, p. 306, Feb., 1854.
Anculotus (sp.), Say, GRAY, Genera, Zool. Proc., pt. 15, p. 153, 1847. REEVE, Monog., April, 1860.
Anculosa (sp.), Say, AUCT.
Melania (sp.), AUCT.

Description. — Shell ovately fusiform or oval, small, smooth. Aperture not so distinctly channelled in front as in the typical *Angitremæ*. Columella with an anterior and posterior callous deposit.

Geographical Distribution. — Like the typical species, we find the *Lithasiæ* inhabiting principally the waters of Tennessee and North Alabama; but one of the species is completely separated from the geographical area of the group, its habitation being confined to the Ohio River and tributaries. This shell, *L. obovata*, is somewhat removed from the general type, but is connected with it, by *L. undosa*, a Kentucky species. Another allied shell, *L. consanguinea*, has heretofore been found in Indiana only.

SYNOPSIS OF SPECIES.

A. Shell large, ovate, inflated.

1. L. FULIGINOSA, Lea, Reeve, sp. 401.
2. L. FLORENTIANA, Lea. Not of Reeve, Anculotus, fig. 4.
3. L. VENUSTA, Lea.
4. L. DILATATA, Lea.
5. L. IMPERIALIS, Lea.

B. Shell small, compact, oval-elliptical, thick.

6. L. VITTATA, Lea.
7. L. SHOWALTERII, Lea, Reeve, Melania, fig. 421.
8. L. NUCLEOLA, Anthony, Reeve, Melania, fig. 348.
9. L. OBOVATA, Say, Reeve, Anculotus, fig. 21. *L. Hildrethiana*, Lea, *L. undosa*, Anthony, Reeve, Melania, fig. 447. *L. rarinodosa*, Anthony (Manuscript), Reeve, Melania, fig. 268. *L. consanguinea*, Anthony, Reeve, Anculotus, fig. 2.

C. Shell obliquely flattened.

10. L. COMPACTA, Anthony, Reeve, Melania, fig. 343.
11. L. NUCLEA, Lea, Reeve, Melania, fig. 423.

D. Shell subcylindrical.

12. L. BREVIS, Lea, Reeve, Melania, fig. 344. *L. solida*, Lea, *non* Reeve, Melania, fig. 454.
13. L. FUSIFORMIS, Lea.
14. L. DOWNIEI, Lea.

A. *Shell large, ovate, inflated.*

1. L. fuliginosa, LEA.

Melania fuliginosa, LEA, Philos. Proc. Philos. Trans., viii, p. 170, t. 5, f. 17. Obs., iii, p. 8. DEKAY, Moll. N. Y., p. 94. TROOST, Cat. WHEATLEY, Cat. Shells U. S., p. 25. BINNEY, Check List, No. 113. CATLOW, Conch. Nomenc., p. 186. BROT, List, p. 40. REEVE, Monog. Melania, sp. 401.
Leptoxis fuliginosa, Lea, ADAMS, Genera, i, p. 307.

Description. —Shell smooth, fusiform, somewhat inflated, rather thick, dark brown; spire obtuse; sutures impressed; whorls six, somewhat convex; aperture large, at the base angular and channelled.

Fig. 65.

Habitat. —Big Bigby Creek, Maury Co., Tenn.

Diameter, ·50; length, ·85 of an inch.

Observations. —In general form this species resembles the *M. Duttoniana* (nobis), but differs in being less elevated in the spire, in being without tubercles, and of a very dark color; the substance of the shell is disposed to be purple. The epidermis is thick and very dark. Mr. Dutton found it rare. —*Lea.*

I was at first disposed to consider this the same as *L. Florentiana*, Lea; but it appears to be always colored differently, being darker, with, generally, broad brown bands, and sometimes the general surface is brilliant green ornamented with the bands, while *Florentiana* is of uniform color. This species also differs from *Florentiana* in being more inflated.

2. L. Florentiana, Lea.

Melania Florentiana, LEA, Philos. Proc. Philos. Trans., viii, p. 188, t. 6, f. 53. Obs., iii, p. 26. DeKAY, Moll. N. Y., p. 99. WHEATLEY, Cat. Shells U. S., p. 23. BINNEY, Check List, No. 110. CATLOW, Conch. Nomenc., p. 186. BROT, List, p. 40.
Io Florentiana, Lea, H. & A. ADAMS, Genera, i, p. 299.

Description.— Shell tuberculate, elliptical, ponderous, pale; spire obtuse; sutures impressed; whorls six, slightly convex; aperture elongated, whitish.

Fig. 66.

Habitat.— Tennessee River, Florence, Alabama.

Diameter, ·47; length, ·87 of an inch.

Observations.— An elliptical species resembling the *M. olivula*, Conrad. Its aperture is so much elongated as to be more than half the length of the shell. Three of the specimens are without bands, a fourth has several very indistinct ones. The whorls are somewhat flattened on the superior part and are disposed to be tuberculated below the sutures. In the young the tubercles are more distinct. In some of the adult specimens they are entirely wanting.—*Lea.*

This species is well represented now, in our cabinets, and very seldom exhibits the tuberculation which appears to have faintly characterized Mr. Lea's first specimens. Reeve's fig. 4, of *Anculosa Florentiana*, more properly represents *L. Tuomeyi*, Lea.

3. L. venusta, Lea.

Melania venusta, LEA, Philos. Proc. Trans., viii, p. 187, t. 6, f. 52. Obs., iii, p. 25. DeKAY, Moll. N. Y., p. 99. JAY, Cat. 4th edit., p. 275. TROOST, Cat. WHEATLEY, Cat. Shells U. S., p. 27. BINNEY, Check List, No. 285. CATLOW, Conch. Nomenc., p. 189. BROT, List, p. 40. REEVE, Monog. Melania, sp. 315.

Description.— Shell disposed to be tuberculate, fusiform, somewhat thin, yellowish above; spire rather obtuse; sutures roughly impressed; whorls six, convex; aperture elongated, at the base angulated and channelled, within whitish.

Habitat.— Tennessee.

Diameter, ·43; length, ·80 of an inch.

Observations.— Dr. Troost sent me a single specimen of this species which is very distinct, the columella is very much thick-ened, particularly above, in which it resembles the genus *Melanopsis.* The aperture is rather more than half the length of the shell. In this specimen a single obscure band may be observed within, close to the base of the columella.— *Lea.*

Fig. 67.

This species is more narrowly cylindrical than *L. Florentiana;* besides, it is lighter colored, heavier in tex-ture, with the two deposits of callus on the columella more prominent and the canal narrower and better developed. It is a rather rare species.

4. L. dilatata, LEA.

Lithasia dilatata, LEA, Proc. Acad. Nat. Sci., p. 55, 1861. Jour. Acad. Nat. Sci., v, pt. 3, p. 260, t. 35, f. 69. Obs., ix, p. 82.

Description.— Shell smooth, subglobose, rather thick, grayish-green, yellowish below the sutures, obscurely banded; spire obtusely conical; sutures irregularly impressed; whorls five, the last one large and ventricose; aperture large, subrhomboidal, brownish within and angular at the base; columella thick-ened above and below, incurved; outer lip sharp and much dilated.

Fig. 68.

Habitat.— Tennessee; Dr. Troost.

Diameter, ·45; length, ·73 of an inch.

Observations.— This is a well-characterized species, nearly allied to two species which I described some years since, before *Lithasia* was established, under the names of *Melania Florentiana* and *M. venusta,* both of which must be removed to the well recognized genus *Lithasia.* It is nearest to the former, but is more globose, more glaucous and darker inside, and has a larger callus above. The bands on this species are very obscure, and are, indeed, simply the general color interrupted by light, transverse, fine lines. On the upper part of the body-whorl there are several low tubercles, which may not be found in all the individuals of this species. The callus above is tinted with brown. The outer lip is bordered with white. The length of the best specimen is nearly three-quarters of an inch, and the aperture is more than half the length of the shell. — *Lea*

The type of Mr. Lea's description I have figured. It is, I

think, a good species, although very close to *L. fuliginosa*. It appears to be a more solid shell than that species, however, and the aperture is narrower below, with a more distinct fuse.

5. L. imperialis, LEA.

Lithasia imperialis, LEA, Proc. Acad. Nat. Sci., p. 55, 1861.   Jour. Acad. Nat. Sci., ·
v, pt. 3, p. 258, t. 35, f. 67. Obs., ix, p. 80.

Description.— Shell tuberculate, fusiform, rather thick, dark horn-color; spire raised, conoidal; sutures irregularly and much impressed; whorls six, the last rather large, irregularly tuberculate above,

Fig. 69.

rather inflated; aperture rather small, elongately rhomboidal, whitish within, furnished with brown hair-like lines, channelled at the base and recurved; columella sigmoid, slightly thickened above; outer lip somewhat expanded, acute at the margin.

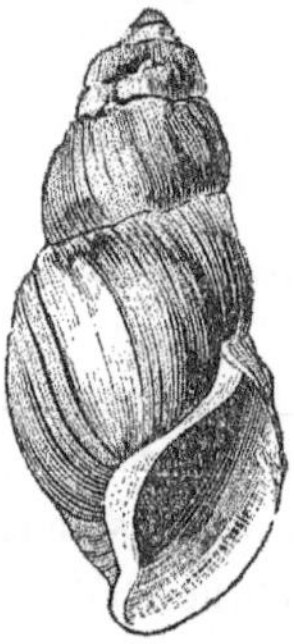

Operculum rather small, very dark brown, rhomboidal, with the polar point on the left edge near the base.

Habitat.—North Alabama; Prof. Tuomey.

Diameter, ·70 of an inch; length, 1·55 inches.

Observations.— This is much the largest *Lithasia* I have seen. Although several of the whorls of the vertex are eroded off, still it measures one and a half inches in length. A single specimen only was received, and this without the operculum. The tubercles are large and irregular, and not much raised. The capillary brown lines in the interior are numerous and rather obscure, but this may not be the case with more perfect specimens. They seem to replace the usual bands. They do not reach the edge, which is bordered with white. Below the sutures there is a stricture which nearly amounts to a furrow. It more nearly resembles *Melania* (*Lithasia*) *Duttonia* (nobis), than any other known species, but is a larger, more ponderous species, and has not the numerous small tubercles, nor the bands of that species.—*Lea.*

B. *Shell small, compact, oval-elliptical.*

6. L. vittata, LEA.

Lithasia vittata, LEA, Proc. Acad. Nat. Sci., p. 273, 1862. Jour. Acad. Nat. Sci.,
v, pt. 3, p. 354, t. 35, f. 67. Obs., ix, p. 176.

Description.— Shell smooth, cylindrical, rather thin, dark horn-color, four-banded; spire short, decollate; sutures irregularly impressed; whorls flattened, the last very large; aperture large, rhomboidal, whitish within and much banded; outer lip acute; columella thickened, white, incurved.

Operculum ovate, thin, light brown, with the polar point on the inner edge near to the base.

Habitat. — Coosa and Cahawba Rivers, Alabama; E. R. Showalter. Diameter, ·40; length, ·88 ? of an inch.

Observations. — This is a beautifully banded species, which is so near to *brevis* (nobis) in size and outline that I considered it at first as a strongly marked variety of that species. From examination now of about a dozen specimens before me, sent by Dr. Showalter and Dr. Lewis, I am perfectly satisfied that this is a distinct species. All the specimens I have seen have four well expressed dark brown bands, which are strongly exhibited within. All the specimens are so much worn at the apex that it is impossible to say how many whorls they naturally have. There is a great difference in the form of the apertures of the specimens before me, — some have quite an angular base, while others are rounded almost like a *Melania.* The aperture is probably two-thirds the length of the shell.— *Lea.*

Fig. 70.

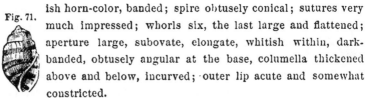

7. L. Showalterii, Lea.

Lithasia Showalterii LEA, Proc., Acad. Nat. Sci., p. 188, 1850. Jour. Acad. Nat. Sci., v, pt. 3, p. 202, t. 35, f. 72. Obs. ix, p. 84.
Melania Showalterii, Lea, REEVE, Monog., sp. 423. BROT, List, p. 33.

Description. — Shell smooth, ovately cylindrical, rather thick, yellowish horn-color, banded; spire obtusely conical; sutures very much impressed; whorls six, the last large and flattened; aperture large, subovate, elongate, whitish within, dark-banded, obtusely angular at the base, columella thickened above and below, incurved; outer lip acute and somewhat constricted.

Fig. 71.

Habitat.— Cahawba River, at Centreville, Alabama; E. R. Showalter. Diameter, ·38; length, ·70 of an inch.

Observations.— This species presents a number of varieties, but the character of the flattened enlarged side, frequently producing quite a large shoulder, is generally preserved. Sixteen out of nineteen specimens before me have very much the same character of bands, viz.:

three broad, nearly equal, distant, revolving ones.   The other three lose all the yellowness of the epidermis, and present an intensely deep purplish brown hue inside and out.   The largest of these three has a more constricted aperture than any of the others, and it has revolving striæ more distinct towards the base, which I have not observed in the others.   The aperture is also quite channelled below, which is indistinct in the others.   Another of these three dark specimens has a higher spire and a shorter aperture, leaning towards the form of a *Melania*.   The shoulder in many of the specimens is large and well pronounced, while in others it is small.   The aperture is about two-thirds the length of the shell.   This species reminds one, as to its outline, of *Melania undosa*, Anth., from Kentucky.   It is, however, larger, more cylindrical and has the callus on the columella, which *undosa*, of course, has not.   *Undosa* is also much paler and has a higher spire.   I have great pleasure in dedicating this species to Dr. Showalter, who is doing so much for the natural history of his adopted state.— *Lea.*

This species resembles the preceding, but is less cylindrical, with the aperture wider, and the outer lip more curved.   The spire is shorter and more rapidly acuminate.

8. L. nucleola, ANTHONY.

Melania nucleola, ANTHONY, Proc. Bost. Soc. Nat. Hist., iii, p. 360, Dec., 1850. BIN-
NEY, Check List, No. 181. BROT, List, p. 40. REEVE, Monog., sp. 348.

Description.— Shell small, thick, eroded, subglobose or subcylindrical, smooth, greenish, encircled by two bands; whorls 2-3, ventricose, the last at length cylindrical; aperture

Fig. 72.   Fig. 73.

semilunar; lip dilated in front, thickened behind; columella with a copious callous deposit.

Habitat.— Tennessee.

Longitude, ½; latitude, ⅜ of an inch.

Observations.— This species, which resembles closely *L. nuclea*, Lea, may be distinguished by being rather larger; differently colored, being light brown; while *nuclea* has a tinge of green; by having two chestnut-colored bands in place of the four dark ones of Mr. Lea's species; and by the columella being not so much thickened.   It is a rare species, whilst *nuclea* appears to be rather an abundant one.

Belongs to a group of solid, ellipsoidal species peculiar to the region of Lower Tennessee and Alabama. It has a very sparing development of the spire, and a remarkable flattening about the middle of the last whorl.—*Anthony.*

9. L. obovata, SAY.

Melania obovata, SAY, New Harmony Dissem., No. 18, p. 276, Sept. 9, 1829; Reprint, p. 18, 1840. BINNEY'S Reprint. p. 143. DEKAY, Moll. N. Y., p. 98. WHEATLEY, Cat. Shells U. S., p. 26. CATLOW, Conch. Nomenc., p. 188. JAY, Cat., 2d edit., p. 45.
Anculotus obovatus, Say, JAY, Cat., 4th edit., p. 276. REEVE, Monog. Mel., f. 21.
Leptoxis obovata, Say, HALD., Monog. Lept., p. 2, t. 1, f. 27-34. BINNEY, Check List, No. 374. BROT, List, p. 25.
Lithasia obovata, Say, CHENU, Manuel, i, f. 2056-8. ADAMS, Genera, i, 308.
Anculosa obovata, Say, WHEATLEY, Cat. Shells U. S., p. 26.
Melania Hildrethiana, LEA, Philos. Proc. Philos. Trans., viii, p. 164, t. 5, f. 1. Obs., iii, p. 2, t. 5, f. 1. DEKAY, Moll. N. Y., p. 92. WHEATLEY, Cat. Shells U. S., p. 25. BINNEY, Check List, No. 138. CATLOW, Conch., Nomenc., p. 187.
Leptoxis Hildrethiana, Lea, ADAMS, Genera, i, p. 307.
Melania undosa, ANTHONY, Ann. N. Y., Lyc., vi, p. 124, t. 3, f. 25, March, 1854. BINNEY, Check List, No. 280. BROT, List, p. 39. REEVE, Monog. Mel., sp. 447.
Melania rarinodosa, Anthony, MSS., REEVE, Monog., sp. 268. BROT, List, p. 39.
Melania consanguinea, ANTHONY, Ann. N. Y. Lyc., vi, p. 125, t. 3, f. 26, March, 1854. BINNEY, Check List, No. 66. BROT. List, p. 39.
Anculotus consanguineus, Anthony, REEVE, Monog. Anc., sp. 2.

Description.— Shell subobovate, dark brown or blackish, volutions nearly five; spire remarkably rounded, short; body-whorl with a very obtuse, slightly indented band or undulation, a little above the middle; aperture more than twice the length of the spire, narrow; labium polished, with a callus above; labrum not projecting near the base, subrectilinear from the shoulder to the basal curve, very convex at the shoulder; base rounded and without indentation.

Fig. 74.

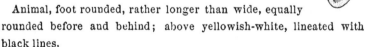

Animal, foot rounded, rather longer than wide, equally rounded before and behind; above yellowish-white, lineated with black lines.

Habitat.— Kentucky River, and some other tributaries of the Ohio.

Length, three-fourths; breadth, nearly half an inch.

Var. A. Indented band almost obsolete.

Observations.— The spire, and even a part of the body-whorl in all old shells, are sometimes remarkably eroded, as in the *M. (Anculotus) prærosa,* nob., and indeed, the general appearance is such, that at a little distance, and without particular observation, it might be readily mistaken for that shell, but the form is less globular, and the aperture is altogether different. I found it very abundant in Kentucky River in company with that shell and other species of *Melania*. I also observed it at the falls of the Ohio.. Lesueur and Troost obtained specimens in Fox River of the Wabash. When young, the undulation is hardly visible, and the shell is often of a dull yellowish color, which on the larger volutions becomes gradually of the characteristic color.—*Say.*

Melania Hildrethiana, Lea, is the half grown stage of this species, as I have verified, by an examination of Mr. Lea's original specimens, one of which he kindly presented to me (see figure). In uniting it with *obovata*, it is proper to say that Prof. Haldeman and Dr. Jay have preceded me.

The following is Mr. Lea's description of

Melania Hildrethiana.—Shell smooth, fusiform, rather thick, horn-color; spire short, pointed at the apex; sutures deeply impressed; whorls five, convex; aperture large, angular at base, ovate, white or purple.

Fig. 78.

Habitat.—Ohio River, near Marietta; Dr. Hildreth.

Diameter, ·25; length, ·37 of an inch.

Observations.—The aperture of this little species is nearly two-thirds the length of the shell. In outline it is allied to *M. fusiformis*, herein described. It may be distinguished by the sutures being more impressed, and the base being more angular. One of the specimens is purple on the columella and at the base. I dedicate it to Dr. Hildreth, to whose kindness I owe several specimens.—*Lea.*

This is nothing more than a small variety of *L. obovata*, Say. I have not seen many specimens, but they all appear to be of stunted growth, and I should not be surprised if future research proves them to be living in circumstances unsuited to their full development.

The following description is of a not entirely full grown shell, retaining the spire complete to the apex. It is a rare state, several whorls being generally lost by truncation.

The remarkably shouldered whorls and smaller size   Fig. 76.
of *M. undosa* will scarcely distinguish it as a variety of this species. Its description here follows :—

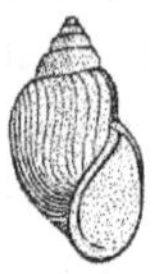

Melania undosa.—Shell ovate, smooth, olivaceous, moderately thick; whorls 6-7, rapidly converging to the apex, convex; body-whorl ample, with a distinct, but somewhat rounded shoulder; suture impressed; aperture irregularly ovate; outer lip waved; inside of the aperture whitish or brownish, often with obscure bands; columella rounded, extending into a broad, shallow sinus.

Habitat.— Nolin River, Kentucky.

Diameter, ·38 (10 millim.); length, ·66 inch (17 millim.) Length of aperture, ·35 inch (9 millim.). Breadth of aperture, ·19 inch (5 millim.).

Observations.— A somewhat variable species; the remarkably shouldered body-whorl will, however, readily distinguish it; differs from *M. obovata*, Say, by its more distinct spire, its greater proportionate breadth, and by the form of the aperture; it is also much less ponderous; many specimens are obscurely banded on the body-whorl; this is more distinctly visible in the young shell.— *Anthony.*

The shell figured and described by Mr. Reeve as *rarinodosa* is evidently the same as the above. The description is

Melania rarinodosa.— Shell ovately turbinated, olive, obscurely broad-banded; whorls 5–6, flatly convex, obtusely swollen and obsoletely noduled round the upper part; aperture ovate; columella twistedly effused.

Fig. 77.

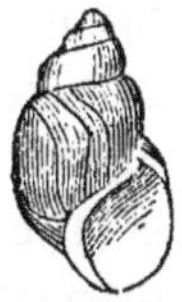

Habitat.— United States.

Anthony, Manuscript in Mus. Von dem Busch.

Observations.—Rather a doubtful species, received by Dr. Busch from Mr. Anthony with the above name in manuscript.— *Reeve.*

Melania consanguinea.— Shell ovate, smooth, thick, brownish-olive; spire short, acuminate; whorls eight, the upper ones nearly flat, the last two or three much shouldered; body-whorl very large, slightly constricted in its upper portion, and very faintly banded; sutures deeply impressed; aperture regularly ovate, within livid, approaching to purple far within; columella rounded, with scarcely a perceptible sinus, tinged with purple at base.

Fig. 79.

Habitat.— Indiana.

Diameter, ·40 inch (10 millim.); length, ·75 inch (20 millim.). Length of aperture, ·40 inch (10 millim.); breadth of aperture, ·20 inch (5 millim.).

Observations.— Allied to, but perfectly distinct from, *M. undosa;* its greater solidity, more elongated spire, and greater number of whorls will at once distinguish it· the whorls of the spire are much more convex, and there is no prominent angle formed by the shoulder on the body-whorl as in *M. undosa.*—*Anthony.*

C.   *Shell obliquely flattened.*

10.   L. compacta, ANTHONY.

Melania compacta, ANTHONY, Ann. N. Y. Lyc., vi, p. 122, t. 3, f. 22, March, 1854.
  BINNEY, Check List, No. 62. BROT, List, p. 32. REEVE, Monog., sp. 343.
Lithasia nuclea, LEA, Proc. Acad. Nat. Sci., p. 188, 1860. Jour. Acad. Nat. Sci., v,
  pt. 3, p. 263, t. 35, f. 73. Obs., ix, p. 85. BINNEY, Check List, No. 301.
Melania nuclea, Lea, REEVE, Monog., sp. 423. BROT, List, p. 33.

Description.—Shell ovate-conic, smooth, thick yellowish-green; spire
obtusely elevated; whorls about five, nearly flat; body-whorl large,

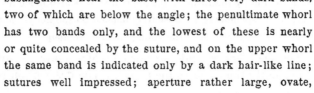

Fig. 80.

subangulated near the base, with three very dark bands,
two of which are below the angle; the penultimate whorl
has two bands only, and the lowest of these is nearly
or quite concealed by the suture, and on the upper whorl
the same band is indicated only by a dark hair-like line;
sutures well impressed; aperture rather large, ovate,
within whitish and banded; columella strongly indented, base regularly rounded, without any sinus.

Habitat.—Alabama.

Diameter, ·38 inch (10 millim.); length, ·60 inch (15 millim.).
Length of aperture, ·30 inch (7½ millim.); breadth of aperture, ·18
inch (4½ millim.).

Observations.—A short, thick, compact species, with seldom. more
than three perfect whorls remaining, other two whorls being indicated on the abruptly decollate spire; the whorls are slightly shouldered, and the lines of growth are curved and prominent; compared
with *M. fusiformis*, Lea, it is less fusiform, more ponderous, has the
spire less acute, and an aperture entirely different; from *M. proteus*,
Con., it differs in its totally different spire and aperture, and its want
of the tuberculous shoulder of that species; the bands in the interior
are very dark and well defined.—*Anthony.*

The following appears to be a synonyme, judging from the
comparison of type specimens of each.

Fig. 81.

Lithasia nuclea.— Shell smooth, elliptical, yellowish-olive,
thick, solid, three-banded; spire obtuse-conical; sutures impressed; whorls five, the last large and slightly inflated;
aperture rather small, ovately rounded, white and three-
banded within, recurved at the base; columella thickened above and
below, incurved; outer lip sharp.

Habitat.— Coosa River, Alabama: E. R. Showalter, M.D.

Diameter, ·34; length, ·60 of an inch.

Observations.— I have nine specimens before me of this little species, which has much the aspect of an *Anculosa*, as well also of some *Melaniæ*. But the callus on the lower and upper parts of the columella naturally places it in *Lithasia*. The longest of these specimens is not more than half an inch, and all are banded precisely alike, the three bands being nearly of equal size and equidistant. It would appear then that these bands are more constant than usual in the *Melanidæ*. Four out of the nine have a light purple spot on the middle of the columella, the others are entirely white. Without being at all like *Melania obovata*, Say (*consanguinea*, Anth.), in outline or general appearance, the columella is very much the same, both being thick with an incipient channel at base. Indeed, *M. obovata* properly belongs to the genus *Lithasia*. In form, color and bands, *nuclea* reminds one of *M. basalis* (nobis), but it is more rotund, has a thicker columella, has a less brilliant epidermis and is a more solid shell. The aperture is about one-half the length of the shell. Dr. Showalter says in his letter that "this is the most uniform species in my collection."— *Lea.*

D. *Shell subcylindrical.*

11. L. brevis, LEA.

Melania brevis, LEA, Philos. Proc., ii, p. 242. Philos. Trans., ix, p. 6. Obs., iv, p. 26. WHEATLEY, Cat. Shells U. S., p. 24. BINNEY, Check List, No. 38. BROT, List, p. 32. REEVE, Monog., sp. 344.

Anculosa solida, LEA, Philos. Proc., ii, p. 243. Philos. Trans., ix, p. 29. Obs., iv, p. 29. WHEATLEY, Cat. Shells U. S., p. 28.

Leptoxis solida, Lea, BINNEY, Check List, No. 384. BROT, List, p. 25.

Melania trivittata, REEVE, Monog., sp. 420.

Description.— Shell striate, subcylindrical, somewhat solid, yellow; spire rather short; sutures impressed; whorls flattened; columella thickened above; aperture ovate, white.

Fig. 82.

Habitat.—Alabama.

Diameter, ·41; length, ·60 of an inch.

Observations.— A single specimen only of this species is before me. The apex being eroded, the number of whorls cannot with certainty be ascertained; there appear to be about five. On this specimen there are eight indistinct impressed striæ, and several low, irregular folds on the body-whorl, which may be more distinct on the superior whorls when found perfect. The aperture is about half the length of the shell. — *Lea.*

The following is Mr. Lea's description of

Lithasia solida.— Shell smooth, elliptical, rather thick, yellowish-brown; spire somewhat drawn out; sutures impressed; whorls flattened; columella incurved, thickened above and below; aperture elongated, elliptical, white.

Habitat.— Alabama.

Diameter, ·38; length, ·60 of an inch.

Observations.— Three specimens only were sent to me by Dr. Foreman. They differ very little from each other, except that one exhibits a few indistinct, elevated, revolving striæ. Other specimens may present this character more strongly. Neither of the specimens has a perfect spire, the apices being eroded. The number of whorls I should think, however, were five. The aperture seems to be rather more than half the length of the shell. The columella is remarkable for its callus near the base as well as having another above.— *Lea.*

Until the possession of more specimens will enable naturalists to distinguish *L. brevis* and *L. solida,* they had probably better remain united as one species. Reeve's figure of the latter appears to have too long a spire, and to be differently formed in the aperture.

Mr. Reeve has not recognized the genus *Lithasia,* and accordingly changes the name to *trivittata,* Reeve, because Mr. Lea had already used *brevis* for a Melanian.

12. L. fusifórmis, Lea.

Lithasia fusiformis. Lea, Proc. Acad. Nat. Sci., p. 54, 1861. Jour. Acad. Nat. Sci., v, pt. 3, p. 261, t. 35, f. 71. Obs., ix, p. 71.

Description.— Shell sulcate, fusiform, rather thin, obscurely furrowed, reddish-brown, four-banded, conical; sutures irregularly impressed; whorls six, the last large and somewhat inflated;

Fig. 83.

aperture elongately rhomboidal; whitish within and four-banded, channelled and recurved at the base; columella with double curve, thickened above; outer lip somewhat constricted, with an acute margin.

Operculum small, ovate, dark brown, serrate around the base and outer margin, with the polar point inside the left edge about one-third above the basal margin.

Habitat.— Coosa River, Alabama; E. R. Showalter, M. D.

Diameter, ·30; length, ·52 of an inch.

Observations.—Six specimens are before me. Neither, I think, quite full grown. This species differs materially from *Showalterii* (nobis) from the same river. It is not quite so large, is not inflated, but more constricted on the body-whorl, and has rather distant, low, longitudinal folds, which in some specimens are scarcely observable. It differs in having four brown bands, the *Showalterii* having but three. The most remarkable character of *fusiformis* is the long, recurved channel which brings it close to the genus *Io*. All the specimens have transverse furrows, which are more strongly developed in some of them than in others. The *operculum* is very remarkable, having the margin from near to the polar point round the upper part of the outer margin completely *serrate*. Fortunately, two of the specimens were found to have the operculum adhering to the desiccated parts within, and both were found to possess this peculiar character, which I have never observed in any other species of the *Melanidæ*. The aperture is nearly two-thirds the length of the shell.— *Lea.*

It is not improbable that this may eventually prove to be the young of some other species—*Showalterii*,—or even *Downiei*.

13. L. Downiei, LEA.

Lithasia Downiei, LEA, Proc. Acad. Nat. Sci., p. 273, 1862. Jour. Acad. Nat. Sci., v, pt. 3, p. 354, t. 39, f. 227. Obs., ix, p. 176.

Description.— Shell sparsely nodulous, subcylindrical, chestnut-colored; spire obtusely conoidal, somewhat raised; sutures irregularly impressed; whorls seven, flattened, the last rather large, rhomboidal, white or banded within; outer lip sharp, sinuous; columella white and incurved.

Fig. 84.

Habitat.— Cumberland River; Major T. C. Downie.

Diameter, ·44; length, ·98 of an inch.

Observations.— This is an unusual form of *Lithasia* and cannot be confounded with any known species. The spire is exserted like most of the *Melanidæ*, but the aperture has all the characteristics of the true *Lithasiæ*. Its most remarkable character is the formation of the few low, elongate tubercles which it possesses. These are formed by an enlargement on the middle of the edge of the outer lip at each stage of growth,—a character I have not observed in any other species of *Melanidæ*. I suspect that this species will generally be found to be banded. One of the two specimens before me has six well-defined bands, which are indistinct on the outside, but are well marked on the inside. The other has only one band, and this is visible only on the upper whorls, the aperture being whitish, with a brown, indistinct band at the base. The upper callus is well marked, and the channel below is well defined. The aperture is more than one-third the length of the shell. I have great pleasure in naming this fine species after Major T. C. Downie, to whom I owe the acquisition of many new and rare mollusks.—*Lea.*

Subgenus STREPHOBASIS, Lea.

Strephobasis, LEA, Proc. Acad. Nat. Sci., p. 96, April, 1861. Jour. Acad. Nat. Sci., v, pt. 3, pp. 264 and 355. Obs., ix, pp. 86, 177.
Megara (sp.), H. & A. ADAMS, Genera, i, p. 306, Feb., 1854.

A.   *Shell ovate-conical.*

1. S. curta, HALDEMAN.

Melania curta, HALDEMAN, Monog. Limniades, No. 3, p. 3 of Cover. BINNEY, Check List, No. 80. BROT, List, p. 32. REEVE, Monog., sp. 345.
Melania solida, LEA, Philos. Trans., t. 9, f. 27. Obs., iv, p. 57. BINNEY, Check List, No. 245. BROT, List, p. 31. REEVE, Monog. Melania, f. 454.
Strephobasis solida, LEA, Jour. Acad. Nat. Sci., v, pt. 3, p. 266, t. 35, f. 77. Obs., ix, p. 88.
Megara solida, Lea, ADAMS, Genera, i, p. 306.

Description.—Shell short, conical, smooth; spire plane, nearly twice
Fig. 85.   as long as the aperture, which is narrow and quadrate with a narrow anterior sinus; color green or chestnut.

   Habitat.— Ohio River.

   Length, ⅝ of an inch.

   Observations.—Resembles *M. conica*, Say, but the whorls increase more rapidly in size.— *Haldeman.*

The above description is not a satisfactory one, but the shell is recognized as identical with *solida* by authenticated types in the collection of Mr. Anthony, one of which is here figuŕed. It is a mistake to assign the Ohio River as the habitat of this species.

Mr. Lea's descriptions and copy of his last figure here follow :—

Melania solida.— Shell smooth, obtusely conical, thick, solid, dark horn-color; spire rather short; sutures much impressed; whorls convex; aperture small, rhomboidal, twisted at the base, white within; columella inflected.

Habitat.— Tennessee.

Diameter, ·5; length, ·9 of an inch.

Observations.— This species in form somewhat resembles *M. alveare*, Conr., on one side, and *M. canaliculata*, Say, on the other. It has not, however, either furrows or tubercles. The three specimens before me have all mutilated apices, and therefore the number of whorls cannot be correctly ascertained. There may be seven or eight. The aperture is about one-third the length of the shell. There is no appearance of bands in these. This is one of those species which have a twisted aperture, being auger-shaped, the outer lip being spread out, and the edge having a line of a double curvature. The columella is very much twisted.

Strephobasis solida.—Shell smooth, subcylindrical, thick, solid, dark horn-color or olive; spire obtusely conical; sutures impressed; whorls slightly convex, the last slightly constricted; aperture rather large, nearly quadrate, whitish within; outer lip acute, very sinuous; columella sinuous, thickened below and channelled backwards.

Fig. 86.

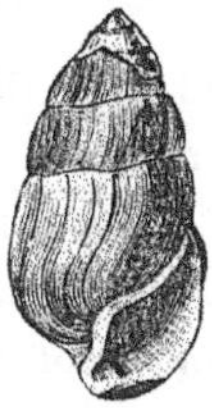

Operculum subovate, very dark brown, with the polar point near the middle of the base.

Habitat.— Tennessee; E. Foreman, M.D. : East Tenn.; President Estabrook: Pulaski Creek, Kentucky; Joseph Lesley.

Diameter, ·50 of an inch.

Observations.— I described and figured an imperfect specimen of this species in the Trans. Am. Phil. Soc., May 2, 1845, under the name of *Melania solida*. The figure shows the specimen to have been very imperfect in the aperture. Having subsequently received a number of perfect specimens (except in the apex), and finding its proper place to be in the genus *Strephobasis*, I have made a new description, and propose to give a more perfect figure. The specimens before me, more than a dozen, vary much in outline, some being more cylindrical than others. One of them has two obscure bands, visible inside and out. Another has an indistinct band inside at the base of the columella; others are white. Two from Kentucky have two broad dark bands, and two are of an olive color, with a purple spot at the base of the columèlla. In mature specimens the inner edge of the outer lip is thickened. Some of the mature specimens have a broad furrow round the body-whorl. The length of the aperture is usually about the third of the length of the shell.—*Lea.*

Messrs. Haldeman and Anthony both agree with me in considering *curta* and *solida* to be identical.

2. S. pumila, LEA.

Melania pumila, LEA, Philos. Proc., iv, p. 166, Aug., 1845. Philos. Trans., x, p. 60, t. 9, f. 36. Obs., iv, p. 60. BINNEY, Check List, No. 223. BROT, List, p. 33. REEVE, Monog.. sp. 446.
Megara pumila, Lea, ADAMS, Genera, i, p. 306.

Description.— Shell smooth, obtusely conical; rather thick, dark horn-color; spire depressed; sutures much impressed; whorls slightly convex; aperture elongate, contracted, twisted at the base, within whitish.

Fig. 87.

Habitat.—Tuscaloosa, Alabama.

Diameter, ·27; length, ·53 of an inch.

Observations.— The two specimens before me are, in form and size, the same. They differ in one having two broad, purple bands, and the other being entirely without. On the inferior part of the whorl one has five rather distinct striæ, the other has these less distinct. The apex of each of these is eroded, and therefore the number of whorls cannot be ascertained. This species is closely allied to *M. alveare*, Conrad, but is a much smaller shell, and in the two individuals before me there is no appearance of the tubercles which usually exist on the carina of the lower whorl of that species. —*Lea.*

This is a very distinct species. The Smithsonian collection contains a number of specimens, labelled "Tennessee." They are very uniform in size, color and markings.

S. pumila is more nearly allied to *P. productum*, Lea (*glossum*, Anth.), than to *alveare;* but it is very much smaller, heavier and differs in the form of the aperture.

3. S. carinata, LEA.

Strephobasis carinata, LEA, Proc. Acad. Nat. Sci., p. 273, 1862. Jour. Acad. Nat. Sci., v, pt. 3, p. 355, t. 39, f. 228. Obs., ix, p. 177.

Description. — Shell carinate, subfusiform, inflated, rather thin, greenish, four-banded; spire obtuse; sutures very much impressed; whorls six, flattened, carinate at the apex, the last one inflated; aperture rather large, rhomboidal, whitish and banded within; outer lip sharp, somewhat sinuous; columella thickened, bent back and much twisted.

Fig. 88.

Habitat. — Tennessee River; W. Spillman, M.D.

Diameter, ·20; length, ·37 of an inch.

Observations. — A single specimen, no doubt young, and somewhat fractured on the outer lip, is the only one received among the shells from Dr. Spillman. The spire is perfect, and all the whorls but the lowest one are carinate. It is, perhaps, nearest to *S. Clarkii* (nobis), but may be at once distinguished by the inflated form, the size and the bands. The aperture is about half the length of the shell.— *Lea.*

The figure is a copy of Mr. Lea's. It is doubtless a distinct species although the adult will probably differ much.

B. *Shell cylindrical.*

4. S. olivaria, LEA.

Strephobasis olivaria, LEA. Proc. Acad. Nat. Sci., p. 273, 1862. Jour. Acad. Nat. Sci., v, pt. 3, p. 356, t. 39, f. 229. Obs., ix, p. 178.

Description. — Shell smooth, elliptical, thick, banded, dark olive; spire obtusely conical; sutures very much impressed; whorls about

Fig. 89.

seven, convex, the last one large; aperture large, rhomboidal, white within and banded; outer lip acute, slightly sinuous; columella thickened below and twisted backwards.

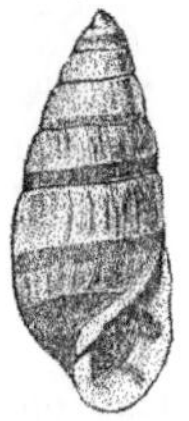

Habitat. — Knoxville, Tennessee; J. Clark.

Diameter, ·42; length, ·99 of an inch.

Observations. — Some twenty specimens are before me, all having very much the same size, form and general appearance. Generally there are two broad, well-characterized bands, strongly marked on the inside and observable on the outside. Two of the specimens have no bands, one has a single band, two have

four bands, and three are purple inside. This species is nearest to
solida, herein described, but it is more elliptical, less ponderous and
of quite a different color,— that species being light horn-color. The
aperture is about four-tenths the length of the shell.—*Lea.*

5. S. plena, ANTHONY.

Melania plena, ANTHONY, Ann. Lyc. N. H. New York, vi, p. 121, t. 3, f. 21, March,
1854. BINNEY, Check List, No. 210. BROT, List, p. 33. REEVE, Monog. Mel.,
sp. 450.
Strephobasis Spillmanii, LEA, Proc. Acad. Nat. Sci., p. 96, 1861. Jour. Acad. Nat.
Sci., v, pt. 3, p. 264, t. 35, f. 74. Obs., ix, p. 86.

Description.— Shell oblong ovate, smooth, thick, dark olive-green;
spire abruptly decollate, not elevated; whorls 4-5, convex; body-

Fig. 90.

whorl large, a little constricted in the centre, having two
very faint, distant bands, more distinct in the interior;
sutures irregularly and distinctly impressed; aperture
large, subrhomboidal, within livid and banded; columella
strongly indented and twisted, with a strong sinus at
base.

Habitat.— Alabama.

Diameter, ·45 inch (11 millim.); length, ·80 inch (21 millim.).
Length of aperture, ·42 inch (11 millim.); breadth of aperture, ·20
inch (5 millim.).

Observations.— A strong, corpulent shell, of a dark livid color,
which cannot well be confounded with any other; its most prominent characters are, its full broad form, the paucity of its whorls, and
its strongly indented columella.— *Anthony.*

Mr. Anthony's shell above described was figured from a
specimen not mature; for comparison another specimen from
the cabinet of that gentleman is here figured. It will
be seen to be the same, evidently, as Mr. Lea's, which
is copied from his plate. *Spillmanii* is thus described :—

Fig. 91.

Strephobasis Spillmanii.— Shell smooth, cylindrical, somewhat thick, dark brown or greenish, shining, very much
banded; spire obtuse, short, carinate at the apex; sutures
irregularly impressed; whorls slightly convex above, the last one
constricted; aperture rather large, somewhat square, bluish and
much banded within; outer lip acute, sinuous; columella sinuous,
thickened at the base and channelled backward.

Habitat.—Tennessee River, four miles above Chattanooga; Wm. Spillman, M. D.

Diameter, ·41; length, ·95 of an inch.

Observations.—I owe to the kindness of Dr. Spillman a number of this remarkable shell, to which he gave the habitat of Tennessee River, but did not designate from what part. Fortunately, there were some young specimens which, with those approaching maturity, gave us the advantage of tracing the great difference between the old and young. The old are decollate, and present, by the body-whorl being flattened, an almost perfect cylindrical form, while the young, which have the spire entire or nearly so, are almost perfectly oval and do not present a quadrate aperture, but an ovato-rhombic one. The callus at the base of the columella is strong, and amounts nearly to a fold, below which the channel suddenly turns backwards. The upper portion of the whorl, immediately below the suture, is tumid, and hence it has a bulbous appearance. This portion is usually lighter colored than the other parts of the whorl. The color differs in some of the specimens, some being more disposed to being dark brown, while others again are greenish. All which I have seen are more or less banded, some of them so thickly as to make the specimen almost black. These bands are all apparent on the inside. The length of the aperture is naturally, I presume, about half the length of the shell, but none of the mature specimens before me have perfect spires, and therefore the proportion cannot be correctly ascertained. There are six or seven whorls.

I have great pleasure in dedicating this interesting species to Dr' Spillman, to whom I am not only indebted for this, but for very many of the mollusks which he has so successfully discovered in the streams which flow through other districts as well as his own.— *Lea.*

6. S. cornea, Lea.

Strephobasis cornea, Lea, Proc. Acad. Nat. Sci., p. 96, 1861. Jour. Acad. Nat. Sci., v, pt. 3, p. 265, t. 35, f. 75. Obs., ix, p. 87.

Description.—Shell smooth, cylindrical, thick, horn-color; spire obtuse; sutures irregularly impressed; whorls slightly convex above, the last one constricted; aperture rhombo-quadrate, yellowish-white within; outer lip acute, sinuous; columella sinuous, thickened and channelled backward at its base.

· *Operculum* small, ovate, spiral, dark brown, with the polar point near the base.

Habitat.— Tennessee River, four miles above Chattanooga; William Spillman, M.D.

Diameter, ·41; length, ·88 of an inch.

Observations.— Among the previously described species from **Dr.** Spillman were two of this, which, while it has a close resemblance, still may easily be distinguished from it. They totally

Fig. 92.

differ in the color of the epidermis and the *cornea* is without any bands. The substance of the shell is stouter and the channel below not quite so well pronounced. There is al o a disposition to thickening on the upper part of the columella which the other has not. In both of the specimens before me there is a thickening following the inner edge of the outer lip. The lines of growth in both are well marked, and in all cases they begin below the antecedent one. The length of the aperture would, I presume, be rather less than half the length of the shell, but both specimens being decollate, the true length of the shell cannot be ascertained, nor can the character of the apical whorls be observed.—*Lea.*

7. S. Lyonii, Lea.

Strephobasis Lyonii, Lea, Proc. Acad. Nat. Sci. 5, 1864. Obs., xi, 107.

Description.— Shell smooth, subcylindrical, thick, dark horn-color or olive, rarely banded; spire obtusely conical; sutures impressed; whorls eight, somewhat convex; aperture somewhat constricted, rhomboidal, whitish within, rarely banded;   Fig. 93. outer lip acute, somewhat sinuous; columella thickened below and channelled and drawn back at the base.

Habitat.— Holston River at Knoxville, East Tennessee.

Diameter, ·48; length, ·92 of an inch.

Observations.— I have about a dozen, of various ages, of this well characterized species, which is nearly allied to *Spillmanii* (nobis). It differs in having a shorter aperture, in being rather larger, and in not being so cylindrical. In the young of the two there is a marked difference in outline, *Lyonii* being much more conical. Some of the less cylindrical specimens approach *olivaria* (nobis), but that is a smaller species, of a darker color, and almost always having two bands; *Lyonii* is usually without bands. Among the specimens before me two have a single band, one has two bands, one has four bands, and another has five bands. Four have a dark purple mark round the base of the columella. In those before me the color of the epidermis is very variable; several are light horn-color, one young one is almost a cinnamon-brown, and three are olivaceous. The old specimens are much eroded at the apex, and this causes a more cylindrical outline. The aperture is about four-tenths of the length of the shell.—*Lea.*

8. S. corpulenta, ANTHONY.

Melania corpulenta, ANTHONY, Ann. Lyc. N. H., vi, p. 127, t. 3, f. 28, March, 1854. BINNEY, Check List, No. 70. BROT. List, p. 32.

Description.—Shell ovate, smooth, yellowish, banded; whorls 6–7, convex; body-whorl very full, with two distant dark brown bands quite broad, which are nearly concealed on the upper

Fig. 94.

whorls by the revolutions of the spire; sutures impressed; aperture narrow ovate, broadest at base, banded within; columella much curved below the middle, white, and thickened at base, with a broad and distant sinus in that region.

Habitat.—Alabama.

Diameter, ·42 inch (10 millim.); length, ·80 inch (20 millim.).

Length of aperture, ·40 inch (10 millim.); breadth of aperture, ·17 inch (4 millim.).

Observations.—Its most prominent character is the corpulence of the body-whorl, and its regular oval form. May be compared with *M. bitæniata,* Conr., but its body-whorl is much more rounded or oval, it is less banded, and the bands are more distinct; the spire is more elevated and less abrupt.—*Anthony.*

In the shape of the aperture this resembles *S. cornea,* Lea, but it appears to differ in the superior portion of the body-whorl being swelled out.

9. S. bitæniata, CONRAD.

Melania bitæniata, CONRAD. New Fresh Water Shells, p. 52, t. 8, f. 6, 1834. DEKAY, Moll. N. Y., p. 94. WHEATLEY, Cat. Shells U. S., p. 24. BINNEY, Check List, No. 34. BROT, List. p. 32. HANLEY, Conch. Misc., t. 8. f. 73.
Anculotus bitæniatus, Conrad, REEVE, Monog. Anculotus, t. 3, f. 25.
Strephobasis Clarkii, LEA. Proc. Acad. Nat. Sci., p. 66, 1861. Jour. Acad. Nat. Sci., v, pt. 3, p. 262, t. 35, f. 76. Obs., ix, p. 87.

Description.—Shell conic, with convex whorls; spire short; one whorl entire, very convex; apex eroded; color olive, with two broad

Fig. 95.

purple bands on the body-whorl; one on the contiguous whorl; columella with a callus above and another near the base; aperture half the length of the shell; labrum regularly arcuated; within bluish, with purple bands

Habitat.—Black Warrior River.

Observations.—It is a rare species, remarkable for its broad, purple bands and convex whorls.

There can be no doubt of the identity of *bitæniata* and *Clarkii*. I give a good figure of the former from an authenticated specimen in Coll. Anthony. The number of bands on the body-whorl varies from two to five. Mr. Lea's description of *Clarkii* and a copy of his figure follow:—

Strephobasis Clarkii.—Shell smooth, cylindrical, rather thin, yellowish horn-color, trebly banded; spire very obtuse, short; sutures irregularly impressed; whorls five, slightly convex above, the last one constricted; aperture rather large, squarish, whitish and much banded within; outer lip acute; columella sinuous, white at the base, thickened and channelled backward.

Habitat.—Tennessee River, at Chattanooga, Tenn.; Joseph Clark. Diameter, ·38; length ·72 inch.

Observations.—Several specimens of this shell were long since sent to me by my deceased friend, Mr. Clark, and it is with peculiar pleasure that I dedicate it to him who, during a long life,

Fig. 96.

devoted his best energies to the investigation of the fauna and flora of Ohio, and other Western States. This species differs from the other two, herein described (*cornea* and *Spillmanii*), in being more regularly cylindrical; in being shorter and in having three regularly revolving brown bands, one of which only ˋis observable on the upper whorls. The aperture is more than one-half the length of the shell. There is a thickening in the interior of the upper part of the whorls, which in some specimens is irregular and oblique, and is observable from the outside. It gives a yellowish appearance to this part of the whorl under the suture.—*Lea.*

Subgenus PLEUROCERA, RAFINESQUE.

Pleurocera, RAFINESQUE, Jour. de Phys. Bruxelles, tome 88, p. 423,
   1819. BLAINVILLE, Dict. Sc. Nat., xxxii, p. 236, 1824, xli, p. 376,
   1826. Man. Malacologie, p. 441, 1825. Rang, Man. Conchyl., p.
   374, 1829. MENKE, Syn. Method, 2d edit., p. 43, 1830. FERUSSAC,
   Bull. Zool., p. 93, 1835. SOWERBY, Conch. Man., 2d edit., p. 231,
   1842. HERMANNSON, Indicis Gen. Malacoz., i, p. 296, 1846.
   HALDEMAN, Iconog. Encyc., p. 84.
Ceriphasia, SWAINSON, Malacol., pp. 204, 342, 1840. GRAY, Syn.
   Brit. Mus., 1844. HERMANNSON, Indic. Gen. Mal., i, p. 208, 1846.
   GRAY. Zool. Proc., pt. 15, p. 153, 1847. H. and A. ADAMS, Gen-
   era Recent Moll., i, p. 297, 1854. CHENU, Manuel de Conchyl.
   i, p. 288, 1859.
Telescopella, GRAY, Proc. Zool. Soc., pt. 15, p. 153, 1847.
Elimia (part), H. und A. ADAMS, Genera, i, p. 300, 1854. CHENU,
   Man. de Conchyl., i, p. 290, 1859.
Megara (part), H. and A. ADAMS, Genera, i, p. 306, 1854. CHENU,
   Man. de Conchyl., i, p. 293, 1859.
Trypanostoma, LEA, Proc. Acad. Nat. Sci., p. 169, April, 1862. Jour.
   Acad. Nat. Sci., 2d ser., v, pt. 3, p. 268, March, 1863. Obs., ix,
   p. 90, March, 1863.
Melania (sp.), of authors. BINNEY, Check List. REEVE, Monog,
   Mel., Nov., 1859, to June, 1861. BROT, Cat. Syst., p. 30, 1862.

Description.—Shell generally lengthened conical or cerithiform, aperture moderate, prolonged into a short spout or
canal in front. Columella not callously thickened.

Geographical Distribution.—The species contained in this
subgenus are inhabitants of the valleys of the Ohio, Tennessee and Alabama rivers. Two or three species are found
as far north as the Great Lakes, but none, so far as I am
aware, have been found in any of the rivers of the Atlantic
seaboard, or west of the Mississippi.

The species generally have a wide distribution within the
limits referred to and are numerously represented in individuals.

Mr. Lea has described several of the species as *Io's*, but I
restrict the typical form of *Io* to the fusiform, ventricose
species, in which the canal and spire are subequal.

A. *Tuberculate.*

1. P. alveare, CONRAD.

Melania alveare, CONRAD, New Fresh-Water Shells, p. 54, t. 4, f. 7, 1834. DEKAY,
   Moll. N. Y., p. 94. WHEATLEY, Cat. Shells, U. S., p. 24. JAY, Cat. 4th edit., p.
   272. BINNEY, Check List, No. 11. BROT, List, p. 30. HANLEY, Conch. Misc.
   t. 8, f. 74. MÜLLER, Synopsis, p. 46, 1836.
Megara alveara, Conrad, CHENU, Manuel, i, f. 2022. ADAMS, Genera. i, p. 306.
Melania torquata, LEA, Philos. Proc., ii, p. 242, Dec., 1842. Philos. Trans., ix, p.
   27. Obs., iv, p. 27. WHEATLEY, Cat. Shells U. S., p. 27. BINNEY, Check List,
   No. 271. ADAMS, Genera, i, 306.
Melania pernodosa, LEA, Philos Proc., iv, p. 105, Aug., 1845. Philos. Trans., x, p.
   66, t. 9, f. 49. Obs., iv. p. 66, t. 9, f. 49. BINNEY, Check List, No. 202.
Io pernodosa, Lea. ADAMS, Genera, i, p. 229.
Melania nupera, SAY (young), American Conchol., pt. 1, t. 8, middle figure.
*Melania producta,** LEA, Philos. Proc., ii, p. 243, Dec., 1842. Philos. Trans., ix, p.
   28. Obs., iv, p. 28. WHEATLEY, Cat. Shells U. S., p. 26. BINNEY, Check List,
   No. 217. BROT, List, p. 36.
*Melania grossa,** ANTHONY, Proc. Acad. Nat. Sci., p. 59, Feb., 1860. BROT, List,
   p. 40. REEVE, Monog., f. 411.

Description.— Shell short, conical, ventricose; whorls flattened,
with a line of wide compressed tubercles at the base of the penultimate whorl; body-whorl angulated; angle armed with prominent
tubercles; base hardly convex, with about five prominent lines;
aperture obliquely elliptical; less than half the length of the shell.

Observations.— Inhabits with the preceding species (*M. lima*) Elk
River, Alabama.   The spire is very regularly conical and the base
strongly ribbed.—*Conrad.*

The figure (No. 99) is from a type specimen in the collection
of my friend Mr. Haldeman, who very kindly placed in my
Fig. 97.   Fig. 98.   hands his entire valuable series of Conrad's,
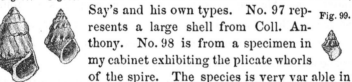
Say's and his own types.   No. 97 rep-   Fig. 99.
resents a large shell from Coll. Anthony.   No. 98 is from a specimen in
my cabinet exhibiting the plicate whorls
of the spire.   The species is very var able in
length.   No. 101 represents an elongated specimen from Cumberland River, Tennessee; this variety Mr. Lea has described
as *M. torquata.*

The following are the descriptions of *pernodosa* and *torquata.*

---

* *M. producta* and *grossa* are the young of a large variety of *alveare.*

Melania pernodosa.— Shell tuberculate, conical, rather thick, horn-color, striate below; spire elevated, ribbed on the apex; sutures undulated; whorls eight, flattened, tuberculate on the in-   Fig. 100. ferior portion; aperture small, angular and canaliculate at the base, within white.

Habitat.— Cypress Creek, Florence, Alabama.

Diameter, ·4; length, ·68 of an inch.

Observations.— This is a very remarkable species, having numerous, somewhat oblique tubercles, thickly set in a single row on the middle of the whorls. In the specimen before me, the only one I have seen, there is a dark spot between each of the tubercles. Towards the apex, the tubercles are more elongate and closely set, so as absolutely to become ribs across the whole of the whorl. The aperture is rather more than one-third the length of the shell. The striæ on the inferior half of the whorls are very regular and distinct, and number eight in this specimen.— *Lea.*

Fig. 101.

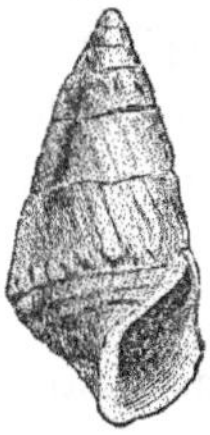

Melania torquata.— Shell tuberculate, subfusiform, shining, rather thin, yellow; spire rather elevated; sutures impressed; whorls seven, somewhat convex; aperture elongated, angular at the base, within whitish.

Habitat.— Tennessee.

Diameter, ·42; length, ·80 of an inch.

Observations.— This is a very beautiful species, of which I have only one specimen before me. The necklace-like row (whence its name) of small closely set tubercles, gives it an attractive appearance. Each successive whorl covers up these tubercles as well as several striæ below them, leaving the whole spire smooth. The aperture is rather contracted, and nearly half the length of the shell. The outer lip is sharp, and very much curved. It has some resemblance to *M. alveare* (Conr.) but is a larger shell, less solid, and more fusiform.—*Lea.*

The young of the large specimen figured, having attained to the full size of the ordinary adults and still differing from them, has been described as distinct by both Messrs. Lea and Anthony. Copies of their descriptions are given below. Having examined numerous specimens I have no doubt of their identity with *alveare*.

As already mentioned, *Strephobasis pumila*, Lea, is closely allied in general appearance to *alveare*.

Mr Lea believes *alveare* to be a *Lithasia*, but I do not find the callous deposits on the columella sufficiently well marked to place it in that genus.

Melania producta. — Shell folded, subfusiform, rather thin, horn colored; spire obtusely conical; sutures impressed; whorls eight, flattened; aperture elliptical, whitish.

Habitat. — Tennessee.

Diameter, ·57; length, ·70 of an inch.

Observations. — This species has rather distant folds on the first six whorls, and a disposition to tuberculation on the middle of the lower whorl, the superior part being disposed to be striate. The base of the columella is twisted, and the channel well impressed. The aperture is quite one-half the length of the shell.— *Lea.*

Melania grossa. — Shell ovate, folded, thick; spire obtusely ele-

Fig. 102. vated, composed of about eight convex whorls rapidly

attenuating to an acute apex; whorls folded, except the last two; body-whorl tumid, smooth; color of epidermis light greenish olive; aperture elliptical, whitish inside; columella rounded; outer lip much curved, with a well marked sinus at the base.— *Anthony.*

Habitat. — Tennessee.

Observations. — A short, thick species whose chief characteristics are its bulbous form, and short but prominent ribs on the upper whorls. All the whorls but the last are remarkably narrow and crowded, lines of growth prominent, four or five striæ revolve around the base of the shell. Resembles *M. glandula*, nob., in form, but its different color and texture, with its prominent ribs, will at once distinguish it. — *Anthony.*

The figure is from Mr. Anthony's type.

2. P. Foremanii, LEA.

Melania Foremanii, LEA, Philos. Proc., ii, p. 242. Philos. Trans., ix, p. 27. Obs., iv, p. 27. BINNEY, Check List, No. 111. BROT, List, p. 30. REEVE, Monog., f. 432. WHEATLEY, Cat. Shells U. S., p. 25.

Description. — Shell tuberculate, pyramidal, rather thick, yellowish-brown; spire elevated; sutures irregularly lined; whorls nine, flattened; aperture elongated, angular and channelled at the base, within whitish.

Habitat. — Alabama.

Diameter, ·52 of an inch; length, 1·28 inches.

Observations.—A fine, large, symmetrical species, furnished with a row of closely-set tubercles on the middle of the whorl, and several irregular transverse striæ disposed to be tuberculate. The seven or eight specimens before me are very similar, differing but little in form or color. The oldest one is rather browner. It is remarkable for its regular pyramidal form. The aperture is contracted, and rather more than one-third the length of the shell. I have great pleasure in dedicating it to Dr. Foreman to whose kindness I owe the specimen in my cabinet.—*Lea.*

Fig. 103.   Fig. 103*a*.

This species differs from other tuberculate *Pleurocera* in the oval form of the base of the body-whorl and in possessing several instead of one row of tubercles. Figure 103*a* is from a specimen in my cabinet, from Coosa River, Alabama, authenticated by Mr. Lea.

I have been much puzzled by the resemblance of this shell to *P. prasinatum*, Conr. and *P. Anthonyi*, Lea, and it would not surprise me if the three should be found to be but one species, as the forms of the shell and aperture are similar, and specimens of *Foremanii* in Coll. Haldeman are scarcely tubercled, while in one of the Smithsonian types of *Anthonyi* a disposition to tuberculation is evident.

2a. P. Lesleyi, LEA.

Trypanostoma Lesleyi, LEA, Proc. Acad. Nat. Sci., p. 4, 1864. Jour. Acad. Nat. Sci., vi, p. 146, t. 23, f. 59, 1867.

Description.— Shell tuberculate, pyramidal, dark horn-color; spire exserted; sutures irregularly impressed; whorls about eight, somewhat impressed; aperture rather small, rhomboidal, white and sometimes banded within; outer lip acute, very sinuous; columella thickened.

Operculum ovate, dark brown, rather thin, with the polar point near the base.

Habitat.— Smith's Shoals, Cumberland River, East Tennessee; Pulaski County, Kentucky.

Diameter, ·80; length, 1·2 inches.

Observations. — This species is closely allied to *T. undulatum*, Say, but may at once be distinguished by its lower spire and proportionately wider base, where it is flatter. The undulations on Mr. Say's shell are low, while in *Lesleyi* these are replaced by well defined tubercles, which are disposed to be compressed and incline to the left. There is only a single row of these tubercles, but those of the row above cause swellings on the upper part of the whorls. In the young state they differ totally, the *undulatum* being entirely smooth, while the *Lesleyi* has tubercles to the apex, except that on the first two or three whorls they change into folds. In the multiplicity of nodules it resembles *Lithasia pernodosa* (nobis). In the spire it also resembles *L. armigera*, Say, and *L. Jayana* (nobis), but differs in the aperture being Trypanostomose and of course not belonging to the same genus. I have ten specimens before me. Those from Prof. Troost I have had for a long time and believed they might be a variety only of *undulatum*, but the young sent by Mr. Lesley and Major Lyon convinced me at once that the species was new and distinct. The aperture is more square than in *undulatum* and the fuse is less. The young are striate on the under part of the whorls, which is never the case with *undulatum*. The aperture is about one-third the length of the shell. I have great pleasure in naming this after Mr. Joseph Lesley, Civil Engineer, to whose kindness I am indebted for many Kentucky species. — *Lea.*

Fig. 104.

A second specimen, kindly furnished by Mr. Lea, is more elongated than his type. The species bears the same relation to *undulatum* that *filum* does to *canaliculatum;* and it is strikingly like Say's *armigera*.

3. P. undulatum, SAY.

Melania undulata, SAY, New Harmony Dissem., p. 261; Reprint, p. 17; BINNEY'S edit., p. 142. REEVE, Monog., f. 307. HALDEMAN, Am. Jour. Sci., xlii., p. 216, Dec., 1841. ANTHONY'S List, 1st and 2d edits. DEKAY, Moll. N. Y., p. 92. WHEATLEY, Cat. Shells U. S., p. 27. JAY, Cat., 4th edit., 275. BINNEY, Check List, No. 281. BROT, List, p. 31. HANLEY, Conch. Misc., t. 1, f. 10. CATLOW, Conch. Nomenc., p. 189. BROT, Mal. Blatt., ii, p. 106, July, 1860;

Megara undulata, Say, CHENU, Man. Conchyl., i, f. 2025. ADAMS, Genera, i, p. 306.

Description. — Shell large, elevated, conic, brownish, with a broad, equally impressed band; inferior boundary of the band elevated and deeply crenate; superior boundary elevated and some- Fig. 105. times nodulous; volutions at least eight, not convex; suture not impressed, hardly obvious, undulated by revolving on the inferior crenate boundary of the impressed band; labrum near the base, much protruded; sinus very obtuse.

Habitat.— Ohio River.

Length one inch and four-tenths.

Observations.— I observed this large species to be abundant in Kentucky River, when travelling in that state two years since with Mr. Maclure. It seems to approach nearest in character to the *canaliculata*, nob., but its rough appearance will distinguish it even at first sight.—*Say.*

A fine specimen from Mr. Anthony's collection is the original of our figure.

The various species of this general type, described by Mr. Lea, *nobilis, moniliferum, nodosum,* are not sufficiently distinct. This shell may (for the present) remain separated from them on account of the sulcate band encircling the periphery and its being wider.

This species extends through Ohio, Indiana, Illinois, Kentucky, Tennessee, Alabama, and West Georgia and presents great variation of contour. The number of nodules on the periphery varies, and also the development of the canal. Many of the large specimens, broadly banded, are very beautiful.

4. P. excuratum, CONRAD.

Melania excurata, CONRAD, New Fresh-Water Shells, p. 49. t. 4, f. 6, 1834. ANTHONY, List, 1st and 2d edits. JAY, Cat., 4th edit., p. 273. DEKAY, Moll. N. Y., p. 96. BINNEY, Check List. No. 103. MÜLLER, Synopsis, p. 43, 1836.
Melania excurvata, Conrad, WHEATLEY, Cat. Shells U. S., p. 25.
Melania rorata, REEVE, Monog. Mel., sp. 306. BROT, List, p. 31.
Io Spillmanii, LEA, Proc. Acad. Nat. Sci., p. 394, 1861. Jour. Acad. Nat. Sci., v, pt. 3, p. 348, t. 39, f. 215. Obs., ix, p. 170.

Description.— Shell subulate, with a spiral band of slightly oblique subcompressed tubercles on the base of the inferior whorls; above this is a prominent line with slight intervening channel, volutions towards the apex nearly entire; base with three prominent lines, the superior one largest; the third hardly prominent and approximate to the middle one.

Observations.— A large and beautiful species, common in the Tennessee River at Florence. It is perhaps most nearly allied to *M. Sayi*

Fig. 107.   Fig. 106.   Fig. 108.

(*M. canaliculata,* Say), but the elevated line and form of the tubercles will distinguish it from that species. The epidermis is reddish-brown or black.— *Conrad.*

Mr. Conrad's figure not being a very good one I have had a figure drawn from a fine specimen from the original locality, kindly furnished to me by Mr. Lea. I have included *rorata,* Reeve and *Spillmanii,* Lea, in the synonymy of this species, finding no characters by which to distinguish them. I have already expressed a doubt whether any of the species immediately following *undulata* are really distinct from it.

The figures of the accompanying descriptions are copies of those of Messrs. Reeve and Lea.

Fig. 109.

Melania rorata.—Shell pyramidally conical, brownish-olive, spire raised, whorls 10-11, slopingly convex, corded throughout with rather close-set ridges, some of which are beaded; aperture ovate, columella callous, twisted, effusely channelled.

Habitat.— Alabama.— *Reeve.*

The following species may be regarded as an immature form of *excuratum* rather than as a distinct species.

Io Spillmanii.— Shell smooth, attenuately conical, pale horn-color; spire regularly conical, striate above; sutures slightly impressed; whorls about ten, flattened, obtusely angular in the middle; aperture small, rhomboidal; outer lip sharp and sinuous; columella white and very much twisted; canal short and subeffuse.

Habitat.—Tennessee River, Alabama? Wm. Spillman, M. D.

Diameter, ·46; length, 1·25 inches.

Observations.— This species is nearly allied to *modesta,* herein described, but may be distinguished by its longer and more attenuate spire, the upper whorls being covered with regular close transverse striæ. The channel is also rather longer and more twisted. One only of four specimens received is full grown. This has, above the angle of the last whorl, a few undefined tubercles. Below this angle there are five or six well defined transverse striæ. None of the specimens have bands. Should adults generally be found with tubercles, then this species should be placed in the tuberculate group and not in the smooth one, where I have now placed it in the above description. The aperture is nearly one-third the length of the shell. I have great pleasure in dedicating the species to Dr. Spillman, who has done so much for the natural history of his own and other Southern States.

Fig. 110.

The typical *excuratum* differs widely enough from *undulatum*, Say, but there exist intermediate forms of a nature to perplex the naturalist. Among these may be mentioned *P. ponderosum*, Anth. (*dux*, Lea), with the tubercles and canal nearly obsolete and the revolving striæ very faint, so that the surface of the shell appears at first sight flat and smooth; also *annuliferum*, Conr., in which the revolving lines are more strongly developed. These shells all partake of one general type and form a natural group of closely related species, at the least.

5. P. moniliferum, Lea.

Trypanostoma moniliferum, Lea, Proc. Acad. Nat. Sci., p. 172, 1862. Jour. Acad. Nat. Sci., v, pt. 3, p. 295, t. 36, f. 125, March, 1863. Obs., ix, p. 117.

Io nodosa, Lea, Proc. Acad. Nat. Sci., p. 393, 1861. Jour. Acad. Nat. Sci., v, pt. 3, p. 346, t. 39, f. 212, March, 1863. Obs., ix, p. 168.

Io variabilis, Lea, Proc. Acad. Nat. Sci., p. 393, 1861. Jour. Acad. Nat. Sci., v, pt. 3, p. 347, t. 39, f. 214, March, 1863. Obs., ix, p. 169.

Description. — Shell tuberculate, thick, pyramidal, yellowish or greenish, banded or without bands; spire high, pyramidal; sutures irregularly impressed; whorls about ten, flattened, striate below, sometimes obscurely sulcate, tuberculate on the periphery; aperture rather large, rhomboidal, within either white or salmon and generally double-banded; outer lip acute, very sinuous; columella thickened below and very much twisted.

Operculum ovate, very dark brown, with the polar point near the base.

Habitat. —Tennessee; Prof. Troost and Mr. Anthony: Florence, Alabama; Rev. G. White, Mr. Pybas and Mr. Thornton: Cumberland River; Dr. Powell: Ohio River, near the mouth in Illinois; J. Ronaldon: New Harmony, Indiana; Mr. Carley and Mr. Sampson: Warrior River, Alabama; Prof. Brumby.

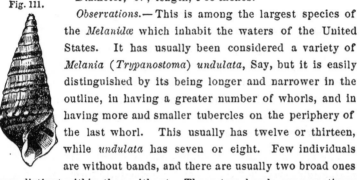

Fig. 111.

Diameter, ·67; length, 1·53 inches.

Observations.— This is among the largest species of the *Melanidæ* which inhabit the waters of the United States.   It has usually been considered a variety of *Melania* (*Trypanostoma*) *undulata*, Say, but it is easily distinguished by its being longer and narrower in the outline, in having a greater number of whorls, and in having more and smaller tubercles on the periphery of the last whorl.   This usually has twelve or thirteen, while *undulata* has seven or eight.   Few individuals are without bands, and there are usually two broad ones more distinct within than without.   These two bands are sometimes separated into four.   The first three or four whorls are usually carinate. The tubercles, which are usually beautifully defined, are highly ornamental, but usually do not exist above the ultimate and penultimate whorls.   This species seems to be widely distributed, and few or none of our species are more beautiful.   There is usually a revolving raised line above, and parallel with, the row of tubercles.   The color of the epidermis varies much.   Some specimens are of a rich straw yellow, and others are greenish, while others again are of a deep olive-brown, with a fine natural polish.   Some have the upper band so broad that a single whitish line is visible under the suture.   This may be remarked more particularly in the specimens from the vicinity of New Harmony. The aperture is about one-third the length of the shell.— *Lea.*

Fig. 112.

Io nodosa.— Shell tuberculate, raised, conical, greenish horn-color, banded; spire irregularly conical; sutures very much impressed; whorls about ten, flattened, tuberculate on the middle, striate below; aperture rather small, rhomboidal, banded within; outer lip sharp and sigmoid; columella white and very much twisted; canal rather short.

Operculum ·pyriform, spiral, dark chestnut-brown, with the polar point near to the basal margin.

Habitat. — Tennessee River, Alabama? Wm. Spillman, M. D.

Diameter, ·57; length, 1·58 inches.

Observations.—This is one of those species of *Melanidæ* which we have considered to belong to the group with a regular channel at the base, like the genus *Fusus*, but which really belongs to the genus *Io*, having other characters differing from *Melania*. It is nearly allied to the species which I described as *Melania nobilis** in the Trans. Am. Phil. Soc., vol. x, pl. 9, fig. 48, from a single imperfect specimen. It is a smaller species, and is not so fusiform, having a shorter channel, which is not quite so much twisted, and the nodules are not so large. The aperture is more than one-third the length of the shell. — *Lea.*

Io variabilis.— Shell smooth, raised, conical, subfusiform, banded, deep purple or greenish; spire regularly conical; sutures slightly impressed; whorls about nine, flattened, angular in the middle; aperture elongately rhomboidal; outer lip sharp and sinuous; columella white or purple and very much twisted; canal long and narrow.

Fig. 113.

Habitat.—Tennessee River, Alabama? Wm. Spillman, M.D. Diameter, ·40; length, ·88 of an inch.

Observations. — A number were received from Dr. Spillman, but they are generally young, and the older specimens were much injured in the delicate fuse and outer lip. It is a small, thin species, with a well developed, nearly straight, channel. It seems to be a very variable species, some individuals being of intense purple, nearly black, while others are yellowish, with numerous bands; others again are greenish, without bands. Some are carinate towards the apex, while others are free from carination. There is a disposition in several to be tuberculate along the angle on the middle of the lower whorl. Generally there is a light line along the upper part of the whorls. The aperture is nearly one-half the length of the shell.—*Lea.*

The four species *undulatum, excuratum, moniliferum* and *robustum* are mainly distinguished by the following differences :—

Undulatum is a stout, broadly conical shell, strongly angled on the periphery and having large tubercles. The base is much flattened.

Robustum, with much the same general outline, is not much angled on the periphery, with the inferior portion of the whorl

---

* In transferring this to the genus *Io*, I think it may properly be considered the type of a group of the genus.

longer and more convex. It bears the same general relation to *undulatum* that *Troostii* does to *canaliculatum;* and these shells may prove to be only tuberculate varieties of the others.

Excuratum is a much longer, narrower species than either of the above, with the whorls almost flat, and the upper ones thickly striate. This feature is most apparent in the young shell (*Spillmanii*, of Lea).

Moniliferum is not so narrow in its proportions as *excuratum*, and is generally beautifully banded. It differs from *excuratum* in the young shells being smooth instead of striate on the spire.

6. P. nobile, LEA.

Melania nobilis, LEA, Philos. Proc., iv, p. 165, Aug. 1845. Philos. Trans., x, p. 65, t. 9, f. 48. Obs., iv, p. 65. BINNEY, Check List, No. 179.
Io nobilis, Lea, ADAMS, Genera, i, p. 299.

Description.— Shell tuberculate, conical, rather thick, yellowish horn-color; spire elevated; sutures irregularly undulate; whorls flattened, in the middle tuberculate; aperture rather large, elongated, angular, and channelled at the base, within yellowish; columella twisted.

Habitat.— Alabama.

Diameter, ·72; length, 1·7 inches.

Observations.— This is among the finest of our American species. It is remarkable for its large size and extended sinus, which allies it

Fig. 114.

to the genus *Io*, in which it might, with no great impropriety, be placed. The specimen before me has eight whorls, and the broken apex would probably present about three more. The central ones have a dark band below, and are of a rather bright horn-color above. In this specimen there is a rather coarse stria above the row of tubercles, and two smaller ones below. The margin of the outer lip is quite sinuous. It has some resemblance to *M. excurata*, Conr., but may be distinguished by having a larger fuse, and in the position of the tubercles, which are not oblique, as described in that shell. When other specimens shall be observed it may be found to differ in some of the characters described above. Aperture rather more than one-third the length of the shell.— *Lea.*

Chiefly distinguished by the narrow lengthened canal which terminates the aperture. Mr. Lea's figure being imperfect I have figured a specimen in Mr. Anthony's collection.

7. P. robustum, LEA.

Io robusta, LEA, Proc. Acad. Nat. Sci., p. 393, 1861. Jour. Acad. Nat. Sci., v, pt. 3, p. 346, t. 39, f. 213, March, 1863. Obs., ix, p. 168.

Description.— Shell canaliculate, slightly tuberculate, raised, conical, pale horn-color, obscurely banded below; spire regularly conical; sutures very much impressed; whorls about ten, flattened about the apex, channelled below; aperture rather small, rhomboidal, banded within; outer lip sharp and sigmoid; columella pale salmon color; channel rather short.

Fig. 115.

Operculum ovately angular, spiral, very dark brown, with the polar point near to the basal margin.

Habitat.— Tennessee River, Alabama? Wm. Spillman, M. D.

Diameter, ·76; length, 1·49 inches.

Observations.— There are two specimens before me. Both have tubercles below the sulcate channel, but one has them much better developed than the other. The aperture within is pale salmon in both specimens, but this may not be constant. It is rather shorter in the channel than *nodosa*, herein described, and the spire is also shorter. The aperture is more than one-third the length of the shell.— *Lea.*

This species is exceedingly closely allied to *undulatum* but appears to be rather wider, more obtusely conical and more robust. The aperture is produced into a somewhat longer canal at the base than that species usually exhibits.

The figure is a copy of that of Mr. Lea.

B. *Sul*

8. P. canaliculatum, SAY.

Melania canaliculata, SAY, Jour. Acad. Nat. Sci., ii, p. 175, January, 1821. BINNEY'S
Reprint, p. 65. BINNEY, Check List, No. 45. DEKAY, Moll. N. Y., p. 94.
WHEATLEY, Cat. Shells U. S., p. 24. RAVENEL, Cat., p. 11. JAY, Cat , 4th edit.,
p. 273. ANTHONY, List, 1st and 2nd edits. KIRTLAND, Report Zool. Ohio, p.
174. CATLOW, Conch. Nomenc., p. 185. BROT, List, p. 30. REEVE, Monog.
Mel., sp. 304.
Io canaliculata, Say, MORCH, Yoldi Cat., p. 56.
Ceriphasia canaliculata, Say, CHENU, Manuel, Conchyl. i, f. 1959.
Ceriphasia canaliculata, Say, ADAMS, Genera, i, p. 297.
Melania conica, SAY, Jour. Acad. Nat. Sci., ii, p. 176, January, 1821. BINNEY'S
Reprint, p. 70. BINNEY, Check List, No. 65. REEVE, Monog. Mel., sp. 252.
DEKAY, Moll. N. Y., p. 95. RAVENEL, Cat., p. 11. HALDEMAN, Monog.
Limniades, No. 7, p. 4 of Cover. BROT, List, p. 30. KIRTLAND, Rep. Zool.
Ohio, p. 174. ANTHONY, List, 1st and 2nd edits. JAY, Cat., 4th edit., p. 273.
WHEATLEY, Cat. Shells U. S., p. 24. CATLOW, Conch. Nomenc., p. 186.
SOWERBY, Mollusca, Fauna Boreali Americana, iil, p. 316, 1836.
Melania substricta, HALDEMAN, Suppl. to Monog. of Limniades.
Pirena plana (Jan.), BROT, Mel., p. 60, note.
Strombus Sayi, WOOD, Index Testaceol. Suppl., t. 4, f. 24.
Melania Sayi (Wood), SHORT and EATON, Notices, p. 82. ANTHONY, List, 1st and
2nd edits.
Melania Sayi, Ward, WHEATLEY, Cat. Shells U. S., p. 27.
Melania Sayi, Ward, KIRTLAND, Rept. Zool. Ohio, p. 174. JAY, Cat., 4th edit., p.
274. HIGGINS, Cat., p. 7.
Melania Sayi, Deshayes, CATLOW, Conch. Nomenc., p. 188.
Melania Sayi, DESHAYES, Encyc. Meth. Vers., ii, p. 427, 1830.
Melania exarata, MENKE, Syn. Meth., p. 135, 1830. BINNEY, Check List, No. 100.
Mel.mia ligata, MENKE, Syn. Meth., p. 236, 1830. BINNEY, Check List, No. 162.
Melania auriscalpium, MENKE, Syn. Meth., p. 133, 1830. BINNEY, Check List, No. 25.
Gyrotoma conica, Say, ADAMS, Genera, i, p. 305.

Description.— Shell tapering, horn-color; volutions about seven,
slightly wrinkled; spire towards the apex much eroded, whitish;
body, with a large obtuse groove, which is obsolete upon the whorls
of the spire in consequence of the revolution of the suture on its inferior margin; this arrangement permits the superior margin of the
groove only, to be seen on the spire, in the form of an obtuse carina
on each of the volutions; aperture bluish-white within with one or
two obsolete revolving sanguineous lines; labrum slightly undulated
by the groove and with a distinct sinus at the base of the columella.

Habitat.— Ohio River.

Breadth, three-fifths of an inch; length, one inch and one-tenth.

Greatest transverse diameter more than two-fifths. Very common
at the Falls of the Ohio River. It is probably the largest species
of this genus in the United States, and may be readily distinguished
from its congeners by its broad groove.—*Say.*

The deep sulcus which distinguishes Mr. Say's *Mel. canalicu-lata*, in its typical form, shades off so gradually into a smooth, flattened surface, that not only is it difficult to arrange the species of this group, but it is even doubtful whether many of the species which are placed in other groups are really distinct. Especially, may it be doubted whether the small shells recently described by Mr. Lea under the names of *bivittatum*, *pami-lum*, *simplex*, etc., are distinct from the young of *canalicula-tum*.

Mr. Say describes the young shell of *canaliculata* as *Melania conica*. It is differently formed from the adult shell and does not possess the sulcated body-whorl. The illustrations of this species, all drawn from specimens, exhibit the various stages of growth, etc.

Fig. 116 is a tall, slender form from the Ohio River, scarcely sulcate. No. 117 represents a stunted specimen also from the Ohio. No. 119 is from Tennessee River. No. 120 is a quite young shell from the Falls of the Ohio. No. 122 is a heavy northwestern form; the specimen probably came from the interior of Ohio. No. 121, a beautiful sharply sculptured form, is from Tennessee. Nos. 117, 118, 119 represent the *M. conica* of Say. It will be seen that there is much variation of form in this species; so the color also varies from a light green and yellow to a dark brown or nearly black and is either uniform or banded. The area of geographical distribution is very great, extending from the interior of Ohio to Alabama and through Indiana and Illinois.

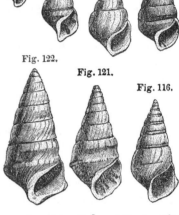

Fig. 120. Fig. 118. Fig. 117. ig. 119.

Fig. 122.

Fig. 121.

Fig. 116.

The following is Mr. Say's description of

M. conica. — Shell conic, rapidly attenuating to an acute apex, very slightly wrinkled, olivaceous; suture not deeply impressed; volutions seven or eight; aperture oblique, equalling the second, third, and fourth whorls conjunctly.

Var. A. With from one to three revolving, rufous or blackish lines.

Habitat. — Ohio River.

Length, nearly three-fifths inch; of the aperture, one-fourth inch.

Observations. — May be readily distinguished from *M. Virginica* by the much more rapid attenuation of the spire, and in the proportional difference in the length of the aperture, which in the *Virginica* is not more than equal to the length of the second and third whorls. — *Say.*

Melania substricta was proposed by Prof. Haldeman instead of *conica* under the impression that the latter name was pre-occupied. He afterwards used the name for a new species.

The following species, described by Menke, are all synonymes of *canaliculatum :*

Melania exarata. — Shell conically turreted, acute; apex eroded; striate, greenish-brown; last whorl encircled by two transverse sulci, plane between; the other whorls carinate in the middle; aperture obliquely ovate; lips alate, arcuate, margined within, extreme margin subreflected.

Habitat. — Ohio River, at Cincinnati.

Long., 13 lin. ; lat., 6 lin. — *Menke.*

Melania ligata. — Shell turreted, apex eroded, truncate, with transverse acute striæ, below sulcate, corneous; whorls seven, convex, the last bifasciate, the others singly banded.

Habitat. — Ohio River, at Cincinnati.

Long., 9 lin. ; lat., 3½ lin. — *Menke.*

Melania auriscalpium. — Shell turreted, apex truncately eroded, smooth, corneous, whorls six, convex, the last doubly banded, the others singly banded; lip arcuate, sub-alate, produced in front.

Habitat. — Ohio River, near Cincinnati.

Long., 10; lat., 3¼ lin. — *Menke.*

It is questionable whether *P. canaliculatum* is really distinct from *P. undulatum;* indeed, the transition between the smooth and tubercled surface is so gradual, and the range and

Fig. 123.

development of the two species in different localities so exactly similar that I am inclined to think them identical, but like Mr. Lea and Prof. Haldeman, who entertain the same views, I do not feel at liberty to unite them as yet.

As an illustration of the great difficulty attending the determination of species in this family, I figure (fig. 123) a depauperate specimen of *canaliculatum* furnished me by Prof. Haldeman.

9. P. filum, Lea.

Melania filum, Lea, Philos. Proc., iv, p. 165. Philos. Trans., x, p. 62, t. 9, f. 41.
Obs., iv, p. 62. Binney, Check List, No. 109. Brot, List, p. 30. Reeve, Monog.
Mel., sp. 402?
Elimia filum, Lea, Chenu, Man. Conchyl., i, f. 1980. Adams, Genera, i, p. 300.

Description.—Shell carinate, conical, rather thin, dark horn-color;
spire elevated; sutures impressed; whorls flattened, carinate in the

Fig. 124.   Fig. 125.   Fig. 125a.   Fig. 126.   Fig. 127.

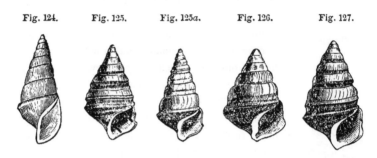

middle; aperture small, rhomboidal, angular at the base, within
whitish, columella twisted.

Habitat.—Alabama.

Diameter, ·47; length, 1·06 inches.

Observations.—A single specimen only of this species was submitted to me by Major LeConte. It is very nearly allied to *M. elevata*,
Say, but may be distinguished by its thread-like carina on the middle
of the whorls, which, on the superior ones, presents a mere simple
line. The outer lip is remarkably patulous, presenting the auger-
shaped lip which belongs to a certain group of the *Melaniæ*. The
apex being imperfect, the number of whorls cannot be ascertained.
There are eight visible on this specimen, and it probably possesses
ten in a perfect state. The aperture is about one-third the length of
the shell.—*Lea.*

This species has by many been considered a variety of
canaliculatum; my impression is, that it is well distinguished
by its more strictly conical shape, flattened whorls, and more
elevated carina on the periphery. It almost entirely replaces
canaliculatum in the waters of Tennessee (I have seen numerous specimens from all portions of the state), and, if specifically
identical with the latter species, must at least be distinguished
as a local variety. The type figure which I have copied (fig.

124) is very poor, and in fact looks much like the young of *P. ponderosum.*

10. P. ponderosum, SAY.

Melania ponderosa, ANTHONY, Proc. Acad. Nat. Sci., Feb., 1860, p. 59. BINNEY, Check List, No. 213. BROT, List, p. 59.
Trypanostoma dux, LEA, Proc. Acad. Nat. Sci., p. 170, 1862. Jour. Acad. Nat. Sci., v, pt. 3, p. 283, t. 36, f. 105. Obs., ix, p. 105.

Description.— Shell conic, broad, smooth, olivaceous, thick; spire considerably but not acutely elevated; whorls 7–8, subconvex; lines

Fig. 128.

of growth curved and strong; sutures distinct; aperture rhombic, rather small, whitish within; columella indented, outer lip much curved forward, forming a broad, well marked sinus at base.

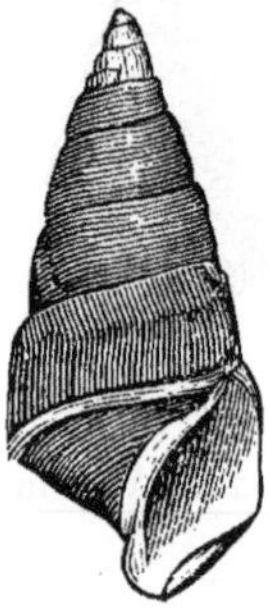

Habitat.— Tennessee.

Observations.— One of the most ponderous of the genus. In form it resembles *M. canaliculata*, Say, but has not the channel of that species, and differs also in the aperture. The body-whorl is strongly keeled about the middle, and has another and less clearly defined carina about midway between the first and the suture above. The lines of growth are very strong and occasionally varicose. A strong deposit of white callus is found upon the columella, which is much thickened near the base.— *Anthony.*

At a meeting of the New York Lyceum of Natural History held in June, 1860, Dr. Budd referred this species to Mr. Conrad's *excurata.* I have already remarked upon the resemblance in the description of the latter species. There can be no doubt that Mr. Lea's *T. dux* is a synonyme. Mr. Lea's description here follows. The figure of *ponderosa* is from the original type, that of *dux* is copied from Mr. Lea's.

Trypanostoma dux. — Shell carinate, pyramidal, thick, reddish-brown; spire much raised; sutures slightly impressed; whorls about nine, flattened; aperture rather large, rhombic, pale salmon-color within and very much twisted.

Operculum subpyriform, dark brown, with polar point near to the basal line.

Habitat.— Tennessee River; Dr. W. Spillman: Fox River, Illinois; J. Sampson; Oostenaula: Rev. G. White: Tuscumbia; D. Pybas.

Diameter, ·75; length, 1·80 inches.

Observations.— This is the largest species of *Trypanostoma* of our country which I have seen. It is nearly two inches long and is athletic. It is closely allied to *Melania* (*Trypanostoma*) *canaliculata* and *undulata*, Say, which two may indeed be only varieties of each other. It has a carina like each of them, and this is sometimes slightly nodulous like the latter, and there is a slight furrow-like impression above the carina which reminds one of the former. The whorls are remarkably flat and the color of the epidermis is more brownish. Three specimens out of six before me are more or less banded inside. The specimen from Tuscumbia is whitish inside and has two indistinct bands. It is an imperfect specimen, and may really not belong to this species. The aperture is more than one-fourth the length of the shell.— *Lea.*

Fig. 129.

11. P. Troostii, Lea.

Trypanostoma Troostii, Lea, Proc. Acad. Nat. Sci., p. 171, 1862. Jour. Acad. Nat. Sci., v, pt. 3, p. 285, t. 36, f. 107. Obs., ix, p. 107.
Trypanostoma viride, Lea, Proc. Acad. Nat. Sci., p. 172, 1862. Jour. Acad. Nat. Sci., v, pt. 3, p. 291, t. 36, f. 119. Obs., ix, p. 113.
Trypanostoma ligatum, Lea, Proc. Acad. Nat. Sci., p. 171, 1862. Jour. Acad. Nat. Sci., v, pt. 3, p. 288, t. 36, f. 114. Obs., ix, p. 110.

Description.— Shell carinate, conical, very much inflated, yellowish horn-color or greenish, banded or without bands; sutures irregularly and very much impressed; whorls about nine, rather impressed, sometimes channelled; aperture large, rhomboidal, whitish and sometimes banded within; outer lip acute, sinuous; columella thickened below and very much twisted.

Fig. 130. Fig. 131.

Habitat.— Tennessee; Prof. G. Troost: Florence, Alabama; Rev. G. White: Oostenaula River, Georgia; Bishop Elliot: Fox River, near New Harmony, Indiana; J. Sampson.

Diameter, ·64; length, 1·29 inches.

Observations.— I have five specimens before me; that from the late Prof. Troost (after whom I have great pleasure in naming it), I have had for a long time. It is one of the largest species we have in the

Fig. 132.

United States. It is perhaps nearest to *Melania* (*Trypa-nostoma*) *canaliculata*, Say. It is, however, more inflated, the aperture is larger and the columella more extended. All the specimens are not channelled, but all are more or less carinate at the periphery. Two of the specimens are obscurely banded inside, and one very much banded inside and out. The old specimens are thickened inside the edge of the lip. The aperture is more than one-third the length of the shell.— *Lea*.

Without making a *positive* decision in this matter I am *in-clined* to believe that *T. Troostii* is distinct from *canaliculatum*. It appears to be more inflated in its form, not so flatly conical, with a longer, rounded base.

The specimens before me convince me, however, that *T. vir-ide* and *ligatum* are only young shells of the same species.   I give Mr. Lea's descriptions of the latter two.   The figures are copied from his plate.

Trypanostoma viride.— Shell subsulcate, somewhat thick, subfusi-form, olivaceous; spire obtusely conical; sutures much impressed; whorls seven, convex, the last slightly canaliculate; aperture rather large, rhomboidal, purple or whitish within; outer lip acute, sinuous; columella thickened below and slightly twisted.

Fig. 133.

Habitat.— Tennessee; Prof. Troost.

Diameter, ·48; length, ·89 of an inch.

Observations.— I have about a dozen specimens before me, all of which have the same olive-green hue.   They have been in my possession a long time, and I had put them among the young of *Melania* (*Trypanostoma*) *canali-culata*, Say. I have now no doubt but that they are distinct from that large species.   None of them are half the size, the color is darker and they are wider in proportion.   The revolving furrow above the periphery of the last whorl is hardly observable in some specimens. Every one of my specimens has a purplish-brown spot at the base of the columella, and in some specimens this color pervades the whole of the interior.   The aperture is more than a third of the length of the shell.— *Lea*.

Trypanostoma ligatum.— Shell carinate, subfusiform, rather thick, inflated, shining, with or without bands, yellowish-olive; spire ob-

tusely conical; sutures impressed; whorls seven, slightly convex,
the last very large, corded on the periphery; aperture large, rhomboidal, obscurely banded within; outer lip acute, sinuous; columella
thickened below, with reddish spots at the base, and much contorted.

Habitat.— Tennessee; Prof. Troost: Cumberland River; C. T.
Downie: North Alabama; Prof. Tuomey: Ohio River, at Cincinnati;
U. P. James.

Diameter, ·38; length, ·71 of an inch.

Observations.— This is a short thick species with a fine natural
olivaceous polish. A specimen from Prof. Troost has been in my
possession many years, and is the most perfect. It has two obscure
bands inside. Another I recently obtained from Dr.
Hartman, who received it from Prof. Tuomey. A third is
an old eroded specimen, quite brown, sent by Mr. Downie.
After the above description was made, I received from Mr.
James four specimens, neither of them entirely mature,
which he took in the Ohio River at Cincinnati. Two only
have the ligatures round the periphery of the last whorl.

Fig. 134.

Two have four bands, one has two well-defined bands and two are without. One of the two without bands is of very dark brown, and the
other very light brown. The aperture is nearly one-half the length of
the shell. The obsolete bands within are dark brown, but the spot at
the base of the columella is of a bright reddish color. The upper part
of the whorls, which are slightly rounded, is of a yellowish color.
Very different from the description of *Melania ligata*, described by
Menke, Synopsis, 82.— *Lea.*

12. P. affine, Lea.

Trypanostoma affine, Lea, Proc. Acad. Nat. Sci., p. 4, 1864. Jour. Acad. Nat. Sci.,
vi, p. 145, t. 23, f. 57, 1867.

Description.— Shell channelled, pyramidal, horn-color; spire very
much raised; sutures regularly impressed; whorls about nine, channelled, flattened above; aperture subrhomboidal, whitish or banded
within; outer lip acute, sigmoid; columella thickened and very much
twisted.

Habitat.— Smith's Shoals, Cumberland River, East Tennessee.

Diameter, ·60; length 1·35 inches.

Observations.— This species is allied to *Thorntonii* (nobis), and belongs to the group of which *canaliculatum*, Say, may be considered

the type. It differs from that species in having a longer fuse or basal
channel, in which character it approaches the genus *Io*.

Fig. 135.

It is closely allied to *moniliferum* (nobis), but differs in
having a shorter spire; being channelled on the periphery and having no nodules. There is usually a well
defined channel above the periphery, the middle of the
lower whorl being carinate. Below the carina there
is usually a single stria. Two specimens of the four
before me have a broad single band on the upper whorls
and several bands in the interior. The base of the columella is very much twisted backwards, and the edge of
the outer lip is disposed to be thickened. The aperture is rather more
than one-third the length of the shell.— *Lea*.

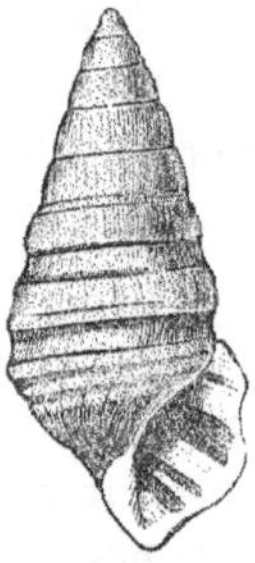

13. P. moriforme, Lea.

Trypanostoma moriforme, Lea, Proc. Acad. Nat. Sci., p. 172, 1862. Jour. Acad. Nat.
Sci., v, pt. 3, p. 290, t. 36, f. 118. Obs., ix, p. 112.

Description.— Shell sulcate, subcylindrical, solid, single banded,
horn-color; spire obtusely conical; sutures impressed; whorls about
nine, impressed canaliculate; aperture rather small, rhombic, white
within, with a single band; outer lip acute, very sinuous; columella
thickened below and very much twisted.

Habitat.— Oostenaula River, near Rome, Georgia; Rev. G. White:
Tennessee River; Dr. Spillman: Tuscumbia, Alabama;
B. Pybas.

Fig. 136.

Diameter, ·52; length, 1·08 inches.

Observations.— This is a well characterized species. I
have nearly forty specimens from different habitats before
me. It is nearly allied to *Melania* (*Trypanostoma*) *infrafasciata*, Anthony, but it differs in being more solid and
being subcylindrical as well as having a more contracted
aperture. It has very much the same kind of fine line near the base.
It is not quite so angular. The aperture is not quite one-third the
length of the shell. It belongs to the group of which *Melania* (*Trypanostoma*) *canaliculata*, Say, may be considered the type.— *Lea*.

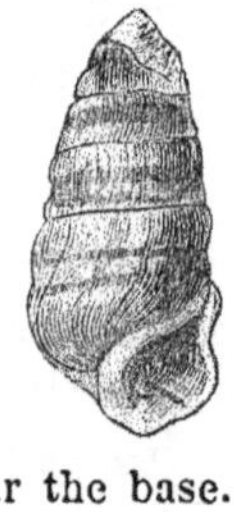

The figure is a copy of Mr. Lea's. The peculiar features of
this species appear to be well preserved in several specimens
before me. Partaking of the general features of *canaliculatum*,
it is yet distinguished by its more cylindrical, elongated form.

14. P. Pybasii, Lea.

Trypanostoma Pybasii, Lea, Proc. Acad. Nat. Sci., p. 172, 1862. Jour. Acad. Nat
Sci., v, pt. 3, p. 289, t. 36, f. 115. Obs., ix, p. 111.

Description.— Shell obtusely carinate, obtusely conical, solid, double-
banded, greenish-brown; spire obtuse; sutures much impressed;
whorls about eight, slightly convex; aperture small, rhombic, white
and banded within; outer lip acute and very sinuous; columella thickened below and very much twisted.

Fig. 137.

Habitat.— Tuscumbia, Alabama; B. Pybas.

Diameter, ·46; length, 1·05 inches.

Observations.— Quite a number of specimens were sent
by Mr. Pybas, which are all very nearly alike. Some are
darker than others. The angle on the periphery of the
whorls is obtuse, and in many specimens obsolete. The
lower whorl is usually flattened, sometimes impressed, making quite a
channel. It is near to *T. moriforme* herein described, but is not so
turgid, is of a darker color and has usually two dark bands inside;
moriforme usually has a thin band but sometimes none. The length
of the aperture is not quite one-third the length of the shell. I name
this after Mr. B. Pybas, to whom I am indebted for it and many fine
species from this vicinity.— *Lea.*

15. P. Showalterii, Lea.

Trypanostoma Showalterii, Lea, Proc. Acad. Nat. Sci., p. 172, 1862. Jour. Acad.
Nat. Sci., v, pt. 3, p. 293, t. 36, f. 122. Obs., ix, p. 115.

Description.— Shell striate, sometimes smooth, much drawn out,
subcylindrical, thick, horn-color or brown, sometimes banded below;
spire much raised; sutures much impressed: whorls nine, somewhat
flattened; aperture small, rhomboidal, whitish or salmon-color within;
outer lip sharp, somewhat sinuous; columella thickened below and
very much twisted.

Operculum ovate, dark brown, with the polar point near to the
base.

· *Habitat.*— Cahawba River, Alabama; Dr. E. R. Showalter: Tuscaloosa, Alabama; Dr. Budd: Oostenaula River, Georgia; Rev. G.
White and Bishop Elliott.

Diameter, ·46; length, 1·38 inches.

Observations.— This is a very remarkable species, having a high subcylindrical spire and a small aperture. Six from the Oostenaula are all more or less striate, two of them having a well defined revolv-

Fig. 138.

ing band near the base on the inside, one has an obsolete band, and the remaining three are without a band. Three of these specimens are of a bright horn-color, the others are dark brown, and one has indistinct bands above the dark one. The thickened part of the columella in three specimens is of a light salmon. Three of the four from Cahawba River are slightly striate, the fourth smooth. These have no bands and are all white on the columella. The aperture is about one-fourth the length of the shell. I have great pleasure in naming this after Dr. Showalter, who has done so much in the development of the Mollusca of his State.

This species is closely allied to *Melania* (*Trypanostoma*) *Ordii* (nobis), but it is more attenuate and more cylindrical.— *Lea.*

C. *Angulate, striate below the periphery.*

16. P. Thorntonii, LEA.

Trypanostoma Thorntonii, LEA, Proc. Acad. Nat. Sci., p. 170, 1862. Jour. Acad.
    Nat. Sci., v, pt. 3, p. 284, t. 36, f. 106. Obs., ix, p. 106.

Description.— Shell carinate, pyramidal, rather thick, horn-color, banded or not banded; spire regularly elevated; sutures somewhat impressed; whorls about ten, flattened; aperture rather small, rhombic, white within; outer lip acute, very sinuous; columella thickened below and very much twisted.

Fig. 139.

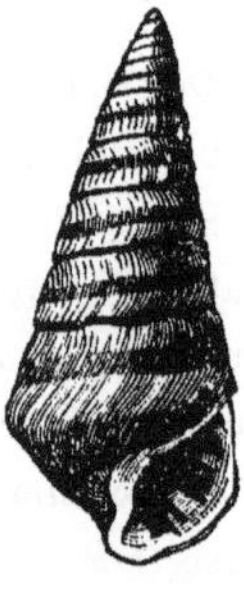

Operculum ovate, dark brown, with the polar point near to the base.

Habitat.—Tuscumbia, Alabama; L. B. Thornton, Esq. and Rev. G. White: Chattanooga, Tennessee; J. Clark.

Diameter, ·62; length, 1·37 inches.

Observations.— This appears to be a common species about Tuscumbia and up the Tennessee River. I have about sixty specimens before me. They came with a large number mixed up with *Mel.* (*Trypanostoma*) *undulata*, Say, but were easily separated from that species. They are always smaller, and none have undulations. Like *undulata* they are usually banded; only eight are without bands entirely. Some specimens have a single broad revolving band on all the whorls, some have several bands, and others again have a capillary line visible on the inside only. Four are dark purplish-green, the color being caused by the broad bands on the inside. It is nearly allied to *T. moriforme* herein described, but is not cylindrical. The specimens are usually of a very regular pyramid with a short base. The carina of the periphery is usually strong, but not always so. In this it is near to *Melania* (*Trypanostoma*) *filum* (nobis), but it is more slender than that species. The aperture is about one-third the length of the shell. Most of the specimens are slightly channelled on the lower whorl. I name it after L. B. Thornton, Esq., to whom I am indebted for many fine specimens of this and other shells.— *Lea.*

This species is shorter in the canal, possesses wider bands and wants the tubercles of *moniliferum* which it otherwise much resembles.

17. P. trivittatum, Lea.

Trypanostoma trivittatum, Lea, Proc. Acad. Nat. Sci., p. 175, 1862. Jour. Acad. Nat. Sci., v, pt. 3, p. 282, t. 36, f. 102. Obs., ix, p. 104.

Description.— Shell smooth, subfusiform, rather thin, shining, olivaceous, three-banded; spire conical, pointed, carinate at the apex; sutures line-like; whorls, eight, flattened, the last one being large; aperture rather large, rhombic, banded within; outer lip acute, sinuous; columella slightly thickened and incurved.

Fig. 140.

Operculum ovate, dark brown, with the polar point near the base.

Habitat.— Tombigbee River, Mississippi; Wm. Spillman, M. D.

Diameter, ·39; length, ·78 of an inch.

Observations.— I have examined about twenty specimens of this species and find them differing very slightly. Every one has three bands, the lower two of which are more distinct on the outside than the upper one, while inside they are well defined and much alike. Three of the specimens are very dark, almost purple, but the bands are distinguishable inside. There is a white line immediately below the sutures. In some specimens there is a disposition to be somewhat angular on the periphery, below which there are transverse striæ in some individuals. The aperture is about three-eighths the length of the shell.— *Lea.*

Very closely allied to *P. Thorntonii*, but a little more convex, with longer canal.

18. P. infrafasciatum, ANTHONY.

Melania infrafasciata, ANTHONY. Proc. Acad. Nat. Sci., p. 57. Feb., 1860. BINNEY, Check List, No. 148. BROT, List, p. 30. REEVE, Monog. Melania, sp. 301.

Description.—Shell conical, smooth, solid, of a pale brown color, form moderately slender and elevated; whorls 8–9, decollate, slightly concave; sutures distinct; lines of growth curved and very distinct;

Fig. 141.

body-whorl decidedly concave, with a well marked ridge revolving near the summit of the aperture, so as to make a tolerably sharp angle near the middle of the body-whorl; two or three coarse striæ revolve parallel with it; below this is a dark brown band, continued around the base of the shell; aperture rhombic, ovate, livid and banded within; columella strongly incurved, with a callous deposit its whole length and well defined sinus at base.

Observations.— Compared with *M. gradata*, nobis, it is more elongate, more solid and has not the carina and regularly graded whorls so characteristic of that species; less conical than *M. canaliculata*, Say, and less broad.   Like *M. annulifera*, Con., in form, but has not the revolving costæ of that species.— *Anthony.*

The figure above is from Mr. Anthony's type.

18a. P. fastigiatum, ANTHONY.

Melania fastigiata, ANTHONY. Ann. N. Y. Lyc., vi, p. 113, t. 3, f. 13, March, 1854. BINNEY, Check List, No. 108. REEVE, Monog. Melania, sp. 302.

Description.— Shell conical, smooth, moderately thick; of a pale yellowish-green color, ornamented with two distinct, distant, reddish-brown bands on each whorl, except those near the apex, which are carinate; spire elevated, rising from the broad body-whorl with regularly decreasing volume in a pyramidal form to the acute apex; whorls ten, not convex, with rather indistinct sutures in a furrowed channel; lines of growth curved and strong, particularly on the penult and body-whorl, where they are almost folds; body-whorl distinctly carinated, having one carina at the middle, another short distance below, with a broad band immediately above the carinæ, and

Fig. 142.

another far within, near the base. Aperture small, subrhomboidal, whitish within, three bands visible in the interior; columella nearly straight, a little thickened, outer lip very much curved, auger-like; sinus narrow, recurved.

Habitat.—Tennessee.

Diameter, ·38 of an inch (10 millim.); length, ·80 of an inch (20 millim.). Length of aperture, ·32 of an inch (8 millim.); breadth of aperture, ·16 of an inch (4 millim.).

Observations.—A fine symmetrical species, which is, perhaps, most nearly allied to *M. vestita*, Conr.; from that shell it differs in being less ponderous, more acute in its outline, and in its flat whorls, the *M. vestita* being angulated below the middle; it has also a double band, while *vestita* has a single one. From *M. elevata*, Say, it differs by its less slender outline, its want of "thread-like carinæ" on the whorls, and its lines of growth are more curved, more elevated and more distant; differs from *M. spinalis*, Lea, by not having carinated whorls, by its more delicate color, and it has not the superior part of the whorl darker than below, as described in *M. spinalis.—Anthony.*

Figured from the type. This species is very close to *Thorntonii*, Lea, but its outline is narrower. It may also be compared with *infrafasciatum*, but differs in having more acutely carinated whorls and a·longer, more distinct fuse. The two narrow bands are present in all the specimens I have examined.

19. P. Postellii, Lea.

Trypanostoma Postellii, Lea, Proc. Acad. Nat. Sci., p. 171, 1862. Jour. Acad. Nat. Sci , v, pt. 3, p. 286, t. 36, f. 110. Obs., ix, p. 108.

Description.— Shell carinate, pyramidal, rather thick, horn-color; spire regularly conical; whorls eight, flattened, the last rather small; aperture very small, rhomboidal, whitish within; outer lip acute, very sinuous; columella thickened below and very much twisted.

Fig. 143.

Habitat.—Tennessee River; J. Postell: North Alabama; Prof. Tuomey.

Diameter, ·35; length, ·85 of an inch.

Observations.—I have from Mr. Postell eight specimens, and from Professor Tuomey, five. They vary very little, but most of them are imperfect at the apex or outer lip. This species very closely resembles *Thorntonii* herein described, but is a much smaller species, with a smaller aperture and compressed whorls. All the specimens before me are more or less angulate on the periphery. None have bands. The aperture is about two-ninths the length of the shell. I name this after Mr. Postell, to whom I am indebted for specimens of this and many other new species of Mollusca.— *Lea.*

This species is closely allied to *infrafasciatum* but may be distinguished by its whorls being more flattened, and by its narrower form.

20. P. incurvum, LEA.

Trypanostoma incurvum, LEA, Proc. Acad. Nat. Sci., p. 171, 1862. Jour. Acad. Nat. Sci., v, pt. 3, p. 286, t. 36, f. 109. Obs., ix, p. 168.

Description.— Shell carinate, conical, rather thin, horn-color; spire somewhat elevated; sutures regularly impressed; whorls eight, flattened, obscurely striate below; aperture rather small, rhombical, whitish within; outer lip acute, extremely sinuous; columella very much twisted.

Fig. 144.

Habitat.— Florence, Alabama; Rev. G. White.

Diameter, ·37; length, ·89 of an inch.

Observations.— Among the *Melanidæ* sent to me by Mr. White, I found three specimens of this species which, being near to *Thorntonii*, herein described, evidently was supposed to be the same species. It is, however, a smaller, thinner and more slender species, and the remarkable sinuous edge of the outer lip at once marks the difference. The inward curve, starting at once in that direction from the suture, turns forward before it reaches the periphery of the whorl and again curves to the base, making a complete sigmoid curve. The aperture is about one-third the length of the shell.— *Lea.*

This species resembles the last but is very distinct in the incurved tip. It differs from *infrafasciatum* by the same characters as *Postellii*.

21. P. Alabamense, Lea.

Trypanostoma Alabamense, Lea, Proc. Acad. Nat. Sci., p. 171, 1862. Jour. Acad. Nat. Sci., v, pt. 3, p. 288, t. 36, f. 113. Obs., ix, p. 110.

Description.— Shell carinate, somewhat thick, subfusiform, dark horn-color; spire somewhat attenuate; sutures regularly impressed; whorls about eight, flattened, striate below; aperture rather small, rhomboidal, whitish within; outer lip acute, sinuous; columella blackened below and very much twisted.

Fig. 145.

Habitat. — North Alabama; Prof. Tuomey: Florence, Alabama; Rev. G. White.

Diameter, ·46; length, 1·11 inches.

Observations.—This species is allied to *Florencense*, herein described in outline, but is a much smaller species, less exserted in the spire, of a much lighter color and with fewer whorls. The three specimens before me differ but little in size or color, neither has a perfect apex, and therefore the character or the exact number of the upper whorls cannot be ascertained. They all have a few indistinct revolving striæ below the periphery of the last whorl. The aperture is about one-third the length of the shell.— *Lea.*

Very distinct from the preceding two species in the longer spire and canal. A variety with a light line below the sutures and yellowish-brown within occurs in Powell's River, Cumberland Gap, E. Tennessee.

21a. P. Florencense, Lea.

Trypanostoma Florencense, Lea, Proc. Acad. Nat. Sci., p. 171, 1862. Jour. Acad. Nat. Sci., v, pt. 3, p. 287, t. 36, f. 112. Obs., ix, p. 109.

Description.— Shell subcarinate, turreted, rather thick, dark brown or yellowish horn-color; spire very much raised; sutures slightly impressed; whorls about eleven, slightly convex; aperture rather small, rhombic, within bluish-white; outer lip acute, sinuous; columella whitish and very much twisted.

Habitat.—Florence, Alabama; Dr. Spillman: Tuscumbia; L. B. Thornton, Esq.

Diameter, ·59; length 1·65 inches.

Observations.— This is a large, rather slim species. Among eight

specimens, the longest is one inch and six-tenths. It is nearly allied
to *Melania* (*Trypanostoma*) *elongata* (nobis), but is not carinate

Fig. 146.

like that species, nor are the whorls so flat. The two
specimens from Florence are larger, and very dark
brown. Of the six from Tuscumbia, four are yellowish,
and two are banded and greenish. Two of the yellowish
ones are disposed to salmon-color inside. There is a
slight disposition above the periphery to flatness or indentation. The aperture is more than the fourth of the
length of the shell.—*Lea*.

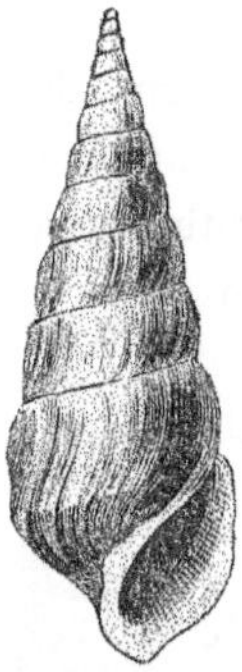

I have seen some specimens from Coosa River,
Alabama, in which the whorls are more convex than
Mr. Lea's figure. The species has a more extended
distribution than the above localities would indicate, Mr. Lea
having specimens from New Harmony, Indiana.

The preceding species (*Alabamense*) may prove to be the
young of this shell.

22. P. olivaceum, LEA.

Trypanostoma olivaceum, LEA, Proc. Acad. Nat. Sci., p. 172, 1862. Jour. Acad. Nat.
  Sci., v, pt. 3, p. 2:0, t. 3:, f. 117. Obs. ix, p. 112.

Description.—Shell carinate, subfusiform, rather thick, olivaceous;
spire rather obtuse; sutures impressed; whorls about eight, flatttened;
aperture rather large, rhomboidal, whitish within; outer   Fig. 147.
lip sharp, sinuous; columella thickened below and very
much twisted.

Operculum ovate, dark brown, with polar point near to
the base.

Habitat.—Tombigbee River, Mississippi; W. Spillman,
M. D.

Diameter, ·50; length, 1·06 inches.

Observations.— Dr. Spillman sent me quite a number of this species.
In outline and size it is very near to *Strephobasis olivaria* (nobis), but it
differs in the base of the columella, which separates it from the genus
Strephobasis, and it is more flattened on the whorls, and is not banded;
except in rare cases it has an obscure small band near the base. The
olive-green hue of the epidermis is very constant. The carina generally leaves a thread-like line along the suture. The aperture is about
one-third the length of the shell.—*Lea*.

This shell is very nearly allied to *P. ponderosum*, Anthony (*P. dux*, Lea). The figure is from Mr. Lea's plate but differs in the form of the aperture, in color and in size.

22a. P. canalitium, LEA.

Trypanostoma canalitium, LEA, Proc. Acad. Nat. Sci., p. 175, 1862. Jour. Acad. Nat. Sci., v, pt. 3, p. 292, t. 36, f. 121. Obs., ix, p. 114.

Description.— Shell canaliculate, conical, rather thick, horn-color, obscurely banded; spire regularly conical, somewhat raised, double-banded towards the point; sutures impressed; whorls about Fig. 118. seven, flattened, the last canaliculate; aperture small, rhomboidal, white or salmon, and banded within; outer lip sharp and sigmoid; columella twisted, recurved at the base.

Habitat.— Yellowleaf Creek, Alabama; E. R. Showalter, M. D.

Diameter, ·43; length, ·99 of an inch.

Observations.— Three specimens are before me all of the same size, and having the appearance of half-grown *Melania* (*Trypanostoma*) *canaliculata*, Say, but they are mature and evidently distinct. The channel above the middle of the whorl is smaller, but well characterized. In the form of the aperture they are very much the same, being auger-shaped like *Cerithium*. It is very nearly allied to *Melania* (*Trypanostoma*) *infrafasciata*, Anth., from Tennessee, but may be distinguished by its channel above the middle of the whorls, and in having three bands visible in the interior, while the *infrafasciata* has but one, as described by Mr. Anthony, and none on the superior whorls, as all our three have. The aperture is about three-tenths the length of the shell.— *Lea*.

This figure is a copy of Mr. Lea's. In specimens of this shell, from Columbus, Miss., the canal is much better developed than in the above figure.

23. P. Clarkii, LEA.

Trypanostoma Clarkii, LEA, Proc. Acad. Nat. Sci., p. 171, 1862. Jour. Acad. Nat. Sci., v, pt. 3, p. 285, t. 36, f. 108. Obs. ix, p. 107.

Description.— Shell obtusely carinate, conical, rather thick, dark olive; spire raised; sutures very much impressed; whorls about eight, flattened; aperture rather small, rhomboidal, within whitish; outer lip acute, sinuous; columella white and twisted.

Fig. 149.   Fig. 150.

Operculum ovate, dark brown, with the polar point near the basal margin.

Habitat. — French-broad and Tellico Creeks, Tennessee; J. Clark and Prof. Christy: Florence, Alabama; Rev. G. White: Noxubee River, Mississippi; Dr. Spillman: Clinch River, Tennessee; Dr. Warder: and Coosa, Cahawba and Alabama Rivers, Alabama; Dr. Showalter.

Diameter, ·46; length, 1·13 inches.

Observations. — This species has the color of *Spillmanii*, herein described, but it is a smaller and thicker species, and has a distinct carina. It is also less attenuate. The specimen from Clinch River is pale horn-color. Those from Tellico Creek are nearly all furnished with 2–4 bands. Two or three from French-broad are of a deep purple. The aperture is about one-third the length of the shell.

I have great pleasure in naming this after my deceased friend, Joseph Clark, to whom I am indebted for many species brought by Prof. Christy.— *Lea.*

I doubt whether this species is really distinct from *P. canalitium*. It appears, however, to be rather a broader shell proportionally, with a better developed carina and recurved canal. Both are common species.

24. P. Anthonyi, Lea.

Trypanostoma Anthonyi, Lea, Proc. Acad. Nat. Sci., p. 172, 1862. Jour. Acad. Nat. Sci., v, pt. 3, p. 293, t. 36, f. 123. Obs., ix, p. 115.

Description. — Shell rugosely striate, pyramidal, thick, yellowish, olive; spire raised; subrugosely impressed; whorls about nine, flattened; aperture rather large, rhomboidal, white within; outer lip acute, sinuous; columella thickened below and very tortuous.

Fig. 151.

Operculum subovate, dark brown, with the polar point near to the base on the left.

Habitat. — Tennessee; J. G. Anthony: Warrior River and Yellow Leaf Creek, Alabama; Dr. Showalter: Fox River, Indiana; J. Sampson.

Diameter, ·63; length, 1·43 inches.

Observations. — A number of specimens of this fine large species are before me from various habitats. It is allied to *Melania* (*Trypanostoma*) *canaliculata*, Say, but it may easily be distinguished from it by the absence of a regular canal, and being a less ponderous shell. The color, too, is more of a yellow-green; usually there are three or four rather coarse striæ about the middle of the whorl, which form irregular canals. The canal at the base is wide and much recurved. Some specimens are almost entirely smooth, and some are 1¾ inches long. The aperture is about one-third the length of the shell. I name this after Mr. J. G. Anthony, to whom I am indebted for several fine specimens, and many other species from Tennessee.— *Lea.*

Fig. 152.

The first figure is from a Tennessee specimen, and is a copy of that given by Mr. Lea. The shells quoted from " Fox River, Indiana, J. Sampson," are more closely allied to *Florencense*, and are probably identical with that species.

This shell appears to be distinct from its congeners, but approximates closely to *Florencense* on one side and *Troostii* on the other side. It is a common species.

25. P. prasinatum, Conrad.

Melania prasinata, Conrad, Am. Jour. Sci., 1st ser., xxv, p. 342, t. 1, f. 14, January, 1834. Jay, Cat., 4th edit., p. 274. Binney, Check List, No. 216. Brot, List, p. 33. Catlow, Conch. Nomenc., p. 188. DeKay, Moll. N. Y., p. 98. Reeve, Monog. Melania, sp. 403.

Fig. 153.   Fig. 154.

Description. — Shell subulate, slightly turreted, whorls seven or eight, flattened, aperture elliptical, a little oblique; about one-third of the length of the shell; body-whorl sub-angulated at base; epidermis green-olive.

Var. A. With broad revolving costæ, those on the body-whorl crenulated. Inhabits Alabama River, adhering to limestone rocks. Cabinet of the Academy of Natural Sciences of Philadelphia.—*Conrad.*

L. F. W. S. IV.

25a. P. incrassum, ANTHONY.

Melania incrassata, ANTHONY, Ann. Lyc. N. Y., vi, p. 99, t. 2, f. 17, March, 1854.
  BINNEY, Check List, No. 144. BROT, List, p. 34.
Trypanostoma Hartmanii, LEA, Proc. Acad. Nat. Sci., p. 173, 1862. Jour. Acad.
  Nat. Sci., v, pt. 3, p. 270, t. 36, f. 80. Obs., ix, p. 92.
Trypanostoma bivittatum, LEA, Proc. Acad. Nat. Sci., p. 175, 1862. Jour. Acad
  Nat. Sci., v, pt. 3, p. 279, t. 36, f. 97. Obs., ix, p. 191.

Description.—Shell conical, smooth, thick; spire elevated; whorls
8-9, very convex, somewhat biangulated; sutures deeply impressed;
body-whorl striated, with a constriction about the middle, which
also extends to the penultimate whorl; aperture ovate, within
reddish; columella not indented, reflected, sinus deep.

Habitat.———?

My Cabinet.

Diameter, ·45 of an inch (12 millim.); length, 1·12 inches (29 millim.). Length of aperture, ·37 inch (9 millim.); breadth of aperture,
·18 inch (4½ millim.).

Observations.—Only one specimen has come under my notice, which,
however, is so unlike any other that I cannot hesitate to consider it
new.—*Anthony.*

Fig. 155.

It is a thick, ponderous species, with narrow
convex or biangulated whorls, faintly banded on
the angulations.

Trypanostoma Hartmanii.— Shell smooth, sometimes obscurely channelled, solid, greenish, or reddish-brown, regularly conical, banded or without bands; spire pyramidal;
sutures regularly impressed; whorls about nine, slightly
convex; aperture small, rhombic, white or salmon-color
within; outer lip acute, sinuous; columella thickened
below and very much twisted.

Habitat.—Cahawba and Coosa Rivers; Dr. Showalter: Warrior
River, Alabama; Dr. Budd: Knoxville; J. Clark: Tennessee River,
Alabama; Dr. Spillman.

Diameter, ·50; length, 1·25 inches.

Observations.—Two or three specimens of this fine species have been
in my collection for a long time, and were given to me under the name
of *Melania pyrenella,* Con., but Mr. Conrad's shell is not so solid, has
flatter whorls and is carinate. Some of the specimens of *Hartmanii*
are furnished with two broad bands, which are usually well marked

inside, others are without bands, and these are usually salmon-colored within. Three of the specimens out of some thirty before me are of a rich dark brown, which arises from the interior nacre being purplish. The aperture is more than one-third the length of the shell. I have great pleasure in naming this after my friend W. D. Hartman, M.D., who has furnished me with a number of fine specimens.*— *Lea.*

P. bivittatum.— Shell smooth, conical, rather thick, yellow, double-banded; spire obtusely conical; sutures much impressed; whorls seven, rather convex, the last one large; aperture rather large, somewhat rhomboidal, white and double-banded within, outer lip acute, somewhat sinuous; columella thickened below and very much twisted.

Fig. 156.

Habitat.— Tennessee; Prof. Troost.

Diameter, ·34; length, ·68 of an inch.

Observations.— This is a small robust species. Five specimens came many years since from Prof. Troost, mixed with many young specimens of *M. canaliculata*, Say, to which it has some resemblance, but it may easily be distinguished by its shorter spire, and larger body-whorl. All the specimens have two regular deep brown bands. The aperture is about two-fifths the length of the shell. Two or three of these specimens were mixed with some young shells from Cincinnati, I think by accident, but still it is possible that they may have come from Cincinnati.— *Lea.*

Figured from Mr. Lea's plate. There can be no doubt that this is the young of Mr. Lea's *Hartmanii*.

25b. P. Jayi, Lea.

Trypanostoma Jayi, Lea, Proc. Acad. Nat. Sci., p. 173, 1862. Jour. Acad. Nat. Sci., v, pt. 3, p. 270, t. 36, f. 81. Obs., ix, p. 92.

Description. — Shell smooth, pupæform, thick, shining, reddish-brown; spire obtusely conical; sutures very much impressed; whorls eight, rather swollen, the last rather large; aperture small, rhom-

---

*Since the above was written, a letter received from Dr. Hartman says, that Dr. Showalter informed him that "the orange color of the animal is remarkable." Dr. Hartman also mentions that he and Dr. Showalter had distributed this shell under the name of *Melania pyrenella*, Con., which mistake Dr. Hartman corrected by reference to the type specimen, which is in the collection of the Academy of Natural Sciences.— *Lea.*

boidal, rather narrow, pale brown within; outer lip acute, sinuous; columella thickened below and twisted.

Fig. 157.

Habitat.— Alabama? J. C. Jay, M.D.

Diameter, ·46; length, 1·16 inches.

Observations.— A single specimen was given to me many years since by Dr. Jay under the name of *Melania prasi-nata*, Con., but it is a very different shell from the type of that species in the collection of the Academy of Natural Sciences, that being of a greenish color, having a few nodes round the periphery, which is angulated, neither of which characters belongs to *Jayi*. Indeed, our shell is much nearer to *clausa* (nobis) in outline, but it is not so pupæform, and it has a more twisted columella, the spire being more conical.

It is to be regretted that a single specimen only should be under observation, as others may be different in color. The interior as well as the columella is of a dull salmon, and the darkness is occasioned by obscure bands which do not extend quite to the edge, which is slightly thickened. The aperture is not quite one-third the length of the shell. I name this species after Dr. Jay, to whom I owe the possession of it, and who has done so much to advance a knowledge of our conchology.— *Lea.*

26. P. tortum, LEA.

Trypanostoma tortum, LEA, Proc. Acad. Nat. Sci., p. 174, 1862. Jour. Acad. Nat. Sci., v, pt. 3, p. 275, t. 36, f. 89. Obs., ix, p. 97.

Description.— Shell smooth, conical, horn-color, rather thick; spire rather obtusely conical; sutures very much impressed; whorls seven, flattened; aperture rather large, subrhomboidal, white or brownish within; outer lip acute, scarcely sinuous; colu-

Fig. 158.

mella very much incurved, slightly thickened above, more thickened below and very much twisted.

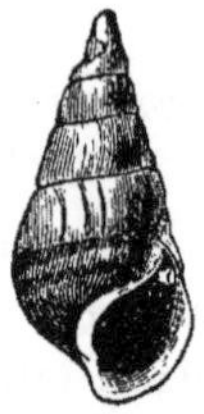

Habitat.— Little Uchee, below Columbus, Georgia; G. Hallenbeck.

Diameter, ·44; length, ·96 of an inch.

Observations.— Several specimens of this species are before me. In one of the specimens there are three or four obscure striæ about the periphery. It is probable that others may be found with this character more developed. On the upper whorls there is a raised line revolving immediately above the suture, which causes the suture to be more impressed. The columella is more than usually twisted, whence the name of the species. Two of the specimens are of a dull brown within, but have a whitish margin. The aperture is rather more than the third of the length of the shell.— *Lea.*

27. P. dignum, LEA.

Trypanostoma dignum, LEA, Proc. Acad. Nat. Sci , p. 273, 1862. Jour. Acad. Nat. Sci., v, pt. 3, p. 350, t. 39, f. 219. Obs., ix, p. 172.

Description.— Shell slightly noduled, subfusiform, somewhat thick, honey-yellow, single-banded spire raised, regularly conical; sutures impressed; whorls about eight, flattened, the last rather large; aperture ovately rhombic, salmon or white within, single-banded within; outer lip acute, sinuous; columella bent in, twisted, obtusely angular at the base.

Fig. 159.

Habitat.— Yellowleaf Creek, Shelby County, Alabama; E. R. Showalter, M. D.

Diameter, ·52; length, 1·06 inches.

Observations.— I have two specimens of this beautiful species before me. The smaller has a well-defined row of small tubercles on the middle of the whorls. The larger has an ill-defined, obscure row, which is partly made up by a raised line. Below this is a well-marked capillary, brown band, which is distinct outside and in. The clear, bright, smooth epidermis is of a honey-yellow, inclining to brown. In outline it is near to *Melania (Goniobasis) Vanuxemiana* (nobis), but it cannot be confounded with that species. The aperture is more than one-third the length of the shell.— *Lea.*

D. *Carinate, striate Pleuroceræ.*

28. P. unciale, HALDEMAN.

Melania uncialis, HALD., Monog. Limniades, No. 4, p. 3 of Cover, Oct. 5, 1841. JAY, Cat., 4th edit., p. 275. BINNEY, Check List, No. 279. BROT, List, p. 37. REEVE, Monog. Mel., sp. 435.

Melania oblita, LEA, Philos. Trans., x, p. 298, t. 30, f. 6. Obs., v, p. 54. BINNEY, Check List, No. 182. BROT, List, p. 36.

Melania bicostata, ANTHONY, Proc. Acad. Nat. Sci., p. 56, February, 1860. BINNEY, Check List, No. 33. BROT. List, p. 30. REEVE, Monog. Melania, sp. 246.

Melania rigida, ANTHONY, Proc. Acad. Nat. Sci., p. 62. February, 1860. BINNEY, Check List, No. 229. REEVE. Monog. Melania, sp. 270.

Melania sugillata, REEVE, Monog. Mel., sp. 319, September, 1860. BROT, List, p. 31.

Description.— Shell pale olivaceous, turreted, with eight or ten slightly convex whorls, the earlier ones of which are strongly carinated; lines of growth curved; aperture ovate, with a sinus anteriorly. One-inch long.

Habitat.— Beaver Creek, N. E. Tennessee.

Observations.— Bears a general resemblance to *M. Virginica*. As far as I can judge from the description, it must be somewhat like *M. Warderiana*, Lea.— *Haldeman.*

Fig. 160.   Fig. 161.

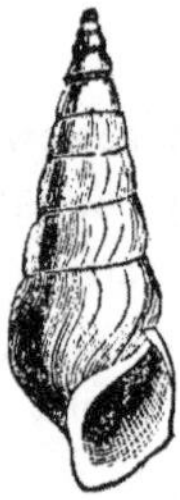

The figure is from Prof. Haldeman's type specimen. It is a common species, and inhabits also West Virginia.

The following appear to me to be synonymes :

M. oblita.— Shell very much carinated, turreted, screw-shaped, rather thin, horn-colored; spire drawn out; sutures linear; whorls twelve, acutely carinate; aperture small, elliptical, within whitish ; columella white and twisted.

Habitat.— Tennessee?

Diameter, ·30; length, ·96 of an inch.

Observations.— I have about a dozen of this species, which is very distinct from any with which I am acquainted. The locality I am uncertain about, the label being by some accident lost. I believe it comes from Tennessee, but am not certain. Its very marked character of a screw, or rather of a gimlet, strikes one at once. In most species there is a thread-like line above the carina and several below. The carina is not usually persistent on the body-whorl. It is nearest in form and size to *M. percarinata*, Con., but may be easily distinguished by the absence of granules between the carinæ, the length o the spire, having three or four more whorls, and in being less shiny. The aperture is not quite one-third the length of the shell.— *Lea.*

Fig. 162.

Fig. 163.

Melania bicostata.— Shell conical, light horn-color, rather thick; spire elevated, acute ; whorls 11–12, strongly carinate near the apex and decidedly so on each succeeding whorl, not excepting even the body-whorl in most cases, though sometimes obsolete there; carinæ often in pairs, near to and parallel with each other; sutures deeply impressed, often with a decided furrow at that point, caused by the carinæ. Aperture broadly elliptical, or subrhombic; within dirty-white or obscurely banded; columella deeply rounded, with a well marked sinus at base.

Habitat.— Tennessee, near Athens.

Observations.—Appears to be a very abundant and rather variable species. Several hundred individuals have come under my notice. It cannot well be confounded with any other species, though of a form by no means uncommon. The sharp double carinæ will at once generally determine it. Occurs abundantly near Athens, in small streams. —*Anthony.*

The figure illustrates one of Mr. Anthony's type specimens. The following is the young of *bicostatum.*

M. rigida.— Shell conic, elevate, carinate, rather thin; whorls 8-9, carinate and banded; sutures distinctly marked; aperture small, elliptical, whitish within; columella indented; sinus small Figs. 164, 165. but very distinct.

Habitat.— Tennessee.

Observations.— This is one of those sharply keeled Melaniæ of which *M. bella*, Conr., *M. carino-costata* and *M. oblita*, Lea, may be considered good examples. The whorls of the spire have each two carinæ with generally a dark band between them though this is sometimes wanting; the body-whorl has four or five of these carinæ and generally two bands, one of which revolves within the aperture. To the touch this species has a peculiarly rough feel.— *Anthony.*

Figure 165 is from Mr. Anthony's type.

Fig. 166.

M. sugillata.— Shell acuminately turreted, livid gray, whorls ten to eleven, the first few encircled with a very sharp keel, the rest smooth; aperture rotundately ovate, columella twisted, sinuately reflected at the base.

Habitat.— Alabama.

Observations.— Of a smooth, livid, bruised aspect, encircled towards the apex with a particularly prominent fine keel, which soon disappears.— *Reeve.*

The above figure is copied from Reeve. *Generally*, but little dependence can be placed in the correctness of the localities given for American species of *Strepomatidæ* in the Cumingian collection—and in the present instance, the locality may be questioned, as the species is rather of the Tennessee type.

29. P. subulare, LEA.

Melania subularis, LEA, Philos. Trans., iv, p. 100, t. 15, f. 30. Obs., i, p. 110, t. 15 f. 30. RAVENEL, Cat., p. 11. DEKAY, Moll. N. Y., p. 92, t. 7, f. 138. WHEATLEY, Cat. Shells U. S., p. 27. JAY, Cat., 4th edit., p. 275. BINNEY, Check List, No. 257. BROT, List, p. 35. REEVE, Monog. Melania, sp. 428. WHITEAVES, Canad. Naturalist, viii, p 102, April, 1863.
Ceriphasia subularis, Lea, ADAMS, Genera, i, p. 287.

Description.—Shell elevated and acutely turreted, horn-color; apex acute; whorls about twelve, flat, carinate on the middle of the body-whorl; base angulated; aperture white and one-fourth the length of the shell.

Fig. 167.

Habitat.—Niagara River.

Diameter, ·4; length, 1·3 inches.

Observations.—I took this species at the Falls of Niagara, and being unable to refer it to any described species, have given it a place here. It resembles the *Virginica* (Say), but differs greatly in elevation, the *Virginica* having about seven whorls only. The carina causes the whorls to be flatter in the *subularis*. In some specimens the columella is purple.—*Lea.*

This is one of our most beautiful species; the clear, polished surface is quite translucent, banded below the sutures by yellow and light blue. It appears to be a common species in the great lakes and their tributaries.

Fig. 167 is a copy of Mr. Lea's.

The species is reported from St. Lawrence River, by Mr. Whiteaves.

29 a. P. intensum, ANTHONY.

Melania intensa, Anthony, REEVE, Monog. sp. 371. BROT, List, p. 30.

Description.—Acuminated, purple-black, whorls ten, flatly convex, encircled with a keel above the sutures, last whorl slightly angled and ridged at the base; aperture rather small, purple-black.

Anthony. MSS. in Mus. Cuming.

Habitat.—United States.

A very characteristic purple-black shell, encircled by a keel so near to the suture as to give them an appearance of being more than usually excavated.— *Reeve.*

I have seen specimens of this shell, but without locality attached to the label. It much resembles *subulare*, Lea, and may be a variety of that species, but I have seen no specimens of the latter species which at all resemble this in color.

Fig. 167 *a*.

The specimens before me and also Mr. Reeve's specimen, as exhibited by his figure, are ornamented by a narrow yellowish band below the sutures.

30. P. subulæforme, Lea.

Trypanostoma subulæforme, Lea, Proc. Acad. Nat. Sci., p. 174, 1862. Jour. Acad. Nat. Sci., v, pt. 3, p. 289, t. 36, f. 116. Obs., ix, p. 111.

Description.—Shell carinate, subulate, rather thin, horn-color; spire attenuately conical; sutures very much impressed; whorls ten, flattened below and carinate above; aperture small, subrhomboidal, whitish within; outer lip acute, sinuous; columella slightly thickened and twisted.

Operculum ovate, dark brown, with the polar point near the base slightly on the left.

Habitat.—Knoxville, Tennessee; Prof. Troost and W. Spillman, M.D.

Fig. 168.

Diameter, ·39; length, 1·07 inches.

Observations.—This species is nearly allied to *Melania* (*Trypanostoma*) *bicostata*, Anth., and in outline and size very close to *Melania* (*Trypanostoma*) *Ocoeénsis* (nobis). From *bicostata*, it may be distinguished by the difference in the aperture, in being more subulate and in having the carina less marked. The channel of *bicostata* is more retrorse and more angular at the point. The aperture is about one-fourth the length of the shell. Two of the three specimens before me are without any bands, the third has a well-defined brown band within the aperture. It is nearly the same in outline as *attenuatum* herein described, but differs in the form of the aperture and in being carinate.

I doubt whether this is more than the adult form of *P. Henryanum*, Lea.

31. P. Henryanum, LEA.

Trypanostoma Henryanum, LEA. Proc. Acad. Nat. Sci., p. 272, 1862. Jour. Acad.
Nat. Sci., v, pt. 3, p. 351, t. 39, f. 222. Obs., ix, p. 173.

Description.—Shell carinate, attenuate, sharp-pointed, thin, semi-
transparent, pale horn-color, without bands; spire regularly
attenuately conical; sutures regularly impressed; whorls ten,
flattened, the last one regularly carinate and striate in the
middle; aperture, small, subrhomboidal, whitish within; outer
lip very sharp and sinuous; columella bent in and very much
twisted.

Fig. 169.

Habitat.—Tennessee? Smithsonian Institution.

Diameter, ·29; length, ·80 inch.

Observations.—Among the *Melanidæ* sent to me by Prof. Henry,
Secretary of the Smithsonian Institution, were a few of this species,
which I at first regarded as a variety of *Melania* (*Trypanostoma*)
uncialis, Hald., but it is certainly a distinct species. In the spire it is
very much the same, but the color is paler, and in the form of the
aperture it is quite different,—*uncialis* having a retrorse channel at the
base while our species curves towards the front and has a more delicate columella, and is altogether more fragile. All the specimens
before me have six revolving striæ on the lower whorl, below the
periphery. The aperture is not quite one-third the length of the
shell.

I have sincere pleasure in dedicating this species to my friend Prof.
Joseph Henry, Secretary of the Smithsonian Institution, who liber-
ally has placed the fresh-water mollusca of that admirable Institution
under my examination.— *Lea.*

32. P. Lewisii, LEA.

Trypanostoma Lewisii, LEA, Proc. Acad. Nat. Sci., p. 172, 1862. Jour. Acad. Nat.
Sci., v, pt. 3, p. 292, t. 36, f. 120. Obs., ix, p. 114.

Description. — Shell sulcate, somewhat thin, high, conical, dark
brown or horn-color, banded; spire very much drawn out; sutures
slightly impressed, whorls about eleven, flattened; aperture small,
subrhomboidal, banded within; outer lip acute, slightly sinuous;
columella slightly thickened below and very much twisted.

Habitat.—Peoria, Illinois; J. Lewis, M. D.

Diameter, ·47; length, 1·12 inches.

Observations.—I have three specimens before me, all of which differ slightly. Two are dark brown and they are purple within. The third is light horn-color, with light brown bands covering the greater part of the whorls. The upper whorls of all three are carinate. It is allied to *Melania (Trypanostoma) annulifera*, Con., but it is a smaller shell, more attenuate, and the aperture is more rounded at the base. The aperture is about one-fourth the length of the shell. I have great pleasure in calling this after my friend Dr. Lewis, of Mohawk, New York, who has aided me greatly by sending me very many new shells from our fresh waters.—*Lea.*

Fig. 170.

This species may be only a striate form of *elevatum*, Say.

33. P. annuliferum, CONRAD.

Melania annulifera, CONRAD, New Fresh Water Shells, p. 51, t. 8, f. 2, 1834. JAY, Cat., 4th edit., p. 272. BINNEY, Check List, No. 17. DEKAY, Moll. N. Y., p. 94. WHEATLEY, Cat. Shells U. S., p. 24. BROT, List, p. 30. CATLOW, Conch. Nomenc., p. 185. REEVE, Monog. Melania, sp. 308. MÜLLER, Synops. 44.
Melania annulata, Conrad. JAY, Cat., 2nd edit., p. 455.
Melania Ordiana, LEA, Philos. Proc., ii. p. 242. Dec., 1842. Philos. Trans. ix, p. 26. Obs., iv, p. 26. WHEATLEY, Cat. Shells U. S., p. 26. BINNEY, Check List, No. 191. BROT. List, p. 30.
Ceriphasia annulifera, Conr., ADAMS, Genera, i, p. 297.
Ceriphasia Ordiana, LEA, ibid., p. 297.

Description.— Shell elevated, subconical, with flattened whorls and elevated, distant ribs, alternately smaller; about five on the body whorl and three on the adjoining one; suture obsolete; color generally blackish exteriorly and dark purple within.

Fig. 171. Fig. 172. Fig. 173. Fig. 174.

Observations. — Inhabits with the preceding species, from which it differs in being less ventricose, and having the ribs plain; the aperture is shorter than in the preceding The three specimens figured are from Alabama; it will be noticed that in one of them, the central striæ are tuberculate, thus forming a connection with *Foremanii*, Lea. — *Conrad.*

The following is regarded as a synonyme :—

Mel. Ordiana.—Shell striate, pyramidal, dark brown; spire drawn out; sutures deeply impressed; whorls flattened; aperture rhombic; small, whitish.

Habitat.—Alabama.

Diameter, ·52; length, 1·25 inches.

Observations.—A single specimen only of this species is before me,

Fig. 175.

and that unfortunately is decollate, in having lost, probably, four or five whorls : the four lower whorls are perfect. The outer lip is much curved, giving the aperture an auger-like appearance and causing the channel to be much impressed. On the body-whorl there are four rather distant elevated striæ, three of which are large; the whorls above exhibit two. The aperture is about one-fourth the length of the shell. This species resembles *M. canaliculata* (Say), and *M. annulifera* (Conr.). It has not the channel of the former, and differs from the latter in having deeply impressed sutures, in the form of the aperture, in the outer lip and in the striæ. I dedicate it to my old friend, Geo. Ord, Esq.—*Lea.*

The description of *Mel. Ordiana* quoted above answers exactly to a variety of *P. annuliferum*, which varies much in outline and in the development of the canal. In the Smithsonian Collection are preserved fine specimens of a variety of this species in which the shell is much broader than usual, with the periphery sharply angulated.

34. P. Brumbyi, LEA.

Melania Brumbyi, LEA, Philos. Trans., x, p. 298, t. 30, f. 5. Obs., v, p. 54. BINNEY, Check List, No. 40. BROT, List, p. 30. REEVE, Monog. Melania, sp. 277.

Description.—Shell striate, pyramidal, rather thick, reddish-brown; spire very much elevated, carinate at the apex; sutures but slightly impressed; whorls flattened; aperture rather large, rhomboidal, within rubiginose; columella twisted.

Habitat.—Coosa River, Ala.; Huntsville, Ala.

Diameter, ·53; length, 1·72 inches.

Observations.—This is a very remarkable species, and among the largest of our Melaniæ. In form and size it is allied to *annulifera*,

Conr., but may easily be distinguished by its more numerous striæ, its reddish color and the form of its aperture, which is more open. In the 'Brumbyi there is an angle in the middle of the whorl, which gives the aperture a. rhomboidal form. The columella is rufous and the channel whitish. The apex of each of them being broken, the number of whorls cannot be correctly ascertained. I should suppose there were at least ten. Some of the specimens here are beautifully granulate between the striæ. The aperture is not quite one-fourth the length of the shell. Along the suture, on the upper part of the whorl, there is a line of a lighter color than the other part. I dedicate this species to Prof. R. T. Brumby, who has done so much in bringing to light the interesting shells of Alabama. — Lea.

Fig. 176.

35. P. Currierianum, Lea.

Trypanostoma Currierianum. Lea, Proc. Acad. Nat. Sci., p. 155, May, 1863. Jour. Acad. Nat. Sci., vi, p. 147, t. 23, f. 61, 1867.

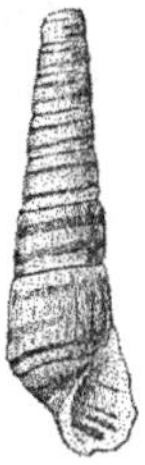

Description.—Carinate, very attenuate, with dark brown bands spire very much drawn out; sutures linear, scarcely impressed; whorls about ten, flattened; aperture small, rhomboidal, banded within; outer lip acute, very sinuous; columella whitish and very much twisted.

Fig. 177.

Operculum ovate, reddish-brown, rather thick, with the polar point near the base towards the left margin.

Habitat.—Florence, Alabama.

Diameter, ·31; length, 1·26? inches.

Observations.—I have seven specimens before me for examination, none of which are perfect at the apex, and therefore the number of whorls is somewhat uncertain. It is a well-characterized shell, all the specimens being without any variation except in age. There are five dark brown bands, the upper and lower being the broadest. The lower two of the three in the middle are on two revolving striæ. The whorls above the body-whorls exhibit two of the five bands all the way to the apex. In old individuals the outer lip is much expanded and slightly thickened inside of the edge. It is allied to *Melania Trypanostoma elongata* (nobis), but may easily be distinguished by being more attenuate, smaller, thinner and in having five bands.

The aperture is about one-fifth the length of the shell. I name this after Mr. A. O. Currier, to whom I am indebted for it. — *Lea.*

E. *Plicate Species.*

36. P. Sycamorénse, Lea.

Trypanostoma Sycamorénse, Proc. Acad. Nat. Sci., p. 175, 1862. Jour. Acad. Nat. Sci., v, pt. 3, p. 283, t. 37, f. 104. Obs., ix, p. 105.

Description. — Shell plicate, conical, yellowish horn-color, rather thick; spire attenuate, pointed; sutures impressed; whorls eleven, somewhat convex, carinate above, plicate in the middle; aper-

Fig. 178.

ture rather small, rhomboidal, whitish within; outer lip acute, sinuous; columella incurved, thickened below and twisted.

Habitat.—Sycamore, Claiborne County, East Tennessee; J. Lewis, M. D.

Diameter, ·36; length, ·92 inch.

Observations.—A single specimen only is before me. It is a rather small, very symmetrical species.   The seven upper whorls are carinate, the three middle ones are furnished with numerous rather obscure folds, the lower whorl is smooth.   In outline it resembles *labiatum*, herein described, but cannot be confounded with that species which is not plicate nor yellowish, and the form of the lower part of the aperture is very different.   The aperture is little more than the fourth of the length of the shell. — *Lea.*

The figure is copied from Mr. Lea's plate.

37. P. plicatum, Tryon.

Pleurocera plicatum, Tryon, Proc. Acad. Nat. Sci., Oct., 1863.

Description.—Shell ovate-conical, spire attenuate, the upper whorls closely plicate, the lower ones smooth or obsoletely con-

Fig. 179.

centrically striate.   Whorls but slightly convex, sutures well impressed.   Color light green, with usually a lighter band below the sutures and ornamented with narrow or broad brown bands.   Aperture canaliculately produced. The outer lip and columella twisted.

Diameter, ·35; length, ·7 inch.

Habitat?—Nashville, Tenn.

Observations.—I owe to Dr. Gould the opportunity of describing this beautiful little species. It differs from *P. grossa*, Anth. (young of *alveare*) in being more slender, different in color and in having bands; the aperture is not nearly so large proportionally and the plicæ are finer. — *Tryon.*

F. *Smooth, Angulate Pleuroceræ.*

38. P. elevatum, SAY.

Melania elevata, SAY, Jour. Acad. Nat. Sci., ii, p. 176, Jan., 1821. BINNEY, Reprint. p. 70. BINNEY, Check List, No. 97. JAY, Cat., 4th edit., p. 273. LAPHAM, Cat. Moll. Wisconsin, p. 308. DEKAY, Moll. N. Y., p. 96. WHEATLEY, Cat. Shells U. S., p. 25. CATLOW, Conch. Nomenc., p. 186. BROT, List, p. 30, REEVE, Monog. Melania, sp. 442.

Ceriphasia elevata, Say, CHENU, Manuél, i, f. 1961.

Melania elongata, LEA, Philos. Trans., iv, p. 121, t. 15, f. 29. Obs., i, p. 130. TROOST, Cat. BINNEY, Check List, No. 99. WHEATLEY, Cat. Shells U. S., p. 25. BROT, List, p. 30.

Ceriphasia elongata, Lea, CHENU, Manuél, i, f. 1959.

Elimia elevata, Lea, ADAMS, Genera, i, p. 300.

Melania tracta, ANTHONY, Bost. Proc., iii, 361, 1850. REEVE, Monog. 429, 1861.

Description.—Shell gradually attenuating to the apex, slightly and irregularly wrinkled, olivaceous; suture not deeply impressed; volutions nine or ten, with several more or less elevated revolving lines, of which one being more conspicuous gives the shell a carinated appearance; aperture oblique, equalling the length of the second, third and fourth volutions conjunctly.

Fig. 180. Fig. 181.

Length, one inch; breadth, two-fifths.

Habitat.—Ohio River.

Observations.—Distinct from our other species, by the elevated revolving lines.— *Say.*

It may be doubted whether *elevatum* and *Lewisii* will not eventually prove to be the same species; I am much inclined to doubt their specific distinction.

The present shell inhabits the waters of Ohio, Indiana and Illinois, the Ohio River, Kentucky and West and Middle Tennessee.

Mr. Say and other conchologists have considered Mr. Lea's *elongatum* to be a synonyme of *elevatum*, in which opinion I concur. The following is the description and copy of the figure of

Melania elongata.—Shell elevated and acutely turreted, dark horn-color with purple bands; apex acute; whorls about ten and slightly

Fig. 182.

depressed; base angulated, aperture bluish-white and about one-fourth the length of the shell.

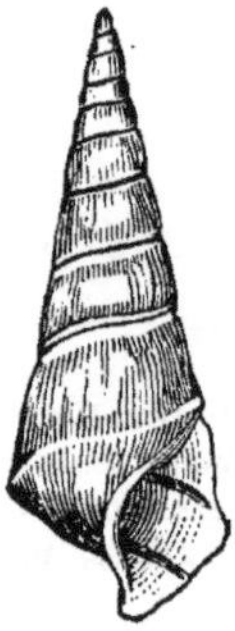

Habitat.—West Tennessee; John Lea.

Diameter, ·5; length, 1·5 inches.

Observations.—This fine *Melania* seems most to resemble the *subularis* (nobis). It differs from it in being wider, in being darker colored and in having a less number of whorls. The bands in some specimens are scarcely visible.—*Lea.*

Reeve figures a shell under the name of *elongata* (Monog. sp. 305) which certainly does not represent this species—it may represent a very fine specimen of *T. annulifera*, Conrad.

The species varies very much in form, and a very long narrow variety has been described as distinct by Mr. Anthony, as follows:

Fig. 183.

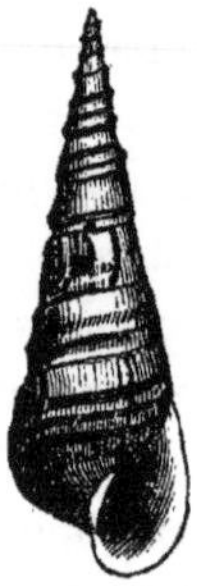

Melania tracta. — Shell ovately-lanceolate, gracile, brownish-green, longitudinally varicosely-plicate and encircled with elevated lines; whorls 7, very convex; sutures profound; aperture contorted, narrowly oval lip produced in front; columella white, mouth livid.

Long. 1⅛; lat. ⅝ poll.

Habitat.—Ohio.

Observations.—General form like *M. Virginica*, but with the whorls more rounded. The delicate raised lines which surround it are among its more obvious characters. —*Anthony.*

39. P. gradatum, ANTHONY.

Melania gradata, ANTHONY, Ann. Lyc., N. Y., vi, p. 112, t. 3, f. 12, March, 1854.
    BINNEY, Check List, No. 130. BROT, List, p. 30. REEVE, Monog. Melania,
    sp. 261.
Melania eximia, ANTHONY, Ann. Lyc. N. Y., vi, p. 107, t. 3, f. 7, March, 1854.
    BINNEY, Check List, No. 106. BROT, List, p. 58. REEVE, Monog. Melania, sp.
    408.
Trypanostoma curtatum, LEA, Proc. Acad. Nat. Sci., p. 155, May, 1863.

Description.—Shell conical, smooth, solid, greenish horn-color; spire not much elevated; whorls 7-8, slightly concave, with a distinct, elevated ridge, closely overlying the suture and the projecting shoulder of the succeeding whorl, so as to form a series of steps to the subacute apex; body-whorl large, generally angulated or distinctly ribbed at base, which is not much rounded; sutures impressed; aperture subrhomboidal, whitish within; outer lip much bent forward towards the base; columella straight, produced into a narrow deep sinus, which is slightly recurved.

Length, ·85 inch (22 millim.); diameter, ·42 inch (11 millim.). Length of aperture, ·30 inch (8 millim.); breadth of aperture, ·20 inch (5 millim.).

Habitat. — Alabama.

Observations. — Belongs to the group of which *M. canaliculata* may be considered the type. It is, however, much less elevated than *M. canaliculata*, has not the conspicuous grooving on the body-whorl as in that species, and its spire has the whorls flat instead of exhibiting an obtuse carina, as described by Mr. Say; a sharp elevated carina at the base of the whorls closely overlies the suture beneath; the extreme upper whorls having this more distant from the suture become distinctly carinated. The regular gradation of the whorls is its most distinctive character. — *Anthony.*

Fig. 184.

Very closely allied to *T. arata*, Lea. The figure is from Mr. Anthony's original type. The shell described as *eximia* by Mr. Anthony is the young of *gradatum*, and the latter name is retained as being more characteristic of the species. For a complete suite of young and old specimens, I am indebted to Prof. Haldeman, who collected them in Holston River, Washington Co., S. W. Virginia. I suspect that Mr. Anthony's locality, "Alabama," for *gradatum*, is incorrect.

Mr. Lea has recently described the same species as *Trypanostoma curtatum*, his shells being rather shorter and more obese than Mr. Anthony's type of *gradatum*. Some of the varieties of this species are finely banded, and others sharply carinate. The following is the description of

Melania eximia. — Shell deeply sulcate and carinate, ovate; of a beautiful, light, apple-green color, ornamented with two dark-green bands, and an elevated, prominent carina of a light color revolving between them; spire not remarkably elevated, but acute, of a rather convex outline; whorls 8-9, somewhat convex, and with sutures not prominent, but channelled; body-whorls with about four carinæ, the lowest one being indistinct; aperture small, subrhomboidal, with two bands in the interior, distant from each other and from the edge of the outer lip; outer lip much twisted, auger-like, causing the sinus, which is small, to curve backwards.

Fig. 185.

Diameter, ·28 inch (7 millim.); length, ·60 inch (15 millim.). Length of aperture, ·25 inch (6 millim.); breadth of aperture, ·13 inch (3 millim.).

Habitat. — Tennessee.

Observations. — A beautiful little shell, of a singularly bright, lively appearance; the colors are well contrasted, very distinct, and the prominent carinæ add to the general effect. On the upper whorls, but one band is visible, the lower one being concealed, or nearly so, by the revolutions of the spire. It cannot well be compared with any other species. — *Anthony.*

Mr. Anthony's type is figured. The following is Mr. Lea's description of

Trypanostoma curtatum. — Shell smooth, pyramidal, yellowish, thick; whorls seven, flattened, the last one impressed; aperture rhomboidal, whitish within; outer lip acute, expanded, very sinuous; columella thickened, bent in, and very much twisted.

Fig. 186.   Fig. 187.

Operculum ovate, dark brown, with polar point near the base on the left.

Habitat. — Powell's River, near Cumberland Gap, East Tennessee.

Diameter, ·41; length, ·75 inch.

Observations. — Quite a number of this species were sent to me by Major Lyon. It is a short thick species, with a well-characterized aperture, the columella being much thickened, drawn back and twisted. It is allied to *T. pumilum* and *minor* (nobis), but differs from both in having the sides flattened and being angular about the middle of the body-whorl. Very few of *curtatum* are banded, while all I have seen of the above two species are banded, and the epidermis polished. The aperture is about one-third the length of the shell. — *Lea.*

40. P. aratum, LEA.

Melania aratum, LEA, Philos. Proc. ii, p. 242, Dec., 1842. Philos. Trans. ix, p.24.
Obs., iv., p. 24. DEKAY, Moll. N. Y., p. 98. BROT, List, p. 30.
Melania exarata, LEA, Philos. Proc. ii, p. 14, Feb., 1841. Philos. Trans., viii, p. 183,
t. 6, f. 44. Obs., iii, p. 21. TROOST, Cat. BINNEY, Check List, No. 101. CAT-
LOW, Conch. Nomenc., p. 186.
Ceriphasia exarata, Lea, ADAMS, Genera, i, p. 297.
Trypanostoma cinctum, LEA, Proc. Acad. Nat. Sci., p. 112, 1864. Jour. Acad. Nat.
Sci., vi, p. 147, t. 23, f. 60, 1867.

Description.—Shell carinate, conical, rather thick, black; sutures rather deeply grooved; whorls flattened, carinate; aperture small, at the base angular and channelled, dark within.

Fig. 188. Fig. 189.

Habitat.— Tennessee.

Diameter, ·28; length, ·57 of an inch.

Observations.— I received only two specimens of this species, both of which are decollated. It is perfectly distinct, and remarkable for its jetty hue, its carina and its deeply impressed sutures, which are caused by the carina. — *Lea.*

First described as *exarata*, which was preoccupied by Menke. I suspect that this species is identical with *Pl. gradatum*, Anthony, the latter being the adult form. The following is no doubt identical.

Trypanostoma cinctum. — Carinate, subfusiform, somewhat thick, dark horn-color; spire somewhat raised; suture impressed; whorls about seven, flattened; aperture rather small, rhomboidal, whitish within; outer lip acute and sinuous; columella thickened and twisted below.

Fig. 190.

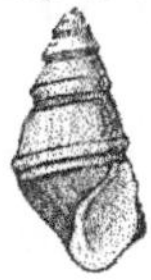

Habitat. — North Alabama.

Diameter, ·32; length, ·65 inch.

Observations.—A single specimen only was received, and it was among several specimens of *Alabamense* (nobis), to which it is allied; but it is evidently a smaller species, with a comparatively shorter spire and with a more developed angle on the periphery, which is accompanied by a furrow. The angle on the lower whorl is cord-like, while on the upper whorls it is sharper and has the furrow deeper above. There are no colored bands on this specimen, and I suspect that it will be found to be generally if not always

without them.   The aperture is rather more than one-third the length
of the shell. — *Lea.*

41. P. carinatum, Lea.

Trypanostoma carinatum, Lea, Proc. Acad. Nat. Sci., p. 4, 1864. Jour. Acad. Nat.
  Sci., vi, p. 148, t. 23, f. 62, 1867.

Shell carinate, acutely conical, reddish horn-color, thin, transparent;
spire acutely conical and sharp at the point; sutures very much
impressed; whorls about nine, carinate and striate above;
aperture rather small and rhomboidal; outer lip acute, sinuous;
columella somewhat thickened and twisted.

Fig.191.

Habitat.—Bull Run, tributary to Clinch River, East Tenn.
Diameter, ·19; length, ·44 inch.

Observations.—Two specimens only were received, having somewhat the aspect of young shells, but I suspect they are nearly if not
quite mature.   It is evidently a delicate species.   It has rather a wide
channel, with the outer lip not much produced.   In outline it resembles *Melania* (*Goniobasis*) *sculptilis* (nobis), but differs from it gener-
ically as well as in being shorter in the spire and in not having deep
striæ over the whole of the whorls.   The aperture is more than
one-third the length of the shell.— *Lea.*

That this species *is* very young is evident, and I have a
conviction that it will be found to be the quite young of
P. aratum.

42. P. lativittatum, Lea.

Trypanostoma lativittatum, Lea, Proc. Acad. Nat. Sci., p. 273, 1862. Jour. Acad.
  Nat. Sci., v, pt. 3, p. 352, t. 39, f. 223. Obs., ix, p. 174.

Description.— Shell carinate, subattenuate, rather thin, shining,
dark, broadly banded; spire conical; sutures linear; whorls
about seven, flattened above, yellow at the base; aperture
small, subrhomboidal, broadly banded within; outer lip sharp,
sinuous; columella bent in, thickened below.

Fig. 192.

Habitat.— Chikasaha River, Alabama; W. Spillman, M. D.
Diameter, ·26; length, ·62 inch.

Observations.— This is a small, gracefully formed species, with a
very broad, intensely brown band around the middle of the whorl.
There is a second narrow band immediately under the suture.   The

angle forming the carina is continued, is well defined on all the whorls, and immediately below it is a hair-like elevated line parallel to it. The area at the base of the columella is of a fine yellow, and contrasts sharply with the dark-brown band above. It is allied to *Chikasahaensis* (nobis), but differs in being more gracefully slender, having different bands and less impressed sutures. The aperture is about one-third the length of the shell. — *Lea.*

42 a. P. strictum, LEA.

Trypanostoma strictum, LEA, Proc. Acad. Nat. Sci., p. 272, 1862. Jour. Acad. Nat. Sci., v, pt. 3, p. 352, t. 39, f. 224. Obs., ix, p. 174.

Description.—Shell carinate, rather attenuate, thin, semi-transparent, pale horn-color, single banded; spire regularly conical; sutures linear; whorls about six, flattened above; aperture rather small, rhomboidal, whitish and single banded within; outer lip sharp, slightly sinuous; columella slightly bent in and twisted.

Fig. 193.

Habitat.—South Carolina; Prof. L. Vanuxem.

Diameter, ·24; length, ·60 inch.

Observations.—Among the numerous mollusca brought from the South long since by my friend, the late Prof. Vanuxem, I found a single specimen of this species, which is different from all others brought by him. I do not know from what part of South Carolina it came, but probably from Spartanburg District, as many of his specimens were from there. This is a small, very regularly formed species, in general outline near to *lativittatum*, herein described, but totally different in the band, that species having it broad and dark while this is hair-like and pale. It is also more fusiform. The aperture is more than one-third the length of the shell. — *Lea.*

P. lativittatum has a line below the angle which this shell has not.

43. P. modestum, LEA.

Io modesta, LEA, Proc. Acad. Nat. Sci., p. 394, 1861. Jour. Acad. Nat. Sci., v, pt. 3, p. 348, t. 39, f. 216. Obs., ix, p. 170.

Description.—Shell smooth, conical, greenish horn-color; spire regularly conical; sutures impressed; whorls nine, flattened, angular in the middle; aperture small, regularly rhomboidal; outer lip sharp and sinuous; columella white and very much twisted; canal short and effuse.

Habitat.—Tennessee River, Alabama? Wm. Spillman, M. D.

Fig. 194.

Diameter, ·39; length, ·88 inch.

Observations.—I have about a dozen of various ages before me. There is no variation in them, either in color or form, but some are slightly carinate towards the apex. None have bands. The channel is short and the outer lip flattened out, so that this species closely impinges on the auger mouthed *Melanidæ*. None before me have the least appearance of colored bands. It is allied to *Spillmanii*, herein described, but is a shorter shell and not so attenuate. The aperture is more than one-third the length of the shell. — *Lea.*

This is evidently a young shell, but whether a distinct species or not I cannot say.

44. P. Leaii, TRYON.

Io viridula, LEA, Proc. Acad. Nat. Sci., p. 394, 1861. Jour. Acad. Nat. Sci., v, pt. 3, p. 349, t. 39, f. 218. Obs., ix, p. 171.

Description.—Shell smooth, cylindrico-conoidal, greenish; spire somewhat raised; suture slightly impressed; whorls about nine, flattened, obtusely angular in the middle; aperture rather small, rhomboidal; outer lip sharp, sinuous; columella purple at the base, slightly twisted; canal short and dilate.

Fig. 195.

Habitat.—Coosa River, Alabama; Wm. Spillman, M. D.

Diameter, ·40; length, ·98 inch.

Observations.—There are three adult specimens before me. Neither has a perfect spire, but the upper whorls show slight carination. There are a few obscure transverse striæ below the angle of the last whorl. The general color is of a faded dark olive-green. Along the sutures the color is light. Within the aperture the color is dull purple in two specimens; in the third, there are four obscure, broad bands. The aperture is a little more than one-fourth the length of the shell. This species has so short a channel and so dilated an outer lip, that it is little removed from the group of *Melanidæ*, which has the auger-shaped aperture, and which I have called *Trypanostoma*. — *Lea.*

Figured from Mr. Lea's plate. The name *viridula* being preoccupied by Mr. Anthony, I gladly avail myself of the present opportunity to dedicate this species to a gentleman who by his immense labors conducted during a period of nearly forty years, has done more for the science of conchology than any other American naturalist. It is closely allied in form to *P. Tuomeyi*, Lea, but differs in the striate spire and in the form of the aperture strikingly. In the latter respect it presents rather an unusual type among the *Pleurocerœ*.

45. P. Tuomeyi, Lea.

Trypanostoma Tuomeyi, Lea, Proc. Acad. Nat. Sci., p. 171, 1862. Jour. Acad. Nat. Sci., v, pt. 3, p. 287, t. 36, f. 111. Obs., ix, p. 109.

Description. — Shell carinate, somewhat thick, high conical, dark brown; spire attenuate conical; sutures scarcely impressed; whorls about ten, flattened; aperture small, rhomboidal, very dark within; outer lip sharp, sinuous; columella a little thickened below and very much contorted.

Fig. 196.

Habitat. — North Alabama; Prof. Tuomey : Florence, Alabama; Rev. G. White.

Diameter, ·45; length, 1·23 inches.

Observations. — I have about a dozen specimens before me from the two habitats. In outline and size it is perhaps nearest to *Melania* (*Trypanostoma*) *elongata* (nobis) from West Tennessee, but it is easy to distinguish it from that species, by its being rather more slender and its being darker. In outline and color it is very close to *Melania* (*Trypanostoma*) *Brumbyi* (nobis), but it differs in the form of the mouth and in not being striate. The aperture is rather more than one-fourth the length of the shell. I have great pleasure in dedicating this species to my deceased friend, Prof. Tuomey, to whom I am greatly indebted for many new and interesting species collected by himself while engaged in his geological survey of the State of Alabama.— *Lea.*

Closely allied to *pyrenellum*, Conr., but differing in the better developed canal, etc.

46. P. gracile, LEA.

Io gracilis, LEA, Proc. Acad. Nat. Sci., p. 394, 1861. Jour. Acad. Nat. Sci., v, pt. 3,
  p. 349, t. 30, f. 217. Obs., ix, p. 171.

Description.— Shell smooth, conical, pale purple; spire regularly
conical; sutures regularly impressed; whorls about nine, flattened.
angular in the middle; aperture rather small, rhomboidal;
outer lip acute and sinuous; columella pale purple, very much
twisted and bent out; canal short and widely effuse.

Fig. 197.

Habitat.— Coosa River, Alabama; Wm. Spillman, M. D.

Diameter, ·36; length, ·90 inch.

Observations.— I have two adults before me. They are
precisely alike, except that one has an obscure band visible
in the inside. It is a graceful, symmetrical species, with a
slight purplish tint which is stronger at the base than at the apex.
It is allied to *Io Spillmanii* on one side and to *Io viridula* on the other,
both herein described. The epidermis is rather more shining than
usual, and the channel is short and wide. The upper part of the
whorls, below the line of the suture, is lighter. The aperture is about
one-third the length of the shell.—*Lea.*

The figure is from Mr. Lea's plate.

47. P. Spillmanii, LEA.

Trypanostoma Spillmanii, LEA, Proc. Acad. Nat. Sci., p. 173, 1862. Jour. Acad.
  Nat. Sci., v, pt. 3, p. 271, t. 36, f. 82. Obs., ix, p. 86.

Description.— Shell smooth, regularly conical, dark olive; spire
much raised; sutures regularly impressed; whorls about
nine, flattened; aperture rather small, rhomboidal, white
within, sometimes banded; outer lip acute, sinuous; columella white and very much twisted.

Fig. 198.

Operculum ovate, reddish-brown, rather thin, with the
polar point near the base.

Habitat.— Noxubee River, Mississippi; Wm. Spillman,
M. D.: and Tennessee; J. Clark.

Diameter, ·46; length, 1·20 inches.

Observations.— Six specimens are before me, one of them
is slightly carinate. In some there is a disposition to put on a whit-

ish line below the suture. The aperture is about one-third the length of the shell.

I have great pleasure in naming this species after my friend Dr. Spillman. — *Lea.*

This species appears to me to be very closely allied to *pyrenellum* on one side and to *elevatum* on the other side.

48. P. planogyrum, ANTHONY.

Melania planogyra, ANTHONY, Ann. Lyc. N. Y., vi, p. 111, t. 3, f. 11, March, 1854. BINNEY, Check List, No. 207. BROT, List, p. 30. REEVE, Monog. Melania, sp. 382.

Description.— Shell conical, rather smooth, thick; of a dull, dark horn-color, unrelieved by any other except a rather indistinct, brown band, revolving near the base of each whorl, immediately below

Fig. 199.

which a raised, rounded, subcrenulated ridge revolves between it and the suture below; spire much but not acutely elevated, with a nearly rectilinear outline; whorls ten to eleven, flat or concave, and with a well-impressed, channelled suture; aperture small, rhomboidal, diaphanous, exhibiting the dark band of the exterior through its substance very faintly, far within; columella deeply curved, not indented, thickened at base; outer lip angularly curved, extended forwards; sinus rather broad, not deep.

Diameter, ·46 inch (12 millim.); length, 1·37 inches (34 millim.). Length of aperture, ·40 inch (10 millim.); breadth of aperture, ·24 inch (6 millim.).

Habitat.—Alabama.

My cabinet.

Observations.—A stout species which most resembles *M. regularis,* Lea, in general appearance, from which, however, its concave whorls, elevated carina, and dark band will readily distinguish it. It has not the channelled body-whorl of *M. canaliculata,* Say, nor the convex, subangulated upper whorls which distinguish that species.

The lines of growth are very coarse and prominent, and extending over the raised line near the base of the whorls, give the latter an interrupted or subcrenulated appearance.—*Anthony.*

The figure is from the original type.

49. P. pyrenellum, CONRAD.

Melania pyrenella, CONRAD, New Fresh Water Shells, p. 52, t. 8, f. 5, 1834. DEKAY, Moll. N. Y., p. 99. WHEATLEY, Cat. Shells U. S., p. 26. BINNEY, Check List, No. 226. BROT, List, p. 30. REEVE, Monog. Melania, sp. 303. MÜLLER, Synopsis, p. 45.

Description.— Shell elevated, with flattened whorls, having an obso-

Fig. 200.

lete spiral line on each; suture impressed; body-whorl angulated; angle defined by a prominent line; base hardly convex, labrum angulated near the centre; aperture patulous; columella obtusely rounded at the base.

Observations.—Inhabits streams in North Alabama. The aperture is remarkably patulous, and the labrum profoundly angulated.—*Conrad.*

The figure is that of Conrad's type in the collection of the Academy of Natural Sciences.

50. P. Conradii, TRYON.

Description.— Shell narrow, lengthened, with nine flattened whorls, which are angulated in the middle of the body and just above the suture of the spire. Dark brown, smooth, apical whorls, slightly carinate. Aperture small, not produced below, fuse short, scarcely perceptible.

Fig. 201.   Fig. 202.

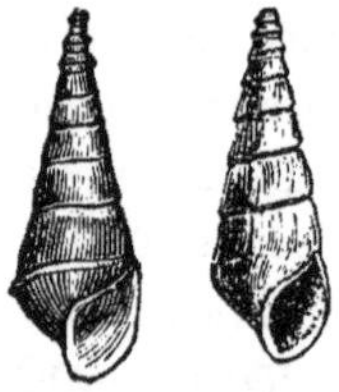

Diameter, ·36; length, 1 inch.

Habitat.—Tennessee. — *Tryon.*

This shell has been distributed very extensively in cabinets under the name of *Melania pyrenella*, Conrad. It is, however, a much narrower species and darker in color.

51. P. regulàre, LEA.

Melania regularis, LEA, Philos. Proc., ii, p. 12, Feb., 1841. Philos. Trans., viii, p. 170, t. 5, f. 16. Obs., iii, p. 8. DEKAY, Moll. N. Y., p. 94. HIGGINS, Cat. TROOST, Cat. JAY, Cat., 4th edit, p. 274. WHEATLEY, Cat. Shells U. S., p. 26. BINNEY, Check List, No. 227. CATLOW, Conch. Nomenc., p. 188. BROT, List, p. 30.
Ceriphasia regularis, Lea, CHENU. Manuél, i, f. 1956. ADAMS, Genera, i, 297.

Description. — Shell smooth, conical, rather thick, dark horn-colored; spire elevated; sutures somewhat impressed; whorls flat; aperture small, whitish.

Fig. 203.

Habitat.—Oconee District, Tennessee; Dr. Troost. Diameter, ·40; length, 1·22 inches.

Observations.—This species has a regularly increasing and elevated spire. Neither of the three before me has perfect tip. The number of whorls must be about ten. The aperture is about one-fourth the length of the shell.— *Lea.*

Apparently very closely related to *pyrenella*, Conrad, but appears to be a heavier shell and not so strongly angulated.

The figure is a copy of that of Mr. Lea.

52. P. validum, ANTHONY.

Melania valida, ANTHONY, Proc. Acad. Nat. Sci., p. 59, Feb., 1860. BINNEY, Check List, No. 282. BROT, List, p. 33. REEVE, Monog. Melania, sp. 317.

Description. — Shell ovate-conic, smooth, olivaceous, thick; spire obtusely elevated, decollate; whorls flat, only about six remaining; sutures distinct; lines of growth very strong, amounting to varices on the body-whorl; aperture ovate, bluish-white within; columella strongly curved or indented about the middle, white; sinus well developed at base; body-whorl obscurely, concentrically striate, the striæ forming faint nodules where they intersect the varices.

Fig. 204.

Habitat. —Tennessee.

Observations.—This species may be compared with *M. tenebro-cincta* herein described; from that species it may be distinguished by its more robust form, uniform, dark, olivaceous color and the absence of the dark bands so conspicuous in that species. It has a very solid, compact form, and this with its regular, uniform size up to the point of decollation, may serve to distinguish it from all others — *Anthony.*

Figure 204 is from Mr. Anthony's original type specimen.

52 a. P. cylindraceum, LEA.

Trypanostoma cylindraceum, LEA, Proc. Acad. Nat. Sci., p. 4, 1864. Jour. Acad.
    Nat. Sci., vi, p. 142, t. 73, f. 57, 1867.

Description.— Shell smooth, cylindrical, rather thick, banded or
without bands; spire rather raised; sutures irregularly impressed;
whorls flattened, slightly impressed, swollen below the sutures;
aperture rather small, rhomboidal; outer lip acute, somewhat sinuous; columella thickened, incurved and twisted.

Fig. 205.

Habitat.— Roane County, East Tennessee.

Diameter, ·41; length, 1·4 inches.

Observations.— I have three specimens of this pupæform
species before me. Two of them are of a light horn-color;
the third has a dark-brown band over more than two-thirds
of the whorls, above which along the sutures it is yellow.

In this specimen, the base of the columella is purple and
the interior is purplish. In all the three specimens the body-whorl is
impressed; above the periphery, amounting almost to a channel. It
is allied to *parvum* and *moriforme* (nobis) but is larger and more cylindrical than the first, and smaller and less pyramidal than the latter.
The aperture is about one-third the length of the shell. The apices
were too much eroded to ascertain the number of whorls, but there
are probably about eight.— *Lea.*

52 b. P. Roanense, LEA.

Trypanostoma Roanense, LEA, Proc. Acad. Nat. Sci., p. 4, 1864. Jour. Acad. Nat.
    Sci., vi, p. 142, t. 23, f. 52, 1867.

Description.—Shell smooth, obtusely conical, thick, banded or without bands; spire obtuse; sutures impressed; whorls flattened, swollen
below the sutures; aperture rather small, rhomboidal; outer
lip acute, sinuous; columella whitish, thickened and very
much twisted.

Fig. 206.

Habitat.—Roane County, East Tennessee.

Diameter, ·41; length, ·80? inch.

Observations.—This species is allied to *cylindraceum,* but
differs in being shorter and wider in proportion. It differs also in the
form of the bands where they exist. Two of the six specimens before
me have a single narrow band below the middle, and one has a second

band above the middle. All the specimens have apices so much eroded that the number of whorls cannot be correctly ascertained. There may be six or seven. The aperture is probably more than one-third the length of the shell.—*Lea*.

Notwithstanding the differences pointed out by Mr. Lea, I suspect that this and *cylindraceum* will prove to be one species.

G. *Smooth species, not angulated.*

53. P. glandulum, ANTHONY.

Melania glandula, ANTHONY, Proc. Acad. Nat. Sci., p. 60, Feb., 1860. BINNEY, Check List, No. 124. BROT, List, p. 39. REEVE, Monog. Melania, sp. 393.
Melania glans, ANTHONY, Ann. N. Y. Lyc., vi, p. 123, t. 3, f. 23, March, 1854.

Description.— Shell ventricose-conic, smooth, thick, dark-olive; spire acuminate, but not elevated; whorls eight, convex, rapidly converging to the apex; body-whorl very large, rounded beneath; sutures well defined, white; aperture not large, elliptical, within dark-purple; columella indented near the base; sinus well developed.

Diameter, ·38 inch (10 millim.); length, ·75 inch (19 millim.). Length of aperture, ·34 inch (9 millim.); breadth of aperture, ·16 inch (4 millim.).

Fig. 207.

Habitat.—Tennessee.

Observations.—A plain sombre-looking species with no very remarkable distinguishing characters except its large, bulbous form, and dark, purple mouth. It cannot be compared with any other species. The whorls are slightly shouldered, with a very narrow, whitish, sutural region.—*Anthony*.

The specific name "*glans*," first used by Mr. Anthony, being preoccupied, he changed it to *glandula*. It is a curious species, resembling *Jayi*, Lea, in the channel of the aperture, but is much more inflated.

The figure is from Mr. Anthony's type specimen.

53 a. P. subrobustum, Lea.

Trypanostoma subrobustum, LEA, Proc. Acad. Nat. Sci., p. 4, 1864. Jour. Acad.
    Nat. Sci., vi, p. 141, t. 23, f. 50, 1867.

Description. — Shell smooth, pyramidal, dark horn-color, thick;
spire pyramidal and elevated; sutures impressed; whorls about nine,

Fig. 208.

flattened; aperture small, rhomboidal; outer lip sharp
and very sinuous; columella thickened and very much
twisted.

Operculum ovate, dark-brown, with polar point near
the base on the left side.

Habitat. — Holston River, at Knoxville, East Tennessee.

Diameter, ·61; length, 1·25 inches.

Observations.—A single specimen only, with an imperfect outer lip and much eroded spire, was received.
This is greatly to be regretted, as such a fine large species ought to be
well represented. This specimen has no bands and is without striæ.
It belongs to the group of which *Hartmanii* may be considered the
type, but may be distinguished by its being a larger and more robust
species, with a much larger body-whorl. The aperture is about one-
third the length of the shell.—*Lea.*

54. P. Christyi, Lea.

Trypanostoma Christyi, LEA, Proc. Acad. Nat. Sci., p. 173, 1862. Jour. Acad. Nat.
    Sci., v, pt. 3, 272, t. 36, f. 83. Obs., ix, p. 94.

Description. — Shell smooth, elongately conical, somewhat thick,
horn-color, rarely banded; spire very much elevated; sutures regularly impressed; whorls about ten, slightly convex; aperture small, subrhomboidal, whitish within; outer lip acute,

Fig. 209.

sinuous; columella white and twisted.

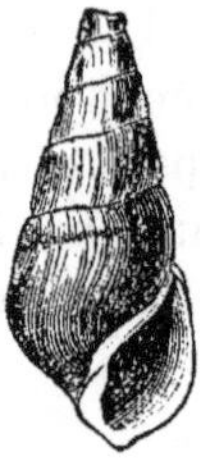

Operculum subovate, dark-brown, with polar point near
to the basal margin.

Habitat.—Cane Creek, Tennessee; Prof. D. Christy.

Diameter, ·48; length, 1·12 inches.

Observations.—I am indebted to the late Joseph Clark
for many specimens from the above habitat, brought by
Prof. Christy.   It is allied to *Estabrookii*, herein described, but it is a
larger and heavier shell, has a larger aperture, a much more twisted

columella and is of a darker horn-color. One of the specimens is somewhat carinate on the body-whorl, and has a more developed channel. The form of the channel is very like to *Melania* (*Trypanostoma*) *regularis* (nobis) but it is not so cylindrical nor so green. The aperture is about the third of the length of the shell. I name this after Prof. David Christy, Hamilton, Butler Co., Ohio, who collected many fine shells in East Tennessee and North Carolina, which he kindly gave to Mr. Clark.—*Lea.*

This species may be distinguished from *labiatum* principally by its more ponderous proportions and more flattened volutions.

55. P. labiatum, LEA.

Trypanostoma labiatum. LEA, Proc. Acad. Nat. Sci., p. 173, 1862. Jour. Acad. Nat. Sci., v, pt. 3, p. 272, t. 36, f. 84. Obs., ix, p. 94.

Description.—Shell smooth, acutely conical, rather thick, shining, greenish horn-color; spire attenuate, sharp-pointed; sutures regularly impressed: whorls about ten, somewhat convex, carinate towards the beak, the last rather large; aperture rather small, rhomboidal, whitish within; outer lip sharp, thickened towards the margin, very much dilated and very sinuous; columella whitish, thickened below and much twisted.

Fig. 210. Fig. 211.

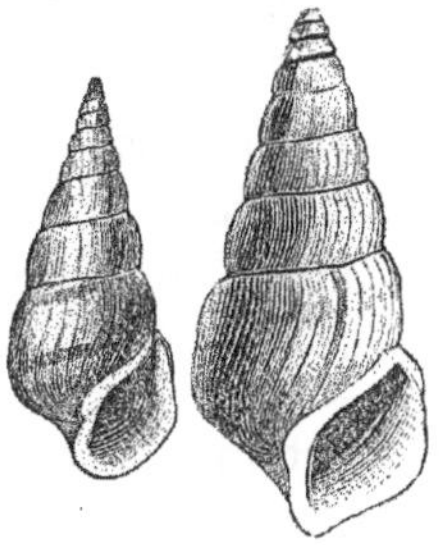

Operculum subovate, dark brown, rather thin, with the polar point near the middle towards the base.

Habitat.—Big Miami River, Ohio; J. Clark.

Diameter, ·43; length, ·98 inch.

Observations.—A number of these were sent to me some years since, by Mr. Clark. They were supposed to be *Melania neglecta*, Anth., but they are not very closely allied to the species which Mr. Anthony sent to me under that name, nor are they like his figure, nor will they answer to his description. This species has a remarkably expanded outer lip, unusually thickened inside of the edge. It is nearly allied to *Whitei* herein described, but may be distinguished by being not quite so attenuate, having rather more convexity in the whorls, having a larger outer lip and slightly differing in the cut of the open channel at the base. The aperture is three-tenths the length of the shell.—*Lea.*

55 a. P. univittatum, LEA.

Trypanostoma univittatum, LEA, Proc. Acad. Nat. Sci., p. 112, 1864. Jour. Acad. Nat. Sci., vi, p. 145, t. 23, f. 58, 1867.

Description.—Shell obtusely carinate, pyramidal, somewhat thick, pale olive, shining, with a single band; spire elevated; sutures impressed; whorls flattened; aperture rather small, rhomboidal, whitish within, obscurely single-banded; outer lip acute, much curved; columella thickened below and very much twisted.

Fig. 212.

Habitat.—Cahawba River, Alabama.

Diameter, ·45; length, 1·2 inches.

Observations.—A single specimen was received by Dr. Hartman from Dr. Showalter and kindly lent to me for description. It seems to be most nearly allied to *T. Anthonyi* (nobis), but it is a smaller species, without the striæ and obscure sulcations of that species, and it has a band which I have never observed in *Anthonyi*, and probably a less number of whorls. It is also somewhat allied to *Hartmanii* (nobis), but not so elevated, and it is smaller. When *Hartmanii* is banded, it always has, I believe, two. This specimen of *univittatum* has a single band above the periphery which is observable on all the whorls above. The apex being eroded, I cannot state the number of whorls, but they seem to be about eight. The aperture is about one-third the length of the shell.—*Lea.*

Certainly very closely allied both to *subrobustum* and *Christyi*.

55 b. P. pallidum, LEA.

Trypanostoma pallidum, LEA, Proc. Acad. Nat. Sci., p. 174, 1862. Jour. Acad. Nat. Sci., v, pt. 3, p. 275, t. 36, f. 90. Obs., ix, p. 97.

Description.—Shell smooth, attenuately conical, rather thick, pale horn-color; spire very much raised; sutures very much impressed; whorls eleven, slightly convex, somewhat geniculate above; aperture rather small, subrhomboidal, white within; outer lip sharp, sinuous; columella white and very much twisted.

Operculum subovate, light chestnut-brown, with the polar point on the left near the basal margin.

Habitat.—Niagara Falls, New York, St. Lawrence at Montreal; E. Billings, Esq.

Diameter, ·46; length, 1·36 inches.

Observations.—Many years since I found two specimens of this species above the Falls, on the New York side. They were accompanied with *Melania* (*Trypanostoma*) *Niagarensis* and *subularis* (nobis). I hesitated when I described the above two, whether this was a new species. There is no doubt in my mind now. It is nearest allied perhaps to *Melania* (*Trypanostoma*) *Sayi*, Ward, but it is a more slender species and has a higher spire and more whorls. The aperture is rather more than the fourth of the length of the shell.—*Lea.*

Fig. 213. Fig. 214.

Fig. 213 is a copy of Mr. Lea's figure; the banded shell (fig. 214) is from an Ohio specimen named by Mr. Anthony "*M. neglecta.*"—See remarks on that species.

The "*Melania Sayi*, Ward" quoted above by Mr. Lea is doubtless intended to be *Melania* (*Strombus*) *Sayi*, Wood (Index Testaceologicus), as Dr. Ward never published a species under that name. Mr. Lea has, however, entirely mistaken the characters of this species, *his* shell being the *neglecta* of Anthony, while the true *M. Sayi* is a *canaliculatum*, as will appear by reference to the Index Testaceologicus, Supplement, t. 4, f. 24.

· 56. P. neglectum, ANTHONY.

Melania neglecta, ANTHONY, Ann. Lyc. N. Y., p. 128, t. 3, f. 29, March, 1854. BINNEY, Check List, No. 173. BROT, List, p. 34. CURRIER, Shells of Grand River Valley, Mich., 1859. REEVE, Monog. Melania, sp. 247.

Description.—Shell conical, rather thin, light yellow; whorls ten, upper ones nearly flat, with a slight ridge revolving just above the suture. This ridge disappears as it approaches the penult whorl, but two of them become visible on the last whorl, which is subangulate. Sometimes the last whorl is encircled by two dark brown bands, of which the uppermost is also visible throughout the upper whorls, covering the ridge above mentioned; sutures impressed; aperture ovate, of a delicate rosy hue within; outer lip waved; columella nearly straight, twisted, roseately recurved into a deep sinus.

Diameter, ·38 inch (10 millim.); length, ·90 inch (23 millim.). Length of aperture, ·33 inch (8 millim.); breadth of aperture, ·18 inch (4½ millim.).

Habitat.—Great Miami River, near Dayton, Ohio.

Fig. 215.

Observations.—A fine large species, which seems to exhibit considerable variation, both in form and coloring. The banded varieties are among our most beautiful species, while we also find those which are of a plain, delicate horn-color, or with bands but faintly indicated by an almost imperceptible difference of color in the interior of the mouth, which in these specimens is generally, and in the banded specimens occasionally, tinged with a delicate rosy hue.—*Anthony.*

The light horn-colored variety alluded to by Mr. Anthony has since been separated by Mr. Lea as *T. labiatum.* It is certainly distinct as the whorls are more swollen, shell larger, color different, as is also the aperture. The two figures are from Mr. Anthony's types.

57. P. vestitum, Conrad.

Melania vestita, Conrad, New Fresh Water Shells, p. 57, t. 8, f. 12, 1834. DeKay, Moll. N. Y., p. 101. Wheatley, Cat. Shells U. S., p. 27. Binney, Check List, No. 287. Brot, List, p. 31. Reeve, Monog. Melania, sp. 322. Müller, Synopsis, p. 47.
Melania mucronata, Lea, Proc. Acad. Nat. Sci., p. 117, 1861.
Trypanostoma mucronatum, Lea, Jour. Acad. Nat. Sci., v, pt. 3, p. 277, t. 36, f. 93. Obs., ix, p. 99.

Description.—Shell subulate, subturreted; volutions nine, each angulated below the middle; suture deeply impressed; epidermis smooth, polished, horn-colored, with a dark band revolving below the angle of each whorl; whorls near the apex acutely carinated.

Fig. 217.

Observations.—Inhabits small streams in Greene County, Alabama, among the grass which grows on the rocks. The shell is always coated with a deposit which obscures its characters.—*Conrad.*

The following is the description of

T. mucronatum.—Shell smooth, awl-shaped, thin, diaphanous, straw-yellow; spire extended, pointed; sutures slightly impressed; whorls six, flattened above; aperture rather small, ovately rhombic, yellowish-white within; outer lip acute, sinuous; columella slightly thickened at the base, subeffuse and somewhat recurved.

Operculum ovate, spiral, light brown, with the polar point on the inner side near to the base.

Fig. 218.

Habitat. — Big Prairie Creek, Alabama; E. R. Showalter, M. D.

Diameter, ·36; length, ·98 inch.

Observations. — This is an acuminate species with about eight, regular, graceful whorls, which are towards the apex usually carinate. There are five specimens before me, all without bands. One of them has on the upper whorls, a disposition to take on a brownish color. This species is allied to *Melania (Goniobasis) Ocoëensis* (nobis). It is not quite so subulate, has not quite so many whorls and the aperture is not so quadrate. The aperture is not quite three-tenths the length of the shell.—*Lea.*

Mr. Lea's description and figure refer to this species not quite fully grown. It is curious that in his description he mentions six whorls, in his observations he gives it eight, while his figure exhibits ten.

I have before me a suite of over one hundred specimens from North Alabama, collected by Dr. Showalter, and presented to the Smithsonian Institution by Dr. Jas. Lewis. About half of them are banded. I have also author's types from Haldeman's collection and collection of the Academy of Natural Sciences. Fig. 217 is one of the latter. Fig. 218 represents Mr. Lea's original figure.

57 a. P. lugubre, LEA.

Melania lugubris, LEA, Philos. Proc. iv, p. 166, August, 1845. Philos. Trans. x, p. 58, t. 9, f. 29. Obs. iv, p. 58. BINNEY, Check List, No. 164. BROT, List. p. 31.
Melania spurca, LEA, Philos. Proc. iv, p. 166, Aug., 1845. Philos. Trans., x, p. 59, t. 9, f. 31. Obs. iv, p. 59. BINNEY, Check List, No. 248. BROT, List, p. 31.
Melania modesta, LEA, Am. Philos. Trans., x, p. 83, t. 9, f. 34, 1847.

Description.—Shell smooth, rather acutely conical, rather thick, dark-brown; spire rather elevated; sutures widely impressed; whorls flattened; aperture small, rhomboidal, within bluish, angular below.

Habitat.—Alabama.

Diameter, ·37; length, ·85 of an inch.

Observations.—A single specimen only of this species was received by Major LeConte. There are no strong characters to separate it, but it is certainly different from any with which I am ac-

Fig 219.

quainted. Like the *canaliculata*, Say, it is auger-shaped on the right lip, but it is a much smaller shell, and without the sulcations of that species. There is an angle on the middle of the whorl which causes the sutures to be rather wide and marked. The apex being eroded, the number of whorls cannot be ascertained — probably eight. The aperture is about one-third the length of the shell.—*Lea.*

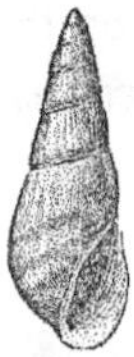

The following, described at the same time as the above, is an undoubted synonyme.

Melania spurca. — Shell smooth, pyramidal, somewhat thick, dark brown; spire somewhat elevated; sutures slightly impressed; whorls eight, flattened; aperture small, rhomboidal, angular at the base, within white.

Fig. 220.

Habitat.— Alabama.

Diameter, ·43; length, ·98 of an inch.

Observations. — This species, of which only a single one was received by Major LeConte, has no striking character, but cannot be placed with any other with which I am acquainted. It is very regular in its form, with a patulous, auger-shaped outer lip, the margin of which is quite sinuous. The aperture is nearly one-third the length of the shell. It more nearly resembles *M. regularis* (nobis), than any other species, but is not so large or solid a shell.—*Lea.*

Mr. Reeve's figure does not represent this species at all. I give a copy of Mr. Lea's figure.

I also place in the synonymy of this species

Melania modesta.—Shell smooth, conical, somewhat fusiform, rather thin, black, spire rather elevated; sutures linear; whorls flattened, the last angular in the middle; aperture elliptical, rather large, within dark.

Habitat.—Chattahoochee River at Columbus, Georgia.

Diameter, ·28; length, ·67 of an inch.

Observations.—A single specimen of this species came from Dr. Boykin, with some others which I published some years since. This one was deferred in the hopes of getting more for comparison. In outline and color, it is very closely allied to a shell I described, from Tennessee, under the name of *tenebrosa*. It differs from it in having the aperture less distended, in having an angle on the middle Fig. 221. of the whorl and in being more fusiform. The apex being eroded, the number of whorls cannot be ascertained; there are about seven. The aperture is nearly one-half the length of the shell. The bands are so broad and dark as to give, in this specimen, a black appearance to the whole shell, except at the termination of the whorl, where the outer lip is yellow.—*Lea*.

The figure is copied from Mr. Lea's plate. Reeve's figure does not represent this species.

57 b. P. abruptum, LEA.

Melania abrupta, LEA, Philos. Proc., iv, p. 165. Philos. Trans., x, p. 59, t. 9, f. 32. Obs. iv, p. 59, t. 9, f. 32. BINNEY, Check List, No. 2. BROT, List, p. 37. REEVE, Monog. Melania, sp. 397.
Leptoxis abrupta, Lea, ADAMS, Genera, i, p. 307.

Description.— Shell smooth, short, conical, rather thick, yellowish; spire very short; sutures linear, whorls seven, flattened; aperture large, ovate, within whitish.

Fig. 222.

Habitat.—Alabama.

Diameter, ·3; length, ·64 of an inch.

Observations. — This species in size and form is somewhat allied to *M. Nickliniana* (nobis), but has the spire more elevated and is not reddish. The two specimens before me, have each two purple bands. This character may be frequent without being constant. The aperture is nearly half the length of the shell.—*Lea*.

Figured from Mr. Lea's plate.

57 c. P. tortum, LEA.

Melania torta, LEA, Philos. Proc. iv, p. 165, Aug., 1845. Philos. Trans., x, p. 58, t. 9, f. 30. Obs. iv, p. 58. BINNEY, Check List, No. 272. BROT, List, p. 39. REEVE, Monog. Melania, sp. 377.

Description.— Shell smooth, club-shaped, rather thick, dark brown; spire obtuse; sutures impressed; whorls convex; aperture large, elliptical; columella twisted.

Habitat.— Big Creek, Lawrence County, Tennessee.

Diameter, ·36; length, ·73 of an inch.

Observations.— There were eight specimens of this species submitted to my examination by Mr. Clark, of Cincinnati. In general outline and size, it very closely resembles *M. Warderiana* (nobis), but differs from the specimens of that species which have come

Fig. 223.

under my notice in not being carinate, and in having a more twisted columella. The apices of the individuals now before me are slightly eroded, and the number of the whorls may be seven or eight. One of the specimens has small folds near the apex, decussating striæ. The inside is bluish-white, one of the specimens having a brown mark at the columella. The aperture is nearly one-half the length of the shell. Over the whole surface there are small, irregular ridges. The body-whorl is very long.—*Lea.*

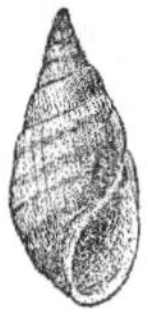

This species differs from all the others of this group in the great acumination of the upper part of its spire. In young shells (in which state only, the spire is perfect) the spire is narrowly subulate for the first few whorls, then suddenly expands into a bulbous form.

58. P. strigosum, Lea.

Melania strigosa, Lea, Philos. Proc., ii, p. 13, Feb., 1841. Philos. Trans., viii, p. 175, t. 5, f. 24. Obs. iii, p. 131. DeKay, Moll., N. Y., p. 95. Troost, Cat. Binney, Check List, No. 250. Wheatley, Cat. Shells U. S., p. 27. Catlow, Conch. Nomenc., p. 188. Brot, List, p. 38. Reeve, Monog. Melania, sp. 320.

Description.—Shell smooth, acutely turreted, thin, pale yellow, striate above; spire drawn out; sutures impressed; whorls nine, flattened; aperture small, elliptical, angular at the base, within bluish.

Fig. 224.

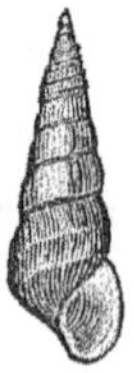

Habitat.— Holston River, Tennessee.

Diameter, ·27; length, ·85 of an inch.

Observations.—This species is somewhat like the *teres* herein described. It may be distinguished, however, at once, by its flattened whorls and darker color.—*Lea.*

The figure is a copy of Mr. Lea's.

59. P. pictum, LEA.

Melania picta, LEA, Philos. Proc., ii, p. 82, Oct., 1841. Philos. Trans. ix, p. 19. Obs. iv, p. 19. WHEATLEY, Cat. Shells U. S., p. 26. BINNEY, Check List, No. 205. REEVE, Monog. Melania, sp. 290.
Melania picturata, REEVE, Errata to Monog. Melania. BROT, List, p. 38.

Description.— Shell smooth, obtusely conical, thick, subfusiform, greenish, banded; spire rather elevated; sutures impressed, above furrowed; whorls eight, flattened; aperture elongated, trapezoidal; columella incurved.

Fig. 225.

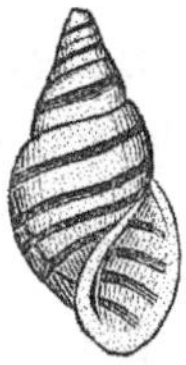

Habitat.—Holston River, East Tennessee.

Diameter, ·30; length, ·70 of an inch.

Observations.— The four specimens before me have each three bands, which with the yellowish tint below the sutures give the shell a lively appearance. The superior whorls are disposed to be bicarinate, and the lower carina being covered with the whorl below, causes a furrow along the suture. The aperture is more than one-third the length of the shell, angular at the base, with rather a large sinus.—*Lea.*

The figure is copied from Reeve.

Mr. Anthony has placed specimens in my cabinet with the habitat Alabama, affixed.

60. P. spinalis, LEA.

Melania spinalis, LEA, Am. Philos. Trans., x, p. 89, t. 9, f. 42, 1847.

Description.— Shell carinate, acutely conical, rather thin, yellow, double-banded; spire elevated; sutures ploughed out; whorls flattened; aperture small, ovate, angular at the base, white within.

Fig. 226.

Habitat.— Alabama.

Diameter, ·33; length, ·96 of an inch.

Observations.— A single specimen only was submitted to me, and this not very perfect. It is a peculiar shell in its general appearance, the color being of an unusually bright yellow, with two broad, distinct bands, one immediately above the middle of the whorl, and the other below. The superior part of the whorl is darker than that below. The number of whorls cannot be given, the apex being broken. There were probably nine or ten. The aperture is about one quarter the length of the shell. — *Lea.*

If an opinion founded on a single specimen, such as Mr. Lea has described, be admissible, I would suggest the too close resemblance of this shell to Conrad's *vestitum* (Lea's *mucronatum*).

61. P. tenebrocinctum, ANTHONY :

Melania tenebrocincta. ANTHONY, Proc. Acad. Nat. Sci., p. 58, Feb., 1860. BINNEY, Check List, No. 266. BROT, List, p. 31. REEVE, Monog. Melania, sp. 271.
Trypanostoma parvum, LEA, Proc. Acad. Nat. Sci., p. 174, 1862. Jour. Acad. Nat. Sci., v, pt. 3, p. 276, t. 36, f. 91. Obs., ix, p. 98.

Description.—Shell conic ovate, smooth, rather thick; spire rather obtusely elevated; whorls 6–7, nearly flat, but with an obtuse carina below the middle of each, and one more decided between that

Fig. 227.

and the suture; aperture well marked, and with a pale band near it; lines of growth decided; aperture linear, ovate, within dusky, and having two dark bands there; sinus very decided.

Habitat.— Tennessee.

Observations. — Compared with *M. valida* (nobis), it is smaller, less robust, more slender, and may also be distinguished from that plain species by its more lively exterior. The dark brown band or bands contrast finely with the general color of the shell, and with a light band near the sutures.—*Anthony.*

The following is Mr. Lea's description.

T. parvum.—Shell smooth, somewhat thick, conical, horn-color, banded or without bands; spire conoidal; sutures regularly impressed; whorls eight, flattened; aperture small, rhomboidal, within whitish; outer lip acute, somewhat sinuous; columella slightly thickened below and twisted.

Fig. 228.

Habitat. — Knoxville; President Estabrook: and French Broad River, Tennessee; J. Clark.

Diameter, ·34; length, ·94 inch.

Observations.—I have three specimens of this small species from French Broad River, and one from Knoxville. ·They are all perfect, and have two bands, one broad and well defined, the lower one obsolete. It is disposed to be slightly angular on the periphery. The aperture is about one-third the length of the shell. This is among the few small species of this genus. In outline and general appearance it is allied to *T. Hartmanii,* herein described, but it is a very much smaller species and cannot be easily confounded with it.—*Lea.*

Figure 227 is from Mr. Anthony's type specimen. Figure 228 is a copy of Mr. Lea's figure quoted above.

62. P. Vanuxemii, LEA.

Trypanostoma Vanuxemii, LEA, Proc. Acad. Nat. Sci., p. 175, 1862. Jour. Acad. Nat. Sci. v, pt. 3, p. 280, t. 36, f. 98. Obs. ix, p. 102.

Description.—Shell smooth, conical, yellowish, double-banded or without bands; spire obtusely conical; sutures impressed; whorls six, somewhat convex; aperture rather small, subrhomboidal, whitish within; outer lip acute, sinuous; columella thickened below and much twisted.

Habitat.—South Carolina; Prof. L. Vanuxem.

Diameter, ·28; length, ·69 inch.

Observations. — Among other species of the *Melanidæ* given to me a long time since by my friend, the late Prof. Vanuxem, were four specimens of this. Three of them are double-banded inside Fig.229. and out. The fourth has no appearance of bands. One of them is about half grown and perfect to the apex. The outer lip is somewhat thickened and expanded. It is somewhat like *bivittatum,* herein described, but it differs in having a higher spire, is not so wide proportionally, and is not highly polished or so yellow as that species. The aperture is more than one-third the length of the shell.—*Lea.*

Figured from Mr. Lea's plate. Too closely allied to the preceding.

63. P. Chakasahaense, LEA.

Trypanostoma Chakasahaense, LEA, Proc. Acad. Nat. Sci., p. 175, 1862. Jour. Acad. Nat. Sci., v, pt. 3, p. 280, t 36, f. 99. March, 1863. Obs., ix, p. 102.

Fig. 230. *Description.*— Shell smooth, conical, brownish-green, rather thin, double-banded; spire somewhat attenuate; sutures very much impressed; whorls about eight, convex, carinate above; aperture small, rhomboidal, white and banded within; outer lip sinuous; columella incurved, thickened below and very much twisted.

Habitat. — Chakasaha River, Mississippi; Wm. Spillman, M. D.

Observations.—Of eight specimens received from Dr. Spillman, three of them had transverse striæ on the periphery of the whorls reaching to the last whorl, on which two raised striæ are noticeable. In general outline and size it is near to *parvum*, herein described, but differs in being flatter on the whorls, in the bands being more distant, and in having a less twisted columella. It reminds one of *M. gracilis*, Anth., but has many distinctive characters. The aperture is about one-third the length of the shell.—*Lea.*

The figure is copied from Mr. Lea's plate.

64. P. altipetum, ANTHONY.

Melania altipeta, ANTHONY, Ann. N. Y. Lyc., vi, p. 87, t. 2, f. 5. BINNEY, Check List, No. 442. BROT, List, p. 34. REEVE, Monog. Mel., sp. 280.
Trypanostoma corneum, LEA, Proc. Acad. Nat. Sci., p. 112, 1864. Jour. Acad. Nat. Sci., vi, p. 148, t. 23, f. 63, 1867.

Description.— Shell conical, smooth, horn-colored, thick; spire elevated; whorls about ten, small, convex, the upper ones carinate, or only striate; sutures distinctly impressed; aperture small, elliptical, banded within; a small but distinct sinus, with an acute termination at base.

Fig. 231.

Habitat.—Raccoon Creek, Vinton County, Ohio.

Diameter, ·24 inch (6 millim.); length, ·62 inch (16 millim.). Length of aperture, ·21 inch (5 millim.); breadth of aperture, ·10 inch (2½ millim.).

Observations.—A very graceful, rather slender species, with somewhat of a club-shaped form by its bulbous body-whorl. Two specimens only are before me; one has a narrow band at the base of the body-whorl; the other has an additional band on the penultimate, faintly indicated also on the upper whorls of the spire.

It may be compared with *M. conica*, Say, but is more elevated, the whorls are more narrow and crowded, as well as more numerous than in that species, and the aperture much smaller, being only about one-fourth the length of the shell.

From *M. neglecta* it differs by its more slender form, smaller and more condensed whorls, and by its entirely different aperture. The apical whorls seem to be slightly folded.—*Anthony.*

This species is almost entitled to a place in the striate division of *Pleuroceræ*, the lines being generally crowded on all except the lower whorl.   The figure is from Mr. Anthony's type.

The following is Mr. Lea's description of

Trypanostoma corneum.— Shell striate, exserted, thin, semi-transparent, pale horn-color; spire raised; sutures regularly impressed; whorls eight, somewhat convex; aperture elongate, narrow, elliptical, whitish within; outer lip acute and very sinuous; columella thin and twisted.

Habitat.—Tennessee.

Diameter, ·27; length, ·76 inch.

Observations.—Two specimens were sent to me some years since by Mr. Anthony. I do not know from what part of Tennessee Fig.232, they came.   In these two specimens, all the whorls but the body-whorl have six or ten transverse striæ.   The base is prolonged almost into a channel, and thus approaches the genus *Io*. In outline and color it is allied to *T. venustum*, herein described, but differs in not being fusiform, in having a larger aperture, and in having striæ.   The aperture is more than one-third the length of the shell.—*Lea*.

Either **Mr.** Anthony sent these specimens *before* describing *altipetum*, or else he must have forgotten his own species.

65. P. Ocoëensis, Lea.

Melania Ocoëensis, Lea, Philos. Proc. ii, p. 12, Feb., 1841. Philos. Trans., viii, p. 169, t. 5, f. 13. Obs. iii, p. 7. DeKay, Moll. N. Y., p. 94. Troost, Cat. Shells Tennessee. Brot, List, p. 38. Wheatley, Cat. Shells U. S., p. 26. Catlow, Conch. Nomenc. p. 188.
Melania Ocoëensis, Lea, Binney, Check List, No. 166.
Potadoma Ocoëensis, Lea, Chenu, Man. de Conch., i, f. 1969.
Potadoma Ocoëensis Lea, Adams, Genera, i, p 299.

Description.— Shell smooth, conical, somewhat thick, dark horn-colored; spire obtuse, towards the apex lined; sutures impressed; whorls somewhat convex; aperture small, ovate, bluish.

Fig. 233.

Habitat.—Ocoee District, Tennessee; Dr. Troost.

Diameter, ·32; length, ·92 of an inch.

Observations.—Five specimens are before me, all of which are more or less decollate. None of them have bands.

Oblique, irregular striæ may be observed more or less on all those which 1 have examined.—*Lea*.

Mr. Reeve's figure decidedly does not represent this species. The identity of *Ocoëensis* with *tenebro cinctum*, Anth., is scarcely doubtful.

66. P. hastatum, ANTHONY.

Melania hastata, ANTHONY, Ann. N. Y. Lyc., vi, p. 85, t. 2, f. 3, March, 1854. BIN-
NEY, Check List, No. 136. BROT, List, p. 31. REEVE, Monog. Mel., sp. 394.

Description.—Shell conical, smooth, rather solid, dark chestnut, spire rather obtusely elevated; whorls 8-9 in number, slightly convex, with occasional delicate spiral striæ, the upper ones sub-

Fig. 234.

carinate; body-whorl subcarinate, with a narrow yellowish band beneath the angle; sutures moderately impressed, yellowish; aperture small, pyriform, purple within; columella and outer lip much twisted together, forming a broad, rather deep, reflexed sinus at base.

Diameter, ·30 inch (7½ millim.); length, ·90 inch (23 millim.).   Length of aperture ·30 inch (7½ millim.). Breadth of aperture ·16 inch (4 millim.).

Habitat.—Alabama.

Observations.—A fine symmetrical species, which seems to have no affinities so close as to be easily confounded with any other. Its most prominent characters, perhaps, are the nearly uniform diameter of the two or three lower whorls, while above these the spire curves more rapidly to the rather acute apex, and the dark purple aperture. These two points will readily serve to distinguish it.—*Anthony.*

Figured from Mr. Anthony's type.

The habitat given above is probably erroneous as Mr. Anthony's tablet is marked "Tennessee" and I have a number of specimens collected by Prof. Haldeman in Holston River, S. W. Virginia. I doubt if it be distinct from *aratum*, Lea, also an inhabitant of the Holston.

67. P. Lyonii, LEA.

Trypanostoma Lyonii, LEA, Proc. Acad. Nat. Sci., p. 155, May, 1863.

Description.—Shell smooth, conical, greenish horn-color, without bands; spire somewhat raised; sutures impressed; whorls about six, convex; aperture rather small, rhomboidal, whitish within; outer lip acute, very sinuous; columella white, thickened below and twisted.

Operculum ovate, very dark brown, with the polar point on the basal margin at the left.

Habitat.—Cumberland River near the Ford, north side of the mountain, and Big Creek, south of mountain, at Cumberland Gap, Tenn.

Diameter, ·32; length, ·85 inch.

Observations.—Quite a number of specimens were sent to me by Major Lyon, from both the above habitats. They are all very much the same in color and size, and none are banded. None were perfect at the apex, but the upper whorls, I think, from indica-

Fig. 235.

tions in a few specimens will be found to be carinate. It is between *Christyi* and *modestum* (nobis). From the former it differs in having the base of the columella less twisted, in having a smaller aperture, and having the whorls more convex. From the latter it differs in being a smaller species, being darker and having a less expanded outer lip. The aperture is about one-third the length of the shell. I name this after Major S. S. Lyon, of the Engineer Corps of the U.S. Army, being collected by him during the campaign, last year, to Cumberland Gap, East Tennessee, where he obtained several new *Melanidæ.*—*Lea.*

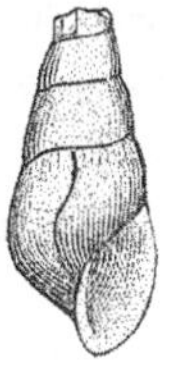

68. P. viridulum, ANTHONY.

Melania viridula, ANTHONY, Ann. Lyc. N. Y., vi, p. 84, t. 2, f. 2, March, 1854. BIN-NEY, Check List, No. 293. BROT, List, p. 31. REEVE, Monog. Mel., sp. 243.

Description.— Shell conical, smooth, rather thick; olive-green; spire much elevated; whorls eight or nine, slightly convex; sutures impressed; aperture elliptical, small, within whitish; outer lip much

Fig. 236. waved or auger-shaped, extending forward at base, and form-

ing a broad sinus in that region.

Diameter, ·35 inch (9 millim.); length, 1 inch (26 millim.). Length of aperture, ·32 inch (8 millim.); breadth of aperture, ·16 inch (4 millim.).

Habitat.—Tennessee.

Observations.—Somewhat like *M. Saffordi*, Lea, but is clearly distinguishable by its more elongated form, its greater number of whorls and size and color of aperture. Differs from *M. regularis*, Lea, by its less number of whorls, and their convexity, as well as by its peculiar green color.— *Anthony.*

This is one of the few species of *Strepomatidæ* which in the absence of all other distinguishing characters rests its specific weight on color alone.   It is a very common species and exceedingly uniform in all of its characters.

The figure is from Mr. Anthony's type.

69. P. striatum, LEA.

Trypanostoma striatum, LEA, Proc. Acad. Nat. Sci., p. 173, 1862. Jour. Acad. Nat. Sci., v, pt. 3, p. 294, t. 36, f. 124. Obs., ix, p. 116.

Trypanostoma rostellatum, LEA, Proc. Acad. Nat. Sci., v, p. 272, 1862. Jour. Acad. Nat. Sci., v, pt. 3. p. 353, t. 39, f. 225. Obs., ix, p. 175.

Description.—Shell striate, subulate, rather thin, horn-color; spire raised; sutures impressed; whorls about eight, somewhat convex, the last rather small; aperture small, subrhomboidal, whitish within; outer lip acute, very sinuous, expanded; columella somewhat thickened and very sinuous.

Fig. 237.

Habitat.—Florence, Alabama; B. Pybas.

Diameter, ·31; length, ·95 inch.

Observations.—Nearly a dozen of this species were received among a number of small shells from Mr. Pybas. It is not an attractive species, being dull horn-color and without bands.

The upper whorls are covered with revolving striæ which rarely extend to the last one, except a single one on the upper part of this whorl. It has much the form and size of *Melania* (*Trypanostoma*) *strigosa* (nobis), but may at once be distinguished by the difference in the form of the aperture, the base of the columella of *striatum* being rounded, while *strigosa* is nearly straight. The length of the aperture is about three-tenths the length of the shell.—*Lea.*   Fig. 238.

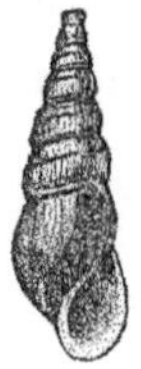

The figure is from Mr. Lea's plate. I can detect no specific difference between this and the following :—

T. rostellatum.—Shell striate, attenuate, rather thin, horn-color, without bands; spire raised; sutures very much impressed; whorls eight, slightly convex; aperture small, rhomboidal, whitish within; outer lip very sinuous; columella bent in and very much twisted.

Operculum ovate, dark brown, with the polar point near the base on the left.

Habitat.—Florence, Alabama; Rev. G. White.

Diameter, ·30; length ·88 inch.

Observations.—Quite a number of this species were among the shells sent to me by Mr. White, collected by him in the northern part of Alabama some years since. It was supposed to be a variety of *Melania* (*Goniobasis*) *proxima*, Say, but the form of the aperture is quite different, having an expanded outer lip. It is also larger, some specimens being nearly an inch long, and it has not a carina, but usually three striæ, the middle one of which rises almost to a carina. In some specimens there is only a single stria, sometimes two, ordinarily three, and rarely four. Usually the upper stria is continued on the lower whorl, extending to the aperture, but rarely any of the others. The aperture is about two-sevenths the length of the shell. It is allied to *Whitei*, herein described, but is a smaller species and differs in color, striæ and in the aperture.—*Lea.*

Figure 238 is a copy of that given by Mr. Lea.

70. P. Knoxvillense, Lea.

Trypanostoma Knoxvillense, Lea, Proc. Acad. Nat. Sci., p. 173, 1862. Jour. Acad. Nat. Sci., v, pt. 3, p. 274, t. 36, f. 87. Obs. ix, p. 96.

Description.—Shell smooth, subulate, rather thin, pale horn-color; spire attenuately conical, sharp pointed; sutures regularly impressed; whorls ten, slightly convex, carinate towards the apex, the last somewhat constricted; aperture small, subrhomboidal, white within; outer lip acute, sinuous; columella thickened below and a little twisted.

Fig.239.

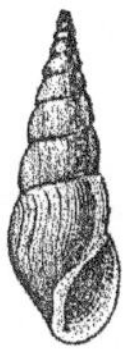

Habitat.—Knoxville, Tennessee; President Estabrook.

Diameter, ·50; length, ·80 inch.

Observations.—A single specimen only of this species was received from President Estabrook. It is closely allied to *Estabrookii*, herein described, but may be distinguished by the form of the inferior part of the columella and the channel being more drawn backwards. It is a smaller species, of rather lighter horn-color and the whorls are rather more bulging. The aperture is less than one-third the length of the shell.—*Lea.*

Figured from Mr. Lea's plate. I doubt whether this is distinct from *Trypanostoma Sycamorense*, Lea, which, like this, is described from one specimen only.

71. P. Whitei, Lea.

Trypanostoma Whitei, Lea, Proc. Acad. Nat. Sci., p. 173, 1862. Jour. Acad. Nat. Sci., v, pt. 3, p. 272, t. 36, f. 85. Obs., ix, p. 95.

Description.—Shell smooth, attenuately conical, somewhat thick, dark horn-color; spire very much raised; sutures regularly impressed; whorls about nine, slightly convex; aperture small, subrhomboidal, whitish within; outer lip acute, sinuous; columella thickened below and twisted.

Fig. 240.

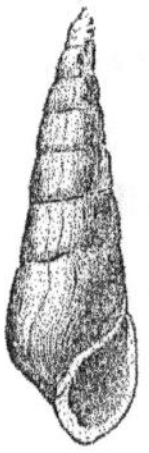

Habitat.—Lafayette County and Marietta, Georgia; Rev. G. White: Farland's Creek, Mississippi; Dr. Spillman: and Tennessee; J. G. Anthony.

Diameter, ·34; length, 1·8 inches.

Observations.—From the four habitats I have sixteen specimens. There is very little difference between them. The tips are either striate or carinate. It is nearly allied to *Estabrookii*, herein described, but it is a smaller species, with a smoother and darker epidermis, and has a smaller aperture and more twist at the base of the columella. The aperture is about three-tenths the length of the shell. I am indebted for many specimens, to the Rev. George White, after whom I name the species.—*Lea.*

72. P. attenuatum, Lea.

Trypanostoma attenuatum, Lea, Proc Acad Nat. Sci., p. 174, 1862. Jour. Acad. Nat. Sci., v, pt. 3, p. 274, t. 36, f. 88. Obs., ix, p. 96.

Description.—Shell smooth, subulate, rather thin, horn-color; spire attenuate; sutures impressed; whorls nine, scarcely convex, the last small, aperture small, rhomboidal, white within; outer lip acute, very sinuous; columella slightly thickened and twisted.

Fig. 241.

Operculum small, ovate, dark brown, with the polar point near the base.

Habitat.—Lafayette, Georgia; Rev G. White: and Tennessee; Dr. Hartman.

Diameter, ·38; length, 1·02 inches.

Observations.—Only two specimens have come under my observation. One is not full grown. In size and general outline this species has a very strong resemblance to *Melania strigosa* (nobis), but it differs much in the aperture and the direction of the base of the columella.

The aperture is quite rhombic, like *Melania Alexandrensis* (nobis). The apical whorls are carinate and the aperture is about one-fifth the length of the shell.—*Lea.*

Figured from Mr. Lea's plate.

73. P. Estabrookii, LEA.

Trypanostoma Estabrookii, LEA, Proc. Acad. Nat. Sci., p. 173, 1862. Jour. Acad. Nat. Sci., v, pt. 3, p. 273, t. 36, f. 86. Obs. ix, p. 95.

Description.—Shell smooth, attenuately conical, rather thin, horn-color; spire very much raised, carinate towards the apex; sutures impressed; whorls about ten, convex; aperture small, sub-rhomboidal, whitish within; outer lip acute, subsinuous; columella white and twisted.

Fig. 242.

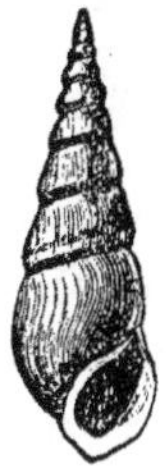

Operculum subovate, dark brown, with polar point near to the basal margin.

Habitat. — East Tennessee; President Estabrook and Bishop Elliott: near Cleveland, Tennessee; Prof. Christy: and Monroe County, Tennessee; J. Clark.

Diameter, ·38; length, 1·11 inches.

Observations.—A number of specimens were received from the above mentioned habitats; all varying very little. It is closely allied to *Christyi* herein described, but while it nearly agrees in color, it is usually smaller and has more convex whorls. These are, in some specimens, more inflated on the lower part. It has a strong resemblance to *M. strigosa* (nobis), but is larger and the aperture is more twisted at the base of the columella. The aperture is about one-fourth the length of the shell. I have great pleasure in naming this species after my deceased friend, President Estabrook of Knoxville, from whom I first received it many years since.—*Lea.*

Figured from Mr. Lea's plate. Allied to *P. subulæforme*, Lea, and to *unciale*, Hald. Indeed, in taking an enlarged view of specific values, all these shells would fall into one species. It is a remarkable and suggestive fact, that the examination of specimens from hitherto unsearched localities generally tends to diminish the number of species, by furnishing connecting links, rather than to increase them.

74. P. modestum, LEA.

Trypanostoma modestum, LEA, Proc. Acad. Nat. Sci., p. 174, 1862. Jour. Acad. Nat.
    Sci., v, pt. 3, p. 276, t. 36, f. 92. Obs. ix, p. 98.
Trypanostoma Knoxense, LEA, Proc. Acad. Nat. Sci., p. 175, 1862. Jour. Acad. Nat.
    Sci., v, pt. 3, p. 281, t. 36, f. 101. Obs. ix, p. 103.

Description.— Shell smooth, conical, rather thin, greenish horn-
color; spire somewhat raised; sutures linear; whorls about seven,
somewhat convex, the last somewhat compressed; aperture rather
small, subrhomboidal, bluish-white within; outer lip acute,

Fig. 243.

sinuous, expanded; columella slightly thickened below and
twisted.

    Habitat.—Chilogita Creek, Blount County, Tennessee, J.
Clarke.

    Diameter, ·32; length, ·80 inch.

    Observations.—I have had a number of this species for some
years and had considered it a variety of *Melania (Goniobasis) dubiosa*
(nobis), but the difference in the outer lip, which is much more
expanded and some other characters, render it specifically different.
The expanded outer lip, which is slightly thickened towards the edge,
resembles that of *Whitei*, herein described, but it has a longer channel
and is not so truncate at the base. It also differs in being a shorter
species with a less number of whorls. None of the specimens before
me have bands. There is a disposition on the apical whorls to be
carinate. None of the specimens were perfect at the apex. Every
one was purplish above. The aperture is about one-third the length
of the shell. It is a very different shell from *Melania (Goniobasis)
modesta* (nobis).—*Lea.*

Figured from Mr. Lea's plate.
The following is evidently the same species.

    T. Knoxense.—Shell smooth, conical, ferruginous or banded, rather
thick, spire rather attenuate, pointed; sutures impressed;
whorls eight, slightly convex, carinate above; aperture small,

Fig. 244.

white or brown within; outer lip sharp, sinuous, expanded;
columella slightly thickened and twisted.

    Habitat.—Flat Creek, Knox County, Tennessee; Prof. D.
Christy.

    Diameter, ·31; length, ·76 inch.

    Observations.—About a dozen of this little species were sent to me some years since by my deceased friend, Joseph Clark. They were collected by Prof. Christy. There is great variety in the color of these specimens. Some are entirely ferruginous, others have a single light line under the sutures, others again have two well defined rather broad brown bands. It is closely allied to *Vanuxemii*, herein described, from South Carolina, but differs in having a larger aperture and a higher spire. The aperture is about one-third the length of the shell.—*Lea.*

The figure is a copy of that given by Mr. Lea.

75. P. luteum, LEA.

Trypanostoma luteum, LEA, Proc. Acad. Nat. Sci., p. 273, 1863. Jour. Acad. Nat. Sci. v, pt. 3, p. 350, t. 39, f. 220. Obs. ix, p. 172.
Trypanostoma Carolinense, LEA, Proc. Acad. Nat. Sci., p. 273, 1862. Jour. Acad. Nat. Sci., v, pt. 3, p. 351, t. 39, f. 221. Obs. ix, p. 173.

Description.—Shell smooth, obtusely conical, rather thick, straw color, without bands, sharp pointed; spire obtusely conical; sutures impressed; whorls eight, somewhat convex; aperture rather small, rhombic, pale straw color within; outer lip sharp, sinuous, thickened near the margin; columella bent in, thickened and twisted below.

Fig. 245.

Habitat.—South Carolina? Prof. L. Vanuxem.

Diameter, ·34; length, ·75 inch.

Observations.—Two specimens of this pretty little species were found among many shells long since given to me by my friend, the late Prof. Vanuxem. It is allied to *Vanuxemii* (nobis), but may at once be distinguished by being without bands, and being a larger and yellow species. The aperture is rather more than one-third the length of the shell.—*Lea.*

Figured from Mr. Lea's plate.

I cannot distinguish specifically the following :—

Trypanostoma Carolinense. — Shell smooth, conical, rather thick, horn-color; spire obtusely conical; sutures impressed; whorls seven, slightly convex; aperture rather small, rhomboidal, whitish or brownish within; outer lip sharp, sinuous; columella bent in, thickened and twisted.

Habitat.—South Carolina; Prof. L. Vanuxem.

Diameter, ·34; length, ·76 inch.

Observations.—Among the mollusca brought long since by my friend, the late Prof. Vanuxem, were about a dozen of this little species.

Fig. 246. The district of the State was not given with the habitat. In

some of the specimens there is a disposition to put on a purplish mark on the inside of the base of the columella. In most of the specimens there is a pale light line immediately below the suture. This species is allied to *simplex*, herein described, but may be distinguished by its being more slender, being a darker horn-color, and in having a more elongated aperture. The aperture is about one-third the length of the shell.—*Lea.*

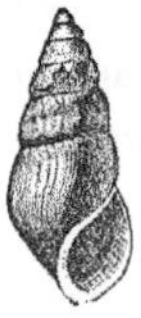

Figured from Mr. Lea's plate.

76. P. curvatum, LEA.

Melania curvata, LEA, Philos. Proc. ii, p. 243. Philos. Trans. ix, p. 23. Obs. ix,
   p. 28. WHEATLEY, Cat. Shells, U. S., p. 25. BROT, List, p. 30. BINNEY, Check
   List, No. 81.
Gyrotoma curvata, Say, ? ADAMS, Genera, i, p. 305.

Description.—Shell obtusely carinate, somewhat pyramidal, rather thick, dark horn-color; spire somewhat elevated; sutures impressed; whorls eight, convex; aperture small, curved, whitish.

Fig. 247.

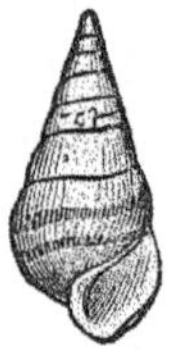

Habitat.—Tennessee.

Diameter, ·40; length, ·73 inch.

Observations.— The two specimens before me vary very little in all their characters. This is a very distinct species, resembling more, perhaps, *M. conica*, Say, than any other. The whorls are close, and about the middle are placed two or three obscure carinæ, which cause a slightly impressed channel. The aperture is small, being a little more than one-third the length of the shell. The outer lip is sharp and very much curved, causing the base of the columella to be twisted. In one of the specimens an obscure band near the base in the interior may be observed.—*Lea.*

77. P. simplex, LEA.

Trypanostoma simplex, LEA, Proc. Acad. Nat. Sci., p. 174, 1862. Jour. Acad. Nat.
   Sci., v, pt. 3, p. 277, t. 36, f. 94. Obs. ix, p. 99.

Description.—Shell smooth, conical, rather thick, yellowish-olive; spire rather elevated; sutures somewhat impressed; whorls eight, somewhat convex, the last somewhat constricted; aperture small, constricted, rhomboidal, whitish within; outer lip acute, sinuous; columella thickened below and twisted.

Habitat.—Cincinnati, Ohio; T. G. Lea.

Diameter, ·33; length, ·76 inch.

Observations.—Among a large number of young *Melania* (*Trypanostoma*) *canaliculata* and *conica*, Say, sent by my brother, long since, I found eight specimens of this small species. All seem to be full grown and are very nearly of the same size. They may be at once distinguished from *canaliculata* by their being much smaller, being much more narrow and having no channel or furrow on the middle of the whorl. The aperture is also much smaller. It differs entirely from *conica* in the whorls, which regularly decrease to the apex, while in that species they decrease rapidly to the apex, which is sharp-pointed. The aperture is about one-third the length of the shell. None of these specimens have bands; one is slightly brownish inside towards the base. This is very different from Mr. Say's *Melania simplex.*—*Lea.*

Fig. 248.

The figure is a copy of that given by Mr. Lea.

78. P. turgidum, Lea.

Melania turgida, Lea, Philos. Proc. ii, p. 82, Oct., 1841. Philos. Trans. ix, p. 18. Wheatley, Cat. Shells U. S., p. 27. Binney, Check List, No. 278. Brot, List, p. 33.

Description.—Shell smooth, obtusely conical, inflated, thick, banded; spire short, pointed at the apex; sutures slightly impressed; whorls seven, flattened; aperture small, trapezoidal; columella thickened, white.

Habitat.—Holston River, East Tennessee.

Diameter, ·35; length, ·55 inch.

Observations.—This is a very short and thick species, having a very large body-whorl disposed to be obtusely angular at the middle. The number of bands varies. One of the specimens has a single one, another has two bands, and five have five bands, there being seven specimens before me. That with a single band is of a bright yellow; the others are of a greenish-yellow. The aperture is nearly one-half the length of the shell, and twisted at the base.—*Lea.*

This species appears to be very closely allied to *T. minor*, Lea.

79. P. minor, LEA.

Trypanostoma minor, LEA, Proc. Acad. Nat. Sci., p. 174, 1862. Jour. Acad. Nat. Sci., v, pt. 3, p. 278, t. 36, f. 95. Obs. ix, p. 100.

Description.—Shell smooth, obtusely conoidal, rather thick, yellowish, banded; spire obtusely conical; sutures much impressed; whorls seven, somewhat convex, the last large; aperture large, subrhomboidal, white and usually banded within; outer lip acute, sinuous; columella incurved, thickened below and slightly twisted.

Fig. 249.

Habitat.—Tennessee; Prof. Troost.

Diameter, ·32; length, ·54 inch.

Observations.—Four specimens were found among a number of young shells from Prof. Troost. It is a modest little species which might easily be taken for a young *Melania conica*, Say. It is most nearly allied to *bivittata*, herein described, but may be distinguished by being wider in proportion, having a shorter spire, being less polished, and not so bright a yellow. It differs also in the brown bands being much less distinctly marked, the upper whorls showing none, while the other is beautifully banded to the apex. The two species differ in columella, *minor* having nearly half of it perpendicular, while *bivittata* has that portion twisted backwards. The bands seem to be uncertain in this species, one having two bands, two having one band and the other having no band. The aperture is nearly half the length of the shell.—*Lea.*

It is very probable that this is the juvenile of some described species.

80. P. pumilum, LEA.

Trypanostoma pumilum, LEA, Proc. Acad. Nat. Sci., p. 174, 1862. Jour. Acad. Nat. Sci., v, pt. 3, p. 279, t. 36, f. 96. Obs. ix, p. 101.

Description.—Shell smooth, shining, conoidal, rather solid, yellowish-green, double-banded; spire obtusely conical; sutures much impressed; whorls seven, somewhat convex, the last very large; aperture rather large, rhomboidal, whitish and double banded within; outer lip acute, sinuous; columella thickened below and very much twisted.

Habitat.—Tennessee; Prof. Troost.

Diameter, ·38; length, ·71 inch.

Observations.—Two specimens of this small species came with *bivittatum*, herein described, mixed with the young of other species. It is rather larger than it and, although very close, may be distinguished by difference of size, being more pyramidal, having a darker epidermis, and in the aperture being more rhombic. Two bands only are visible on the exterior, but the interior of the larger displays a third close to the base of the columella, making a spiral turn round it. The aperture is about three-eighths of the length of the shell. It is very different from *Melania pumila* (nobis) described in Trans. Am. Phil. Soc. v. x, p. 86, which indeed belongs to the genus *Lithasia.*—*Lea.*

Fig. 250.

81. P. opaca, ANTHONY.

Melania opaca, ANTHONY, Proc. Acad. Nat. Sci., p. 58, Feb. 1860. BINNEY, Check List, No. 189. BROT, List p. 38. REEVE, Monog. Melania, sp. 384.

Melania iostoma, ANTHONY, Proc. Acad. Nat. Sci., p. 62, February, 1830. BINNEY, Check List, No. 152. BROT, List, p. 31. REEVE, Monog. Melania, sp. 351.

Melania nigrostoma, Anthony, REEVE, Monog. Melania, sp. 463, 367. BROT, List, p. 38.

Trypanostoma Tennesseénse. LEA, Proc. Acad. Nat. Sci., p. 175, 1862. Jour. Acad. Nat. Sci., v, pt. 3, p. 281, t. 37, f. 100. Obs. ix, p. 103.

Melania iostoma. — Shell ovate conic, smooth; spire obtusely elevated; whorls about six, subconvex; body-whorl exhibiting uncommonly strong lines of growth, curved and varicose; color, greenish-olive, shining; sutures distinct; body-whorl strongly but not sharply angulated on the middle, aperture broad ovate, within light purple, which becomes very deep on the columella, which is regularly rounded; outer lip somewhat produced, and having a well developed sinus at base.

Fig. 251.

Habitat.—Tennessee.

Observations.—This species approaches nearest in form and color *M. glans* (nobis), now changed to *glandula*, from which it differs in being less globular, of a lighter color generally, and by the angulated body-whorl. Compared with *M. pinguis*, Lea, it is less obese, more elongate and has not the rapidly attenuating spire of that species. From all others it is readily distinguished.— *Anthony.*

The following species, which is figured from a type specimen also, will, I am confident, prove to be the young of *iostoma.*

Melania nigrostoma, Anthony.—Shell conically ovate, deep purple-black within and without, whorls five, flatly sloping, smooth, the last

Fig. 252. rather stout, obtusely angled in the middle; aperture ovate.

Anthony, manuscript.

Habitat.———?

Observations.—A dense purple-black species, received from Mr. Anthony with the above name, without habitat.—*Reeve.*

Mr. Reeve first figured this species by mistake (No. 367) as *nigrina*, Lea.

Melania opaca.—Shell ovate, thick, smooth, of a dark brown color; spire short, composed of about six convex whorls; body- Fig. 253. whorl large, subangulated in the centre; sutures indicated by a narrow lighter line, and very distinct; aperture ovate, livid within; columella indented, and tinged with purple; outer lip a little curved; sinus not remarkable.

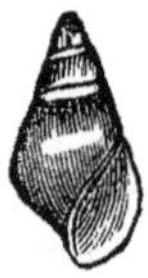

Habitat.—Alabama.

Observations.—A dusky inconspicuous shell of no great beauty. Only two specimens have ever come under my notice, but I am persuaded, nevertheless, that they are distinct—cannot well be compared with any other species. More smooth than *M. athleta* (nobis) and devoid of ribs, which that species has. Its dark, dirty brown color down to about the middle of the body-whorl, and pale olive-green underneath, together with its purple columella, may sufficiently distinguish it.—*Anthony.*

An examination of Mr. Anthony's type specimen of *opaca* convinces me that the species is the same as *iostoma.* Mr. Lea agrees with me that his *Pl. Tennesseénse* described below is a synonyme.

Pl. Tennesseénse. — Shell smooth, obtusely conical, very much

Fig. 254. inflated, rather thick, dark brown; spire short and very obtuse, sutures impressed; whorls about six, convex; aperture large, subrhomboidal, dark within; outer lip acute, much expanded below; inflected and very sinuous: columella very much thickened below, and twisted.

Habitat.—Tennessee; Drs. Troost and Currey: Lebanon County, Tennessee; J. M. Safford.

Diameter, ·47; length, ·84 inch.

Observations.—I have four specimens of this species. The two larger have been in my possession for a long time. They are from Dr. Troost, and are more inflated. While the older part is dark brown, the newer part is dark green, and the interior partakes of these colors. The specimen from Mr. Safford is rather smaller and browner, is purplish within and is thickened on the outer lip near the base. All have a light line under the suture. That from Dr. Currey is about half grown, and has two broad bands. The largest specimen is figured, the lower part of the specimen is more expanded than the others, and is very remarkable in this respect. In outline it is allied to *M. pinguis* (nobis), but differs much in the form of the aperture. The aperture is nearly half the length of the shell.—*Lea.* ·

82. P. trochulus, LEA.

Trypanostoma trochulus, LEA, Proc. Acad. Nat. Sci., p. 175, 1862. Jour. Acad. Nat. Sci., v, pt. 3, p. 282, t. 37, f. 103. Obs. ix, p. 104.

Description.—Shell smooth, top-shaped, very much swollen, yellow, single banded below; spire very obtuse; sutures impressed; whorls six, flattened above and inflated below; aperture large, rhomboidal, whitish and single-banded within; outer lip acute, sinuous; columella thickened below and very much twisted.

Fig. 255.

Habitat.—Holston River, Tennessee; Prof. G. Troost.

Diameter, ·37; length, ·49 inch.

Observations.—A single specimen of this pretty little species was received from Prof. Troost, a long time since, with *Melania turgida* (nobis), but it is a very different species, having a more characteristic auger-shaped mouth, and this specimen has a single band, while four specimens of *turgida* have each five bands. It is also top-shaped while the *turgida* is globose. It is not easily confounded with any other species, being wider for its length than any other *Trypanostoma* with which I am acquainted. The aperture is full one-half the length of the shell, and the body-whorl is nearly two-thirds the length of the whole shell.—*Lea.*

83. P. napoideum, LEA.

Trypanostoma napoideum, LEA, Proc. Acad. Nat. Sci., p. 112, 1864. Jour. Acad. Nat. Sci., vi, p. 143, t. 23, f. 54, 1867.

Description.—Shell smooth, obtusely conical, rather thick, horn-color, without bands; spire short, pointed at the apex; sutures impressed; whorls seven, slightly convex above, the last one very much inflated; aperture large, subrhomboidal, white within; outer lip acute, sinuous; columella thickened below and very much twisted.

Habitat.—Tennessee.

Diameter, ·30; length, ·51 inch.

Observations.—This is one of the many species sent to me long since by my excellent friend the late Prof. Troost. There were but two specimens, and as they had very much the aspect of young *Melania conica*, Say, I refrained from describing them in hopes that others would be received. Feeling satisfied that it is a distinct species, I propose the name from its round, short form, somewhat like a turnip. One of the specimens has a purple spot at the base of the columella; the other is devoid of it. The aperture is quite one-half the length of the shell.—*Lea.*

Fig.256.

Goniobasic Section.

Genus GONIOBASIS, Lea.

Goniobasis, Lea, Proc. Acad. Nat. Sci., p. 262, May, 1862.   Jour. Acad. Nat. Sci., v, pt. 3, p. 217, March, 1863.   Obs. ix, p. 39.

Ceriphasia (sp.), Swainson, H. and A. Adams, Genera, i, p. 298, Feb., 1854.   Chenu, Man. de Conchyl., i, p. 290, 1859.

Pachycheilus (sp.), Lea, H. and A. Adams, Genera, i, p. 298, Feb., 1854.

Potadoma (sp.), Swainson, H. and A. Adams, Genera, i, p. 299, Feb., 1854.   Chenu, Man. de Conchyl., i, p. 290, 1859.

Elimia (sp.), H. and A. Adams, Genera, i, p. 300, Feb., 1854.   Chenu, Man. de Conchyl., i, p. 290, 1859.

Melasma (sp.), H. and A. Adams, Genera, i, p. 300, Feb., 1854.   Chenu, Man. de Conchyl., i, p. 292, 1859.

Hemisinus (sp.), Swainson, H. and A. Adams, Genera, i, p. 302, Feb. 1854.

Juga (sp.), H. and A. Adams, Genera, i, p. 304, Feb., 1854.   Chenu, Man. de Conchyl., i, p. 293, 1859.

Megara (sp.), H. and A. Adams, Genera, i, p. 306, Feb., 1854.   Chenu, Man. de Conchyl., i, p. 293, 1859.

Pleurocera, Rafinesque, Haldeman, Proc. Acad. Nat. Sci., p. 274, 1863.

Melania (sp.), Auct.

SPECIES.

A. *Shell spirally ridged.*

1. G. procissa, ANTHONY.

Melania procissa, ANTHONY, Ann. Lyc. Nat. Hist. N. Y., vi, p. 109, t. 3, f. 9, March, 1854. BINNEY, Check List, No. 218. BROT, List, p. 59. REEVE, Monog. Melania, sp. 342.

Description.—Shell ovate, rather thick, brown; whorls supposed to be about five, rather convex; body-whorl surrounded by about five carinæ, of which two central ones are more prominent; sutures linear; aperture large, ovate, exhibiting the elevated ridges on the body-whorl, as linear, brown bands seen through the substance of the shell; columella rounded, deeply indented, having a small purple spot below the middle, with a slight sinus at the base.

Fig. 257.

Diameter, ·35 inch (9 millim.); length, ·56 inch (14 millim.). Length of aperture, ·28 inch (7 millim.); breadth of aperture, ·18 inch (4½ millim.).

Habitat.—Alabama.

Observations.—The only specimen I have is somewhat mutilated, but seems nevertheless perfectly distinct; the only known species with which I can compare it is *M. sulcosa,* Lea, which is a much thinner and more elevated species. The aperture of the present shell is also proportionally much larger, and the number of whorls less, for, though injured in that part, the rapid diminution of the whorls does not indicate an elevated spire; the number of raised lines on the body-whorl is also less, and they are rather very elevated *costæ* than *striæ* as in Mr. Lea's species.—*Anthony.*

This species, at first sight very distinct, may be only a lengthened variety of Mr. Anthony's *Anculosa canalifera;* and the latter is perhaps a variety of *A. carinata,* Bruguière (*dissimilis,* Say). The locality given is probably incorrect, as the shell has the aspect of the North Carolina *Strepomatidæ* rather than those of Alabama.

B. *Shell tuberculate or nodulous.*

2. G. varians, LEA.

Melania varians, LEA, Proc. Acad. Nat. Sci., p. 120, 1861.
Goniobasis varians, LEA, Jour. Acad. Nat. Sci., v, pt. 3, p. 219, t. 34, f. 2, March,
1863. Obs. ix, p. 41.

Description.—Shell smooth, plicate or striate, raised conical, rather

Fig. 258.   thick, yellowish or pale brown, banded; spire raised; sutures
impressed; whorls seven, flattened above; aperture rather
small, elliptical, whitish and banded within; outer lip acute;
columella whitish, incurved, obtusely angular at the base.

Habitat.—Coosa River, Alabama; Dr. Showalter and Dr.
Budd.

Diameter, ·40; length, 1·4 inches.

Observations.—I have a number of specimens before me,
some of which have been in my possession for several years. They
are allied to *Melania Haysiana* (nobis), and I formerly
thought they were a mere variety of that species; but the

Fig. 259.

numerous and fine specimens sent to me, of various ages
and forms, by Dr. Showalter, satisfy me that the species is
quite distinct. It is very variable, some being smooth and
beautiful, while others are plicate and others again roughly
striate, with a shoulder below the sutures, giving it quite a
different aspect. The aperture is more than one-third the
length of the shell. It usually has four bands, but in some individuals
there are none and others have one, two, three or four.—*Lea.*

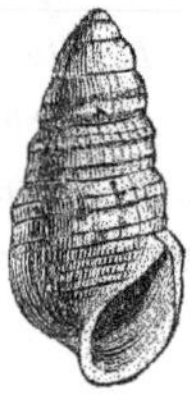

The first figure is a copy of Mr. Lea's; the other figure is
from a specimen belonging to the Smithsonian Institute. This
latter appears to be the typical form of the species.

3. G. Hydeii, CONRAD.

Melania Hydeii, CONRAD, New Fresh Water Shells. p. 50, t. 8, f. 1, 1834. REEVE,
Monog. Melania, sp. 248. DEKAY, Moll. N. York, p. 93. WHEATLEY, Cat.
Shells, U. S. p. 25. BINNEY, Check List, No. 141. Conrad, MÜLLER, Synopsis,
p. 44.
Melania Hydei, Conrad. JAY, Cat. Shells. 4th edit., p. 273. BROT, List, p. 32. HAN-
LEY, Conch. Misc. t. 1, f. 3.
Melania Hydii, Conrad, CATLOW, Conch. Nomenc., p. 187.

Description.—Shell conical, rather elevated; whorls flattened, with spiral acute tuberculated lines, one or two only on each whorl of the spire, and about four on the body-whorl, the inferior one plain; aperture elliptical.

Fig. 260.

Fig. 261.

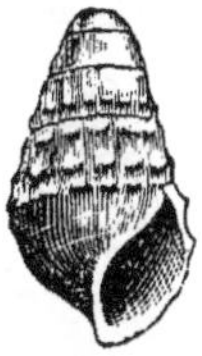

Observations.—Inhabits rocks in the Black Warrior River, south of Blount's Springs, Alabama, and is very abundant. It is remarkable for its distant tuberculated lines. Young specimens are olive, with a purple band on each whorl, and are without tubercles; the body-whorl is angulated, and carinated.

It is named in honor of Mr. William Hyde, an industrious and excellent conchologist.—*Conrad.*

4. G. decorata, ANTHONY.

Melania decorata, ANTHONY, Proc. Acad. Nat. Sci., p. 55, Feb., 1860. REEVE, Monog.
Melania, sp. 251. BINNEY, Check List, No. 86. BROT, List, p. 32.
Goniobasis Tryoniana, LEA, Proc. Acad. Nat. Sci., p. 272, 1862. Jour. Acad. Nat.
Sci., v, pt. 3, p. 342, t. 38, f. 207, March, 1863. Obs., ix, p. 164, t. 38, f. 207.
Goniobasis granata, LEA, Proc. Acad. Nat. Sci., p. 272, 1862.~Jour. Acad. Nat. Sci.
v, pt. 3, p. 343, t. 38, f. 209, March, 1863. Obs., ix, p. 165.

Description.—Shell short, thick, ovate; whorls about five, but truncate as to show only two or three remaining; whorls prominently ribbed and intersected by revolving striæ, forming nodules where they cross each other; dark bands also revolve around the whorls, giving them a highly decorative appearance, columella often thickened by a callous deposit; sinus small.

Fig. 262.

Habitat.—Oostenaula River, Georgia.

Observations.—I collected some two hundred specimens of this species in Oostenaula River, Georgia, in 1853, I then supposed they would prove to be merely the young of *M. cælatura,* Conr.

Fig. 263.

Closer examination and comparison, however, have convinced me that they are not identical. Many of the specimens are decidedly mature, and differ from *cælatura* by the greater regularity of their folds, which are also interrupted by a revolving raised line near the sutures, and by their dark bands and less elongate form; cannot well be compared with any other.—*Anthony.*

The following are the descriptions of the species believed to be synonymes.

Goniobasis Tryoniana.—Shell granulose or striate, subfusiform, yellowish-brown or dark brown, thick, robust, banded, rarely not banded; spire obtusely conical; sutures irregularly im-

Fig. 264.

pressed: whorls about six, the last very large; aperture very large, ovately rhomboidal, much banded within; outer lip subcrenulate, scarcely sinuous; columella slightly bent in and scarcely twisted.

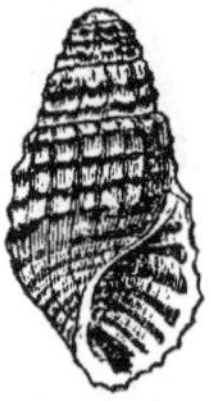

Operculum ovate, rather thick, dark brown, with the polar point near the left margin, above the base.

Habitat. — Oostenaula, near Rome; Bishop Elliott: Etowah River, Georgia; J. Postell: and Oconee River and Tennessee River; Rev. G. White.

Diameter, ·52; length, 1·01 inches.

Observations.—I have a number of specimens from the above various habitats, and they vary very much. Some are more obtuse than others, and some are tuberculate, while others are only transversely striate, close striæ often covering the whole surface. Usually the bands do not show on the outside, often giving the surface a clouded appearance, while in the interior usually the bands are well marked and sometimes number as many as eight, but sometimes the aperture is entirely white; rarely the whole is purple inside, in which case the exterior is very dark brown. The base of the columella is usually yellowish outside. It is somewhat allied to *Melania (Goniobasis) Coosaensis* (nobis), but that species is more constricted and has a narrow aperture! The aperture is nearly one-half the length of the shell. I name this species after Mr. G. W. Tryon, Jr., who has done much to promote the study of malacology.—*Lea.*

Goniobasis granata.—Shell granulose, striate below, fusiform, banded, rather thick, shining, inflated, olivaceous or reddish; spire depressed; sutures irregularly impressed; whorls about

Fig. 265.

five, flattish, the last one very large; aperture large, ovately rhomboidal, much twisted.

Operculum ovate, rather thin, dark brown, with the polar point near to the left margin above the base.

Habitat.—Etowah River, near Canton, Georgia; Bishop Elliott and Rev. G. White.

Diameter, ·36; length, ·70 inch.

Observations.—A number of specimens were sent to me by Bishop Elliott and the Rev. Mr. White; some are much more granulate than others, which are transversely striate with rugose granulations. When perfectly granulate there are three or four rows of beautiful small nodes surrounding the whorls. There are usually seven bands well marked inside, but obscure on the exterior. A single specimen is entirely brownish-purple inside. It is rarely without color; usually there is a small yellowish spot at the base of the columella outside. Those sent by Mr. White are all olive-green and without an iron deposit. Those from Bishop Elliott were all covered with the black oxide of iron, which on being removed exhibit a rubiginose color, and do not show much color in the bands. In outline it is near to *Melania* (*Goniobasis*) *bellula* (nobis), but is more inflated and is striate and granose. The aperture is about one-half the length of the shell. —*Lea.*

This species is a good one but has unfortunately not been properly distinguished from *cœlatura*, Conrad.

Mr. Anthony's description of *decorata* applies to the juvenile form only, but his name has priority and must be adopted. Mr. Anthony has misunderstood the range of characters of the species, and some of the specimens labelled *decorata* by him are the young of *cœlatura*. Mr. Lea's type figure of *Tryoniana*, which is here copied, exhibits the mature form, but he has made his description to cover both this roughly granose species and the smoother *cœlatura*. Indeed, some of the shells which he has presented to me are really *cœlatura*.

Mr. Lea's *granata* is a young shell and is in all respects identical with Mr. Anthony's species. The original figure is copied. Luckily in the present instance a number of lots of specimens, numbering several hundred individuals in all, have enabled me to make the above decisions with confidence.

There is a wide range of variation in color, form, texture and ornamentation in this species.

5. G. cælatura, CONRAD.

Melania cœlatura, CONRAD, Proc. Acad. Nat. Sci., iv, p. 154, Feb., 1849. Jour. Acad. Nat. Sci., i, pt. 4, p. 278, t. 38, f. 3, Jan., 1850. BINNEY, Check List, No. 58. BROT, List, p. 32. REEVE, Monog. Melania, sp. 245.
Goniobasis Tryoniana, LEA, Description in part.

Description.—Ovate-oblong, turreted; volutions six, with longitudinal ribs and unequal prominent revolving lines, subnodulous where they cross the ribs; the ribs on the body-whorl do not reach the middle; the color ochraceous and brown; aperture narrow, elliptical;

Fig. 266.   Fig. 266a.

labium with interior brown bands; superior part of columella somewhat callous.

Habitat.—Savannah River.—*Conrad.*

Mr. Lea's description of *Tryoniana* includes this species. Fig. 266 is a copy from Conrad's plate. It is readily distinguished from the preceding species by being narrower, more fusiform and closely nodulously striate; the tuberculations not being so well developed as in *decorata.* As mentioned before, Mr. Anthony has distributed the young of this species under the latter name.

6. G. Stewardsoniana, Lea.

Goniobasis Stewardsoniana, Lea, Proc. Acad. Nat. Sci., p. 272, 1862. Jour. Acad. Nat. Sci., v, pt. 3, p. 344, t. 38, f. 210, March, 1863. Obs. ix, p. 166.

Description.— Shell granulate, transversely striate, subfusiform, thick, shining, inflated, green or brown, without bands; spire very obtuse; sutures impressed; whorls slightly convex; aperture very large, ovately rhomboidal, white within; outer lip sharp, slightly sinuous; columella bent in, thickened above and below and twisted.

Fig. 267.

Habitat.—Knoxville, Tennessee; B. W. Budd, M.D.

Diameter, ·42; length, ·70 inch.

Observations.—Two specimens, one perfect, the other with little more than the body-whorl, were given to me long since by Dr. Budd, to whom I am indebted for many fresh water *mollusca* of our Western and Southwestern States, one of which, properly belonging to this genus, I called *Melania Buddii.* Of the two specimens before me, the younger is almost entirely perfect, and presents a fine, smooth, dark green epidermis with transverse striæ, which on the upper part of the whorls are broken up into granulations. These striæ are raised and rounded, and are darker than the ground. The old specimen is of a rusty color, having been covered with oxide of iron. The aperture is more than half the length of the shell. There is some resemblance of this shell to *Melania (Gonio-*

Fig. 268.

basis) *Hydei*, Con., but that is conical, having a high granular spire. I name this after my friend Thomas Stewardson, M.D., to whom I am indebted for many fine specimens of our Southern mollusca.—*Lea.*

I at first considered this shell the young of *cœlatura*, but have finally concluded that it is distinct. The surface is ridged around, the ridges being fretted, disposing to tuberculation ; the shell is very solid and generally dark green and polished. A figure of the adult satisfactorily exhibits the differences between it and *cœlatura*.

7. G. flavescens, Lea.

Goniobasis flavescens, Lea, Proc. Acad. Nat. Sci., p. 271, 1862. Jour. Acad. Nat. Sci., v, pt. 3, p. 339, t. 38, f. 202, March, 1863. Obs. ix, p. 161.

Description.—Shell striate, sometimes granulate and folded, subcylindrical, yellowish, thick; spire obtusely conical; sutures irregularly impressed; whorls slightly convex, the last very large; aperture large, subrhomboidal, banded or white within; outer lip sharp, scarcely sinuous; columella bent in, very much thickened above and twisted.

Fig. 269.

Operculum ovate, rather thick, brown, with the polar point near the left margin above the base.

Habitat.— Oconee and Tennessee Rivers, Tennessee; Rev. G. White.

Diameter, ·43; length ·97 inch.

Observations. — Quite a number of specimens were sent to me by Mr. White, and among them there is great variation. They are allied on one side to *Tryoniana*, herein described, and on the other to *Melania (Goniobasis) brevis* (nobis.) It is a larger species than the latter, and smaller and more cylindrical than the former. Brown bands are more or less observable in the interior of about half the specimens before me. The callus above is usually thick and often colored. One specimen only is entirely brown inside. The aperture is more than one-third the length of the shell, none have the apex sufficiently perfect to ascertain the number of whorls. There are probably about six. There is a close affinity between this and *Melania (Goniobasis) Holstonia* (nobis), which, however, is more robust, of a different color and more granulate.—*Lea.*

8. G. occata, HINDS.

Melania occata, HINDS, Ann. and Mag. Nat. Hist. xiv, p. 9. Zool. Voy. Sulphur.
    Moll. ii, p. 56, t. 15, f. 5. CATLOW, Conch. Nomenc., p. 188. BROT, List, p. 34.
    LEA, Proc. Acad. Nat. Sci., p. 81, April, 1856. REEVE, Monog. Mel., sp. 267.
Juga occata, Hinds, CHENU, Man. de Conchyl., i, f. 2016.
Melania Shastaensis, Lea, REEVE, Monog. Mel. sp. 318.

Description.—Shell ovate, elongate, lutescent; whorls few, rounded,
grooved, intermediate ridges narrow, acute; spire eroded
above the fourth whorl; aperture cærulescent.

Fig. 270.

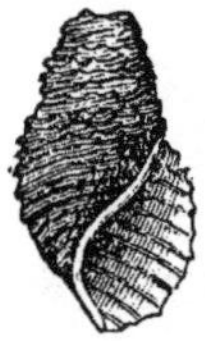

Habitat.—River Sacramento, California.

Observations. — The rounded whorls are ploughed into
numerous furrows and the intervening ridges are comparatively narrow and keel-shaped; the lower part of
the apérture is somewhat dilated, and slightly disposed
to elongate in the manner of *Io.* — *Reeve.*

Mr. Reeve, and Dr. Brot following him, have fallen into the
error of quoting *Shastaensis* as a synonyme through that prolific source of error " an authentic specimen." The figure of
" *Shastaensis* " given by Reeve from a specimen in the collection of Mr. Cuming is finer than any specimen of *occata* that
I have seen. The species varies in form very much.

9. G. catenaria, SAY.

Melania catenaria, SAY, Jour. Acad. Nat. Sci., ii, p. 379, Dec. 1822. BINNEY, Reprint,
    p. 111. BINNEY, Check List, No. 52. REEVE, Monog. Melania, sp. 336. DEKAY,
    Moll. N. York, p. 93. WHEATLEY, Cat. Shells U. S. p. 24. GIBBES, Rep't.
    S. Carolina, p. 19. JAY, Cat. 4th edit., p. 273. CATLOW, Conch. Nomenc. p. 185.
    BROT, List, p. 34.
Elimia catenaria, Lea, ADAMS, Genera, i, p. 300.
Melania sublirata, CONRAD, Jour. Acad. Nat. Sci., 2nd ser. i, pt. 4, p. 277, t. 38, f. 1.
    Jan. 1850. BROT, List, p. 37. REEVE, Monog. Melania, sp. 339.

Description.—Shell conic, black; whorls seven or eight, slightly
undulated transversely, and with eight or nine revolving,
elevated lines, of which the four or five superior ones are
almost interrupted between the undulations.

Fig. 271.

Length less than half an inch.

Habitat.—South Carolina.

Observations.—The essential specific character resides in the cate-
nated appearance of the superior revolving lines of the whorls, result-
ing from their being more prominent on the undulations which they

cross, than between them, where they are often obsolete. This species was sent to me by Mr. Stephen Elliot, who obtained it in Limestone Springs, St. John's, Berkley.—*Say.*

The shell described by Mr. Say is a quite young one — as is evident from an inspection of the figure, which is drawn from the original type, now in the possession of Jno. G. Anthony. Mr. Lea described under the same name a species from Georgia, but Prof. Haldeman (Monog. Limniades, Cover No. 6) called attention to the fact that the name was preoccupied by Say, and Mr. Lea subsequently changed his name to *catenoides*.

Fig. 272. Fig. 273.

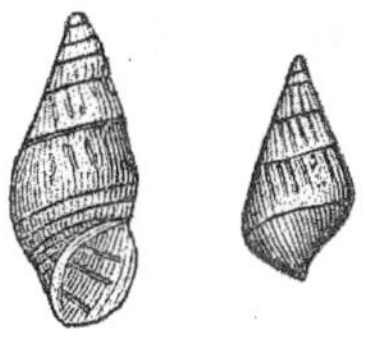

That the following is the adult of this species cannot be doubted.

Melania sublirata. — Elongate-conoidal; volutions six, the sides flattened above; whorls of the spire with a carinated angle near the base of each, and longitudinally ribbed; ribs not prominent; upper whorls with two distant revolving lines on each; base of the body-whorl striated, the upper portion of body-whorl obscurely ribbed; color olivaceous with obscure brown bands.

Habitat.—Savannah River.—*Conrad.*

10. G. Floridensis, Reeve.

Melania Floridensis, Reeve, Monog. Melania, sp. 334. Brot, List, p. 34.

Fig. 274. Fig. 275.

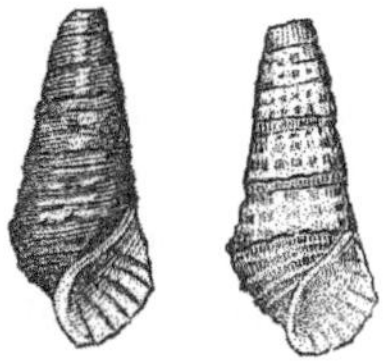

Description.—Shell somewhat pyramidally turreted, blackish-olive, whorls seven to nine, broadly sloping, then slightly angled, longitudinally indistinctly plaited, corded throughout with fine noduled ridges; aperture ovate, a little effused at the base.

Habitat.—Florida.

Observations. — Sculptured throughout with fine corded ridges which are noduled on crossing the rather obscure longitudinal plaits. — *Reeve.*

11. G. catenoides, LEA.

Melania catenaria, LEA, Proc. Philos. Soc. i, p. 289, Oct. 1840 (preoc.).
Melania catenoides, LEA, Philos. Trans. viii, p. 228, t. 6, f. 60. Obs., iii, p. 66.
    DEKAY, Moll. N. Y. p. 101. WHEATLEY, Cat. Shells U. S. p. 24. JAY, Cat.
    4th edit., p. 273. BINNEY, Check List, No. 53. CATLOW, Conch. Nomenc. p. 185.
    BROT. List. p. 34. REEVE, Monog. Melania, sp. 298.
Elimia catenoides, Lea, CHENU, Man. de Conchyl. i, f. 1982. ADAMS, Genera, l,
    p. 300.

Description.—Shell granulate, elevated, conoidal, livid; apex folded;
sutures small; aperture ovate.

Fig. 276.   Fig. 277.   *Habitat.* — Chattahoochee River,   Fig. 278.    Fig. 279.
Columbus, Geo.

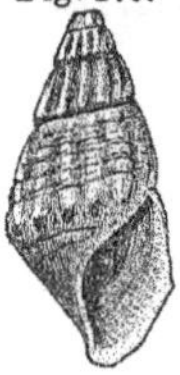

Diameter, ·43; length, ·93 of an
inch.

Observations. — This species differs from the *M. Boykiniana,* in
being without tubercles and carina.

The colored revolving hair-like lines are numerous and, being pitted,
present the appearance of a chain. Some of the old specimens are
quite black, while the younger ones are green or yellow. In some
cases where the apex is eroded or worn off and the shell black and
old, it looks like *M. Virginica* (Say), as no grains can be observed.—
Lea.

12. G. Etowahensis, LEA.

Melania Etowahensis, Lea, REEVE, Monog. Mel. sp. 426, May, 1861.
Goniobasis Canbyi, LEA, Proc. Acad. Nat. Sci., p. 271, 1862. Jour. Acad. Nat. Sci.,
    v, pt. 3, p. 340, t. 38, f. 204, March, 1863. Obs., ix, p. 162.

Description.—Shell tuberculate, plicate, transversely striate below,
turreted, thin, brown or pale brown, maculate; spire turreted; sutures,
irregularly impressed; whorls seven, carinate, with com-
Fig. 280.
pressed tubercles on the periphery; aperture small, rhomboidal, spotted within; outer lip crenulate, sinuous; columella
bent in and very much twisted.

Habitat. — Lake Monroe, Florida; W. Canby: and Etowah
and Tennessee Rivers, Georgia; J. Postell.

Diameter, ·35; length, ·76 inch.

Observations.—Several bleached specimens were collected by Mr.
Canby of Wilmington, Delaware, from Enterprise, on Lake Monroe.
Mr. Postell sent me two perfect specimens from Etowah River,

Georgia, and a bleached one from the Tennessee River. All these specimens are without variation. There are usually five revolving striæ below and two above that round the periphery, which make compressed tubercles where they are crossed. These folds are bright brown, nearly red on their left side, and give a maculate appearance to the whole shell. These maculations are visible on the inside. The compressed, sharp tubercles almost constitute spines, and, on first looking at this shell, one is reminded of *Melania spinulosa*, Lam., but it cannot be confounded with that species. In outline and in most of its characters it is allied to *Hallenbeckii*, herein described, but it is much smaller, and differs in being maculate instead of banded. The aperture is about one-third the length of the shell. I dedicate this to my friend, Mr. Canby, who kindly brought me some specimens.—*Lea*.

I presume it was Mr. Lea's first intention to describe this species under the name of *Etowahensis*, as specimens are before me, which that gentleman sent to Mr. Anthony under the latter name. Mr. Reeve's description, which it is unnecessary for us to reproduce here, is drawn up from Mr. Anthony's specimen. The figure, which is copied from the original one, gives but a faint idea of this beautifully variegated species, which for gracefulness of contour stands unrivalled.

It is doubtful whether this species is really distinct from *papillosa*, Anth. In the young shells, particularly, it is extremely difficult to draw a line of distinction between the two.

13. G. Hallenbeckii, Lea.

Goniobasis Hallenbeckii, Lea, Proc. Acad. Nat. Sci., p. 271, 1862. Jour. Acad. Nat.
   Sci., v, pt. 3, p. 339, t. 38, f. 203, March, 1863. Obs., ix, p. 161.
Melania Hallenbeckii, Lea, Reeve, Monog. Melania, sp. 332.

Description.—Shell tuberculate, transversely striate below, turreted, rather thin, pale horn-color or olivaceous, banded, or without bands; spire elevately turreted; sutures very much impressed; whorls eight, carinate, with compressed tubercles at the periphery; aperture large, ovately rhomboidal, whitish within; outer lip crenulate, sinuous; columella bent in, slightly thickened, and very much twisted.

Fig. 281.

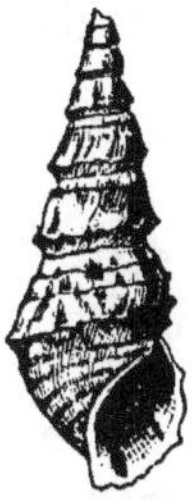

Habitat. — Randall's Creek, near Columbus, Georgia; G. Hallenbeck.

Diameter, ·47; length, 1·24 inches.

Observations.—This is a very beautiful species, having some resemblance in outline to *Melania (Goniobasis) Boykiniana* (nobis), but it is larger, has more tubercles, and a more elevated spire.   Many speci-

Fig. 282.

mens are disposed to be plicate, and on the periphery where these folds traverse the raised striæ, a compressed tubercle is caused.   These are sometimes repeated obscurely by the inferior striæ.   Most of the specimens before me are banded, but many are entirely free from bands.   Usually, there are four bands, rarely five, two being visible on the upper whorls.   The lower band near to the base of the columella is usually well defined.   The aperture is about one-third the length of the shell.   I have great pleasure in dedicating this fine species to Mr. Hallenbeck, who has done much to develop the natural history of Georgia.—*Lea.*

Dr. Brot makes this species a synonyme of *Boykiniana*, but I cannot, from the material that has passed under my inspection, coincide in this decision, although the two are closely allied, and *may* be the same.

14. G. Boykiniana, LEA.

Melania Boykiniana, LEA, Proc. Philos. Soc., i, p. 289, Oct., 1840.   Philos. Trans. viii, p. 228, t. 6, f. 59.   Obs., iii, p. 66.   DEKAY, Moll. N. Y., p. 100.   WHEATLEY, Cat. Shells U. S., p. 24.   REEVE, Monog. Melania, sp. 77.   JAY, Cat. Shells, 4th edit., p. 273.   BINNEY, Check List, No. 37.   CATLOW, Conch. Nomenc., p. 185.   BROT, List, p. 34.
Elimia Boykiniana, Lea, CHENU, Man. de Conchyl., i, f. 1978.   ADAMS, Genera, i, p. 300.
Juga Troostiana, Lea, CHENU, Man. de Conchyl., i, f. 2017.

Description.—Shell granulate, elevated, somewhat turreted, at the carina tuberculate; sutures impressed; aperture long, ovate.

Fig. 283.   Fig. 284.

Habitat. — Chattahoochee River, Columbus, Georgia.

Fig. 285.

Diameter, ·38; length, ·94 of an inch.

Observations. — This is a very distinct and remarkable species.   Although many of the individuals differ, the prevailing character is to have the whole of the whorls covered with numerous granulate, revolving lines, generally bearing a purple or brown line.   In some the tubercles of the carina assume the character of folds.—*Lea.*

Figure 283 is a copy of the original figure. 284 and 285 are from specimens in the Smithsonian collection. Like *Hallenbeckii*, this species is numerous in individuals. Many specimens are light green with raised, revolving lines of very dark color, giving them a strikingly handsome appearance.

So great are the variations of form in this shell and in *catenaria*, that I should not be surprised if the latter proved to be a younger stage of the former.

15. G. Bentoniensis, Lea.

Goniobasis Bentoniensis, LEA, Proc. Acad. Nat. Sci., p. 271, 1862. Jour. Acad. Nat. Sci. v, pt. 3, p. 336, t. 38, f. 198, March, 1863. Obs. ix, p. 158.

Description.—Shell carinate, folded, striate, conical, greenish horn-color, without bands; spire raised, conical; sutures very much impressed; whorls seven, slightly convex; aperture rather small, ovately rhomboidal, whitish within; outer lip acute, scarcely sinuous; columella bent in, s o m e w h a t twisted.

Fig. 286.

Habitat.—Benton County? North Alabama; G. Hallenbeck. My cabinet and cabinet of Dr. Hallenbeck.

Diameter, ·39; length, ·93 inch.

Observations.—There are two specimens before me sent by Mr. Hallenbeck. He is not positively certain that they were found in Benton County. Both these have revolving striæ over all the whorls. The upper whorls have folds which, where they cut the striæ, are raised into obtuse nodes. The larger striæ on the body-whorl are represented on the inside by white lines. It is rare that any species is carinate, plicate and striate at the same time. It is allied to *Melania (Goniobasis) Boykiniana* (nobis), but is not tuberculate, nor is it so large. The aperture is about one-third the length of the shell.—*Lea.*

Doubtfully distinct from *papillosa*, Anthony.

16. G. papillosa, Anthony.

Melania papillosa, Anthony, REEVE, Monog. Mel., sp. 467, May, 1861. BROT, List, p 34.
Goniobasis Downieana, LEA, Proc. Acad. Nat. Sci., p. 272, 1862. Jour. Acad. Nat. Sci., v, pt. 3, p. 341, t. 38, f. 206, March, 1863. Obs. ix, p. 163.

Description.—Shell somewhat pyramidally ovate, fulvous-olive;

whorls five, slopingly convex, then keeled, longitudinally faintly plicated, transversely nodulosely
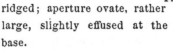
Fig. 287. Fig.287a. Fig. 288.   Fig. 290. Fig.289.

ridged; aperture ovate, rather
large, slightly effused at the
base.

Habitat.—Florida.

Observations. — Distinguished
by a papillose sculpture though being crossed with transverse ridges,
passing over oblique longitudinal folds.—*Reeve.*

The following is a copy of the description of

Goniobasis Downieana.— Shell tuberculate, subturreted, clathrate
and subcarinate above, transversely striate below, thin, pale brown;
spire conical, clathrate; sutures irregularly impressed; whorls seven,
subcarinate; compressed tuberculate on and above the periphery;
aperture rather large, ovately rhomboidal, whitish within; outer lip
crenulate, sinuous; columella bent in and twisted.

Habitat.—Etowah River; J. Postell.

Diameter, ·33; length, ·71 inch.

Observations.—Two specimens only of this beautiful species are
before me, neither of them being entirely perfect. These two are
without bands, but one has in the interior slight lines of color,
which indicate that other individuals may be well banded.   Fig. 291.
The striæ below the periphery are six, and they are thick
enough to cause corresponding white lines in the interior
The three lines above the periphery are cut by close folds on
ribs and these make the upper parts beautifully clathrate.
This species is closely allied to *Canbyi* herein described but
it is shorter and wider, and the tubercles are more numerous and
smaller, having about twenty on the periphery while *Canbyi* has about
thirteen. These three ornamented little species—*Canbyi, Couperii,
Downieana*—form a distinct group among American species, which
one would hardly expect to find existing here. The aperture is rather
more than one-third the length of the shell. I name this species
after T. C. Downie, Esq., civil engineer, who has done much to develop the natural history of Georgia.—*Lea.*

17. G. Couperii, Lea.

Goniobasis Couperii, Lea, Proc. Acad. Nat. Sci., p. 271, 1862. Jour. Acad. Nat. Sci., v, pt. 3, p. 341, t. 38, f. 205, March, 1863. Obs. ix, p. 163.

Description.—Shell tuberculate, plicate, striate above and below, turreted, thin, dark brown, banded at the base; spire turreted; sutures very much impressed; whorls seven, subcarinate, with compressed tubercles on and above the periphery; aperture very small, subrhomboidal, dark and single-banded within; outer lip crenulate, very sinuous; columella bent in, twisted and purple.

Fig. 292.

Habitat.—Etowah River; Mr. Couper by J. Postell.

Diameter, ·27; length, ·72 inch.

Observations. — This ornamented little species was sent by Mr. Postell with the *Canbyi*, which he found also in Etowáh River. They are closely allied, but *Couperii* is slimmer, has more striæ above the periphery, which are all cut by the folds, thus filling the spire with small, compressed tubercles. It differs also in being much darker, in not being maculate and in having a broad band near the base which is well marked inside. Below the periphery there are six well-defined, raised revolving striæ. The aperture is not quite one-third the length of the shell. Mr. Postell informs me that this species, as well as *Canbyi* and *Downieana*, from Etowah River, were brought some years since by Mr. Couper, son of James Hamilton Couper, Esq., of Hopeton, near Darien, and I have great pleasure in naming this species after him.—*Lea.*

18. G. inclinans, Lea.

Goniobasis inclinans, Lea, Proc. Acad. Nat. Sci., p. 267, 1862. Jour. Acad. Nat. Sci., v, pt. 3, p. 318, t. 37, f. 166, March, 1863. Obs. ix, p. 140.

Description.—Shell very much folded, somewhat drawn out, rather thin, obscurely banded; spire subattenuate, sharp pointed; sutures furrowed; whorls eight, flattened, covered with oblique folds; aperture small, rhomboidal, pale brown within; outer lip acute, sinuous; columella very much bent in, brownish-red and very much twisted.

Fig. 293.

Operculum ovate, very thin, light brown, with the polar point nearer to the centre than usual.

Habitat.—New Albany, Georgia; Rev. G. White: Etowah; J. Postell: Tuscumbia, Alabama; B. Pybas.

Diameter, ·27; length, ·68 inch.

Observations.—A large number of this species was sent to me by Mr. White and Mr. Pybas. They were generally incrusted with carbonate of lime, which was easily removed. It has some resemblance to *Melania (Goniobasis) Deshaysiana*, but it is a smaller species, with numerous folds much inclining to the left, and generally covering all the whorls. These folds are crossed by revolving striæ which form numerous nodes, giving a general rough appearance to the surface, Below the suture there is generally a light line. There is usually a dark band at the base of the columella, more distinct inside, and sometimes several indistinct ones may be observed above. It reminds one of *Melania (Goniobasis) Edgariana* (nobis), but that is a much larger species, and different in color and folds. The aperture is about one-fourth the length of the shell.—*Lea.*

Figured from Mr. Lea's plate.

19. G. Postellii, Lea.

Melania Postellii, Lea, Proc. Acad. Nat. Sci., p. 166, July, 1858. Binney, Check List, No. 214. Brot, List, p. 34.
Melania Portellii, Lea, Reeve, Monog. Melania, sp. 427.
Goniobasis Postellii, Lea, Jour. Acad. Nat. Sci., v, pt. 3, p. 343, t. 38, f. 208, March, 1863. Obs., ix, p. 165.

Description.—Shell granulate, attenuate, rather thin, yellowish-olive, transversely striate below; spire raised; sutures irregularly impressed;

Fig. 294.

whorls rather flattened, about eight; aperture small, elliptical, white or banded within; outer lip sharp; columella twisted.

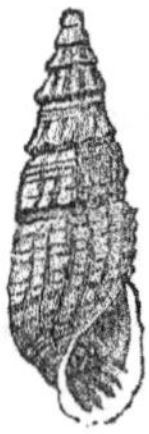

Habitat.—Altamaha River, Georgia; James Postell.

Diameter, ·36; length, 1·06 inches.

Observations.—Some dozen specimens were received from Mr. Postell, which were all more or less covered with a black deposit of oxide of iron, but underneath the epidermis was quite perfect, and of a light horn-color. Most of the specimens have four or five brown bands, but others are entirely without them, while others, again, are altogether deep purple inside. It has a very close resemblance to *Melania (Goniobasis) caliginosa* (nobis), but that species is cancellate, the cancellation not amounting to granulations as in *Postellii*. It is also near to *catenaria*, Say, from South Carolina, but that shell is quite cancellate. I name this after James Postell, Esq., of St. Simon's Island, to whom I owe the acquisition of many fine *mollusca*, from Georgia. Fine specimens were subsequently sent to me by Dr. Wilson, of St. Simon's Island, procured in Lewis' Creek.—*Lea.*

This is a beautiful and rather common species—easily distinguished from all others belonging to this group.

20. G. arachnoidea, ANTHONY.

Melania arachnoidea, ANTHONY, Ann. Lyc. Nat. Hist. N. Y., vi, p. 95, t. 2, f. 14, March, 1854. BINNEY, Check List, No. 19. BROT, List, p. 34. REEVE, Monog. Melania, sp. 83.

Melania intertexta, ANTHONY, Proc. Acad. Nat. Sci., p. 62, February, 1860. BINNEY, Check List, No. 151. BROT, List, p. 34. REEVE, Monog. Melania, sp. 296.

Description.—Shell conic, rather thin, horn-colored; spire slender and much elevated; whorls twelve, very strongly striated and ribbed, particularly the upper ones; the ribs extend only to a prominent, acute carina on each whorl, situated below the middle, between which and the suture below, one or two coarse striæ alone are visible, sutures deeply impressed; aperture very small, ovate, purplish within; columella regularly curved, without indentation, and with but a small, very narrow sinus at base.

Fig. 295.

Diameter, ·28 inch (7 millim.); length, 1 inch (26 millim.). Length of aperture ·22 inch (2½ millim.); breadth of aperture, ·15 inch (4 millim.).

Habitat.—A small stream emptying into the Tennessee River, near London, Tennessee.

Observations.—This is one of the slenderest and most elevated of the genus; more than forty specimens are before me, and they are very constant in all their characters; it comes nearest to *M. striatula*, Lea, by its folds and striæ, but should not be confounded with it, being different in every other particular; the number of whorls is greater by one-half, the *striatula* having only eight; its proportions are altogether more slender, the *striatula* standing as 21 to 49, while this is 28 to 100. The present species is also much more folded and rough than the *striatula*, which is essentially a *striate* shell. Upon the older specimens the folds are nearly obsolete on the two lower whorls, being *there* coarsely striate only. About twelve striæ on the body-whorl and six on the penultimate; more elevated in the centre, which renders these whorls subangulated; lines of growth strong, by reason of which the last two whorls have quite a varicose appearance.—*Anthony*.

The following is the description of

Melania intertexta.—Shell conical, acute and highly elevated; whorls about ten, each strongly ribbed longitudinally and furnished also with revolving striæ which, becoming more elevated near the suture, arrest the ribs at that point; sutures decidedly impressed; aperture pyriform, not' large, whitish within; columella slightly rounded, not indented; sinus distinct but small.

Fig. 296.

Habitat.—Tennessee.

Observations.—A very abundant species. About two hundred specimens are now before me, and present characters remarkably uniform. May be compared with *M. bella*, Conrad, but differs by its more elongate, and sharply elevated form; its ribs are more decided, and it has not the bead-like prominences, so common in *M. bella*, and kindred species. From *M. arachnoidea* (nobis), it may be distinguished by its less elongate but more acute form, difference of aperture and less number of whorls; the striæ revolve around the whorls and over the folds without being arrested by them, giving the shell a woven appearance; hence its name.— *Anthony*.

I cannot distinguish the two species indicated by the synonymy at the commencement of this article; I therefore reprint the descriptions in full and figure the types. The examination of a great many specimens has convinced me that this shell varies much in its proportions, although very distinct from the other species of the genus.

21. G. Conradi, Brot.

Melania Conradi, BROT, List, p. 36.
Melania symmetrica, CONRAD, Proc. Acad. Nat. Sci., iv, p. 155, Feb., 1849. Jour. Acad. Nat. Sci., i, pt. 4, p. 278, t. 38, f. 5, Jan., 1850. BINNEY, Check List, No. 260.

Description.—Subulate, whorls nine, slightly convex, with longitudinal, slightly curved, narrow ribs, interrupted near the suture by a revolving granulated line; ribs on the body-whorl not extending as far as the middle; margin of labrum profoundly rounded; color ochraceous and black.

Habitat.—Savannah River.

Observations.—Near the apex, two or three volutions have a fine granulated or carinated line.— *Conrad.*

Fig. 297.

Dr. Brot proposes the name *Conradi* for this species as *symmetrica* is preoccupied by Prof. Haldeman. I doubt whether this species is distinct from *carinifera*, Lam.

22. G. carinifera, LAMARCK.

Melania carinifera, LAMARCK, Anim. sans Vert. DESHAYES, Anim. sans Vert., 2d edit., viii, p. 433. WHEATLEY, Cat. Shells U. S., p. 24. BINNEY, Check List, No. 48. CATLOW, Conch. Nomenc., p. 185. BROT, List, p. 36. REEVE, Monog. Melania, sp. 273.

Melania bella, CONRAD, New Fresh Water Shells, Appendix, p. 6, t. 9. f. 4, 1834. BINNEY, Check List, No. 29. BROT, List, p. 36. REEVE, Monog. Melania, sp. 269.

Elimia bella, Conrad, ADAMS, Genera, i, p. 300.

Melania perangulata, CONRAD, Proc. Acad. Nat. Sci., iv, p. 154, Feb., 1849. Jour. Acad. Nat. Sci., i, pt. 4, p. 278, t. 38, f. 6. BINNEY, Check List, No. 199. BROT, List, p. 36. REEVE, Monog. Melania, sp. 285.

Melania percarinata, CONRAD, Proc. Acad. Nat. Sci., iv, p. 155, Feb., 1849. Jour. Acad. Nat. Sci., 2d ser., i, pt. 4, p. 278, t. 38, f. 10. BINNEY, Check List, No. 200. BROT, List, p. 36.

Melania nebulosa, CONRAD, Proc. Acad. Nat. Sci., iv. p. 155, Feb., 1849. Jour. Acad. Nat. Sci., i, pt. 4, p. 278, t. 38, f. 9. BINNEY, Check List, No. 172. BROT, List, p. 36.

Melania bella-crenata, HALDEMAN, Monog. Limniades, No. 4, p. 3 of cover, Oct. 5, 1841. JAY, Cat., 4th ed., p. 273. BINNEY, Check List, No. 30. BROT, List, p. 36.

Melania monilifera, Anthony, JAY, Cat., 4th ed., p. 474.

Description.—Shell ovate-oblong, longitudinally subrugose, brownish-black; whorls carinated in the middle; spire more strongly carinate.

Fig. 298.　Fig. 299.

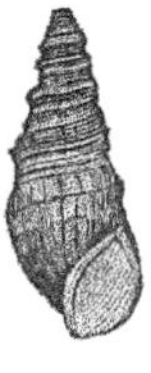

Habitat.—Cherokee County (Georgia).

Length, 7½ lignes.

Observations.—The spire is longer than the last whorl; its carinæ are very prominent and its sutures are plainly granulose.—*Lamarck.*

Fig. 300. Fig. 301.

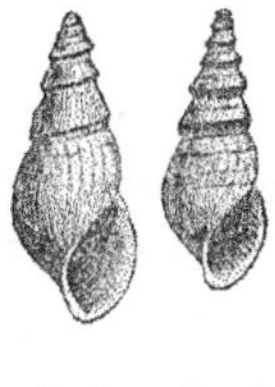

Melania bella.— Shell subulate, with carinated whorls, and a prominent crenulated line near the summit of each; aperture elliptical.

Habitat.—Streams in North Alabama.— *Conrad.*

Melania perangulata.— Subulate; volutions nine or ten, with an acutely carinated angle on all except the body-whorl, which is subcarinated; on each whorl of the spire is a revolving granulated line above the carina; color olive-brown.

Habitat.— Savannah River.—*Conrad.*

Melania percarinata.—Elongate-conoidal; volutions of the spire with a carinated line below the middle, and a revolving granulated line above; body-whorl with a granulated revolving line near the suture, and three carinated lines, the superior one largest, the lower one fine; color dark olive-brown.

Fig. 302.

Habitat.—Savannah River.—*Conrad.*

Melania nebulosa.—Elongate-conoidal; volutions six or seven with revolving raised lines; whorls of the spire carinated below the middle, above which they are longitudinally ribbed, and have two or three revolving granulated lines; granules compressed; aperture widely elliptical; color ochraceous, with brownish-black stains.

Fig. 303.

Habitat.—Savannah River.—*Conrad.*

The figure of *carinifera* is copied from Delessert and represents the original specimen of Lamarck's description.   That of *percarinata* is from Mr. Conrad's plate.   *G. bella* (fig. 301) is from the type specimen in possession of Prof. Haldeman.   Dr. Brot was the first author on Melanidæ to recognize the identity of all these species.   The following description also belongs to this species, which exhibits many varieties, but may be known through them all by its encircling row of beadlike elevations.

Fig. 304.

Melania bella-crenata. —Shell reddish, subulate, whorls eleven, marked with a strong carina and a crenulated line posterior to it.

Habitat.—Alabama.

Length ¾ of an inch.

Observations.—Differs from *M. bella,* Con., by having an oval aperture.—*Haldeman.*

Melania monilifera, Anthony, unpublished, but quoted in Jay's Catalogue, belongs here, as I have ascertained by a specimen so labelled by Mr. Anthony, in Coll. Gould.

I have seen specimens of *carinifera* from Yadkin River, S. C., and from North Alabama, but in Georgia it is exceedingly numerous in the Savannah and other rivers.

23. G. vittata, ANTHONY.

Melania vittata, ANTHONY, Ann. Lyc. Nat. Hist. N. Y., vi, p. 89, t. 2, f. 7, March 1854. BINNEY, Check List, No. 294. BROT, List, p. 37. REEVE, Monog. Melania, sp. 262.

Description.—Shell conic, nearly smooth; spire elevated; whorls about nine, flat, with two fine, distant, brown lines on each, the lower one revolving upon an angle near the suture; lines obsolete on the extreme upper whorls and increased to four or five on the body-whorl visible also within the aperture; sutures deeply impressed; aperture ovate, within whitish, but exhibiting also the brown lines of the epidermis; columella curved, sinus inconspicuous.

Fig. 305.

Habitat.—Alabama.

Diameter, ·32 of an inch (8 millim.); length, ·86 of an inch (22 millim.). Length of aperture, ·33 of an inch (8 millim.); breadth of aperture, ·16 inch (4 millim.).

Observations.—May be compared with *M. Taitiana*, Lea, but may be distinguished by its flat subangulated whorls. It also exhibits somewhat coarse striæ (amounting nearly, if not quite, to ribs in some specimens) upon all the whorls; even the body-whorl is no exception. The sutures also are deeply impressed, the contiguous whorls shelving towards each other to form quite a furrow there. Upper whorls carinate. It is a very beautiful species, the distinct reddish-brown, hair-like bands contrasting finely with the yellowish-brown color of the general shell.—*Anthony.*

24. G. abbreviata, ANTHONY.

Melania abbreviata, ANTHONY, Bost. Proc., iii, p. 360, Dec., 1850. BINNEY, Check List, No. 4. REEVE, Monog. Melania, sp. 424.
Melania elegantula, ANTHONY, Ann. Lyc. Nat. Hist. N. Y.. vi, p. 103, t. 3, f. 2. March, 1854. BINNEY, Check List, No. 96. BROT, List, p. 32. REEVE, Monog. Melania, sp. 346.
Melania coronilla, ANTHONY, Ann. Lyc. Nat. Hist. N. Y., vi, p. 126, t. 3, f. 27, March, 1854. BINNEY, Check List, No. 69. BROT, List, p. 32. REEVE, Monog. Melania, sp. 418.
Melania chalybœa, Anthony, BROT, List, p. 37.
Melania curvilabris, ANTHONY, Ann. N. Y. Lyc. Nat. Hist., vi, p. 102, t. 3, f. 1, Mar. 1854. BINNEY, Check List, No. 82. BROT, List, p. 31. REEVE, Monog. Melania, sp. 378.

Melania coronilla.—Shell ovate, moderately thick; of a dark, dull, horn-color, sometimes decorated with two or three linear revolving bands at, and below, the upper part of the aperture; spire short, with a rather convex outline to the truncated apex; whorls 5–6, convex, one of which seems to have been lost by truncation; obtusely shouldered and shelving, with about ten, short, thick, elevated, rather distant, longitudinal ribs on each which, on the body-whorl, are nearly obsolete,

Fig. 306.

rarely extending below the shoulder; sutures distinctly impressed, but rendered irregular by the interruptions of the longitudinal folds; aperture not large, ovate, reddish or banded within; columella much curved, with an indentation below the middle, and thickened by a calcareous deposit along its whole length, more prominent near the upper angle of the aperture.

Habitat.—Tennessee.

Diameter, ·22 of an inch (5½ millim.); length, ·50 of an inch (13 millim.). Length of aperture, ·24 of an inch (6 millim.); breadth of aperture, ·13 of an inch (3 millim.).

Observations.—I know no species with which the present one can easily be confounded; its short, rather broad outline, with its thick, prominent, longitudinal ribs on the short whorls of the spire, will readily distinguish it. Six specimens only are before me, three of which are banded, and three are plain; the specimens are otherwise very uniform in appearance.—*Anthony.*

The figure is from Mr. Anthony's original type. Other specimens exhibit slight folds on the body-whorl.

An examination of the types of *coronilla, elegantula* and *abbreviata,* together with other specimens, convinces me that they are all varieties of one species, which does not always develop the folds on the spire. It is a very remarkable species in the form of the shell, tubercles and aperture, and particularly in the broad band of a lighter color than the general hue of the shell.

The following is the description of

Melania elegantula.—Shell obtusely conical, smooth; whorls 5-6, irregularly shouldered and angulated; body-whorl dark olive-green color, with two or three darker bands, which are visible also within the aperture; upper whorls of a very light green color, with *one* light brown sub-central band, and *another* so near the upper part of the whorl as to be almost concealed by the suture; sutures rather obscure; aperture rather large, irregularly ovate; columella much indented near its base, outer lip sinuous.

Fig. 307.

Habitat.—Kentucky.

Diameter, ·25 of an inch (6 millim.); length, ·60 of an inch (15 millim.). Length of aperture, ·28 of an inch (7 millim.); breadth of aperture, ·16 of an inch (4 millim.).

Observations.—A singularly ornamented species, of which only two specimens are before me, and which cannot be compared with any described species. The apex is eroded in the specimens under observation, and only five whorls are visible, but it evidently has one more when perfect. The whorls form a shelving shoulder from the suture, and are then nearly flat, the body-whorl being, perhaps, slightly concave. Altogether it presents a remarkable and beautiful appearance, and no one need be at a loss to recognize it after once having seen a specimen. Three bands are visible in the interior.—*Anthony.*

Melania curvilabris.—Shell conical, smooth, rather thick, greenish horn-color; spire elevated; whorls 7-8, convex or subangulated; body-whorl angulated, with a depression broad, but not deep; sutures deeply and irregularly impressed; aperture very irregular, by the twisted columella and the sinuous curving of the outer lip, within whitish; outer lip deeply and singularly curved, so as to give this part of the shell almost a *pleurotomose* character; columella very much curved and indented, leaving a small, umbilical indentation, and having a distinct sinus at base.

Fig. 308.

Habitat.—Tennessee.

Diameter, ·30 of an inch (8 millim.); length, ·72 of an inch (19 millim.). Length of aperture, ·25 of an inch (6 millim.); breadth of aperture, ·15 of an inch (4 millim.).

Observations.—May be compared with *M. elegantula* in general form, but its peculiarly curved outer lip will at once distinguish it from all others.—*Anthony.*

Figured from Mr. Anthony's original type.

Melania abbreviata.—Shell small, ovately conical, turreted, somewhat solid, corneous, acuminate; whorls five, flattened, the last compressed; aperture rotundately-ovate, contorted, lip dilated in front, widely sinuated behind.

Fig. 309.

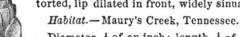

Habitat.—Maury's Creek, Tennessee.

Diameter, ¼ of an inch; length, ½ of an inch.

Observations.—A peculiar shell, though not easily characterized. Its abbreviated form, shouldered whorls and the compression of the last whorl, are among its peculiarities.—*Anthony.*

25. G. vesicula, Lea.

Melania vesicula, Lea, Proc. Acad. Nat. Sci., p. 118, 1861.
Goniobasis vesicula, Lea, Jour. Acad. Nat. Sci., v, pt. 3, p. 242, t. 35, f. 45, March,
1863. Obs. ix, p. 64.

Description.—Shell obscurely folded, elliptical, yellow, without spots,
rather thin; spire very short and obtuse; sutures rather impressed;
Fig. 310. whorls three, somewhat convex; aperture large, regularly

ovate, pale salmon within; outer lip sharp; columella thickened, incurved, rounded at the base.

Habitat.—Alabama; E. R. Showalter, M.D.

Diameter, ·18; length, ·37 inch.

Observations.— A single specimen of this very small species was
found among others of a different species from Dr. Showalter. It is a
small, regularly oval, inflated species. In this specimen there is a disposition on the upper part of the whorls to plication, and this produces obscure spots round this part of the whorls. Other specimens
may not have this character. The aperture is very large, being two-
thirds the length of the shell. It is nearly allied to *Melania* (*Goniobasis*) *auriculæformis* (nobis), but is not so large and has a wider
aperture, which is not so elongate. The color is nearly the same, but
the tint is rather brighter. It cannot be confounded with *Melania*
(*Goniobasis*) *corneola*, Anth., although of the same size and color, that
shell being fusiform, with a conical spire and an aperture only half
the length of the shell.—*Lea.*

C.   *Shell plicate.*

26. G. obesa, Anthony.

Melania obesa, Anthony, Reeve, Monog. Melania, sp. 469, May, 1861. Brot, List.
p. 33.

Description.— Shell globosely ovate, solid, fulvous,   Fig. 311.
obscurely banded with olive-green; spire short, rather
immersed; whorls five, slopingly rounded, longitudinally, obsoletely, rudely plicated, last whorl spirally
ridged and striated round the lower part; aperture
ovate, a little effused at the base.

Anthony, manuscript.

Habitat.—Alabama, United States.—*Reeve*

This species, which I have not seen, does not appear to be closely related to any other plicate species.

27. G. Leai, TRYON.

Melania blanda, LEA, Proc. Acad. Nat. Sci., p. 122, 1861.
Goniobasis blanda, LEA, Jour. Acad. Nat. Sci., v, pt. 3, p. 242, t. 35, f. 44, March, 1863. Obs., ix, p. 64, t. 35, f. 44.

Description.—Shell plicate, obtusely fusiform, obtusely conical above, rather thin, dark horn-color; spire very obtuse; sutures impressed; whorls five, flattened above, the last large and subangular; aperture rather large, elliptical, yellowish-white within; outer lip acute; columella thickened, inflected, subangular below.

Fig. 312.

Habitat.—Yellowleaf Creek, Alabama; Dr. E. R. Showalter.
Diameter, ·37; length, ·73 inch.

Observations.—A single specimen only was received from Dr. Showalter. I think it is not entirely mature. The folds are low, somewhat distant and vertical. The aperture is about half the length of the shell. In outline it is near to *Lithasia Duttoniana*, which I described as a *Melania*, but it has not the callus above and below on the columella, which constitute that genus, nor has it any tubercles, being covered above by folds.—*Lea.*

The name *blanda* is preoccupied by Mr. Lea himself in a species of Goniobasis published by him over twenty years ago.

The shell is a very variable one, being generally more dilated than the figure, with impressed, distinct striæ below the periphery, which is sometimes tuberculate. The young shell is very sharply angulate. Except in being plicate, this species is very nearly related to *G. straminea*, Lea.

28. G. æqualis, HALDEMAN.

Melania æqualis, HALDEMAN, Monog. Limniades, No. 4, p. 3 of cover, Oct. 5, 1841.
JAY, Cat. 4th ed., p. 272. BINNEY, Check List, No. 7.

Description.—Shell thick, short, conical; with five flat whorls ornamented with longitudinal ribs; texture thin, surface smooth, aperture narrow, elliptic, as long as the spire. Color brown.

Habitat.—Nolachucky River.

Length, ½ of an inch.

Observations.—Closely resembles the young of *Io spinosa*, and differs from the young of *Melania nupera* as figured by Say (Am. Conch., pl.

Fig. 313. Fig. 314.

3), by the want of the concentric elevated lines on the anterior slope. This figure, as I am informed by Mrs. Say, does not represent the young of the principal figures (*Lithasia nupera*), but another species which, if distinct, will retain the name of *M. nupera*, as it appears to be a true Melania.—*Haldeman.*

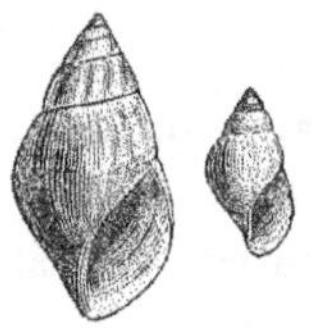

The two figures, representing a young and adult shell, are from Prof. Haldeman's types. The peculiar form of the aperture distinguishes all the specimens I have seen. Somewhat allied to *carinocostata*, Lea, but in that species the plicæ are terminated by a rib or angle on the body-whorl and the spire is angled or carinate. The largest specimen I have seen attains $\frac{4}{5}$ inch.

29. G. semigradata, Reeve.

Melania semigradata, Reeve, Monog. Melania, sp. 472, May, 1861. Brot, List. p. 33.

Description.—Shell pyramidally conical, fulvous-olive, encircled with a green band; whorls 5–6, flatly sloping, sharply   Fig. 315.
keeled around the lower part, first few whorls longitudinally plicated, last whorl double-keeled; aperture ovate, a little effused at the base.

Habitat.—Alabama, United States.

Observations.—A striking new species, in which the whorls are double-keeled at the periphery, the lower keel being hid in all but the last whorl by the overlapping of one whorl upon another.—*Reeve.*

Very closely related to *G. Gerhardtii.*

30. G. carinocostata, Lea.

Melania carinocostata, Lea, Philos. Proc. iv, p. 165, 1845. Philos. Trans., x, p. 62, t. 9, f. 40. Obs., iv, p. 62. Binney, Check List, No. 49. Brot, List, p. 35. Reeve, Monog. Melania, sp. 333.
Elimia carinocostata, Lea, Adams, Genera, i, p. 300.
Goniobasis strenua, Lea, Proc. Acad. Nat. Sci., p. 267, 1862. Jour. Acad. Nat. Sci., v, pt. 3, p. 316, t. 37, f. 161, March, 1833. Obs., ix, p. 138.
Goniobasis Leidyana, Lea, Proc. Acad. Nat. Sci., p. 268, 1862. Jour. Acad. Nat. Sci., v, pt. 3, p. 322, t. 38, f. 173, March. 1833. Obs., ix, p. 144.
Melania scabrella, Anthony, Reeve, Monog. Melania, sp. 388.
Melania scabriuscula, Brot, List, p. 36.

Description.— Shell plicate, carinate, conical, rather thin, yellow or chestnut-colored; spire somewhat elevated; sutures sulcate; whorls flattened; aperture small, elliptical; columella smooth.

Fig. 316.   Fig. 317.

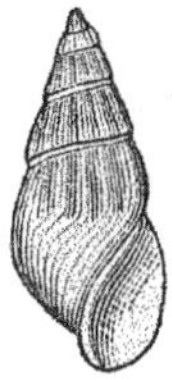

Habitat.—Alabama. Tennessee.

Diameter, ·36; length, ·98 of an inch.

Observations.— This is a species not easily confounded with any other known to me. The character of the ribs or folds is peculiar; they being arrested near the sutures by an abrupt carina, which has a smaller parallel one between it. The folds and the carinæ are conspicuous, being perfectly pronounced. Two of the six specimens before me are of a dark chestnut-brown, with the nacre of the interior quite rufous. One is more horn-colored, having four bands and the nacre whitish. The three others, all from Dr. Budd, are wax-yellow, the ribs less expressed, and the interior yellowish. The apex of each being broken, the number of whorls cannot be determined. I should think there were about eight. The inferior part of the whorl is smooth. The aperture is rather more than one-third the length of the shell.—*Lea.*

Fig. 316 is copied from Mr. Lea's figure. The following figure, from a shell in Mr. Anthony's collection, determined by Mr. Lea, locality Georgia (?), is much broader in outline and constitutes a well marked variety, if not distinct species.

The following are synonymes :—

Melania scabrella.—Shell somewhat fusiformly conoid, dull-chestnut, whorls 5-6, slopingly convex, concentrically, closely, plicately ridged, keeled above and below; sutures impressed; aperture oblong, ovate, canaliculately produced at the base.

Fig. 318.   Fig. 319.

Habitat. —Georgia, U. S.

Fig. 320.

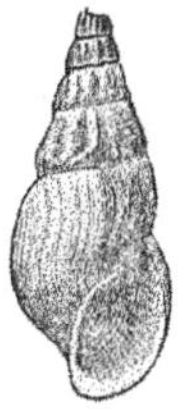

Observations. — Distinguished by a characteristic sculpture of arched, concentric ridges, interrupted by a keel, which gives a peculiarly impressed aspect to the sutures.— *Anthony.*

Goniobasis strenua.—Shell folded, subfusiform, brownish-olive, rather thin, without bands; spire somewhat raised; sutures very much impressed; whorls about seven, flattened; aper-

ture rather large, ovately rhomboidal, whitish within; outer lip subsinuous; columella bent in and twisted.—*Lea.*

Habitat.—Benton County, northeast Alabama; G. Hallenbeck.

Diameter, ·44 of an inch; length, 1·01 inches.

Observations.—Two specimens only were procured by Mr. Hallenbeck, and these are before me. The smaller one is rather lighter in
color and inclines to be more brown. It is allied to *Melania (Goniobasis) athleta,* Anth., but is a shorter shell, with two or three less
number of whorls. It also differs in being of a greenish color, and
in having fewer and more distant folds. It also differs in the base
of the columella being more direct. In our shell the folds are lost in
a carinate edge above the suture. In the body-whorl there are
minute venations. Immediately below the suture there is a line of
lighter color. The aperture is four-tenths the length of the shell.—
Lea.

Goniobasis Leidyana.—Shell folded, fusiform, rather thin, yellowish

Fig. 321. horn-color, without bands; spire obtusely conical; sutures

linear; whorls six, flattened; aperture very large, ovately
rhomboidal, whitish within; outer lip acute, thin; columella
bent in, twisted at the base.

Operculum ovate, thin, brown, with the polar point close
on the left margin, near to the base.

Habitat.—Benton County? northeast Alabama; G. Hallenbeck.

Diameter, ·39; length, ·80 of an inch.

Observations.—Two specimens were sent by Mr. Hallenbeck for
my examination. Both have imperfect plicæ on the spire which is
very obtuse, and both are evidently adults. The upper whorls are
carinate, but the inferior whorl closes on the angle so as to obliterate
the carination. On the body-whorl this angulation is nearly obsolete. It has nearly the outline of *Melania (Goniobasis) abrupta*
(nobis), but that species is not plicate and is a thicker shell. The
aperture is one-half the length of the shell. I dedicate this species
to my friend, Joseph Leidy, M.D., who has done so much for American zoology and comparative anatomy.—*Lea.*

31. G. perstriata, LEA.

Melania perstriata, LEA. Philos. Trans., x, p. 296, t. 30, f. 2. Obs., v, p. 52. BIN-
NEY, Check List, No. 203. BROT, List, p. 36.

Description.—Shell striate, acutely conical, rather thin, cinnamon-brown; spire elevated, somewhat attenuate, at the apex carinate and granulate; sutures impressed; whorls seven, convex; aperture small, elliptical, angular at the base, reddish within; columella smooth.

Fig. 322.

Habitat.—Coosa River, Alabama: Huntsville, Tennessee.

Diameter, ·28; length, ·83 of an inch.

Observations.—Among the numerous *Melaniæ* sent to me long since by my late friend, Prof. Troost, were several specimens of the young of this species. I could not satisfactorily place them in any known species, and I put them temporarily with *striatula* (nobis), which is strongly allied to the species which I have described above. Recently, I have received from Prof. Brumby and from Mr. J. Clark several adult specimens, which leave the younger in my possession no longer in doubt; they were recognized at once to belong to those more recently received. All the specimens before me, some dozen, are reddish; the *striatula* is horn-colored, with a white aperture. The latter is also flatter in the whorls, and not so carinate above, nor are the sutures so deeply impressed. Some of the specimens are quite smooth on the body-whorl. Aperture about one-third the length of the shell.—*Lea.*

32. G. Lecontiana, LEA.

Melania Lecontiana, LEA, Philos. Proc., ii, p. 13, Feb., 1841. Philos. Trans., viii, p. 177, t. 5, f. 29. DEKAY, Moll. N. York, p. 96. WHEATLEY, Cat. Shells, U. S., p. 26. BROT, List, p. 35. JAY, Cat., 4th edit., p. 274. BINNEY, Check List, No. 160. CATLOW, Conch. Nomenc., p. 187.
Melasma Lecontiana, Lea, CHENU, Man. Conchyl., i, f. 2002. ADAMS, Genera, i, p. 300.

Description.—Shell folded, conical, thick, horn-color; spire obtusely elevated; sutures small; whorls six, flattened; aperture large, elliptical, bluish.

Fig. 323.

Habitat.—Georgia; Major Le Conte.

Diameter, ·35; length, ·80 of an inch.

Observations.—The folds of this species extend over the whole shell, except the inferior half of the body-whorl. The aperture is large, and somewhat dilated, being nearly one-half the length of the shell. I owe the possession of several specimens to the kindness of Major Le Conte, to whom I dedicate it.—*Lea.*

Mr. Reeve's figure does not represent this species, it approaches nearer to *decorata*, Anthony. The outer lip in this species is not so expanded as in *carinocostata*, and the body-whorl is not angulate as in that species.

33. G. obtusa, LEA.

Melania obtusa, LEA, Philos. Proc., ii, p. 13, Feb., 1841. Philos. Trans., viii, p. 176, t 5, f. 28. Obs., iii, p. 14. DEKAY, Moll. New York, p. 96. BINNEY, Check List, No. 183. TROOST, Cat. Shells, Tennessee. WHEATLEY, Cat. Shells, U. S., p. 26. CATLOW, Conch. Nomenc., p. 188. BROT, List, p. 59.

Goniobasis cadus, LEA, Proc. Acad. Nat. Sci., p. 272, 1862. Jour. Acad. Nat. Sci., v, pt. 3, p. 345, t. 38, f. 211, March, 1863. Obs., ix, p. 167.

Melania substricta, HALDEMAN, Monog. Limniades, vii, p. 4 of cover, Jan., 1844. WHEATLEY, Cat. Shells, U. S., p. 27. BINNEY, Check List, No. 256. BROT, List, p. 36.

Description.— Shell folded, fusiform, rather thick, horn-color; spire obtuse; sutures impressed; whorls four, the last semi-plicate; aperture large, whitish.

Fig. 324.

Habitat.— Tennessee.

Diameter, ·27; length, ·55 of an inch.

Observations.— A fusiform species with costæ or folds half way down the last whorl.— *Lea.*

The following are believed to be synonymes :—

G. cadus.— Shell cancellate, subfusiform, somewhat thick, inflated, yellowish, without bands; spire very obtuse; sutures irregularly impressed; whorls five, slightly convex, cancellate above; aperture very large, ovately rhomboidal, white within; outer lip sharp, slightly sinuous; columella bent in, thickened and twisted.

Habitat.— Georgia; Major Le Conte.

My cabinet.

Diameter, ·33; length, ·63 of an inch.

Observations.— A single specimen has been in my possession for many years. The description was delayed in the hope of other specimens being found. It was a single one among many species, brought by our late lamented vice president from Georgia, which he placed in my hands. This species reminds one of *Melania* (*Goniobasis*) *Deshayesiana* (nobis), but it is entirely different in the outline and number of its whorls, being a very short shell with a very different size of aperture. The aperture is more than half the length of the shell.—*Lea.*

Fig. 325.

Melania substricta.— Brown, lengthened conical, upper whorls flattened, with numerous folds; body-whorl slightly convex, suture impressed; aperture pyriform, purple, obtusely rounded before, five-eighths of an inch.

Habitat.— Tennessee; Mr. Anthony.

Observations.— Bears some resemblance to *M. decora*, Lea. I formerly proposed the name *substricta* for *M. conica*, Say, supposing the name to have been previously applied to the *M. conica*, Gray. A subsequent examination of the dates has satisfied me that Say's name has priority, so that Mr. Gray's species now requires a new name, unless the citation of the author presents a sufficient distinction.—*Haldeman.*

34. G. amœna, LEA.

Goniobasis amœna, LEA, Proc. Acad. Nat. Sci., p. 268, 1862. Jour. Acad. Nat. Sci., v, pt. 3, p. 323, t. 38, f. 175, March, 1863. Obs., ix, p. 145, t. 38, f. 175.

Description.— Shell folded, subfusiform, thick, pale chestnut-color, without bands; spire obtusely conical; sutures irregularly impressed; whorls about six, somewhat convex; striate at the apex; aperture large, ovately rhomboidal, whitish within; outer lip acute, slightly sinuous; columella thickened, incurved and twisted.

Operculum ovate, thin, light brown, with the polar point on the left margin near the base.

Habitat.— North Alabama; Prof. Tuomey.

Diameter, ·29; length, ·70 of an inch.

Observations.— A number of these species were sent to me by the late Prof. Tuomey, but the older ones are very imperfect, being generally decollate. Most of them are young. The largest is nine-tenths of an inch long, but it is too imperfect to figure. The folds are close, regular and are oblique to the right. On the upper whorls there are one or two striæ which cut the folds as in *Melania (Goniobasis) Deshayesiana* (nobis). The aperture is nearly half the length of the shell.—*Lea.*

Fig. 326.

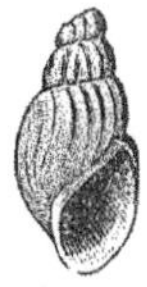

35. G. Tuomeyi, LEA.

Goniobasis Tuomeyi, LEA, Proc. Acad. Nat. Sci., p. 266, 1862. Jour. Acad. Nat. Sci., v, pt. 3, p. 311, t. 37, f. 153, March, 1863. Obs., ix, p. 133.

Description.—Shell smooth, fusiform, slightly thick, yellowish-olive, banded or without bands; spire obtusely conical, minutely plicate at the apex; sutures impressed; whorls about six, flattened above, the last one ventricose; aperture large, rhomboidal, whitish within; outer lip acute, somewhat sinuous; columella thickened, bent in and twisted.

Habitat.— North Alabama; Prof. M. Tuomey.

Diameter, ·35; length, ·70 of an inch.

Observations.— My friend, the late Prof. Tuomey, sent to me during his geological survey of the state of Alabama, many new *Mollusca*, most of which I described at the time. Some were laid over for more leisure and further examination. Among them were a number

Fig. 327.

of this species which I now dedicate to his memory with peculiar gratification. He was an ardent student of nature, and warm and generous in his friendships. This species varies very much. None of the specimens have perfect tips, but some are nearly so, and display on the apical whorls very minute and close plicæ. Some have minute venations on the body-whorl. They are generally without bands, yet some have two bands, but more frequently only one, which is about one-third of the whorl below the suture. It is rather broad and distinct inside and out. In outline and size it is closely allied to *Melania* (*Goniobasis*) *gracilis* (nobis), but it is not so high in the spire, nor is it so yellow. The aperture is about one-half the length of the shell.—*Lea.*

Differs from *G. strenua* in being more ventricose and in the aperture being narrower below. This species is allied in form to *G. Leidyana*, but in that species the body-whorl is plicate.

36. G. interveniens, LEA.

Goniobasis interveniens, LEA, Proc. Acad. Nat. Sci., p. 268, 1862. Jour. Acad. Nat. Sci., v, pt. 3, p. 320, t. 38, f. 169, March, 1863. Obs., ix, p. 142.

Description.— Shell folded, conical, rather thin, dark horn-color or brown, double-banded or without bands; spire obtusely conical; sutures irregularly and very much impressed; whorls about six, flattened, with slightly bent folds; aperture rather large, rhomboidal, white, brown or banded within; outer lip acute, sinuous; columella bent in and somewhat twisted.

Habitat.— North Alabama; Prof. Tuomey.

Diameter, ·32; length, ·74 of an inch.

Observations.— Some half dozen specimens were among the shells received from Prof. Tuomey obtained during his geological survey. This is rather a small species between *Melania* (*Goniobasis*) *costulata* (nobis), and *Melania* (*Goniobasis*) *Edgariana* (nobis). It has a less number of folds than the former, and about the same number as the latter, but these folds differ in not being so much raised and protruded above as in *Edgariana*, nor is the spire so high. The interior is usually white, sometimes double-banded, and one of the specimens is dark brown. The aperture is nearly half the length of the shell.—*Lea.*

Fig. 328.

Resembles *G. Curreyana*, Lea, but differs in being shorter and wider.

37. G. olivella, Lea.

Goniobasis olivella, Lea, Proceed. Acad. Nat. Sci., p. 269, 1862. Jour. Acad. Nat. Sci.,v, pt. 3, p. 327, t. 38, f. 182, March, 1863. Obs., ix, p. 149.

Description.— Shell folded, fusiform, rather thick, olivaceous, shining, without bands; spire obtusely conical; sutures irregularly and very much impressed; whorls about five; somewhat convex; aperture large, rhomboidal, whitish; outer lip acute, scarcely sinuous; columella bent in and twisted.

Fig. 329.

Habitat.—Tennessee; Prof. Troost.

My cabinet.

Diameter, ·31; length, ·60 of an inch.

Observations.—I have two specimens before me varying little but in size. It is a well characterized species, having folds, more or less distinct on all the whorls. These folds are rather close, and incline to the left. In one of the specimens there are two lines which cut the folds immediately under the suture. In outline it is near to *ornatella*, herein described, but it cannot be confounded with that species, which is of a different color and banded. The aperture is nearly the half of the length of the shell.—*Lea.*

38. G. interrupta, Haldeman.

Melania interrupta, Haldeman, Supplement to No. 1, Monog. Limniades, Oct., 1840. Wheatley, Cat. Shells, U. S., p. 25. Jay, Cat., 4th edit., p. 274. Brot, List, p. 34. Reeve, Monog. Melania, sp. 398.
Goniobasis Christyi, Lea, Proc. Acad. Nat. Sci., p. 269, 1862. Jour. Acad. Nat. Sci., v. pt. 3, p. 328, t. 38, f. 185, March, 1863. Obs., ix, p. 150.
Goniobasis instabilis, Lea, Proc. Acad. Nat. Sci., p. 269, 1862. Jour. Acad. Nat. Sci., v, pt. 3, p. 329, t. 38, f. 186, March, 1863. Obs., ix, p. 151.

Description.—Shell conical, with four flat whorls, which are crossed

Fig. 330.   Fig.331. Fig. 332. Fig. 333.

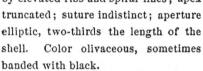

by elevated ribs and spiral lines; apex truncated; suture indistinct; aperture elliptic, two-thirds the length of the shell.   Color olivaceous, sometimes banded with black.

Length, ⅓ of an inch.

Habitat.—Tennessee.—*Haldeman.*

The following are synonymes.

Goniobasis Christyi.— Shell folded, striate or granulate, fusiform, rather thick, inflated, yellowish-olive, banded; spire obtusely conical; sutures impressed; whorls five, slightly convex; aperture very large, ovately rhomboidal, banded within; outer lip sharp, scarcely sinuous; columella thickened, slightly twisted.

Operculum ovate, thin, brown, with the polar point well removed from the left margin and the base.

Habitat.—Valley River, Cherokee City, N. C.; Prof. David Christy.

Diameter, ·37; length, ·C7 of an inch.

Observations.—I have about a dozen of this species from Mr. Clark, collected by Prof. Christy in North Carolina. All the specimens are nearly of the same size and outline, and have the same

Fig. 334.

bands, usually four, but they differ much in the exterior. Some have no striæ, but those which have cut the irregular folds and form granules. Usually, there are four bands indistinct on the outside, but well marked within, the two middle ones being approximate.   The upper band is the largest, and the callus above is often purple. Some specimens have five or six bands. It reminds one of *Melania* (*Goniobasis*) *basalis* (nobis), but that shell is not so much inflated, nor has it folds, striæ or granules like this. The aperture is more than half the length of the shell. I name this after Prof. David Christy, who collected it, with many interesting shells, while in the northwestern part of North Carolina.—*Lea.*

This and *instabilis* are adult forms.

Goniobasis instabilis.— Shell folded or smooth, fusiform, thick, somewhat inflated, banded or not banded, olivaceous; spire conical; sutures impressed; whorls about five, slightly convex; aperture large, ovately rhomboidal, banded within; outer lip acute, scarcely sinuous; columella thickened, somewhat bent in and twisted.

Operculum ovate, thin, light brown, with the polar point well removed from the left margin and the base.

Habitat.—Twenty-one miles north of Murphy, and other places in Cherokee County, N. C.; Prof. David Christy.

Diameter, ·32; length, ·64 of an inch.

Observations.—I have a number of these from several habitats in Cherokee County, North Carolina. From the different habitats there is a great variety of character, about half seem to be plicate, the others perfectly smooth; the folds not being on the upper whorls, but commencing on the body-whorls or the penultimate, and these folds are on the shoulder, and somewhat curved and close. Some are lighter green and white inside being without bands. The bands are usually four in number, with the two middle ones approximate. The smooth, green, elongate varieties look very much like *Melania (Goniobasis) Saffordii* (nobis), but it cannot be confounded with that species. The dark banded varieties might be mistaken for the *Melania (Goniobasis) subangulata,* Anth. The aperture is about half the length of the shell.—*Lea.*

Fig. 335.

39. G. crispa, Lea.

Goniobasis crispa, Lea, Proc. Acad. Nat. Sci., p. 269, 1862. Jour. Acad. Nat. Sci., v, pt. 3, p. 326, t. 38, f. 180, March, 1863. Obs., ix, p. 148.

Description.—Shell folded and transversely striate, fusiform, rather thick, yellowish, crispate, without bands; spire obtuse; sutures irregularly impressed; whorls about six; somewhat convex; aperture large, ovately rhomboidal; whitish within; outer lip acute, scarcely sinuous; columella slightly bent in and twisted.

Fig. 336.

Habitat.—Florence, Alabama; Rev. G. White.

Diameter, ·30; length, ·62 of an inch.

Observations.—A single specimen only was found among the numerous shells kindly sent to me some years since by Mr. White. The folds are rather close, well-defined, and incline to the left, reaching half way down the body-whorl, and are crossed by transverse striæ, which cover the whole surface, and cause the upper portion to be clathrate. The aperture is nearly half the length of the shell.—*Lea.*

More convex than *nassula*, Con., with more regular striæ, and is altogether a handsomer species.

40. G. formosa, Conrad.

Melania formosa, Conrad, New Fresh-Water Shells, Appendix, p. 5, t. 9, f. 3, 1834.
WHEATLEY, Cat. Shells, U. S., p. 23. BINNEY, Check List, No. 112.
Melania formosa, Anthony, REEVE, Monog. Melania, sp. 387. BROT, List, p. 35.
Goniobasis ornatella. LEA, Proceed. Acad. Nat. Sci., p. 269, 1862. Jour. Acad. Nat.
Sci., v, pt. 3, p. 326, t. 38, f. 181, March, 1863. Obs., ix, p. 148.

Description.—Shell with distant, robust, rounded ribs, and six convex whorls, with two approximate, prominent lines at the summit of each; base profoundly striated; color olivaceous, with distant, brown bands.

Fig. 337.

Habitat.—Inhabits streams in North Alabama.—*Conrad.*

The figure is from an authentic specimen in the collection of Mr. Anthony. Prof. Haldeman also possesses an author's type. It is a very beautiful species and apparently very constant in its characters. *G. nassula,* Conrad, is an allied species, but is striate and more rounded in the form of the aperture and in the whorls.

The following is a synonyme.

Goniobasis ornatella.— Shell folded, fusiform, rather thick, yellowish horn-color, banded; spire obtusely conical; sutures irregularly and very much impressed; whorls about six, convex; aperture large, ovately rhomboidal, whitish and obscurely banded; outer lip acute, scarcely sinuous; columella slightly bent in and.twisted.

Habitat.— Tennessee; Coleman Sellers.

Diameter, ·27; length, ·53 of an inch.

Observations.— A single specimen was among a number of *Melanidæ* kindly given to me by Mr. Sellers a long time since, one of which I then named after him. This pretty little species is ornamented with regular folds, which are slightly curved, and incline to the right. These folds cease at the middle of the body-whorl, being cut by an indented line below the suture, causing a granulation. In this specimen are five bands which are indistinct. It has nearly the same outline as *crispa,* herein described, but it is smaller, is not clathrate above, and the folds are not so strong. The aperture is about half the length of the shell.—*Lea.*

41. G. mediocris, LEA.

Goniobasis mediocris, LEA, Proceed. Acad. Nat. Sci., p. 269, 1862. Jour. Acad. Nat. Sci., v, pt. 3, p. 326, t. 38, f. 179, March, 1863. Obs., ix, p. 148.

Description.— Shell folded, subfusiform, rather thin, ash-color, shining, banded; spire conical; sutures irregularly impressed; whorls six, flattened; aperture somewhat large, rhomboidal, whitish and banded within; outer lip sinuous; columella bent in, thickened and twisted.

Fig. 338.

Habitat.— Tennessee; Dr. Edgar, and President Lindsley.

Diameter, ·23; length, ·57 of an inch.

Observations.— A single specimen was among a number of shells simply labelled, "Tennessee." This is a well characterized little species, which cannot be confounded with any I know. It has two obscure bands, one of which shows on the whorls of the spire, which is covered with rather distant folds, which curve to the right. The spire, embellished with folds and a colored band, reminds one of some of the small *Mitræ.* The aperture is nearly one-half the length of the shell.—*Lea.*

42. G. Duttonii, LEA.

Goniobasis Duttonii, LEA, Proceed. Acad. Nat. Sci., p. 266, 1862. Jour. Acad. Nat. Sci., v, pt. 3, p. 314, t. 37, f. 158, March, 1863. Obs., ix, p. 136.

Description.— Shell folded, conoidal, pale reddish-yellow, thick, double-banded; spire conoidal; sutures irregularly impressed; whorls about seven, somewhat convex; aperture ovately rhomboidal, white and double-banded within; outer lip acute, sinuous; columella bent in, thickened and very much twisted.

Fig. 339.

Habitat.— Maury County, Tennessee; T. R. Dutton: Grayson County, Kentucky; S. S. Lyon.

Diameter, ·38; length, ·80 of an inch.

Observations.— This is a well marked species, allied to *Pybasii,* herein described, and to *Melania (Goniobasis) laqueata,* Say. It is a stouter shell than either, and may at once be distinguished from them by its two well defined brown bands, the upper one of which is the larger. The folds are rather indistinct, close, not curved, and inclining to the right. The specimen from Maury County, Tennessee, is more robust, and has a shorter spire than that from Kentucky. The aperture is about three-eighths the length of the shell. I name this after Mr. T. R. Dutton, who sent it to me long since with other mollusca from Tennessee. This must not be confounded with the shell which I called *Melania Duttoniana*, Trans. Am. Phil. Soc., vol. 8, pl. 6, which is really a *Lithasia.— Lea.*

Differs from *G. Tuomeyi* in the form of the aperture. The specimens before me are not all double banded, some of them being without bands and of a light yellow-color. It is a remarkably fine species.

43. G. laqueata, SAY.

Melania laqueata, SAY, New Harmony Disseminator, p. 275, September, 1829. SAY's Reprint, p. 17. American Conchology, No. 5, t. 47, f. 1. BINNEY's edition, pp. 143 and 200. BINNEY, Check List, No. 158. DEKAY, Moll. New York, p. 97. WHEATLEY, Cat. Shells, U. S., p. 25. JAY, Cat, 4th ed., p. 274. REEVE, Monog. Melania, sp. 281, 288? BROT, List, p. 35. CATLOW, Conch. Nomenc., p. 187.
Melasma laqueata, Say, ADAMS, Genera, i, p, 300.
Melania monozonalis, LEA, Philos. Proc., ii, p. 13, February, 1841. Philos. Trans., viii, p. 178, t. 6, f. 31. Obs., iii, p. 16. DEKAY, Moll. New York, p. 96. BINNEY, Check List, No. 168. TROOST, Cat. Shells, Tennessee. WHEATLEY, Cat. Shells, U. S., p. 26. CATLOW, Conch. Nomenc., p. 187. BROT, List, p. 40.

Description.— Shell oblong, conic; spire longer than the aperture,

Fig. 340. Fig. 341. Fig. 342. elevated, acute at tip; volutions moderately

convex, with about seventeen, regularly elevated, equal, equidistant costæ on the superior half of each volution, extending from suture to suture and but little lower on the spire, and becoming obsolete on the body-whorl; suture moderately impressed; labrum and columella a little extended at base.

Observations.— This species was found by Dr. Troost in Cumberland River. The elevated costæ, without any revolving lines, distinguish this shell from the other species of our country.— *Say.*

Figure 340 is a copy of Mr. Say's, which is drawn from a poor specimen. Shells somewhat like it are before me. The species being very variable in outline and marking, two other figures are given.

Melania monozonalis.— Shell folded, fusiform, rather thick, banded, light colored; spire obtuse; sutures linear; whorls five, rather convex; aperture large, elliptical, angular at base, white.

Habitat.—Tennessee.

Diameter, ·21; length, ·42 of an inch.

Observations.—But a single specimen of this was sent to me by Dr. Troost. It is a very distinct species, and remarkable for Fig. 343. a single broad band on the upper part of the whorl. In other specimens this band may not always be found to present the same character; and the number of bands in others again may even be increased. The aperture is about one-half the length of the shell.—*Lea.*

G. monozonalis is an unusually wide juvenile *laqueata*, as I have ascertained from the inspection of numerous specimens.

44. G. Pybasii, Lea.

Goniobasis Pybasii, Lea, Proc. Acad. Nat. Sci., p. 266, 1862. Jour. Acad. Nat. Sci., v, pt. 3, p. 313, t. 37, f. 157, March, 1863. Obs., ix, p. 135, t. 37, f. 157.

Description.—Shell folded, very much drawn out, yellowish, thin, banded; spire attenuate, sharp-pointed; sutures impressed; whorls seven, flattened; aperture ovately rhomboidal, whitish and Fig. 344. banded within; outer lip acute, sinuous; columella slightly bent in, somewhat thickened and twisted.

Habitat.—Tuscumbia, Alabama; B. Pybas.

Diameter, ·31; length, ·82 of an inch.

Observations.—I found four specimens among numerous *Melanidœ* sent to me by Mr. Pybas. It is allied to *Melania (Goniobasis) Deshayesiana* (nobis), but it is more slender, has bands, and has not the granulations of that shell on the upper part of the whorls. It differs from *Lyonii* herein described, in having a longer aperture, being thicker, not being striate, and in having bands. It is evident that this species usually has four well marked revolving bands, the two middle ones being approximate. The broadest is at the bottom. In this character it is very like to *Melania (Goniobasis) grata*, Anth., and it reminds one of *Melania Goniobasis laqueata*, Say. In one of the specimens an indistinct fifth band is observable. The folds are not very strongly marked and do not extend to the body-whorl. They are not very close, are slightly curved and incline to the left. The aperture is more than one-third the length of the shell.

I dedicate this species with great pleasure to Mr. B. Pybas, of Tuscumbia, who has sent me many new mollusca from his vicinity.—*Lea.*

45. G. versipellis, ANTHONY.

Melania versipellis, ANTHONY, Proc. Acad. Nat. Sci., p. 60, February, 1860. BINNEY, Check List, No. 286. BROT, List, p. 59. REEVE, Monog. Melania, sp. 436.

Description.—Shell small, ovate, folded, rather thin; spire not elevated, but acute composed of about seven flat whorls; whorls of the

Fig. 345.

spire all more or less folded, penult and body-whorl smooth; body-whorl bulbous, subangulated, concentrically striate; color olivaceous, ornamented with dark brown bands, of which four are on the body-whorl, and one only on the spiral ones, located upon or near the shoulder of each volution; aperture elliptical, about half the length of the shell, banded within.

Habitat.—Tennessee.

Observations.—A small and somewhat variable species as to coloration, though very constant in other characters; it is sometimes very dark both as to bands and general color, and often very light, with bands scarcely distinguishable, and many varieties between. It seems not to be a very common species.—*Anthony.*

Fig. 345 is from Mr. Anthony's type specimen. This shell is more frequently *not* striate. It resembles in form a young, bulbous *G. laqueata*, but is a rather heavy shell, although small.

46. G. gracilis, LEA.

Melania gracilis, LEA, Philos. Proc., ii, p. 12, Feb., 1841. Philos. Trans., viii, p. 168, t. 5, f. 11. Obs., iii, p. 6. DEKAY, Moll. N. York. p. 94. TROOST, Cat. Shells, Tenn. WHEATLEY, Cat. Shells, U. S., p. 25. BINNEY, Check List, No. 128. CATLOW, Conch. Nomenc., p. 187. BROT, List, p. 38.
Potadoma gracilis, Lea, CHENU, Manuel de Conchyl., i, f. 1968. H. and A. ADAMS, Genera, i, p. 299.

Fig. 346.

Description.—Shell smooth, club-shaped, rather thin, horncolored; spire acute; sutures impressed; whorls eight, convex; aperture small, ovate, whitish.

Habitat.—Tennessee; Dr. Troost.

Diameter, ·32; length, ·75 of an inch.

Observations.—This resembles the *clavata* in form, but is rather more robust. It differs also in color. The aperture is rather more than one-third the length of the shell.—*Lea*

The figure, which is a copy of Mr. Lea's, does not represent the plicate upper whorls of the spire; and Mr. Lea, it will be perceived, supposed it to be a smooth species and described it as such. In a number of specimens before me the upper whorls are slightly ribbed.

47. G. paucicosta, ANTHONY.

Melania paucicosta, ANTHONY, Proc. Acad. Nat. Sci., p. 57, February, 1860. BINNEY, Check List, No. 198. BROT, List, p. 36. REEVE, Moncg. Melania, sp. 255.

Description.— Shell conical, nearly smooth, of a dark greenish horn-color; spire obtusely elevated; whorls nearly flat, with a few distinct, longitudinal ribs on the upper ones; body-whorl entirely smooth; sutures well marked; aperture ovate, within livid or purple; columella rounded; sinus small.

Fig. 347.

Habitat.— Tennessee.

Observations.— Belongs to a group of which *nitens* may be considered the type. From that species it differs, however, by its more robust form and stronger ribs. There is also a marked peculiarity in this species not often observed in the genus; the spire being acute at the apex, increases regularly for the first four or five turns, and then suddenly expanding, becomes as it were distorted in appearance. The ribs are distant from each other and very strongly expressed, differing in this respect from *M. athleta*, which it otherwise resembles. It is a beautiful, and appears to be an abundant, species.— *Anthony.*

48. G. tenebrosa, LEA.

Melania tenebrosa, LEA, Philos. Proc., ii, p. 13, February, 1841. Philos. Trans., viii. p. 176, t. 5, f. 26. Obs., iii, p. 14. DEKAY, Moll. N. Y., p. 95. TROOST, Cat. Shells, Tenn. WHEATLEY, Cat. Shells. U. S., p. 27. BINNEY, Check List, No. 267. CATLOW, Conch. Nomenc., p. 189. REEVE, Monog. Melania, sp. 443. BROT, List. p. 39.

Fig. 348. Fig. 349.

Description.— Shell smooth, conical, rather thick, nearly ovate; spire rather elevated; sutures impressed; whorls flattened; aperture rather large, elliptical, at the base angular, within bluish.

Habitat.— Tennessee.

Diameter, ·30; length, ·72 of an inch.

Observations.— Two specimens of this species were sent to me by Dr. Troost, both of which are decollated. On one there is a slight

disposition to striæ on the upper remaining whorl. In general outline it resembles a small *Virginica*, Say.— *Lea.*

The first specimens received by Mr. Lea being decollate,
he was not aware that it is a plicate species. I have copied
Mr. Lea's figure, but give also a figure of a more perfect
specimen.

49. G. coracina, ANTHONY.

Melania coracina, ANTHONY, Bost. Proc., iii, p. 361, Dec., 1850. BINNEY, Check
    List, No. 67. BROT, List, p. 58.
Melania Sellersiana, LEA, Philos. Trans., x, p. 299, t. 30, f. 8. Obs., v, p. 55. BIN-
    NEY, Check List, No. 239.

Description.— Shell small, thin, conically turreted, piceous, shining,
whorls 6-7, flattened above, generally, plicately ribbed, the

Fig. 350.

last ventricose and subangulate; aperture rotundately-ovate,
rounded in front, columella narrow, blackish.

    Observations.— The peculiar, dark, purplish-black color of
this prettily sculptured species is a very decisive character.
It is allied to *M. decora* and *M. costulata.*— *Anthony.*

The figure is from the original type. Mr. Anthony writes
to me that the shells described by Mr. Lea as *Sellersiana* had
first been submitted to himself, when he selected specimens
and described them as *M. coracina.* An inspection of the
copy of Mr. Lea's figure, which is here given, will show the
identity of the two species. Mr. Anthony has considerable
priority in the publication.

The following is the description of

Melania Sellersiana.— Shell folded, small, conical, rather thick,
very dark brown; spire rather short; sutures linear; whorls slightly
convex; aperture large, elliptical, rounded at the base, within purple;
columella very much incurved.

Habitat.— Caney Fork, Tennessee.

Diameter, ·16; length, ·38 of an inch.

Observations.— This is an interesting little species, somewhat like
M. Nickliniana (nobis), in its general appearance and size,
but is less inflated, and of a darker color. It might be supposed that its being a plicate shell would at once distinguish
it; but the Sellersiana seems to be very variable in the character of its folds, some of the specimens really having none
remaining. These may have had folds near the apex, which is now

Fig. 351.

eroded. Some of those before me are beautifully folded down to the last half of the body-whorl, the folds being rather large and straight. The surface varies very much; some of the specimens being beautifully malleate, while on others no such marks can be observed. The outer lip is broken. The apex being eroded in all the specimens, I am not sure of the number of the whorls; there may be about six. The aperture is about one-half the length of the shell. I dedicate this species to Mr. Coleman Sellers of Cincinnati.—*Lea.*

50. G. intersita, HALDEMAN.

Melania intersita, HALDEMAN, Monog. Limniades, No. 4, p. 4 of cover, Dec., 28, 1841. BINNEY, Check List, No. 150. BROT, List, p. 35. REEVE, Monog. Melania, sp. 376.

Description.—Shell conic, plicated, with four convex whorls; aperture elliptical; color olivaceous.

Habitat.— Swan Creek, Indiana; Mrs. Say.

Length, ½ of an inch.

Observations.— Allied to *M. comma.*— *Haldeman.*

Mr. Reeve's figure does not well represent this species and his description does not accord with that given by Haldeman. He seems to have obtained a poor specimen, which does not exhibit the longitudinal folds. The above figure illustrates Prof. Haldeman's type. The species is interesting as being one of the few species of the present group inhabiting north of the Ohio River.

Fig. 352.

51. G. columella, LEA.

Melania columella, LEA, Philos. Proc., ii, p. 13, Feb., 1841. Philos. Trans., viii, p. 179, t. 6, f. 33. Obs., iii, p. 17. DEKAY, Moll. N. Y., p. 96. BINNEY, Check List, No. 60. TROOST, Cat. Shells, Tenn. WHEATLEY, Cat. Shells, U. S., p. 24. CATLOW, Conch. Nomenc., p. 186. BROT, List, p. 35. REEVE, Monog. Melania, sp. 441.

Description.— Shell obscurely plicate, conical, rather thin, horn-color; spire rather elevated, striate towards the apex; sutures impressed; whorls six, somewhat convex; aperture small, elliptical, angular at base, whitish.

Fig. 353.

Habitat.—Tennessee.

Diameter, ·26; length, ·63 of an inch.

Observations.—This species is remarkable for the impressed curve on the columella. In its general character it resembles the

M. blanda herein described.   The aperture is about one-third the length of the shell.—*Lea.*

52. G. blanda, LEA.

Melania blanda, LEA, Philos. Proc., ii, p. 13, Feb., 1841. Philos. Trans., viii, p. 79, t. 6, f. 34. Obs., iii, p. 17. DEKAY, Moll. N. Y., p. 97. BINNEY, Check List, No. 36. TROOST, Cat. Shells, Tennessee. WHEATLEY, Cat. Shells, U. S., p. 24. CATLOW, Conch. Nomenc., p. 185. BROT, List, p. 35.
Melasma blanda, Lea, ADAMS, Genera, i, p. 300.

Description.— Shell folded, conical, rather thin, shining, horn-color;

Fig. 354.

spire rather elevated; towards the apex, striate; sutures impressed; whorls seven, rather flattened; aperture small, elliptical, angulated at the base, whitish.

Habitat.— Tennessee.

Diameter, ·26; length, ·69 of an inch.

Observations.— A single specimen of this species was received from Dr. Troost.   The folds are obscure and the striæ small. The aperture is not quite one-third the length of the shell.—*Lea.*

53. G. nitens, LEA.

Melania nitida, LEA, Philos. Proc., ii, p. 14, February, 1841.
Melania nitens, LEA, Philos. Trans., viii, p. 182, t. 6, f. 40. Obs., iii, p. 20. DEKAY, Moll. N. Y., p. 98. BINNEY, Check List, No. 178. TROOST, Cat. Shells, Tennessee. WHEATLEY, Cat. Shells, U. S., p. 26. CATLOW, Conch. Nomenc. p. 187, BROT, List, p. 36.

Description.— Shell folded, somewhat thick, dark brown; Fig. 355.
spire obtuse; sutures impressed; whorls seven, somewhat convex; aperture small, elliptical, angular at the base, reddish within.

Habitat.— Tennessee.

Diameter, ·30; length, ·76 of an inch.

Observations.— This is a shining, dark brown species, with rather regular ribs on the superior whorls.   The aperture is about one-third the length of the shell.   A single specimen only was received.—*Lea.*

This species very much resembles the last.   Closely allied to *Deshayesiana,* but without the subsutural striæ which characterize that species.

54. G. mutata, Brot.

Melania Deshayesiana, Reeve, Monog. Melania, sp. 278, September, 1860.
Melania mutata, Brot, List, p. 37.

Description.— Shell acuminately ovate, raised at the apex, dull olive; whorls slopingly tumid, the first few longitudinally plicated plaits soon disappearing, transversely ridged; ridges obso-

Fig. 356.

lete towards the aperture; aperture ovate, rather contracted, at the upper part; columella thinly effused at the base.

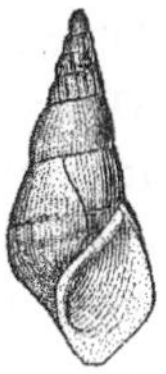

Habitat.—Tennessee, United States.

Observations.— The whorls of this species are swollen in a sloping manner towards the upper part, and the spire is acuminately raised at the apex. The first few whorls are decussately sculptured, but the sculpture soon becomes obsolete.—*Reeve.*

Changed by Dr. Brot to *mutata* because *Deshayesiana* is preoccupied by Mr. Lea. This species is closely allied to *difficilis*, Lea.

55. G. suturalis, Haldeman.

Melania suturalis, Haldeman, Supplement to Monog. Limniades, No. 1, Oct., 1840.
WHEATLEY, Cat. Shells, U. S., p. 27. Jay, Cat., 4th ed., p. 275.
Goniobasis mutabilis, Lea, Proc. Acad. Nat. Sci., p. 270, 1862. Jour. Acad. Nat.
Sci., v, pt. 3, p. 331, t. 38, f. 189, March, 1863. Obs., ix, p. 153.

Description.— Shell lengthened, conical, composed of six quite flat

Fig. 357.

whorls, which are separated by a well marked angular suture, bordered on each edge by an elevated, revolving line, which is double upon the body-whorl; aperture narrow, elliptic, one-half the entire length, bluish-white and banded; color dark olivaceous or black.

Habitat.— Ohio.

Length, ¾ of an inch.—*Haldeman.*

An examination of the original and only specimen of *suturalis* convinces me that it is the same as *G. mutabilis;* and that it is not found in Ohio will, I think, be admitted. Prof. Haldeman has probably mistaken its habitat.

The following is the description and figure of

Goniobasis mutabilis.— Shell carinate, plicate or striate, subfusi-

form, somewhat thick, yellowish-green, four-banded, or without
bands; spire obtusely conical; whorls six, slightly flattened; aperture rather large, rhomboidal, whitish within; outer lip acute,

Fig. 358.

scarcely sinuous; columella bent in, thickened, somewhat
twisted.

Operculum ovate, thin, dark brown, with the polar point
well removed from the left margin.

Habitat.—Butts County, Georgia; Rev. G. White.

Diameter, ·31; length, ·65 of an inch.

Observations.—This is a most variable species, most are carinate,
but many are striate, and some are plicate, and on a few neither of
these characters can be observed, the surface being entirely smooth.
All are disposed to carination on the apical whorls. Many are without bands, but most are four-banded, having the two medial bands
approximate. All were more or less covered with the black oxide of
iron. In outline it is nearly allied to *Melania* (*Goniobasis*) *Lecontiana*
(nobis), but it is not so fusiform, nor so large, nor is it always plicate, as that species is. Some of the specimens are entirely white
inside, and thickened, but usually they are four-banded. In several
instances there is an indistinct fifth band. The aperture is more
than one-third the length of the shell.—*Lea.*

56. G. Viennaënsis, Lea.

Goniobasis Viennaënsis, Lea, Proc. Acad. Nat. Sci., p. 267, 1862. Jour. Acad. Nat.
  Sci. v, pt. 3, p. 315, t. 37, f. 160, March, 1863. Obs., ix, p. 137.

Description.—Shell folded, subfusiform, olivaceous, rather thin,
without bands; spire regularly conical; sutures irregularly impressed; whorls seven, flattened; aperture rather large, rhomboidal,
bluish·white within; outer lip acute, sinuous; columella
bent in, thickened and somewhat twisted below.

Fig. 359.

Habitat.—Near Vienna, Dooly County, Georgia, in a small
stream, tributary to Flint River; Rev. G. White.

Diameter, ·36; length, ·90 of an inch.

Observations.—A number of this species came with *Doolyensis*, herein described, but it is quite a different species. It
is regularly conical, while the other is subcylindrical, and the
ribs are more numerous and closer, and are not quite so much curved.
The aperture is also larger. It is allied to *Melania* (*Goniobasis*)
Deshayesiana (nobis), but while it is nearly of the same outline it

differs in being wider, also in color, and it has no decussating revolving striæ. The aperture is more than one-third the length of the shell.—*Lea.*

57. G. Curreyana, LEA.

Goniobasis Curreyana, LEA, Philos. Proc., ii, p. 13, Feb,. 1841. Philos. Trans., viii, p. 180, t. 6, f. 36. Obs., iii, p. 18. WHEATLEY, Cat. Shells, U. S., p. 25. BINNEY, Check List, No. 79. DEKAY, Moll. N. Y., p. 97. REEVE, Monog. Melania, sp. 286. TROOST, Cat. Shells, Tennessee. CATLOW, Conch. Nomenc., p. 186. BROT, List, p. 35.
Melasma Curreyana, Lea, CHENU, Man. de Conchyl., i, f. 2003. ADAMS, Genera i, p. 300.

Description.— Shell folded, conical, rather thick, horn-color; spire somewhat elevated; sutures irregularly impressed; whorls seven, rather convex; aperture small, angular below, purplish within.

Habitat.— Barren River, Kentucky.

Diameter, ·27; length, ·73 of an inch.

Fig. 360.

Observations. — Two specimens of this species are before me, which I owe to the kindness of Dr. Currey of Nashville, after whom I name it. It is remarkable for its large and strong folds. It is without striæ, and the body-whorl is smooth, except near to the suture. The aperture is about one-third the length of the shell. One of the specimens has quite a dark purple aperture, and the lip is thickened and reflexed. In these two specimens the ribs seem disposed to alternate in size.— *Lea.*

58. G. costifera, HALDEMAN.

Melania costifera, HALDEMAN, Monog. Melania, No. 2, p. 3 of cover, Jan., 1841. BINNEY, Check List, No. 72. BROT, List, p. 34. REEVE, Monog. Melania, sp. 440.

Description.— Shell lengthened, composed of eight, slightly convex turns, having numerous, spiral, elevated lines, crossing a series of curved ribs, on all the whorls; spire twice the length of the aperture; suture well marked; aperture ovate.

Fig. 361.

Habitat.— Hennepin, Illinois.

Length, 1 inch.

Observations.— The aperture is wider in the allied species, and the costæ are better developed.—*Haldeman.*

The plicæ are more numerous (though not so prominent in this species) than in *Curreyana*, the aperture more rounded below and the spire more acuminate.

59. G. Deshayesiana, Lea.

Melania plicatula, LEA, Proc. Philos. Soc., ii, p. 14, Feb., 1841. Philos. Trans., viii, p. 182, t. 6, f. 41. Obs., iii, p. 20. TROOST, Cat. Shells, Tenn. JAY, Cat., 4th Edit., p. 274. CATLOW, Conch. Nomenc., p. 188. BROT, List, p. 34.
Melasma plicatula, Lea, CHENU, Man. de Conchyl., i, f. 1998. ADAMS, Genera, i, p. 300.
Melania Deshayesiana, LEA, Philos. Proc., ii, p. 242, Dec. 1842. Philos. Trans., ix, p. 24. DEKAY, Moll. N. Y., p. 98. WHEATLEY, Cat. Shells, U. S., p. 25, TROOST, Cat. Shells, Tennessee. JAY, Cat. Shells, 4th Edit., p. 273. BINNEY, Check List, No. 88. BROT, List, p. 34.
Melania Deshayesii, Lea, REEVE, Monog. Melania, sp. 330.
Melasma Deshayesiana, Lea, ADAMS, Genera, i, p. 300.

Description.— Shell folded, conical, thin, dark horn-color; spire rather elevated; sutures impressed; whorls

Fig. 362. Fig. 363. Fig. 364.

eight, rather convex, striate above; aperture rather small, elliptical, at the base somewhat angular, within whitish.

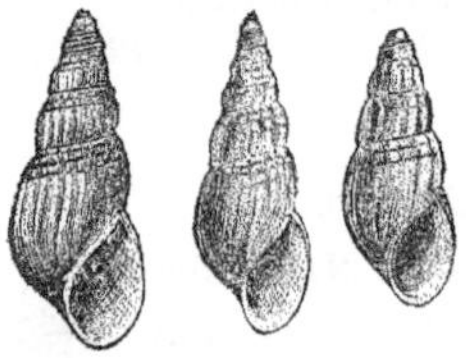

Habitat.— Tennessee.

Diameter, ·35; length, ·85 of an inch.

Observations.— Dr. Troost and Mr. Edgar both procured this species from Tennessee, but their labels do not state the district. The ribs are numerous and close, and most individuals have two striæ above, which, crossing the ribs, produce a granulation. The mouth is about one-third the length of the shell.— *Lea.*

This species was described as *plicatula*, but that name having been preoccupied by Deshayes, Mr. Lea changed it to *Deshayesiana*. It is very closely allied to *crebricostata* and *tenebrosa*.

60. G. Abbevillensis, Lea.

Goniobasis Abbevillensis, LEA, Proc. Acad. Nat. Sci., p. 208, 1862. Journ. Acad. Nat. Sci., v, pt. 3, p. 323, t. 38, f. 174, Mar., 1863. Obs., ix, p. 145.

Description.— Shell folded, conical, rather thick, chestnut-color, shining, without bands; spire conical, sutures linear; whorls seven, somewhat convex, nearly flat, carinate and striate at the apex; aperture slightly large, ovately rhomboidal somewhat ochraceous within; outer lip acute, scarcely sinuous; columella thickened and twisted.

Fig. 365.

Habitat.— Abbeville District, South Carolina; J. P. Barratt, M.D.

Diameter, ·30; length, ·63 of an inch.

Observations.— This is a pretty species with very regular spire and folds. It is allied to *Melania (Goniobasis) Deshayesiana* (nobis), but is a smaller species. Its chestnut-brown color reminds one of *Melania (Goniobasis) castanea* (nobis), but it is not so elongate and is thicker. The aperture is more than one-third the length of the shell.—*Lea.*

61. G. Doolyensis, LEA.

Goniobasis Doolyensis, LEA, Proc. Acad. Nat. Sci., p. 266, 1862. Jour. Acad. Nat. Sci., v, pt. 3, p. 315, t. 37, f. 159, Mar., 1863. Obs., ix, p. 137.
Goniobasis induta, LEA, Proc. Acad. Nat. Sci., p. 267, 1862. Jour. Acad. Nat. Sci., v, pt. 3, p. 319, t. 37, f. 166, March, 1863. Obs., ix, p. 141.

Description.— Shell folded, subcylindrical, dark horn-color or somewhat ash-gray, thin, without bands; spire drawn out; sutures irregularly impressed; whorls about nine, slightly convex; aperture small, ovately rhomboidal, whitish within; outer lip acute, sinuous; columella very much bent in, impressed in the middle and very much twisted.

Fig. 366.

Habitat.— Tennessee, Prof. Troost; near Vienna, Dooly County, Georgia, in a small stream tributary to Flint River; Rev. George White.

Diameter, ·32; length, ·91 of an inch.

Observations.— I have a number of specimens from Mr. White, and one a long time since from Prof. Troost. It belongs to the group of which *Melania (Goniobasis) costulata* (nobis) may be considered the type, but it is more cylindrical and has more distant folds. It is also allied to *Melania (Goniobasis) decora* (nobis), but is more cylindrical, has more distant folds and has no cancellate striæ. The folds are curved and incline slightly to the left. The aperture is not quite one-third the length of the shell. Some specimens are disposed to be slightly brownish inside.—*Lea.*

Goniobasis induta.—Shell very much folded, conical, rather thin, polished, dark, four-banded; spire conoidal, sharp-pointed; sutures very much impressed; whorls eight, flattened, clothed with erect folds; aperture small, rhomboidal, whitish and four-banded within; outer lip acute, sinuous; columella bent in and twisted.

Fig. 367.

Operculum ovate, thin, light brown, with the polar point well inside of the margin.

Habitat.—Near Vienna, Dooly County, Georgia; Rev. G. White.

Diameter, ·31; length, ·76 of an inch.

Observations.—This is a very ornate little species, being covered with close, perpendicular ribs and four, dark brown, revolving bands, which give the shell a dark appearance, although the ground is yellow. The two middle bands are approximate, and the lowest band is the strongest. Immediately below the suture there is usually a light line. It belongs to the group of which *Melania* (*Goniobasis*) *Deshayesiana* (nobis) may be considered the type, but is nearest allied to *inclinans*, herein described. It is nearly of the same size and outline, but the regular perpendicular folds and the distinct bands distinguish it at once. The apical whorls are disposed to be carinate. The aperture is one-third the length of the shell. The specimens were all incrusted with black oxide of iron, which, being removed, the epidermis was found to be smooth and polished. One or two revolving striæ immediately under the suture decussate the folds.—*Lea.*

62. G. inconstans, LEA.

Goniobasis *inconstans*, LEA. Proc. Acad. Nat. Sci., p. 269, 1862. Jour. Acad. Nat.
    Sci., v, pt. 3, p. 325, t. 38, f. 178, Mar., 1863. Obs., ix. p. 147.

Description.— Shell folded, subfusiform, rather thin, horn-color, olivaceous or dark brown, banded or without bands; spire obtusely conical; sutures impressed; whorls six, somewhat convex, folded above; aperture somewhat large, subrhomboidal, whitish within, pale purple or banded; outer lip acute, slightly sinuous; columella bent in and twisted.

Fig. 368.

Habitat.— Etowah River; J. Postell.

Diameter, ·26; length, ·60 of an inch.

Observations.—This is a small and very variable species, varying from light horn-color to dark brown, a few having two broad bands. The folds rarely reach to the body-whorl, but they cover the upper whorls, and are somewhat distant and nearly straight. Some of the specimens closely resemble *proletaria*, herein described, in form, but this has a more pointed apex, and is more fusiform. The aperture is not quite one-half the length of the shell.—*Lea.*

63. G. continens, LEA.

Goniobasis *continens*, LEA, Proc. Acad. Nat. Sci., p. 268, 1862. Jour. Acad. Nat.
    Sci., pt. 3, p. 324, t. 38, f. 176, March, 1863. Obs., ix, p. 146.
Goniobasis *proletaria*, LEA. Proc. Acad. Nat. Sci , p. 268, 1862. Jour. Acad. Sci.,
    v, pt. 3, p. 325, t. 38, f. 177, March, 1863. Obs., ix, p. 147.

Description.—Shell folded, conical, rather thin, yellowish horn-color, without bands; spire irregularly conical; sutures impressed; whorls about seven, somewhat convex, with folds slightly bent; aperture rather small, ovately rhomboidal, bluish-white within; outer lip acute, scarcely sinuous; columella somewhat bent in and twisted.

Operculum ovate, thin, light brown, with the polar point well removed from the margin and towards the base.

Habitat.—North Alabama; Prof. Tuomey.

My cabinet and cabinet of Dr. Hartman.

Diameter, ·29; length, ·79 of an inch.

Observations.—I have eight specimens before me of this modest little species. They were taken by Prof. Tuomey during his geological survey of Alabama many years since. The folds are not on the body-whorl; they incline to the left. It is allied to *Melania* (*Goniobasis*) *acuta* (nobis), but is not so small nor so pointed, and it is more of a horn-color. The aperture is about one-third the length of the shell.—*Lea.*

Goniobasis proletaria.—Shell folded, obtusely conical, rather thin, horn-color, without bands; spire obtusely conical; sutures impressed; whorls about six, slightly convex, folded above; aperture somewhat large, subrhomboidal, whitish within; outer lip acute, sinuous; columella bent in, thickened and twisted.

Fig. 370.

Habitat.—Florence, Alabama River; Rev. G. White.

Diameter, ·31; length, ·65 of an inch.

Observations.—A single specimen only was received, and that far from being perfect. The epidermis of it is very thin and most of it removed. It is nearly of the size and somewhat like *paupercula*, herein described, but is more conical and has larger and more distant folds, which are very slightly inclined to the left. The aperture is more than one-third the length of the shell.—*Lea.*

Appears to be the young of *continens*.

64. G. viridicata, Lea.

Goniobasis viridicata, Lea, Proc. Acad. Nat. Sci., p. 268, 1862. Jour. Acad. Nat. Sci., v, pt. 3, p. 322, t. 38, f. 172, March, 1863. Obs., ix, p. 144.

Description.—Shell folded, somewhat drawn out, thin, greenish, without bands; spire conical, exserted; sutures impressed; whorls about seven, flattened, with rather close folds; aperture very small, rhomboidal, bluish-white within; outer lip acute, somewhat sinuous; columella bent in, yellowish above, whitish below and twisted.

Habitat.— Grayson County, Kentucky; S. S. Lyon.

Diameter, ·24; length, ·64 of an inch.

Observations.— Three specimens were sent to me by Mr. Lyon, taken on his geological survey of Kentucky. It is a graceful, greenish little species with the folds inclining to the left, and with a paler line below the suture. The body-whorl has no folds, but is in two of the specimens covered with minute irregular veins. The middle whorls are plicate, while the apical whorls are carinate and striate. It is about the size of *cerea*, herein described, but differs in outline and other characters· In outline it is near *Doolyensis*, herein described, but is a much smaller species, and differs in the folds and the aperture. The aperture·is about one-third the length of the shell.— *Lea.*

Fig. 371.

65. G. purpurella, LEA.

Goniobasis purpurella, LEA, Proc. Acad. Nat. Sci., p. 269, 1862. Jour. Acad. Nat. Sci., vi, pt. 3, p. 327, t. 38, f. 183, March, 1863. Obs., ix, p. 149.

Description.— Shell folded, conical, thin, purplish, shining, banded or without bands; spire conical; sutures impressed; whorls about seven, flattened; aperture somewhat large, rhomboidal, dark within; outer lip acute, scarcely sinuous; columella bent in and twisted.

Habitat.— Caney Fork River, Tennessee; J. Lewis, M.D.

Diameter, ·22; length, ·48 of an inch.

Fig. 372.

Observations.— Several specimens were sent to me by Dr. Lewis for examination, nearly all more or less imperfect. They are usually without bands, but when banded the number is four, the two middle being approximate. An impressed line under the suture cuts the folds, forming a row of granules. The folds are close, inclining a little to the right. Below the suture some specimens have a light line. This species is nearly allied to *Melania* (*Goniobasis*) *Sellersiana* (nobis), but differs in being more pointed, in having bands and especially in having granules along the sutures. The aperture is more than one-third the length of the shell.—*Lea.*

66. G. semicostata, CONRAD.

Melania semicostata, CONRAD, New Fresh-Water Shells, **App. p. 7,** t. 9, f. 6, 1834.
BINNEY, Check List, No. 241. Brot, List, p. 59.

Description.— Shell elevated; longitudinally ribbed; whorls
convex, with fine, spiral striæ; body-whorl without ribs, ob- Fig. 373.
scurely striated above, subangulated in the middle; aperture
large, obliquely elliptical; within bluish, with brown bands.
Habitat.— Inhabits streams in North Alabama.— *Conrad.*

The figure is from the author's type specimen in the collection of the Academy of Natural Sciences of Philadelphia.

67. G. dislocata, RAVENEL.

Melania dislocata, RAVENEL, Cat. Shells, p. 11, 1834. BINNEY, Check List, No. 90.
BROT, List, p. 35. REEVE, Monog. Melania, sp. 380.
Goniobasis Lindsleyi, LEA, Proc. Acad. Nat. Sci., p. 267, 1862. Jour. Acad. Nat.
Sci., v, pt. 3, p. 319, t. 37, f. 167, March, 1863. Obs., ix, p. 141.

Description.— Shell ovately turreted, yellowish; whorls convex,
Fig. 374. longitudinally, plicately ribbed; ribs obsolete towards the
base; aperture ovate, rather small, a little effused at the base.
Habitat. — Dan River, North Carolina.— *Reeve.*

Mr. Reeve's publication of this species was made
the year previous to that of *Lindsleyi* by Mr. Lea.
I give a figure from Ravenel's type, which is in possession of Mr. Anthony.

Goniobasis Lindsleyi.— Shell folded, cylindrico-conical, rather thin,
yellowish horn-color, without bands; spire conoidal; sutures irregularly and very much impressed; whorls flattened; clothed with erect
folds; aperture rather small, rhomboidal, bluish white within; Fig. 375.
outer lip acute, sinuous; columella bent in and twisted.

Habitat.— Tennessee; President Lindsley and Dr. Edgar.
Diameter, ·31; length, ·80 of an inch.

Observations.— A few, imperfect specimens only are before
me, and the number of whorls cannot be ascertained, probably
eight. It is allied to *Melania* (*Goniobasis*) *costulata* (nobis),
but it is more cylindrical, and has the folds further apart. The aperture is probably one-third the length of the shell. It has two or
three decussating striæ immediately under the suture which make

small nodes. I dedicate this species to my friend, President Lindsley of Nashville, who sent it to me with many other shells from the streams of Tennessee.— *Lea.*

68. G. paupercula, Lea.

Goniobasis paupercula, Lea, Proc. Acad. Nat. Sci., p. 268, 1862. Jour. Acad. Nat.
Sci., v, pt. 3, p. 324, t. 38, f. 176, March, 1863. Obs., ix, p. 146.

Description.— Shell folded, subcylindrical, rather thin, chestnut-color or dark olive, without bands; spire rather short, sutures impressed; whorls somewhat convex, folded above and striate at the apex; aperture small, ovately rhomboidal, whitish within; outer lip acute, slightly sinuous; columella bent in and slightly twisted.

Fig. 376. Fig. 377.

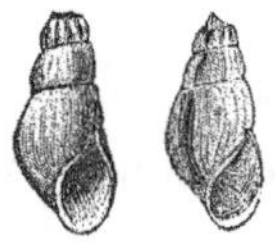

Operculum ovate, thin, light brown, with the polar point well in from the margin and above the base.

Habitat.— North Alabama; Prof. Tuomey.

Diameter, ·27; length, ·63 of an inch.

Observations.— I have quite a number of this small species sent many years since by Prof. Tuomey, not a single one with an entirely perfect apex, being usually decollate at the second whorl from the base. Most of them, therefore, do not exhibit the folds, which are only on the upper whorls; there they are pretty close and perpendicular. They were all covered with black oxide of iron, which on being removed exhibits a smooth, brown or greenish epidermis. The aperture is probably not one-third the length of the shell.— *Lea.*

69. G. corneola, Anthony.

Melania corneola, Anthony, Proc. Acad. Nat. Sci., p. 61, Feb., 1860. Binney.
Check List, No. 68. Brot, List, p. 35. Reeve, Monog. Melania, sp. 456,

Description.— Shell small, conical, rather thin; spire short Fig. 378.
and not very acute, composed of five or six subconvex whorls; whorls all more or less folded and with revolving raised striæ, which give them a subnodulous appearance; the body-whorl has four or five faint bands, which appear also within the aperture; aperture small, ovate; sinus small.

Habitat.— Alabama. My cabinet.

Observations.— This is a small and not very remarkable species nor can it well be compared with any other. One is at first view forcibly reminded of *Columbella avara*, Say, which it resembles, both in size and general appearance. The bands alluded to are often interrupted and never very fully expressed; body-whorl subangulated below the middle; does not seem to be a very abundant species. Only six individuals are before me.— *Anthony.*

Fig. 378 is from Mr. Anthony's type. The shell is not entirely adult, probably, but I cannot assimilate it to any other species. A number of specimens are before me, which are very uniform in character; in one, however, the bands are three in number, broad and dark. This shell inhabits Black Warrior River, Alabama,— *teste* Showalter.

70. G. nassula, CONRAD.

Melania nassula, CONRAD, New Fresh-Water Shells, p. 55, t. 8, f. 9, 1834. BINNEY, Check List, No. 171. DEKAY, Moll. New York, p. 97. JAY, Cat. 4th edit., p. 274. WHEATLEY, Cat. Shells, U. S., p. 26. BROT. List, p. 34. REEVE, Monog. Melania, sp. 412. CATLOW, Conch. Nomenc., p. 187.
Melania Edgariana, LEA, Philos. Proc., ii, p. 14, Feb., 1841. Philos. Trans., viii, p. 180, t. 6, f. 37. Obs., iii, p. 18. DEKAY, Moll. N. Y., p. 97. JAY, Cat. 4th edit., p. 273. BINNEY, Check List, No. 94. TROOST, Cat. Shells, Tenn. REEVE, Monog. Melania, sp. 430. WHEATLEY, Cat. Shells, U. S., p. 25. CATLOW, Conch. Nomenc., p. 186.
Melasma Edgariana, Lea, CHENU, Man. de Conchyl, i, f. 1997.

Description.— Shell elevated; whorls convex or subangulated, with longitudinal ribs, crossed by numerous, spiral, elevated lines, about seven on the penultimate whorl, and about eleven on ^{Fig. 379.} the body-whorl; suture impressed; apex much eroded.

Habitat.— Inhabits the limestone spring at Tuscumbia, Ala.

Observations.— Immense numbers of this pretty species congregate on the rocks where Spring Creek finds a passage through a cavern of the carboniferous limestone.— *Conrad.*

The figure is from an author's example in collection of Anthony. I have also examined author's examples in collections of Haldeman and Gen. Totten, which are shorter in consequence of the erosion of the apices. This shell is allied to *G. formosa*, Con., but has no bands.

Mr. Lea agrees with me that his *Edgariana* is a synonyme of *nassula*. The following is his description:—

Melania Edgariana— Shell folded, conical, rather thin, striate, yel-

13

lowish-brown; spire elevated; sutures irregularly impressed; whorls eight, rather flattened; aperture small, elliptical, angular below, bluish.

Habitat.—Cany Fork, Tennessee.

Diameter, ·29; length, ·77 of an inch.

Observations.—I owe to Mr. Edgar's kindness, several specimens

Fig. 380.   Fig. 381.

of this pretty species, which I name after him. It is remarkable for being folded and transversely striate on all the whorls, except the lower part of the body-whorl, which is striate only. The crossing of the folds and striæ give it a cancellated appearance. The aperture is rather more than one-fourth the length of the shell. The number of striæ on the body-whorl is about ten.—*Lea.*

This species is by no means uncommon in cabinets, and some specimens attain to noble proportions.

71. G. rugosa, Lea.

Melania corrugata, Lea, Philos. Proc., ii. p. 13, Feb., 1841. Philos. Trans., viii, p. 177, t. 5, f. 30. Obs., iii, p. 15. Troost, Cat. Shells, Tenn. Wheatley, Cat. Shells, U. S., p. 24.

Melania rugosa, Lea, Philos. Proc., ii, p. 237. Dec., 1842. Philos. Trans. viii. p. 248. Obs., iii, p. 86. DeKay, Moll. New York, p. 96. Binney, Check List, No. 235. Catlow, Conch. Nomenc., p. 188. Brot, List, p. 34.

Description —Shell folded, conical, rather thin, translucent, transversely striated, horn color; spire rather elevated; sutures very much impressed; whorls seven, convex, cancellated above; aperture rather large, elliptical, angular below, whitish.

Fig. 382.

Habitat.— Tennessee.

Diameter, ·22; length, ·50 of an inch.

Observations.—This is a small, folded species of which a single specimen was received from Dr. Troost. The superior whorls are carinated. The folds extend to the body-whorl. The aperture is rather more than one-third the length of the shell.— *Lea.*

I have not seen this species, but it is evidently a young shell. It was first described as *M. corrugata,* but as that name was preoccupied by Lamarck it was changed to *rugosa.*

72. G. costulata, LEA.

Melania costulata, LEA, Philos. Proc., ii, p. 14, Feb., 1841. Philos. Trans., viii, p. 181, t. 6, f. 39. Obs., iii, p. 19. BINNEY, Check List, No. 73. DEKAY, Moll. N. Y., p. 98. JAY, Cat. 4th edit., p. 273. TROOST, Cat. Shells, Tennessee. WHEATLEY, Cat. Shells, U. S., p. 24. REEVE, Monog. Melania, sp. 272, 360. BROT, List, p. 35.
Melasma costulata, Lea, ADAMS, Genera, i, p. 300.

Description.— Shell folded, conical, rather thin, yellow, above carinate; spire rather elongated; sutures impressed; whorls nine, rather convex; aperture small, subovate, within bluish.

Fig. 383.

Habitat.— Barren River, Kentucky: Tennessee.

Diameter, ·20; length, ·82 of an inch.

Observations.— In its general characters this species resembles *M. laqueata*, Say. It may be distinguished in its being of less diameter and being more slender. The specimens received from both Dr. Troost and Dr. Currey were covered with a deposit from the oxide of iron, giving them a black hue. Under this the epidermis is yellow. The aperture is about one-third the length of the shell.— *Lea.*

73. G. cinerella, LEA.

Goniobasis cinerella, LEA, Proc. Acad. Nat. Sci., p. 269, 1862. Jour. Acad. Nat. Sci., v, pt. 3, p. 328, t. 38 f. 184, March, 1863. Obs., ix, p. 150.

Description.— Shell folded, subfusiform, thin, pale, ash-color, without bands; spire obtusely conical; sutures irregularly impressed; whorls six, slightly convex; aperture somewhat large, ovately

Fig. 384.

rhomboidal, whitish within; outer lip acute, scarcely sinuous; columella bent in and slightly twisted.

Habitat.— Tennessee; Coleman Sellers.

Diameter, ·23; length, ·49 of an inch.

Observations.— A single specimen only was received from Mr. Sellers. It came with two young *Melania* (*Goniobasis*) *rugosa* (nobis), which it resembles, but this little species is not clathrate over the whole of the upper whorls, having only two transverse striæ, which cut the folds below the suture, forming granules. The folds are close and thick, and nearly straight. The aperture is nearly half the length of the shell.— *Lea.*

74. G. caliginosa, LEA.

Melania caliginosa, LEA, Philos. Proc., ii. p. 15, Feb., 1841. Philos. Trans., viii,
   p. 189, t. 6, f. 56. Obs., iii, p. 27. WHEATLEY, Cat. Shells, U. S., p. 24. REEVE,
   Monog. Melania, sp. 293. DEKAY, Moll. New York, p. 100. BINNEY, Check
   List, No. 44. TROOST, Cat. Shells, Tenn. JAY, Cat. 4th edit., p. 273. CATLOW,
   Conch. Nomenc., p. 185. BROT. List, p. 34.
Elimia caliginosa, Lea, ADAMS, Genera, i, p. 300.

Description— Shell cancellate, conical, somewhat thick, transversely
striated; very dark brown; spire elevated; sutures irregularly im-
Fig. 385. pressed; whorls eight, rather convex; aperture small, elliptical, purplish within.

Habitat.— Tennessee.

Diameter, ·34; length, ·91 of an inch.

Observations.— A fine, cancellate species with ten or twelve
revolving striæ on the body-whorl, crossing the folds. The
aperture is about one-third the length of the shell. It nearly
answers to Mr. Conrad's description of *M. nassula*, but has
five striæ on the penultimate whorl, while the *nassula* has seven. It
differs from *M. catenaria*, Say, in having a more elevated spire, and
in having two or three more revolving striæ. In some individuals
the aperture is bluish-white.— *Lea.*

75. G. nodulosa, LEA.

Melania nodulosa, LEA, Philos. Proc., ii. p. 15, Feb., 1841. Philos. Trans., viii,
   p. 190, t. 6, f. 57. Obs., iii, p. 28. DEKAY, Moll. N. Y., p. 100. BINNEY, Check
   List, No. 180. TROOST, Cat. Shells, Tennessee. WHEATLEY, Cat. Shells,
   U. S., p. 26. CATLOW, Conch. Nomenc., p. 188. BROT, List, p. 34. REEVE,
   Monog. Melania, sp. 276.
Elimia nodulosa, Lea, ADAMS, Genera, No. 300.

Description.— Shell cancellate, conical, thick, dark brown; sutures
irregularly impressed; whorls somewhat convex; aperture rather
large, elliptical, subangular below, within bluish.
Fig. 386.   Fig. 386a.

Habitat.— Tennessee.

Diameter, ·34; length, ·82 of an inch.

Observations.— Two imperfect specimens only
were received from Dr. Troost, and both are
much eroded at the apex, consequently the number of whorls could not be ascertained. The
body-whorl has about twenty well defined, raised

striæ, which on the superior part are crossed by folds, giving the
whole of the upper part of the shell a granulate appearance. It is

somewhat like *M. catenaria*, Say, but may be distinguished at once by the number of striæ.— *Lea*.

This beautiful species being poorly represented by Mr. Lea's figure I have had drawn a specimen named by Mr. Lea in museum of Mr. Anthony and also a younger shell in museum of Mr. Haldeman.

76. G. difficilis, Lea.

Goniobasis difficilis, Lea, Proc. Acad. Nat. Sci., p 267, 1862. Jour. Acad. Nat. Sci.,
v, pt. 3, p. 317, t. 37, f. 163, March, 1863. Obs., ix, p. 139.

Description.—Shell folded, somewhat attenuate, dark olive or brownish, rather thin, without bands; spire attenuate, sharp pointed; sutures regularly impressed; whorls about eight, slightly convex; aperture rather small, ovately rhomboidal, whitish within; outer lip acute, subsinuous; columella bent in, thickened and twisted.

Fig. 387. Fig. 388.

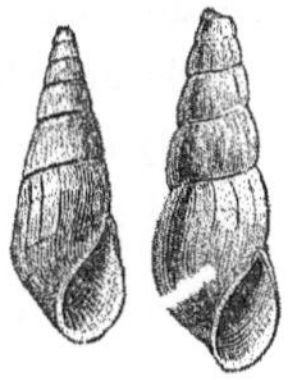

Habitat.— Tennessee; Dr. Edgar.

Diameter, ·31; length, ·82 of an inch.

Observations.— This is one of the *Melania* (*Goniobasis*) *Deshayesiana* group, and is nearly allied to *sparus*, herein described, but may at once be distinguished from that species by being flatter on the whorls, and by being of a darker color. There is but a single adult specimen before me, the apical whorls of which are eroded. Some of the young specimens are perfect to the apex, and the upper whorls present close folds slightly curved and decussate, with revolving striæ. These are hardly perceptible on the adult specimen. In outline it resembles *Melania* (*Goniobasis*) *columella* (nobis), but differs in the color and in the form of the lower part of the columella. The aperture is about one-third the length of the shell.— *Lea*.

This shell is somewhat like *G. glauca*, but the whorls are more convex. Except in the shell being more cylindrical, *baculum* is closely related to it.

77. G. sparus, Lea.

Goniobasis sparus, Lea, Proc. Acad. Nat. Sci., p. 267, 1862. Jour. Acad. Nat. Sci.,
v, pt. 3, p. 316, t. 37, f. 162, March, 1863. Obs., ix, p. 138.
Goniobasis cerea, Lea, Proc. Acad. Nat. Sci., p. 268, 1862. Jour. Acad. Nat. Sci.,
v, pt. 3, p. 321, t. 38, f. 171, March, 1863. Obs., ix, p. 143.

Description.— Shell folded, somewhat drawn out, pale yellow, somewhat thick, without bands; spire attenuate, sharp-pointed; sutures
irregularly impressed; whorls eight, slightly convex; aperture rather
large, ovately rhomboidal, white within; outer lip acute, sin-

Fig. 389.

uous; columella somewhat bent in, yellow above and white
below, twisted.

    Habitat.— Tennessee; Dr. Currey and President Lindsley.
Diameter, ·28; length, ·74 of an inch.

    Observations.— This is a graceful, sharp-pointed species,
closely allied to *Deshayesiana* (nobis), but is rather more
slender, is a little more inflated below the sutures and is rather
more solid in its structure. It has the same striæ along the upper
part of the whorls which decussate the folds. It is more ovate in
the aperture, the base not being so angular. The folds on the upper
whorls are close and well defined, but disappear below. They are
slightly curved, and the aperture is about one-third the length of the
shell.— *Lea.*

The following is a younger shell.

Goniobasis cerea.— Shell folded, conical, rather thin, wax-colored,
without bands; spire conical; sutures impressed; whorls six, somewhat convex, with small folds; aperture rather large, elongately
rhomboidal, whitish within; outer lip acute, sinuous; columella bent in and twisted.

Fig. 390.

    Habitat.— Tennessee; Prof. Troost: and Duck Creek, Ten-
nessee; J. Clark.

Diameter, ·26; length, ·64 of an inch.

    Observations.— Two specimens only are before me. That
from Mr. Clark, which I believe was collected by Prof. Christy, is of
a lighter color than the other, which is brownish and may even prove
to be a distinct species, as it is slimmer and is rather smaller in the
aperture. The folds are delicate, inclining to the right, and do not
reach to the body-whorl. There are indistinct striæ on the upper
part of the whorls decussating the folds. It is about the size and
nearly the same outline as *inosculata*, herein described, but that is a
carinate species with a somewhat differently formed aperture. The
aperture is more than one-fourth the length of the shell.—*Lea.*

78. G. Thorntonii, Lea.

Goniobasis Thorntonii, Lea, Proc. Acad. Nat. Sci., p. 268, 1862. Jour. Acad. Nat.
Sci., v, pt. 3, p. 320, t. 38, f. 168, March, 1863. Obs., ix, p. 142.

Description.— Shell roughly folded, conical, rather thin, horn-color,
without bands; spire conical; sutures irregularly and very much
impressed; whorls slightly convex, clothed with distant bent folds;
aperture rather large, rhomboidal, white within; outer lip
acute, sinuous; columella somewhat bent in and twisted.

Fig 391.

Operculum ovate, thin, brown, with the polar point one-
third from the base on the left of the centre.

Habitat.— Tuscumbia; L. B. Thornton, Esq.: Florence,
Alabama; Rev. G. White.

Diameter, ·38; length, ·87 of an inch.

Observations.— Some dozen specimens, most of them imperfect are
before me. The number of whorls could not be ascertained— probably eight. The folds are large, distant and curving to the right;
about the middle of a whorl there is a line which decussates the fold,
making a node. It belongs to the group of which *Melania* (*Goniobasis*) *costulata* (nobis), may be considered the type, and it closely
resembles *Lindsleyi,* herein described, but differs in not being cylindrical, in having larger and more distinct ribs and and a larger
aperture. The aperture is rather more than one-third the length of
the shell. I name this after L. B. Thornton, Esq., Attorney at Law,
Tuscumbia, who very kindly has sent to me many fine specimens from
his vicinity.— *Lea.*

79. G. cancellata, Say.

Melania cancellata, Say, New Harmony Disseminator, p. 260, Aug., 1829. Say's
Reprint, p. 16. Binney's edit., p. 141. Binney, Check List, No. 46. DeKay,
Moll., N. Y., p. 93. Wheatley, Cat. Shells, U. S., p. 24.
Elimia cancellata, Say, Adams, Genera, i, No. 84

Description.— Shell rather slender, attenuated; volutions convex,
with about twenty-six, reclivate, longitudinal, elevated lines, crossed
by about eighteen revolving ones, the eight or nine towards the
base crowded.

Length, more than four-fifths of an inch.

Habitat.— Florida.

Observations.— For this shell I am indebted to Captain Le Conte,
who informed me that he obtained it in St. John's River. It differs

from all other species in the numerous, longitudinal and transverse, elevated lines, with the exception of the *catenaria* (nobis), than which it is of a much more elongated and attenuated form.— *Say.*

I have not been able to procure a specimen of this shell. Does it = *curvicostata*, Anthony?

80. G. circincta, Lea.

Melania circincta, Lea, Philos. Proc., ii, p. 15, Feb., 1841. Philos. Trans., viii, p. 187, t. 6, f. 51. Obs., iii, p. 25. DeKay, Moll., N. Y., p. 99. Troost, Cat. Shells, Tenn. Wheatley, Cat. Shells, U. S., p. 24. Catlow, Conch. Nomenc., p. 186. Brot, List. p. 31. Reeve, Monog. Melania. sp. 289.
Melania circinnata, Lea, Binney, Check List, No. 54.
Juga circinnata, Lea, Chenu, Man. de Conchyl., i, f. 2015. Adams, Genera, i, p. 294.

Description.— Shell striate above, turreted, rather thin, pale yellow, banded; spire drawn out; sutures small; whorls nine, slightly con-

Fig. 392.

vex, carinate in the middle; aperture rather small, ellip-tical, angular at the base, and white within.

Habitat.— Tennessee.

Diameter, ·35; length, ·90 of an inch.

Observations.— This beautiful species is peculiar for its pale yellow ground and broad band, which is placed immediately upon the carina. A very indistinct band may be observed below the carina, where in some indi-viduals may also be observed a few striæ. In some, the striæ on the superior part of the shell are accompanied by indistinct ribs. —*Lea.*

Except in the development of the carina, and in being longer, this species resembles *G. laqueata*, Say.

81. G. athleta, Anthony.

Melania athleta, Anthony, Ann. Lyc. Nat. Hist. N. Y., vi, p. 83, t. 2, f. 1, March, 1854. Binney, Check List No. 23. Brot, List, p. 34. Reeve, Monog. Mel., sp. 258.
Melania glauca, Anthony, Proc. Acad. Nat. Sci., p. 57, Feb., 1860. Binney, Check List, No. 125. Brot, List, p. 35. Reeve, Monog. Melania, sp. 389.
Goniobasis Lyonii, Lea, Proc. Acad. Nat. Sci., p. 263. Jour. Acad. Nat. Sci., v, pt. 3, p. 313, t. 37, f. 156, March, 1863. Obs., ix, p. 135.

Description.— Shell conical, nearly smooth, dark horn-color; spire much elevated; whorls ten, nearly flat, with faint, longitudinal ribs, most distinct on the upper part of the whorls; sutures well marked;

aperture small, ovate, within whitish, tinged near the base with rose; columella rounded, and forming a slight sinus at base.

Fig. 393.

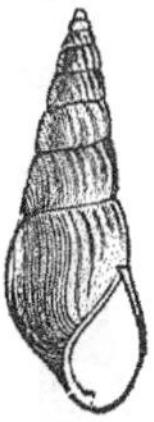

Diameter, ·40 of an inch (10 millim.); length, 1·35 inches 32 (millim). Length of aperture, ·40 (10 millim.); breadth of aperture, ·23 of an inch (6 millim.).

Habitat.— Tennessee.

Observations.—A stout species, and one of the most beautiful with which I am acquainted. The ribs are not strongly expressed, and on the lower whorls are nearly obsolete, having there the appearance of striæ of growth merely; body-whorl a little angulated at base.— *Anthony.*

Figured from the type specimen.

Melania glauca.— Shell conical, folded, of a green color in the lower whorls, often modified by a brown tinge on the upper ones; whorls ten, slightly convex, with prominent longitudinal ribs, obso-

Fig. 394.

lete on the body-whorl; sutures well defined, but not deeply marked; aperture ovate, livid within and with occasionally a faint, rosy tinge there; columella angulated at the middle; sinus well defined.

Habitat.— Tennessee.

Observations.— A stout species, with prominent, curved ribs on all the upper whorls, those on the body-whorl being less clearly defined or else absolutely wanting. Color a beautiful apple-green, relieved by a broad, yellow band near the suture; and this color often passes into a yellowish-brown on the upper whorls. Near the apex, the folds are often traversed by four or five prominent striæ, which pass over without being interrupted by the longitudinal ribs. May be compared with *M. viridula* (nobis) as to color, but is less slender, and the ribs at once distinguish it.— *Anthony.*

The figure, which is a very poor one, represents the type specimen. The species is better illustrated by the figure of *G. Lyonii,* which is a synonyme. The following is a description of the latter

Goniobasis Lyonii. — Shell folded, striate above, carinate at the apex, yellowish, very thin, very much drawn out; spire attenuate, sharp-pointed; sutures irregularly impressed; whorls nine, slightly convex; aperture rather small, subrhomboidal, whitish within; outer lip acute, sinuous; columella bent in, thickened and slightly twisted.

Habitat.— Grayson County, Kentucky; S. S. Lyon.

Diameter, ·30; length, ·92 of an inch.

Fig. 395.

Observations.— A single specimen of this species was among the *Melanidæ* collected by Mr. Lyon in the geological survey of Kentucky. It was accompanied by *Melania (Goniobasis) Deshayesiana* (nobis), to which it is closely allied in some of its characters. It differs in having two or three more whorls, in being more cancellate above, by the striæ decussating the longitudinal ribs, and particularly in the lower part of the columella being nearly straight, while that part in *Deshayesiana* is oblique to the right. The ribs are pretty close and slightly curved, the inner margin of the outer lip is slightly thickened. The aperture is rather less than one-third the length of the shell. I dedicate this with great pleasure to Mr. Lyon, civil engineer and state geologist.— *Lea.*

82. G. curvicostata, ANTHONY.

Melania curvicostata, ANTHONY, MSS. REEVE, Monog. Melania, sp. 462. BROT, List, p. 35.
Melania densecostata, REEVE, Monog. Melania, sp. 465. BROT, List, p. 35.

Description.— Shell ovately turreted, livid olive, encircled towards the apex with a reddish line, whorls convex, longitu-

Fig. 396.

dinally, plicately ribbed, ribs curved, gradually fading towards the aperture; aperture ovate, slightly effused at the base, interior tinged with purple.

Habitat.— Florida, United States.— *Reeve.*

Fig. 397.

The following appears to me to be the same species.

Melania densicostata. — Shell subulately turreted, burnt olive, whorls eight to nine, rather flat, longitudinally, densely plicately ribbed, the last obtusely angled; aperture rather small, ovate, interior very faintly tinged with purple.

Habitat.— Florida, United States.

This interesting little species is of the same type as *M. curvicostata*, just described, but the ribs are stout and comparatively straight, ending abruptly on an obtuse angle of the last whorl.— *Reeve.*

83. G. striatula, LEA.

Melania striata, LEA, Philos.- Proc., ii, p. 15, Feb., 1841. Philos. Trans., viii,
  p. 186, t. 6, f. 49. Obs., iii, p. 24. TROOST, Cat. Shells, Tenn. WHEATLEY,
  Cat. Shells, U. S., p.
Juga striata, Lea, CHENU, Man. de Conchyl., i, f. 2018. ADAMS, Genera, i, p. 304.
Melania striatula, LEA, Philos. Proc., ii, p. 237, Dec., 1842. Philos. Trans.. viii,
  p. 248, Obs., iii, p. 83. DEKAY, Moll. New York, p. 99. JAY, Cat. 4th edit.,
  p. 275. BINNEY, Check List, No. 249. CATLOW, Conch. Nomenc., p. 158.
  REEVE, Monog. Melania, sp. 463. BROT, List, p. 35.

Description.— Shell striate, conical, rather thin, dark brown, carinate above; spire somewhat elevated; sutures impressed; whorls
eight, convex; aperture small, elliptical, within reddish.

Habitat.— Tennessee.

Diameter, ·21; length, ·49 of an inch.

Observations.— Rather a small species of a dark reddish-brown.
In some individuals the folds are numerous. In others
the striæ predominate and cover nearly all the whorls.
The aperture is rather more than one-third the length
of the shell.— *Lea.*

Fig. 398. Fig. 399.

This shell was originally described under the
name of *striata,* but finding that name to be preoccupied, Mr.
Lea subsequently changed it to *striatula.* Mr. Reeve's figure
is not a good representation of the shell.

84. G. tripartita, REEVE.

Melania tripartita, REEVE, Monog. Melania, sp. 364, Dec., 1860. BROT, List, p. 37.

Description.— Shell acuminated, olive; whorls eight to nine, somewhat rounded, spirally, distantly ridged, the first few strongly keeled,
then longitudinally, plicately ribbed, afterwards smooth;
aperture small, semilunar.

Fig. 400.

Habitat.— ——?

Observations.— This is without doubt, a United States
species, but I know of none with which it can be satisfactorily identified.— *Reeve.*

The figure is copied from Mr. Reeve's plate. I
do not recognize the species, although it approaches closely
to several others of the present group.

85. G. decora, LEA.

Melania decora, LEA, Philos. Proc., ii, p. 14, Feb., 1841. Philos. Trans., viii, p 181,
t. 6, f. 38. Obs., iii, p. 19. DEKAY, Moll., N. Y., p. 98. BINNEY. Check List,
No. 85. TROOST, Cat. Shells, Tenn. WHEATLEY, Cat. Shells, U. S., p. 25.
REEVE, Monog. Melania, sp. 202. CATLOW, Conch. Nomenc , p. 186. BROT,
List, p. 35.

Description.— Shell folded, acutely turreted, rather thin, horn-color,
above striate; spire acute, elevated; sutures impressed; whorls nine,
rather flattened; aperture small, elliptical, whitish.

Fig. 401. *Habitat.*— Tennessee : Green River, Kentucky.

Diameter, ·26; length, ·82 of an inch.

Observations.— This species resembles *M. costulata*, herein
described. It is, however, more elevated in the spire, and
the folds are closer. On the two lower whorls the folds
become obsolete.— *Lea.*

Reeve's figure is either a very poor one or it does not represent this species. It is scarcely necessary to add that his
locality " Niagara " is entirely wrong, as no plicate species is
found there.

86. G. crebricostata, LEA.

Melania crebricostata, LEA, Philos. Proc., ii, p. 13, Feb., 1841. Philos. Trans., viii,
p. 179, t. 6, f. 35. Obs., iii, p. 17. DEKAY, Moll., New York, p. 97. JAY, Cat.
4th edit., p. 273. TROOST, Cat. Shells, Tenn. WHEATLEY, Cat. Shells, U. S.,
p. 24. REEVE, Monog. Melania, sp. 374. BINNEY, Check List, No. 74. BROT,
List, p. 35.
Melasma crebricostata, Lea, CHENU, Man. de Conchyl., i, f. 1099. ADAMS, Genera,
i, p. 300.

Description.— Shell closely folded, conical, rather thick, horn-color;
spire elevated; sutures linear; whorls seven, flattened; aper- Fig. 402.
ture small, elliptical, below angular, bluish.

Habitat.— Robinson County, Tennessee.

Diameter, ·28; length, ·90 of an inch.

Observations.— This is rather a slender shell, and is peculiar
for its numerous folds, which are slightly curved and parallel.
They extend over the whole shell, except the inferior half of
the body-whorl. The aperture is about one-third the length of the
shell.— *Lea.*

The species is a common one. Dr. Brot suggests that this
species should, perhaps, be united with *M. costulata*; I think,
however, that they are sufficiently distinct.

87. G. comma, CONRAD.

Melania comma, CONRAD, New Fresh-Water Shells, p. 53, t. 8, f. 7, 1834. WHEATLEY, Cat. Shells, U. S., p. 24. REEVE, Monog. Melania, sp. 107. BINNEY Check List, No. 61. DEKAY, Moll., New York, p. 95. JAY, Cat. 4th edit., p. 273. BROT, List, p. 35. CATLOW, Conch. Nomenc., p. 186. MÜLLER, Synopsis, p. 45.
Melasma comma, Conrad, ADAMS, Genera. i, p. 300.

Description.— Shell subulate, much elongated, slender; whorls eight or nine, flattened, indented at the sutures, with longitudinal, distant, slightly arcuated ribs, disappearing on the lower volutions; labrum thin; aperture elliptical, produced at base; color olive, with a dark band above the middle of each whorl.

Fig. 403. Fig. 404.

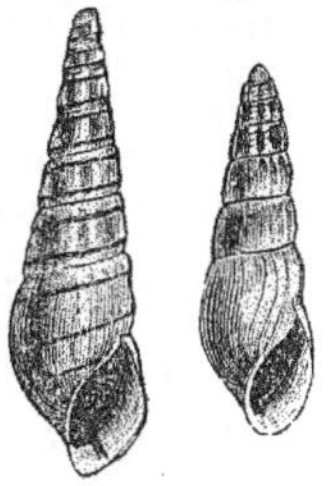

Habitat.— Inhabits rivulets which are tributary to the Black Warrior in mountain districts in Alabama.

Observations.— It is greatly elongated, and the ribs are separated by an indented space at the sutures.— *Conrad.*

A slender variety, which we have figured, occurs in Tennessee. The first figure is from the type in collection of Acad. Nat. Sci., Philadelphia. Mr. Haldeman possesses an author's example.

88. G. acuta, LEA.

Melania acuta, LEA, Philos. Trans., iv, p. 101, t. 15, f. 32. Obs., i, p. iii. TROOST. Cat. Shells, Tennessee. WHEATLEY, Cat. Shells, U. S., p. 24. BINNEY. Check List, No. 4. BROT, List, p. 3. REEVE, Monog. Melania, sp. 274.
Ceriphasia acuta, Lea, ADAMS, Genera, i, p. 297.

Description.— Shell acutely turreted, thin, horn-colored; apex acute; whorls eight, carinate immediately above the suture, longitudinally undulated and transversely lineated; base angulated: aperture white, and one-fourth the length of the shell.

Fig. 405.

Habitat.— Tennessee River; Prof. Vanuxem.

Diameter, five-twentieths; length, thirteen-twentieths of an inch.

Observations.—I have seen no described species to which this bears a close resemblance. Its delicate form, furnished with undulations and transverse lines, will easily distinguish it.— *Lea.*

Mr. Say (cover of No. 6 Am. Conch.) says this equals his
Melania semicarinata, but I can see no good reason to unite
them, as that shell has not the longitudinal folds of *acuta*.
The specimen figured by Mr. Lea, and here copied, is evidently not mature. A shell closely allied to this species
inhabits the Great Lakes, etc., and Mr. Lea and other conchologists labeled it *acuta*. It is *never* plicate and I have
described it under the name of *Haldemani*.

89. G. subcylindracea, Lea.

Melania subcylindracea, Lea, Philos. Proc., ii, p. 12, Feb., 1841. Philos. Trans.,
viii, p. 169, t. 5, f. 14. Obs., iii, p. 7. DeKay, Moll., New York, p. 94. Troost,
Cat. Shells, Tenn. Binney, Check List, No. 253. Wheatley, Cat. Shells,
U. S., p. 27. Catlow, Conch. Nomenc., p. 188. Brot, List, p. 39, Reeve,
Monog. Melania, sp. 399.
Potadoma subcylindracea, Lea, Adams. Genera, i, p. 299.

Description— Shell smooth, subcylindrical, somewhat thick, horn-
color; spire obtusely elevated; sutures impressed; whorls convex;
Fig. 406. Fig. 407. aperture small, ovate, whitish.

Habitat.— Tennessee; Dr. Troost.

Diameter, ·32; length, ·85 of an inch.

Observations.— This is a club-shaped species with
an aperture about the third of the length of the shell.
All the specimens sent by Dr. Troost are more or
less decollate.— *Lea.*

Figured from Mr. Lea's plate. Some specimens are more
lengthened and cylindrical than the type specimen.

90. G. baculum, Anthony.

Melania baculum, Anthony, Ann., N. Y. Lyc. Nat. Hist., vi, p. 98, t. 2, f. 16, March
1854. Binney, Check List, No. 27. Brot, List, p. 34. Reeve, Monog. Mela
nia, sp. 431.

Description.— Shell conical, thick; of a dull, reddish-brown color,
with a lighter shade near the upper part of each whorl. Spire much
elevated, not diminishing rapidly as it ascends, and with nearly a rectilinear outline; whorls eight remaining, and with an appearance
of having lost several by truncation; hardly convex and with a
deeply impressed suture; aperture small, broadly ovate, light red
within; columella rounded, indented, with a small sinus.

Diameter, ·48 of an inch (12 millim.); length, 1·28 inches (33 millim.). Length of aperture, ·35 of au inch (9 millim.); breadth of aperture, ·20 of an inch (5 millim.).

Fig. 408.

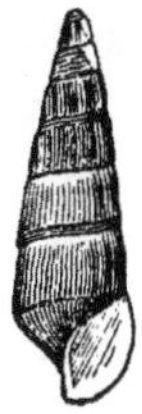

Observations.— The most striking characteristic of this species is its robust, cylindrical form, combined with its pale sutural region; compared with *M. teres*, Lea, it is much less slender and turreted, much more plicate, and the whorls are less inflated. *M. rufa* is not folded, and is a more acutely elevated species. The curve in the columella resembles that of *M. columella*, Lea, but that shell is much less elongated, has only six whorls, and is destitute of distinct folds.— *Anthony.*

91. G. concinna, Lea.

Melania concinna, Lea, Philos. Proc., ii, p. 14, Feb., 1841. Philos. Trans., viii, p. 183, t, 6, f. 42. Obs., iii, p. 21. DeKay, Moll., New York, p. 98. Troost, Cat. Shells, Tennessee. Wheatley, Cat. Shells, U. S., p. 24. Catlow, Conch. Nomenc., p. 186. Binney, Check List, No 63. Brot, List, p. 34.
Melasma concinna, Lea, Adams, Genera, i, p. 300.

Description.— Shell folded, acutely turreted, thin, brown; spire drawn out; sutures impressed; whorls nine, carinate, flattened; aperture small, elliptical, angular at base, whitish.

Fig. 409.

Habitat.— Tennessee.

Diameter, ·25; length, ·75 of an inch.

Observations.— A single individual only was received from Dr. Troost. Its mouth is about one-fourth the length of the shell. It is remarkably flattened on the whorls, and the superior part is transversely striate.— *Lea.*

This species resembles *baculum*, but is narrower, smaller, and the plications are closer. It has been extensively distributed by Mr. Anthony as a variety of *comma*. Allied to *eliminata*, but differs in the plicæ, being smaller, also in the form of the mouth: the shell is rather stouter and the body-whorl more angular.

92. G. eliminata, Anthony.

Melania eliminata, Anthony, Ann. New York. Lyc. Nat. Hist., vi, p. 97, t. 2, f. 15, Mar., 1854. Binney, Check List, No. 98. Brot, List, p. 34.

Description.— Shell conic, thin, brownish; spire slender, elevated; whorls about eight, convex, with transverse folds and spiral striæ, both of which, however, disappear towards the lower portion of each whorl, and are hardly visible on the last whorl; sutures deeply impressed; aperture small, ovate, within translucent, exhibiting the exterior coloring through its substance; columella but little rounded, except near its base, where with the much curved lip it forms a sharp, narrow sinus.

Fig. 410.   Fig. 411.

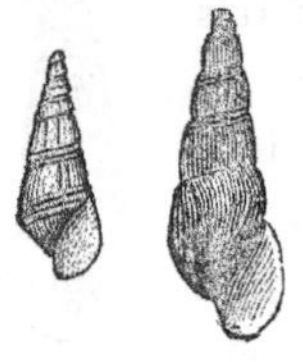

Diameter, ·24 of an inch (6 millim.); length, ·80 of an inch (21 millim.). Length of aperture, ·26 of an inch (7 millim.); breadth of aperture, ·15 of an inch (4 millim.).

Habitat.— Kentucky, near Owenborough.

Observations.— This is a very slender and elevated species, resembling in this respect *M. comma*, Con., from which it differs very materially by the character of its folds and striæ, which are more decided, being nearly as prominent, though less distant than in *M. curreyana*, Lea; the striæ revolve round the whorls and over the ribs without being interrupted by them; differs from *M. Edgariana*, Lea, by its brown color, more slender form, less convex whorls, and thinner texture; it is more slender than *M. decora* or *costulata*, and less acute, the whorls tapering more gradually to the apex; on the upper whorls there are about five striæ, the lowest of which is much more elevated than the others, and the folds are arrested by it near the suture.   The penultimate whorl is often subangulated at its base.—*Anthony.*

93. G. teres, LEA.

Melania teres, LEA, Philos. Proc., ii, p. 13, Feb., 1841. Philos. Trans., viii, p. 176, t. 5, f. 27. Obs., iii, p. 14. DEKAY, Moll., New York, p. 96. TROOST, Cat. Shells, Tenn. WHEATLEY, Cat. Shells, U. S., p. 27. BINNEY, Check List, No. 269. JAY, Cat. 4th edit., p. 275. CATLOW, Conch. Nomenc., p. 189. BROT, List, p. 35.

Melania terebralis, LEA, Philos. Proc., ii, p. 13, Feb., 1841. Philos. Trans., viii, p. 178, t. 6, f. 32. Obs., iii, p. 16. DEKAY, Moll., New York, p. 96. TROOST, Cat. Shells, Tenn. WHEATLEY, Cat. Shells, U. S., p. 27. BINNEY, Check List, No. 268. CATLOW, Conch. Nomenc., p. 189. BROT, List, p. 36.

Description.— Shell folded, acutely turreted, thin, horn-colored; spire drawn out; sutures impressed; whorls nine, convex; aperture small, elliptical, whitish within.

Habitat.— Tennessee.

Diameter, ·25; length, ·87 of an inch.

Observations.— This is a remarkably elevated species, with the whorls much inflated, and the last whorl very small. Some of the specimens before me are but obscurely folded.— *Lea.* Fig. 412. Fig. 413.

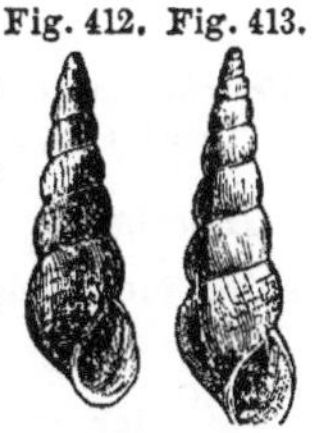

Figured from Mr. Lea's plate. This is a very distinct species, on account of the great convexity of the whorls.

The following description and figure represents half grown specimens :—

Melania terebralis. — Shell folded, acutely turreted, rather thin, shining, reddish-brown; spire much elevated; sutures much impressed; whorls nine, convex, carinate above; aperture small, elliptical, whitish.

Fig. 414.

Habitat.— Tennessee.

Diameter, ·24; length, ·67 of an inch.

Observations.— This species differs in the form of the folds from any which have come under my notice. These folds are from each other, but slightly raised, and give the shell a distant varicose appearance. The mouth is about the fifth part of the length of the shell.— *Lea.*

94. G. gracillima, ANTHONY.

Melania gracillima, ANTHONY, Proc. Acad. Nat. Sci., p. 62, Feb., 1860. BINNEY, Check List, No. 129. BROT, List, p. 36. REEVE, Monog. Melania, sp. 437.

Description.— Shell conic, thin, brownish; spire very slender, elevated, composed of eight, convex whorls, the upper ones folded and striate, the lower ones smooth, the striæ being replaced by indistinct, slender, brown lines; sutures very deeply impressed, a sharp carina on the lower portion of each whorl, rendering them Fig. 415. Fig. 416. quite distinct; aperture small, ovate, banded inside; columella indented; sinus small.

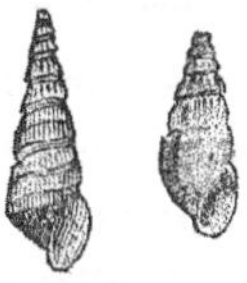

Habitat.— South Carolina.

Observations.—A peculiarly slender, graceful species, in form somewhat like *M. strigosa*, Lea, but more folded and more slender. The striæ on the upper whorls are very distinct where they intersect the folds, and give the shell a tuberculous appearance; the folds are arrested by the carina which is elevated. The brown lines on the body-whorl are often slightly elevated, but nevertheless, indistinct and are about four in number. A faint line or band of a yellow color revolves around the upper portion of the two lower whorls.— *Anthony.*

95. G. Clarkii, LEA.

Melania Clarkii, LEA, Philos. Trans., x, p. 297, t. 30, f. 4. Obs., v, p. 53. BINNEY, Check List, No. 56. BROT, List, p. 34. REEVE, Monog. Melania, sp. 356.

Description.— Shell folded, club-shaped, rather thin, dark brown; spire elevated, drawn out; sutures somewhat impressed; whorls flattened; aperture small, rather elliptical, at the base angu-
Fig. 417. lar, within dark; columella twisted.

Habitat.— Duck Creek, Tennessee.

Diameter, ·23; length, ·73 of an inch.

Observations.— The form of this species is more attenuate than usual, with the clavate forms. It has about ten whorls; those above the body-whorl being disposed to be both plicate and striate. Towards the apex they are all thickly striate. On all the specimens before me, on the lower whorls, there are irregular, oblique striæ, somewhat similar to those on the *M. Ocoeĕnsis* (nobis), which give them a malleate character. On the upper margin of the whorls, along the sutures, there is usually an indistinct, light line. The outer lip is broken.— *Lea.*

Figured from Mr. Lea's plate. Specimens before me differ somewhat in the closeness of the plicæ. Some are even more attenuately lengthened than Mr. Lea's figure. This is the narrowest species inhabiting North America. In collection of Mr. Gould are specimens from Lee County, Georgia.

96. G. De Campii, LEA.

Goniobasis De Campii, LEA, Proc. Acad. Nat. Sci., p. 154, May, 1863.

Description.— Shell plicate, striate below, greatly attenuated, thin, corneous, without bands; spire subulate; sutures linear, Fig. 418. impressed; whorls fully ten, subconvex, above with slightly bent plicæ; aperture very small, subrhomboidal, whitish within; lip acute, somewhat sinuous; columella whitish, in-curved and twisted.

Habitat.— Huntsville, Alabama; Wm. H. De Camp, M.D., surgeon, United States army.— *Lea.*

97. G. plicifera, LEA.

Melania plicifera, LEA, Philos. Trans., vi, p. 93, t. 23, f. 90. Obs., ii, p. 93. WHEATLEY, Cat. Shells, U. S., p. 26. JAY, Cat., 4th edit., p. 274. BINNEY, Check List, No. 211. REEVE, Monog. Melania, sp. 284. COOPER, Report, p. 374. BROT, List, p. 36. GOULD, Moll. Expl. Exped., p. 143, f. 165. TROSCHEL, Archiv, fur Naturgesch., ii, p. 227.

Melasma plicifera, Lea, CHENU, Manuel, i, f. 2001. ADAMS, Genera, i, p. 300.

Description.— Shell acutely turreted, rather thick, nearly black; spire full of folds; apex truncate; whorls some-what convex, the last being smooth above and striate below; aperture white.

Fig. 419.   Fig. 420.

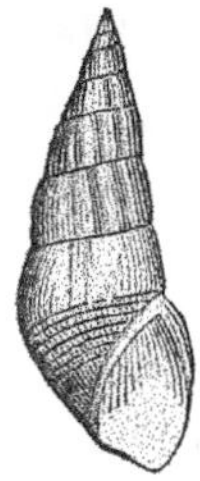

Fig. 421.   Fig. 422.

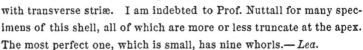

Habitat.— Wahlamat, near its junction with the Columbia River; Prof. Nuttall. Diameter, ·4 of an inch; length, 1·1 inches.

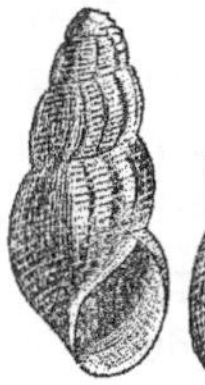

Observations. — Among the fine shells brought by Prof. Nuttall from beyond the Rocky Mountains, was this single species of *Melania*. It is remarkable for its numerous folds, or ribs, which fill the superior whorls. The inferior whorl is entirely without these ribs, but the inferior portion is furnished with transverse striæ. I am indebted to Prof. Nuttall for many specimens of this shell, all of which are more or less truncate at the apex. The most perfect one, which is small, has nine whorls.— *Lea.*

This is an exceedingly common and variable species, and I give several figures of its most usual forms. Occasionally the shell is thickly striate, with folds on the upper whorls only.

Dr. Gould, in the Mollusca of the United States Exploring Expedition says of this species:—

" This shell seems to be subject to great variety, or else these are several allied species. The typical shell has the spire elongated, pointed, and the whorls flattened, with coarse, longitudinal folds. Others are surrounded by numerous, raised lines, and are nearly destitute of folds. A variety from Lake George (Oregon) must be very corpulent. It is much decollated, and is light and thin. Whorls convex; aperture rounded, ovate; lip very flexuous, having a sinus posteriorly, and a very deep one at the point of the columella; color pale

olive-green. Even the little *M. siliqua* may be only a starved specimen of the Nisqually variety. All have a varix half a volution from
the mouth."

Fig. 422, Lake George specimen.

98. G. silicula, GOULD.

Melania silicula GOULD, Bost. Proc., ii, p. 224, June, 1847. Otia Conchologica, p.
    46. Moll. Expl. Exped., p. 141, f. 164, 164a. COOPER, Report, p. 374. BINNEY,
    Check List, No. 243. BROT, List, p. 52.
Juga silicula, Gould, ADAMS, Genera, i, p. 304.
Melania Shastaensis, LEA, Proc. Acad. Nat. Sci., viii, p. 80, April, 1856. BINNEY,
    Check List, No. 242. COOPER, Report, p. 374.
Goniobasis Shastaensis, LEA, Jour. Acad. Nat. Sci., v, pt. 3, p. 337, t. 38, f. 199,
    March, 1863. Obs., ix, p, 159.
Melania rudens, REEVE, Monog. Melania, sp. 224, May, 1860. BROT, List, p.

Description.—A small, slender, nearly cylindrical species, covered
with a somewhat clouded, dark chestnut epidermis. There are about
four, entire whorls, several others being lost from the tip; they are
Fig. 423. well rounded, and marked with numerous, fine, revolving
threads, and all but the two largest ones are longitudinally
plaited. The aperture is small, rounded-ovate, scarcely produced in front, and about one-fourth the length of the shell.
The throat has a pale violet tint. The last whorl has a dark,
narrow band around it, just at the junction of the lip of it.
Length, one-half; breadth, one-fifth of an inch.

Habitat.— Nisqually, Oregon.

Observations.— It resembles *M. proxima,* Say, which is less cylindrical and without folds.— *Gould.*

Melania silicula.— Shell small, graceful, subcylindrical, truncated;
epidermis chestnut-brown; spire of four remaining whorls, rounded,
spirally lirate, the upper longitudinally plicate; the last whorl banded
with brown; sutures well impressed; aperture roundly ovate, scarcely
produced in front; pale violaceous.

Longitude one-half; latitude, one-fifth poll.

Habitat.— Nisqually, Oregon.— *Gould.*

This species differs much from *plicifera* in being more narrowly cylindrical, the whorls, generally, but not always, more
convex, and especially in the broad band. It is a beautiful
and numerous species, extending to all parts of Oregon and
California. Dr. Gould's description refers to a young shell,
of which *G. Shastaensis,* Lea is the adult. *Melania rudens* of

Reeve is a more rugose variety of the same species. The *M. Shastaensis* of Reeve, sp. 318, is a good figure of *G. occata*, Hinds.

Melania Shastaensis. — Shell striate, subcylindrical, rather thin, dark horn-color, banded; spire elevated, folded at the apex; sutures very much impressed; whorls convex; aperture small, ovate, white within; columella smooth, incurved and recurved.

Operculum ovate, the polar point being near the left side and below the middle.

Habitat.— Shasta and Scott Rivers, California; Dr. Trask: and Fort Umpqua, O. T., Smithsonian Institution.

Diameter, ·34 of an inch; length, 1·05 inches.

Observations.— Nearly thirty specimens of this species were kindly sent to me by Dr. Trask. The form and size of this species is very much the same as *Melania (Goniobasis) Virginica*, Say. It Fig. 424. differs in the form of the aperture, in having but a single, revolving, wide band, and in being more cylindrical. The *Shastaensis* varies like the *Virginica*, in being very uncertain as to striation. Some of the specimens are covered with minute, revolving striæ, while others are almost entirely destitute of them. In every specimen before me there is a broad, revolving, brown band on the middle of the whorls, more or less distinct, and always with more intense color on the superior whorls. This band often becomes obsolete on the inferior whorls, but when that is not the case it may be seen within the aperture also. A few of the specimens have the columella slightly purple. Every specimen in my possession has the apex eroded, so that the number of whorls cannot be with certainty stated. I should suppose the number to be nine or ten. Some of them are sufficiently perfect to show several upper whorls with regular folds. The aperture is probably rather more than one-fourth the length of the

Fig. 425.

shell.— *Lea.*

Melania rudens.— Shell narrowly turriculated, dull olive; whorls rounded, constricted at the sutures, spirally ridged, striated, the first strongly, concentrically plicated; aperture small, rounded.

Habitat.— ——— ?

Observations.— Strongly characterized by the constricted sutures and by the rib-like plications of the earlier whorls.—*Reeve.*

99. G. nigrina, Lea.

Melania nigrina, Lea, Proc. Acad. Nat. Sci., p. 80, April, 1856.
Goniobasis nigrina, Lea, Jour. Acad. Nat. Sci., v, pt. 3, p. 302, t. 37, f. 137, March,
1863. Obs., ix, p. 124. Binney, Check List, No. 177.

Description.— Shell smooth, small, conical, rather thin, nearly black,
polished; spire somewhat elevated; sutures impressed; whorls regularly convex; aperture small, ovate, angular above, dark purple
within; columella incurved, purple.

Operculum dark brown, the polar point being low down and near to
the left margin.

Habitat.— Clear Creek, Shasta County, California; Dr. Trask.

Diameter, ·23; length, ·67 of an inch.

Observations.— A number of good specimens, with their opercula,
were sent to me by Dr. Trask. In form, size and color this species
is very like to *Melania semicarinata*, Say, from Georgia and
South Carolina. It may be distinguished at once by not having the carination of that species, which is usually strongly
marked. It is not quite so high in the spire, and the aperture
is more rounded at the base. In all the specimens of *nigrina*,
which I received, the apex is worn off. In the half grown ones I
can see no disposition to carination or plication in the upper whorls.
I should suppose that in perfect specimens the number of whorls
would be found to be about seven, and that the aperture would be
about the third of the length of the shell. In some of the specimens
there is a disposition to put on a few, fine striæ, and in most of them
there is a very small, angular line running below the suture. I am
not acquainted with Dr. Gould's *Melania silicula* and *bulbosa* from
Oregon, described in the Proc. Boston Soc. Nat. Hist., July, 1847;
but from the descriptions I have no doubt that they are different from
both species therein described.— *Lea.*

The upper whorls of this species are sometimes plicate. The
shell is like *silicula* in form, but is rather more cylindrical, of
a darker color, shaded with red internally. It is particularly
distinguished by the carinated upper whorls.

This is not the *nigrina* of Reeve's Conch. Icon., that species
being the *nigrocincta*, Anth., as Mr. Reeve states in his "errata."

Fig. 426.

100. G. rubiginosa, Lea.

Goniobasis rubiginosa, Lea, Proc. Acad. Nat. Sci., p. 270, 1862. Journ. Acad.
Nat. Sci., v, pt. 3, p. 333, t. 38, f. 193, March, 1863. Obs., ix, p, 155.

Description.— Shell carinate, somewhat awl-shaped, rather thin,
shining, reddish, obscurely banded; spire subattenuate; sutures very
much impressed; whorls about six, convex; aperture very small,
subrhomboidal, pale reddish and obscurely double-banded within;
outer lip acute, sinuous; columella slightly bent in and twisted.

Operculum broadly ovate, dark brown, with the polar point near
the left margin above the base.

Habitat.— Oregon; W. Newcomb, M.D.

Diameter, ·27; length, ·74 of an inch.

Observations.— Two specimens only were sent to me by Dr. W.
Newcomb. The four upper whorls are carinate, and a Fig. 427. Fig. 428.
small, thread-like line below runs parallel with the more
raised one. The two obscure bands are near to each
other and are in the middle of the whorl. In outline
it is near to *Melania* (*Goniobasis*) *nigrina* (nobis), but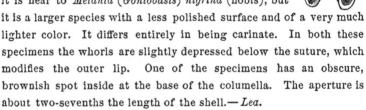
it is a larger species with a less polished surface and of a very much
lighter color. It differs entirely in being carinate. In both these
specimens the whorls are slightly depressed below the suture, which
modifies the outer lip. One of the specimens has an obscure,
brownish spot inside at the base of the columella. The aperture is
about two-sevenths the length of the shell.— *Lea.*

Mr. Lea's figure, of which the accompanying one (Fig. 427)
is a copy, does not exhibit plicæ on the spire, nor does his
description mention their existence, still I am convinced that
when specimens with more perfect spires are discovered, they
will, in common with the other lengthened species, exhibit this
character. Except in the character of the carinated upper
whorls this shell is allied to *Shastaensis.*

101. G. Bairdiana, Lea.

Goniobasis Bairdiana, Lea, Proc. Acad. Nat. Sci., p. 267, 1862. Jour. Acad. Nat.
Sci., v, pt. 3, p. 317, t. 37, f. 164, March, 1863. Obs., ix, p. 139, t. 37, f. 164.

Description.— Shell folded, somewhat drawn out, dark brown, rather
thick, single-banded; whorls subattenuate, sharp-pointed; sutures im-

pressed; whorls eight, slightly convex; aperture rather small, ovately rhomboidal, whitish within and single-banded; outer lip scarcely sinuous; columella bent in, somewhat thickened and very much twisted.

Habitat.— Columbia River at Fort George, Oregon; J. Drayton.

Diameter, ·26; length, ·66 of an inch.

Observations.— In size, color and outline this is nearly allied to *Draytonii*, herein described, but may at once be distinguished by that Fig. 429. species having no folds, and in being more convex in the

whorls. It cannot be confounded with *Melania* (*Goniobasis*) *Newberryii* (nobis), which is shorter, more inflated, and has two bands. The *Bairdiana* has five or six apical whorls, furnished with close, regular, well formed, perpendicular folds. The lower whorls have two or three very minute, revolving striæ immediately below the suture, where the color is lighter. There is a disposition to thickening on the inner margin of the outer lip, and along this edge a little coloring of brown is observable. The aperture is nearly the third of the length of the shell. I have great pleasure in dedicating this interesting little species to my friend, Prof. Spencer F. Baird of the Smithsonian Institution, to whom I am greatly indebted for many kind services, and who has done so much for the advancement of the Natural Sciences of our country.— *Lea.*

This species differs very much in form from the others inhabiting the Pacific States.

D.   *Shell angulate.*

102. G. trochiformis, CONRAD.

Melania trochiformis, CONRAD, New Fresh-Water Shells, p. 56, t. 8, f. 11, 1834. DEKAY, Moll., New York, p. 100. WHEATLEY, Cat. Shells, U. S., p. 27. BINNEY, Check List, No. 275. BROT, List, p. 31. MÜLLER, Synopsis, p. 47.

Description.— Shell short, conical, ventricose, turreted;   Fig. 430. two spiral, prominent lines on each whorl, the intervening spaces concave; summit of the whorls flattened, angulated; body-whorl angular, with the periphery carinated; base flattened; aperture small; labrum angulated in the middle.

Habitat.— Streams in North Alabama.

Observations.— A species easily recognized by its strong ribs, or by its sulci, and its trochiform shape.— *Conrad.*

The figure is a copy of that in Mr. Conrad's work. It is evidently a very poor one, however. It is probable this will prove to be identical with Mr. Anthony's *T. cristata.*

103. G. cristata, ANTHONY.

Melania cristata, ANTHONY, Ann. Lyc. Nat. Hist. N. Y., vi, p. 108, t. 3, f. 8, March, 1854. BINNEY, Check List, No. 77. BROT, List, p. 32. REEVE, Monog. Melania, sp. 413.

Description.— Shell carinate on the body-whorl, rhomboidal; thin, horn-colored; upper whorls not carinate, but somewhat shouldered; whorls five, flat, slightly concave, diminishing rapidly to the apex; sutures not remarkable; body-whorl with a strong, well developed carina, extending from the upper part of the aperture, and revolving round so as to be at its centre when it reaches the mouth again. The carina and a smaller one below it are indicated in the interior by a grooved channel with a dark band running through it; aperture rhomboidal, banded within; columella straight, with an acute sinus at base.

Fig- 431.   Fig. 432.

Fig. 433.    Fig. 434.

Habitat.— Alabama.

Diameter, ·34 (9 millim.); length, ·50 of an inch (13 millim.). Length of aperture, ·30 (8 millim.); breadth of aperture, ·16 of an inch (4 millim.)

Observations.— Only one specimen of this remarkable species has come under my notice, but it is so widely different from all others that no one can for a moment doubt its distinctive character. The upper whorls are obscurely banded near the suture.— *Anthony.*

Fig. 434 is from the type specimen. It is not an adult, and is also a malformation. The succeeding figures represent different varieties and ages. The carination appears to be lost in an obscure angle on the periphery of the adult shell.

104. G. cruda, LEA.

Goniobasis cruda, LEA, Proc. Acad. Nat. Sci., p. 270, 1862. Jour. Acad. Nat. Sci.,
    v, pt. 3, p. 332, t. 38, f. 190, March, 1863. Obs., ix, p. 151.

Description.— Shell carinate, subfusiform, rather thin, shining, dark
brown, obscurely banded; spire obtuse; sutures slightly impressed;
whorls flattened above, the last one large; aperture rather large,
rhomboidal, dark within; outer lip acute, scarcely sinuous; columella slightly incurved, scarcely thickened.

Habitat.—Tennessee River; Dr. Spillman.

Diameter, ·38; length, ·68 of an inch.

Observations.— Only two specimens were received from Dr. Spillman, both much worn at the apex. Two of the lower whorls
only are perfect. The bands on both are imperfect and obscure.
They may be considered to be three, one being on the periphery of the whorl. One is much darker in the interior than
the other, and has a dark purple mark at the base of the columella. It has very much the form of *Melania* (*Goniobasis*) *perfusca*
(nobis), but differs in size, in aperture and in carination. The character of the upper whorls cannot be ascertained by these specimens,
nor the proportion of the aperture, but it must be nearly one-half the
length of the shell.— *Lea.*

Fig. 435.

105. G. Whitei, LEA.

Goniobasis Whitei, LEA, Proc. Acad. Nat. Sci., p. 266, 1862. Jour. Acad. Nat. Sci.,
    v, pt. 3, p. 310, t. 37, f. 151, March, 1863. Obs., ix, p. 132.

Description.— Shell smooth, fusiform, thick, very much inflated,
yellowish-brown, bright, three-banded; spire very obtuse, sutures
somewhat impressed; whorls five, flattened above, the last being
ventricose; aperture very large, widely rhomboidal; outer lip acute,
straight; columella bent in, thickened and twisted.

Habitat.— Georgia; Rev. George White.

Diameter, ·35; length, ·61 of an inch.

Observations.— Two specimens were received among Mr. White's
shells, but the part of Georgia was not designated from where he
obtained them, probably towards the north. In outline it closely
resembles *Nickliniana*, as well as *Vauxiana*, herein described. It is
rather more obtuse in the apex than *Nickliniana*, and not so round

at the base, and it has bands which the other has not. Both the specimens are furnished with three, equidistant, brown bands, Fig. 436. obscure outside, but well defined inside. The older of these two has a thickening inside of the outer lip, and the bands do not extend to the margin. The aperture is more than the half the length of the shell. I dedicate this species to the Rev. George White, who has done so much to elucidate a knowledge of the mollusca of his State. — *Lea.*

The figure copied does not represent the three bands referred to ; but they are present on all the specimens before me.

106. G. expansa, LEA.

Melania expansa, LEA, Trans. Am. Philos. Soc., ix. p. 28.

Description.— Shell smooth, somewhat fusiform, rather thick, yellowish; spire obtusely conical; sutures somewhat impressed; whorls five, slightly convex; aperture large, expanded, whitish.

Habitat.— Alabama.

Diameter, ·43; length, ·63 of an inch.

Observations.— A solitary specimen of this was among the shells sent by Dr. Foreman. In form it resembles *M. variabilis* (nobis), but may be distinguished from that species in being larger, and having a larger proportionate aperture, which is more expanded. The aperture is full one-half the length of the shell. The specimen under examination has four bands, and the yellow epidermis is nearly covered with a deposit of the oxide of iron.— *Lea.*

This shell has not been figured. The species is unknown to me.

107. G. casta, ANTHONY.

Melania casta, ANTHONY, Ann. N. Y. Lyc. Nat. Hist., vi, p. 100. t. 2, f. 19, March, 1854. BINNEY, Check List, No. 50. BROT, List, p. 32. REEVE, Monog. Melania, sp. 381.

Description.— Shell conical, nearly smooth, thick; spire obtusely elevated; whorls 6–7, nearly flat; sutures well impressed; upper whorls smooth, or only modified by the lines of growth, which are coarse and distinct; body-whorl with five prominent striæ below the middle, of which the lower three also revolve within the aperture on the columella; aperture small, elliptical, within whitish, subnacreous; columella not indented; sinus small.

Habitat.— Alabama.

Fig. 437.   Fig. 438.

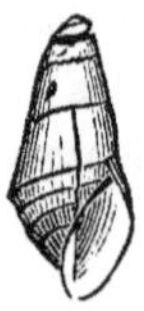

Diameter, ·30 (8 millim.); length, ·75 of an inch (19 millim.). Length of aperture, ·33 (8 millim); breadth of aperture, ·17 of an inch (4 millim.).

Observations.—A singularly pale, greenish-white species, the distinguishing marks of which are its regular, subcylindric form, and the smooth spire, combined with the prominent striæ at the base of the shell. These are characters which I do not recognize on any other species so combined. There is also a distinct carina on the penultimate whorl, near the top of the aperture, above which may be observed a faint, interrupted line.— *Anthony.*

Another specimen in Mr. Anthony's collection has not the angulation so well developed and is covered with slight striæ. The type specimen is figured, figure 438.

108. G. rhombica, ANTHONY.

Melania rhombica, ANTHONY, Ann. N. Y. Lyc. Nat. Hist., vi, p. 116, t. 3, f. 16, March, 1854. BINNEY, Check List, No. 228. BROT, List, p. 38. REEVE, Monog. Melania, sp. 347.

Description.—Shell conic, rather thin, brown; spire regularly pyramidal; not elevated; whorls about six, flat, regularly and very distinctly striate; body-whorl angulated about the middle, nearly smooth, except as modified by the lines of growth, which are quite distinct, the concentric striæ being nearly obsolete on the body-whorl; sutures inconspicuous; aperture rather large, ovate, whitish within; columella very slightly rounded, with little or no sinus.

Fig. 439.

Habitat.— Alabama.

Diameter, ·22 (5½ millim.); length, ·43 of an inch (11 millim.). Length of aperture, ·20 (5 millim.); breadth of aperture, ·12 of an inch (3 millim.).

Observations.—This cannot well be confounded with any known species; its short spire, flat, striated whorls, regularly and rapidly decreasing to the apex, the prominent, acute carina, which encircles it near the top of the aperture, beneath which the striæ, so prominent above are hardly discernible, and its rather broad form, will readily distinguish it from *M. striatula*, Lea, to which it might seem allied by form and color; it has somewhat the form of *M. vicina* (nobis), but that shell is more slender, less distinctly carinated, and has not the striation of the present species.—*Anthony*.

A very distinct and not uncommon species, remarkably uniform in form and ornamentation. One of Mr. Anthony's types is figured. In younger specimens the striæ are more strongly developed.

109. G. angulata, ANTHONY.

Melania angulata, ANTHONY, Ann. N. Y. Lyc. Nat. Hist., vi, p. 117, t. 3, f. 17, March, 1854. BINNEY, Check List, No. 14. BROT, List, p. 37. REEVE, Monog. Melania, sp. 386.
Melania cinnamomea, Anthony, REEVE, Monog. Melania, sp. 379. BROT, List, p. 35.
Goniobasis intercedens, LEA, Proc. Acad. Nat. Sci., p. 265, 1862. Journ. Acad. Nat. Sci., v, pt 3, p. 305, t. 37, f. 143. Obs., ix, p. 127.

Description.— Shell acutely conic, smooth, brown, rather thick; spire not remarkably elevated, but tapering regularly with a rectilinear outline to the apex, which is entire and acute; whorls eight, nearly flat, upper ones carinate, and with a well defined suture; body-whorl with a distinct angle, more distinct where it revolves near the top of the aperture; below this the base is rather concave on the columella side; aperture moderate, narrow, ovate, whitish or faintly tinged with red within; columella slightly curved, not indented; sinus slight, but well defined.

Fig. 440.

Habitat. — Tennessee.

Diameter, ·25 (6 millim.); length, ·56 of an inch (14 millim.). Length of aperture, ·25 (6 millim.); breadth of aperture, ·13 of an inch (3 millim.).

Observations.— A singularly neat, precise looking shell. Its trim appearance, its pale color, unornamented by any band, and its sharp, well defined angle, amounting almost to a carina, will serve to distinguish it from all others.— *Anthony*.

The above description is that of the juvenile shell. In the adult state it has been described by both Mr. Anthony and Mr. Lea as follows:

Melania cinnamomea.— Shell ovately conoid, cinnamon-brown, with a narrow, chestnut zone at the sutures; whorls 6-7, slopingly ven-

tricose, longitudinally wrinkle striated, last whorl irregularly transversely wrinkled; aperture ovate, effused at the base.

Fig. 441.

Habitat.— Alabama.

Observations— An obese, cinnamon-colored shell, encircled by a narrow, chestnut band at the sutures. The surface is sculptured with longitudinal, close-set striæ and
transverse, interrupted, keel-like wrinkles.— *Reeve.*

Goniobasis intercedens. — Shell smooth, fusiform, rather
thin, yellow, honey-bright without bands; spire conoidal, sharp-
pointed, carinate at the apex; sutures linear; whorls eight, flattened,
varicose; aperture rather large, rhomboidal, whitish within; outer lip
acute, scarcely sinuous; columella slightly bent in, somewhat thickened, nearly straight below.

Habitat.— Cahawba River, Alabama; E. R. Showalter, M.D.

Diameter, ·30; length, ·69 of an inch.

Observations.— This species is very closely allied to *Melania* (*Goniobasis*) *mellea* and *Bridgesiana*, herein described. It is the
same color, but may be distinguished by its being more slender and having a higher spire. It has also a less twisted
columella. In the interior there is a slight disposition to
yellowness. Neither of the two specimens before have any
appearance of bands. The larger of the two is not complete
on the outer lip, but the smaller one is perfectly so, and shows a disposition to thickening on the inner edge. The aperture is about one-
half of the length of the shell.— *Lea.*

Fig. 442.

110. G. Bridgesiana, LEA.

Goniobasis Bridgesiana, LEA, Proc. Acad. Nat. Sci., p. 265, 1862. Jour. Acad. Nat.
Sci., v, pt. 3, p. 305, t. 37, f. 142, March, 1863. Obs., ix, p. 173, t. 37, f. 142.

Description— Shell smooth, fusiform, somewhat inflated, rather thin,

Fig. 443.

honey-yellow, without bands; spire obtusely conical, carinate
at the apex; sutures linear; whorls about seven, flattened;
aperture large, subrhomboidal, whitish within; outer lip
acute, scarcely sinuous; columella somewhat bent in, thickened above and below and slightly twisted.

Habitat.— Cahawba River, Alabama; E. R. Showalter, M.D.

Diameter, ·40; length, ·83 of an inch.

Observations.— A single specimen only was received from Dr. E. R.
Showalter. It was considered by him to be *Melania gravida*, Anth.,

but it does not answer to his description. It is allied to *Melania
(Goniobasis) mellea* (nobis), but differs in being more regularly fusiform, in not being so much inflated, nor having so sharp an apex, and
the whorls are flatter. The interior of this specimen is slightly disposed to yellowness. There is no appearance of bands on this specimen, and I doubt if it will be found banded. The aperture is nearly
one-half the length of the shell. I dedicate this species to my friend,
R. Bridges, M.D., who has done so much to promote the knowledge
of our zoology.— *Lea.*

I doubt whether this is more than an adult form of *angulata*,
Anth.

111. G. cubicoides, ANTHONY.

Melania cubicoides, ANTHONY, Proc. Acad. Nat. Sci., p. 60, Feb., 1860. BINNEY,
Check List, No. 78. BROT, List, p. 39. REEVE, Monog. Melania, sp. 445.

Description.— Shell ovate, smooth, thick; whorls 6–7, flat, the upper
ones rapidly enlarging to the body-whorl, which is broad and acutely
angulated; sutures distinct, rendered more so by a sharp carination
on the lower part of each whorl; aperture broadly ovate, Fig. 444.
within whitish; columella deeply indented; sinus small.

Habitat.— Wabash River, Indiana.

Observations.—One of the short, thick species, in form not
unlike *M. cuspidata* (nobis), but differing by its sharp, carinated body-whorl and imbricated spire; the body-whorl is
also strongly striate and obscurely ribbed; these longitudinal ribs are
very faint, but sufficiently distinct at the sharp carina, near the
summit of the aperture to modify its outline into a waving, subnodulous line.— *Anthony.*

Figured from Mr. Anthony's type. The longitudinal ribs
alluded to by Mr. Anthony are very indistinct in his type
specimen, and do not exist in other specimens; both old and
young, before me.

112. G. Spillmanii, LEA.

Goniobasis Spillmanii, LEA. Proc. Acad. Nat. Sci., p. 264, 1862. Journ. Acad. Nat.
S i., v, pt. 3. p. 302, t. 37, f. 198, March, 1863. Obs. ix, p. 134.

Description — Shell smooth, fusiform, thin, greenish horn-color,
shining, without bands; spire obtusely conical; sutures linear;

whorls about six, flattened, somewhat impressed below the sutures; aperture large, rhomboidal, diaphanous˙ within; outer lip acute, slightly sinuous; columella slightly bent in and thin.

Fig. 445.

Habitat.— Tennessee River; W. Spillman, M.D.

Diameter, ·39; length, ·94 of an inch.

Observations.— Only three specimens were received from Dr. Spillman, two of which are little more than half grown. In outline it is near to *Melania* (*Goniobasis*) *gracilis* (nobis), but it is more *fusiform*, rather larger and not so thick. The color is very nearly the same. There is a slight disposition to angulation on the periphery of the whorls. The aperture is about four-tenths the length of the shell. I dedicate this species to Dr. Spillman, who has done so much to elucidate the natural history of the Southern States.— *Lea.*

113. G. pallidula, ANTHONY.

Melania pallidula, ANTHONY, Ann. N. Y. Lyc. Nat. Hist., vi, p. 115, t. 3, f. 15, March, 1854. BINNEY, Check List, No. 197. BROT, List, p. 38. REEVE, Monog. Melania, sp. 417.

Description.— Shell elongate-ovate, smooth, moderately thick; of a pale, horn-color, with a faint, brown, narrow band on the penult whorl, increased to two on the body-whorl, and obsolete on the apical ones; spire obtusely elevated, with a rather convex outline and a well defined suture; whorls four remaining, with indications of two more lost by truncation; body-whorl angulate, and rather coarsely striate longitudinally; aperture rather large, ovate, pale within, ornamented with the two bands of the body-whorl, which do not reach the outer edge, a broad, plain area intervening; columella curved, with a very slight sinus at base.

Fig. 446.

Habitat.— Tennessee.

Diameter, ·25 (6 millim.); length, ·50 of an inch (12 millim.). Length of aperture, ·27 (7 millim.); breadth of aperture, ·15 of an inch (4 millim.).

Observations.— This is a very neat, pretty species, whose affinity with any other is not so strong as to endanger its being easily confounded; from *M. angulata* (nobis) it differs in being broader, less angulated, paler in color, less elongated, and by its brown bands, that species being entirely plain.— *Anthony.*

114. G. vicina, ANTHONY.

Melania vicina, ANTHONY, Ann. N. Y. Lyc. Nat. Hist., vi, p. 114, t. 3, f. 14, March, 1854. BINNEY, Check List, No. 288. BROT, List, p. 39. REEVE, Monog. Melania, sp. 291.

Description.—Shell conical, smooth, rather thick, yellowish-brown; spire short; whorls six, upper ones subconvex, with a brown Fig. 447. band immediately above the suture; body-whorl a little shouldered beneath the suture, and angulated in the middle, surrounded by two narrow bands, one above and the other below the angle; sutures impressed; aperture ovate, banded within; columella much curved, with hardly a perceptible sinus at base.

Habitat.— Alabama.

Diameter, ·21 (5 millim.); length, ·45 of an inch (11 millim.). Length of aperture, ·20 (5 millim.); breadth of aperture, ·12 of an inch (3 millim.).

Observations.— A small, not inelegant species, which may be compared with *M. ovoidea*, Lea, and *M. depygis*, Say, as its nearest congeners. The former species I have never seen, but judging from the description this differs from it in many particulars; its form is proportionately broader, the bands are more distinct; the body-whorl has a distinct angle, which is also apparent on the penultimate whorl, amounting there to a carination. The aperture also is much smaller. The same particulars apply with equal force to *Melania depygis*, Say, the two being so nearly alike in description that the *M. ovoidea* may prove to be only a variety of Mr. Say's *depygis.—Anthony.*

Except in the striæ not being present, the shell resembles *G. rhombica*, Anth. All the specimens before me are labelled "Kentucky" by Mr. Anthony.

115. G. Spartenburgensis, LEA.

Goniobasis Spartenburgensis, LEA, Proc. Acad. Nat. Sci., p. 205, 1862. Jour. Acad. Nat. Sci., 2d ser., v, pt. 3, p. 307, t. 37, f. 147, March, 1863. Obs. ix, p. 129.

Description.— Shell smooth, fusiform, rather thin, greenish horn-color, bright, banded or without bands; spire acutely conical, carinate at the apex; sutures impressed; whorls eight, flattened, aperture rather large, elongately rhomboidal, white within; outer lip acute, scarcely sinuous; columella slightly bent in, thickened below.

Operculum ovate, thin, brown, with the polar point near to the base on the left margin.

Habitat.—Spartanburg District, South Carolina; Prof. L. Vanuxem: Marietta, Ohio; Dr. Hildreth: Wabash River, Ind.; H. C. Grosvenor.

Diameter, ·23; length, ·54 of an inch.

Observations.—I have seven specimens from Spartanburg, seven from Marietta and two from the Wabash. This small, graceful spe-

Fig. 448.

cies has a wide, geographical distribution. I can see very little difference between the specimens of the different habitats. The two from the Wabash are very much smaller and thinner, and may be much younger, but they differ in having a purplish columella which is not observable in the others. One of them has a remarkable row of brown spots under the sutures on the body-whorl. The other is without spots or bands. Usually this species has two bands; six of the seven from Marietta are two-banded. Of the seven from Spartanburg two only are double-banded. The others are without bands. The species is very nearly allied to *Melania* (*Goniobasis*) *ovoidea* (nobis), but it is more elongate and the aperture is less effuse. The aperture is not quite half the length of the shell.—*Lea.*

I fear the specimens mentioned as from Marietta, Ohio, and Wabash River, Ind., are not distinct from *depygis*, Say.

116. G. Gerhardtii, LEA.

Goniobasis Gerhardtii, LEA, Proc. Acad. Nat. Sci., p. 270, 1862. Journ. Acad. Nat. Sci., v, pt. 3, p. 330, t. 38, f. 187, March, 1863. Obs., ix, p. 152.
Goniobasis infuscata, LEA, Proc. Acad. Nat. Sci., p. 270, 1862. Journ. Acad. Nat. Sci., v, pt. 3, p. 330, t. 38, f. 188, March, 1863. Obs. ix, p. 152.

Description.—Shell carinate, fusiform, thin, shining, yellowish-green, four-banded; spire regularly conical; aperture small, rhomboidal, whitish and banded within; outer lip acute, slightly sinuous; columella bent in, slightly thickened below.

Operculum ovate, thin, dark brown, with the polar point on the left above the base.

Habitat.—Chattanooga River, Georgia; Alexander Gerhardt: Coosa River, Alabama; Dr. Spillman.

Diameter, ·36; length, ·72 of an inch.

Observations.—From the two habitats I have a number of specimens, nearly all of which are young. The largest, one of which will be figured, were from the Smithsonian Institution, kindly sent to me by Prof. Henry, the Secretary, having been received from Mr. Gerhardt. Those from Dr. Spillman were smaller, and gen-

Fig. 449.

erally much darker. It is a beautiful, regular and graceful species. The young are very acutely angular, having on the periphery a very dark, raised line. There are four bands which are remarkably uniform, being nearly the same in every specimen. The two middle ones are close together, the upper of the two being the larger. The upper one is near to the suture above; the lower one is broad and near the base. At the base of the columella the area is usually quite yellow. A few young ones from the Coosa are without bands. In the number and position of the bands we are reminded of *Melania* (*Goniobasis*) *suavis* (nobis) and *Melania* (*Goniobasis*) *grata*, Anth., but this is a much thinner and a carinate species. The aperture is about half the length of the shell. I name this after Mr. Alexander Gerhardt, who has done much to elucidate the zoology of his district in North Georgia.— *Lea.*

The following is the description of the adult form of this species :—

Goniobasis infuscata.—Shell carinate, fusiform, rather thin, shining, dark, nearly black, three-banded; spire conical, sutures impressed; **Fig. 450.** whorls about six, flattened above, the last one large; aperture rather large, rhomboidal, whitish or brown, and three-banded within; outer lip acute, slightly sinuous; columella bent in, slightly thickened below.

Habitat.— Georgia; Rev. G. White : Coosa River, Alabama; Dr. Spillman.

Diameter, ·37 ; length, ·82 of an inch.

Observations.— A single specimen only from each of the habitats was received. That from Mr. White is the larger and is not so dark, the epidermis being olive-brown, and the interior being whitish with the three bands well defined. That from Dr. Spillman is of so dark a brown that it has the appearance of being entirely black, but in the inside, the three bands may be distinguished, but the exterior is totally and intensely dark. In outline it is nearly the same with *Gerhardtii*, herein described, but differs in the number and character of the bands. The aperture is not quite half the length of the shell.— *Lea.*

E.   *Whorls very strongly carinated.*

117. G. acutocarinata, Lea.

Melania acutocarinata, Lea, Philos. Proc., ii, p. 14, Feb., 1841. Philos. Trans., viii,
   p. 184, t. 6, f. 46. Obs. iii, p. 22. DeKay, Moll. N. Y., p. 99. Troost, Cat.
   Shells, Tenn. Wheatley, Cat. Shells, U. S., p. 24. Binney, Check List, No. 5.
   Catlow, Conch, Nomenc., p. 185. Brot, List, p. 36.
Elimia acutocarinata, Lea, Chenu, Manuel de Conchyl., i, f. 1979. Adams, Genera,
   i, p. 300.
Melania pagodiformis, Anthony, Ann. N. Y. Lyc. Nat. Hist., vi, p. 106, t. 3, f. 6,
   March, 1854. Binney, Check List, No. 195. Brot, List, p. 36. Reeve, Monog.
   Melania, sp. 260.
Melania torulosa, Anthony, Ann. N. Y. Lyc. Nat. Hist., vi, p. 110, t. 3, f. 10, March
   1854. Binney, Check List, No. 273. Brot, List, p. 37. Reeve, Monog. Mela-
   nia, sp. 370.

Description.— Shell carinate, conical, rather thick, shining, dark
brown; spire obtusely elevated; sutures impressed; whorls six; ap-
Fig. 451. Fig. 452. erture rather large, elliptical, angular at base, purplish

within.

Habitat.— Tennessee.

Diameter, ·30; length, ·66 of an inch.

Observations.— I received a single specimen only of
this species. It seems to be distinct in its large carina, which extends over all the whorls, but is scarcely distinct on the last. The
columella is remarkably indented. The aperture is nearly one-half
the length of the shell.— *Lea.*

This shell is believed by Prof. Haldeman to be a variety
of *simplex*, but I doubt if they are the same, as this species is
acutely carinate in some specimens, smooth in others, but as
it appears to me *always* narrowly lengthened.

The following is the description of:—

Melania pagodiformis.— Shell conical, thin, brownish-olive; spire
obtusely elevated; whorls 7–8, smooth; the upper ones are Fig. 453.
surrounded by a sharp, elevated keel just above the suture;
the body-whorl is angulated in the middle by two keels, of
which the upper is the more prominent; sutures deeply impressed; aperture ovate, ending in an acute angle below,
whitish within; columella rounded, produced into a narrow, but
slight sinus.

Habitat.— Battle Creek, Tennessee.

Diameter, ·28 (7 millim.); length, ·50 of an inch (13 millim.). Length of aperture, ·26 (7 millim.); breadth of aperture, ·14 of an inch (3½ millim.).

Observations.—Bears some resemblance to *M. acuto-carinata*, Lea, but differs from it in many particulars. It is of a much lighter color, has the carina on every whorl, the body-whorl not excepted, its columella is not remarkably indented as in that species, and it is altogether a thinner and broader shell. The aperture is generally uncolored, but some specimens present a faint tinge of violet there.— *Anthony.*

M. torulosa, Anth., is only a variety of the above, a number of specimens before me exhibiting every gradation between the two species. The following is the description:—

Melania torulosa.—Shell conic, chestnut-colored, rather thick; spire little elevated, acute; whorls 7–8, strongly carinated a little above the suture; sutures linear; aperture not large, broad, ovate, purplish within; columella regularly but not remarkably curved, with a small sinus.

Habitat.—Tennessee.

Diameter, ·28 (7 millim.); length, ·58 of an inch (15 millim.). Length of aperture, ·23 (6 millim): breadth of aperture, ·15 of an inch (4 millim.).

Observations.—But a single specimen of this species is before me, but it differs so much from all others that I cannot hesitate to place it among well established species. *M. acuto-carinata*, Lea, is the only one with which it may be compared, but that species has the carina obsolete on the body-whorl, the very point where it is most remarkably developed in this; the whorls also in the *M. torulosa* diminish much more rapidly to an acute apex, which in *M. acuto-carinata* is said to be *obtusely* elevated; the *M. torulosa* is remarkable for its acute elevation from the broad base of the carina on the body-whorl. In the columella too of the present species there is no indentation, while in *M. acuto-carinata* it is "remarkably indented."—*Anthony.*

Fig. 454.

F. *Body whorl by-multiangulated.*

118. G. tabulata, ANTHONY.

Melania tabulata, ANTHONY, Ann. N. Y. Lyc. Nat. Hist., vi, p. 118, t. 3, f. 18, March, 1854. BINNEY, Check List, No. 262. BROT, List, p. 39.

Description.— Shell ovate-conic, smooth, thin, of a dark brown color externally; spire not remarkably elevated, with a rather con-cave outline; whorls about five, upper ones convex, penult whorl flat, body-whorl subangulated into several planes, with a distinctly

Fig. 455. Fig. 455a.   impressed suture; aperture rather large, ovate,

within of a beautiful, reddish-purple; columella slightly curved, indented, and with a narrow, re-curved sinus at base.

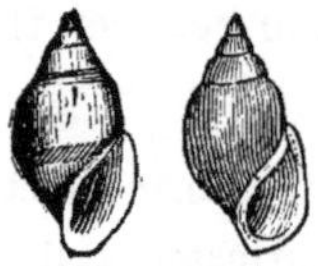

Habitat.— Tennessee.

Diameter, ·34 (8½ millim.); length, ·62 of an inch (16 millim.). Length of aperture, ·31 (8 millim.); breadth of aper-ture, ·17 of an inch (4 millim.).

Observations.— I know of no species with which this is liable to be confounded; its ample body-whorl, the broad, angular, and shelving shoulder on the body and penult whorls, while the upper ones are wanting in this character, and above all the tabulation of the penult whorl are its most striking characteristics, and will at once distin-guish it from all other species; the lines of growth are rather coarse, curved and approximate.— *Anthony.*

119. G. pulcherrima, ANTHONY.

Melania pulcherrima, ANTHONY, Proc. Acad. Nat. Sci., p. 58, Feb., 1860. BINNEY Check List, No. 222. BROT, List, p. 37. REEVE, Monog. Melania, sp. 336.

Description.— Shell conical, carinate, elevated, acute; whorls 6-8, flat, upper ones obscurely ribbed, longitudinal; body-whorl sharply angulated, with a dark brown band directly upon the carina, and two or three below it, of which one is very near the carina; upper whorls with two bands each, widely separated; sutures distinct, rendered more so by the neighboring carina; aperture ovate, within three or four banded; columella rounded and indented; sinus small.

Habitat.— North Carolina.

Observations. — A small, but remarkably beautiful species; its bright yellow ground, and conspicuous, dark lines give, by contrast, a lively and pleasant character to the shell. Compared with *M. nigrocincta* (nobis) it is a larger species, its colors are more Fig. 456. decided, and its carina is also a prominent mark of difference. *M. clara* (nobis) is a larger and more globose species its bands are broader, and it has no carina. It seems to be an abundant species, varying occasionally in some of its characters, but always easily recognized. More than one hundred specimens are before me.— *Anthony*.

120. G. subangulata, ANTHONY.

Melania subangulata, ANTHONY. Ann. N. Y. Lyc. Nat. Hist., vi, p. 91, t. 2, f. 9, March, 1854. BINNEY, Check List, No. 252. BROT, List, p. 37. REEVE, Monog. Melania, sp. 242.

Description. — Shell conical, smooth, rather thick; spire obtusely elevated; whorls about six, convex, subangulated below the middle, brown banded; sutures deeply impressed, and situated in a deep furrow formed by the inclination of two whorls towards each other at that part; lower band below the angulation, upper one midway between it and the suture above; body less angulated, with about six, reddish-brown bands, the upper and lower of which are distinct and distant, the central ones confluent, more distinct in the interior, Fig. 456a. aperture small, long-ovate, within reddish and banded; columella regularly curved, purplish, no sinus at base.

Habitat.—Alabama.

Diameter, ·30 (7½ millim.); length, ·62 of an inch (17 millim.) Length of aperture, ·30 (7½ millim.); breadth of aperture, ·17 of an inch (4 millim.).

Observations.— Somewhat allied to *M. rufescens*, Lea in general form, but that species has regularly, convex whorls and no bands, and has at least two more whorls. The number of whorls in this species cannot, however, with certainty be determined, since in all my specimens, seventy or eighty in number, every one is decollate, but the form does not indicate the loss of more than two whorls at most, and only four are present. *M. rufescens* is described as having eight. A few of the specimens are irregularly and strongly striate on the body-whorl.— *Anthony*

121. G. paula, LEA.

Melania paula, LEA, Proc. Acad. Nat. Sci., p. 122, 1861. Jour. Acad. Nat. Sci., v,
pt. 3, p. 244, t. 35, f. 48, March, 1863. Obs., ix, p. 66.

Description.—Shell carinate, conical, thin, diaphanous, reddish horn-
color; spire subelevated; sutures slightly impressed; whorls six,
acutely carinate above, the last subcarinate; whorls rather small,
widely elliptical, whitish within; outer lip acute; columella either
whitish or reddish, obtusely angular at the base.

Habitat.— Cahawba River, Alabama; E. R. Showalter, M.D.

Diameter, ·27; length, ·66 of an inch.

Observations.—A very small species, about two-thirds of an inch
long. Four specimens are before me, nearly all of the same size and
Fig. 457. color. This species is very closely allied to *Melania* (*Goniobasis*) *bicincta,* Anth., but it is not much more than half the
size, and the carina below that on the middle of the whorl is
more indistinct. In the aperture they also differ, the *bicincta*
having it larger and more disposed to be rhombic, and having
indistinct bands within, which this has not. In all the specimens the
carina is sharp. The aperture is about two-fifths the length of the
shell. It reminds one also of *Melania* (*Goniobasis*) *rhombica,* Anth.,
being about the same length, but that species has a single sharp
carina, with a less exserted spire and a larger mouth.— *Lea.*

Differs from *vittata,* Anth., in the more rounded aperture
and outer lip.

122. G. symmetrica, HALDEMAN.

Melania symmetrica, HALDEMAN, Monog. Lim., No. 4, p. 2 of cover, October 5, 1841.
    BINNEY, Check List, No. 261. JAY, Cat. 4th ed., p. 275. BROT, List, p. 35.
    REEVE, Monog. Melania, sp. 328.
Ceriphasia symmetrica, Haldeman, ADAMS, Genera, i, p. 297.
Melania imbricata, ANTHONY, Ann. N. Y. Lyc. Nat. Hist., vi, p. 105, t. 3, f. 5, March,
    1854. BINNEY, Check List, No. 142. BROT, List, p. 36. REEVE, Monog. Mela-
    nia, sp. 259.
Melania bicincta, ANTHONY, Proc. Acad. Nat. Sci., p. 56, Feb., 1860. BINNEY,
    Check List, No. 31. BROT, List, p. 36. REEVE, Monog. Melania, sp. 327.
Melania assimilis, ANTHONY, Proc. Acad. Nat. Sci., p. 60, Feb., 1860. BROT, List,
    p. 36. REEVE, Monog. Melania, sp. 464.
Melania assimilis, Lea (mistake), BINNEY, Check List, No. 22.
Goniobasis Ucheénsis, LEA, Proc. Acad. Nat. Sci., p. 270, 1862. Journ. Acad. Nat.
    Sci., v, pt. 3, p. 334, t. 38, f. 194, March, 1863. Obs. ix, p. 156.
Goniobasis Barrattii, LEA, Proc. Acad. Nat. Sci., p. 271, 1862. Journ. Acad. Nat.
    Sci., v, pt. 3, p. 335, t. 38, f. 196, March, 1863. Obs. 9, p. 57.
Goniobasis Catabœa, HALDEMAN, Amer. Jour. Conch., vol. 1, No. 1, Feb. 25, 1865.

Description.— Shell olivaceous, turreted, with eight or nine convex whorls, separated by a deep suture; apex carinated anterior to the middle of the whorls; aperture ovate.

Habitat.— Roanoke River, Virginia.

Length, ¾ of an inch.

Observations.— Less ponderous than the preceding species, *M. uncialis*, and distinguishd from *M. Virginica* by the carinated apex.— *Haldeman.*

Fig. 458.

This is a variable species inhabiting from Virginia to Georgia, Alabama and Tennessee. In some localities the carinæ of the body-whorl are better developed, and the color differs from light to dark brown, which has caused the species ·to be described several times. The largest *symmetrica* I have seen attains to over one inch.

The following is the description of

Melania imbricata. — Shell conical, nearly smooth, rather thick light horn-colored; spire elevated, but not acutely so; whorls 8-9, flat; lines of growth distinct, having almost the appearance of ribs; two lines, distant, slightly visible, surround each whorl, and from these the whorls incline towards each other to form a broad groove between them; sutures well impressed; aperture small, narrow, ovate, within whitish; columella much indented and curved, forming a slight sinus at base.

Habitat.—Alabama.

Diameter, ·30 (8 millim.); length, ·88 of an inch (23 millim.). Length of aperture, ·33 (8 millim.); breadth of aperture, ·21 of an inch (5 millim.).

Observations.—A fine, symmetrical shell, some of its varieties approaching *M. sordida*, Lea, in form, but differing in every other respect. The whorls enlarge regularly, and the lower raised line on the whorls being consequently more prominent; the spire has somewhat an imbricated appearance, giving rise to its specific name. The specimens before me, twelve in number, are all decollate. The upper whorls are often rather prominently ribbed, and the concentric lines thereby rendered crenulous.— *Anthony.*

It is doubtful whether this species came from Alabama, as stated above, or Georgia, as Mr. Anthony's specimens have the latter locality attached to the label.   I do not observe the ribs mentioned by Mr. Anthony, in the numerous suite of specimens before me.

Melania bicincta. — Shell conical, elevated; spire very acute; whorls seven, upper ones bicarinate, and body-whorl encircled by three or four carinæ, the upper two of which are carinate, while the lower two are of ten striæ merely; color dark olive-brown, very shining, Fig. 458*a.* and relieved by a faint or yellow, narrow band near the suture; sutures distinct; aperture ovate, and brown within; columella deeply indented.

Habitat. — Tennessee.

Observations. — A beautifully distinct and well marked species of that group which *M. bella*, Conrad may be considered most fitly to represent.   May be distinguished from *M. bella* by its broader and more acute form, more distinct carination and absence of the beaded line so characteristic of that species; lines of growth conspicuous and crowded.   Differs from *M. bicostata* (nobis) by its less robust form, darker color and by the form of its spire, which diminishes more rapidly towards the apex. — *Anthony.*

All the specimens of *bicincta* before me, including Mr. Anthony's type, are labelled by him "North Carolina," and this shell certainly belongs to a group of species characterizing that State.

Melania assimilis. — Shell small, short, conic, not thick; spire acute composed of about seven, flat whorls; sutures very distinct, of a light horn-color; aperture small, ovate, dusky within; columella indented; body-whorl angulated; sinus not broad, but well formed.

Fig. 458*b.*

Habitat. — Tennessee.

Observations. — A small, delicate species; compared with *M. pallidula* (nobis) it is more slender and elevated, has a greater number of whorls, and is devoid of bands.   From *M. angulata* (nobis) it differs in being more slender, more carinate and having a more elevated spire. — *Anthony.*

The above description applies, of course, to young shells of *symmetrica*, in which the carinæ are well developed.

Goniobasis Uchéénsis.— Shell carinate, obtusely conical, rather thin, horn-color, without bands; spire obtuse; sutures impressed; whorls about six, flattened; aperture rather large, ovately rhomboidal, whitish within; outer lip acute, somewhat sinuous; columella bent in and somewhat twisted.

Operculum ovate, light brown, with the polar point near to the left margin above the base.

Habitat.— Little Uchee River, below Columbus, Ga.; G. Hallenbeck. Diameter, ·24; length, ·58 of an inch.

Observations.— This is a very small species, nearly allied to *Melania* (*Goniobasis*) *proxima*, Say, but may be distinguished by its smaller size, its lighter color, its shorter spire, and its having a raised line above and below the carina on the upper whorls. The aperture is rather more than one-third the length of the shell.—*Lea.*

Fig. 459.

Goniobasis Barrattii.— Shell carinate, subfusiform, rather thin, greenish or reddish horn-color, obscurely banded, or without bands; spire obtusely conical; sutures very much impressed; whorls seven, slightly convex, folded at the apex; aperture rather large, subrhomboidal, whitish or obscurely banded within; outer lip acute, scarcely sinuous; columella somewhat bent in and twisted.

Habitat.— Abbeville District, South Carolina; J. P. Barratt, M.D. Diameter, ·25; length, ·53 of an inch.

Observations.— A number of specimens were sent to me by Dr. Barratt many years since. In outline all the specimens are very much the same, but they differ in some having the apical whorls obscurely plicate, while others are only carinate. All the specimens are carinate down to the last whorl. In very few specimens can the bands be seen on the outside, but usually two bands are visible on the inside near the middle. In some specimens four bands are observable. Usually the four apical whorls are obscurely plicate. The aperture is more than one-third the length of the shell. It is nearly allied to *Melania* (*Goniobasis*) *tenebrosa* (nobis), but it is more slender, has higher carinæ and is plicate. I dedicate this to the late Dr. Barratt, from whom I have formerly received many interesting specimens of the mollusca of South Carolina and Georgia.— *Lea.*

Fig. 460.

Goniobasis Catawbæa. — Shell short, conic, inflated; the whorls flat, the body convex, bright green polished; sutures well impressed; whorls five or six, encircled in the middle with two raised lines;

aperture ovate, bluish and translucent within, acuminate below; columella nearly straight. Some of the specimens are marked in the Fig. 460a. centre of the body-whorl with two, very narrow, dark, approximate bands.

Habitat.— Catawba River, near Morgantown, N. Carolina.

Length, ·63; width, ·34 of an inch. Length of aperture, ·3; width of aperture, ·17 of an inch.

Observations.— This species is nearest related to *G. proxima*, Say, which inhabits the same river. It is, however, a wider, more inflated species than G. *proxima.*— *Haldeman.*

123. G. iota, ANTHONY.

Melania iota, ANTHONY, Ann. Lyc. Nat. Hist., vi, p, 86, t. 2, f. 4, March, 1854. BROT, List, p. 36. BINNEY, Check List, No. 153.

Description.— Shell conical, smooth, greenish horn-colored; spire acutely elevated; whorls about ten, lower ones convex, upper with a strong carina below the middle; sutures impressed; aperture pyriform, small, within whitish; columella but little rounded, Fig. 460b. not indented; sinus very small.

Habitat.— ——?

Diameter, ·25 (6 millim.); length, ·78 of an inch (20 millim.). Length of aperture, ·26 (7 millim.); breadth of aperture, ·15 of an inch (4 millim.).

Observations.— A beautiful, slender, graceful species, in form not unlike *M. percarinata*, Con., and *perangulata*, Con., but differs from both in coloring, in the want of a crenulated or beaded line on the volutions, and in other respects. The upper whorls are often obscurely folded down to the carina on each, where they are arrested; below the carina the whorls shelve towards the suture, which thus becomes situated in a deep furrow. It cannot be confounded with *M. elevata*, Say, which has flat whorls, a dark epidermis, and a totally different aperture. The columella of the present species is faintly tinged with purple. I am not quite sure as to the habitat of this species, but think it an Ohio shell.—*Anthony.*

124. G. nigrocincta, ANTHONY.

Melania nigrocincta, ANTHONY, Ann. N. Y. Lyc. Nat. Hist., vi, p. 90, t. 2, f. 8, March, 1854. BROT, List, p. 36. BINNEY, Check List.

Description.— Shell conical, smooth, not much or acutely elevated;

thin, brown; whorls about six, subconvex, often slightly angulated near the suture below; sutures impressed; body-whorl not large, a little angulated, ornamented with four very dark bands, the upper and lower of which are distant, and the central ones approximate or confluent; aperture somewhat large, elliptical, banded within; columella regularly but not remarkably curved or indented, with a small sinus.

Habitat.— Tennessee.

Diameter, ·27 (7 millim.); length, ·58 of an inch (15 millim.). Length of aperture, ·27 (7 millim.); breadth of aperture, ·15 of an inch (4 millim.).

Observations.— A rather small species, which when once seen, will readily be recognized afterwards. Compares with *M. sub-* Fig. 461. *angulata* (nobis); it is less robust, more acute, and the bands are of a totally different character; the texture is quite thin, and the dark bands are distinctly seen in the aperture, through the substance of the shell. It has somewhat of the club-shaped form of that group of shells of which *M. clavæformis*, Lea, and *M. castanea*, Lea, are members, but is more angular, and its dark bands and thin texture are prominent differences.— *Anthony.*

◆This may equal *quadricincta*, Lea, young.

125. G. tecta, ANTHONY.

Melania tecta, ANTHONY, Ann. N. Y. Lyc. Nat. Hist., vi, p, 105, t. 3, f. 4, March, 1854. BINNEY, Check List, No. 265. BROT, List, p. 37. REEVE, Monog. Melania, sp. 253.
Goniobasis macella, LEA, Proc. Acad. Nat. Sci., p. 270, 1862. Jour. Acad. Nat. Sci., v, pt. 3, p. 333, t. 38, f. 192, March, 1863. Obs., ix, p. 155.

Description.— Shell conical, thin, brown; spire elevated; whorls 7-8, flat, with a distinct, but not elevated carina on each at its lower Fig. 462. edge, near the suture; sutures very deeply impressed; aperture oval, within reddish and lightly banded; columella curved, sinus small.

Habitat.— Ohio.

Diameter, ·26 (6½ millim.); length, ·60 of an inch (15 millim.). Length of aperture, ·23 (6 millim.); breadth of aperture, ·14 of an inch (3½ millim.).

Observations.— May be compared with *M. pulchella*, Anth., but is

readily distinguishable by its more slender proportions, thinner texture, lighter color, and above all by its peculiarly shaped whorls,
which, increasing regularly, and being carinate at their bases, have
somewhat the appearance of the roof of a house, hence its name.
Lines of growth distinct; one or two indistinct, narrow bands are
often visible on the shell; a very neat and graceful species.—*Anthony.*

The following is the description of *macella*, which, notwithstanding the wide difference of habitat, appears to be the same
in every respect as *tecta* :—

Goniobasis macella.— Shell carinate, awl-shaped, thin, olivaceous,
without bands; spire subattenuate; sutures very much impressed;
Fig.463. whorls seven, somewhat convex; aperture very small, sub-

rhomboidal, whitish within; spotted at the base; outer lip
acute; slightly sinuous; columella bent in and slightly twisted.

Operculum ovate, thin, light brown, with the polar point
well in from the left of margin.

Habitat.— Coosa River, Alabama; Prof. Brumby.

Diameter, ·22; length, ·62 of an inch.

Observations. — This is a little species received from Professor
Brumby a long time since. It is closely allied to *rubella*, herein
described, but differs in being somewhat smaller, in color, in having
rather flatter whorls and in having a brown, elongate spot at the
base of the columella inside. The few specimens before me are
minutely veined on the lower whorl. The upper whorls are carinate
and substriate. The aperture is about one-fourth the length of the
shell.— *Lea.*

126. G. hybrida, ANTHONY.

Melania hybrida, ANTHONY, Proc. Acad. Nat. Sci., p. 60, Feb., 1860. BINNEY, Check
    List, No. 140. BROT, List, p. 36.
Melania subcarinata, Anthony, REEVE, Monog. Melania, sp. 282.

Description.— Shell conical, elevated, nearly smooth, horn-colored;
whorls 8–9, upper ones carinated deeply, lower ones entirely smooth;
color reddish-brown, or dark horn-color; sutures distinctly impressed;
aperture small, ovate, tinged with rose-color or violet within; columella rounded, but not deeply indented; sinus small.

Habitat. — Tennessee.

Observations.— A neat, pretty species, with no very strong, distinctive characters; from *intertexta* (nobis), which it somewhat resembles, it may be distinguished by its less acute form, less numerous whorls, and by its want of reticulated surface so peculiar to that species. Bears some resemblance to *M. bella*, Con., but differs in form of outline and aperture, and has no beaded line; is also more elevated than *M. bella.*— *Anthony*.

Fig. 464.

This species differs from *symmetrica* in being more cylindrical, with the whorls more flattened.

127. G. fuscocincta, ANTHONY.

Melania fuscocincta, ANTHONY, Ann. N. Y. Lyc. Nat. Hist., vi, p. 120, t. 3, f. 20, March, 1854. BINNEY, Check List, No. 118. BROT, List, p. 40. REEVE, Monog. Melania, sp. 415.

Description.— Shell ovate, smooth, moderately thick; spire very short, consisting of 4-5 nearly flat whorls, with a broad, dark brown band revolving in the centre of each; body-whorl large, with one band above the middle, and another at base, subangulated; sutures irregularly impressed, distinct; columella well rounded, indented and reflected at the middle so as partially to conceal a small, umbilical opening; aperture large, broad, ovate, within banded.

Fig. 465.

Habitat. —Alabama.

Diameter, ·30 (7½ millim.); length, ·44 of an inch (11 millim.). Length of aperture, ·25 (6 millim.); breadth of aperture, ·17 of an inch (4 millim.).

Observations.—A short shell almost like an *anculosa;* a single specimen only is before me, but is too remarkable to be confounded with any known species. The uncommonly broad, dark band, surrounded by the generally yellow epidermis, gives it a lively appearance.— *Anthony*.

Figured from Mr. Anthony's type.

128. G. congesta, CONRAD.

Melania congesta, CONRAD, Amer. Jour. Sci , 1st ser. xxv, p. 343, Jan.. 1834. DEKAY, Moll. N. Y., p. 96. WHEATLEY, Cat. Shells U. S., p. 24. BINNEY, Check List, No. 64. JAY, Cat., 4th edit , p. 273. BROT, List, p. 36. MÜLLER, Synopsis, p. 43.

Description.— Shell subulate, with about nine volutions, the lower ones obscurely angulated, those of the spire becoming acutely carinated towards the apex; suture well defined; body-whorl obscurely subangulated; aperture longitudinal, elliptical.— *Conrad.*

This species is unknown to me and has not been figured.

G.   *Short clavate, smooth species.*

129. G. auriculæformis, LEA.

Melania auriculæformis, LEA, Philos. Proc., iv, p. 166. Philos. Trans. x, p. 62,
t. 9, f. 39. Obs., iv, p. 62, t. 9, f. 39. BINNEY, Check List, No. 24. BROT, List,
p. 32. REEVE, Monog. Melania, sp. 409.
Megara auriculæformis, Lea, ADAMS, Genera, i, p. 306.

Description.— Shell smooth, elliptical, rather thin, yellow; sutures impressed; whorls six, slightly convex; aperture elongate, contracted, at the base rounded, within whitish.

Fig. 466.

Habitat.— Tuscaloosa, Alabama.

Diameter, ·24; length, ·45 of an inch.

Observations.— This species has very much the aspect of an *auricula*. It is a very regularly formed and pretty shell, with a smooth, yellow, polished epidermis. The aperture is about two-thirds the length of the shell, regularly rounded below and angular above, where there is a good deal of nacreous matter deposited.— *Lea.*

This shell reminds one of a small *olivula*, Con., but it differs in texture from that species. The figure is copied from Mr. Lea's plate.

130. G. Nickliniana, LEA.

Melania Nickliniana, LEA, Philos. Proc., ii, p. 12, Feb., 1841. Philos. Trans., viii
p. 171, t. 5, f. 18. Obs. iii, p. 9. DEKAY, Moll. N. Y., p. 95. REEVE, Monog.
Melania, sp. 375. WHEATLEY, Cat. Shells U. S., p. 26. CATLOW, Conch.
Nomenc., p. 187.
Leptoxis Nickliniana, Lea, BINNEY, Check List, No. 371. ADAMS, Genera, i, p. 307.

Description. — Shell smooth, obtusely conical, solid, very dark; sutures impressed; whorls six, slightly convex; aperture large, somewhat rounded, within purple.

Habitat.—Bath County, Virginia; P. H. Nicklin.

Diameter, ·27; length, ·45 of an inch.

Observations.—This is a robust, small species which seems not to have been before noticed. It was found by Mr. Nicklin Fig.467. Fig.468. in a small stream of cold water at the Hot Springs in Virginia. It is amongst the smallest species I have seen. The purple color of the interior of most of the specimens gives the shell a very dark appearance. I owe to the kindness of Mr. Nicklin, to whom I dedicate it, the possession of several specimens of this species.—*Lea.*

131. G. aterina, Lea.

Goniobasis aterina, Lea, Proc. Acad. Nat. Sci., p. 155, May, 1863.

Description. — Shell smooth, subfusiform, black or greenish-black, Fig. 469. thick; spire obtuse; sutures regularly impressed; whorls six, convex; aperture rather large, subovate, within purple, *aliquanto?* white; lip acute, *vix?* sinuated; columella inflected, purple, thickened and contorted.

Habitat.— Gap Spring, Cumberland : Gap and Rogers' Spring, west of Fincastle, East Tennessee; Capt. S. S. Lyon, U. S. Army.—*Lea.*

Resembles *ebenum*, Lea, in color and texture, but is a smaller, narrower species, more angulate at the periphery. It is not an uncommon species.

132. G. Binneyana, Lea.

Goniobasis Binneyana, Lea, Proc. Acad. Nat. Sci., p. 266, 1862. Jour. Acad. Nat. Sci., v, pt. 3, p. 310, t. 37, f. 152, March, 1863. Obs., ix, p. 132.

Description.— Shell smooth, obtusely fusiform, rather thin, very much inflated, dark olive, obscurely banded; spire depressed; sutures impressed; whorls five, flattened above, the last one ventricose; aperture very large, subovate, dark within; outer lip acute, slightly sinuous; columella thickened, spotted at the base.

Habitat.—Coosa River, Alabama; Wm. Spillman, M.D.

Diameter, ·29; length, ·53 of an inch.

Observations.— Only two specimens were received from Dr. Wm. Spillman. The smaller one is rather the thicker. It has very Fig. 470. much the outline of *Lithasia Showalterii* (nobis), and at first I thought it was only a variety of that species, but the absence of a callus above and below on the columella, and a channel at the base preclude its being a *Lithasia*. It is nearly allied to *Melania*

(*Goniobasis*) *fusiformis* (nobis), but differs in being more ovate, in having a shorter spire, larger aperture, and in being of a darker color. The aperture is more than half the length of the shell. I dedicate this species to Mr. W. G. Binney, who has done so much to elucidate American conchology.—*Lea.*

This species may be distinguished from the following by its more oval form, and by the lip being less expanded.

133. G. ebenum, Lea.

Melania ebenum, LEA, Philos. Proc., ii, p. 12, Feb., 1841. Philos. Trans., viii, p. 166, t. 5, f. 7. Obs., iii, p. 4. DEKAY, Moll. New York, p. 93. JAY, Cat., 4th edit., p. 273. BINNEY, Check List, No. 93. TROOST, Cat. Shells Tenn. WHEATLEY, Cat. Shells U. S., p. 25. REEVE, Monog. Melania, sp. 350. CATLOW Conch. Nomenc., p. 186. BROT, List, p. 31.
Anculotus ebenum, Lea, REEVE, Monog. Anculotus, t. 4, f. 31.
Nitocris ebena, Lea, ADAMS, Genera, i, p. 308.

Description.— Shell smooth, obtusely conical, thick, black; spire ob-
Fig. 471. Fig. 472. tuse; sutures small; whorls somewhat convex; aper-

ture rather large, ovate, subangular at base, within purplish.

Habitat.—Robinson County, Tennessee; Dr. Currey.
Diameter, ·20; length, ·47 of an inch.

Observations. — A very dark colored and rather robust species. It resembles *M. tenebrosa*, herein described, but differs in having the whorls rather more convex, and in the outer lip being more curved. All the specimens received had the apex eroded, the number of whorls is therefore not ascertained; the aperture is more than one-third the length of the shell. It is usually purplish on the whole of the inside of the aperture. Some of the specimens are, however, bluish.— *Lea.*

134. G. Vauxiana, Lea.

Goniobasis Vauxiana, LEA, Proc. Acad. Nat. Sci., p. 265, 1862. Jour. Acad. Nat. Sci., v, pt. 3, p. 309, t. 37, f. 150, March, 1863. Obs., ix, p. 131.

Description.— Shell smooth, fusiform, rather thin, green; spire very obtuse; sutures somewhat impressed; whorls five, flattened, carinate above; aperture very large, widely rhomboidal; outer lip acute, straight; columella somewhat bent in.
Habitat.—Coosa River, Alabama.

Diameter, ·31; length, ·58 of an inch.

Observations.— Two specimens were sent to me many years since by Prof. Brumby, and I then considered them to be a variety of *Melania (Goniobasis) Nickliniana* (nobis). They differ, however, in being more angular at the base of the aperture, in being thinner, and in having the upper whorls carinate. The two specimens before me are different in the color and markings. The one from which the diagnosis is made is of a darker green and has not four well defined bands like the other, but it has two broad, indistinct ones above and below, and the lower half of the columella is purplish. The aperture is more than half the length of the shell. I dedicate this species to my friend, W. S. Vaux, Esq., who has done so much to promote the objects of our Academy.—*Lea.*

Fig. 473.

135. G. larvæformis, LEA.

Melania larvæformis, LEA, MSS. REEVE, Monog. Melania, sp. 357, Dec., 1860. BROT, List, p. 38.

Description. — Shell conically ovate, olive; whorls six to seven, smooth, the first few minutely keeled; aperture ovate.

Fig. 474.

(Lea, manuscript in Museum Cuming.)

Habitat.— United States.

Observations.— Of few whorls, convex and smooth, but yet minutely keeled near the apex.— *Reeve.*

This species is certainly *very* closely allied to *ebenum* or *Vauxiana*, but I am unable to decide whether it is identical with either of them or not.

136. G. auricoma, LEA.

Goniobasis auricoma, LEA, Proc. Acad. Nat. Sci., p. 265, 1862. Jour. Acad. Nat. Sci., v, pt. 3, p. 308, t. 37, f. 148, March, 1863. Obs., ix, p. 130.

Description.— Shell smooth, fusiform, rather thin, honey-yellow, banded; spire very obtuse; sutures linear; whorls five, scarcely convex; aperture very large, subrhomboidal, yellowish within; outer lip acute, scarcely sinuous; columella bent in and slightly thickened.

Habitat.— Tennessee River; Wm. Spillman, M.D.

Diameter, ·25; length, ·46 of an inch.

Observations.— A single specimen only of this little species was

received among a large number of mollusca from Dr. Spillman. It
réminds one of *Melania* (*Goniobasis*) *corneola*, Anth., but it is a large
Fig. 475. and more robust species, and has not the plicæ of that species.
It has also affinities to *Melania* (*Goniobasis*) *fusiformis* (nobis),
but differs in color, has a higher spire and a less incurved columella.   The specimen of *auricoma* before me has four bands,
the three lower ones are broad, equidistant and not very distinct.
The upper one is more distant and very indistinct.   Under the microscope may be observed in this specimen numerous, very minute, impressed revolving lines.   The aperture is little more than half the
length of the shell.— *Lea.*

137. G. glabra, LEA.

Melania glabra, LEA, Proc. Acad. Nat. Sci., ii, p. 82, Oct., 1841. Philos. Trans., ix,
    p. 18. Obs., iv, p. 18. WHEATLEY, Cat. Shells U. S., p. 25. BINNEY, Check List,
    No. 123. BROT, List, p. 38. REEVE, Monog. Melania, sp. 439.

Description. — Shell smooth, conical, rather thin, shining, dark
chestnut color; spire rather elevated; sutures slightly impressed;
whorls rather flattened; aperture elongated, trapezoidal, purplish
within; columella incurved.

Habitat.— Holston River, East Tennessee.

Diameter, ·32; length, ·70 of an inch.

Observations.— The apex in all the specimens before me is slightly
eroded, and therefore the number of the whorls cannot be Fig. 476.
accurately ascertained; it may be six or seven.   The aperture
is more than one-third the length of the shell.   The superior
whorls are disposed to be carinate, and below the sutures the
color is lighter.   The columella is much incurved.   Within
the aperture, indistinct, confluent bands may be observed.   These are
scarcely observable without, but give the shell a very dark aspect,
somewhat like *M. rufa* (nobis).   It is very different, however, in form
from that species.— *Lea.*

138. G. gibbosa, LEA.

Melania gibbosa, LEA, Philos. Proc., ii, p. 34, April, 1841. Philos. Trans., x, p. 301,
    t. 30, f. 12. Obs., v, p. 57, t. 30, f. 12. BINNEY, Check List, No. 121. BROT, List,
    p. 40.

Description.— Shell smooth, obtusely conical, gibbous, subfusiform,
rather thin, greenish horn-color; spire obtuse; sutures irregularly

impressed; whorls five, somewhat convex; aperture large, elliptical, within double banded; columella rubiginose, thickened, flattened, impressed and much curved.

Habitat.— Scioto River, Ohio.

Diameter, ·25; length, ·43 of an inch.

This is a small, very remarkable species. There is a slight depression above the middle of the whorl, which gives it a somewhat gibbous form. The most unusual character pertaining to this species is, however, the very flat and impressed columella, more im- Fig. 477. pressed at the point of the umbilical region. The columella on the upper part of these two specimens is not thickened, but it is of a dark brown color, and being also dark below the color extends to the outer side of the whorl, and there makes two rather indistinct bands. In outline it is allied to *M. fusiformis* (nobis), but they differ entirely in the columella and in the length of the aperture. The aperture is rather more than one-half the length of the shell. I have had some doubts of the Scioto being the real habitat of this shell; but Mr. Wheatley says it was sent from thence to him. It seems to have a more southern aspect.— *Lea.*

139. G. graminea, HALDEMAN.

Goniobasis graminea, HALD., American Journ. Conch., i, 37, t. 1, f. 4, 1865.

Description. — Shell fusiform, short, inflated; spire very obtuse; surface smooth, polished, brilliant green, with a light yellow, sutural band; spire brownish; whorls five, somewhat convex; aperture large, Fig. 478. rhomboidal, somewhat angular below, bluish within; columella somewhat curved, tinged with brown.

Habitat.— Unknown.

Diameter, ·3; length, ·56 of an inch. Aperture, ·3; diameter, ·2 of an inch.

Observations.— This shell is very closely allied to *G. Vauxiana*, Lea; but that species is banded, and the spire is carinated; it has not the light sutural band which distinguishes *graminea.*— *Haldeman.*

140. G. cognata, ANTHONY.

Melania cognata, ANTHONY, Proc. Acad. Nat. Sci., p. 60, Feb., 1860. BINNEY, Check List, No. 59. BROT, List, p. 39. REEVE, Monog. Melania, sp. 458.

Description.— Shell ovate, short, smooth, moderately thick; spire obtusely elevated, consisting of 5–6 convex whorls; color brownish-yellow, with three dark brown bands about the middle of the body-whorl, and one very obscure one at the suture; suture deeply impressed; aperture broad, ovate, not large, exhibiting the bands inside; columella deeply rounded, indented and callous; sinus none.

Habitat.— Tennessee.

Observations.— A short, pretty species with no very marked characters, though easily recognized as distinct on examination; in form and coloring somewhat like *M. compacta* (nobis), but far less solid and heavy than that species; the spire is more elevated and acute and the surface smooth. It most nearly resembles, perhaps, *M. coronilla* (nobis), but is less elevated and has not the peculiar crowning ribs of that species, which is sufficient at once to distinguish it. It is also more robust.— *Anthony.*

Fig. 479.

Figured from Mr. Anthony's type specimen. Much more inflated and shorter than *G. Georgiana*, Lea. It also differs from that species in possessing two bands only.

141. G. Georgiana, Lea.

Goniobasis Georgiana, Lea, Proc. Acad. Nat. Sci., p. 265, 1862. Jour. Acad. Nat. Sci., v, pt. 3, p. 308, t. 37, f. 148. Obs., ix, p. 130.

Description.— Shell smooth, fusiform, inflated, rather thick, yellowish, bright, banded; spire very obtuse; sutures impressed; whorls five, convex; aperture large, subrhomboidal, whitish and banded within; outer lip acute, straight; columella bent in, thickened and somewhat twisted.

Operculum subovate, dark brown, with the polar point near to the base on the left margin.

Habitat.— North Georgia.

Diameter, ·26; length, ·57 of an inch.

Observations.— Among a number of *Melanidæ* from the Smithsonian Institution, were two small specimens which have the same outline and same form of aperture, but which differ much in color. Fig. 480. That which is described above seems to me to be the normal character and will serve as the type. This has three well defined bands, the middle one of which is the broadest, and it has a character which I have not seen in any of our *Melanidæ*, that is, longitudinal, whitish maculations, which are dispersed over the body-whorl, and seem under the microscope to be slightly raised on the surface. The second specimen is horn-color and has no bands. In outline this species is closely allied to *Melania (Goniobasis) Nickliniana* (nobis), but is not so pointed at the apex. is not so inflated in the body-whorl, and differs in color. The aperture is quite half the length of the shell.— *Lea.*

142. G. depygis, SAY.

Melania depygis, SAY, New Harmony Disseminator, p. 291. SAY'S Reprint, p. 19. Am. Conchology, Part 1, t. 8, f. 4, 5. BINNEY'S Reprint, pp. 145 and 157, t. 8. BINNEY, Check List, No. 87. LAPHAM, Cat. Moll. Wisconsin. KIRTLAND, Am. Jour. Sci. KIRTLAND, Rep. Zool., Ohio, p. 174. SHAFFER, Catalogue. HIGGINS, Catalogue. ANTHONY, List, 1st and 2d edit. SAGER, Rept. Michigan Moll., p. 15. WHEATLEY, Cat. Shells U.S., p. 25. DEKAY, Moll. N. Y., p. 89, t. 7, f. 135. STIMPSON, Shells of New England, p. 32. JAY, Cat. Shells, 4th edit., p. 273. ADAMS, Am. Jour. Sci., xl, p. 366. Adams, THOMPSON'S Hist. Vermont, p. 152. CATLOW, Conch. Nomenc., p. 186. BROT, List, p. 37. DESHAYES, Lamark, Anim. saus. Vert., viii, p. 441. REEVE, Monog. Melania, sp. 373.
Potadoma depygis, Say, ADAMS, Genera i, p. 298.
Melania occulta, ANTHONY, Proc. Acad. Nat. Sci., p. 5, Feb., 1860. BINNEY, Check List, No. 185. BROT, List, p. 38. REEVE, Monog. Melania, sp. 254.

Description.— Shell oblong, conic-ovate, not remarkably thickened; spire as long as the aperture, or rather longer, often much eroded, with a broad, revolving, rufous line near the suture, occupying a considerable portion of the surface; whorls about five, Fig. 481. Fig. 482. hardly rounded; suture moderately impressed; body-whorl yellowish, with two rufous, revolving lines equidistant from the suture, base and each other, the superior one broader, and its locality a little flatter than the general curvature; aperture ovate, acute above, moderately dilated; labium with calcareous deposit, particularly above; labrum not projecting near the base, nor arquated near its junction with the second volution; base regularly rounded.

Observations.— I found this species, in great abundance, on the rocky flats at the Falls of the Ohio, where they were left by the subsiding of the river, in company with numerous other shells. In old specimens the spire is very much eroded, exhibiting a white, irregular surface. It varies a little in color, and a few occurred, of which the color is fuscous, the bands being obsolete.—*Say.*

The following description is founded on elongated specimens of *depygis,* of which it is undoubtedly a synonyme.

Melania occulta.— Shell conic, smooth, rather thin; color lemon-yellow, inclining to brown, with a darker brown band on each whorl, increasing to two on the body-whorl; whorls 7–8, rather convex; suture deeply impressed; aperture ovate, within dusky-white, with

Fig. 482a.

the outer bands seen faintly through its substance; columella beautifully rounded; outer lip produced, a small sinus at base.

Habitat.— Wisconsin.

Observations.—A very beautiful and lively species. Bears some resemblance to *M. pulchella* (nobis), but is elongate, more delicately colored, and of a less solid texture; the bands are often obsolete, and never so distinctly expressed as in *pulchella;* its spire is also more acute, and the whorls more rounded. Compared with *M. brevispira* (nobis), which in form it resembles, it is more attenuate, has a greater number of whorls, and its bands also distinguish it. Its delicate yellow color also is not a common character in the genus, and forms a prominent mark for determination.—*Anthony.*

143. G. livescens, MENKE.

Melania livescens, MENKE, Syn. Meth., p. 135, 1830. BINNEY, Check List, No. 163. GOULD, Lake Superior, p. 245. JAY, Cat., 4th edit., p. 274. REEVE, Monog. Melania, sp. 229. BROT, List, p. 38. CURRIER, Shells of Grand River Valley, Mich., 1859.
Melania Niagarensis, LEA, Philos. Proc., ii, p. 12, Feb., 1841. Philos. Trans., viii, p. 173, t. 5, f. 21. Obs., iii, p. 11. DEKAY, Moll., N. Y., p. 90. WHEATLEY, Cat. Shells U. S., p. 26. BINNEY, Check List, No. 175. CATLOW. Conch. Nomenc., p. 187. BROT, List, p. 38. CURRIER, Shells of Grand River Valley, Mich. BELL, Canad. Naturalist, iv, pt. 3, p. 213, June, 1859.
Potadoma Niagarensis, Lea, ADAMS, Genera, i. p. 299.
Melania napella, ANTHONY, Bost. Proc., iii, p. 362, Dec., 1850. BINNEY, Check List, No. 170. BROT, List. p. 59.
Melania cuspidata, ANTHONY, Bost. Proc. iii, p. 362, Dec., 1850. BINNEY, Check List, No. 83. REEVE, Monog. Melania, sp. 283.
Melania correcta, BROT, List, p. 39.

Description.—Shell ovately oblong, smooth, bluish flesh-color; spire conically acute; lip horn-color, produced in front, border Fig. 483. purple; columella thinly callous, purplish.

Longitude, ·7; latitude, ·3½ lin.

Habitat.—Lake Erie, New York; sent by my friend, Hæninghaus.— *Menke.*

The following are the descriptions of the species which I consider synonymes.

Melania Niagarensis.— Shell smooth, obtusely conical, thick, horn-colored; spire short; sutures linear; whorls rather flat; aperture rather large, elliptical, within purple.

Habitat.— Falls of Niagara.

Diameter, ·25; length, ·55 of an inch.

Observations.— I obtained this shell many years since at the foot of the Falls of Niagara, where it exists in abundance. It may generally have been confounded with *M. depygis*, Say. When I procured it I placed it in my cabinet under that name with a mark of doubt. It is a smaller shell than the *depygis*, has a shorter spire and a narrower aperture. This species has a purple columella and interior, which in some cases are very dark. The specimens procured were all more or less eroded, and the apex removed. The number of whorls is either six or seven. The aperture is nearly half the length of the shell.— *Lea.*

Fig. 484.

Melania napella.— Shell small, ovate, acuminate, smooth, light corneous; whorls seven, the upper ones conical and carinate at the sutures; aperture one-half the length of the shell, narrowly lunate; lip dilated in front, sinuate posteriorly.

Fig. 485.

Longitude, $\frac{1}{2}$; latitude, $\frac{1}{4}$ poll.

Habitat.— Ohio.

Observations.— A pale, rather singular species, from its bulbous form. Some immature specimens of *M. simplex* are often much like it.— *Anthony.*

Melania cuspidata.— Shell small, short, ovate, acuminate, smooth, greenish-purple, lighter on the sutures; whorls six, convex, sometimes flattened, apical ones carinate, the last ventricose; aperture large and equalling half the length of the shell; lip dilated in front, posteriorly scarcely sinuate.

Fig. 486.

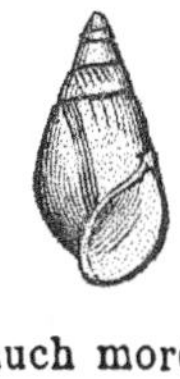

Habitat.— Maumee River, Ohio.

Longitude, three-fifths; latitude, three-tenths poll.

Observations.— Allied to *M. napella*, having the same peculiar bulbous form and produced lip. It is, however, much more elongated. It resembles *M. Warderiana*, Lea.— *Anthony.*

The identity of these species has long been conceded by most of our best conchologists. They all possess in common the short, bulbous form and conical spire, frequently slightly carinate; and are readily known by the very convex, outer lip, salmon-purple interior and dark purple-tinged columella. The epidermis is corneous in fresh specimens, but most of them are without epidermis and then present a livid bluish-white appearance. Considerable variation may be noticed in the form of the shell and in its texture. It is an exceedingly numerous species inhabiting the waters of the Northwestern States. Dr. Brot proposed the name *correcta* instead of *cuspidata*, Anth., preoccupied in *Melania*.

144. G. Milesii, LEA.

Goniobasis Milesii, LEA, Proc. Acad. Nat. Sci., p. 154, May, 1863.

Description.— Shell smooth, subfusiform, olivaceous, without bands; spire subelevated; sutures irregularly impressed; whorls six, sub-

Fig. 487.

inflated; aperture rather large, subrhomboidal, brownish within; lip acute, scarcely sinuate; columella purplish, slightly incurved.

Habitat. — Tuscola County, Michigan; M. Miles, State zoologist.— *Lea.*

This species is certainly very closely allied to *livescens* in many respects but appears to be more convex in the whorls, and to attain a larger size. I am by no means satisfied that it is distinct, however.

145. G. simplex, SAY.

Melania simplex, SAY, Jour. Acad. Nat. Sci., v, p. 126, Sept. 1825. BINNEY's edition, p. 115. BINNEY, Check List, No. 244. DEKAY, Moll. N. Y., p. 100. WHEATLEY, Cat. Shells U.S., p. 27. REEVE, Monog. Melania, sp. 148. JAY, Cat., 4th edit., p. 275. BROT, List, p. 38.
Pachycheilus simplex, Say, ADAMS, Genera, i, p. 298.
Melania Warderiana, LEA, Philos. Proc., ii, p. 14, Feb., 1841. Philos. Trans., viii, p. 185, t. 6, f. 47. Obs., iii, p. 23. DEKAY, Moll. N. Y., p. 90. CATLOW, Conch. Nomenc., p. 189. BINNEY, Check List, No. 297. BROT, List, p. 39. REEVE, Monog. Melania, sp. 353.
Melania Wardiana, Lea, WHEATLEY, Cat. Shells U.S., p. 27.
Potadoma Warderiana, Lea, CHENU, Manuel de Conchyl., i, f. 1972. ADAMS, Genera, i, p. 299, CHENU. Manuel, i, f. 1972.
Melania densa, ANTHONY, Bost. Proc., iii, p. 360., Dec., 1850. BINNEY, Check List, No. 89. BROT, List, p. 31. REEVE, Monog. Melania, sp. 250.
Melania subsolida, Philos. Proc., ii, p. 12, Feb., 1841. Philos. Trans., viii. p. 168, t. 5, f. 12. Obs., iii, p. 6. TROOST, Cat. Shells Tenn. BINNEY, Check List, No. 255. WHEATLEY, Cat. Shells U.S., p. 27. DEKAY, Moll. N. Y., p. 94. CATLOW, Conch. Nomenc., p. 188. BROT, List, p. 39.
Potadoma subsolida, Lea, H. and A. ADAMS, Genera, i, p. 299.
*Goniobasis Vanuxemii,** LEA, Proc. Acad. Nat. Sci., p. 265, 1862. Jour. Acad. Nat. Sci., v, p. 307, t. 37, f. 146. Obs., ix, p. 129.

---

* Changed to *G. Prestoniana*, LEA, Proc. Acad., 1864, p. 3.

Description.— Shell conic, blackish, rather rapidly attenuated to an acute apex; suture not deeply impressed; volutions about eight, but little rounded; aperture longitudinal; within dull reddish; labrum with the edge not undulated, or but very slightly and obtusely so near the superior termination.

Length, three-fifths; greatest breadth, three-tenths of an inch.

Observations.— For this species we are indebted to Prof. Vanuxem, who presented several specimens to the Academy. He informs me that he obtained them in Virginia, in a stream running from Fig. 488. Abington to the salt works, and from the stream on which General Preston's grist-mill is situated, near the salt works, as well as in a brook running through the salt water valley, and discharging into the Holston River. Near the summit the whorls are marked by an elevated line near their bases. It cannot be mistaken for the *conica* (nobis) for in that species the aperture is obviously oblique.— *Say.*

The synonymy of the species indicated by the above table is due to the investigations of Professor Haldeman, whose fine suite of self-collected specimens demonstrates their entire identity. Figure 488 is from an author's example of *simplex* in Museum Anthony. I have specimens of the same form, but of much larger size. *Warderiana* is figured from Mr. Lea's plates.

The following are the descriptions of the synonymes:—

Melania Warderiana. — Shell carinate, club-shaped, rather thick, very dark; spire conical; sutures linear; whorls eight, convex; aperture ovate, rather large, within flesh-color.

Habitat.— Cedar Creek, a branch of Clinch River, Russell County, Virginia.

Diameter, ·37; length, ·76 of an inch.

Observations.— I have two specimens before me. The two lowest Fig. 489. whorls are smooth, the superior ones are carinate, with a small, intermediate stria, the upper whorls diminish very rapidly. The exterior of the shell is very black and shining, and its color appears to arise from a deposit of ferruginous matter, as the substance of the shell is reddish. The aperture is rather more than one-third the length of the shell. I name it after Dr. Warder of Cincinnati, to whom I owe the possession of this and other interesting specimens.— *Lea.*

Melania subsolida. — Shell smooth, subfusiform, somewhat solid, horn-colored; spire acute; sutures impressed; whorls somewhat con-
Fig. 490. vex; aperture somewhat elongated, within purple.

Habitat. — Tennessee; Dr. Troost.

Diameter, ·32; length, ·82 of an inch.

Observations. — This species has a strong resemblance to *M. simplex*, Say. It is, however, more elevated in the spire. It is purplish within, but white towards the margin of the lip.— *Lea.*

Melania densa. — Shell solid, elongately ovate, acuminate, light olivaceous; spire produced; whorls 6–7, ventricose, angulated below, the upper ones small, the last subcylindrical, equalling two-   Fig. 490a. thirds the length of the shell; aperture narrowly ovate, scarcely effused, rounded in front; columella quite callous; within yellowish.

Habitat. — Maury's Creek, Tennessee.

Longitude, ⅞; latitude, ⅜ poll.

Observations. — Somewhat like *M. basalis*, Lea.   The shelving of the whorls towards the suture and the acumination of the spire are among its most striking characters.— *Anthony.*

Goniobasis Vanuxemii. — Shell smooth, fusiform, rather thick, horn-color; spire obtusely conical; sutures impressed; whorls seven,
Fig. 491. slightly convex; aperture large, subrhomboidal, white or purple within; outer lip acute, slightly sinuous; columella bent in, thickened above and below.

Operculum ovate, very thin, light brown, with the polar point near to the base on the left.

Habitat. — North Fork of the Holston River, Virginia; Prof. L. Vanuxem.

Diameter, ·27; length, ·54 of an inch.

Observations. — Many years before the decease of my lamented friend, Prof. Vanuxem, he gave me a number of mollusca collected during his journeys in South Carolina and Western Virginia. Among them was quite a number of this little species which I now dedicate to him. It is nearly allied to *Melania* (*Goniobasis*) *Niagarensis* (nobis), but is a small species with a shorter spire, and is straighter at the base of the columella. The aperture is rather more than one-third the length of the shell.—*Lea.*

146. G. Potosiensis, LEA.

Melania Potosiensis, LEA, Philos. Proc., ii, p. 14, Feb., 1841. Philos. Trans., viii, p. 184, t. 6, r. 45. Obs., iii, p. 22. DEKAY, Moll. N. Y., p. 99. WHEATLEY, Cat. Shells U.S., p. 26. BINNEY, Check List, No. 215. CATLOW, Conch. Nomenc., p. 188. BROT, List, p. 36. REEVE, Monog. Melania, sp. 295.
Elimia Potosiensis, Lea, H. and A. ADAMS, Genera, i, p. 300.

Description.— Shell carinate, conical, rather thin, brown; spire obtusely elevated; sutures much impressed; whorls eight, convex; aperture large, ovate, purplish.

Habitat.— Potosi, Missouri.

Diameter, ·28; length, ·62 of an inch.

Observations.— The rotundity of the outer lip in this is different from the species generally, with the same elevation of spire. Fig. 492. The aperture is more than one-third the length of the shell, and is entirely purple, in the only two specimens before me. In one specimen the carina is distinct on all the whorls but the last; in the other it is not visible on the last·two whorls.—*Lea.*

Were it not for the wide difference of locality I should suspect this to be identical with *simplex*. I have not seen specimens, but the figure and description are certainly very close to that species.

147. G. Saffordi, LEA.

Melania Saffordi, LEA, Philos. Trans., x, p. 300, t. 30, f. 10. Obs., v, p. 56. BINNEY, Check List, No. 236. BROT, List, p. 38. REEVE, Monog. Melania, sp. 365.
Melania virens, ANTHONY, Ann. N. Y. Lyc. Nat. Hist., vi, p. 93, t. 2, f. 11, March, 1854. BINNEY, Check List, No. 289. BROT, List, p. 40.

Description. — Shell smooth, obtusely conical, thick, subfusiform, dark green; spire rather short; sutures linear; whorls a little convex, the last large; aperture rather large, ovately elongated, within purple; columella purple and twisted.

Fig. 493.

Habitat.— Lebanon, Wilson County, Tennessee.

Diameter, ·37; length, ·85 of an inch.

Observations.— This is a very distinct species, with a not uncommon form. The green color is unusual. On the upper part of the whorl, and on the line of the suture there is a light or brownish band. The body-whorl is rather suddenly enlarged in the middle, which gives it a slight gibbous appearance, and it is irregularly, transversely striate. The apex of each of the three specimens under my examination being eroded, the number of whorls cannot be exactly ascertained, but I think there must be about six. The aperture is quite one-half the length of the shell. It is allied to *M. sordida* (nobis) in outline, but may easily be distinguished in color and the gibbous swelling on the whorls. I name this after Mr. Safford, to whose kindness I owe this and some other fine specimens from Tennessee.— *Lea.*

The following shell appears to be in every respect identical with the above :—

Melania virens.— Shell ovate-conic, smooth, rather thick; spire

Fig. 494.   Fig. 494a. rather obtusely elevated, with a somewhat

 convex outline, and with sutures decidedly impressed; color light uniform green, paler towards the summit; whorls five only remaining, and indications of one lost by truncation, convex; aperture rather large, elliptical, bluish within; columella well rounded, not perceptibly indented, and with a small, recurved sinus at base.

Habitat.— Alabama.

Diameter, ·40 (10 millim.) : length, ·87 of an inch (22 millim.). Length of aperture, ·42 (10 millim.); breadth of aperture, ·21 of an inch (5 millim.).

Observations.— A broad species with an outline and proportions not unlike a *Paludina*, to which genus its pale, uniform green color seems to ally it. I am not sure that it should not be referred to that genus. It cannot be compared with any known species.—*Anthony.*

148. G. Newberryi, Lea.

Goniobasis Newberryi, Lea, Proc. Acad. Nat. Sci., March, 20, 1860. Jour. Acad. Nat. Sci., v, pt. 3, p. 300, t. 37, f. 135, March, 1863. Obs., ix, p. 122. Binney, Check List, No. 174. Brot, List, p. 38.

Description.—Shell smooth, ovately conical, rather thin, dark brown, triple-banded, yellow below the sutures; spire somewhat raised; sutures much impressed; whorls six, inflated; aperture rather small, ovately rounded, whitish and banded within; outer lip inflated; columella whitish, incurved.

Operculum ovate, rather thin, dark brown, with the polar point near the inner inferior edge.

Habitat.—Upper Des Chutes River, Oregon Territory; J. S. Newberry, M.D.

Diameter, ·30; length, ·64 of an inch.

Observations.— This is a rather small species, very nearly allied to *Melania* (*Goniobasis*) *Taitiana* (nobis), from Claiborne, Alabama, but differs in being rather more inflated, of a darker color, and having three dark bands instead of four. The bands in *Newberryi* are broad and dark, sometimes running into each other, while the *Taitiana* has thinner ones of a lighter color. In some specimens of the latter the bands are absent, but I have seen no specimen of the former without bands. These give a dark appearance to the shell, which is well relieved by the yellow margin under the sutures. I have great pleasure in naming it after Dr. Newberry, the discoverer of it.— *Lea.*

Fig. 495.

149. G. bulbosa, GOULD.

Melania bulbosa, GOULD, Bost. Proc., ii, p. 225, July, 1847. Otia Conchologica, p. 46. Moll. Expl. Exped., p. 142, f. 163, 163a, 1852. BINNEY, Check List, No. 43. BROT, List, p. 58.

Description.— Shell small, conically oblong, shining, eroded, greenish-brown; spire of 2-3 rounded whorls, remaining; sutures profound; aperture ovately-rounded, scarcely effused.

Fig. 496.

Habitat.— Columbia River.

Longitude, one-half; latitude, nine-twentieths poll.

Observations. — The whorls are very cylindrical, so as to appear like a succession of bulbs. It is much like *M. perfusca*, Anth.; but in that the whorls slope gently to the suture. A broken specimen shows that it often attains a considerable size.— *Gould.*

This species is exactly similar in outline to Mr. Lea's *Newberryi*, but none of the specimens before me, including Dr. Gould's types, exhibit the slightest indications of bands, while Mr. Lea declares his species to be always banded.

150. G. Lithasioides, LEA.

Goniobasis Lithasioides, LEA, Proc. Acad. Nat. Sci., May, 1863. Obs., xi, p. 89, t. 23, f. 37.

Description.— Shell smooth, subfusiform, horn-color, without bands; spire conoidal; sutures impressed; whorls six, somewhat constricted, flattened above; aperture rather large, rhomboidal, white within;

outer· lip acute, somewhat sinuous; columella white, bent in and somewhat twisted.

Habitat.— Ohio; J. P. Kirtland, M.D.

Diameter, ·28; length, ·65 of an inch.

Observations.—A single specimen was received many years since from Dr. Kirtland with *Melania* (*Goniobasis*) *depygis*, Say, but while it agrees with it in color and size, it is quite different in the body-whorl, and in the form of the aperture. The aperture is very much like *Lithasia*, and is slightly thickened above on the columella, but there is neither a channel nor a callus below. In the whole outline and form of the aperture it is very like *Lithasia Downiei* (nobis), but it is a much smaller shell, a much lighter color, has no tubercles and has no channel at the base. It is among the few species which are impressed on the body-whorl, but it is not so much so as *G. informis*, herein described, and is a larger and stouter species. The aperture is not quite half the length of the shell. Dr. Kirtland did not state from what part of Ohio it came.— *Lea.*

Fig. 496a.

151. G. infantula, LEA.

Goniobasis infantula, Proc. Acad. Nat. Sci., May, 1863. Obs., xi, p. 91, t. 23, f. 39.

Description.—Shell smooth, fusiform, dark horn-color, much banded; spire short; sutures slightly impressed; whorls five, flattened above; aperture rather large, ovate, banded within; outer lip acute, slightly sinuous; columella purple, thickened and twisted.

Operculum ovate, reddish-brown, rather thin, with the polar point near the base on the left edge.

Habitat.—Falls of the Ohio at Louisville, Ky.; W. H. DeCamp, M.D.

Diameter, ·20; length, ·38 of an inch.

Observations.— This is a pretty little species, usually with four well marked, rather broad, brown bands. In one of the six speci- mens before me there are only three indistinct bands. It is closely allied to *Melania* (*Goniobasis*) *cognata*, Anthony, and near to *Georgiana* (nobis). It differs from *cognata* in being more drawn out in the spire and having less inflation of the body-whorl. The aperture is about one-half the length of the shell.— *Lea.*

Fig. 496b.·

152. G. Louisvillensis, LEA.

Goniobasis Louisvillensis, LEA, Proc. Acad. Nat. Sci., May, 1863. Obs., xi, p. 89, t. 23, f. 36.

Description. — Shell smooth, fusiform, dark horn-color, without bands; spire short; sutures irregularly impressed; whorls about five, somewhat convex; aperture rather large, long elliptical, white within; outer lip acute, slightly sinuous; columella white, thickened above and twisted.

Operculum ovate, reddish-brown, rather thin, with the polar point on the left, near the base.

Habitat. — Falls of the Ohio at Louisville, Ky.; W. H. DeCamp, M.D. Diameter, ·25; length, ·56 of an inch.

Observations. — Two specimens only were received, neither perfect at the apex. It is a simple species with an unusually thick-

Fig. 496c.

ened columella, approaching indeed to *Lithasia*. It is near to *Spartanbergensis* and *ovoidea* (nobis) and is somewhat like *depygis*, Say, but cannot be confounded with this last species, from the same habitat, being much shorter in the spire, and having a differently formed aperture. Neither of the two specimens has any appearance of bands, but they may exist on other specimens. The aperture is about one-half the length of the shell.— *Lea.*

H. *Smooth, elevated species.*

153. G. pulchella, ANTHONY.

Melania pulchella, ANTHONY, Bost. Proc., iii, p. 361, Dec., 1850. HIGGINS, Cat., p. 7. REEVE, Monog. Melania, sp. 257. BINNEY, Check List, No. 221. BROT, List, p. 38. CURRIER, Shells of Grand River Valley, Mich.

Description. — Shell small, thin, elongately conical, brownish horn, banded with brown; spire conical; whorls 7-8, convex; aper-

Fig. 497.

ture large, equalling one-third the length of the shell, elongately ovate.

Habitat. — ———?

Longitude, seven-tenths; latitude, one-fourth poll.

Observations. — A pretty species, ornamented by dark, rather broad bands, somewhat like *M. Taitiana* and some varieties of *M. Virginica.* — *Anthony.*

An exceedingly common species in various parts of Ohio, extending into Michigan. It varies considerably in form and size, but is larger and more elevated than *depygis*, which it resembles in color and ornamentation. From *gracilior* it is distinguished by its lighter color and convexity of the superior half of the lip, which in the latter species is incurved or flattened.

154. G. cinerea, LEA.

Goniobasis cinerea, LEA, Proc. Acad. Nat. Sci., p. 265, 1862. Jour. Acad. Nat. Sci., v, pt. 3, p. 306, t. 37, f. 145. Obs., ix, p. 128.

Description. — Shell smooth, conical, thin, ash-gray, bright: spire obtusely conical, sharp-pointed, carinate at the apex; sutures very Fig. 498. much impressed; whorls eight, somewhat convex; aperture rather large, subrhomboidal, bluish-white within; outer lip acute, somewhat sinuous; columella bent in, slightly thickened and purplish.

Habitat. — South Carolina; Prof. L. Vanuxem.

Diameter, ·25; length, ·60 of an inch.

Observations. — A single specimen, of this gracefully formed species was among a number of shells given to me by my friend, the late Prof. Vanuxem. The exact habitat was not given. It is a thin, sub-diaphanous species, of an ashen gray, with a remarkably thin epidermis. There is an obscure appearance of a band towards the upper portion of the whorls and a purple oblique marking at the interior of the base of the axis. It is allied to *Ohioensis*, herein described, but it is more slender, thinner, and has a more elongate aperture. The aperture is six-sixteenths the length of the shell.— *Lea.*

This species is so nearly allied to *G. pulchella* that I much doubt whether it is distinct.

155. G. gracilior, ANTHONY.

Melania gracilis, ANTHONY, Cover of No. 4, HALDEMAN'S Monog. Limniades. Dec., 28, 1841. Shells of Cincinnati, 1st edit. NEWBERRY, Proc. American Association for Adv. of Science, v, p. 105. JAY, Cat., 4th edit., p. 273.
Melania gracilior, ANTHONY, Ann. N. Y. Lyc. Nat. Hist., vi, p. 129, t. 1, f. 5, 1854. HIGGINS' Cat., p. 7. BINNEY, Check List, No. 127. REEVE, Monog. Melania, sp. 244.
Melania gracilis, Lea, REEVE, Monog. Melania, sp. 369.

Description. — Shell conical, smooth and shining, color dark brown, texture light; whorls about eight, upper ones nearly flat, the last is usually slightly constricted beneath the suture, and beneath this stricture on the periphery of the last whorl revolve one or two broad bands of yellowish-green; sutures impressed, and of paler color than the rest of the shell; aperture small, pyriform, and inwardly ornamented with alternate bands of a dark ruby color and translucent white, which render this part of the shell peculiarly lively and beautiful; outer lip sinuate; columella dark brown, arcuate, and produced into a distinct sinus.

Habitat.— Congress and Springfield Lakes, Stark County, Ohio.

Diameter, ·28 (7 millim.); length, ·75 of an inch (19 millim.). Length of aperture, ·25 (6 millim.); breadth of aperture, ·17 of an inch (5 millim.).

Observations.— This is a very distinct and beautiful species, remarkable for its long, slender form, its polished surface, and for a profound stricture on the body-whorl of many of the specimens, though this last character is not always present; when it is present it furnishes a mark by which this species can be readily distinguished from any other. It is seldom that any of our *Melaniæ* are found inhabiting waters so still as those of the small lakes so numerous in Stark and the neighboring counties in Ohio; nearly all the family are denizens of rapid streams abounding with rocks, to which they adhere, often in great numbers. Occasionally, however, they attach themselves to the dead bivalve shells which pave many of the rivers in our Southern and Western States, or cling to the long grass which grows in them. This species was first published on the cover of Haldeman's Monograph of the Fresh-water Shells of North America, No. 4, December 28, 1841. A short time previous Mr. Lea had published a species from Tennessee under the same name, which publication I had not then seen. It becomes expedient, therefore, to change its name to one not preoccupied, and I propose in redescribing the species to confer upon it that of *gracilior*, which seems even *more* appropriate than the name originally given to it.— *Anthony.*

Fig. 499.

156. G. Canbyi, TRYON.

Goniobasis Etowahensis, LEA,* Proc. Acad. Nat. Sci., p. 264, 1862. Jour. Acad. Nat.
    Sci., v, pt. 3, p. 299, t. 37, f. 133, March, 1863.

Description. — Shell smooth, conoidal, thin, dark, double-banded;
spire somewhat raised; sutures impressed; whorls seven, slightly
convex; aperture rather large, subrhomboidal, dark and broadly
banded within; outer lip acute and sinuous; columella bent in and
very much twisted.

Habitat. — Etowah River, Georgia; J. Postell.

Diameter, ·30; length, ·74 of an inch.

Observations. — A single specimen only was sent to me by Mr.

Fig 500.

Postell. At first sight it would be taken for *Melania* (*Goniobasis*) *gracilior*, Anth., having the same dark hue, made so by
the two, broad, dark brown bands. It differs from it in being
less conical, in having a larger aperture which is more angular at the basal margin. The two broad bands cover nearly
two-thirds of the last whorl, leaving a yellowish interspace.
In this specimen there is a brown, elongate spot at the base of the
columella. The aperture is about three-eighths the length of the
shell.— *Lea.*

157. G. ovoidea, LEA.

Melania ovoidea, LEA, Philos. Proc., iv, p. 167, Aug., 1845. Philos. Trans., x, p. 61,
    t. 9, f. 38. Obs., iv, p. 61. BINNEY, Check List, No. 193. BROT, List, p. 38.
Potadoma ovoideus, Lea, ADAMS, Genera, i, p. 299.

Description. — Shell smooth, elliptical, rather thick, horn-color;
spire short; sutures slightly impressed; whorls six, slightly convex;
aperture large, nearly ovate, within white.

Habitat. — Alexandria, Louisiana.

Diameter, ·2; length, ·44 of an inch.

Observations. — A single specimen only of this little species was
found among the shells sent by Dr. Hale. It differs entirely   Fig. 501.
from the other two species, and approaches Mr. Say's *depygis*,
but is smaller, and has a proportionately larger aperture.
The aperture is quite one-half the length of the shell. The
columella is somewhat thickened on the superior portion. In the

---

**G. Etowahensis*, Lea, being preoccupied by Mr. Reeve, who described and figured *G.
Canbyi*, Lea, under that name in advance of Mr. Lea's description, we apply the latter's
name to this species.

specimen before me there are two, broad, rather indistinct, brown bands.—*Lea.*

Mr. Reeve's figure represents a species of *Lithasia.*

158. G. translucens, ANTHONY.

Goniobasis translucens, ANTHONY, Am. Journ. Conch., i, 36, t. 1, f. 1, 2, 1865.

Description. — Shell ovately bulbous, consisting of five convex whorls, or the upper ones sometimes flattened. Aperture ovate, slightly angular at the base; columella curved to the right inferiorly; color light horn, thin, translucent, ornamented with two Fig. 502. dark brown bands, of which one is apparent on the whorls of the spire; columella sometimes tinged with brown.

Habitat.— Canada.

Length, ·7; breadth, ·35 of an inch.

Observations.— This beautiful species is distinguished by its coloration and thin texture from *G. livescens,* which it otherwise greatly resembles.—*Anthony.*

159. G. grata, ANTHONY.

Melania grata, ANTHONY, Proc. Acad. Nat. Sci., p. 61, Feb., 1860. BINNEY, Check List, No. 131. BROT, List, p. 34. REEVE, Monog. Melania, sp. 433.
Goniobasis Prairiensis, LEA, Proc. Acad. Nat. Sci., p. 264, 1862. Jour. Acad. Nat. Sci., v, pt. 3, p. 299, t. 37, f. 132, March, 1863. Obs., ix, p. 121.

Description. — Shell conic, elevated, smooth, thick; whorls nine, flat, terminating in an acute apex, the first three or four whorls being carinated; color light greenish-yellow, ornamented by a single dark band on the spiral whorls, and four similar bands on the body-whorl, giving the shell a truly lively and beautiful appearance; sutures very distinct; aperture ovate, banded within; columella deeply

Fig. 502a.

indented and curved at base, where there is a small but rather broad sinus.

Habitat.— Alabama.

Observations.—The colors in this species are finely contrasted, and the general appearance is very lively and pleasing; the bands on the body-whorl are not uniformly distributed, the upper and lower ones being widely separated, while the central ones are very close together and less distinct. Altogether it is one of our most beautiful species.— *Anthony.*

Goniobasis Prairiensis. — Shell smooth, elongately fusiform, thin, olivaceous, shining, four-banded; spire raised, sharp-pointed; sutures regularly impressed; whorls nine, flattened; aperture rather large, Fig. 503. subrhomboidal, whitish and four-banded within; outer lip acute and sinuous; columella bent in and twisted.

Operculum ovate, dark brown, with the polar point on the left, one-fourth above the basal margin.

Habitat.—Big Prairie Creek, Alabama; E. R. Showalter, M.D. Diameter, ·35; length, ·85 of an inch.

Observations.— Among some twenty specimens before me there is no difference in form or markings, except that some have the bands slightly broader than others. The two middle bands are rather closer together and the under one of these two is generally the smaller. It was sent to me by Dr. Showalter under the name of *M. grata*, Anth., but while it has the four bands like that species, it is more slender, is not yellow, has a less aperture and one more whorl, and is more fusiform. The aperture is rather more than one-third the length of the shell.—*Lea.*

Mr. Anthony's types of *M. grata* are before me, and do not represent the shell, which Mr. Lea distinguishes in the above description by that name, but are identical in every respect with *G. Prairiensis.* The shell which Mr. Lea mistook for *M. grata*, he has since described as *quadricincta.*

160. G. quadricincta, LEA.

Goniobasis quadricincta, LEA, Proc. Acad. Nat. Sci., Apr., 1864, p. 112. Obs., xi, 87, t. 23, f. 33.

Description.—Shell smooth, or obscurely folded, somewhat fusiform, somewhat thick, yellow, four-banded; spire conical; Fig. 504. Fig. 505. sutures regularly impressed; whorls about eight, flattened, angular towards the apex; aperture rather large, ovate and four-banded within; outer lip acute and somewhat sinuous; columella thin and somewhat twisted.

Operculum ovate, rather thin, light brown, with the polar point near the left edge.

Habitat.— Coosa and Cahawba Rivers, Alabama; Dr. Showalter; East Tennessee and North Georgia; Bishop Elliott.

Diameter, ·37; length, ·93 of an inch.

Observations.—I have about two dozen specimens before me from the different habitats. Those from East Tennessee are shorter and not so well characterized, having less marked bands, some even being without them. The best developed are from the Coosa River. Two specimens from Fannin County, Georgia, have a bright yellow epidermis without bands, and may belong to a distinct species. The four bands are remarkably regular in this species. The two middle ones are near to each other and the lower of the two is smaller than the upper. It is allied to *grata*, Anth. The aperture is rather more than one-third the length of the shell.— *Lea.*

161. G. flava, LEA.

Goniobasis flava, LEA, Proc. Acad. Nat. Sci., p. 264, 1862. Jour. Acad. Nat. Sci., v, pt. 3, p. 303, t. 37, f. 139, March, 1863. Obs., ix, p. 125.

Description.— Shell smooth, obtusely conical, rather thin, yellow, three-banded; spire obtusely conical; sutures very much impressed; whorls about six, somewhat convex; aperture rather small, ovate, white and three-banded within; outer lip acute, slightly sin- Fig. 506. uous; columella bent in and thickened.

Operculum ovate, dark brown, with the polar point near to the edge and above the basal margin. ʻ

Habitat.—Benton County? N. E. Alabama; G. Hallenbeck. Diameter, ·35; length, ·88 of an inch.

Observations.— A single specimen, only, of this pretty species, was sent to me by Mr. Hallenbeck. It cannot be confounded with any other species known to me. It reminds one of *Melania grata*, Anth., but it has a rounder base, is not fusiform, and has but three bands, which are well marked inside and out. The three bands are equidistant and of equal size. The upper part of the columella is thickened, and in this specimen the color of the upper band is extended over part of this callus. The aperture is rather more than one-third the length of the shell.— *Lea.*

162. G. tenebrovittata, LEA.

Goniobasis tenebrovittata, LEA, Proc. Acad. Nat. Sci., p. 264, 1862. Jour. Acad. Nat. Sci., v, pt. 3, p. 301, t. 37, f. 136, March, 1863. Obs., ix, p. 123.

Description.— Shell smooth, high conical, rather thin, yellowish, banded or without bands; spire somewhat raised; sutures slightly impressed; whorls flattened; aperture rather large, subrhomboidal, whitish within; outer lip acute, slightly sinuous; columella somewhat bent in.

Operculum ovate, dark brown with the polar point near the edge above the basal margin.

Habitat.— Coosa River; W. Spillman, M.D.

Diameter, ·43 of an inch; length, 1·07 inches.

Observations.— This species is allied to *Melania (Goniobasis) grata,*

Fig. 507. Anth., which puts on many phases. It may be at once distinguished, however, by *grata* being more pointed, having a more yellow epidermis and narrower bands. Two out of ten specimens before me have a greenish epidermis and are without bands. One specimen has a purplish interior. The prevailing character of the bands is, two being proximate in the middle, and two, one above the other below, being more removed. The two middle ones are sometimes closed, forming a single broad band. The aperture is more than one-third the length of the shell.— *Lea.*

163. G. tenera, ANTHONY.

Melania tenera, Anthony, REEVE, Monog. Melania. sp. 407, Apr., 1861. BROT, List,
   p. 39.

Description.—Shell elongately ovate, subcylindrical, yellowish-olive, encircled with narrow, distant, red-brown bands; whorls slopingly convex, the first few keeled next the suture; aperture ovate, narrowly effused at the base; columella thinly reflected, rather produced.

Habitat.— Alabama, United States.

Observations.—Chiefly distinguished by its encircling pattern of red-brown linear bands upon a pale yellowish-olive ground.—*Anthony.*

I at first thought this to be the same as *G. Brumbyi,* Lea, but the latter species grows larger and is of a narrower form.

164. G. Brumbyi, LEA.

Goniobasis Brumbyi, LEA, Proc. Acad. Nat. Sci., p. 263, 1862. Jour. Acad. Nat. Sci.,
   v, pt. 3, p. 296, t. 37, f. 127, March, 1863. Obs., ix, p. 118.

Description.— Shell smooth, attenuate, rather thin, ash-gray, four-banded; spire drawn out, carinate at the apex; sutures very much impressed; whorls about eight, slightly convex; aperture small, sub-rhomboidal, whitish and four-banded within; outer lip acute; columella bent in, obtusely angular at base.

Habitat.— Alabama; Prof. Brumby.

Diameter, ·32; length, ·74 of an inch.

Observations.— Two specimens were sent to me among other species, by the late Prof. Brumby of Columbia, South Carolina. One is but little more than half grown, and is more perfect in the epidermis and in the aperture. It is very closely allied to *Melania* (*Goniobasis*) *Kirtlandiana* (nobis), but it is more attenuate and has bands which I have never seen on *Kirtlandiana.* Both the specimens before me have four bands, the two middle ones being nearer to each other. The aperture of the mature specimen is not quite one-third the length of the shell, while that of the younger is more than the third, and it is also more angular at the base, the older one not being entirely perfect. I dedicate this species to the late Prof. R. T. Brumby, to whom I am indebted for it.—*Lea.*

Fig. 508.

The shell figured is the half grown specimen; the other one is much longer.

165. G. Elliottii, Lea.

Goniobasis Elliottii, Lea, Proc. Acad. Nat. Sci., p. 271, 1862. Jour. Acad. Nat. Sci., v, pt. 3, p. 338, t. 38, f. 201, March, 1863. Obs., ix, p. 160.

Description.— Shell obscurely striate, rather obtusely conical, somewhat thick, yellowish or brownish, without bands; spire rather obtuse; sutures very much impressed; whorls about six, slightly convex; aperture large, ovately rhomboidal, whitish or brown within; outer lip sharp, slightly sinuous; columella slightly bent in, thickened and somewhat twisted.

Fig. 509.

Operculum subovate, thin, dark brown, with the polar point on the edge near the base.

Habitat.— Fannin County, Ga.; Bishop Elliott: Uchee and Little Uchee Rivers, Alabama: G. Hallenbeck and Dr. Gesner.

Diameter, ·41; length, ·94 of an inch.

Observations.— I have quite a number of this species. It is well marked, and not easily confounded with any other I know. The interiors of some specimens are dark brown, with a white thickened margin on the outer lip; others are light brown, inclining to obscure bands, while about one-half of all are white. The apical whorls are usually carinate. The body-whorl has generally two or three obscure, transverse striæ about the periphery, below which, towards the base, they are closer and coarser. There is a strong disposition in some specimens to a depression below the suture. The aperture is about three-eighths the length of the shell. I dedicate this to the Right Reverend Stephen Elliott, who has done so much to develop the zoology of Georgia.— *Lea.*

166. G. pallescens, LEA.

Melania pallescens, LEA, Philos. Proc., iv, p. 166, August, 1845.   Philos. Trans., x,
    p. 63, t. 9, f. 43.   Obs., iv, p. 63.   BINNEY, Check List, No. 196.   BROT, List, p. 31.
Goniobasis inosculata, LEA, Proc. Acad. Nat. Sci., p. 270, 1862.   Jour. Acad. Nat.
    Sci., v, pt. 3, p. 334, t. 38, f. 195, March, 1863.   Obs., ix, p. 153.
Goniobasis parva, LEA, Proc. Acad. Nat. Sci., p. 264, 1862.   Jour. Acad. Nat. Sci.,
    v, pt. 3, p. 297, t. 37, f. 129, March, 1863.   Obs., ix, p. 119.

Description.— Shell carinate, rather acutely conical, somewhat thin, yellow; spire somewhat elevated; sutures impressed; whorls nine, rather convex; aperture small, ovate, angular at the base, within whitish.

Habitat.— Chester District, South Carolina.

Diameter, ·34; length, ·87 of an inch.

Observations.— Many years since, I was not satisfied that it was not merely a variety of *semicarinata,* Say, but I am disposed to think it Fig. 510. differs too much to be considered merely a variety. It is a larger shell, with more whorls and more distinct carinations. The color also differs, in being much lighter. A single specimen was among the shells sent from Major LeConte, which, I suspect, is from Georgia, the locality not being certain. Those from Professor Vanuxem are from Major Green's farm. The aperture is less than one-third the length of the shell. All the specimens are without bands but one, which has four, large, distinct ones.— *Lea.*

Figured from Mr. Lea's plate. The following is the description of a half grown shell of this species.

Goniobasis inosculata.— Shell carinate, conical, rather thin, yellowish horn-color, without bands; spire somewhat raised; sutures impressed; whorls about seven, a little convex; aperture rather large, rhomboidal, whitish within; outer lip acute, sinuous; columella somewhat bent in and thickened below.

Operculum subrotund, thin, light brown, with the polar point on the left near the edge.

Habitat.— Little Uchee River, below Columbus, Ga.; G. Hallenbeck.

Diameter, ·30; length, ·74 of an inch.

Observations.— Nearly a dozen of this species were mixed up with the *Uchéensis*, herein described. It is closely allied, but may Fig. 511. be distinguished by the form of the aperture, which is much more rhombic. It is also of a lighter color, and the outer lip is more sinuous. The aperture is more than one-third the length of the shell.— *Lea.*

The following is a still younger form of *pallescens :* —

Goniobasis parva.— Shell smooth, conical, thin, horn-color, without bands; spire somewhat raised, sharp-pointed; sutures impressed; Fig. 512. whorls seven, flattened; aperture rather small, whitish within, subrhomboidal; outer lip acute and sinuous; columella bent in and somewhat thickened.

Habitat.— Georgia; Right Rev. Stephen Elliott.

Diameter, ·27; length, ·66 of an inch.

Observations.— This is a small species of which I received only three specimens, neither of them entirely perfect. It is very near to *Melania (Goniobasis) lœvis* (nobis), but it is more attenuate, having a higher spire and rather smaller aperture. The aperture is about two-fifths the length of the shell.—*Lea.*

167. G. Anthonyi, LEA.

Goniobasis Anthonyi, LEA, Proc. Acad. Nat. Sci., p. 264, 1862. Journ. Acad. Nat. Sci., v, pt. 3, p. 303, t. 37, f. 140, March, 1863. Obs., ix, p. 125.

Description.— Shell smooth, obtusely conical, rather thin, shining, dark chestnut brown, without bands; spire obtuse; sutures impressed; whorls about six, somewhat convex; aperture rather large, elongately rhombic, brownish within; outer lip acute, white towards the margin and slightly thickened; columella bent in and very much twisted.

Habitat.— Tennessee; J. G. Anthony.

Diameter, ·33; length, ·77 of an inch.

Observations.—A single specimen of this species was sent to me some years since by Mr. Anthony, who collected it in Tennessee, but

Fig. 512a.

I am not aware in what part. I then thought it might be a variety of *Melania* (*Goniobasis*) *perfusca* (nobis), but it is a smaller species with a longer aperture. It has the smooth, dark chestnut-brown and polished epidermis of *Melania* (*Goniobasis*) *nitens* (nobis), but is larger and has a longer aperture. In the specimen before me there is a line of light brown below the suture. On the inside are two, obscure, brownish bands, but none are apparent on the outside. The aperture is nearly half the length of the shell. I name this after Mr. J. G. Anthony, who kindly sent it to me with other specimens.— *Lea.*

168. G. Cahawbensis, LEA.

Melania Cahawbensis, LEA, Proc. Acad. Nat. Sci., p. 121, 1861.
Goniobasis Cahawbensis, LEA, Jour. Acad. Nat. Sci., v, pt. 3, p. 223, March, 1863.
  Obs., ix, p. 45.

Description.— Shell smooth, somewhat fusiform, raised conical, pointed, rather thin, dark horn-color, obscurely banded; spire somewhat raised; sutures line-like; whorls eight, flattened above, the last rather large; aperture rather small, ovate, whitish or yellowish within; outer lip acute; columella arcuate, somewhat rounded at the base.

Fig. 512b.

Habitat.— Cahawba River, Alabama; E. R. Showalter, M.D. Diameter, ·42; length, ·84 of an inch.

Observations.— This is a regularly formed, graceful species, with very obscure bands. In three of the specimens these bands are scarcely noticeable, but the fourth, which is the youngest, has three bands well defined within the aperture. It is nearly allied to *Melania germana*, Anth., but it is more elongate and has not the carination of the middle of the whorl, nor the rhomboidal aperture. The aperture is more than one-third the length of the shell. The apical whorls are carinate.— *Lea.*

169. G. Gabbiana, LEA.

Goniobasis Gabbiana, LEA, Proc. Acad. Nat. Sci., p. 265, 1862. Jour. Acad. Nat. Sci., v, pt. 3, p. 304, t. 37, f. 141, March, 1863. Obs. ix, p. 126.

Description.— Shell smooth, subfusiform, rather thin, horn-color, without bands; spire slightly elevated, sharp-pointed; sutures impressed; whorls about eight, convex and varicose; aperture rather small, subrhomboidal, whitish within; outer lip acute, slightly sinuous; columella bent in and twisted.

Habitat.—Tennessee; Prof. G. Troost: Alabama; Prof. Tuomey. Diameter, ·25; length, ·54 of an inch.

Observations.—I have only seen two specimens and indeed I have some doubts if that from Alabama be not specifically distinct. That from the late Prof. Troost I consider the type. It has been in my possession many years. They are very much the same in outline and size, and both have veiny lines on the bodywhorl. That from Alabama is, however, slightly more inflated, is of a darker color, and has plicæ on the apical whorls with striæ beneath. It also has a less number of whorls by two. When more specimens shall be found from both habitats, and these differences be found to be persistent, I would consider them as distinct species. The aperture is about one-half the length of the shell. I name this after my young friend, Mr. W. M. Gabb, who has done much to advance the conchology of our country.— *Lea.*

Fig. 513.

170. G. sordida, LEA.

Melania sordida, LEA, Philos. Proc., ii. p. 12, Feb., 1841. Philos. Trans., viii, p. 170, t. 5, f. 13. Obs., iii, p. 8. DEKAY, Moll. N. Y., p. 94. REEVE, Monog. Melania, sp. 449. JAY, Cat. 4th edit., p. 275. TROOST, Cat. Shells Tennessee. CATLOW, Conch. Nomenc., p. 188. WHEATLEY, Cat. Shells U. S., p. 27. BINNEY, Check List, No. 246. BROT, List, p. 33.

Potadoma sordida, Lea, CHENU, Manuel de Conchyl., i, f. 1971. H. and A. ADAMS, Genera, i, p. 299.

Melania perfusca, LEA, Philos. Proc., ii, p. 82, Oct., 1841. Philos. Trans., ix, p. 18. Obs., iv, p. 18. WHEATLEY, Cat. Shells U. S. p. 26. JAY, Cat., 4th edit., p. 274. BINNEY, Check List, No. 201. BROT, List, p. 31. REEVE, Monog. Melania, sp. 354.

Melania incurta, Anthony, REEVE, Monog. Melania, sp. 300. BROT, List, p. 38.

Melania plebeius, ANTHONY, Bost. Proc., iii, p. 362, Dec., 1850. REEVE, Monog. Melania, sp. 414. BINNEY, Check List, No. 209.

Melania plebeia, Anthony, BROT, List, p. 38.

Melania brunnea, ANTHONY, Ann. N. Y. Lyc. Nat. Hist., vi, p. 92, t. 2, f. 10, March, 1854. BINNEY, Check List, No. 41. BROT, List, p. 39. REEVE, Monog. Melania, sp. 319.

Melania Paula, Anthony, BROT, List, p. 40.

Description.—Shell smooth, conical, somewhat thick, dark horncolored; sutures impressed; whorls somewhat convex; aperture rather large, somewhat rounded, within bluish.

Habitat.— Tennessee; Dr. Troost.

Diameter, ·40 of an inch; length, 1·02 inches.

Observations.— The whole. of five individuals before me have the apex decollate. This species closely resembles the *Ocoeensis*, herein

Fig. 514.   described. It is, however, larger in the aperture, which is more rotund, and the species seems to be larger.— *Lea.*

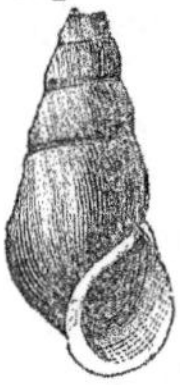

The following are synonymes :—

Melania plebeius.— Shell small, rather solid, plain, truncated ovate-conical, reddish-brown; whorls three, flattened, the last large, ventricose, subangulated; sutures well impressed, aperture large, ovate; lip dilated anteriorly,   Fig. 515. Fig. 516. scarcely sinuated posteriorly; columella white or

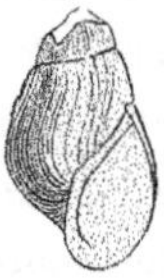

Fig. 517. Fig. 518.   stained with red.

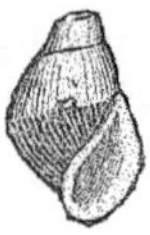

Habitat.— Saline Co., Arkansas.

Observations.— A small, apparently variable species, without any attractive characters. The angle around the last whorl is more or less marked, or even wanting. Small specimens appear to be much like *M. Nickliniana.*—*Anthony.*

The figures are from type specimens.

Melania brunnea, Anth., is characterized from thinner and better grown specimens of this shell. *M. paula,* Anth., (unpublished) is the young, not yet half grown. The species resembles somewhat *M. iostoma,* Anth., and Mr. Lea believes them to be identical, but as it appears to me *iostoma* is darker, 'and a little more angulate at the periphery. *M. Nickliniana* is smaller, wider, with spire more truncate. The following is the description of

Melania brunnea.— Shell elongate-ovate, smooth, thin, brown; spire obtusely elevated; whorls six, nearly flat; body-whorl convex, sometimes three-banded; sutures irregularly but decidedly impressed; aperture large, broad, elliptical, within whitish,   Fig. 519. or tinted with reddish; columella somewhat indented below the middle, and forming a very small sinus at base.

Habitat.— Alabama.

Diameter, ·32 (8 millim.); length, ·76 of an inch (20 millim.) Length of aperture, ·37 (9 millim.); breadth of aperture, ·23 of an inch (6 millim.).

Observations.— A smooth, fine species, with no very prominent characters. May be compared with *M. perfusca,* Lea, but is less cylindrical, and much less ponderous; the whorls are also more convex, and the sutures more distinctly impressed; it is altogether a broader and thinner shell. Some specimens are finely banded, the lower band being often concealed partially by the revolutions of the succeeding whorl. The body-whorl has three bands in the *variety*, and these also appear within the aperture. All the specimens before me, some fifty in number, are more or less decollate, and only two or three are banded.— *Anthony.*

Melania perfusca.— Shell smooth, conical, rather thick, dark brown; spire exserted; sutures linear; whorls rather flattened; aperture large, inflated, ovate, within pale purple.

Fig. 520.

Habitat.— Calf-killer Creek, Tennessee.

Diameter, ·50 of an inch; length, 1 inch.

Observations.— A single specimen, with the spire truncate, is before me. The lower portion is perfect. The apex being destroyed the number of whorls cannot be ascertained. The aperture is, I presume, rather more than one-third the length of the shell. The lower part of the margin protrudes considerably. It seems to be nearly allied to *M. ebenum* (nobis), but is a larger shell, more inflated, and has a larger aperture, being less elliptical.— *Lea.*

Melania incurta.— Shell somewhat pyramidally conical, yellowish-olive; whorls smooth, slopingly contracted round the upper part, then rounded; aperture ovate; columella reflected, slightly sinuated at the base.

Fig. 521.

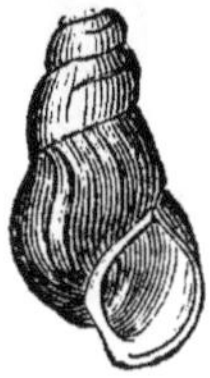

(Anthony, manuscript in Museum Cuming).

Habitat.— United States.

Observations.—All I can say of this shell is, that it is in Mr. Cuming's cabinet with the above name in manuscript, alleged to have been received from Mr. Anthony.— *Reeve.*

An extensive suite of specimens, which I have had before me, through the kindness of Messrs. Gould and Haldeman proves the identity of the above described species, the variation of form being very great.

171. G. castanea, Lea.

Melania castanea, Lea. Philos. Proc., ii, p. 11. Philos. Trans., viii, p. 164, t. 5, f. 2. Obs., iii, p. 2. DeKay, Moll. N. Y., p. 92. Troost, Cat. Shells Tennessee. Wheatley, Cat. Shells U.S., p. 24. Reeve, Monog. Melania, sp. 337.

Description.—Shell smooth, club-shaped, rather thin, dark brown;

spire elevated, carinate towards the apex; sutures small; whorls eight, somewhat convex; aperture small, elliptical, purple.

Habitat.— Maury County, Tennessee; Thomas R. Dutton.

Diameter, ·25; length, ·67 of an inch.

Observations.— This species is remarkable for its club-shaped form.

Fig. 522.

It differs from the *clavæformis* herein described, in having a less pointed apex, in being a smaller species, and in being of a darker color. The first three or four whorls are carinate, and disposed also to be striate and plicate. The aperture is about one-third the length of the shell. The three individuals before me are entirely purple inside, and this gives a very dark appearance to the shell.— *Lea.*

172. G. clavæformis, Lea.

Melania clavæformis, Lea, Philos. Proc., ii, p. 12, Feb., 1841. Philos. Trans., viii, p. 168, t. 5, f. 10. Obs., iii, p. 6. DeKay, Moll. N. Y., p. 93. Jay, Cat., 4th edit., p. 273. Troost, Cat. Shells Tennessee. Wheatley, Cat. Shells U.S., p. 25. Reeve, Monog. Melania, sp. 396. Binney, Check List, No. 57. Catlow, Conch. Nomenc., p. 186. Brot, List, p. 37.

Description.—Shell smooth, club-shaped, rather thin, chest-   Fig. 523. nut-brown, shining; spire acute; sutures somewhat impressed; whorls eight, convex; aperture elongated, light purple.

Habitat.— Ocoee District and Clinch River, Tennessee.

Diameter, ·27; length, ·67 of an inch.

Observations.— The aperture is about one-third the length of the shell. In color it differs from most species.— *Lea.*

173. G. adusta, Anthony.

Melania adusta, Anthony, Proc. Acad. Nat. Sci., p. 55, Feb., 1860. Binney, Check List, No. 2. Brot, List, p. 37. Reeve, Monog. Melania, sp. 338.
Melania funebralis, Anthony, Proc. Acad. Nat. Sci., p. 56, Feb., 1860. Binney, Check List, No. 114. Brot, List, p. 38. Reeve, Monog. Melania, sp. 372.
Goniobasis Cumberlandiensis, Lea, Proc. Acad. Nat. Sci., p. 155, May, 1863.

Description. — Shell conical, smooth, shining; color dark brown, with a pale line near the sutures; whorls 7–8, flat; body-whorl rather large, subangulated, and with somewhat coarse lines of growth; sutures distinct, but not remarkable; aperture ovate, dark purple within; outer lip curved; columella deeply rounded, a broad sinus at base.

Habitat.— Tennessee.

Observations.— A neat, pretty species, of rather plain appearance. Compared with *M. gracilior* (nobis), it is broader, shorter, and of darker color; the broad, deep cincture on the body-whorl and beautiful red bands in the interior, so conspicuous in Fig. 524. Fig. 525.

M. gracilior, are also wanting. From *athleta* it differs by its shorter, more acute form, and by the absence of folds. It is less slender than *M. viridula.* — *Anthony.*

Melania funebralis.— Shell conic, smooth, solid, of a dark chestnut color; spire elevated and generally abruptly truncate; whorls from 3–5 only remaining, slightly convex; aperture ovate, within bluish; columella white, tinged occasionally with purple; sinus small.

Habitat.— Tennessee.

Observations.— A very neat, pretty species with no very decided

Fig. 527. Fig. 526. character to distinguish it from allied species. May

be compared with *M. brevispira* (nobis), but is far more solid in its texture, of a darker color, and its surface is more polished and shining; much less slender too than *brevispira*, and that species is never so abruptly decollate. It appears to be an abundant species. — *Anthony.*

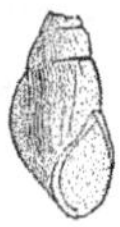

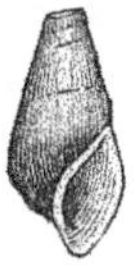

This species is narrower and more elongated than the typical form, *M. adusta*, and has not the yellowish, sutural band of that species.

Goniobasis Cumberlandiensis. — Shell smooth, acuminately conoidal, rather thin, reddish-brown; spire somewhat elevated; sutures Fig. 528. regularly impressed; whorls eight, slightly convex; aperture small, subrhomboidal, white or purple within; lip acute, slightly sinuous; columella white or purple, inflated and contorted.

Habitat. — Gap Spring, Cumberland Gap, Tennessee; Capt. Lyon: and Knoxville, Tennessee; William Spillman, M.D. — *Lea.*

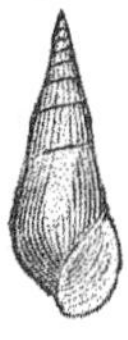

174. G. furva, Lea.

Melania furva, Lea, Philos. Trans., x, p. 299, t. 30, f. 7. Obs., v, p. 55. Binney, Check List, No. 115. Brot, List, p. 38.

Description.— Shell smooth, conical, rather thick, dusky; spire rather elevated; sutures furrowed; whorls flattened: aperture small, subrhomboidal, at the base angular, within purplish; columella purple and twisted.

Habitat.— Branch of Coosa River, Alabama.

Diameter, ·30; length, ·84 of an inch.

Observations.— A single specimen of this species was received from
Prof. Brumby. It has the apex so much eroded as to present

Fig. 528a

only a little more than three whorls, which are, however, perfect, and enable me to distinguish it from its allied species, the nearest of which is *M. arata* (nobis). The sutures have the same furrowed line, and the sides of the whorl are alike flattened. The aperture, however, differs in form and color. In the *arata* the columella is straight down to the channel at the base; in the *furva*, it is curved to the right and the channel is less marked. The length of the aperture, in perfect specimens, must be about one-third the length of the shell. The *Alexandrensis* (nobis) from Louisiana, is very closely allied to this species, and when perfect specimens of both shall be obtained, they may possibly be found to be the same.—*Lea.*

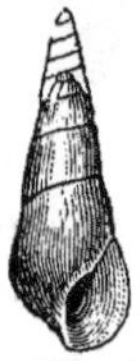

175. G. dubiosa, Lea.

Melania dubia, LEA, Philos. Proc., ii, p. 11, Feb., 1841.
Melania dubiosa, LEA, Philos. Trans., viii, p. 166, t. 5, f. 6. Obs., iii, p. 4. DEKAY,
    Moll. N. Y., p. 93. BINNEY, Check List, No. 91. TROOST, Cat. Shells Tennessee.
    WHEATLEY, Cat. Shells U. S., p. 25. JAY, Cat. 4th edit., p. 273. CATLOW,
    Conch. Nomenc., p. 183. BROT, List, p. 37.
Goniobasis Estabrookii, LEA, Proc. Acad. Nat. Sci., p. 264, 1862. Jour. Acad. Nat.
    Sci., v, pt. 3, p. 298, t. 37, f. 131, March, 1863. Obs., ix, p. 120.

Description.— Shell smooth, conical, rather thin, horn-color; spire rather elevated; sutures linear; whorls seven, somewhat convex; aperture elliptical, small, subangular at the base, whitish. Fig. 529.

Habitat.— Tennessee; Dr. Troost.

Diameter, ·30; length, ·75 of an inch.

Observations. — This is a rather small species, somewhat like *M. simplex*, Say, but seems to me to differ, in having a more elevated spire, and a smaller aperture. The aperture is rather more than one-third the length of the shell. — *Lea.*

Figured from Mr. Lea's plate. One or two specimens of this species are plicate on the first two or three whorls, but the plicæ are by no means characteristic of the species.

The following is a synonyme :—

Goniobasis Estabrookii.— Shell smooth, conical, rather thin, reddish horn-color, without bands; spire attenuately conical, sharp-pointed; sutures impressed; whorls ten, somewhat convex; aperture rather small, ovate, whitish within; outer lip acute, slightly sinuous; columella bent in.

Operculum ovate, light brown, with the polar point to the left of the centre, towards the basal margin.

Habitat.—Knoxville, Tennessee; President Estabrook.

Diameter, ·34; length, ·89 of an inch.

Observations.— I received from President Estabrook nine specimens of this species. They were all covered with a black deposit of oxide of iron. This being removed, the epidermis was found to be smooth and shining, and of a reddish horn-color, inclining to yellow. It is very closely allied to *Melania* (*Goniobasis*) *dubiosa* (nobis), but differs in the aperture being slightly more constricted and in being rather longer, having one more whorl. It is also near to *castanea* (nobis), but is larger and not chestnut-brown. The aperture is about one-third the length of the shell. I dedicate this species to the late President Estabrook of Knoxville, Tennessee.— *Lea.*

Fig. 530.

176. G. interlineata, ANTHONY.

Goniobasis interlineata, ANTHONY, Am. Jour. Conch., vol. i, p. 36, t. 1, f. 3, Feb. 25, 1865.

Description.— Shell thin, elongate, slender, of a grayish horn-color, alternating with narrow, brown, hair-like lines, longitudinally and closely arranged; whorls 7–8, subconvex, smooth; sutures distinct; aperture small, elliptical, ashen gray within; columella regularly rounded, much curved at base, and with a faint indentation or notch where the outer lip meets it.

Fig. 531.

Habitat.— Christy Creek, Indiana.

Length of shell, ·62 of an inch. Length of aperture, ·25; breadth of aperture, ·15 of an inch.

Observations.— A most beautifully delicate, slender species, whose most prominent characteristic is indicated by its specific name. Upon a light grayish horn-colored surface we find narrow, brown, longitudinal lines, distinctly drawn. These are very conspicuous under the microscope, and appear to be slightly raised. It presents a general resemblance to *G. elata* (nobis) and *G. bicolorata* (nobis), but its peculiarly varied exterior will at once distinguish it from either. I know of no other American species so marked.— *Anthony.*

I am pretty well satisfied that this is only a local variety of *semicarinata*, the thickened, deeper colored, longitudinal lines indicate periods of arrested growth.

177. G. lævigata, LEA.

Melania lævis, LEA, Philos. Proc., ii, p. 237, Dec., 1842. Philos. Trans., viii, p. 248.
  Obs., ii, p. 86.
Melania lævigata, LEA, Proc. Philos. Soc., ii, p. 237. Philos. Trans., vii, p. 165, t. 5,
  f. 3. Obs., iii, p. 3. WHEATLEY, Cat. Shells U. S., p. 23. CATLOW, Conch.
  Nomenc., p. 187. REEVE, Monog. Melania, sp. 459.
Potadoma lævigata, Lea, H. and A. ADAMS, Genera, i, p. 299.
Melania Leaii, BROT, List, p. 34.

Description.— Shell smooth, obtusely conical, rather thin, shining, yellowish; spire rather short, carinate towards the apex; sutures Fig. 532. linear; whorls seven, rather convex; aperture rather large,

elliptical, angular at base, whitish.

Habitat.— Alabama River at Claiborne; Judge Tait.

Diameter, ·25; length, ·55 of an inch.

Observations.—With the *M. Taitiana* herein described, came two specimens of this species, which differ from the *Taitiana* in the elevation of the spire, and the form and size of the aperture. In the most perfect specimen the columella and base are purplish. The aperture is more than one-third the length of the shell. The upper whorls are slightly carinate on their lower portions.—*Lea.*

Originally described as *lævis*, which was preoccupied. Dr. Brot proposed the name *Leaii* for this species, because *lævigata* is preoccupied in *Melania*, but in *Goniobasis* that name has not been previously used, and consequently stands good.

The figure is a copy of that given by Mr. Lea. I doubt whether this is more than an immature shell of *dubiosa*, Lea.

178. G. Ohioensis, LEA.

Goniobasis Ohioensis, LEA, Proc. Acad. Nat. Sci., p. 265, 1852. Jour. Acad. Nat.
  Sci., v, pt. 3, p. 306, t. 37, f. 144. Obs., ix, p. 128.

Description.— Shell smooth, conical, somewhat thin, without bands; spire obtusely conical, sharp pointed, carinate at the apex; sutures very much impressed; whorls about nine, convex; aperture small, somewhat rounded, white within; outer lip acute, scarcely sinuous; columella bent in, very much thickened.

Habitat.—Yellow Springs, Ohio.

Diameter, ·31; length, ·65 inch.

Observations. — Many years since two specimens of this species were brought by a member of my family from the Yellow Springs of Ohio, a much frequented watering place. They are both dead specimens, but are well preserved in form, while the epidermis has been Fig. 533. entirely removed. The columella is remarkably thick, and the edge stands off from the whorls, displaying an impression at the axis amounting nearly to an umbilicus. It is nearly allied to *Grosvenorii* herein described, but may be distinguished in having a shorter spire, less impressed sutures, a thicker columella, and having an umbilical impression. The outer lip also is not so sinuous and the whorls are not so attenuate. It has its affinities to *Melania (Goniobasis) varicosa*, Ward, but has a different aperture and has no veins. The aperture is about two-sevenths the length of the shell.—*Lea.*

This species is probably not distinct from *semicarinata*, Say.

179. G. brevispira, ANTHONY.

Melania brevispira, ANTHONY, Bost. Proc., iii, p. 361, Dec., 1850. BINNEY, Check List, No. 39. JAY, Cat., 4th edit., p. 474. BROT, List, p. 37. REEVE, Monog. Melania, ep. 263.

Melasma brevispira, Anthony, ADAMS, Genera, i, p. 300.

Description.— Shell small, elongate, ovate, truncate, rather solid, Fig. 534. Fig. 535. plain, shining, brownish-green, paler at the sutures;

  whorls 4–5, convex, somewhat declining at the sutures: aperture ovate; lip dilated before, sinuated behind.

Habitat.— Ohio.

Longitude, three-fifths; latitude, three-tenths poll.

Observations.— A small, plain species, with no very obvious, distinctive marks. It is allied to *M. plebejus*, but is rather more slender. It is usually much eroded.— *Anthony.*

180. G. semicarinata, SAY.

Melania semicarinata, Say, New Harmony Disseminator, p. 261. Reprint, p. 16. American Conchology, Part 5, t. 47, f. 4. BINNEY'S Reprint, p. 142, 200. BINNEY, Check List, No. 240. DEKAY, Moll. N. Y., p. 100. REEVE, Monog. Melania, sp. 368. WHEATLEY, Cat. Shells U. S., p. 27. JAY, Cat. Shells, 4th edit., p. 275. CATLOW, Conch. Nomenc., p. 188. BROT, List, p. 38. KENNICOTT, Trans. Ills. State Ag. Soc. p. 595.

Melania angustispira, ANTHONY, Proc. Acad. Nat. Sci., p. 55, Feb., 1860. BINNEY, Check List, No. 16. BROT, List, p. 37.

Melania angusta, Anthony, REEVE, Monog. Melania, sp. 359.

Melania exilis, HALDEMAN, Suppl. to No. 1 Monog. Limniades, Oct., 1840.

Juga exilis, Haldeman, ADAMS, Genera, i, p. 304.

Melania rufula, HALDEMAN, Monog. Limniades, No. 2, p. 3 of Cover, January, 1841, BINNEY, Check List, No. 234. BROT, List, p. 39.

Melania Kirtlandiana, LEA, Philos. Proc., ii, p. 11, Feb., 1841. Philos. Trans., viii, p. 165, t. 5, f. 4. Obs., iii, p. 3. ANTHONY, Cat., 1st edit. HIGGINS, Cat. DEKAY, Moll. N. Y., p. 92. WHEATLEY, Cat. Shells U. S., p. 25. REEVE, Monog. Melania, sp. 361. BINNEY, Check List, No. 155. BROT, List, p. 36. CATLOW, Conch. Nomenc., p. 187.

Ceriphasia Kirtlandiana, ADAMS, Genera, i, p. 297.

Melania Kirtlandia, LEA, Philippi, Beschreib, Neuer, Conchyl. Melania, t. 3, f. 8.

Melania elata, ANTHONY, Bost. Proc., iii, p. 362, Dec., 1850. BINNEY, Check List, No. 95. BROT, List, p. 37. REEVE, Monog. Melania, sp. 331.

Melania bicolorata, ANTHONY, Bost. Proc., iii, p. 361, Dec., 1850. BINNEY, Check List, No. 32. BROT, List, p. 58.

Melania bicolor, Anthony, REEVE, Monog. Melania, sp. 265.

Melania inornata, ANTHONY, Bost. Proc., iii, p. 360. Dec., 1850.

Potadoma inornatus, ADAMS, Genera, i, p. 299.

Melania succinulata, ANTHONY, Bost. Proc., iii, p. 363, Dec., 1850. BINNEY, Check List, No. 258. BROT, List, p. 59.

Melana varicosa, Ward, HALDEMAN, Monog. Limniades, Part iii, p. 3 of Cover, March 13, 1854. ANTHONY, List, 1st and 2d editions. JAY, Cat., 4th edit., p. 275. BINNEY, Check List, No. 284. CATLOW, Conch. Nomenc., p. 189.

Melania livida, REEVE, Monog. Melania, sp. 434. BROT, List, p. 30.

Goniobasis Grosvenorii, LEA, Proc. Acad. Nat. Sci., p. 263, 1862. Jour. Acad. Nat. Sci., v, pt. 3, p. 297, t. 37, f. 128, March, 1863. Obs., ix, p. 119.

Melania Babylonica, LEA, Philos. Proc., ii, p. 14, Feb., 1841. Philos. Trans., viii, p. 183, t. 6, f. 43. Obs., iii, p. 21. DEKAY, Moll., N. Y., p. 98. WHEATLEY, Cat. Shells U. S., p. 24. BINNEY, Check List, No. 23. CATLOW, Conch. Nomenc., p. 185. BROT, List, p. 36.

Description.— Shell small, conic, turreted; spire acute at the apex, Fig. 538. Fig. 537. Fig. 536. the four apical volutions carinate below; volu-

tions about eight, somewhat convex; suture moderately impressed; surface, especially of the body-whorl, slightly wrinkled; labrum a little prominent near the base; within slightly tinged with reddish-brown.

Observations.— This pretty little species occurred in great numbers in a small stream in Kentucky. It may be distinguished from our other species by its small size, combined with the existence of a carinated line only formed in its immature state; having increased to four or five volutions the carina is no longer formed.— *Say.*

The following are synonymes :—

Melania exilis.— Shell long and slender, composed of about eight convex whorls; apex pointed; suture deep; aperture narrow, elliptic, equally curved on both sides; labrum much advanced anteriorly.

Habitat.— Kentucky and Ohio.

Length, ¾ of an inch.

Observations.— More slender than *M. simplex,* Say.— *Haldeman.*

Melania rufula.— Shell lengthened, conical, composed of eight whorls, the four anterior of which are convex, and those of the apex flat; suture well marked; spire twice the length of the aperture; apex suddenly tapered to a point; aperture ovate, elliptic.

Fig. 539.

Habitat.— Lake Pepin.

Length, 1 inch.

Observations.—Distinguished from *M. simplex* by having the peritreme level, and from *M. Virginica* by the flattened apex.— *Haldeman.*

Melania Kirtlandiana.— Shell smooth, acutely conical, rather thick, shining, horn-colored; spire elevated towards the apex, carinate; sutures impressed; whorls nine, rather convex; aperture small, elliptical, whitish.

Fig. 540.

Habitat.— Richmond, Indiana: Duck Creek near Cincinnati and Miami, Ohio: Little Miami.

Diameter, ·30; length, ·87 of an inch.

Observations.— This is a finely formed, graceful species, with an indistinct carina on the lower part of the whorls, near the apex. The aperture is nearly one-third the length of the shell. I name it after Professor Kirtland of Poland, Ohio.—*Lea.*

Fig. 541.

Melania inornata.— Shell moderate in size, rather solid, ovately lanceolate, simple, yellowish-green, deeper below, and paler at the sutures; whorls eight, the apical ones carinate, the last equal to two-fifths the length of the shell. Aperture a third of the total length, narrowly lunate, subacute before produced; columella narrow, white, with a callus in front.

Habitat.— Lorrain County, Ohio.

Longitude, seven-eighths; latitude, three-tenths poll.

Observations.—A simple species like *M. simplex* and *M. gracilis.* Its pale, sutural region is perhaps its most obvious character.—*Anthony.*

Melania bicolorata.— Shell small, slender, brownish-green, at the sutures flavescent; whorls 6–7, flattened, encircled above with nar-

Fig. 542.   row lines, the last expanded in front. Aperture ovate; lip dilated in front, sinuate behind; tinged with pink.

Habitat.— Camp Creek, near Madison, Indiana.

Longitude, ⅛; latitude, ¼ poll.

Observations.— An unadorned species, rather remarkable for its elongated, slender form, and well rounded whorls. It comes near *M. exilis* and *M. terebralis* having the lip threaded as in these species.— *Anthony.*

Melania elata.— Shell thin, gracile, elongate, light horn-color, paler at the sutures; whorls 8–9, rather flat, carinate above; aper- Fig. 543. ture ovate, effused before; columella thin.

Habitat.— Maumee River, Ohio.

Longitude, one; latitude, three-tenths poll.

Observations.— A plain, slender species of an unusually pale color. The whorls vary much in obliquity and convexity. It is similar in many respects to *M. bicolorata.*— *Anthony.*

Melania succinulata.— Shell elongate, acuminate, ovately conical, thin, plain, pinkish, horn-colored; whorls 7–10, rather convex, the apical ones carinate at the sutures, the last equalling two-thirds the length of the shell, subattenuate in front; aperture narrow, ovate, contorted, somewhat dilated in front.

Habitat.— Ohio.

Length, ⅜; width, ¼ of an inch.

Observations.— A smooth, delicate species, much thinner than usual, and when well cleaned nearly as transparent and amber-colored as a *succinea.* It may be compared with *M. clavæformis.*—*Anthony.*

Fig. 544.   *Melania varicosa.*— Shell olivaceous, conical, with seven convex whorls, flattened at the apex; later whorls marked with thick, varicose lines; aperture elliptic.

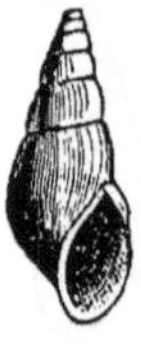

Habitat.— Ohio.

Length, ¾ of an inch.

Observations.— Allied to, but less slender than, *M. exilis.* It may prove to be a variety of *M. rufula,* Hald.— *Haldeman.*

Melania angustispira.— Shell thick, elongate, very slender; color reddish-brown, with a narrow, pale line at the suture; whorls 9–10, lower ones subconvex, smooth, upper ones flattened and carinate near their bases; sutures slight; aperture narrow, ovate, within pale purple; columella regularly curved; sinus not remarkable. Fig. 545.

Habitat.— Tennessee.

Observations.— May be compared with *M. exilis*, Hald., than which it is more slender, more attenuate and of more solid texture; its color is also entirely different, being more like *M. Warderiana*, Lea, but wanting the peculiar, bulbous form of that species. The carinations do not extend to the three lower whorls; upon these they are entirely wanting. It is a peculiarly slender and graceful species.— *Anthony.*

Goniobasis Grosvenorii.— Shell smooth, subattenuate, thin, horn-color, bright without bands; spire subattenuate, pointed, carinate at the apex; sutures regularly and very much impressed; whorls eight, Fig. 546. convex; aperture small, subrotund, white within; outer lip acute, slightly sinuous; columella bent in, thin and contorted.

Habitat.—Fox River, Illinois; H. C. Grosvenor: and Quincy, Ohio; J. Clark.

Diameter, ·29; length, ·79 of an inch.

Observations.—I have about a dozen specimens from Quincy, and one from Fox River. The former are fresh, and of a dark horn-color. The latter is whitish and probably bleached, being evidently a dead shell. It is allied to *M. varicosa*, Ward, and is very much the same outline and size, but it has no veins and has no light line below the sutures. The aperture is not quite one-third the length of the shell. I name it after Mr. Grosvenor, to whom I am indebted for the specimen from Fox River, and many other species.— *Lea.*

Messrs. Anthony and Haldeman's species, described above, are all figured from their types. Mr. Lea's are copies from his plates. The shells indicated by the above several descriptions embrace very great variety in form and convexity of the whorls, still I cannot, with several thousand specimens before me, ascertain the dividing line, they all seem to merge together.

With regard to *exilis*, Hald., there is no doubt of the type belonging to this species, but a very narrow, elongated form, of many flattened whorls, has received the name *exilis* in most of our collections, although it does not at all resemble the type, but is a new species, *G. Haldemani* (nobis). *G. semicarinata* is found in Kentucky, Tennessee and in all the Northwestern States and is everywhere within their limits, a very abundant species.

I also add the following to the synonymy of this species; the description is drawn up from a single specimen, a scalariform monstrosity :—

Melania Babylonica.— Shell carinate, turreted, rather thick; spire rather elevated, striate at the apex; sutures impressed; whorls seven, Fig. 547. angular above; aperture rather large, elliptical, white.

Habitat.— Yellow Springs, Green Co., Ohio.

Diameter, ·36; length, ·78 of an inch.

Observations.— A single specimen only of this shell has come under my notice. If the prominent character of this specimen, the large carina on the superior part of the whorls, be persistent, it marks a very distinct species. On the first four whorls the striæ are well defined. On the remaining three the carina alone exists. The aperture is more than one-third the length of the shell.— *Lea.*

181. G. Haldemani, Tryon.

Goniobasis Haldemani, Tryon, Am. Journ. Conch., i, p. 38, t. 1, f. 8, Feb. 25, 1865.
Melania acuta, Lea, Bell, Canadian Nat., iv, pt. 3, p. 213. Lewis, Bost. Proc., vi, p. 2.
Melania exilis, Haldeman, Adams, Moll. Vermont.

Description.— Shell narrowly elongated; whorls nine, smooth, flat, the last subangulated at the periphery; aperture small, subrhomboidal; lip slightly sinuous; columella incurved; color light horn, not banded, yellowish within.

Habitat.— Lake Erie; Lake Champlain.

Diameter, ¼ of an inch; length, 1 inch.

Observations.— Resembles *P. elevatum,* Say, but differs in the aperture, is still more narrowly elongated, and the whorls more Fig. 547a. flattened, and is entirely without striæ. In this last respect it differs widely from that species, and much resembles *P. Conradi* (nobis). This species has long been known in our cabinets as *G. exilis,* of Haldeman, but does not resemble that species in the remotest degree, as *exilis* is wider, with more convex whorls, and a larger aperture.— *Tryon.*

182. G. informis, LEA.

Goniobasis informis, LEA, Proc. Acad. Nat. Sci., p. 154, May, 1863. Obs., xi, p. 92, t. 23, f. 41.

Description.— Shell smooth, cylindrico-conical, dark horn-color, without bands; spire somewhat elevated; sutures irregularly impressed; whorls about seven, impressed in the middle; aperture rather small, nearly ovate, whitish within; outer lip acute, very sinuous; columella white and very much twisted.

Habitat.— Fall of the Ohio at Louisville, Ky.; W. H. DeCamp, M.D.

Diameter, ·19; length, ·60 of an inch.

Observations.— Only two specimens were sent to me by Mr. Currier, one of which is only about half-grown. It is very different Fig. 547b. from any species I have seen, having the appearance of being deformed by the impressed or constricted middle of the whorl. The bulging of the shoulder immediately below the suture has a corresponding thickening within. The outer lip is very much incurved above the middle of the whorl at the impressed portion of it. The aperture is nearly one-third the length of the shell.— *Lea.*

183. G. vittatella, LEA.

Goniobasis vittatella, LEA, Proc. Acad. Nat. Sci., p. 155, May, 1863. Obs., xi, t. 23, f. 38.

Description.— Shell smooth or subcarinate, conical, dark brown, single-banded; spire somewhat acuminate; sutures linear; whorls eight, flattened; aperture small, subrhomboidal, dark within; outer Fig. 548. lip acute, somewhat sinuous; columella bent in and twisted.

Habitat.— Cumberland Gap, East Tennessee; Major S. S. Lyon, U. S. A.

Diameter, ·20; length, ·55 of an inch.

Observations.— This is a pretty little species when perfect, but most of the specimens sent were imperfect, and covered with vegetable and mineral substances difficult to remove. There is a small, light band on the upper part of the whorls immediately below the suture, which is more or less visible on all the specimens before me, some of which have a carina on the upper terminal whorls. In outline and size it is near to *Melania* (*Goniobasis*) *glabra* (nobis), but

it is more slender, and that species has no band. The aperture is
about three-tenths the length of the shell.— *Lea.*

184. G. Alexandrensis, Lea.

Melania Alexandrensis, Lea. Philos. Proc., iv, p. 167. Philos. Trans., x, p. 61, t.
    9, f. 37. Obs., iv, p. 61. Binney, Check List, No. 8. Brot, List, p. 37.
Ceriphasia Alexandrensis, Lea, Adams, Genera, i, p. 297.

Description.— Shell smooth, rather acutely conical, rather thin,
dark horn-color; spire rather elevated; sutures somewhat impressed;
whorls rather flattened; aperture small and somewhat trapezoidal;
within whitish.

Habitat.— Alexandria, Louisiana.

Diameter, ·22; length, ·58 of an inch.

Observations.— There were only two of this species which came
Fig. 549. from Dr. Hale. It closely resembles the *Haleiana*, herein
described, but has a less elevated spire, and the aperture differs in being somewhat auger-shaped, the outer lip being more
sinuous. The apex of each being broken, the number of
whorls cannot be ascertained. The aperture is rather more
than a fourth of the length of the shell.— *Lea.*

Figured from Mr. Lea's plate.

185. G. Haleiana, Lea.

Melania Haleiana, Lea, Philos. Proc., iv, p. 167, Aug., 1845. Philos. Trans., x, p.
    60, t. 9, f. 35. Obs., iv, p. 60. Binney, Check List, No. 134. Reeve, Monog.
    Melania, sp. 406.
Ceriphasia Haleiana, Lea, Adams, Genera, i, p. 297.

Description.— Shell smooth, acutely conical, rather thin, yellowish
horn-color, polished; spire elevated; sutures impressed; whorls nine,
convex; aperture small, ovate, at the base angular, within Fig. 550.
whitish.

Habitat.— Alexandria, Louisiana.

Diameter, ·17; length, ·64 of an inch.

Observations.— Among some fifty specimens of small *Melaniæ* sent by Dr. Hale, I found three species, nearly the whole, however, being of the above described. It has no very distinctive
character, but cannot be placed with any species with which I am

acquainted. It resembles some of the young varieties of *M. Virginica*, Say, but has the whorls more convex, and the aperture smaller. Four or five specimens are banded, and these have uniformly two bands, the inferior one being larger and much more distinctly marked. The first few whorls of the apex are carinate. The aperture is about one-fourth the length of the shell.— *Lea.*

The figure given by Reeve is perhaps the same as *Haleiana*, but differs considerably.

186. G. rubella, Lea.

Goniobasis rubella, Lea, Proc. Acad. Nat. Sci., p. 270, 1832. Jour. Acad. Nat. Sci., v, pt. 3, p. 332, t. 38, f. 191, March,.1863. Obs., ix, p. 154.

Description.— Shell carinate, awl-shaped, rather thin, reddish, without bands; spire attenuate; sutures very much impressed; whorls eight, somewhat convex; aperture very small, subrhomboidal, whitish or reddish within; outer lip acute, sinuous; columella slightly bent in and twisted.

Habitat.— Near Murphy, Cherokee County, North Carolina; Prof. Christy.

Diameter, ·23; length, ·68 of an inch.

Observations.— I have eight specimens before me, sent some years since by my late friend, Mr. Clark, being part of the collection Fig. 551. made by Professor Christy. In form and size this species is very near to *Melania (Goniobasis) teres* (nobis), but it differs in being carinate, and having striæ which in all the specimens reach more than half way down from the apex. *Teres* is not striate. In the aperture there is also a difference. The aperture is about two-sevenths the length of the shell.— *Lea.*

187. G. spinella, Lea.

Goniobasis spinella, Lea, Proc. Acad. Nat. Sci., p. 264, 1862. Jour. Acad. Nat., Sci., p. 264, 1862. Jour. Acad. Nat. Sci., v, pt. 3, p. 298, t. 37, f. 130, March, 1863. Obs., ix, p. 120.

Description.— Shell smooth, very much attenuate, thin, dark olive, without bands; spire very much raised, sharp-pointed; sutures regularly impressed; whorls about nine, flattened; aperture very small, ovate, whitish within; outer lip acute, slightly sinuous; columella bent in and slightly thickened below.

Habitat.— Sycamore, Claiborne County, Tennessee; J. Lewis, M.D. Diameter, ·20; length, ·67 of an inch.

Observations.—A single specimen only was received from Dr. Lewis. It is nearly of the size of *Melania* (*Goniobasis*) *terebralis* (nobis), Fig. 552. but is a slimmer and darker colored species. It is very nearly of the same outline of *Melania* (*Goniobasis*) *strigosa* (nobis), but much smaller, slimmer and darker color. The specimen before me has neither folds nor angle on the apical whorls. Below the sutures there is a line of a lighter green. The aperture is about one-fifth the length of the shell.— *Lea.*

Of a large number before me many specimens have folds and the upper whorls angular.

188. G. Draytonii, Lea.

Goniobasis Draytonii, Lea, Proc. Acad. Nat. Sci., p. 264, 1862. Jour. Acad. Nat. Sci., v, pt. 3, p. 300, t. 37, f. 134, March, 1863. Obs., ix, p. 122.
Goniobasis nigrina, Lea, Proc. Acad. Nat. Sci., p. 263, 1862. Jour. Acad. Nat. Sci., v, pt. 3, p. 299, t. 37, f. 133. Obs. ix, p. 121.

Description.— Shell smooth, conoidal, somewhat thick, dark chestnut-brown, without bands, or obscurely banded; spire somewhat raised; sutures very much impressed; whorls about six, convex; aperture small, ovate, dark brown within; outer lip acute, slightly sinuous; columella very much bent in and twisted.

Operculum subrotund, thin, light brown, with the polar point well towards the middle on the left.

Habitat.—Fort George, Oregon; J. Drayton: also at Walla.

Diameter, ·27; length, ·68 of an inch.

Observations.— A number of these specimens were sent to me by Professor J. Henry, Secretary of the Smithsonian Institution, having been collected by the late Mr. Drayton, and to his memory Fig. 553. I dedicate it. It is allied to *Melania* (*Goniobasis*) *nigrina* (nobis), but it is not so polished and is a much thicker shell. Some of the specimens before me have a thickened outer lip, with a lighter margin. The deep color within is made by broad, obscure bands. Some of the specimens have a white thickening in the interior at the base, and some have a lighter brown mark on the exterior at the base of the axis.—*Lea.*

Goniobasis nigrina.—Shell smooth, small, conical, rather thin, nearly black, polished; spire somewhat elevated; sutures impressed;

whorls regularly convex; aperture small, ovate, angular above, dark purple within; columella incurved, purple.

Operculum dark brown, the polar point being low down and near to the left margin.

Habitat.— Clear Creek, Shasta County, California; Dr. Trask.

Diameter, ·23; length, ·67 of an inch.

Observations.— A number of good specimens with their opercula were sent to me by Dr. Trask. In form, size and color, this species is very like to *Melania semicarinata*, Say, from Georgia and South Carolina. It may be distinguished at once by not having the carination of that species which is usually strongly marked. It is not quite so high in the spire, and the aperture is more rounded at the base. In all the specimens of *nigrina* which I received, the apex is worn off. In the half grown ones I can see no disposition to carination or plication in the upper whorls. I should suppose that in perfect specimens, the number of whorls would be found to be about seven, and that the aperture would be about the third of the length of the shell. In some of the specimens there is a disposition to put on a few, fine striæ, and in most of them there is a very small angular line running below the suture. I am not acquainted with Dr. Gould's *Melania silicula* and *bulbosa* from Oregon, described in the Proc. Boston Soc. Nat. Hist., July, 1847; but from the descriptions, I have no doubt that they are different from both species herein described.—*Lea.*

189. G. proxima, SAY.

Melania proxima, SAY, Jour. Acad. Nat. Sci., p.126, Sept., 1825. BINNEY's edit. of Say, p. 115. BINNEY, Check List, No. 220. DEKAY, Moll. N. Y., p. 99, WHEATLEY, Cat. Shells U. S., p. 26. GIBBES' Report, p. 19. JAY, Cat., 4th edit., p. 274. BROT, List, p. 38.

Juga proxima, Say, ADAMS, Genera, i. p. 304.

Melania carinata, RAVENEL, Cat., p. 11, 1834. WHEATLEY, Cat. Shells U. S., p. 24. BINNEY, Check List, No. 47.

Melania Taitiana, LEA, Philos. Proc., ii, p. 11, Feb., 1841. Philos. Trans., viii, p. 165, t. 5, f. 5. Obs., iii, p. 3. DEKAY, Moll. N. Y., p. 92. WHEATLEY, Cat. Shells U. S., p. 27. JAY, Cat., 4th edit., p. 275. BINNEY, Check List, No. 274. CATLOW, Conch. Nomenc., p. 180. REEVE, Monog. Melania, sp. 444. BROT, List, p. 37.

Melania approximata, HALDEMAN, Monog. Limniades, No. 4, p. 4 of Cover, Dec., 28, 1841. JAY, Cat., 4th edit., p. 272. BINNEY, Check List, No. 18. BROT, List, p. 36.

Melania abjecta, Haldeman, REEVE, Monog. Melania, sp. 341. BROT, List, p. 34.

Goniobasis rubricata, LEA, Proc. Acad. Nat. Sci., p. 271, 1862. Jour. Acad. Nat. Sci., v, pt. 3, p. 325, t. 38, f. 107. Obs., ix, p. 157, t. 38, f. 107.

Description.— Shell conic, rather slender, black, gradually attenuated to the truncated apex; suture moderately impressed; aperture longitudinal within, milk-white; labrum with the edge not undulated, or but very slightly and obtusely so near the superior termination.

Length to the truncated apex, nearly three-fifths; greatest breadth less than ¼ of an inch.

Observations.— Professor Vanuxem obtained this species in a small brook, which discharges into the Catawba River, near Landsford, Chester district, South Carolina, and also in the Warm Springs, Buncombe County, North Carolina, and in the French Broad River of the same County. It resembles the preceding very closely (*simplex*, Say), but is decidedly more slender, and like that shell it has two elevated lines on the inferior margin of the terminal whorls. The interior of the aperture in many specimens is of a dull reddish color, and in some the same part exhibits the appearance of two or three obsolete bands. Another variety, which Mr. Vanuxem obtained from a limestone spring near Broad River, Spartanburg district, South Carolina, is of a pale horn color. In a stream of the Saluda range of mountains near Mill Gap in Rutherford County, he found another variety of a somewhat smaller size, tinged with reddish-brown, and generally distinctly banded within the aperture; one of these specimens is very remarkably truncated, presenting only about one whorl and a quarter. The same variety also inhabits a brook near the Table Rock. A variety, which seems to differ from the latter only in size, was found by Mr. Vanuxem, near Douthard's Gap of the Saluda mountains; the largest specimen he sent from that locality is only about three-tenths of an inch long.— *Say.*

Fig. 555.

Dr. Jay quotes *carinata*, Rav., as a variety, and I therefore include it in the synonymy of *proxima*. *Carinata* has not been described, nor have I seen an authentic specimen.

All of the following species are believed to be synonymes, giving this species a very wide range; I doubt, however, whether *abjecta* really inhabits Arkansas. The species does not vary much in form and is easily recognizable. It will be seen that the color and ornamentation, however, vary considerably.

The following are the descriptions of the synonymes :—

Melania approxima.— Shell lengthened, conical, tapering gradually to the truncated apex; upper whorls carinated; aperture ovate, tinted with pink; color light brown, with two dark reddish, Fig. 556. approximate, narrow, revolving lines.

Fig. 557. 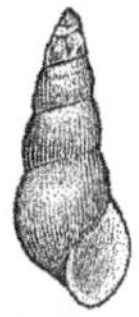 Fig. 558.

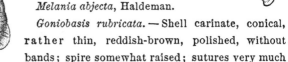

Habitat.— Tennessee.

Length, ½ an inch.—*Haldeman.*

Melania abjecta, Haldeman.

Goniobasis rubricata. — Shell carinate, conical, rather thin, reddish-brown, polished, without bands; spire somewhat raised; sutures very much impressed; whorls about seven, convex; aperture rather large, rhomboidal, pale reddish within; outer lip acute, scarcely sinuous; columella bent in, somewhat thickened.

Operculum ovate, dark brown, with the polar point near the base on the left.

Habitat.— Tennessee; Professor Troost.

Diameter, ·29; length, ·71 of an inch.

Observations.— These specimens sent to me long since by the late Professor Troost are nearly all truncate. I formerly considered them a variety of *Melania (Goniobasis) proxima*, Say, but it is a larger species, more exserted, and has a peculiar appearance in the Fig. 559. whorls of the spire assimilating to a coiled rope. Several young specimens are perfect to the apex, which shows that all are more or less carinate, but very obtusely so. The decollate specimens have no appearance of a carina on the lower whorls. All the specimens were covered with the black oxide of iron, which being removed, the epidermis is found to be smooth, polished and bright reddish-brown. Usually the upper part of the whorl is slightly impressed, which gives to the curve of the whorl a peculiar form. The columella is usually light brown, and some specimens have a whiteness about the middle portion. The aperture is about two-sevenths the length of the shell.— *Lea.*

Fig. 560.  *Melania Taitiana.*— Shell smooth, conical, rather thin, shining, horn-color; spire truncate, carinate towards the apex; sutures impressed; whorls rather convex; aperture small, elliptical, subangular at base, whitish.

Habitat.— Alabama River, Claiborne; Judge Tait.

Diameter, ·25; length, ·80 of an inch.

Observations.— Several years previously to the death of my friend, Judge Tait, he sent me a number of this species, which in form resembles *M. blanda*, described herein. Most of them are without bands; some, however, are finely banded, and all are mutilated at the apex. I dedicate this species to my lamented friend, to whose kindness I owe so many beautiful and interesting objects in the natural history and geology of Alabama.— *Lea.*

190. G. rufescens, LEA.

Melania rufa, LEA, Philos. Proc., ii. p. 12, Feb., 1841.   Philos. Trans., viii, p. 167, t. 5, f. 8.   Obs., iii, p. 5.   TROOST, Cat. Shells Tennessee.   WHEATLEY, Cat. Shells U. S., p. 26.   CATLOW, Conch. Nomenc., p. 188.
Melania rufescens, Lea, DEKAY, Moll. N. Y., p. 93.   JAY, Cat., 4th edit., p. 274. BINNEY, Check List, No. 233.   BROT, List, p. 37.
Potadoma rufescens, Lea, ADAMS, Genera, i, p. 299.

Description.— Shell smooth, turreted, rather thin; shining, dark red; spire elevated; sutures impressed; whorls convex, towards the Fig. 561. apex carinate; aperture small, elliptical, subangular below, within purplish.

Habitat.— Mamma's Creek, Tennessee; S. M. Edgar.

Diameter, ·30; length, ·85 of an inch.

Observations. — In form this species resembles *M. teres*, herein described. It differs in the color being red, and in being carinate on the superior whorls. The most perfect specimen in my possession has the first few whorls broken; I should suppose a perfect one would have eight whorls, and the aperture be one-fourth the length of the shell.— *Lea.*

This species is longer, narrower and darker colored than Tennessee specimens of the preceding species.

I.   *Striate species, spire elevated.*

191. G. Virginica, GMELIN.

Buccinum Virginica, Gmelin, Syst. Nat. 3505.   GREEN, Trans. Alb. Inst., i, p. 135 WOOD, Index Test., t. 24, f. 154.   Schröter, Einleit., i, p. 414, 1783. MARTINI, Berlin Mag., iv, p. 348, t, 10, f. 48.   SCHREIBERS, Einleit. Conchyl., t. 113, f. 7.
Paludina Virginica, SAY, Nicholson's Encyc., iii, t. 2, f. 4.
Melania Virginica, SAY, Am. Conch., pt. 5, t. 47, f. 2.   App. to Long's Exped., ii, p. 265.   BINNEY's edit., p. 131 and 199.   BINNEY, Check List, No. 291.   CATLOW Conch. Nomenc., p. 189.   PHILIPPI, Neüer Conchylien, Melania, t. 2, f. 12, HILDRETH, Am. Jour. Science, xxxi, p. 53.   SAGER, Rept. Zool. Mich., p. 15, CONRAD, Am. Jour. Science, N. S., i, p. 407.   HALDEMAN, Rupp's Hist. Lan-

caster County, Pa., p. 479. HALDEMAN, Am. Jour. Sci., xli, p. 22. DEKAY, Moll. N. Y., p. 90, t. 7, f. 141. WHEATLEY, Cat. Shells U. S., p. 27. HARTMAN, Catalogue Shells, Chester Co., Pa. BROT, List, p. 35. GIRARD, Proc. National Inst., i, No. 2, p. 82. JAY, Cat., 4th edit., p. 275. REEVE, Monog. Melania, sp. 321. VILLA., Cat., Syst. p. 36, 1841.

Io Virginica, Say, MORCH, Yoldi Cat. p. 56.

Ceriphasia Virginica, Gmel., ADAMS, Genera, i, p. 297.

Juga Virginica, Say, CHENU, Man. de Conchyl., i, f. 2019. ADAMS, Genera, i, p. 304.

Melania multilineata, Say, Jour. Acad. Nat. Sci., ii, p. 380, Dec., 1822. Am. Conch., pt. 5, t. 47, f. 2. BINNEY's edit., pp. 111 and 199. BINNEY, Check List, No. 169. DEKAY, Moll. Rept. to Regents, p. 32. Moll. N. York, p. 97. WHEATLEY, Cat. Shells U. S., p. 26. HARTMAN, Cat. Shells Chester Co., Penn. CATLOW, Conch. Nomenc., p. 187. GIRARD, Proc. Nat. Inst., i, No. 2, p. 82, March, 1856. PHIL-IPPI, Neüer Conchyl. Melania, t. 2, f. 13.

Juga multilineata, Say, ADAMS, Genera, i, p. 304.

Melania auriscalpium, MENKE, Syn., Meth., p. 136, 1830.

Melania curta, MENKE, Syn. Meth., p. 136, 1830.

Melania fasciata, MENKE, Syn. Meth., p. 136, 1830.

Melania bizonalis, DEKAY, Moll. N. Y., p. 91, t. 7, f. 140, a. b. 1843. BINNEY, Check List, No. 35.

Melania Buddii, DeKay, WHEATLEY, Cat. Shells U. S., p. 24.

Melania gemma, DEKAY, Moll. N. Y., p. 91, t. 7, f. 142, 1843. BINNEY, Check List, No. 119. BROT, List, p. 38.

Melania strigillata, MUHLFELDT, MSS.

Melania inemta, ANTHONY, Bost. Proc., iii, p. 362, Dec., 1850. BINNEY, Check List, No. 145. BROT, List, p. 58.

Description.— Shell turreted, usually truncate, eroded at tip, olivaceous or blackish-brown; whorls about six, but little rounded, crossed by obvious wrinkles; a dull reddish line revolves near the

Fig. 565a.   Fig. 564.   Fig. 563.   Fig. 565.   Fig. 562.   Fig. 566.

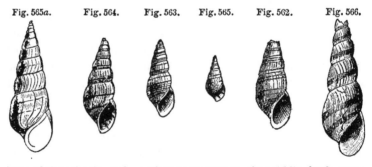

base of the whorls, and another near or upon the middle, both sometimes obsolete or wanting; labrum a little prominent towards the base. Animal bluish-white beneath, with orange clouds each side of the mouth: above pale orange, shaded with dusky and banded with numerous black interrupted lines; mouth advanced into a rostrum as long as the tentacula, which are darker at base, and setaceous; foot with an undulated outline. Var. A. Shell destitute of the rufous bands.

Observations.— This species is very abundant in the Delaware and

Schuylkill Rivers. The basal portion of the labrum in Lister's figure of plate 113, fig. 7, above quoted, is deficient, nevertheless I have no doubt that the figure was intended for this species, and that his lower figure on plate 109 is intended to represent the variety.—*Say.*

The above description applies only to the smooth variety, between which and *multilineata*, every grade occurs. Several of these have been described by DeKay and Menke as distinct species. Were it not for these intermediate stages, and the long continued observations upon this species, in consequence of its favorable habitat, the two extremes would certainly be considered distinct, as Say classed them.

The following are the descriptions referred to.

Melania multilineata.— Shell gradually tapering; apex generally much eroded; whorls about seven, a little convex, with numerous,

Fig. 567.

filiform, elevated, subequal lines, which are from ten to twenty in number on the body-whorl.

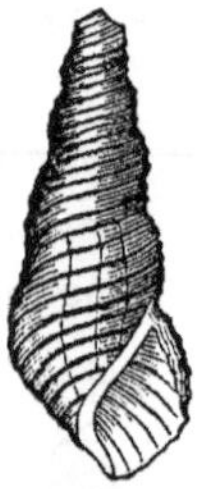

Habitat.— Tributaries to the Delaware.

Length, nineteen-twentieths; greatest width, two-fifths of an inch.

Observations.— I found several specimens of this shell in Frankford Creek, and Professor Vanuxem presented me with others which he obtained from a creek in New Jersey. The *M. elevata* (p. 95 of this work), from its attributed specific characters, might be supposed to be nearly related to this shell, but it differs in being of a more accurate conic form, the whorls being flattened, and not convex as in this species; its raised lines are also few in number.

Synonyme.—*M. curta*, Menke, Synop., Mollusc., p. 81.—*Say.*

Melania curta.— Shell ovately oblong, subturreted; apex cariously truncated, transversely, sulcately striate, brownish-black; aperture oval; lip produced in front.

Habitat.—Philadelphia; Bescke.

Longitude, 7 lin.; latitude, 4 lin.— *Menke.*

Melania fasciata.— Shell conically oblong, turreted; apex eroded, greenish, semipellucid, with a few obsolete sulci, last whorl doubly brown-banded, the others with a single band; lip marginal, rounded, produced in front.

Habitat.— Philadelphia; Bescke.

Longitude, 11; latitude, 4½ lin.—*Menke.*

Melania bizonalis.— Shell tapering, elongated; whorls seven or eight, flattened; the upper whorls with a revolving, strongly cari- nated line just above the suture, and above this two Fig. 568. slightly, but distinctly, elevated, revolving lines; all the volutions with sinuous, vertical, elevated lines becoming obsolete towards the tip; aperture subovate, angular above, and uniting with a broad, white callus on the pillar lip; tip rarely perfect; color olivaceous-brown; epider- mis with two and rarely three dark reddish, revolving lines on the body-whorl, often indistinct, but may be traced.

Length, ·7· Length of aperture, ·23; width of aperture, ·16·

Observations.— For this species I am indebted to Dr. Evans who found it abundantly in Lake Champlain. It approaches *Melania Virginica*, but is, as I view it, very distinct by its flattened whorls and deep, angular sutures.— *DeKay.*

Melania gemma.— Shell moderately large, oblong; spire attenuated, acute; the whole surface covered with waved, vertical wrinkles; whorls eight, all distinctly carinate near the middle, and very acutely so on the apical whorls; on the lower whorls this carina is below the middle, but becomes medial above, in some specimens the lower whorls are bicarinate, or rather the carina is slightly furrowed on its edge; suture deep, occasionally cancellate; the body-whorl has one or more rounded grooves on each side of the carina, which produces corresponding minute, elevated ridges; lip fragile, its margin convex, rarely perfect; color variable from straw-yellow to amber and deep reddish-brown; columella often purple; lower sutures opaque, white.

Length, ·7-1·2 inches. Length of aperture, ·23 of an inch.

Observations.— This species was obtained from Mud Creek, Onon- daga County by Dr. Budd, and was at first referred to the *semicarinata* Fig. 569. of Say, hitherto supposed to be an exclusively western 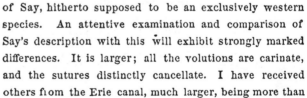 species. An attentive examination and comparison of Say's description with this will exhibit strongly marked differences. It is larger; all the volutions are carinate, and the sutures distinctly cancellate. I have received others from the Erie canal, much larger, being more than an inch long. In these the revolving groove, in descending, gradually approaches nearer the suture, and is continued on the body-whorl, which is vertically rugose. In my catalogue of species, I had named this species after its discoverer, but the practice has been so much abused, it is becoming daily obsolete. I trust that the name proposed will suggest that of the gentleman to whom I have been under many obligations in this department.—*DeKay.*

Melania inemta.—Shell elongate, turreted; apex eroded, unicolored, brownish-green; whorls 3-4, very convex. The last gibbose, constricted behind; sutures impressed; aperture broadly lunate, scarcely effuse; lip brownish.

Habitat.— Virginia.

Observations.—Possibly this may be a largely truncated specimen of *M. Virginica,* which it resembles in its aperture. The form of the ultimate whorl is unusual. — *Anthony.*

Philippi (Neuer Conchyl.) is very much mistaken in his remarks relative to the wide distribution of this species, as it certainly has never been found near Cincinnati nor in Central America. This shell is the only *Melania* inhabiting the eastern portion of the Middle States and is nowhere found in the tributaries of any of the western rivers. As the striate and smooth varieties are frequently observed in conjunction, and as the young shells appear indifferently smooth or striate, there can be no doubt that they all form one species.

Philippi figures the following *varieties* of *multilineata:*—

a. *Sulcosa* equally transversely striate; last whorl one-banded.

b. *Ligata* transverse striæ unequal, two-banded.

c. *Fasciata* rarely-obsoletely sulcate, two-banded.

d. *Concolor* without bands.

The first figures represent specimens from Delaware River. The figures of *gemma* and *bizonalis* are copied from DeKay's work.

192. G. sulcosa, LEA.

Melania sulcosa, LEA, Philos. Proc., ii, p. 14, Feb., 1841. Philos. Trans., viii, p. 185, t. 6, f. 48. DeKay, Moll. N. Y., p. 99. Troost, Cat. Shells Tenn.

Fig. 570.   Catlow, Conch. Nomenc., i, p. 189. Binney, Check List, No. 259.
           Wheatley, Cat. Shells U. S., p. 27. Brot, List, p. 35.
     Ceriphasia sulcosa, Lea, Chenu, Man. de Conchyl., i, f. 1957. Adams,
           Genera, i, p. 297.

Description.—Shell transversely sulcate, conical, thick, yellowish; sutures impressed; whorls flattened; aperture small, ovate, whitish.

Habitat.— Tennessee.

Diameter, ·32; length, ·75 of an inch.

Observations. — A single specimen only, and that imperfect, is before me. The body-whorl has seven or eight distinctly marked striæ. On the penultimate there are three, and these give a sulcate appearance to the shell.—*Lea.*

When perfect specimens are obtained this shell may be found to be a species of *Pleurocera* instead of *Goniobasis*.

193. G. Buddii, LEA.

Melania Buddii, LEA, Philos. Proc., iv, p. 165. Philos. Trans., x, p. 64, t. 9, f. 44. Obs., iv, p. 64. BINNEY, Check List, No. 42. JAY, Cat., 4th edit., p. 273. REEVE, Monog. Melania, sp. 324.
Juga Buddii, Say, H. and A. ADAMS, Genera, i, p. 304.

Description. — Shell striate, cylindrical, rather thin, horn-color; spire attenuated; sutures impressed; whorls flattened; aperture small, elliptical, within whitish.

Habitat.— Tennessee.

Diameter, ·32 of an inch; length, 1·07 inches.

Observations.—I have two specimens before me, both of which have seventeen revolving striæ on the lower whorl. They have also a single small band immediately below the middle of the body-whorl, which is hidden on the superior whorls. Each of the specimens under examination has the apex broken, but I presume the number of whorls may reach to ten. Eight may be counted in one of these. Dr. Budd mentions, in a note, that "out of six, five have a band." The aperture is about one-fourth the length of the shell. This species is nearly allied to the striate variety of Mr. Say's *M. Virginica*, which he called *multistriata* (*multilineata*, G. W. T., Jr.). The *Buddii* may be distinguished by its being flattened on the whorls, in being more angular on the superior part of the whorls, and in being more attenuate.—*Lea.*

Fig. 571.

Figured from Mr. Lea's plate. This shell is so very closely allied to *Virginica* that Dr. Brot has placed it in the synonymy of that species.

194. G. Troostiana, Lea.

Melania Troostiana, Lea, Philos. Proc., ii, p. 34, April, 1841. Philos. Trans., p. 92, t. 23, f. 86. Obs. ii, p. 92. DeKay, Moll. New York, p. 100. Wheatley, Cat. Shells U. S., p. 27. Binney, Check List, No. 276. Troost, Cat. Shells Tenn. Jay, Cat., 4th edit., p. 275. Catlow, Conch. Nomenc., p. 189. Brot, List, p. 35. Reeve, Monog. Melania, sp. 339. Troschel, Archiv fur Naturgesch., ii, p. 227.

Juga Troostiana, Lea, Adams, Genera, i, p. 304.

Description.— Shell elevated, brown, thickly striated; apex acute; whorls ten, above carinate; aperture oval.

Habitat.— Mossy Creek, Jefferson Co., Tennessee.

Diameter, ·5 of an inch; length, 1·2 inches.

Observations.— I owe to Professor Troost this interesting species.

Fig. 571a.

It differs from any American species with which I am acquainted, in having a sharp carina, which is placed on the superior part of the inferior whorls. In its numerous striæ it resembles the *M. multilineata,* Say, which is now I believe conceded to be only a variety, much striated, of *M. Virginica* of the same author. Most of the specimens, which have come under my notice, are white inside, with a purple spot on the columella, and an indistinct, light band along the inferior part of the suture. Some individuals are, however, entirely purple inside, and this gives the epidermis quite a black appearance.— *Lea.*

195. G. latitans, Anthony.

Melania latitans, Anthony, Ann. Lyc. Nat. Hist., New York, vi, p. 88, t. 2, f. 6, March, 1854. Binney, Check List, No. 159. Brot, List, p. 34.

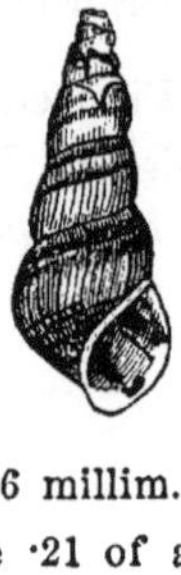

Description.—Shell conical, obscurely striate, greenish-brown, rather thin; spire elevated; whorls 8–9, convex or subangulated, with three or four transverse striæ above the angle, which become obsolete below it, and one or two brown bands at and above the middle of each turn; sutures distinct; lines of growth coarse, amounting almost to ribs on the lower whorls; aperture not large, subrotund or very broad ovate, reddish within and banded; columella very much curved and twisted, with a small sinus at base.

Fig. 572.

Habitat.— Mammoth Cave, Kentucky.

Diameter, ·39 of an inch (10 millim.); length 1 inch (26 millim.). Length of aperture, ·34 (9 millim.); breadth of aperture ·21 of an inch (5 millim.).

Observations.—Bears no very strong resemblance to any known species; but is perhaps more nearly allied to *M. rufa*, Lea, and *M. teres*, Lea, in its elevated spire and convex whorls. It wants, however, the smooth whorls of the former, its dark red color, and elliptical aperture. From the latter it may be distinguished by its striated whorls, its less slender proportions, the absence of folds, its obscure bands, and white aperture. This species is unusually interesting from the fact that it is the first species in conchology known to have been procured from the subterranean river flowing through Mammoth Cave.—*Anthony.*

196. G. porrecta, LEA.

Goniobasis porrecta, LEA, Proc. Acad. Nat. Sci., p. 155, May, 1863.

Description.—Shell striate, attenuate, blackish-brown, one-banded; spire attenuated, acuminate; sutures slightly impressed; whorls nine, flattened; aperture small, ovate, white or blackish within; lip Fig. 573. acute, scarcely sinuous; columella inflected and contorted.

Habitat.—Gap Creek and Spring, Cumberland Gap, East Tennessee; Captain S. S. Lyon, U. S. Army.—*Lea.*

A very distinct, and apparently abundant species, at its locality. I possess a number of specimens, most of which are not banded. They are generally covered with raised striæ, and the sutures almost canaliculate.

197. G. sculptilis, LEA.

Melania sculptilis, LEA, Philos. Trans., x, p. 297, t. 30, f. 3. Obs., v, p. 53, t. 30, f. 3. BINNEY, Check List, No. 238. BROT, List, p. 38.

Description.—Shell thickly striate, conical, rather thin, horn-color; spire pointed, towards the apex carinate and granulate; sutures irregularly impressed; whorls ten, rather flattened; striæ close, and
Fig.574. between them sculptured; aperture small, elliptical, angular at base, white within; columella incurved and twisted.

Habitat.—Tennessee.

Diameter, ·24; length, ·55 of an inch.

Observations.—Two specimens are before me, which are precisely alike. It is a very remarkable species, having regular and close striæ over the whole of the lower whorls, between which striæ there is a double row of minute, indented marks, very close to each other, and only visible with a lens. I have seen no such marks on any other species. In outline it is closely allied to *striatula* (nobis), but it is a smaller species, and has not the cancellation of that species. The aperture is rather more than one-third the length of the shell. The outer lip is broken.— *Lea.*

The specimen figured by Mr. Lea being imperfect, I give a figure from a shell' in Coll. Smithsonian Institute. This species is evidently described from an immature specimen.

198. G. crenatella, Lea.

Melania crenatella, LEA, Proc. Acad. Nat. Sci., v, pt. 3, p. 268, t. 35, f. 79, March, 1863. Obs., ix, p. 90. BINNEY, Check List, No. 76. BROT, List, p. 34. REEVE, Monog. Melania, sp. 457.

Description.— Shell transversely striate, high turreted, subcostate, somewhat folded, rather thin, dark brown, almost black; spire elevated, closely folded at the apex; sutures very much impressed; whorls seven, flattened, covered with transverse ribs; aperture small, oval, banded within; columella whitish, incurved; outer lip somewhat contracted and very crenulate.

Habitat.—Coosa River, Uniontown, Alabama; E. R. Showalter, M.D. Diameter, ·16; length, ·50 of an inch.

Observations.— Five specimens of this very beautiful little species are before me, all of which I owe to the kindness of Dr. Showalter. Fig. 575. Most of these have eleven closely-set, thread-like, transverse

ribs on the last whorl, which are very dark brown, while the interspace is yellowish. On the next whorl above there are usually six, and above these the number diminishes to three. There appear to be about seven whorls. Within the aperture of four out of the five specimens there are brown bands accompanying the lines of the outer ribs, and these terminate in little furrows at the edge, which cause the outer lip to be beautifully and regularly crenulate. One of the specimens has the ribs without color, and therefore it is without bands inside. It is allied to *Melania* (*Goniobasis*) *striatula* (nobis), but is a much smaller species, more cylindrical, of a darker color, and has stronger rib-like striæ.— *Lea.*

J. *Heavy, pupæform or cylindrical species.*

199. G. cylindracea, CONRAD.

Melania cylindracea, CON., New Fresh-water Shells, p. 55, t. 8, f. 10, 1834. MÜLLER, Synopsis, p. 47, 1836. BINNEY, Check List, No. 84.
Melania cylindrica, Con., WHEATLEY, Cat. Shells, U. S., p. 25. REEVE, Monog. Melania, sp. 311. BROT, List, p. 32.
Melania oppugnata, LEA, Philos. Trans., x, p 300, t. 30, f. 9. Obs., v, p. 56. BINNEY, Check List, No. 190.

Description.— Shell subcylindrical, smooth, with a short spire, the whorls of which are small; apex eroded; body-whorl angulated, obtusely rounded above, and at base; aperture Fig. 576. Fig. 577. more than half the length of the shell, narrow, much contracted above.

Observations.— This species is remarkable for the rude, almost deformed, whorls of the spire. It inhabits the Tombigbee River on the soft limestone banks, and is generally coated with a calcareous deposit.— *Conrad.*

Fig. 576 is a copy of Mr. Conrad's original figure. Fig. 577 is from an excellent specimen in Coll. Smithsonian. The numerous shells before me vary in color from brown to dull light green. Whilst most of them are unadorned, a few are banded with dark green. The identity of *cylindracea* and *oppugnata* is conceded by most American conchologists. The following is the description of the latter with a copy of the figure.

Melania oppugnata.— Shell smooth, truncate, cylindrical, very thick, yellowish horn-color; spire cut off; sutures large and very irregularly impressed; whorls very much compressed, geniculate above; aperture very long, very much narrowed, above callous; within white; columella twisted, and very much thickened above.

Fig. 578.

Habitat.— Alabama River.

Diameter, ·41; length,———?

Observations.— This is a very remarkable species. The two specimens before me are both cut off, leaving little more than the body-whorl. When taken they were evidently living and healthy specimens, but the eroded and fractured spires give them the appearance of old and diseased shells, which is by no means the case. The upper part of the whorl, along the suture, is irregularly fractured round the whole circle. This arises from the fact that the animal having filled up the channel with calcareous deposit, suddenly recommences at a new line of growth, some distance below, leaving open and bare of epidermal matter this upper portion of the channel, which, consequently having a sharp edge, becomes more or less fractured. The whorls are so much flattened that the two sides are nearly parallel. One of the specimens has a small spot of brown in the aperture above and below; the other has none. This species is allied to *auriculaformis* (nobis) on one side, and *olivula*, Conrad, on the other, but it may be easily distinguished from both of them. The former is a smaller shell and more fusiform; the latter is more conical, less thickened on the columella, and not irregularly fractured in the suture. The number of whorls or proportional size of the aperture cannot be ascertained on the specimens before me. They have the appearance of having been very much exposed to an attacking enemy, hence the name.—*Lea.*

In Coll. Haldeman are specimens labelled "Kemper County, Mississippi."

200. G. pupoidea, ANTHONY.

Melania pupoidea, ANTHONY, Ann. Lyc. Nat. Hist. N. Y., vi, p. 104, t. 3, f. 3, April, 1854. BROT, List, p. 33. BINNEY, Check List, No. 224. REEVE, Monog. Melania, sp. 249.
Melania propinqua, LEA, Proc. Acad. Nat. Sci., p. 119, 1861.
Goniobasis propinqua, LEA, Journ. Acad. Nat. Sci., v, pt. 3, p. 234, t. 34, f. 29, March, 1863. Obs., ix, p. 56.

Description.— Shell ovate-conic, smooth, rather thick; spire obtusely elevated, with a decidedly convex outline, and a well impressed suture; whorls seven, convex, nearly entire at the apex; color pale green, with one linear band revolving on the spire, and four broader

Fig. 579.

and more distinct bands on the body-whorl; aperture small, narrow ovate, diaphanous, with four distinct, brown bands within; columella rounded, not indented; outer lip curved and extended forward; sinus small.

Habitat.— Alabama.

Diameter, ·35 (9 millim.); length, ·87 of an inch (22 millim.). Length of aperture, ·38 (10 millim.); breadth of aperture, ·17 of an inch (4 millim.).

Observations.— This belongs to that group of which *M. olivula*, Conrad, may be considered the type. From that shell it differs, how-

ever, in being more elongate, and less ornamented with bands, as
well as by its paler and less varnished epidermis. Compared with
M. proteus, Lea, it is even more elongate and less acute; the aperture
is entirely different, and it wants the tuberculous shoulder which distinguishes that species. Its resemblance to the pupæ of some of the
insect tribes has suggested its characteristic.—*Anthony.*

The following is a synonyme.

Goniobasis propinqua. — Shell smooth, subcylindrical, somewhat
thick, yellowish, four-banded; spire somewhat raised, conical; sutures very much impressed; whorls six, flattened above; aperture
elliptical and rather small, whitish and banded within; outer Fig.580.
lip acute; columella slightly thickened and rounded below.

Habitat. — Coosa and Cahawba Rivers, Alabama; E. R.
Showalter, M.D.

Diameter, ·33; length, ·90 of an inch.

Observations.—This species is very closely allied to *Melania*
(*Goniobasis*) *pupoidea*, Anthony, but it differs in being more cylindrical, in being smaller, and in having the base of the aperture more
rounded. Most of the specimens are decollate. One has a few raised
striæ. In some there is a disposition to have a shoulder under the
sutures.— *Lea.*

Without the large series of specimens before me I should
have acquiesced in the institution of *propinqua* as a distinct
species, but I find every grade of form between the two. The
shorter forms become very close to *olivula*, Conrad, with which
indeed, they have been confounded. They are distinguished
by difference of color, and principally of texture, *olivula* being
much heavier.

201. G. lita, Lea.

Melania lita, Lea, Proc. Acad. Nat. Sci., 1861, p. 121.
Goniobasis lita, Lea, Jour. Acad. Nat. Sci., v, pt. 3, p. 239, t. 34, f. 40, March, 1863.
    Obs., ix, p. 61.

Description.— Shell rugosely striate, subfusiform, rather large, four-
banded, variegated, shining; spire obtusely elevated; sutures irregularly impressed; whorls six, convex above, the last elongate; aperture
somewhat constricted, elongately ovate, purplish and banded within;
outer lip acute, thickened; columella incurved and purple below,
rounded at the base.

Habitat.— Cahawba River, Alabama; E. R. Showalter, M. D.

Diameter, ·31; length, ·78 of an inch.

Observations.—I have seen but a single specimen of this species. It is remarkable for the several greenish and brownish tints of the exterior and its purple aperture. The apical whorls are pli-

Fig. 581.

cate. The two lower whorls have rather rugose striæ. Other individuals may differ from the characters given above. The aperture is about two-fifths the length of the shell. It is one of the pupoid group and is nearly allied to *fallax*, herein described, but it is not so cylindrical and the aperture is longer. It differs also in color.—*Lea.*

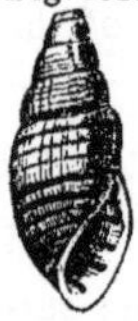

I am doubtful whether this is distinct from *Haysiana*. I have before me two or three specimens which appear to occupy an intermediate position between the two species. In the specimens I have examined, except Mr. Lea's type, the aperture is white within, instead of purple.

202. G. fallax, Lea.

Melania fallax, Lea, Proc. Acad. Nat. Sci., 1861, p. 120.
Goniobasis fallax, Lea, Journ. Acad. Nat. Sci., v, pt. 3, p. 231, t. 34, f. 24, March, 1863. Obs., ix, p. 53.

Description.—Shell smooth, pupæform, somewhat cylindrical, rather thick, either dark brown or dark horn-color, obscurely banded or without bands; sutures impressed; whorls seven, slightly convex, the last small; aperture small, very much constricted, elon-

Fig. 582.

gate elliptical; outer lip sharp; columella a little inflected, obtusely angular at the base.

Habitat.—Coosa River, Alabama; E. R. Showalter, M.D.

Diameter, ·34; length, ·96 of an inch.

Observations.—This species is nearly allied to *clausa*, herein described, but it is a smaller species, rather more cylindrical and with a smaller aperture. The dark specimens are four-banded, the bands being well defined inside, but obscure exteriorly. These dark ones have a light line below the suture. The aperture is not quite one-third the length of the shell.—*Lea.*

203. G. inosculata, Lea.

Goniobasis inosculata, Lea, Proc. Acad. Nat. Sci., p. 263, 1862. Journ. Acad. Nat. Sci., v, pt. 3, p. 206, t. 87, f. 126, March, 1863. Obs., ix, p. 118.

Description. — Shell smooth, pupæform, somewhat raised, rather thick, yellowish-brown, four-banded; spire somewhat raised; sutures very much and irregularly impressed; whorls seven, somewhat convex; aperture small, constricted, subelliptical, whitish within Fig. 583. and banded; outer lip acute; columella white, bent in, twisted and subangular at the base.

Operculum small, ovate, thin, dark brown, with the polar point near the base.

Habitat. — Coosa River, Alabama; E. R. Showalter, M.D. Diameter, ·37; length, ·89 of an inch.

Observations.—A species very closely allied to *Melania* (*Goniobasis*) *Alabamensis* (nobis), but it may be distinguished by its being smaller, more constricted, and being slightly more cylindrical. The bands are smaller and not quite so well expressed. When I received the first specimen, I considered it a small variety of *Alabamensis*, but having received others from Dr. Showalter, I cannot but consider it a distinct species inosculating on the other. The aperture is about one-third the length of the shell.— *Lea.*

This species is nearly related to *G. fallax*, Lea, and at first I united it with that species, but I am now convinced that it is distinct. Among the points of difference may be mentioned the greater convexity of its whorls, brighter color, and the constant ornamentation of four distinct, dark bands, the upper of which is the broadest. A single band appears on the whorls of the spire. *G. fallax* is a more cylindrical species.

204. G. Alabamensis, Lea.

Melania Alabamensis, LEA, Proc. Acad. Nat. Sci., 1861, p. 121.◆
Goniobasis Alabamensis, LEA, Jour. Acad. Nat. Sci., v, pt. 3, p. 232, t. 34, f. 27, March, 1863. Obs., ix, p. 54.

Description. — Shell smooth, pupæform, subelevated, rather thick, yellowish, four-banded; spire raised; sutures very much impressed; Fig. 584. whorls about seven, convex; aperture small, rather constricted, subelliptical, whitish and banded within; outer lip sharp; columella inflected, whitish, obtusely angular at base.

Habitat. — Coosa River, Alabama; E. R. Showalter, M.D. Diameter, ·38; length, ·92 of an inch.

Observations.— This species is allied to *clausa*, herein described, but it is more conical and less cylindrical. One of the two specimens is obscurely banded, while the other has well defined bands, the broadest one being above. The aperture is about one-third the length of the shell.— *Lea.*

205. G. rara, Lea.

Melania rara, Lea, Proc. Acad. Nat. Sci., p. 121, 1861.
Goniobasis rara, Lea, Jour. Acad. Nat. Sci., v, pt. 3, p. 220, t. 34, f. 3, March, 1863.
    Obs., ix, p. 42.

Description.— Shell smooth, high conical, scalariform, rather thick, dark olive, shining; spire raised; sutures irregularly impressed; whorls eight, flattened, angular above; aperture rather small, ellip-

Fig. 585.

tical, dark purple within; outer lip sharp; columella incurved, purple, obtusely angular at the base.

Habitat.— Coosa River, Alabama; E. R. Showalter, M.D.

Diameter, ·38; length, ·90 of an inch.

Observations.— A single specimen only of this species was sent to me by Dr. Showalter. It is remarkable for its fine polish, its dark color and its square shoulder below the sutures. It has a few obscure striæ on the lower part of the whorl. The Babylonic form is unusual. It reminds one of *varians,* herein described, but that species is plicate and not scalariform. The length of the aperture is more than one-third the length of the shell.—*Lea.*

May possibly be a monstrosity of *G. fallax.*

206. G. punicea, Lea.

Melania punicea, Lea, Proc. Acad. Nat. Sci., p. 119, 1861.
Goniobasis punicea, Lea, Jour. Acad. Nat. Sci., v, pt. 3, p. 232, t. 34, f. 27, March,
    1863. Obs., ix, p. 54.

Description.— Shell smooth, somewhat cylindrical, thick, reddish brown; spire elevated, conical; sutures impressed; whorls slightly convex; aperture small, ovately rounded, white within; outer Fig. 586.
lip acute; columella thickened, white, rounded at the base.

Habitat.— Coosa River, Alabama; E. R. Showalter, M.D.

Diameter, ·38; length, ·94 of an inch.

Observations.— All the five specimens before me are decol-late, and have nearly the general outline of *Bulimus decollatus,*
Lam. Some have but two complete whorls, while one has four; probably when complete the number would be seven. Two of the specimens have slight striæ below, and one has a few obscure, capil-

lary bands. The reddish-brown shining epidermis well characterizes the species. The aperture is small, and is probably a little more than the third of the length of the shell.—*Lea.*

Very closely allied to *pudica*, if not identical with that species.

207. G. pudica, Lea.

Melania pudica, Lea, Proc. Acad. Nat. Sci., v, pt. 3, p. 222, t. 34, f. 7, March, 1863. Obs., ix.

Description.— Shell smooth, conical, somewhat thick, olive or reddish; spire conical; sutures irregularly impressed; whorls six, slightly convex; aperture rather small, ovate, bluish-white within; outer lip acute; columella inflected, thickened above, rounded at the base.

Fig. 587.

Observations.— This is a modest little species, with regular, even whorls. One of the specimens has obscure bands, the other has none. It is allied to *æqua*, herein described. The aperture is not quite half the length of the shell.—*Lea.*

This species has been confounded with *olivula*, Conrad; it is a smaller and more solid shell, and appears to be more numerous in individuals.

208. G. fabalis, Lea.

Goniobasis fabalis, Lea, Proc. Acad. Nat. Sci., p. 266, 1862. Jour. Acad. Nat. Sci. v, pt. 3, p. 311, t. 37, f. 154, March, 1863. Obs., ix, p. 133.

Description.— Shell smooth, elliptical, thick, yellow, four-banded; spire very obtuse; sutures irregularly impressed; whorls four, somewhat convex above, the last one very large; aperture large, subrhomboidal, whitish and banded within; outer lip acute, scarcely

Fig. 588.

sinuous; columella thickened above and below.

Habitat.— Tennessee River; W. Spillman, M.D.

Diameter, ·34; length, ·64 of an inch.

Observations.— Among the *Melanidæ* sent by Dr. Spillman, with simply the habitat Tennessee River, were four of this species. I presume they are from that part of the river which is in or near to Alabama. All the three specimens are very similar in color, size and bands. It is one of that group which approaches the genus *Lithasia* by the thickening of the columella above and below, but it has no channel. It is allied to *Melania* (*Goniobasis*) *elliptica* (nobis) and *Melania* (*Goniobasis*) *auriculæformis* (nobis), but differs from the former in being smaller and having a less constricted aperture; from the latter in being larger and having a more obtruded spire, and in the bands. The aperture is about half the length of the shell.— *Lea.*

209. G. Shelbyensis, LEA.

Melania Shelbyensis, LEA, Proc. Acad. Nat. Sci., p. 121, 1861.
Goniobasis Shelbyensis, LEA, Jour. Acad. Nat. Sci., v, pt. 3, p.228, t. 34, f. 18, March, 1863. Obs., ix, p. 50.

Description.— Shell smooth, fusiform somewhat thick, banded or without bands; spire obtusely conical; sutures impressed; whorls flattened above; aperture rather small, subovate, white with-

Fig. 589.

in; outer lip acute; columella inflected, obtusely angular at base.

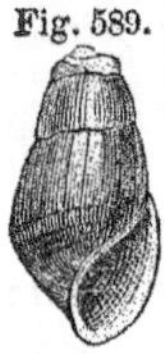

Habitat.—Yellowleaf Creek, Alabama; Dr. E. R. Showalter.

Diameter, ·38; length, ·86 of an inch.

Observations.— This species is allied to *clausa* and to *bellula* herein described. It is more elliptical than either, and smaller than the former. One of the specimens before me has four well defined, though not strong, bands, while another is entirely without any. The aperture is nearly half the length of the shell. Neither of the two specimens before me has a perfect spire, and hence the number of whorls cannot be ascertained.— *Lea.*

This species is closely related to *G. pudica*, but differs somewhat in the base of the aperture and the whorls are flattened.

210. G. fumea, LEA.

Melania fumea, LEA, Proc. Acad. Nat. Sci., 1861, p. 123.
Goniobasis fumea, LEA, Jour. Acad. Nat. Sci., v, pt. 3, p. 222, t. 34, f. 6, March, 1863. Obs., ix, p. 44.

Description. — Shell smooth, conical, rather thin, sooty brown, sometimes obscurely banded; spire somewhat raised; sut-

Fig. 590.

ures irregularly impressed; whorls flattened above, somewhat inflated below; aperture ovately rhombic, whitish within; outer lip acute; columella inflected, slightly thickened above, rounded at the base.

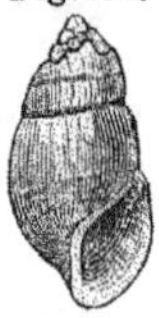

Diameter, ·36; length, ·80 of an inch.

Habitat.— Yellowleaf Creek, Shelby Co., Ala.; Dr. E. R. Showalter.

Observations.— This is an obscure species and is near to *crepera* herein described, but it is more inflated, and reminds one of *bullula* also herein described. But it has not the well marked bands of that species, some individuals being without any bands, while others have a few very obscure ones. In some there are very obscure striæ towards the base of the lower whorl. All the specimens before me being worn at the tips, I cannot make out the character of the apical whorls. The aperture is about one-third the length of the shell.—*Lea.*

Very closely allied to *G. solida.*

211. G. æqua, Lea.

Melania æqua, Lea, Proc. Acad. Nat. Sci., 1861, p. 122.
Goniobasis æqua, Lea, Jour. Acad. Nat. Sci., v, pt. 3, p. 240, t. 34, f. 41, March, 1863. Obs., ix, p. 62.

Description.—Shell substriate, conical, somewhat thick, dark brown; spire somewhat elevated; sutures impressed; whorls about six, flattened above; aperture small, rhomboidal, whitish within; Fig. 591. outer lip acute; columella inflected, slightly thickened, obtusely angular at the base.

Habitat.—Yellowleaf Creek, Alabama; Dr. E. R. Showalter. Diameter, ·37; length, ·34 of an inch.

Observations.—This is a modest looking species near to *pudica* herein described. One of the specimens has a few obscure, transverse striæ on the lower part of the whorls, the other has them nearly over the whole surface. Both specimens are imperfect at the spire. The aperture is about one-third the length of the shell.—*Lea.*

Differs from the previous species of this group in the form of the aperture.

212. G. solidula, Lea.

Melania solidula, Lea, Proc. Acad. Nat. Sci., 1851, p. 121.
Goniobasis solidula, Lea, Jour. Acad. Nat. Sci., v, pt. 3, p. 230, t. 34, f. 23. Obs., ix, p. 52.

Description.—Shell smooth, subfusiform, obtusely conical, somewhat thick, yellowish-green or yellowish-brown, banded; spire raised; sutures impressed; whorls five, above flattened, rounded below, the last large; aperture rather large, ovate, whitish within; outer lip acute; columella arcuate, slightly thickened above, obtusely angular at the base.

Habitat.— Yellowleaf Creek, near its junction with Coosa River, Alabama; E. R. Showalter, M.D.

Fig. 592.   Diameter, ·33; length, ·68 of an inch.

Observations.— Two specimens of this solid little species are before me.   The larger has five well-defined bands, which are visible in the interior as well as the exterior.   The smaller one has obsolete bands on the outside, but none within.   In outline it is very near to *Melania abrupta* (nobis), but it differs in being more solid and less expanded in the aperture.   The aperture is nearly one-half the length of the shell.—*Lea.*

213.   G. olivula, CONRAD.

Melania olivula, CON., Am. Jour. Sci., 1st series, xxv, p. 342, t. 1, f. 13, Jan., 1834.
MÜLLER, Synopsis, p. 42, 1836. WHEATLEY, Cat. Shells U. S., p. 23. DEKAY,
Moll. N. Y., p. 98. JAY, Cat. Shells, 4th edit., p. 274. REEVE, Monog. Melania,
sp. 455. BINNEY, Check List, No. 188. BROT, List, p. 33. HANLEY, Conch.
Miscellany, t. 1, f. 2.
Megara olivula, Con., CHENU, Manuel, i, f. 2027. ADAMS, Fig. 592a. Fig. 593.
Genera, i, p. 306.
Melania olivata, Con., JAY, Cat. 3d edit., p. 45. CATLOW,
Conch. Nomenc., p. 188.

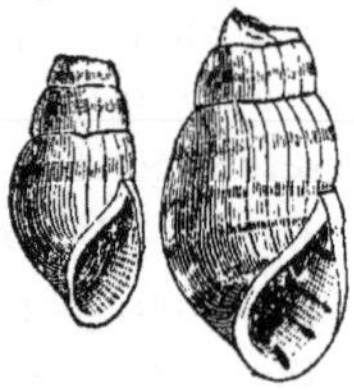

Description.— Shell oblong or narrow, elliptical, smooth and entire; spire conical; volutions five; suture impressed; aperture somewhat elliptical, longitudinal, about half the length of the shell; color green-olive, with strongly marked, brown, revolving bands; about four on the body-whorl.— *Conrad.*

214.   G. fascinans, LEA.

Melania fascinans, LEA, Proc. Acad. Nat. Sci., p. 119, 1861.
Goniobasis fascinans LEA, Jour. Acad. Nat. Sci., v, pt. 3, p. 229, t. 34, f. 20, March,
1863. Obs., ix, p. 51.

Description.—Shell smooth, subfusiform, somewhat thick, yellowish Fig. 594. horn-color, shining; spire high conical; sutures impressed; whorls slightly convex; aperture rather large, white and three-banded within; outer lip acute; columella white and retuse at base.

Habitat. — Yellowleaf Creek, Shelby County, Alabama; E. R. Showalter, M.D.

Diameter, ·38; length, ·92 of an inch.

Observations. — This graceful an beautifully banded species is allied to *Melania pupoidea*, Anth. It is more elongate and has only three bands usually, which are deep brown, well defined and nearly equidistant; but sometimes has a thin additional one below the middle one. Neither of the two specimens before me has a perfect apex, so that the number of whorls might be determined, but a perfect mature specimen would probably exhibit seven. In the penultimate whorl are two bands; on those above only one can be observed. The aperture is more than one-third the length of the shell.—*Lea*.

215. G. Showalterii, Lea.

Melania Showalterii, Lea, Proc. Acad. Nat. Sci., 1861, p. 120.
Goniobasis Showalterii, Lea, Jour. Acad. Nat. Sci., v, pt. 3, p. 220, t. 34, f. 4. Obs., ix, p. 42.

Description.—Shell smooth, raised conical, rather thick, yellowish-brown, four bands; spire obtusely elevated; sutures impressed; whorls about six, flattened above, somewhat inflated below, the last rather large; aperture rather large, ovately rhombic; whitish and banded within; outer lip sharp and slightly sinuate; columella white inflected, slightly thickened above, rounded at the base.

Operculum elongate, tongue-shaped, narrower at the outer end, dark brown, without polar point, having parallel, transverse, slightly curved striæ.

Habitat.—Coosa and Cahawba Rivers, Alabama; Dr. E. R. Showalter.
Diameter, ·42; length, —— of an inch.

Observations.— This remarkable shell was sent to me by Dr. E. R. Showalter last summer who called my attention to the very unusual form of its horny operculum, which in the old specimens is half an inch long, being a quarter of an inch wide at the inner end, grad- Fig. 595. ually diminishing to an angular point at the outer end. It is usually curved, the outer end forming a half circle from the base, the starting or inner end. Thus quite half the length extends outside of the outer lip, the inner half stretching across the aperture of the shell. Dr. Showalter did not observe whether there was any difference in the soft parts of this species from other *Goniobases*, but proposes to examine living specimens. He remarks in his letters that "the operculum is very striking and not observed in any other species, the mouth being remarkably uniform in its shape, as indeed it is in its general form and aspect." "Some of the Coosa *Anculosæ*," he says, "have this peculiar form of operculum," but I have never seen any operculum of the *Melaniæ* take this long tongue-shaped form but in this species.* Having asked Dr. Showalter if he had observed whether the opercula of young individuals were spiral, he very kindly sent me one about one-third grown. This was in no way different from the adults except in size, being rather more than one-third of an inch long. He says that he "finds the young specimens of this species have the same peculiarity in the operculum." Should there be found to exist any difference in the anatomical structure of this mollusk, when the soft parts shall be examined, then it must be eliminated from the *Goniobases*. In which case I propose the name of *Macrolimen* † for it. Among nearly a dozen specimens which I have examined, none have a perfect apex. The length of the shell, therefore, cannot be stated, nor the exact number of whorls, nor the character of the very young. The length of the aperture is probably nearly half the length of the shell. All the specimens I have examined are handsomely adorned with four bands, more or less distinct inside and out. It is nearly allied to *suavis* (nobis) and *bellula* (nobis), and reminds one of *Lewisii* (nobis).—*Lea*.

216. G. clausa, Lea.

Melania clausa, LEA, Proc. Acad. Nat. Sci., 1861, p. 120, v, pt. 3, p. 231, t. 34, f. 25, March, 1863.
Goniobasis clausa, LEA, Jour. Acad. Nat. Sci., Obs., ix, p. 53.

Description.— Shell smooth, subfusiform, thick, olive, banded, or without bands; sutures very much impressed; whorls seven, some-

Fig. 596.

what convex; aperture small, constricted, elliptical, whitish within; outer lip acute; columella slightly inflected, obtusely angular at base.

Habitat.— Coosa River, Alabama; E. R. Showalter, M.D. Diameter, ·42 of an inch; length, 1·2 inches.

Observations.— This species reminds one at once of *Pupa crysalis*, Fér., but the outline is more fusiform. It is nearly allied to *Melania pupæformis*, Anth., but it is a larger and stouter shell and is not so much banded. The aperture is narrow and unusually closed. Some specimens are feebly banded, while others have the usual four bands very broad, which make the interior dark, and

---

*I have several specimens of *A. rubiginosa* (nobis) which have an elongated operculum, but I have never observed it in any other species of *Anculosa*.
† μακροσ, longus; λιμεν, portus.

give the exterior a dark brownish or submaculate appearance. Two of the specimens are entirely without bands. The aperture is about one-third the length of the shell.—*Lea.*

217. G. crepera, Lea.

Melania crepera, Lea, Proc. Acad. Nat. Sci., 1861, p. 123.
Goniobasis crepera, Lea, Jour. Acad. Nat. Sci., v, pt. 3, p. 240, t. 34, f. 42, March, 1863. Obs., ix, p. 62.

Description. — Shell substriate, conical, somewhat thick, sooty-brown; spire somewhat raised; sutures irregularly impressed; whorls six, somewhat convex; aperture ovately rhombic, whitish within; outer lip acute; columella inflected, slightly thickened above, Fig. 597. obtusely angular at the base.

Habitat. — Yellowleaf Creek, Shelby County, Alabama; E. R. Showalter, M.D.

Diameter, ·41; length, ·83 of an inch.

Observations.— This species is closely allied to *Haysiana* (nobis), but is less striate, has a darker epidermis, is rather smaller and not so solid. Some of the specimens have but few and obscure striæ on the lower part of the whorls, while others have them over the whole whorl. None were perfect enough to show the character of the apical whorls. The length of the aperture is more than one-third the length of the shell.—*Lea.*

218. G. abscida, Anthony.

Melania abscida, Anthony, Proc. Acad. Nat. Sci., 1860, p. 56. Binney, Check List, No. 435. Brot, List, p. 32. Reeve, Monog. Melania, sp. 395.

Description.— Shell ovate, smooth, olivaceous, thick; spire obtuse, composed of five low whorls, nearly flat; body-whorl large, occupying nearly the entire length of the shell; aperture not broad, but Fig. 598. long, subrhombic, more than half the length of the shell; columella deeply rounded and indented; outer lip much curved and produced; sinus broad and conspicuous.

Habitat.— Alabama.

Observations.— A ponderous species, whose chief characteristic is its square form and short, truncate spire, resembling in that respect *M. planospira* (nobis). It differs from that species, however, by its more elongate form, narrow, rhombic aperture, and by having several revolving striæ at base. It is a solid shell of compact texture and seems to be rare, as only two specimens have come under my notice.—*Anthony.*

Very closely allied to *G. crepera*, Lea.

219. G. Vanuxemiana, LEA.

Melania Vanuxemiana, LEA, Proc. Philos. Soc., ii, p. 242, Dec., 1842. Philos. Trans. ix, p. 25. Obs., ix, p. 25. REEVE, Monog. Melania, sp. 453. BROT, List, p. 33.
Melania Vanuxemensis, Lea, WHEATLEY, Cat. Shells U. S., p. 27. BINNEY, Check List, No. 283.
Megara Vanuxemiana, Lea, ADAMS, Genera, i, p. 306.

Description.—Shell striate, obtusely conical, solid, yellowish, banded;

Fig. 599.   Fig. 600.   spire rather short; sutures impressed; whorls six, somewhat convex; columella thickened above; aperture ovate, white.

Habitat.—Alabama.

Diameter, ·42; length, ·73 of an inch.

Observations. — A very pretty symmetrical species, having the mouth rather more than one-third the length of the shell. A single specimen only is before me. It has five nearly equidistant, coarse striæ, and four purple bands. It is somewhat like *M. ovalis* herein described, but has a wider aperture, and a higher spire. I name it after my friend, Prof. Vanuxem.—*Lea.*

220. G. Coosaensis, LEA.

Melania coosænsis, LEA, Proc. Acad. Nat. Sci., 1861, p. 118.
Goniobasis coosænsis, LEA, Jour. Acad. Nat. Sci., v, pt. 3, p. 234, t. 34, f. 30, March, 1863. Obs., ix, p. 56.

Description.—Shell striate, fusiform, horn-color, four-banded, rather thick; spire rather raised, conical; sutures very much impressed; whorls seven, slightly convex, sulcate; aperture constricted, Fig. 601. elongate elliptical, whitish and four-banded within; outer lip acute, subcrenulate; columella slightly thickened, incurved and obtusely angular at the base.

Habitat.— Coosa River, Alabama; E. R. Showalter, M.D.

Diameter, ·42 of an inch; length, 1·2 inches.

Observations.— About a dozen specimens of various ages are before me. They all bear the four well marked bands, more distinct from the inside. The transverse striæ are coarse and rounded, cord-like, making well impressed sulcations. This species reminds one of *Melania* (*Goniobasis*) *Vanuxemiana* and *ovalis* (nobis), but it is a more fusiform shell, and has a longer aperture. Some of the young are almost free from striæ, and are disposed to be plicate at the apex.— *Lea.*

Differs from *Haysiana* in the form of the aperture.

221. G. rubicunda, Lea.

Melania rubicunda, Lea, Proc. Acad. Nat. Sci., 1861, p. 118.
Goniobasis rubicunda, Lea, Jour.`Acad. Nat. Sci., v, pt. 3, p. 235, t. 34, f. 32, March, 1863. Obs., ix, p. 57.

Description.— Shell much striate, reddish, subfusiform : spire sub-elevated, conical; sutures impressed; whorls about six, slightly convex; aperture rather constricted, elongate elliptical, reddish Fig. 602. within, obtusely angular at the base; outer lip acute; columella thickened, reddish, incurved.

Habitat.— Coosa River, Alabama; E. R. Showalter, M.D. Diameter, ·43; length, ·96 of an inch.

Observations.— There are five specimens before me, two of them being old and so much eroded as to leave little more than the body-whorl. The other specimens are more perfect, but the apices are worn and their character unascertained. The species is allied to *Melania* (*Goniobasis*) *Haysiana* (nobis), but may be distinguished by its not being cylindrical and by the aperture being longer. Like *Haysiana*, the striæ are coarse and rounded, somewhat cord-like. These striæ number eight to ten. As *Haysiana* is sometimes found without striæ, this species may also be found without them. The aperture is more than one-third the length of the shell.—*Lea.*

222. G. Haysiana, Lea.

Melania Haysiana, Lea, Philos. Proc., ii, p. 242, Dec., 1842. Philos. Trans., ix, p. 25. Obs., iv, p. 25. Wheatley, Cat. Shells U. S., p. 25. Jay, Cat. Shells, 4th edit., p. 273. Binney, Check List, No. 137. Brot, List, p. 32. Brot, Mal. Blatt., ii, p. 108, July, 1860. Reeve, Monog. Melania, sp. 310. Hanley, Conch. Miscel. Melania, t. 1, f. 6.
Megara Haysiana, Lea, Chenu, Manuel, i, f. 1981. Adams, Genera, i, p. 306.

Description.— Shell striate, subcylindrical, solid, yellowish-brown; spire rather elevated; sutures impressed: whorls flattened; aperture small, elliptical.

Habitat.— Alabama.

Diameter, ·43; length, ·90 of an inch.

Observations.— Dr. Foreman submitted many specimens of this species to my examination, and I find them differing very much in form

Fig. 603.   Fig. 604.   Fig. 605.   and color.   Some individuals are so full
of dark purple bands as to give them a
dark hue; others are devoid of bands entirely, and are yellowish.   The aperture is
contracted and about one-third the length
of the shell.   The transverse, raised striæ,
in some, cover nearly all the whorls, while
others are almost or entirely free from them.   In general outline it is
allied to *M. picta* (nobis) all the specimens being more or less eroded
at the beaks.   I am unable to state the number of whorls, but believe
them to be eight or nine.   I dedicate this species to my friend, Isaac
Hays, M.D.—*Lea.*

223. G. arctata, LEA.

Melania arctata, LEA, Philos. Proc., iv, p. 166.   Philos. Trans., x, p. 64, t. 9, f. 46.
    Obs., iv, p. 64.   BINNEY, Check List, No. 20.   BROT, List, p. 32.
Megara arctata, Lea, CHENU, Manuel, i, f. 2024.   ADAMS, Genera, i, f. 306.

Description.—Shell striate, compressed, thick, yellowish horn-color;
spire conical; sutures much impressed; whorls six, flattened; aperture small, rhomboidal, within whitish.

Habitat.— Tuscaloosa, Alabama.

Diameter, ·40; length, ·90 of an inch.

Observations.—Among the seven specimens before me there is a
good deal of difference.   Some are darker than others.   Fig. 606.
Several have the superior portion of the whorl rising into
a ridge, quite nodose, while others are entirely without
it.   This species has more resemblance to *M. Haysiana* than
any other which has come under my notice.   It is not, however, so elliptical a shell, and the aperture is shorter.   The
aperture of the *arctata* is rather more than one-third the
length of the shell; is obtusely angular below, and somewhat acutely
angular above, where it is thickened.—*Lea.*

The nearest affinity of this species is with *G. Coosaensis.*

224. G. ampla, ANTHONY.

Melania ampla, ANTHONY, Ann. N. Y. Lyc., vi, p. 93, t. 2, f. 12, 1854, BINNEY,
    Check List, No. 13. BROT, List, p. 39. REEVE, Monog. Melania, sp. 312.
Melania Hartmaniana, LEA, Proc. Acad. Nat. Sci., 1861.
Goniobasis Hartmanii, LEA, Jour. Acad. Nat. Sci., v, pt. 3, p. 218, t. 34, f. 1, 1863.
    Obs., ix, p. 40.

Description.— Shell ovate conic, smooth, thin; spire obtusely elevated; whorls 5–6, subconvex; body-whorl ample, surrounded with
four dark greenish bands; sutures irregularly and deeply impressed;
aperture large, ovate, within roseate and banded, col-   Fig. 607.
umella rounded, slightly indented, and a little effuse
at base.

Habitat.— Alabama.

Diameter, ·58 of an inch (15 millim.); length, 1·25
inches (32 millim.). Length of aperture, ·58 (15 millim.); breadth of aperture, ·30 of an inch (8 millim.).

Observations.— Compared with *M. olivula*, Conrad,
it is a larger, much less solid species, the epidermis is thinner, less
polished, and has not the fine contrasting colors which render *M. olivula* so lively and pleasing; differs from *M. fuliginosa*, Lea, in being
far less ponderous, with fewer and less distinct bands, by the distinct
angle passing round the shell near the top of the mouth, and by its
capacious aperture, which last two points apply with equal force to
olivula. Although in some points, and particularly in its ample mouth,
it resembles *M. florentina*, Lea, it has not the shouldered whorls and
tubercular armature which distinguish that beautiful species. The
bands within the aperture do not reach its outer edge, but a broad,
plain area is left between.—*Anthony.*

Melania ampla is not a fully grown shell, as will be seen by
reference to the accompanying figure which is copied from Mr.
Anthony's type specimen, but that the species is the same as
Hartmanii cannot be doubted. Some specimens before me
are slightly striate transversely.

The following is Mr. Lea's description of *G. Hartmanii*
together with a copy of his figure.

Description.— Shell smooth, conical, large, dark horn or olive color,
much banded, imperforate; spire obtusely conical; sutures much impressed; whorls somewhat flattened, about seven, the last large;

aperture large, ovately rhomboid, brown, banded within, obtusely angular at the base; outer lip sharp; columella incurved.

Operculum ovate, spiral, dark brown, rather rough, with the polar point on the edge, about ¼ from the base.

Habitat.— Coosa and Cahawba Rivers, Ala.; E. R. Showalter, M.D. Diameter, ·68 of an inch; length, 1·65 inches.

Observations.— This is a fine large species, and among the most

Fig. 608.

robust yet found in the United States. It is much larger than *Melania robusta* (nobis) and cannot be confounded with that species, being entirely smooth and banded. The whorls are also more flattened. The general character of the species is to have four broad, brown bands, very strongly marked on the inside. In some cases these bands are increased in width, and even so combined as to make the fauces nearly black within. These bands do not quite reach the margin. Where the bands are not strong, the exterior is light horn-color. There is a disposition on the upper part of the whorls to geniculation, and this part is there yellowish. The aperture is nearly half the length of the shell. I have great pleasure in naming this fine species after my friend, Wm. D. Hartman, M.D. of Westchester, Pennsylvania, who is always ready to promote the objects of natural history and other branches of science.—*Lea.*

225. G. mellea, Lea.

Melania mellea, Lea, Proc. Acad. Nat. Sci., 1861, p. 120.
Goniobasis mellea, Lea, Jour. Acad. Nat. Sci., v, pt. 3, p. 224, t. 34, f. 10, 1863. Obs., ix, p. 46.

Description. — Shell smooth, subfusiform, conical, rather thick, honey-yellow, sometimes banded; spire very obtuse; sutures regularly impressed; whorls seven, flattened above, the last   Fig. 609.

large and inflated; aperture large, rhomboido-elliptical, yellowish within; outer lip acute; columella thickened, inflected, obtusely angular below.

Operculum ovate, spiral, light brown, with polar point near the edge and base.

Diameter, ·52; length, ·98 of an inch.

Habitat.—Coosa River, at Wetumpka, Alabama; Dr. E. R. Showalter.

Observations.— This is a well marked species with an unusual yellow, smooth epidermis. There are four specimens before me, one being quite young, the others mature or nearly so. One has four somewhat obscure, broad, purplish bands, better defined within. The aperture is about half the length of the shell. In outline it approaches *Lithasia Florentiana* and *L. fuliginosa*, both which were described by me as *Melaniæ*, but it is larger, more yellow, has a higher spire and is not so thickened on the columella as either of those species.— *Lea.*

226. G. ambusta, ANTHONY.

Melania ambusta, ANTHONY, Ann. Lyc. Nat. Hist., vi, p. 94, t. 2, f. 13, 1854. BINNEY, Check List, No. 12. BROT, List, p. 39. REEVE, Monog. Melania, sp. 352.

Description.— Shell ovate, rather thin, smooth, chocolate-colored; spire obtusely elevated; whorls about six, subconvex; body-whorl large, substriate; sutures moderately impressed; aperture large, narrow ovate, reddish within; columella indented, with a broad, not very remarkable sinus at base.

Habitat.— Alabama.

Diameter, ·48 of an inch (12 millim.); length, 1 inch (26 millim.). Length of aperture, ·48 (12 millim.); breadth of aperture, ·23 of an inch (6 millim.).

Observations.— In form not unlike *M. olivula*, Conrad, but its very peculiar plain, dark chocolate-colored epidermis and sombre interior will at once distinguish it from all other species. A few, irregular striæ are visible on the body-whorl, and a very obscure, narrow band may be observed near the sutures; in all of the three specimens before me the columella is slightly reflected over a narrow, umbilical opening near the base, which appears almost disconnected from the outer lip as in *Achatina*. The burnt appearance of the shell has suggested its specific name.— *Anthony.*

Fig. 610.

Figured from Mr. Anthony's type specimen.

227. G. laeta, JAY.

Melania laeta, JAY, Cat. Shells, 3d edit., p. 122, t. 7, f. 11, 1839. JAY, Cat. Shells, 4th
edit., p. 274. WHEATLEY, Cat. Shells U. S., p. 26. BINNEY, Check List, No. 156.
CATLOW, Conch. Nomenc., p. 187. BROT, List, p. 32.
Melania robusta, LEA, Philos. Proc., ii, p. 83, October, 1841. Philos. Trans., ix, p.
19. Obs., iv, p. 19. WHEATLEY, Cat. Shells, U. S., p. 26. BINNEY, Check List,
No. 231.
Melatoma Buddii, Lea, REEVE, Monog. Melatoma, sp. 3.
Melania tæniolata, ANTHONY, Proc. Acad. Nat. Sci., 1860, p. 59. BINNEY, Check
List, No. 263. BROT, List, p. 31. REEVE, Monog. Melania, sp. 392.

Description.— Shell striate, fusiform, thick, yellowish; spire obtuse; sutures rather impressed; whorls six, rather convex; aperture

Fig. 611.   Fig. 612.

elliptical, large, angular at the base, within
white.

Habitat.— Coosa River, Alabama.

Diameter, ·60; length, ·91 of an inch.

Observations.— A single specimen only of
this fine species was obtained by Dr. Griffith.
It presents four rather distant, large, revolving striæ on the body-whorl and two
on the next. In other specimens these may
be found more numerous, or entirely wanting.*   The aperture is nearly half the
length of the shell. In form and size, it
very closely agrees with *M. impressa* herein
described.—*Lea.*

Dr. Jay published merely a name
and figure of his species, without description.   The figure 613 represents
a copy of it.   612 represents Mr. Lea's figure of *robusta* and
611 is from a splendid specimen from Coosa River, while 614
represents a younger shell.

Fig. 613.   Fig. 614.

The following description was drawn up from an immature
specimen; we present a figure from the type :—

Melania tæniolata.— Shell conic, ovate, striate, thick; spire elevated, but not acute, composed of 6-7 nearly flat whorls: sutures
not distinct; aperture subrhombic, small, banded within; columella
indented, callous at its lower portion, and with a small, but distinct,
sinus at base.

---

* In specimens subsequently received, the striæ were found to differ but little.

Habitat.— Alabama.

Observations.— A fine, showy, robust species, of a dark yellow color, enlivened by several dark brown bands, generally two on each whorl; body-whorl angulated, with one band directly upon   Fig. 615. the sharp angle, another in close proximity, and a third quite distant and near the base of the shell. Band obsolete on the first two or three whorls. Surface coarsely striate and obscurely ribbed.— *Anthony.*

This species appears to vary somewhat in form, being only occasionally angulated at the periphery, but the specimens are all covered with alternate, transverse, rounded ribs and sulcations with a few nodules on the former.

228. G. harpa, Lea.

Melania harpa, Lea, Philos. Proc., iv, p. 163, August, 1845. Philos. Trans., x, p.
  64, t. 9, f. 45. Obs:, iv, p. 64. Binney, Check List, No. 135. Brot, List, p. 32.
  Reeve, Monog. Melania, sp. 313, 314.
Megara harpa, Lea, Adams. Genera, i, p. 303.
Melania textilosa, Anthony, Ann. Lyc. Nat. Hist., vi, p. 101, t. 2, f. 20, 1854. Binney, Check List, No. 270. Brot, List, p. 40. Reeve, Monog. Melania, sp. 391.

Description.— Shell striate, conical, rather thick, horn-color; spire rather elevated; sutures rather impressed; whorls somewhat convex; Fig. 616. aperture small, elliptical, angular at the base, within whitish.

Habitat.— Tuscaloosa, Alabama.

Diameter, ·42; length, ·8 of an inch.

Observations.— I am not able to place this with any of the species submitted to me by Dr. Budd, and although a single specimen only is under examination, I have considered it new. It has some resemblance to *M. Haysiana,* but is not so cylindrical, and the aperture is not so narrow. It is transversely striate over the whole whorls. The length of the aperture is about two-fifths the length of the shell. The aperture being eroded the number of whorls cannot be ascertained.—*Lea.*

The following description represents the young of this species :—

Melania textilosa.— Shell conical, thick; color uniform, pale greenish-yellow; spire not acutely elevated; whorls 7-8, nearly flat, obscurely striate and subnodulous; body-whorl coarsely, but not thickly, striate on its upper half; sutures impressed; aperture rather large, ovate, whitish, inclining to roseate.

Habitat.— Georgia.

Diameter, ·40 (10 millim.); length, ·88 of an inch (23 milim.).

Fig. 617.   Length of aperture, ·39 (10 millim.); breadth of aperture, ·20 of an inch (5 millim.).

· Observations. — In form like *M. Duttoniana,* Lea, but without any of the ornamental decorations of that species.   The nodules are not so distinct, appearing more like interrupted folds.   The striæ on the body-whorl are not uniformly distributed, but above the middle there is a plain surface or ground, which becomes more decidedly a furrow on the penultimate whorl.—*Anthony.*

G. harpa is narrower than *laeta* with the mouth more acuminate below.   The striæ are smaller and closer.

229. G. oliva, Lea.

Melania oliva, Lea, Philos. Proc., ii, p. 242, 1842. Philos. Trans., ix, p. 27. Obs., iv, p. 127. Wheatley, Cat. Shells U. S., p. 26. Binney, Check List, No. 187. Brot, List, p. 33.
Megara oliva, Lea, Adams, Genera, i, p. 306.

Description.— Shell striate, elliptical, solid, brown; spire rather short; sutures much impressed; whorls convex; columella incurved, thickened above; aperture ovate, white.

Habitat.— Alabama.

Diameter, ·50 of an inch; length, 1 inch.

Observations.— This is a ponderous and rather large species, with not very distinct striæ on the few specimens before me.   The superior part of the columella is quite callous.   The apex of each is too much eroded to designate the number of the whorls.   The aperture is rather small and contracted.   One of the specimens is rather coarsely plicate.— *Lea.*

This shell is narrower than *laeta,* resembling *harpa* in form, but with the aperture wider and more rounded below.   It is very closely allied to *G. excavata,* which is a smooth species, however.

230. G. grisea, ANTHONY.

Melania grisea, ANTHONY, Proc. Acad. Nat. Sci., 1860, p. 61. REEVE, Monog. Melania, sp. 390. BROT, List, p. 32.

Description.— Shell ovate, smooth, thick, of a dull gray color; whorls seven, convex; sutures very distinct; body-whorl obscurely ribbed, and having two or three inconspicuous bands revolving around it; aperture large, ovate, banded within; columella deeply indented, with a white callus, unusually thickened at the summit   Fig. 618. of aperture; sinus broad but not distinct.

Habitat.— Tennessee River, North Alabama.

My cabinet.

Observations.— A single specimen only of this species has come under my notice, but I cannot consider it referable to any described species. The bands are very obscure, scarcely perceptible, and those within the aperture are arrested before reaching the edge of the lip. The ribs which are inconspicuous on the spire become more decided on the body-whorl, and sometimes appear as varices there; the spire is very obtusely elevated.— *Anthony.*

This species much resembles *G. variata*, Lea.

231. G. culta, LEA.

Melania culta, LEA, Proc. Acad. Nat. Sci., p. 121, 1861.
Goniobasis culta, LEA, Jour. Acad. Nat. Sci., v, p. 13, p. 237, t. 34, f. 36, March, 1863. Obs., ix, p. 59.
Melania suavis, LEA, Proc. Acad. Nat. Sci., p. 169, 1861.
Goniobasis suavis, LEA, Jour. Acad. Nat. Sci., v, pt. 3, p. 228, t. 34, f. 19, March, 1863. Obs., ix, p. 50.

Description. — Shell rugosely striate, subfusiform, inflated, rather thick, greenish-yellow, shining, three-banded; spire very obtuse;

Fig. 619.   sutures irregularly and very much impressed; whorls seven,

carinate above; aperture wide, subrhomboidal, whitish within and banded; outer lip acute; columella incurved, pale rose-color, angular below.

Habitat.— Coosa River, Alabama; E. R. Showalter, M.D.

Diameter, ·42; length, ·79 of an inch.

Observations.— A single specimen only was received from Dr. E. R. Showalter, and this may not be entirely mature. It has six coarse, transverse striæ, which are rather sharp; the two upper ones, being rather distant, cause quite a large furrow between them.   Other specimens may not present these characters, as striæ, whether fine or coarse, vary very much on the *Melanidæ.* The color on the callus of the columella may also vary in other individuals.   The aperture is nearly half the length of the shell.   This species is allied to *Vanuxemiana* (nobis), but it has not so high a spire, and it is wider in proportion.— *Lea.*

Goniobasis suavis.— Shell smooth, subfusiform, rather thick, yellowish-green, polished, four-banded; spire obtusely conical; sutures regularly impressed; whorls six, slightly flattened above; aperture Fig. 620. rather large, elliptical, whitish and banded within; outer lip acute; columella incurved and rounded at the base.

Habitat.— Coosa River, Alabama; E. R. Showalter, M.D.

Diameter, ·33; length, ·68 of an inch.

Observations.— There are two specimens before me of this pretty little species, both of the same size and appearance in every way. The bands are remarkably perfect and well defined, and the two middle ones, in these specimens, are approximate, while they are equidistant from that above and below.   It reminds one of *Melania ovalis* (nobis), but it has a higher spire and is more disposed to be fusiform.   The greenish-yellow tint, its well marked bands and shining surface, give it a very agreeable aspect.— *Lea.*

232. G. luteola, LEA.

Melania luteola, LEA, Proc. Acad. Nat. Sci., p, 119. 1861.
Goniobasis luteola, LEA, Jour. Acad. Nat. Sci. v, pt. 3, p. 230, t. 34, f. 22, March, 1863. Obs., ix, p. 52.
Melania straminea, LEA, Proc. Acad. Nat. Sci., 1861, p. 121.
Goniobasis straminea, LEA, Jour. Acad. Nat. Sci., v, pt. 3, p. 227, t. 34, f. 16, March, 1863. Obs., ix, p. 49.

Description.— Shell smooth, subfusiform, obtusely conoidal, somewhat thick, straw-color; spire raised; sutures impressed; whorls five, the last large and somewhat inflated; aperture large, elongate elliptical, yellowish-white within, outer lip acute; columella arcuate, slightly callous above, obtusely angular at the base.

Operculum ovate, spiral, light brown, with the polar point near the edge towards the base.

Habitat.— Coosa River, Alabama; E. R. Showalter, M.D.

Diameter, ·40; length, ·80 of an inch.

Observations.— The regularly elliptical outline of this species is remarkable among the *Goniobases.* There is no appearance of bands in either of the three specimens sent by Dr. Showalter. One of them has a slight line of brown in the callus of the inte-

Fig. 621.

rior above. The largest specimen has some indistinct striæ towards the base of the whorl. It is nearly allied to *Melania olivula*, Conrad, but it is more inflated and has a shorter spire. The aperture is more than half the length of the shell.—*Lea.*

I cannot detect any difference between the two species included in the above synonymy except that *luteola* appears to be not fully grown. For this reason I give the description of *straminea* first, as being that of the adult shell.

The following is the description of

Goniobasis luteola.— Shell smooth, elliptical, rather thin, pale yellow; spire rather raised, conical; sutures slightly impressed; whorls a little flattened; aperture rather large, whitish within; outer lip acute; columella whitish, incurved, obtusely angular at the base.

Habitat.— Alabama River; E. R. Showalter, M.D.

Diameter, ·28; length, ·62 of an inch.

Observations.— Two specimens of this pale little species are before me. They are nearly allied to *punicea* herein described, but it is a

Fig. 622.

shorter and thinner species and of quite different color in the epidermis. The aperture is more elongate and larger in proportion. Both specimens are decollate, but in one there are four whorls apparent, and I presume the normal number would be six. The larger specimen has an obscure band on the upper part of the whorl, which is well defined inside. The smaller one has none whatever. There is a slight disposition to take on folds on the upper whorls. The aperture is about one-half the length of the shell.—*Lea.*

233. G. gravida, ANTHONY.

Melania gravida. ANTHONY, Proc. Acad. Nat. Sci., p. 59, Feb., 1860. REEVE, Monog. Melania. BROT, List.

Description.— Shell ovate, smooth, thick; spire obtusely elevated; whorls 7-8, nearly flat; sutures well defined; lines of growth fine, but very distinct; body-whorl large, subangulated; aperture oval,

livid inside; columella deeply indented, covered with a white callus;
outer lip curved forward, and with the columella forming a small

Fig. 623.   sinus at base.

Habitat.— Alabama.

Observations.—A stout, heavy shell, in form and color
resembling in some degree *M. solida*, Lea, but is more
ovate than that species; color light brown, smooth but
not very shining; lines of growth very distinct and
curved.   A few indistinct striæ occur at the base of the
shell; the lower part of the columena is often tinged with a golden
hue.— *Anthony.*

Figured from Mr. Anthony's type.

234. G. germana, ANTHONY.

Melania germana, ANTHONY, Proc. Acad. Nat. Sci., p. 61, Feb., 1860.   BINNEY,
    Check List, No. 120.   BROT, List, p. 40.   REEVE, Monog. Melania, sp. 383.

Description.— Shell carinate on the body-whorl; form rhombic;
substance rather thin; varying in color from ash-gray to dark brown;
whorls six, upper ones smooth; suture very distinct; aperture rhombic, within brownish with a white area near the outer   Fig. 624.
edge; columella rounded or angularly indented, slightly
callous; sinus small.

Habitat.—Cahawba River, Alabama.

Observations.—This is another of the short, rhombic
species, which are represented most fitly by *M. rhombica* (nobis), and
includes *M. angulata* (nobis), *M. cubicoides* (nobis), *M. cristata* (nobis)
and many others.   From *M. rhombica*, it differs in being shorter and
less slender, and by wanting the regular concentric striæ so conspicuous on the upper half of that species.   It is also less slender than
M. angulata (nobis) and more solid.   From all other species it may
readily be distinguished.— *Anthony.*

235. G. variata, LEA.

Melania variata, LEA, Proc. Acad. Nat. Sci., p. 119, 1861.
Goniobasis variata, LEA, Jour. Acad. Nat. Sci., v, pt. 3, p. 224, t. 34, f. 11, March,
    1863. Obs., ix, p. 46.

Description.— Shell smooth, subfusiform, somewhat thick, yellowish
or purplish; spire very obtuse; sutures irregularly impressed; whorls

six, flattened above, the last inflated; aperture large, yellowish or purplish within; outer lip sharp; columella arcuate, thickened, obtusely angular at base.

Habitat.— Coosa River, at Wetumpka and Montevallo, Bibb County, Alabama; E. R. Showalter, M.D.

Diameter, ·40; length, ·76 of an inch.

Observations.— Six specimens are before me. Two of them are mature, are yellowish and are somewhat thick. Four are thinner and are purplish inside and out, not disposed to be banded, but are obscurely maculate. The apical whorls have obscure folds. One of the old ones has obscure bands on the inside. The other has none. The aperture is more than half the length of the shell. It is somewhat like *Melania fuliginosa* (nobis) in outline, but it is not so much inflated as that species.— *Lea.*

Fig. 625.

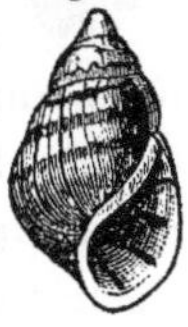

236. G. ovalis, Lea.

Melania ovalis, Lea, Philos. Proc., ii, p. 242, Dec., 1842. Philos. Trans., ix, p. 25. Obs., ix, p. 25. Wheatley, Cat. Shells U. S., p. 26. Binney, Check List, No. 192. Reeve, Monog. Melania, sp. 448 and sp. 309.

Megara ovalis, Lea, Adams, Genera, i, p. 306.

Melania copiosa, Lea, Proc. Acad. Nat. Sci., p. 122, 1861.

Goniobasis copiosa, Lea, Jour. Acad. Nat. Sci., v, pt. 3, p. 239, t. 34, f. 39. Obs., ix, p. 61.

Melania orbicula, Lea, Proc. Acad. Nat. Sci., p. 118, 1861.

Goniobasis orbicula, Lea, Jour. Acad. Nat. Sci., v, pt. 3, p. 238, t. 34, f. 37, March, 1863. Obs., ix, p. 60.

Description.— Shell striate, fusiform, solid, yellow, banded; sutures much impressed; whorls six, rather convex; aperture oval, narrow, whitish within.

Habitat.— Alabama.

Diameter, ·40; length, ·62 of an inch.

Observations.— A number of specimens were kindly sent by Dr. Foreman for my inspection, several of which are young, exhibiting on the first two or three whorls very distinct folds. Those of the larger specimens are worn off. The mature specimens are remarkable for their irregularly elliptical form, generally having transverse striæ over the whole surface. The aperture is very regularly ovate, fully the half of the length of the shell.— *Lea.*

Fig. 626.

This species is not so broadly ovate as *G. laeta* and is also smaller.

Goniobasis copiosa. — Shell striate, broadly fusiform, ventricose, obtusely conical, somewhat thick, yellowish horn-color, obscurely banded; spire very obtuse; sutures irregularly impressed; whorls five, somewhat convex, the last very large; aperture very large, widely

Fig. 627. elliptical, whitish within; outer lip acute, sinuous; columella arcuate, slightly thickened above, rounded at the base.

Habitat. — Coosa River, Alabama; E. R. Showalter, M.D.
Diameter, ·42; length, ·69 of an inch.

Observations. — The single specimen before me seems to be mature. It is allied to *Melania* (*Goniobasis*) *ovalis* (nobis) and to *culta* herein described. It is more inflated than either, and has a more expanded outer lip. In this specimen the upper whorls have a single well defined band, which is obsolete on the lowest whorl. It has ten rather coarse, rounded striæ, which are slightly interrupted by the lines of growth, giving the surface a rugose appearance. These striæ being thickened, cause in the interior whitish lines. The aperture is more than one-half the length of the shell. The apical whorls are plicate.— *Lea.*

Goniobasis orbicula. — Shell striate, globose, somewhat thick, yellowish-green, four-banded; spire short obtuse; sutures very much impressed; whorls five, very much inflated, the last large; aperture large, elliptical, four-banded within; outer lip acute; columella white, incurved, obtusely angular at the base.

Operculum ovate, dark brown, with the polar point near the inner border, one-quarter above the base.

Habitat. — Coosa River, Alabama; E. R. Showalter, M.D.
Diameter, ·31; length, ·54 of an inch.

Observations. — This is a remarkably globose, small species, of which only a single specimen was received. The striæ are Fig. 628. coarse and cord-like, and cover the whole of the body-whorl. It is so nearly like in form and color to *Schizostoma globula* (nobis), that it might easily be taken for that shell, if it were not that there is no appearance of a fissure. The length of the aperture is two-thirds the length of the shell.— *Lea.*

237. G. virgulata, LEA.

Melania virgulata, LEA, Proc. Acad. Nat. Sci., p. 119, 1861.
Goniobasis virgulata, LEA, Jour. Acad. Nat. Sci., v, pt. 3, p. 223, t. 34, f. 9, March, 1863. Obs., ix, p. 45.
Melania glandaria, LEA, Proc. Acad. Nat. Sci., p. 120, 1861.
Goniobasis glandaria, LEA, Jour. Acad. Nat. Sci., v, pt. 3, p. 226, t. 34, f. 14, March, 1863. Obs., ix, p. 48.

Description.— Shell smooth, fusiform, thick, greenish-yellow, four-banded; suture irregularly and much impressed; whorls seven, slightly convex, the last large; aperture long elliptical, subconstricted, whitish within and much banded; outer lip acute, **Fig. 629.** subsinuous; columella arcuate, thickened above and below, slightly canaliculate and twisted.

Habitat.— Coosa River, Alabama; E. R. Showalter, M.D. Diameter, ·42; length, ·86 of an inch.

Observations.— This is a solid species nearly an inch long, and reminds one of the form of an acorn. It is near to some of the forms of *nebula* herein described, but has not the dark maculations of that shell, the four dark brown bands being distinct on the inside. The aperture is half the length of the shell. The upper band is well defined on the upper whorls.— *Lea.*

Goniobasis virgulata.— Shell smooth, fusiform, conical, somewhat thick, shining, yellowish, four-banded; spire conical, sharp-pointed; sutures impressed; whorls seven, constricted above, the last bulbose; aperture rather large, somewhat elliptical, yellowish-white and very **Fig. 630.** much banded within; outer lip sharp; columella inflected, angular at the base and canaliculate.

Operculum ovate, spiral, dark brown, with the polar point on the inner side near the base.

Habitat.— Coosa and Tallapoosa Rivers, Alabama; E. R. Showalter, M.D.

Diameter, ·36; length, ·76 of an inch.

Observations.— This is a beautiful banded species, having the two middle bands more approximate. The four bands are broad and of an intense brown; on the upper whorls a single band only is exhibited. On one specimen this band reaches nearly to the apical whorl, in the other only to the second. Its mucronate spire and inflated body-whorl remind one of *Melania conica*, Say, but it may be distinguished by its having a larger body-whorl and a shorter spire. The aperture is nearly half the length of the shell.— *Lea.*

I think *virgulata* is only the young of *glandaria*. The two figures, which are copies of Mr. Lea's, will assist the reader in forming his judgment of the correctness of my determination.

238. G. clara, ANTHONY.

Melania clara, ANTHONY, Ann. N. Y. Lyc., vi, p. 119, t. 3, f. 19, March, 1854. BIN-
   NEY, Check List, No. 55. BROT, List, p. 32.

Description.—Shell ovate, smooth, thick; spire not elevated; whorls seven, flat, nearly smooth; upper ones rapidly enlarging to the body-whorl, which is very large and ornamented with four conspicuous, brown bands, on a clear and well contrasting yellow ground; the upper band is distant and alone, near the suture, while the others are crowded and below the middle; sutures impressed; aperture large,

Fig. 631.

ovate, banded inside; columella nearly straight, with no remarkable sinus at base.

Habitat.— Alabama.

Diameter, ·38 (10 millim.); length, ·70 of an inch (18 millim.). Length of aperture, ·40 (10 millim.); breadth of aperture, ·20 of an inch (5 millim.).

Observations.—Allied to *M. olivula*, Conrad, in general form, but seems to differ by its body-whorl, which is subangulated at its upper part, near the top of the aperture, and slightly so at the middle; the whorls of the spire have only one band, which is above the middle; lines of growth distinct, giving the upper whorls a slightly varicose character.— *Anthony.*

239. G. inflata, HALDEMAN.

Melania inflata, HALDEMAN, Cover of No. 3, Monog. Limniades, March, 1841.
   BINNEY, Check List, No. 146. BROT, List, p. 40. REEVE, Monog. Melania,
   sp. 410.

Description.— Shell conical, with 3-4 flat turns; lines of growth un-deviating; aperture as long as the spire, very narrow, elliptic, slightly produced, and turned to the left anteriorly; color brown or green, inside banded with reddish.

Fig. 632.

Habitat.—Alabama River; Mr. Conrad.

Length, ½ of an inch.

Observations.— Allied to *M. stygia.*— *Haldeman.*

The above description does not correspond with that of *germana*, Anthony, but if the figure here given (which is copied from Reeve and represents a shell in museum Anthony) is *inflata*, then the two are identical. This species differs from *G. virgulata*, by its obtusely angled whorls and somewhat diamond-shaped aperture.

240. G. fusiformis, LEA.

Melania fusiformis, LEA, Philos. Proc., ii, p. 12, Feb., 1841. Philos. Trans., viii, p. 167, t. 5, f. 9. Obs., iii, p. 5. DEKAY, Moll. N. Y., p. 93. TROOST, Cat. Shells Tenn. WHEATLEY, Cat. Shells U. S., p. 25. BINNEY, Check List, No. 117. CATLOW, Conch. Nomenc., p. 186. BROT, List, p. 40.

Description.— Shell smooth, fusiform, rather thin, yellow, pointed at the apex; spire short; sutures linear; whorls six, the last being large and inflated; aperture ovately elongated, whitish.

Habitat.— Tennessee; Dr. Troost.

Diameter, ·27; length, ·50 of an inch.

Observations.— This is a very remarkable species in regard to its form, resembling as it does the young of some species of *col-* Fig. 633. *umbella.* The aperture is about two-thirds the length of the shell, and is somewhat angular at base above it turns inward. One of six individuals before me has two rather broad bands. On the superior whorls may be observed an indistinct stria.—*Lea.*

The figure is a copy of that of Mr. Lea. Much like *G. ambusta*, when young, but more inflated, and the aperture more broadly rounded below.

241. G. bellula, LEA.

Melania bellula, LEA, Proc. Acad. Nat. Sci., p. 122, 1861.
Goniobasis bellula, LEA, Jour. Acad. Nat. Sci., v, pt. 3, p. 237, t. 34, f. 35, March, 1863. Obs., ix, p. 59.

Description.— Shell striate, subfusiform, somewhat thick, pale horn-color, four-banded; spire obtuse; sutures much impressed; whorls Fig. 634. about five, somewhat convex, the last large; aperture rather large, elliptical, whitish within and spotted; outer lip sharp; columella white, inflected, obtusely angular at the base.

Operculum elliptical, spiral, dark brown, with the polar point near the inner edge about one-fourth from the base.

Habitat.— Yellowleaf Creek, Shelby County, Alabama; E. R. Showalter, M.D.

Diameter, ·43; length, ·78 of an inch.

Observations.— The four bands which are well marked on the three specimens before, seem to be regular and prominent in character. The two middle ones are slightly nearer together than they are to the outside ones. These bands are strongly marked inside and out. The transverse striæ are few, coarse and cord-like. Neither of the specimens is perfect in the apex, and therefore the number of whorls cannot be correctly ascertained. The bands are exhibited on all the whorls. The aperture is nearly the length of the shell. This is a remarkably beautiful species, the deep brown bands forming a contrast to the bright yellowish horn-color of the ground. In outline and general appearance it is closely allied to *Showalterii* herein described, but it is more inflated and has a regularly formed spiral *operculum*, while the *Showalterii* is long tongue-shaped.

The young shell is generally smooth, polished and banded, being very beautiful. This species is smaller than *laeta* and differs in the aperture.

242. G. calculoides, Lea.

Melania calculoides, Lea, Proc. Acad. Nat. Sci., p. 118, 1861.
Goniobasis calculoides, Lea, Jour. Acad. Nat. Sci., v, pt. 3, p. 238, t. 34, f. 38, March, 1863. Obs., ix, p. 60.

Description.— Shell striate, subglobose, thick, horn-color, robust; spire obtusely conical; sutures impressed; whorls six, very much inflated, the last large; aperture rather large, elongately elliptical, whitish within; columella whitish, thickened, arcuate, retuse at the base.

Fig. 635.

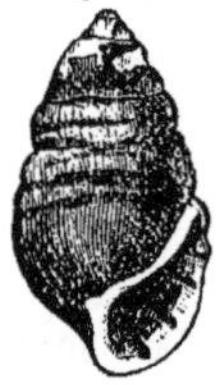

Habitat.— Coosa River, Alabama; E. R. Showalter, M.D. Diameter, ·50; length, ·93 of an inch.

Observations.— Four specimens of different ages were received; two are without bands and two have four bands each. It is not so globose as *orbicula* herein described, and is much larger. It is also higher in the spire. It is nearest to *Melania (Goniobasis) robusta* (nobis), but is not so high in the spire. The two differ in the channel at base of the columella. The aperture is a little more than half the length of the shell. All these specimens are more or less striate, the upper ones being more conspicuous.— *Lea.*

Very closely allied to *G. culta.*

243. G. basalis, LEA.

Melania basalis, LEA, Philos. Proc., iv, p. 166. Philos. Trans., x, p. 59, t. 9, f. 33.
Obs., iv, p. 59. BINNEY, Check List, No. 28. BROT, List, p. 32. REEVE,
Monog. Melania, sp. 471.
Anculotis basalis, Lea, REEVE, Monog. Anculotus, t. 5, f. 40.
Megara basalis, Lea, ADAMS, Genera, i, p. 306.

Description.—Shell smooth, elliptical, rather thick, yellowish-green,
banded; spire short, obtuse; sutures impressed; whorls convex; aperture ovately elongate, at the base acutely angular, within whitish.

Habitat.—Alabama.

Diameter, ·43; length, ·83 of an inch.

Observations.—The elliptical form of this species is very remarkable. The spire is very short and obtuse. The apex of each of
the two specimens before me is eroded, two whorls only being perfect. It has numerous purple bands, and the aperture is   Fig. 636.
rather more than half the length of the shell. The base
of the shell is extended and slightly retuse. One of the
specimens near to the superior part of the whorl is disposed to swell into large tubercles. The epidermis is very
smooth and polished.—*Lea.*

I scarcely think Mr. Reeve's figures represent this species,
as they do not correspond with Mr. Lea's figure, a copy of
which is here given. This species resembles *G. glandaria*, Lea,
but is thinner, the outer lip more expanded and the aperture
rather longer. It is closely allied to *G. fusiformis*, Lea.

244. G. Lewisii, LEA.

Melania Lewisii, LEA, Proc. Acad. Nat. Sci., p. 118, 1861.
Goniobasis Lewisii, LEA, Jour. Acad. Nat. Sci., v, pt. 3, p. 243, t. 35, f. 46, March,
1863. Obs., ix, p. 65.

Description.—Shell striate, somewhat cylindrical, dark green, much
banded; spire somewhat raised, conical; sutures much impressed;
whorls flattened, sulcate, about six; aperture rather small, ovately
rhomboidal, much banded within, obtusely angular at the base; outer
lip acute; columella white and incurved.

Operculum ovate, spiral, nearly black, with the polar point near the
inner edge and close to the base.

Habitat.—Coosa and Tallapoosa Rivers, Alabama; E. R. Showalter, M.D.

Diameter, ·44; length, ·94 of an inch.

Observations.— Several specimens were sent to me by Dr. Lewis and by Dr. Showalter. It is a well marked species, and has some-

Fig. 637.

what the appearance of a *Schizostoma*, but there is no fis-sure. The shoulder below the suture is well marked and like *Schizostoma*, and the suture so wide and deep as to make quite a furrow. There is a disposition to have five to eight coarse, rounded striæ, with sulcations between, but some specimens are nearly smooth. These coarse striæ are cord-like and usually dark colored. The dark brown bands are well defined within, and in each of the eight specimens before me, there are four. On the upper part of the whorls the bands are interrupted with yellowish spots. The aperture is more than one-third the length of the shell. I have great pleasure in dedi-cating this interesting species to my friend, James Lewis, M.D. of Mohawk, N. Y., who has done so much to develop the history of our fresh-water *Mollusks.— Lea.*

The young shell, like most of the species of this group, is sharply angulated on the periphery.

245. G. ellipsoides, Lea.

Melania gracilior,* Lea, Proc. Acad. Nat. Sci., 1861, p. 118.
Goniobasis ellipsoides, Lea, Jour. Acad. Nat. Sci., v, pt. 3, p. 234, t. 34, f. 31, March,
 1863. Obs., ix, p. 56.

Description.— Shell striate, fusiform, greenish-yellow, rather thick; spire rather elevated, conical; sutures irregularly impressed; whorls seven, scarcely convex; aperture somewhat constricted, Fig. 638. elongately elliptical, whitish within; outer lip acute; colu-mella whitish, a little recurved below, rounded at the base.

Habitat.— Coosa River, Alabama; E. R. Showalter, M.D.

Diameter, ·43; length, ·86 of an inch.

Observations.— This species is very near in outline and size to *Coosaensis* herein described. It differs in being without bands except obscure ones on the upper whorls, and in having but few raised striæ. The channel at the base also differs in *ellipsoides* being slightly retuse. The color and whole aspect of the two specimens before me are exactly alike, having a peculiar greenish-yellow epidermis. In

---

* Changed to *ellipsoides*, the name of *gracilior* being preoccupied.

both these specimens there are two raised cord-like striæ above and a few impressed striæ at the base.— *Lea.*

246. G. elliptica, Lea.

Melania elliptica, Lea, Proc. Acad. Nat. Sci., p. 118, 1861.
Goniobasis elliptica, Lea, Jour. Acad. Nat. Sci., v, pt. 3, p. 225, t. 34, f. 13, March, 1863. Obs., ix, p. 47.

Description.—Shell smooth, elliptical, yellowish, four-banded; spire short, obtuse, folded at the tip; sutures impressed; whorls six, sub-convex; aperture rather large, elongate elliptical, four-banded within; obtusely angular at the base; outer lip acute; columella whitish and incurved.

Operculum narrow, elliptical, spiral, light brown, with the polar point near the inner margin above the base.

Habitat.— Coosa River, Alabama; E. R. Showalter, M.D.; and E. Foreman, M.D.

Diameter, ·41; length, ·78 of an inch.

Observations.—This is a remarkably regular elliptical species, pointed at the base and apex. There are five specimens before me. Fig. 639.
One is an old worn one, which I long since received among other species from Dr. Foreman. It looks much like the young or immature of *Melania ovalis* (nobis), but is not so thick, nor has it striæ. It has somewhat the aspect of *Lithasia Showalterii* (nobis), but it has not the callus of that genus, and it is not compressed at the sides, but is regularly convex. All the specimens under examination have four regular bands, and one of them is disposed to be striate. The folds on the upper whorls are represented below by irregularities on the whorls which interrupt the upper band and give it a maculate appearance.—*Lea.*

247. G. bullula, Lea.

Melania bullula, Lea, Proc. Acad. Nat. Sci., p. 121, 1861.
Goniobasis bullula, Lea, Jour. Acad. Nat. Sci., v, pt. 3, p. 221, March, 1863. Obs., ix, p. 43, t. 34, f. 5.

Description.— Shell smooth, conical, inflated, rather thin, greenish-yellow, four-banded; spire raised; sutures impressed; whorls about five, inflated, the last rather large; aperture rather large, widely ovate, whitish and banded within; outer lip acute; columella whitish, thickened above, sinuous, subangular below.

Operculum elliptical, spiral, dark brown with the polar point near the base.

Habitat.—Yellowleaf Creek, Shelby County, Alabama; Dr. E. R. Showalter.

Fig. 640.

Diameter, ·40; length, ·90 of an inch.

Observations.—This is a somewhat inflated species, with four regular brown bands and reminds one of *bullula* herein described. It is not so solid a species, is usually more inflated, higher in the spire and has not usually any striæ, although some specimens have a few. Neither of the specimens before me has a perfect apex, therefore the number of whorls is uncertain. The aperture is not quite half the length of the shell.— *Lea.*

248. G. excavata, ANTHONY.

Melania excavata, ANTHONY, Ann. Lyc. N. Y., vi, p. 99, t. 2, f. 18, March, 1854. BINNEY, Check List, No. 102. BROT, List, p. 32. REEVE, Monog. Melania, sp. 385.

Description.—Shell ovate-conic, smooth, olivaceous, thick; spire obtusely elevated, decollate; whorls 3-4 remaining, flat or concave; sutures distinct; penultimate and body-whorl with a broad, deep, concave excavation, their edges being elevated into an obtuse carina, tipped with a lighter color; lines of growth very strong; aperture not large, ovate, reddish within; columella regularly curved, thickened by a deposit of calcareous matter purplish and white, indented near its base, without any sinus.

Fig. 641.

Habitat.— Alabama.

Diameter (of an eroded example), ·44 (11 millim.); length (of an eroded example), ·84 of an inch (21 millim.). Length of aperture, ·40 (10 millim.); breadth of aperture, ·22 of an inch (5½ millim.).

Observations.—An unadorned species of a dull olive-color, not easily confounded with any of its congeners. Differs from *M. fusiformis*, Lea, by its broad, more elevated spire, its purple mouth, unadorned with bands, but above all, by the peculiar excavation on the lower whorls, which is so peculiar as to distinguish this species from all others. — *Anthony.*

Figured from Mr. Anthony's type, which exhibits so unmistakably the signs of diseased growth that it must not be supposed that the above description will characterize the species in its normal state.

249. G. purpurea, Lea.

Melania purpurea, Lea, Proc. Acad. Nat. Sci., p. 120.
Goniobasis purpurea, Lea, Jour. Acad. Nat. Sci., v, pt. 3, p. 225, t. 34, f. 12, March, 1863. Obs., ix, p. 47.

Description.— Shell smooth, subfusiform, obtusely conical, rather thin, dark brown; spire very obtuse; sutures slightly impressed; whorls five, the last large; aperture rather large, elliptical, dark within; outer lip acute; columella dark and bent inward.

Operculum ovate, spiral, dark brown, with polar point near the inner edge, and one-fourth distance from the base.

Habitat.— Alabama; E. R. Showalter, M.D.

Diameter, ·35; length, ·81 of an inch.

Observations.— There are two specimens before me of this very dark brown shell. The larger one has three bands faintly visible on the inside. It is very possible that it may be found much less intense in color. It is a graceful, well proportioned species. On the upper portion of the whorls, immediately under the suture, there is a disposition to take on a light color, like a thread. The aperture is about one-half the length of the shell. The nearest allied species is *ebenum* (nobis) = *Melania iostoma*, Anth., but it may at once be distinguished by the line of the outer lip, which in *ebenum* is remarkably indented, while in *purpurea* that line is nearly straight. *Ebenum* is also smaller and thicker.—*Lea.*

Fig. 642.

Very nearly related to *G. rara*, Lea.

250. G. quadrivittata, Lea.

Melania quadrivittata, Lea, Proc. Acad. Nat. Sci., 1861, p. 119.
Goniobasis quadrivittata, Lea, Jour. Acad. Nat. Sci., v, pt. 3, p. 226, March, 1863. Obs., ix, p. 48.

Description.—Shell smooth, subelliptical, a little thick, greenish-yellow, shining; spire obtusely conical; sutures very much impressed; whorls eight, somewhat convex; aperture somewhat constricted, ovately rhombic, whitish and four-banded within; outer lip acute, columella incurved, angular at base.

Habitat.—Coosa River, Alabama; E. R. Showalter, M.D.

Diameter, ·33; length, ·84 of an inch.

Observations.—This brilliant species, with its four well defined, dark brown bands on a dark yellow, is allied to *fascinans* herein

Fig. 643.

described, and to *Melania pupoidea*, Anth., but it is shorter and more robust than either. The five specimens before me are very nearly of the same size, and all have four beautiful bands which are somewhat close, and give a darkish color to the whole. The aperture is more than one-third the length of the shell.— *Lea.*

Very closely allied to *G. Alabamense*, **Lea.**

251. G. propria, Lea.

Melania propria, Lea, Proc. Acad. Nat. Sci., p. 118, 1861.
Goniobasis propria, Lea, Jour. Acad. Nat. Sci., v, pt. 3, p. 229, t. 34, f. 21, March, 1863. Obs., ix, p. 52.

Description.— Shell smooth, fusiform, yellowish-olive, four-banded. rather thick; spire obtusely conical; sutures impressed; whorls six, slightly convex; aperture somewhat large, elongately ellip- Fig. 644. tical, whitish within and banded; outer lip acute; columella inflected, white and subangular at base.

Habitat.— Alabama; E. R. Showalter, M.D.

Diameter, ·34; length, ·80 of an inch.

Observations.—This is a regular fusiform species, with an agreeable outline near to that of *gracilior* herein described. It is not so stout a shell and is rather smaller, and having bands cannot be easily confounded with that species. The aperture is more than half the length of the shell, and the apex is quite pointed.— *Lea.*

252. G. negata, Lea.

Goniobasis negata, Lea, Proc. Acad. Nat. Sci., p. 271, 1862. Jour. Acad. Nat. Sci., v, pt. 3, p. 337, t. 38, f. 200, March, 1863. Obs., ix, p. 159.

Description.— Shell striate, elliptical, subconical, thick, yellowish, four-banded; spire obtusely conical; sutures very much and very irregularly impressed; whorls six, somewhat convex, the last large; aperture rather small, ovate, white within, and four-banded; outer lip sharp, slightly thickened; columella bent in, thickened, obtusely angular at the base.

Operculum ovate, rather thin, light brown, with the polar point near to the base.

Habitat.— Coosa River, Alabama; E. R. Showalter, M.D.

Diameter, ·35; length, ·68 of an inch.

Observations.— This species is very nearly allied to *Melania* (*Goniobasis*) *Vanuxemiana* (nobis), having coarse striæ over the whole of the whorls. But it is smaller, rather more elliptical, and has Fig.645. more striæ, the number being about ten. These striæ are rounded, with an intervening groove, and cover the whole of the whorls. The bands are obscure on the outside of both the specimens before me, but are well defined inside. It has some resemblance to *Melania* (*Goniobasis*) *Coosaensis* (nobis), but is a much smaller species, and is more constricted in the whorls and in the aperture. The aperture is nearly half the length of the shell.—*Lea.*

253. G. impressa, LEA.

Melania impressa, LEA, Philos. Proc., ii, p. 83, Oct., 1841. Philos. Trans., ix, p. 19.
 Obs., iv, p. 19. WHEATLEY, Cat. Shells U. S., p. 25. JAY, Cat. Shells, p. 274.
 BINNEY, Chèck List, No. 143. BROT, List, p. 32. REEVE, Monog. Melania, sp.
 316, 349. HANLEY, Conch. Miscel. Melania, t. 8, f. 69.
Megara impressa, Lea, CHENU, Manuel, i, f. 2023. ADAMS, Genera, i, p. 306.
Melania crebristriata, LEA, Philos. Proc., iv, p. 166. Philos. Trans., x, p. 65, t. 9,
 f. 47. Obs., iv, p. 65. BINNEY, Check List, No. 75. CATLOW, Conch. Nomenc.,
 p. 186. BROT, List, p. 32.
Megara crebristriata, Lea, ADAMS, Genera, i, p. 306.

Description.— Shell transversely and thickly sulcate, fusiform, thick, reddish-brown; spire obtuse; sutures impressed; whorls six, flattened; aperture elliptical, rather large, angular at the base, within white.

Fig. 646.

Habitat.— Coosa River, Alabama.

Diameter, ·48; length, ·81 of an inch.

Observations.— Dr. Griffith received a single specimen only of this singularly marked species, and this is not entirely perfect at the spire or aperture. The whole surface of this specimen is covered with very minute, impressed, revolving lines, the body-whorl having twenty-four. They are nearly equidistant and very regular. Its aperture is nearly one-half the length of the shell. On the superior part of the columella, there is quite a large callus.* In form and size, it closely resembles the *M. robusta* herein described.— *Lea.*

My two figures represent an adult and immature specimen.

---

* Other specimens, subsequently received, confirm nearly all the other characters.

It is a beautiful species and occurs not infrequently in the Coosa River.

Melania crebristriata.—Shell transversely and very closely striate, nearly fusiform, thick, yellowish horn-color; spire obtuse; sutures impressed; whorls somewhat convex; aperture small, rather ovate, angular at the base, within whitish; columella inflected and thickened above.

Habitat.—Tuscaloosa, Alabama.

Diameter, ·40; length, ·76 of an inch.

Observations.—This species is nearly allied to *M. impressa* (nobis), but may be distinguished by its color being yellowish, and by its
Fig. 647. coarser striæ. Its aperture also is smaller. The three specimens before me are very differently banded, one having nine, another three, and the last a rather broad one near the upper part of the whorl. These are only seen on the inside. The apex of each being eroded, the number of the whorls could not be accurately counted Perhaps there are six. The striæ are so strong that they cause the edge of the outer lip to be crenate. The aperture is about two-fifths the length of the shell. On the superior whorls there are broad, slightly elevated, somewhat oblique ribs. The number of striæ on the three specimens before me are, respectively, sixteen, eighteen and twenty.— *Lea.*

254. G. pergrata, LEA.

Melania pergrata, LEA, Proc. Acad. Nat. Sci., p. 122, 1861.
Goniobasis pergrata, LEA, Jour. Acad. Nat. Sci., v, pt. 3, p. 243, March, 1863. Obs., ix, p. 65.

Description.—Shell striate, subcylindrical, obtusely conical, somewhat thick, greenish horn-color; spire very obtuse; sutures very much impressed; whorls six, shouldered above, covered with transverse striæ, the last very large and cylindrical; aperture large, elongately ovate, whitish within; outer lip acute; columella arcuate, slightly callous above, somewhat rounded at the base.

Operculum ovate, spiral, dark brown, with the polar point on the edge near to the base.

Habitat.— Coosa River, Alabama; E. R. Showalter, M.D.

Diameter, ·44; length, ·90 of an inch.

Observations.— This species reminds one of *M. crebristriata, M. cap-*

illaris and *M. impressa* (nobis), (all *Goniobases*) by its numerous transverse striæ; but these striæ are neither so numerous, so regular, nor the intervals so deeply impressed, nor do these striæ exist Fig. 648. on the upper whorls, as in those species. The color of the epidermis is also much lighter and brighter. In outline it is near to *impressa*, but the spire is not so elevated, nor has it the bands which are visible on that species. It is to be regretted that a single specimen only was received, as others may be found with different character. This one has an obscure band on the upper whorls, but none whatever on the lower one. The striæ on the outside are represented inside by whitish lines. The aperture is fully half the length of the shell.— *Lea.*

This may be merely a variety of *impressa*, in which the striæ are not so well developed.

255. G. capillaris, Lea.

Melania capillaris, Lea, Proc. Acad. Nat. Sci., p. 122, 1861.
Goniobasis capillaris, Lea, Jour. Acad. Nat. Sci., v, pt. 3, p. 236, t. 34, f. 34, March, 1863. Obs., ix, p. 58.

Description.— Shell thickly striate, subfusiform, somewhat thick, yellowish-brown, covered with close, transverse striæ; spire very obtuse; sutures irregularly impressed; whorls somewhat compressed, the last large; aperture large, widely elliptical, capillary Fig. 649. striæ within; outer lip crenulate; columella whitish, thickened, incurved, obtusely angular at the base.

Operculum ovate, spiral, dark brown, with the polar point near the inner side and near to the base.

Habitat.— Coosa River, Alabama; E. R. Showalter, M.D. and Wm. Spillman, M.D.

Diameter, ·38; length, ·88 of an inch.

Observations.— This species belongs to the group of which *Melania* (*Goniobasis*) *impressa* (nobis) may be considered the type. It is covered with hair-like raised lines, like *impressa* and *Melania* (*Goniobasis*) *crebristriata* from the same river. It may be distinguished from the former by being more cylindrical, being of a slightly lighter brown, and in having more striæ. From the latter by having a less exserted spire, by having finer striæ and being of a darker brown. All three of these species have usually more or less fine brown bands in the interior, but occasionally a specimen may be seen without bands. Among the specimens before me, the *crebristriata* has about fifteen striæ, the *capillaris* about twenty-six, and the *impressa* about twenty-eight. These raised, rounded striæ çause, in all the three species, a beautiful crenated outer lip. The aperture is about half the length of the shell, and the apex is usually decollate. The brown lines of the interior do not reach the edge of the outer lip. In some specimens the columella is so much thickened that it reminds one of the genus *Lithasia.— Lea.*

DOUBTFUL AND SPURIOUS SPECIES.

Melania fuscata, DESHAYES,* Anim. sans. Vert.. viii, p. 435·
Melania ligata, Conrad, BROT, List, p. 33. (Ubi?) Alabama.
Melania ochracea, Cristofori and Jan., BROT, List, p. 59. (In museo deest.)
Melania Buschiana, REEVE,† Monog. Melania, sp. 50. California.
Melania ligata, Cristofori and Jan., BROT, List, p. 58.
Melania oveliana, Lea, WHEATLEY, Cat. Shells, U. S. p. 26. Alabama.
Melania multistriata, Lea, WHEATLEY, Cat. Shells, U. S., p. 26. Alabama.
Melania mutilata, Say,‡ JAY, Cat. Shells. CATLOW, Conch. Nomenc., p. 187. South Carolina.
Melania exigua, CONRAD, = *Amnicolidæ.*
Melania sulculosa, MENKE, Syn. Meth., 2d edit., p. 136. BROT, List, p. 59.
Paludina sulculosa, MENKE, Syn. Meth., 1st edit., p. 80.
Melania costata,§ RAVENEL, Cat., p. 11, 1834. BINNEY, Check List, No. 71. BROT, List, p. 58. Dan River, Virginia.
Melania Wahlamatensis,‖ Lea, BINNEY, Check List. BROT, List, p. 59.
Pleurocera acuta, RAFINESQUE, Enumeration and Account, p. 3, Nov., 1831.
Pleurocera gibbosa, Rafinesque, BINNEY, Check List, No. 122.
Pleurocera gonula, RAFINESQUE, Enumeration and Account, p. 2, Nov., 1831.
Melania marginata, Rafinesque, BINNEY, Check List, No. 165.
Melania (Ambloxus) rugosa, RAFINESQUE, Enumeration and Account, p. 3, Nov., 1831.
Melania viridis, RAFINESQUE. Enumeration and Account, p. 3. Nov., 1831.
Melania vittata, RAFINESQUE, Enumeration and Account, p. 3. BINNEY, Check List, No. 295.
Melania zonalis, Rafinesque, BINNEY, Check List, No. 298. BROT, List, p. 59.

---

* This old species, figured by Born and described in full in Deshayes' edition of Lamarck, is certainly not an American shell, although attributed to Virginia. Its characters are entirely of the East Indian type.
† This shell is evidently of East Indian type.
‡ = *Bulimus decollatus,* L. (*mutilatus,* Say).
§ *Anculosa dissimilis?*
‖ Mr. Lea has not used this name for any of the *Strepomatidæ,* but he has used it for an *Anodonta.*

Genus EURYCÆLON, Lea.

Eurycælon, Lea, Proc. Acad. Nat. Sci., p. 3, Jan., 1864.

Description.— See Preliminary Observations, p. xxx.

Geographical Distribution.—The species of *Eurycælon* are not numerous, and appear to be confined to the waters of East Tennessee and North Alabama.*

1. E. Midas, Lea.

Melania Midas, Lea, Proc. Acad. Nat. Sci., p. 119, 1861.
Goniobasis Midas, Lea, Jour. Acad. Nat. Sci., v, pt. 3, p. 233, t. 34, f. 28, March, 1863. Obs., ix, p. 55.

Description.— Shell smooth, cylindraceo-elliptical, somewhat thick, greenish, obscurely banded; spire very obtuse; sutures irregularly impressed; whorls somewhat compressed, the last very large, obscurely striate below; aperture large, ear-shaped, bluish-white within; outer lip acute; columella bluish-white, thickened and inflected, obtusely angular at the base.

Operculum subelliptical, spiral, dark brown, with polar point near the inner edge and one-fifth from the base.

Habitat.—Coosa and Alabama Rivers, near Wetumpka; Dr. E. R. Showalter.

Diameter, ·48; length, ·98 of an inch.

Observations.— This is a well marked species. There are several specimens before me, differing but little. Two of them have a brown band in the interior of the upper part of the aperture, another has none, but exhibits an obscure row of spots on the upper whorls, which others have also. Two of the specimens have irregular, tuberculous swellings on the upper part of the whorls, which obscure the bands, and thus cause them to take on a maculate character. The increment of growth usually commences below the previous termination, leaving angles on the sutures. In this character one is reminded of *Melania (Goniobasis) oppugnata* (nobis). In these

Fig. 650.

---

* I am now inclined to consider these shells to be distorted *Goniobases* and *Anculosæ*, and in none of them can I find generic characters. They might with advantage to science be relegated to those genera. *April,* 1873.

specimens there is a difference in the form of the base of the aperture, one of them being more rounded; but this may arise from difference of age. In outline this species is allied to *Hartmanii* (nobis), but it cannot be confounded with that shell, which is much larger, more robust, more elevated in the apex, and has more and better developed bands. It is on the other side near to *Melania* (*Goniobasis*) *basalis* (nobis). The aperture is about two-thirds the length of the shell.— *Lea.*

Very closely allied to *G. ambusta.*

2. E. Leai, Tryon.

Eurycælon Leai, Tryon, American Journal of Conchology, vol. 2, No. 1, p. 5, t. 2, f. 3, 1866.

Description.— Shell conical, thick, shining; spire conical, obtusely elevated; suture moderately impressed; whorls about six, slightly convex, everywhere covered with very fine, close, revolving striæ, somewhat shouldered beneath the suture and crimped; body-whorl

Fig. 651.

large, slopingly convex; aperture large, ovate, broad below; wax-yellow or somewhat olivaceous, lighter beneath the suture, white within.

Habitat.— Etowah River, Cartersville, Georgia.

Diameter, 13 mill.; length (eroded), 19 mill.

Observations.— This species is somewhat like *G. luteola*, Lea, in color, striæ and texture, but differs in having tubercles and in the form of the aperture. In *G. pergrata*, Lea, the striæ are coarser and the tuberculations are wanting. It is a very neat species, beautifully marked by the narrowly compressed numerous tubercles under the suture, and its close, waved, revolving striæ.— *Tryon.*

3. E. gratiosa, Lea.

Melania gratiosa, Lea, Proc. Acad. Nat. Sci. p. 122, May, 1861.
Goniobasis gratiosa, Lea, Jour. Acad. Nat. Sci. v, pt. 3, p. 241, t. 35, f. 43, March, 1863. Obs., ix, p. 63.

Description.— Shell tuberculate, sometimes striate, obtusely fusiform, somewhat thick, yellowish-green, banded or without bands; spire very obtuse; sutures impressed; whorls six, flattened above, the last large; aperture rather large, subrhomboidal, whitish within; outer lip acute, slightly sinuous; columella inflected, thickened, subangular at the base.

Operculum ovate, spiral, dark brown, with the polar point near to the base.

Habitat.— Coosa River, Alabama; E. R. Showalter, M.D.

My cabinet and cabinet of Dr. Showalter.

Diameter, ·39; length, ·78 of an inch.

Observations.—This is a very remarkable and beautiful little species. There are three specimens before me, all of them having four somewhat distant, low, obtuse, rather large nodes. I have Fig. 652. never seen any other species with this kind of nodes. The texture of the shell is delicate, the epidermis smooth and shining. Two of the specimens have four well defined, brown bands, which are strongly marked inside and out. The third specimen is without bands, but it is covered with very remarkable transverse striæ, which traverse the nodes as well as the other parts of the surface. The aperture is more than half the length of the shell.— *Lea.*

See remarks on next species (*M. lachryma,* Anthony) with which it is identical.

3a. E. lachryma, Anthony.

Melania lachryma, Anthony, Reeve, Monog. Melania, sp. 473, May, 1861. Brot, List, p. 32.

Description.— Shell conically ovate, thick, fulvous-olive, encircled Fig. 653. Fig. 653a. with numerous black lines; whorls five, slopingly convex round the upper part, then gibbous, and obtusely tubercled, longitudinally, plicately striated throughout; aperture narrowly ovate, rather small, sinuately effused at the base.

Habitat.— United States. (Alabama—label attached to type, G. W. Tryon, Jr.)

Observations.— A prettily painted species of a rude, obtusely tubercled form.— *Reeve.*

The figure is a copy of Mr. Anthony's type. This shell and *gratiosa* are identical, but I am unable to ascertain which has priority. A very beautiful specimen in Mr. Lea's collection is closely and sharply sculptured with transverse striæ.

4. E. lepida, LEA.

Melania propria, LEA, Proc. Acad. Nat. Sci., 1861, p. 123.
Goniobasis lepida, LEA, Jour. Acad. Nat. Sci., v, pt. 3, p. 227, t. 34, f. 17, March,
   1863. Obs., ix, p. 49.

Description.— Shell smooth, subfusiform, rather thin, yellowish
horn-color, obscurely banded, shining; spire raised; sutures very
much impressed; whorls about six, slightly convex above, inflated
below; aperture rather large, ovate, yellowish-white within; outer
lip acute; columella inflected, thickened above and rounded at the
base.

Habitat.— Yellowleaf Creek, Shelby County, Alabama; Dr. E. R.
Showalter.

Diameter, ·42; length, ·98 of an inch.

Observations.— A single specimen was sent to me by Dr. Lewis,
Mohawk, N. Y., who received it from Dr. Showalter. It is allied to
straminea herein described, and to *Melania proteus* (nobis).

Fig. 654.

It was more elongate than the former, and larger and darker
horn-color. It differs from the latter in not being so solid
and in being more oval. The specimen before me is eroded
at the apex, and therefore the apical whorls cannot be described, nor the number correctly ascertained. There is a
slight swelling below the suture, and irregular flattenings
on the bulge of the whorls. A single obscure band is visible on
the upper part of the whorls, and some obscure striæ on the lower
part.— *Lea.*

The shouldered whorls, and irregular flattenings will place
this species in the genus *Eurycœlon*, instead of *Goniobasis*,
where it is put by Mr. Lea. This species was first published
under the name of *propria*, but that name being preoccupied
by Mr. Lea himself, it was subsequently changed to *lepida*.

5. E. proteus, LEA.

Melania proteus, LEA, Philos. Proc., iv, p. 166, 1845. Philos. Trans., x, p. 57, t. 9,
   f. 28. Obs., iv, p. 57. BINNEY, Check List, No. 219. BROT, List, p. 33.
Juga proteus, LEA, Adams, Genera, i, p. 304.

Description.— Shell smooth, subcylindrical, thick, pupæform, yellowish horn-color; spire elevated; sutures impressed; aperture small,
rhomboidal, angular at the base, within whitish.

Habitat.—Tuscaloosa, Alabama.

Diameter, ·5 of an inch; length, 1 inch.

Observations.—There were six specimens submitted to me by Dr. Budd, which I refer to the one species, although they Fig. 655. present considerable difference. Five of the specimens are dead and bleached shells, and are of a light yellow or buff color. The sixth is a fresh and perfect specimen, with four small, purple bands and a tuberculous shoulder, the tubercles being prolonged nearly into folds. Two others are indistinctly banded. Another has a tuberculous shoulder, and is disposed to be granulate. From these varieties arises the name given to it. The aperture is rather contracted, and about two-fifths the length of the shell.—*Lea.*

6. E. gibberosa, Lea.

Goniobasis gibberosa, Lea, Proc. Acad. Nat. Sci., p. 266, 1862. Jour. Acad. Nat. Sci., v, pt. 3, p. 312, t. 37, f. 155, March, 1863. Obs., ix, p. 134, t. 37, f. 155.

Description.—Shell smooth, subfusiform, thick; spire obtuse; sutures irregularly impressed; whorls hump-backed, slightly convex Fig. 656. above, the last one very large; aperture very large rhomboidal, white within; outer lip acute, sinuous; columella bent in, thickened above and below.

Operculum ovate, dark brown, with the polar point near to the base, on the inner edge.

Habitat.—Alabama River; E. R. Showalter, M.D.

Diameter, ·48 of an inch; length, 1·03 inches.

Observations.—Four specimens of this remarkable species are before me. They were sent by Dr. Showalter to Dr. Hartman, who called my attention to them and sent them for examination. The species is singular for the four to six hump-like elevations which exist on the upper half of each of the whorls and which leave flattish spaces between, on one of which spaces the shell will always rest when the specimen is moved on a flat surface. One of the specimens has four distinct bands, one has these obsolete, the two remaining ones are without bands. The only species to which this has close affinities is *Melania (Goniobasis) basalis* (nobis), it having somewhat like irregular elevations, but it is a smaller and thinner species with a greenish epidermis and thick close bands. None of the four specimens before me have more than three-perfect whorls remaining, the upper ones (perhaps six originally) are worn off. The length of the aperture is about one-half that of the shell.— *Lea.*

7. E. nubila, LEA.

Melania nubila, LEA, Proc. Acad. Nat. Sci., p. 118, 1861.
Goniobasis nubila, LEA, Jour. Acad. Nat. Sci., v, pt. 3, p. 235, March, 1863. Obs., ix, p. 57.

Description.— Shell striate, somewhat elliptical, subfusiform, dark green, obscurely spotted, rather thick; spire obtusely elevated; sutures irregularly impressed; whorls six, rather inflated, the last large; aperture rather large, rhomboido-elliptical, four-banded within; outer lip acute; columella arcuate, obtusely angular at the base.

Habitat.— Coosa River, Wetumpka, Alabama; Dr. E. R. Showalter. Diameter, ·45 of an inch; length, 1·1 inches.

Observations.— Several specimens of different ages are before me. The oldest one is about an inch long, the youngest about half an

Fig. 657.

inch. They all bear the same dark nebulous character, but the largest only has the four bands so wide as to combine and give the fauces a dark purple hue, which extends to the callus of the columella. The others have the columella whitish and the bands are distinct within. The oldest has a few coarse striæ on the upper and lower parts of the whorls, but the younger ones in my possession have not these striæ. There is a disposition in all these specimens to have obscure coarse folds, which are yellowish, leaving between them darkish spots. The aperture is nearly one-half the length of the shell.—*Lea.*

8. E. umbonatum, LEA.

Eurycælon umbonatum, LEA, Proc. Acad. Nat. Sci., p. 3, 1864. Obs., xi, 106, t. 23, f. 64.

Description.— Shell nodulous, subfusiform, rather thick, obscurely banded, dark olive; spire very obtuse; sutures very much impressed; whorls with irregular bosses, swollen below the sutures, the last one very large; aperture very large, subelliptical; outer lip acute, slightly sinuous; columella thickened above and somewhat sinuous below.

Habitat. — Smith's Shoals, Cumberland River, East Tennessee; Major S. S. Lyon (U. S. E.).

Diameter, ·48; length, ·80? of an inch.

Observations.—I only received two specimens of this interesting species, and neither being perfect at the apex, the number of whorls cannot be ascertained; probably there are not more than five. Both these specimens have two small, obscure bands on the inside of the upper part of the outer lip. One has dark brown marks inside and is brown at the bottom of the columella. One is much darker on the outside than the other. The large, irregular nodes or bosses are three on the body-whorl of one specimen and five on the other, they are placed on the shoulder of the whorls. The aperture is nearly two-thirds the length of the shell.—*Lea.*

Fig. 658.

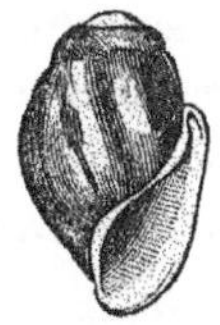

9. E. Anthonyi, Budd.

Anculosa Anthonyi, Budd, Redfield, Ann. Lyc. Nat. Hist., vi, p. 130, t. 1, f. 6, April, 1851.
Leptoxis Anthonyi, Budd, Redfield, Brot, List, p. 23. Binney, Check List, No. 341.
Anculotus Anthonyi, Budd, Redfield, Reeve, Monog. Anc., t. 2, f. 17.

Description.—Shell rhomboidally ovate, covered with an olivaceous-yellowish epidermis, beneath which usually appear two purplish bands encircling the body-whorl; spire short; whorls about four, the upper ones much eroded, the upper portion of the last whorl is shouldered by a series of large, obtuse and irregular tubercles, about four or five in number, there is also a slight tendency towards thickening in the ventral portion of the whorl; aperture ovate, effuse above and below; right lip thin; columella lip usually stained with purple above and below, reflected so as partially to cover a deep, umbilical depression, which, however, is continued towards the base, forming a channel much resembling that of the umbilical region in *Natica.*

Fig. 659.

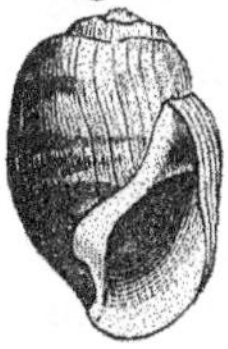

Habitat.—Holstein River, near Knoxville, Tennessee, where it was collected by our associate, O. W. Morris, and also by Mr. Anthony.

Diameter, ·63 (16 millim.); length, ·83 of an inch (21 millim.). Length of aperture, ·61 (16 millim.); breadth of aperture, ·31 of an inch (8 millim.).

Observations.—Allied to *A. salebrosa*, but has the tubercles of its last whorl larger, more obtuse and irregular and fewer in number. In adopting the above name for this species, proposed by Dr. Budd,

I pay a deserved compliment to one of the most industrious and ardent naturalists in our Western States; though in so doing, I reluctantly depart from a wholesome recommendation formally promulgated, first by the Scientific Congress of Great Britain, and afterwards by that of America. It is to be regretted that this recommendation has been so little heeded, but where the recognized *laws* of nomenclature hardly restrain, mere suggestions will be of little avail.— *Redfield.*

This very distinct species attains a large size, ranking in this respect with *E. crassa.* In the collection of Gould are specimens collected in west Georgia.

10. E. crassa, HALDEMAN.

Anculosa crassa, HALDEMAN, Monog. Limniades, No 4, p. 3 of Cover, Oct. 5, 1841.
Anculotus crassus, Haldeman, JAY, Cat., 4th edit., p. 276. REEVE, Monog. Anculotus, t. 2, f. 14.
Leptoxis crassa, HALDEMAN, Monog. Lept., p. 2, t. 1, f. 19-23. BINNEY, Check List, No. 350. BROT, List, p. 24. Haldeman. ADAMS, Genera, i, p. 307.
Leptoxis pisum, HALDEMAN, Monog. Lept., p. 4, t. 3, f. 82. BINNEY, Check List, No. 378. BROT, List, p. 25. Haldeman, ADAMS, Genera, i, p. 307.
Anculosa turbinata, LEA, Proc. Acad. Nat. Sci., 54, 1831. Jour. Acad. Nat. Sci., v, pt. 3, p. 254, March, 1863. Obs., ix, p. 76.

Description.— Shell conical or globose, ponderous; whorls five, flat or slightly convex; spire exserted; aperture ovate, with a well

Fig. 661.   Fig. 660.

marked columellar notch; labium thick; color brown.

Habitat.— Clinch? River, Tennessee.

Length, ¾ of an inch.

Observations.— Differs from *A. prærosa* by the better developed spire and notch.

Haldeman

In his "Monog. of Leptoxis," Professor Haldeman informs us that this species lives in tranquil waters near their margins, and not in rapid currents, like the other species of the genus. This is certainly an unexpected habit in a species so ponderous and it may be doubted whether the species habitually *seeks* such stations. The species appears to be rather common in North Alabama, whence beautiful specimens have been received.

The following is a synonyme :—

Leptoxis pisum.— Shell globular, shining, having the lines of growth effaced; spire very short, decorticated and rounded; mouth widely oval, contracted by the columella in front; columella slightly flattened with an anterior flexure; color shining brown, within white or violet.

Fig. 662.

Habitat.— Tennessee.

Observations.— A species of medium size, remarkable for its exterior and its well developed columellar flexure.— *Haldeman.*

The following is also a synonyme :—

Eurycælon turbinata.— Shell smooth, subrotund, thick, heavy, dark horn color, three-banded; spire obtuse, scarcely exserted; sutures very much impressed; whorls four, the last very large; aperture large, ovate, within white and three-banded, recurved at the base;

Fig. 663.   Fig. 664.

columella incurved, impressed; outer lip acute, expanded and sinuous.

Habitat. — North Alabama; Prof. M. Tuomey and Dr. Lewis: Tuscaloosa; Dr. Budd.

Diameter, ·56; length, ·70 of an inch.

Observations.— I have seen only three specimens of this species. One, that which is figured, I have had for some years. It is not easily confounded with any species I know, being more turbinate than any which has come

Fig. 665.   Fig. 664a.

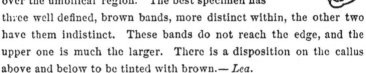

under my notice. It is broad above and pointed below, and has an abrupt curvature near the base of the columella made by the impressed callus over the umbilical region. The best specimen has three well defined, brown bands, more distinct within, the other two have them indistinct. These bands do not reach the edge, and the upper one is much the larger. There is a disposition on the callus above and below to be tinted with brown.— *Lea.*

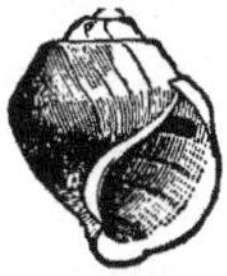

I find that this is only a very much inflated and not fully grown shell of *E. crassa.* I figure a very young specimen (fig. 665), which exhibits a great difference from the adult. In fig. 664*a* the sharp carina of the young shell is disappearing; this is succeeded by the form described by Mr. Lea as *turbinata,* and then follows the mature form.

Genus MESESCHIZA, LEA.

Meseschiza, LEA, Proc. Acad. Nat. Sci., p. 2, Jan., 1864.

Description.— Shell fusiform, imperforate : aperture rhomboidal, below canaliculate ; lip expanded, slit in the middle ; columella smooth, incurved.

Operculum corneous, spiral.— *Lea.**

1. M. Grosvenorii, LEA.

Meseschiza Grosvenorii, LEA, Proc. Acad. Nat. Sci., p. 2, Jan., 1864. Obs., xi, 108, t. 23, f. 67.

Description.— Shell smooth, fusiform, thin, obtusely conical, purple or banded ; spire obtusely conical ; sutures slightly impressed ; whorls Fig.666. about seven, scarcely convex ; aperture large, rhomboidal, generally banded within ; outer lip acute, slightly notched in the middle ; columella slightly thickened and twisted.

Operculum ovate, light brown, rather thin, having several volutions, and with the polar point well removed from the left margin.

Habitat.— Wabash River, Indiana ; H. C. Grosvenor.

Diameter, ·27 ; length, ·43 of an inch.

Observations.— I have thirteen specimens of this remarkable shell. Eight of them have a well defined, though delicate notch, on the edge, at or near to the periphery of the last whorl. In some this notch is a little above the periphery, and in others a little below. Five of the specimens have no notch, which probably arises in four of them from not being fully grown, and in one from having the thin, delicate edge broken off. The specimens vary in color, some being light horn-color with few or many bands, others more or less purple and with or without bands ; others again have obscure, longitudinal thickenings, which being whitish give the specimens the appearance of being folded. In all the specimens there is a light line under the sutures, and some have six or seven brown bands, which are distinctly seen on the inside. The channel at the base is small, but well defined. In outline this species reminds one of *Goniobasis Vauxiana* (nobis) and

---

*Only a single species of this genus has been described, and all the specimens are young shells and from a single locality. I have examined them carefully and I have discovered in every one of those exhibited to me by Mr. Lea, the evidence of diseased growth ; under these circumstances I think the genus may fairly be considered a doubtful one. *April,* 1873.

Melania (*Goniobasis*) *germana*, Anthony. It is a thinner shell than either, and the notch in the lip removes it from that genus. The aperture is about one-half the length of the shell. I have great pleasure in naming this species after Mr. Grosvenor, to whom I am greatly indebted for many of our western mollusca.—*Lea.*

Genus SCHIZOSTOMA, LEA.

Schizostoma, LEA, Philos. Proc., ii, p. 242, Dec., 1842; iv, p. 167, Aug., 1845. Philos. Trans., x, p. 67, 1847. Obs., iv, p. 41, 1847. Proc. Acad. Nat. Sci., May, 1860. Jour. Acad. Nat. Sci., v, pt. 3, p. 245, March, 1863. Obs., ix, p. 67.

Schizocheilus, LEA, Philos. Trans., x, p. 295, 1853. Obs., v, p. 51, 1823.

Gyrotoma, SHUTTLEWORTH, Mittheil. Naturforsch. Bern., p. 88, July 22, 1845. ADAMS, Genera, i, p. 305, Feb , 1854. GRAY, Guide to Mollusca, i, p. 103, 1857. CHENU, Man. de Conchyl., i, p. 293, 1859. ANTHONY, Proc. Acad. Nat. Sci., p. 63, Feb., 1860. BINNEY, Check List, June, 1860. BROT, List, p. 27, 1862.

Melatoma, Anthony, GRAY, Zool. Proc., p. 153, 1847. WOODWARD, Manual, p. 131, 1851. REEVE, Conch. Icon., March, 1860.

Apella, MIGHELS, MSS.

Description.—Shell conical or fusiform; lip fissured above; aperture ovate; columella smooth, incurved.

Geographical Distribution.—The genus appears to be restricted to the waters of the Coosa River, Alabama.

Observations. — The genus *Schizostoma* seems to be capable of being divided into two natural groups in the form of the *fissura*, the cut in the lip. In one group this fissura is deep and direct, that is, parallel with the suture or upper edge of the whorl (fig. 667); in the other it is not deep and is oblique to the suture (fig. 668). Fig. 669 represents the operculum of *S. ovoideum*, Shutt.

Fig. 667. Fig. 668. Fig. 669.

SYNOPTICAL TABLE OF SPECIES.*

FISSURE DIRECT, NARROW AND DEEP.   FISSURE OBLIQUE, SHORT AND WIDE.

1. *Shell striate or ridged.*

A. Shell conical, spire lengthened, sharply carinate.

1. S. CARINIFERUM, Anthony.
    S. Showalterii, Lea.
2   S. CASTANEUM, Lea.

15. S. PAGODUM, Lea.
16. S. PYRAMIDATUM, Shutt.
17. S. WETUMPKAENSE, Lea.
     S. ornata, Anthony.
     S. pagoda, Lea, of Reeve.

B. Shell conic-cylindrical; spire obtuse, not carinate.

3. S. OVOIDEUM, Shuttleworth.

4. S. EXCISUM, Lea.

18. S. ALABAMENSE, Lea.
19. S. ANTHONYI, Lea.
20. S. BABYLONICUM, Lea.
     Spillmanii, Lea.

C. Shell globosely-ovate, spire moderate.

5. S. PUMILUM, Lea.
    Globosum, Lea.
    Alabamense, Lea, of Reeve.
    Showalterii, Lea, of Reeve.

21. S. BUDDII, Lea.
     S. funiculatum, Lea.
     S. pagodum, Lea, of Reeve.

2. *Shell smooth.*

D. Shell elliptic.

6. S. ELLIPTICUM, Anthony.
7. S. LACINIATUM, Lea.

E. Shell quadrately cylindrical.

8. S. AMPLUM, Anthony.
9. S. NUCULUM, Anthony.

10. S. CYLINDRACEUM, Mighels.

22. S. DEMISSUM, Anthony.
     S. Hartmanii, Lea.
23. S. CONSTRICTUM, Lea.
     S. rectum, Anthony.
23a. S. SHOWALTERIANA, Lea.
24. S. SALEBROSUM, Anthony.
     S. robustum, Anthony.
     S. rectum, Anth., of Reeve.

---

* In the above table the opposite species in the two groups are generally exactly similar except in the character of the slit.

F. Shell ovate, whorls obliquely flattened, spire obtuse.

11. S. BULBOSUM, Anthony.   25. S. GLANDULUM, Lea.
    S. ovalis, Anthony.   26. S. INCISUM, Lea.
12. S. CURTUM, Mighels.   *S. virens*, Lea.
         S. quadratum, Anthony.
         S. obliquum, Anthony.
13. S. GLANS, Lea.

G. Shell globose.

14. S. SPHÆRICUM, Anthony.

SPECIES.

1. S. cariniferum, ANTHONY.

Gyrotoma carinifera, ANTHONY, Proc. Acad. Nat. Sci., p. 66, Feb., 1860. BINNEY,
    Check List, No. 310. BROT, List, p. 27.
Melatoma cariniferum, Anthony, REEVE, Monog. Melatoma, t. 2, f. 13.
Schizostoma Showalterii, LEA, Proc. Acad. Nat. Sci., p. 93, March, 1860. Jour.
    Acad. Nat. Sci., t. 35, f. 49, March, 1863. Obs., ix, p. 68.
Gyrotoma Showalterii, Lea, BINNEY, Check List, No. 334. BROT, List, p. 28.

Description.— Shell conic, thick, dark brown; spire obtusely elevated, truncate, though not abruptly so, six whorls remaining, one
or two having apparently been lost by truncation; car-   Fig. 670.
inations elevated, subacute and found on all the whorls,
two on each of the spiral ones and three to four on the
body-whorl; fissure direct, broad, and moderately deep,
extending about one-fifth around the shell; sutures irregular, much modified by the carinæ, and often concealed in part by them; aperture ovate and banded
within; columella much rounded, callous at the lower part only;
outer lip irregularly waved, its outline modified by the carinæ on the
body-whorl; no sinus.

Habitat.— Coosa River, Alabama.

Length of shell, ⅞; breadth of shell, ½ of an inch. Length of aperture, 5½–16 of an inch; breadth of aperture, ¼ of an inch.

Observations.—This species cannot well be confounded with any
other yet described. In general form and in its armature, one is very
forcibly reminded of *Melania annulifera*, Con., from which it differs,
however, not only generally, but by its more ovate base. The carinæ
are lighter in color than the general body of the shell, and are slightly
irregular or subnodulous in outline; it is a stout, heavy species, and

has a smaller aperture, proportionally, than is common in the genus; the bands within the aperture are five in number, very dark, and the three central ones are disposed to be confluent; a dark, broad band revolves around the base of the shell.   Compared with *Schizostoma pagoda*, Lea, it differs in color, in its more elongate form, and by the character of its carinæ, which are more uniform, the main variation being that they are more diffused on the whorl, whereas, in Mr. Lea's species they are particularly conspicuous near the apex.— *Anthony*.

I give below Mr. Lea's description of *Schizostoma Showalterii*, from the Journal of the Academy of Natural Sciences.

Schizostoma Showalterii.— Shell transversely ribbed, subcylindrical, thick, chestnut-color, minutely striate; spire elevated; sutures impressed; whorls flattened; fissure rather large and deep; aperture rather small, elliptical, banded within; columella thick; outer lip slightly crenulate.

Fig. 671.

Operculum ovate, with the polar point near the inner lower edge.

Habitat.— Coosa River, at Uniontown, Alabama; E. R. Showalter, M.D.

Diameter, ·46; length, ·98 of an inch.

Observations.— It is somewhat like *pagoda* (nobis), but is much larger, more robust and subcylindrical.   It also has more and larger ribs, which are very prominent.   The specimens before me have on the last whorl seven ribs, the three lower ones being small, the three middle ones large, looking like cords wrapped round the shell. These are of a lighter brown.   Two ribs only are visible on the upper whorls.   The fissure in the lip is three-tenths of an inch long. The apex being eroded, I am unable to describe that part, nor can I give, consequently, the number of whorls, but they are likely to be seven or eight.— *Lea*.

S. pagoda, Lea, is distinguished from this species, besides the above characters, by its short and oblique slit.   Mr. Reeve figures, in species 23, *Melatoma Showalterii*, which certainly does not apply to this species, but rather to Mr. Lea's *S. pumilum*.

2, S. castaneum, Lea.

Schizostoma castaneum, LEA, Proc. Acad. Nat. Sci., p. 186, May, 1860. Jour. Acad.
Nat. Sci., v, pt. 3, t. 35, f. 50. Obs., ix, p. 69.
Gyrotoma castanea, Lea, BINNEY, Check List, No. 311. BROT, List, p. 27.

Description.— Shell carinate, conical, rather thick, dark brown, imperforate; spire exserted; sutures very much impressed; whorls six,
flattened, with a single carina and four bands; lip-cut straight, narrow
and deep; aperture rather small, elliptical, banded within, rounded
at the base; columella white and thickened; outer lip acute, slightly
sinuous.

Operculum nearly round, light brown, with the polar point below
the middle on the inner side.

Habitat.—Coosa River, Alabama; E. R. Showalter, M.D.

Diameter, ·32; length, ·64 of an inch.

Observations.— Several specimens are before me of nearly the same
size. A single, rather obscure carina follows round the middle of the
lower whorls, and is exhibited on the upper whorls just above Fig. 672.
the suture with more force. The four bands are obscure
on the outside, but well defined on the inside. One specimen
has but three bands, and another has very pale bands. The
impression made by the lip-cut is well defined and forms a
narrow, hem-like line below the suture. The aperture is rather
small, not being quite half the length of the shell, and is rounded at
the base. It is nearest in outline to *pagoda* (nobis), but may at once
be distinguished by the color being usually darker, by being less carinate, in having a deeper lip-cut, and in being rounded at the base,
instead of being angular there, as that species is. The aperture is
rather more than one-third the length of the shell.— *Lea.*

This shell is also closely allied to *Wetumpkaense*, Lea,
which, however, has a short, wide fissure. I have endeavored
in the Synoptical Table of this genus to indicate the close
connection of certain species belonging to the opposite groups,
namely, those with the short, oblique, and those with the narrow, direct fissure. It is curious that almost every species
in the one section has its analogue in the other, with which,
perhaps, it has more affinity than with the nearest of its own
section.

3. S. ovoideum, Shuttleworth.

Gyrotoma ovoideum, Shuttleworth, MITTHEIL, Bern. Nat. Gesell., No. 50, p. 88,
July 22, 1845. H. & A. ADAMS, Genera, iii, t. 32, f. 4.

Description.— Shell conoidal, thick, olivaceous, concentrically stri-

Fig. 673. ate-costate, brown-banded, apex eroded; whorls about
five, thickened at the suture; fissure very narrow, elongate; columella thickened above.

Length, about ·7; breadth, ·4–·4½ of an inch. Length
of aperture, ·3 of an inch. Length of fissure, ·2 of an
inch.

Observations.—Closely approaching *Melania olivula*, Conrad, in form; varied by confluent bands.— *Shuttleworth.*

Figured from H. and A. Adams, "Genera." It appears to
be a more cylindrical and narrower species than the following.

4. S. excisum, Lea.

Melania excisa, LEA, Philos. Proc., p. 242, Dec., 1842. Philos. Trans., ix, 1846.
JAY, Cat., 4th edit., p. 273.
Schizostoma excisa, Lea, WHEATLEY, Cat. Shells U. S., p. 28.
Gyrotoma excisa, Lea, BINNEY, Check List, No. 317. BROT, List, p. 27. Lea,
ADAMS, Genera, i, p. 305.
Melatoma excisum, Lea, REEVE, Monog., sp. 2.

Description.— Shell striate, subfusiform, rather thick, yellowish;
spire ovately conical; sutures impressed; whorls flattened; aperture
cut out above, small, elliptical, white.

Habitat.— Alabama.

Diameter, ·40; length, ·64 of an inch.

Observations.— This shell is very remarkable for the cut in the
superior part of the outer lip, very similar to some species of *Pleurotoma.* This cut extends nearly one-fifth round the
whorl, leaving immediately below the suture an ele-   Fig. 674. Fig. 675.
vated ridge. There are nearly three whorls of this
specimen perfect, and the cicatrix shows the cut to
have extended in due proportion thus far. The aperture is rather small, and rather more than one-
third the length of the shell. On the spire there is a slight disposition to plication. The apex being eroded, the number of whorls is
not certain, perhaps six. This specimen has three revolving, purple
bands.— *Lea.*

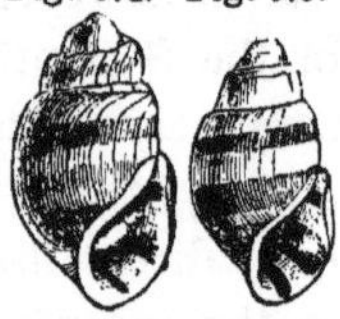

Mr. Reeve, and Dr. Brot following him, place *ovoideum*, Shuttleworth, in the synonymy of this species. As I have no means of comparing specimens of the latter with Mr. Lea's species, I have preferred to separate them in this work.

S. Babylonicum is a larger, wider, more robust species than the one now under consideration.

5. S. pumilum, LEA.

Schizostoma pumilum, LEA, Proc. Acad. Nat. Sci., p. 187, May, 1860. Jour. Acad. Nat. Sci., v, pt. 3, t. 35, f. 57, March, 1863. Obs., ix, p. 74.
Gyrotoma pumila, Lea, BINNEY, Check List, No. 328. BROT, List, p. 27.
Schizostoma globosum, LEA, Proc. Acad. Nat. Sci., p. 187, May, 1860. Jour. Acad. Nat. Sci., v, pt 3, t. 35, f. 58, March, 1833. Obs., ix, p. 74.
Gyrotoma globosa, Lea, BINNEY, Check List, No. 321. BROT, List, p. 27.
Melatoma globosum, Lea, REEVE, Monog. t. 3, f. 18.
Melatoma Alabamense, Lea, of REEVE, Monog. sp. 20.
Melatoma Showalterii, Lea, of REEVE, Monog. sp. 23 ?

Description. — Shell striate, top-shaped; rather thin, pale horn-color, imperforate; spire very obtuse; sutures much impressed; whorls six, ventricose, the last very large; fissure straight and rather short; aperture rather small, ovate, white within, angular at the base and somewhat canaliculate; columella white, twisted and thickened below; outer lip acute and sinuous.

Fig. 676. Fig. 677.

Habitat. — Alabama; B. W. Budd, M.D.

Diameter, ·40; length, ·63 of an inch.

Observations. — This is a rather small, dwarfish looking species, nearly as wide as it is long, which

Fig. 678. Fig. 679.

I have had for a long time from Dr. Budd. One of the specimens has a few obscure bands. It is nearly allied to *glandula* (nobis), but the spire is higher, and it is striate, while the other is not. It is not likely to be confounded with *glans* (nobis), as that is a large species with a higher spire. The hem-like line left by the lip-cut is large and well defined round the whorls. The aperture is about half the length of the shell. One of the specimens before me has three indistinct bands. The other two have none. — *Lea.*

Having before me a number of specimens of Mr. Lea's *S. pumilum* and of his *S. globosum*, I am convinced that the latter is an immature form of the former species. The accompanying figures, the largest of which agrees well with Mr.

Lea's figure of *S. pumilum*, and the smallest with *S. globosum*, with the aid of the intermediate figure (Fig. 678), will exhibit their connection and the mode of growth of the shell. It will be seen that *S. globosum* has attained to four whorls, that the intermediate figure would exhibit (if the loss by erosion were supplied) five, and that the adult has six whorls.

The following is the description of

Schizostoma globosum.— Shell transversely striate, globose, rather thin, yellowish, imperforate; spire short, obtusely conical; sutures impressed; whorls four, three-banded, the last large; lip-cut straight, narrow and short; aperture rather large, elliptical, banded within and angular at the base; columella white, incurved; outer lip sharp and expanded.

Operculum ovate, rather light brown, with the polar point near the inner lower edge.

Habitat.— Alabama; E. R. Showalter, M.D.

Diameter, ·32; length, ·48 of an inch.

Observations.— This is a very small, globose species, more rounded and inflated than any other which has come under my notice, and it is Fig. 680. the smallest which I have seen. The description being made

from two specimens only, it may be found to vary when others are observed. In this specimen the three bands are broad and of a dark brown, the two upper ones having on the outside raised striæ running parallel to the edges. The aperture is large, and is rather more than half the length of the shell. The impression made by the lip-cut is well defined and forms a narrow, hem-like line below the suture. This species is not likely to be confounded with any of the species known, being smaller than all but *laciniatum* (nobis), which is more conical. The aperture is nearly two-thirds the length of the shell.— *Lea.*

The analogue of *S. pumilum* among the obliquely fissured species is *S. Buddii*, Lea, to which it perhaps more nearly approximates than to either *S. glans* or *glandula*, with which Mr. Lea compares it. Although many of the shells in Reeve's Monograph are well figured, their value for the identification of species is seriously impaired by the application to them in several instances of wrong names, and by the insufficiency of the descriptions. This is greatly to be regretted and illustrates the truth of Mr. Brot's remark, that the genus is but little known in Europe.

6. S. ellipticum, ANTHONY.

Melatoma ellipticum, ANTHONY. MSS., REEVE, Monog., t. 3, f. 21, April, 1861.
Gyrotoma elliptica, Anthony, BROT, List, p. 27.

Description. — Shell oblong-ovate, yellowish-olive, encircled with three broad, greenish-black bands; spire rather pro- Fig. 680*a.*
duced, obtuse; whorls flatly convex, smooth, faintly, rudely plicated towards the apex; aperture narrowly ovate; fissure deep.

Habitat. — Coosa River.

Observations. — A well defined species, though partaking of the typical characters of some others. — *Reeve.*

This shell somewhat resembles *S. bulbosum*, Anthony, but is distinguished by its more lengthened form and by the regularly convex outline of the body-whorl and spire.

7. S. laciniatum, LEA.

Schizostoma laciniatum, LEA, Philos. Proc., iv, p. 167, August, 1845. Philos. Trans.
    x, p. 69, t. 9, f. 57, 1853.
Gyrotoma laciniata, Lea, BINNEY, Check List, No. 324. BROT, List, p. 27. ADAMS,
    Genera, i, p. 305.

Description. — Shell smooth, obtusely conical, rather thick, banded, yellowish horn-color; spire obtuse; sutures excavated; whorls convex; fissure deep; aperture elliptical, whitish within; columella smooth, thickened above.

Habitat. — Tuscaloosa, Alabama.

Diameter, ·25; length, ·45 of an inch.

Observations. — This is the smallest species I have seen. The mouth and fissure of this specimen are perfect, but the apex is much eroded, and the number of whorls cannot therefore be ascer-
Fig. 680*b.*
tained. There are four bands very distinctly marked on the inside. The aperture appears to be about one-half the length of the shell. The fissure is very narrow and remarkably deep, extending nearly one-fourth round the whorl. The cicatrix along the suture is of a lighter color. The marks of growth are distinct, and give a laciniate appearance. — *Lea.*

A very neat species which Mr. Reeve seems to have over-looked. The locality given in the above description is prob-ably incorrect. Mr. Lea has recently stated his opinion that this and other species, to which he originally assigned Tusca-loosa as the habitat, were not really found there. Indeed the present state of our knowledge of the species of this genus leads us to believe that they are entirely confined to the waters of the Coosa River. It is wonderful that this group occupies such a restricted space, while others, such as *Lithasia, Pleuro-cera*, etc., extend over nearly the whole of the country between the Mississippi River and the Alleghany Mountains.

8. S. amplum, ANTHONY.

Gyrotoma ampla, ANTHONY, Proc. Acad. Nat. Sci., p. 66, Feb., 1860. BINNEY,
    Check List. No. 306. BROT, List, p. 27.
Melatoma amplum, Anthony, REEVE, Monog., t. 3, sp. 16.

Description.— Shell smooth, ovate, rather thick, olivaceous; spire not elevated, but acute; whorls 6-7, subconvex; sutures well defined; fissure broad, rather deep and waved; aperture moderate, elliptical, flesh-colored and banded within; columella smooth, or slightly thick-

Fig. 681.   Fig. 682.

ened only at the fissure; body-whorl striate and banded; whorls of the spire not banded, but having a thickened, cord-like line near the sut-ure.

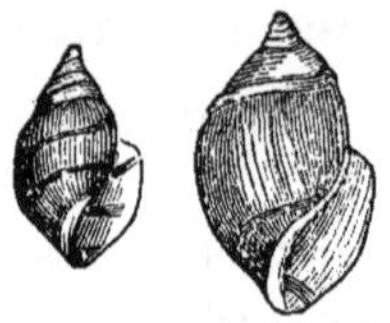

Habitat.— Coosa River, Alabama.

Length, eleven-sixteenths; breadth, seven-sixteenths of an inch. Length of aperture, seven-sixteenths; breadth of aperture, four-sixteenths of an inch.

*Observations.—*A fine, symmetrical species of this interesting genus, which hitherto has not been very productive in species. Compared with *Schizostoma funiculatum*, Lea, which it most nearly resembles, it is smoother, thinner, more acute and has not the double cord-like lines of that species. Most, if not all the species of *Gyrotoma*, have the fissure gradually filled up behind as it is pushed forward in the process of growth, by a cord-like line more or less prominent, often so much so as to produce quite a shoulder at the suture, and this species is so marked, but it has no cord-like line in the middle of the body-whorl, as described in *funiculatum.— Anthony.*

A beautiful species, which may be readily distinguished from all the other deeply fissured *Schizostomæ* by its quadrate form, caused by the flattening of the body-whorl. In its form it approaches closely to *S. salebrosa*, Anthony, which is, however, much larger and belongs, moreover, to the other section of the genus.

9. S. nuculum, ANTHONY.

Melatoma nucula, ANTHONY. MSS., REEVE, Monog. t. 3, f. 19, April, 1861.
Gyrotoma nucula, Anthony, BROT, List, p. 27.

Description.— Shell obtusely conical, fulvous-olive; whorls convex, smooth; aperture narrowly ovate, a little effused at the base; fissure deep.

Fig. 683

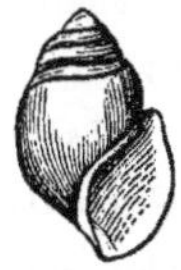

Habitat.— Coosa River, Alabama.

Observations. — Chiefly distinguished by the simplicity of its characters, the shell being neither sculptured nor banded.— *Reeve.*

I have not seen this species. Judging from the figure it appears to me to be the same as *amplum*.

10. S. cylindraceum, MIGHELS.

Schizostoma cylindracea, MIGHELS, Bost Proc., i, p. 189, Oct., 1844.
Gyrotoma cylindracea, Müll., BINNEY, Check List, No. 315. Gould, BROT, List, p. 27. ADAMS, Genera, i, p. 305.

Description.—Shell nearly smooth, cylindrical, thick, with slight, revolving undulations; epidermis olivaceous; spire ovate-conic, eroded; whorls three or four, flattened, shouldered; suture distinct; aperture oval; fissure deep and wide.

Habitat.—Warrior River, Alabama.—*Mighels.*

I can only reprint the original description of this species, the shell being unknown to me.

11. S. bulbosum, ANTHONY.

Gyrotoma bulbosa, ANTHONY, Proc. Acad. Nat. Sci., p. 65, Feb., 1860. BINNEY, Check List, No, 309. BROT, List, p. 27.
Melatoma bulbosum, Anthony, REEVE, Monog., sp. 22.
Gyrotoma ovalis, ANTHONY, Proc. Acad. Nat. Sci., p. 65, Feb., 1860. BINNEY, Check List, No. 325. BROT, List, p. 27.

Description.— Shell striate, ovate, moderately thick, dark olive; spire obtusely elevated, subtruncate, four whorls only remaining; whorls of the spire subconvex; sutures very distinct, rendered more so by the shouldering of the whorls; body-whorl inflated, subangulated a little below the suture, from which angle it shelves towards it, and having two or three dark, broad bands revolving round it; lines of growth curved and very distinct, almost like crowded ribs; fissure perfectly straight, very narrow and not deep; aperture rather

Fig. 684.   Fig. 685.

long, of a dusky color within and ornamented by three broad and distinct bands there; columella smooth, except at the lower part, where it is slightly thickened.

Habitat.— Coosa River, Alabama.

Length of shell, nine-sixteenths; breadth of shell, three-eighths of an inch. Length of aperture, five-sixteenths; breadth of aperture, three-sixteenths of an inch.

Observations.— A short, ovate species resembling in some respects *G. ovalis* (nobis) herein described; it is less elevated than that species, more ventricose, and its surface is rougher; indeed, there seem to be some indications of obscure folds on the body-whorl of this species near the suture, which in very old specimens may be more fully expressed; and thus bring it into close affinity with *M. salebrosa* (nobis).—*Anthony.*

Having compared Mr. Anthony's types of his *S. bulbosum* and *S. ovalis*, together with other specimens, I am convinced that they are the extreme forms of one species. With regard to the striæ of the former being rougher than those of the latter species, some of the specimens of *ovalis* before me have exactly the same striation, disposed somewhat to rise into folds near the suture which distinguishes the typical *bulbosum*. *S. salebrosum* is a larger and more cylindrical species, and *S. bulbosum* is more closely allied to *S. incisum*, Lea.

The description of *S. ovalis* follows, and figures of both that and *bulbosum* are given from Mr. Anthony's types.

Schizostoma ovalis.— Shell smooth, oval, olivaceous, moderately thick; spire obtusely elevated, composed of about 5–6 convex whorls, of which two are generally lost by truncation; sutures deeply impressed; aperture broadly elliptical, banded within; fissure direct, exceedingly narrow and very deep, extending nearly one-half around the shell; columella slightly curved by a callus.

Habitat.— Coosa River, Alabama.

Length of shell, ten-sixteenths; breadth of shell seven-sixteenths of an inch. Length of aperture, seven-sixteenths; breadth of aperture, four-sixteenths of an inch.

Observations.— A fine, symmetrical species remarkable for its regularly oval form and unusually deep, linear fissure; the whorls are somewhat shouldered, though not so much so as in m ny Fig. 686.
of the species; the spiral whorls are furnished with two broad bands, one near the top of each and the other widely separate and near the preceding whorl, being often half concealed by it; there are three bands on the bodywhorl equidistant from each other; compared with *G. bulbosa* (nobis), which it most nearly resembles, it is longer, more linear, and has not the rapidly attenuating spire of that species nor its roughly striate surface.— *Anthony.*

12. S. curtum, MIGHELS.

Schizostoma curta, MIGHELS, Bost. Proc., i, p. 189, Oct., 1844.
Gyrotoma curta, Mighels, BINNEY, Check List, No. 314. Gould, BROT, List, p. 27.
  ADAMS, Genera, i, p. 305.

Description.— Shell short, subglobose, smooth, thick and solid; epidermis dark green, with two or three revolving bands of a darker color; spire short, obtuse, eroded; whorls three or four, flattened in the middle; suture superficial; aperture pear-shaped; fissure distinct.

Habitat.— Warrior River, Alabama.— *Mighels.*

This species is unknown to me except through the description. The locality probably should read Coosa River, instead of Warrior River.

13. S. glans, LEA.

Schizostoma glans, LEA, Proc. Acad. Nat. Sci., p. 186, May, 1860. Jour. Acad. Nat.
  Sci., v, pt. 3, t. 35, f. 52, March, 1863. Obs., ix, p. 70.
Gyrotoma glans, Lea, BINNEY, Check List, No. 320. BROT, List, p. 27.

Description.— Shell smooth, ovately conical, inflated, rather thick, yellowish horn-color or chestnut-brown, striate, imperforate; spire obtusely elevated; sutures regularly impressed; whorls six, obsoletely banded, the last rather large; lip-cut straight, narrow and deep; aperture rather small, elliptical, white within, obtusely angular at the base; columella white, thickened above; outer lip sharp and somewhat sinuous.

Operculum ovate, dark brown, with the polar point near to the inner lower edge.

Habitat.— Coosa River, Alabama; E. R. Showalter, M.D.

Diameter, ·44; length, ·78 of an inch.

Observations.— This is rather a robust species, and judging from the specimens before me, I should presume that there would be much regularity in the species. On one of the specimens there

Fig. 687.

are two obscure, hair-like bands, one on the middle of the body-whorl and another near the base. Other specimens have only a very obscure, thin band near the base. Very probably specimens may be found with a third band near to the suture, and others with better defined bands. Some were chestnut-brown. The upper whorls were rather flattened, and the lines of growth few and obscure. The impression made by the lip-cut is well defined, and forms a strong, narrow, hem-like line below the suture. The outer lip stands close to the body-whorl. The aperture is one-half the length of the shell, and the base is obtusely angular. This species, in general facies, is near to *glandula* herein described, but differs in the form of the lip-cut, which is narrow, deep and straight. It is also a much larger species, and is without the well marked shoulder of *glandula.—Lea.*

This pretty species appears to be allied to *S. bulbosum,* Anthony, but offers the following points of distinction :— it is more inflated and heavier, the color is much lighter, the bands are very narrow and the striation is not so strongly marked. In a very fine individual before me, the body-whorl is disposed to tuberculation below the suture.

14. S. sphæricum, Anthony.

Melatoma sphæricum, Anthony, MSS., Reeve, Monog., sp. 8, April, 1861.

Description.— Shell subglobose, yellowish-olive, encircled with interrupted fillets of greenish-black; spire small, somewhat immersed; whorls convex, smooth, rather inflated; sutural fissure   Fig. 688.

slightly channelled; columella callous.

Habitat.— Coosa River, Alabama.

Observations.— A small, globose shell, with its little spire distinctly immersed, characterized by a copious banding throughout of interrupted fillets of greenish-black, fuscous in the interior.— *Reeve.*

This elegant little species is widely separated in form and ornamentation from any other of the genus. In both these respects it reminds one strongly of Mr. Lea's *Anculosa formosa.*

15. S. pagoda, LEA.

Schizostoma pagoda, LEA, Philos. Proc., iv, p. 167, Aug., 1845. Philos. Trans., x, p. 67, t. 9, f. 52, 1853.
Gyrotoma pagoda, Lea, CHENU, Manuel, i, f. 2,020. BINNEY, Check List, No. 327. BROT, List, p. 27. ADAMS, Genera, i, p. 305.

Description.— Shell carinate, conical, rather thick, dark horn-color; spire rather short; sutures very much impressed; whorls six; fissure small; aperture elliptical, within whitish; columella smooth.

Habitat.— Tuscaloosa, Alabama.

Diameter, ·35; length, ·75 of an inch.

Observations.— Three of this interesting species are before me. They are very distinct, and may easily be known by the carina being very acute on the superior whorls, presenting the appearance of a Chinese pagoda. The lower whorl is slightly and irregularly striate. The fissure is not deep, but rather wide, being about one-fifth the length of the whorl. The columella at the base is rather angular. The aperture is rather more than one-third the length of the shell.— *Lea.*

Fig. 689.

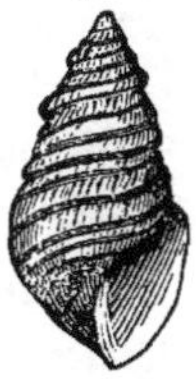

This excellent species in form belongs to that group of which *S. carinifera,* Anthony (*Showalterii,* Lea) may be considered the type. It is not so large a shell as that species, nor is it so strongly carinate. It is also allied to *S. Wetumpkaense,* Lea, but is a more elongated shell. The locality given is extremely doubtful. Mr. Reeve figures two distinct species for *S. pagoda:*—his fig. 1*a* is *S. Wetumpkaense,* Lea, and fig. 1*b* is *S. Buddii,* Lea. It is doubtful whether Mr. Brot has recognized this species, as he refers to Mr. Reeve's figures.

16. S. pyramidatum, SHUTTLEWORTH.

Gyrotoma pyramidatum, SHUTTLEWORTH, Mitt. Bern. Nat. Gesell., No. 50, p. 88, July 22, 1845. BINNEY, Check List, No. 329. BROT, List, p. 27. ADAMS, Genera, i, p. 305.

Description.— Shell pyramidal, thickened, olivaceous or blackish, concentrically, sulcately costate, frequently nodosely geniculate; banded with brown; apex eroded; whorls five or six; fissure wide, short; columella tuberculately thickened above.

Length, ·9; breadth of the ultimate whorl, ·4½–·5 of an inch. Length of aperture, ·3½. Length of fissure, ·1 of an inch.—*Shuttleworth.*

This species is entirely unknown to me, but is evidently closely allied both to the preceding and following.

17. S. Wetumpkaense, LEA.

Schizostoma Wetumpkaense, LEA, Proc. Acad. Nat. Sci., p. 187, May, 1860. Jour.
　Acad. Nat. Sci , v, pt. 3, t. 35, f. 56, March, 1863. Obs , ix, p. 73.
Gyrotoma Wetumpkaensis, Lea, BINNEY, Check List, No. 336. BROT, List, p. 28.
Melatoma Wetumpkaense, Lea, REEVE, Monog., t. 3, f. 17.
Melatoma ornata, ANTHONY, MSS., REEVE, Monog., fig. 11.
Melatoma pagoda, Lea, REEVE, Monog., fig. 1a. (not 1b).

Description.— Shell striate, ovately cylindrical, thick, light brown, umbilicate; spire obtuse, conoidal; sutures very much impressed;

Fig. 690.

whorls six, banded, flattened, the last large; fissure oblique and short; aperture large, ovate, banded within, at the base obtusely angular; columella white, thickened above; outer lip sharp and sinuous.

Operculum spiral, large and long, the polar point being near to the lower left edge.

Habitat.— Coosa River, at Wetumpka, Ala.; E. R. Showalter, M.D. Diameter, ·44; length, ·70 of an inch.

Observations.— Among the specimens from Dr. Showalter were a number of adults and young of this species. Some were eroded so much as to exhibit little more than the body-whorl. The more perfect ones, still slightly eroded at the apex, exhibited six whorls. The half-grown have five whorls, with a cord-like carina on the middle of each, and this carina is raised much above the surface. The quite young have a sharp apex, and carry the carina to near the apex. The suite, which I owe to the kindness of Dr. S., consists of some eighteen specimens, varying from one-fourth to nearly a whole inch in size. In general outline this species approaches *S. Buddii* (nobis), but it is more cylindrical when full grown, and generally has bands. Besides it is umbilicate, while *Buddii* is not. Usually *Wetumpkaënse* is striate and banded, but it is not universally the case. The aperture is less than half the length of the shell. The hem is yellowish and not well marked.—*Lea.*

S. ornata, Anthony, is evidently the young of this species.
I give the original description, and also a figure from the type
specimen.

Melatoma ornatum.— Shell ovate, somewhat pyramidally turreted,
yellowish-olive, neatly, spirally corded with dark green; whorls 5-6,
concavely sloping round the upper part, keeled at the   Fig. 691.
sutures; aperture small; fissure broad, moderately deep;
columella thinly inflected, pinkish-white.

Habitat.— North Carolina, United States.

Observations.— A charming little species, banded in a
most characteristic manner, with raised, dark green, cord-like ridges
upon a clear, yellowish-olive ground.— *Reeve.*

Mr. Anthony's label is marked "Proc. A. N. S. Phil.," but
he never published the species.   Mr. Reeve, misled by this
reference, has quoted *Anculosa ornata,* Anthony, as being the
description referred to, and consequently assigns North Carolina as the habitat.   It is scarcely necessary to repeat that
no species of *Schizostoma* has ever been positively ascertained
to exist in any other waters than those of the Coosa.   I think
it very probable that *pagoda, pyramidatum* and *Wetumpkaense*
are identical, but I have not sufficient data to ascertain the
fact positively.

18. S. Alabamense, LEA.

Schizostoma Alabamense, LEA, Proc. Acad. Nat. Sci., p. 187, May, 1860. Jour. Acad.
    Nat. Sci., v, pt. 3, t. 35, f. 54. Obs., ix, p. 72.
Gyrotoma Alabamensis, Lea, BINNEY, Check List, No. 305. BROT, List, No. 27.

Description.— Shell striate, elliptical, stout, yellowish-olive, imperforate; spire obtusely conical; sutures very much impressed; whorls
Fig. 692.   six, banded, rather inflated, the last very large; fissure
oblique and rather short; aperture rather large, ovate,
banded within and obtusely angular at the base; columella
white, somewhat thickened above and below; outer lip
sharp and sinuate.

Habitat.— Alabama; B. W. Budd, M.D. and Dr. E. R.
Showalter.

Diameter, ·50; length, ·90 of an inch.

Observations.— The specimen from Dr. Budd has been a long time
in my possession, and was considered to be an inflated variety of

excisa, but specimens recently received from Dr. Showalter satisfy
me that it is distinct. It is among the largest of the genus, being
nearly an inch long, and may be distinguished by its robust form and
its regular, elliptical outline. The specimens before me have three
broad, dark purple bands within, which give an indistinct dark green
hue to the outside, and stop short of the edge. The lip-cut stands
well out, and the hem-like margin is distinct and yellowish. The
base of the columella is yellowish. The aperture is half the length
of the shell. The hem is yellow, broad and well marked.—*Lea.*

Mr. Reeve's fig. 20 intended to represent this species, I refer
to *S. pumilum*, Lea. *S. Alabamense* is allied to *Babylonicum*,
Lea, but is, as it appears to me, well distinguished by the regularity of the striæ, which cover the whole surface.

19. S. Anthonyi, Reeve.

Melatoma Anthonyi, Reeve, Monog., sp. 12, April, 1861.
Gyrotoma Anthonyi, Reeve, Brot, List, p. 27.

Description.— Shell conically ovate, rather solid, fulvous-brown;
spire produced; whorls sloping round the upper part, concavely im-

Fig. 693.   Fig. 694.

pressed round the middle, last whorl encircled by a single, dark ridge; aperture
rather narrow, attenuately effused at the
base; columella arcuately twisted.

Habitat.— Alabama.

Observations.— This shell, received from
Mr. Anthony without a name, appears to
me to be distinct, and I am glad to avail
myself of the opportunity of dedicating it to a gentleman to whom
we are so largely indebted beyond all others for his researches after
the *Melaniadæ* of the southern United States of America.— *Reeve.*

Mr. Reeve does not mention the character of the fissure, but
I judge from the figure that it is short and wide. The accompanying woodcuts are copied from Mr. Reeve's.

20. S. Babylonicum, Lea.

Schizostoma Babylonicum, Lea, Philos. Proc., iv, p. 167, Aug., 1845. Philos. Trans.,
  x, p. 68, t. 9, f. 54.
Gyrotoma Babylonicum, Lea, Binney, Check List, No. 307. Chenu, Manuel de
  Conchyl., i, f. 2,021. Brot, List, p. 27. Lea, Adams, Genera, i, p. 305.
Melatoma Babylonicum, Lea, Reeve, Monog., sp. 6.
Schizostoma Spillmanii, Lea, Proc. Acad. Nat. Sci., p. 54, Feb., 1861. Jour. Acad.
  Nat. Sci., v, pt. 3, t. 35, f. 55. Obs., ix, p. 72.
Gyrotoma funiculata, Lea, Adams, Genera, i, p. 305.

Description.— Shell striate, somewhat fusiform, rather thick, chestnut-color; spire obtusely conical; sutures impressed; whorls flattened; fissure small; aperture large, elliptical, somewhat flesh-colored
within; columella smooth, angular at the base, thickened above.

Habitat.— Tuscaloosa, Alabama.

Diameter, ·48 of an inch; length, 1 inch.

Observations.— A single specimen only of this species was submitted to me. It differs from the other described spe-   Fig. 695.
cies in being angular at the superior portion of the
whorl along the lower margin of the fissure, making
quite a shoulder, and giving it the *Babylonic* appearance.
The fissure is wide, but not deep. The apex being
much eroded, the number of whorls could not be ascertained. The aperture is nearly half the length of the
shell. The deposit on the columella in this individual does not cover
the perforation. In others this may differ. The outer lip is quite
patulous.—*Lea.*

S. Babylonicum was described from a single specimen, several years ago, when but few species of the genus were known.
As the description of *S. Spillmanii* appears to be much more
accurate and to apply well to the shell first named, I have
adopted it in this connection. There can be but little doubt
that the two species described by Mr. Lea are identical. I
have before me a splendid suite of this species numbering
about thirty individuals from which the figures of the adult
and young *Spillmanii* are drawn. These were obligingly presented to the Smithsonian Institution by Dr. James Lewis of
Mohawk, N. Y., who received them from Dr. Showalter. Mr.
Reeve's figure 6 intended to represent this shell is too large
and ponderous and must be received with doubt.

The description and figure of *S. Spillmanii* are given below.

Schizostoma Spillmanii.—Shell striate, subcylindrical, rather thick, yellowish-brown, imperforate; spire obtuse, conoidal; sutures impressed; whorls six, very much banded, flattened, the last large; fissure oblique and rather short; aperture large, ovate and banded within, obtusely angular at the base; columella white, thickened above; outer lip sharp and sinuous.

Fig. 696.

Operculum ovate, spiral, rather large, dark brown with the polar points near to the left edge, about one-fifth above the basal margin.

Habitat.— Coosa River, Alabama; Dr. E. R. Showalter. Diameter, ·48; length, ·92 of an inch.

Observations.— I have a number of specimens, chiefly young, from Dr. Spillman, and a fine suite of different ages from Dr. Fig. 698. Showalter. There is much difference among them, some being subcylindrical, while others are disposed to be oval.

Fig. 698.

This species is nearly allied to *Wetumpkaense* (nobis) and closely resembles it in the adult state, but in the young state the two species differ very much. The young of *Wetumpkaense* is remarkably carinate on the middle of the whorl, and this is more marked on the superior whorls, the epidermis being of a light yellowish horn-color, with a distinct brown band on the upper portion of the whorl, and generally two below, sometimes three. The *Spillmanii* has a very obtuse angle along the middle of the whorl, which does not show in the upper whorls, which are dark brown, and the band is interrupted, making the spire somewhat maculate. The aperture is not quite half of the length of the shell. The hem is not well defined. I name this after my friend Dr. Spillman, who sent me a number of fine specimens, old and young.— *Lea.*

Fig. 697.

Fig. 699.

21. S. Buddii, Lea.

Schizostoma Buddii, Lea, Philos. Proc., iv, p. 167, Aug., 1845. Philos. Trans., x, p. 68, t 9, f. 53.
Gyrotoma Buddii, Lea, Binney. Check List, No. 308. Brot, List, p. 27.
Schizostoma funiculatum, Lea, Philos. Proc., iv, p. 167, Aug., 1845. Philos. Trans., x, p. 69, t. 9, f. 56.
Gyrotoma funiculata, Lea, Binney, Check List, No. 318. Brot, List, p. 27.
Melatoma funiculatum, Lea, of Reeve, Monog., sp. 5.
Melatoma pagoda, Lea, of Reeve, Monog., sp. 16.

Description.— Shell striate, subfusiform, thick, dark horn-colored; spire obtusely conical; sutures irregularly impressed; whorls six, rather inflated; fissure small, oblique; aperture large, rhomboidal, whitish within; columella thickened above.

Habitat.— Tuscaloosa, Alabama.

Diameter, ·47; length, ·83 of an inch.

Observations.— This is a robust shell, being thicker and heavier than any other species of this genus which I have observed. The aperture is nearly one-half the length of the shell. Two specimens were sent together by Dr. Budd, presuming they were the Fig. 700. same. One, however, which is not quite a mature shell, has little or no fissure. The other, from which the description is made, has a wide but short fissure, and the margin of it opens obliquely.— *Lea.*

The following is the description of

Schizostoma funiculatum. — Shell striate, elliptical, rather thick, chestnut-colored; spire obtuse; sutures much impressed; whorls

Fig. 701. convex; fissure rather large, oblique; aperture large, elliptical; columella thickened above.

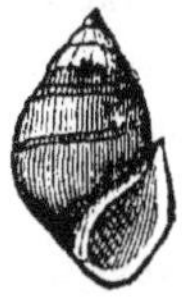

Habitat.— Tuscaloosa, Alabama.

Diameter, ·4; length, ·66 of an inch.

Observations.— A single specimen only was obtained by Dr. Budd of this species. It is short, stout, and almost subrotund. It has two elevated, cord-like lines, revolving on the whorls. One immediately under the suture, the other below that again. The aperture is more than half the length of the shell. The apex is so much eroded as to prevent the number of whorls being ascertained. There are about six.— *Lea.*

Having examined Mr. Lea's original specimens of the above descriptions (both of which are figured) as well as other shells of intermediate forms, I believe that the two should be united. Mr. Reeve's figure 3 of this species is a *Goniobasis læta,* Jay. Mr. Reeve's figure of *funiculatum* quoted above does not so well represent that variety as his figure 1*b,* which he introduced to illustrate Mr. Lea's *S. constrictum* (considered by Mr. Reeve to be a synonyme of *pagoda*). This species is very closely allied to the long-fissured *S. pumilum* of Lea.

22. S. demissum, ANTHONY.

Gyrotoma demissa, ANTHONY, Proc. Acad. Nat. Sci., p. 64, Feb., 1860. BINNEY, Check List, No. 316. BROT, List, p. 27.

Melatoma demissum, Anthony, REEVE, Monog., sp. 9.

Schizostoma Hartmanii, LEA, Proc. Acad. Nat. Sci., p. 187, May, 1860. Jour. Acad. Nat. Sci., v, pt. 3, t. 35, f. 51. Obs., ix, p. 69.

Gyrotoma Hartmanii, Lea, BINNEY, Check List, No. 322. BROT, List, p. 27.

Description.— Shell short, robust, thick, truncate, of a dark horn-color; spire flat by truncation, exhibiting traces of about four whorls; body-whorl cylindrical; fissure broad, waved and rather deep; aperture elliptical, within whitish; columella thickened along its whole extent, but most so at the fissure.

My cabinet.

Length of shell, ten-sixteenths; breadth of shell, seven-sixteenths of an inch. Length of aperture, seven-sixteenths; breadth of aperture, four-sixteenths.

Observations.— A fine, cylindrical species, whose chief character-

Fig. 702.

istics are its very smooth, polished surface, plain russet color and flat, truncate spire; the lines of growth are unusually strong in this species, and the darker lines indicating the terminus of previous mouths are very distinct and numerous, evidencing frequent and many pauses in its growth; the columella is much bent near its base, and a narrow, but distinct sinus is formed at about the middle space between the outer lip and columella. A single specimen only is before me, but seems so very distinct from all others that I have no hesitation in considering it new.—*Anthony.*

Mr. Lea considers that this species = his *S. constrictum.* They are nearly allied, but *constrictum* is a more elongated, narrower shell, and a comparison of Mr. Anthony's types, kindly placed in my hands by that gentleman, has induced me to believe that *constrictum* should rather be united to *S. rectum,* Anthony.

The following is Mr. Lea's description of

Schizostoma Hartmanii.— Shell smooth, subcylindrical, thick, yellowish horn-color, imperforate; spire raised; sutures very much impressed; whorls flattened, the last rather large; fissure straight and rather short; aperture rather small, ovate, white within, obtusely angular at the base; columella white, incurved, somewhat thickened below; outer lip sharp and sinuous.

Habitat.— Coosa River, Alabama; W. D. Hartman, M.D.

Diameter, ·46; length, ·96 of an inch.

Observations.— This specimen, which I owe to the kindness of Dr. Hartman of Westchester, Penn., was no doubt sent to him by Dr. Showalter. It is distinct from any species I have before seen, and is more nearly allied in outline to *Babylonicum* (nobis)

Fig. 703.

than any other species I know. It differs in not being umbilicate, in not having a square shoulder, and in being yellowish horn-color. It is impressed below the hem-like margin of the suture, while the other is not. It is also near to *recta*, Anthony, but is stouter, is of a light color, and has a more twisted columella. The specimen in my possession is nearly an inch in length. With a perfect spire it would exceed an inch. All is imperfect above the second whorl, but there are indications of there being at least six. One specimen has no bands, the other has three obscure ones. The aperture is about half the length of the shell. The hem is rather narrow and is well defined. I have great pleasure in naming this species after my friend Dr. Hartman, who has done so much to promote natural science.— *Lea.*

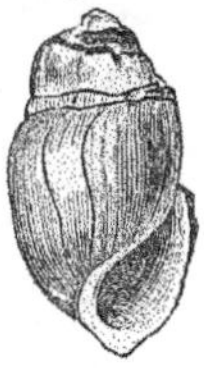

23. S. constrictum, LEA.

Schizostoma constrictum, LEA, Philos. Proc., iv, p. 167, Aug., 1845. Philos. Trans.,
    x, p. 68, t. 9, f. 55.
Gyrotoma constricta, Lea, BINNEY, Check List, No. 302. BROT, List, p. 27. ADAMS,
    Genera, i, p. 305.
Gyrotoma recta, ANTHONY, Proc. Acad. Nat. Sci., p. 64, Feb., 1860. BINNEY, Check
    List, No. 331. BROT, List, p. 27.
Melatoma rectum, Anthony, REEVE, Monog., sp. 10, not sp. 7a.

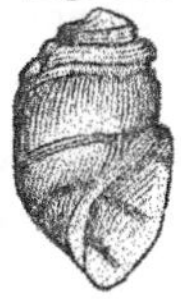

Description.— Shell smooth, cylindrical, yellowish, thick; short,

Fig. 704.

originally furnished with about five low whorls, of which three are nearly lost by truncation; fissure moderately broad, not quite direct and not remarkably deep; sutures lightly impressed; aperture narrow ovate, occupying about three-fifths of the length of the shell; within dusky and obscurely banded; columella callous, thickened abruptly at the fissure.

Habitat.— Coosa River, Alabama.

Length of shell, eleven-sixteenths; breadth of shell, three-eighths of an inch. Length of aperture, seven-sixteenths; breadth of aperture, three-sixteenths of an inch.

Observations.— This is the most cylindrical species I have ever seen in this genus. In its general form and coloring it most nearly resembles *G. demissa* (nobis), but is longer, more elevated, smoother and is ornamented with bands, which on that species are entirely wanting; these bands on the body-whorl are three in number, of which the middle one is the narrowest and least distinct; they are widely distant from each other; the cord-like cincture is very prominent in this species and the fissure is farther removed from the suture than is usual. It is altogether a beautiful and graceful species.— *Anthony.*

Mr. Lea's description, being founded on a single abnormal specimen, is by no means so good as that of Mr. Anthony; I have, therefore, adopted the latter. The types of both are figured.   I have seen other specimens besides Mr. Lea's, which have the constriction of the centre of the whorls, which has given rise to the specific name, but I cannot at present consider this to be a normal character of the species. Mr. Reeve's fig. 7a represents a smooth variety of *salebrosum,* Anthony. Mr. Lea's description and figure are given below.

Schizostoma constrictum.— Shell smooth, somewhat fusiform, rather thin, yellowish horn-color; spire obtuse; sutures impressed; whorls

Fig. 705.   Fig. 706.   constricted; fissure rather large, somewhat oblique; aperture large, elliptical, whitish within; columella smooth, subangular at the base.

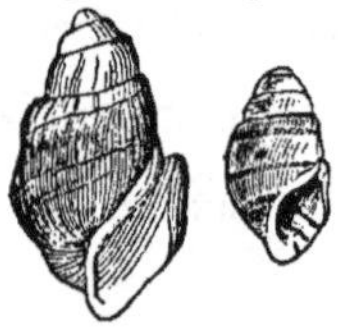

Habitat.— Tuscaloosa; Alabama.

Diameter, ·43; length, ·75 of an inch.

Observations.— A single specimen only of this species was among the shells submitted to me by Dr. Budd. It differs from those I have seen in having a rather broad channel impressed immediately above the centre of the whorl. This character may, however, differ in other individuals. The fissure is rather wide, but not deep. The apex being eroded, the number of whorls could not be ascertained. The aperture is about one-half the length of the shell. There is no appearance of bands about this specimen.— *Lea.*

23a. S. Showalteriana, LEA.

Schizostoma Showalterii, LEA, Proc. Acad. Nat. Sci., 112, 1864. Obs., xi, p. 105, t. 23, f. 56.

Description. — Shell smooth, cylindrical, elevated, thick, honey-yellow, without bands; spire exserted; sutures very much impressed, furnished below with a cord; whorls flattened; fissure rather small; aperture small, elliptical, white within; outer lip acute, somewhat sinuous; columella somewhat thickened above and below.

Operculum elongate, dark brown.

Habitat. — Coosa River, Alabama; E. R. Showalter, M.D

Diameter, ·54 of an inch; length, 1·2 (?) inches.

Observations. — This species, of which I have but a single specimen, is the highest in the spire of any I have seen, and it is to be regretted that. it is not more perfect, the three lower whorls only remaining. These, however, indicate a high spire, which is not common in the genus. The lower whorl reminds one of *constrictum* (nobis), but that species is short, not so thick, has a larger aperture, and the callus is not so thick on the columella. It also has a constriction around the body-whorl which this species has not. It is also devoid of the well marked cord which runs round the sutures of this species, which cord is very remarkable. There are a few iridescent striæ on the upper part of the last whorl in this specimen. Being an imperfect specimen, neither the number of whorls nor the proportion of the aperture can be ascertained. In a former paper I named a fine *Schizostoma* after Dr. Showalter, which he sent to me as new; but I find that Mr. Anthony had very shortly before described the same shell under the name of *carinifera*. Wishing very much that Dr. Showalter's name should be permanent in a genus to which he has so much contributed in bringing so many new species to light, I dedicate this fine species to him, as an acknowledgment of the debt due to him by all students of malacology. — *Lea.*

Fig. 707.

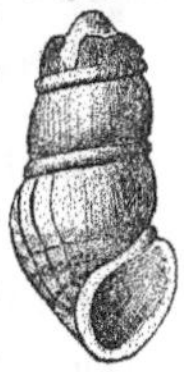

The specific name *Showalterii* having become a synonyme, it cannot be revived by the same author for another species in the same genus. To obviate all difficulty, I have slightly changed the termination of the name.

24. S. salebrosum, ANTHONY.

Gyrotoma salebrosa, ANTHONY, Proc. Acad. Nat. Sci., p. 66, Feb., 1860. BINNEY,
    Check List, No. 333.
Melatoma salebrosum, Anthony, REEVE, Monog., sp. 8 and 15.
Gyrotoma robusta, ANTHONY, Proc. Acad. Nat. Sci., p. 67, Feb., 1860. BINNEY,
    Check List, No. 332. BROT. List, p. 28.
Melatoma robustum, Anthony, REEVE. Monog., sp. 14a, b.
Melatoma rectum, Anthony, of REEVE, Monog., sp. 7a.

Description.— Shell fusiform, robust, thick, nodulous, of a dusky
olive-color; spire truncated, leaving scarcely more than the body-
whorl, but indicating by traces on the truncation the loss of three or
four others; fissure moderately open, waved, not deep; body-whorl
roughly nodulous at the upper part and ornamented by three dark

Fig. 708.   bands below; aperture ample, ovate, dusky within
and banded by three broad bands; columella deeply
rounded, covered with a thick deposit of callus,
white at its lower portion, but tinged with dark brown
at the fissures.

Habitat.— Coosa River, Alabama.

Length of shell, three-fourths; breadth of shell, one-half of an
inch. Length of aperture, nine and one-half sixteenths; breadth of
aperture, five-sixteenths of an inch.

Observations.— This species presents the unusual characteristic of a
nodulous surface, which character has not been observed in any species hitherto described by any American author. These nodules are
very conspicuous and much compressed laterally, so as to present very
much the appearance of coarsely folded ribs.— *Anthony.*

The nodules, or rather folds of *S. salebrosum*, by which
Mr. Anthony distinguishes it from *S. robustum* are caused
by the arrest of growth and indicate the position of former
mouths of the shell.

The type of *S. robustum* (which I figure) is a more than
usually smooth variety, but I have before me a number of
specimens, which exhibit the intermediate stages between it
and the folded *S. salebrosum.*

Mr. Reeve's fig. 7a, intended for *S. rectum*, is, I think, referable to this species.

Mr. Lea's *incisum* is not the same as *salebrosum*, as he supposes, but is quite a different shell in form.

The following is Mr. Anthony's description of

Schizostoma robustum.— Shell fusiform, robust, thick, of a dark olive-color; spire obtuse, consisting of one perfect whorl remaining, with marks of two or three more, lost by truncation; body-whorl broad, curved, not deep, closed behind by a cord-like cincture, very prominent, beneath which and close to it is a narrow depression or furrow; aperture narrow, ovate, banded inside; columella well rounded and covered by callus; lines of growth very distinct and much curved, rendering the shell rough by their prominence.

Habitat.— Coosa River, Alabama.

My cabinet.

Length of shell, seven-eighths; breadth of shell, nine-sixteenths of an inch. Length of aperture, ten-sixteenths; breadth of aperture, five-sixths of an inch.

Observations.— This is a large, robust species, somewhat resembling *Melania ampla* (nobis) in form, and not unlike it in coloring; it is about the largest species I have seen in this genus, and certainly not the least beautiful; compared with *G. salebrosa* (nobis) herein described, it is larger, smoother, more inflated and has not the rib-like prominences so characteristic of that species; the lower part of the columella is somewhat flattened and thickened, and another thickening takes place at the aperture, leaving a thinner space between the two points.— *Anthony.*

Fig. 709.

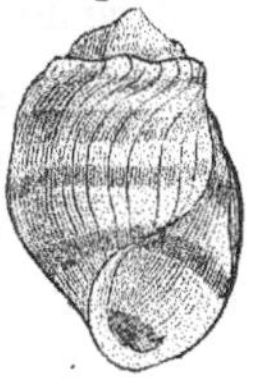

25. S. glandula, LEA.

Schizostoma glandula, LEA, Proc. Acad. Nat. Sci., p. 187, 1860. Jour. Acad. Nat. Sci., v, pt. 3, t. 35, f. 53, March, 1863. Obs., ix, p. 71.
Gyrotoma glandula, Lea, BINNEY, Check List, No. 319. BROT, List, p. 27.

Description.— Shell smooth, short, much inflated, rather thick, yellowish horn-color, minutely striate, imperforate; spire short; sutures much impressed; whorls six, banded, the last large and swollen; lip-cut oblique and short; aperture rather large, elliptical, white within; columella whitish and thickened above; outer lip sharp and somewhat sinuous.

Operculum ovate, brown, with the polar point very close to the inner lower edge.

Habitat.— Coosa River, Alabama; E. R. Showalter, M.D.

Diameter, ·36; length, ·57 of an inch.

Observations.— A single specimen only was received from Dr.

Showalter. The lip-cut in this species is not deep, but it is wider than usual, and, being oblique, presents more of the whorl within than usual.   In the specimen before me there are two small, hair-like bands, one immediately under the shoulder and the other very near

Fig. 710.

to the base, and in the middle there is a slight indication of a band, but these indistinct bands do not become visible in the interior except in a very small degree.   The shoulder is slightly impressed, giving the suture a hem.   In color it is nearly the same with *glans* herein described, but it differs entirely in the lip-cut, and is a much smaller species with a much lower spire.   It is very likely that in other specimens the color may be found to vary.   The outer lip stands well off from the body-whorl, and the base is subangular.   The aperture is more than one-half the length of the shell.   The hem is large and well defined.   It is near to *virens* (nobis) in outline and size, but differs entirely in the color, bands and shoulder.— *Lea.*

Closely allied to the following species (*S. incisum*) it may be distinguished by being heavier, of different color, higher spire and by the body-whorl not being so much flattened around its superior portion.

26. S. incisum, Lea.

Anculosa (Schizostoma) incisa, Lea, Philos. Proc., ii, p. 243, Dec., 1842. Philos. Trans., ix, p. 28, t. 9, f. 28.
Schizostoma incisa, Lea, Wheatley, Cat. Shells U. S., p. 28. Hanley, Conch. Misc. Melania, t. 5, f. 44, 45.
Gyrotoma incisa, Lea, Binney, Check List, No. 323. Brot, List, p. 27. Adams, Genera i, p. 305.
Melatoma incisum, Lea, Reeve, Monog. sp. 4.
Melania incisa, Lea, Jay, Cat., 4th edit., p. 274.
Leptoxis incisa, Lea, Binney, Check List, No. 363. Haldeman, Monog., p. 2, t. 1, f. 24-26.
Gyrotoma quadrata, Anthony, Proc. Acad. Nat. Sci., p. 65, Feb., 1860. Binney, Check List, No. 330.
Melatoma quadratum, Anthony, Reeve, Monog., fig. 7b (not 7a, nor 8).
Schizostoma virens, Lea, Proc. Acad. Nat. Sci., p. 187, 1860. Jour. Acad. Nat. Sci., v, pt. 3, t. 35, f. 59. Obs., ix, p. 75.
Gyrotoma virens, Lea, Binney, Check List, No. 335. Brot, List, p. 28.
Gyrotoma obliqua, Anthony, MSS.

Description. — Shell smooth, ovately gibbous, thick, yellowish-brown: spire short; whorls four, flattened; columella thickened above; aperture large, ovate, white.

Habitat.— Alabama.

Diameter, ·44; length, ·64 of an inch.— *Lea.*

As this shell was one of the first species of the genus described, there did not appear to be so much necessity at that time for an accurate and extended description. That of *S. quadratum*, by Mr. Anthony, will give a better idea of the specific characters.

Fig. 711.

Gyrotoma quadrata.—Shell short, smooth, fusiform, rather thick, olivaceous; spire short, composed of about four very low whorls, the upper two being partially obliterated by erosion; fissure rather broad, waved, but not remarkably deep; sutures distinct; whorls distinctly, but not squarely, shouldered; aperture elliptical, occupying more than half the length of the shell; within three-banded; columella with a light callous deposit.

Habitat.— Coosa River, Alabama.

Length of shell, nine-sixteenths; breadth of shell, seven-sixteenths of an inch. Length of aperture, six-sixteenths; breadth of aperture, three-sixteenths of an inch.

Observations.— The most remarkable characteristic at first view of this species is its short, square form; its color is dark, and the bands,

Fig. 712. Fig. 713.

which are very broad, are not very distinct; hence its general aspect is not so pleasing to the eye as many others; the fissure is broadly separated from the body of the shell; outer lip very sharp and sinuous, forming, with the columella, a small not very distinct sinus at base. In form it approaches most nearly perhaps to *G. salebrosa* (nobis), but is more delicate in texture, thinner and has no armature as in that species.— *Anthony.*

Mr. Lea considers *quadrata*, Anthony, to be a synonyme of his *S. incisum.* An inspection of a number of specimens of both species enables me to agree with him entirely. To these I unite *S. virens*, Lea, recently published, believing it to be a small variety of the same species.

Schizostoma virens.— Shell very slightly nodulous, very much inflated, rather thick, dark green, very minutely striate, imperforate; spire short; sutures impressed; whorls rather flattened and with three bands; lip-cut oblique, short; aperture elongate, nearly pearshaped, within darkly banded; columella whitish and thickened above; outer lip sharp and sinuous.

Operculum ovate, dark brown, with the polar point near to the inner lower edge.

Habitat.— Coosa River, Alabama; E. R. Showalter, M.D.

Diameter, ·32; length, ·50 of an inch.

Observations.—This is rather a small species; at least the specimens before me indicate this. There appear to be about six whorls, the upper ones being disposed to put on indistinct folds. The lower whorl is flattened on the middle, has a distinct shoulder above, the top of which is yellowish. It is furnished with three dark, broad bands. There is no appearance of a hem below the suture. The upper whorls are slightly inflated. The lines of growth are distinctly marked. The aperture is nearly two-thirds the length of the shell, and the base is subangular, and disposed to form a channel like *Lithasia.* The three dark, broad bands are well marked within the aperture. This species is nearer in general outline and color to *bulbosa,* Anthony, than any which have come under my notice, but it does not belong to the deep fissured group and the spire is by no means so high. The aperture is more than half the length of the shell.— *Lea.*

Species unknown to me.

Gyrotoma conica, Shuttleworth (ubi), Brot, List, p. 27.

Genus ANCULOSA, Say.

Anculosa, Say, Jour. Acad. Nat. Sci., ii, p. 178, Nov., 1821. Conrad, Am. Jour. Sci., xxv, p. 342, 1834. Muller, Syn. Test. Viv., p. 39, 1836. Swainson, Manual Malacol., 1840. Haldeman, Suppl. to Monog. Limniades, Oct., 1840. Sowerby, Conch. Manual, 2d edit., p. 66, 1842. Wheatley, Cat. Shells U. S., p. 27, 1845. Lea, Philos. Trans., ix, p. 14, 1846. Anthony, Proc. Acad. Nat. Sci., p. 67, Feb., 1860.

Anculosa, Conrad, Hermannson, Indices Gener. Malac., i, p. 51, 1846.

Anculotus, Say, Jour. Acad. Nat. Sci., v, pt. 1, p. 128, Aug., 1825. Conrad, New Fresh Water Shells, p. 62, 1834. Couthuoy, Bost. Jour., ii, p. 184, Feb., 1839. Anthony, Bost. Jour., iii, p. 278, Jan., 1840. DeKay, Moll. N. Y., p. 101, 1843. Chenu, Bibl. Conch., i, iii. Conrad, p. 26, 1845. Gray, Genera, Zool. Proc., xv, p. 153, 1847. Woodward, Manual, i, p. 131, 1851. Jay, Cat., 4th edit., p. 276, 1852. Reeve, Conch. Iconica, Sept., 1860.

Ancylotus, Say, Hermannson, Indices Gen. Mal., i, p. 51, 1846.

??Leptoxis, RAFINESQUE, Jour. de Phys., lxxxviii, p. 424, 1819. HAL-
DEMAN, Monog. Lept. H. & A. ADAMS, Genera, i, p. 307, Feb.,
1854. CHENU, Man. de Conchyl., i, p. 294, 1859. BINNEY, Check
List, p. 10, June, 1860. BROT, List, p. 23, 1862.

Mudalia, HALDEMAN, Suppl. to Monog. Limn., Oct., 1840.

Nitocris, H. & A. ADAMS, Genera, i, p. 308, Feb., 1854.

1. *Tuberculate species.*

1. A. plicata, CONRAD.

Anculotus plicatus, CONRAD, New Fresh-Water Shells, p. 61, t. 8, f. 18, 1834. DEKAY,
Moll. N. Y., p. 103. JAY, Cat., 4th edit., p. 276. REEVE, Monog., t. 3, f. 22.
MÜLLER, Synopsis, p. 40, 1836.
Anculosa plicata, Conrad, WHEATLEY, Cat. Shells U. S., p. 28.
Leptoxis plicata, Conrad, BINNEY, Check List, No. 379. HALDEMAN, Monog. Lept.,
p. 3, t. 2, f. 35—39. ADAMS, Genera, i, p. 307.
Anculosa bella, LEA, Philos. Proc., ii, p. 83, Oct., 1841. WHEATLEY, Cat. Shells
U. S., p. 28.
Anculosa tuberculata, LEA, Philos. Proc., ii, p. 83, Oct., 1841. Phil. Trans., ix, p. 21.
Obs., iv, p. 21. WHEATLEY, Cat. Shells U. S., p. 28. BINNEY, Check List, No. 392.
Anculotus smaragdinus, REEVE, Monog., t. 3, f. 23, April, 1860.

Description.— Shell suboval, with a short spire, only one whorl
of which is entire, rounded; body-whorl slightly ventricose, with
oblique plaits or lines, which are crenu-

Fig. 715.   Fig. 716.   Fig. 717.

lated on the margins of a slight, spiral
groove near the suture; lines of growth
prominent; epidermis greenish or blackish, with spiral bands; aperture elliptical.

Habitat.— Inhabits tributaries of the
Tennessee River in Alabama, adhering to stones.— *Conrad.*

Here follow descriptions of the synonymes :—

Anculosa bella.— Shell subglobose, rather thin, tuberculate above,
banded, greenish-brown; spire short; sutures linear; whorls three,
convex; aperture subrotund, bluish within; columella maculated.

Habitat.— Warrior River, Alabama; Professor Brumby.— *Lea.*

Figure 715 is a copy of one in Prof. Haldeman's Monograph, which, as he says, is labelled "*bella*" in Mr. Lea's
cabinet. It is a good figure of the original type.

Anculosa tuberculata.—Shell ovate, thick, above tuberculate, brown;
spire short; sutures scarcely impressed; whorls slightly convex: aperture ovate, within flesh-colored; columella thick and spotted.

Habitat.— Warrior River, Alabama.

Diameter, ·28; length, ·50 of an inch.

Observations.— The above description is made from a single specimen, which is truncate at the apex; as the species of this genus

Fig. 718.   usually are.   Three whorls are visible.   In a perfect state,

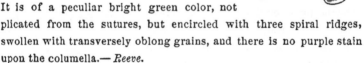

it probably has four.   The specimen before me has two
rows of tubercles.   On the superior part of the whorl, and
below these, are two parallel, indistinct lines, which may,
in other individuals, rise into tubercles.   In the interior,
two purple bands are visible.   On the middle of the columella there
is a large, purple spot.   This may not occur in all specimens.   The
aperture is about two-thirds the length of the shell.— *Lea.*

Anculotus smaragdinus.— Shell ovately turbinate, sometimes rather
solid, bright green; spire tumidly exserted; whorls slopingly convex, smooth, encircled round the upper part
with three ridges; ridges oblong-granuled;
aperture ovate; columella broadly callous.

Fig. 719.   Fig. 720.

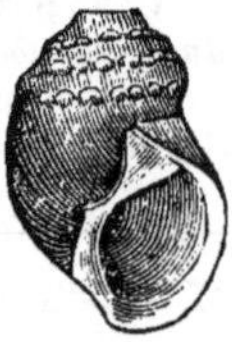

Habitat.— Alabama.

Observations.— This species has been confounded by Mr. Haldeman with the preceding.
It is of a peculiar bright green color, not
plicated from the sutures, but encircled with three spiral ridges,
swollen with transversely oblong grains, and there is no purple stain
upon the columella.— *Reeve.*

As this species is very variable in outline and ornamentation,
four other figures are here given, all of which are from specimens from the Coosa River, Alabama.   Messrs. Haldeman
and Reeve, both make *tuberculata* a synonyme of *plicata;*
M. Brot does the same, and adds *smaragdinus,* Reeve; *bella,*
Lea, is written a synonyme by Prof. Haldeman.

2.   *Sulcate species.*

2.   A. Showalterii, Lea.

Anculosa Showalterii, Lea, Proc. Acad. Nat. Sci., p. 93, 1860.   Jour. Acad. Nat.
    Sci., 2d ser., v, pt. 3, p. 255, t. 35, f. 62, March, 1863.   Obs., ix, p. 77, t. 35, f. 62.
Leptoxis Showalterii, Lea, BINNEY, Check List, No. 385.   BROT, List, p. 25.
Anculotus sulcosus, ANTHONY. MSS., REEVE, Monog. Anculotus, t. 6, f. 44, April,
    1861.
Leptoxis sulcosa, Anthony, BROT, List, p. 26.

Description.— Shell much ribbed, suborbicular, thick, very dark brown, almost black, very finely striate; spire very short; sutures much impressed; whorls inflated, covered with seven transverse ribs; aperture large, nearly round, subangular above, with dark bands inside; columella thick, flattened, dark brown; outer lip very much expanded and very much crenulate.

Operculum ovate, thin, with the polar point on the inner inferior edge.

Habitat.— Coosa River, Uniontown, Alabama; Dr. E. R. Showalter. Diameter, ·37; length, ·40 of an inch.

Observations.— Several specimens of this very remarkable *Anculosa* were sent to me by Dr. Showalter. It differs from all the species I have seen in its peculiar, large ribs which girt it with great strength. The apices being eroded, the number of whorls cannot be Fig. 721. ascertained, but there are probably only three. On the second whorl only three ribs appear above the suture. It reminds us at once of *Paludomus loricata*, Reeve, but the transverse ribs are not beaded like that shell. It is also a diminutive shell compared with that, and has a more depressed spire. The ribs are very large, and sometimes obscurely maculate. They are accompanied on the inside with dark brown bands which terminate at the edge of the lip, each in a small furrow, which produces the crenulations of the lip.— *Lea.*

The following is Mr. Reeve's description :—

Anculosa sulcosus. — Shell ovate, rather thin, inflated, dirty-fulvous; spire very short, flat; whorls spirally keeled; keels very large, rounded, distant, with the interstices broadly excavated; aperture ovate, large, wide, open; columella short.

Habitat. — Alabama.

Observations.— A very remarkable Purpura-shaped species, composed of largely defined, winding, keel-like ribs, broadly excavated in the interstices.—*Reeve.*

As the species is very uniform and Mr. Anthony's types before me do not in the least differ from those of Mr. Lea, I have not considered it necessary to give a figure of *sulcosa.*

3. A. canalifera, ANTHONY.

Anculosa canalifera, ANTHONY, Proc. Acad. Nat. Sci., p. 68, Feb., 1860.
Anculotus canaliferus, Anthony, REEVE, Monog. Anculotus, t. 5, f. 39.
Leptoxis canalifera, Anthony, BINNEY, Check List, No. 345. BROT, List, p. 24.

Description.— Shell ovate, costate, of a brown color, thin; spire acutely elevated, composed of 5-6 sharply carinated whorls; sutures not very distinct; aperture about half the length of the shell, ovate, banded inside; columella deeply indented; sinus none.

Fig. 722. Fig. 723.

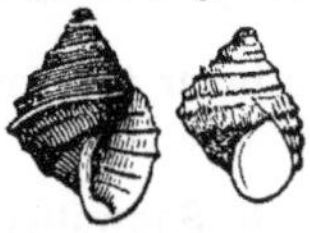

Habitat.— North Carolina, in Dan River.

Observations.— One of our most curious and beautiful species, which no one can easily mistake; the whole shell is crossed with sharp, elevated costæ running round the whorls and corresponding deep grooves between them; about five costæ on the body-whorl; a less number on the spire volutions; these ribs appear as dark bands in the interior of the aperture, and there is a broad, non-elevated band at the base of the shell; differs from *Anculosa costata* (nobis) by the size and prominence of its ribs and by its elevated spire.— *Anthony.*

Figure 722 is from Mr. Anthony's type. This species is very closely allied to *Melania proscissa*, Anthony, from the same locality, and may prove to be a variety of that shell with a shorter spire. It is a very beautiful species.

3. *Striate species.*

4. A. littorina, HALDEMAN.

Anculosa littorina, HALDEMAN, Spec. Number of Monog. Cover of No. 1, Monog. July, 1840.
Leptoxis littorina, HALDEMAN, Monog. Lept., p. 4, t. 4, f. 110. BINNEY, Check List, No. 368. BROT, List, p. 24.
Melania pilula, LEA, Philos. Proc., ii, p. 15, Feb., 1841. Philos. Trans., viii, p. 186, t. 6, f. 50. Obs., iii, p. 24, t. 6, f. 50. DEKAY, Moll. N. Y., p. 99. TROOST, Cat. Moll. Tenn. WHEATLEY, Cat. Shells U. S., p. 26. BINNEY, Check List, No. 204. CATLOW, Conch. Nomenc., p. 188 ADAMS, Genera, i, p. 307.

Fig. 724.

Description. — Shell solid, conical, olivaceous, encircled with transverse lines; whorls four, flattened; apex eroded; sutures scarcely excavated; aperture somewhat rounded, angulated above.

Habitat.— Holston River, Virginia.

Length, ½ of an inch.— *Haldeman.*

The accompanying figure is from Prof. Haldeman's Monography of *Leptoxis.* As Mr. Lea's figure of *Melania pilula* is precisely similar, it is not necessary to reproduce it here; his description is as follows :—

Melania pilula.— Shell striate, subglobose, thick, dark brown; sutures somewhat impressed; whorls convex; aperture ovate, large, angular at the base, within purplish.

Habitat.— Tennessee.

Diameter, ·34; length ·43 of an inch.

Observations.— This is a very distinct species, and is quite as globose as *M. subglobosa,* Say. Two specimens were received, the spires of which are not perfect. I should presume that when perfect they would be found to have four whorls. The raised striæ are very distinct, and consist of eighteen in these two individuals. The aperture is about half the length of the shell. One specimen is dark purple within the aperture. The other is bluish with a tinge of purple on the columella.— *Lea.*

This species resembles somewhat a striate variety of *dilatata.*

5. A. costata, ANTHONY.

Anculotus costatus, ANTHONY, Bost. Jour. Nat. Hist., iii, p. 278, t. 3, f. 1, Jan., 1840.
    DEKAY, Moll. N. Y., p. 102, t. 7, f. 139. REEVE, Monog. Anculotus, t. 5, f. 41.
Anculosa costata, ANTHONY, List of Shells of Cincinnati, 2d edit. WHEATLEY,
    Cat. Shells U. S , p. 28.
Leptoxis costata, Anthony, BINNEY, Check List, No. 349.
Melania occidentalis, LEA, Philos. Proc., ii, p. 12, Feb., 1841. Philos. Trans., viii, p.
    172, t. 5, f. 20. Obs., iii, p. 10, t. 5, f. 20. DEKAY, Moll. N. Y., p. 95. WHEATLEY,
    Cat. Shells, U. S., p. 26. JAY, Cat., 4th edit., p. 274. BINNEY, Check List
    No. 184. CATLOW, Conch. Nomenc., p. 188.
Nitocris costata, Lea, H. and A. ADAMS, Genera, i, p. 308.
Nitocris occidentalis, Lea, ADAMS, Genera, i, p. 308.

Description.— Shell subglobose, with a depressed, convex spire; body-whorl ventricose, with about five costæ revolving Fig. 725. Fig. 726. around it; color olivaceous; aperture obovate; base regularly rounded; purplish within.

Observations.— Found on pebbly shores near the city of Cincinnati.—*Anthony.*

Melania occidentalis. — Shell smooth, subglobose, rather thick, green; spire short, pointed; sutures linear; whorls four, somewhat convex; aperture ovate, large, within, purple or white.

Habitat.— Vicinity of Cincinnati, Ohio.

Diameter, ·30; length, ·37 of an inch.

Observations.— This is a fine species about the size of *Melania subglobosa*, Say (*Anculosa*), and it has been confounded with it. I have specimens of *subglobosa* which were brought by Prof. Vanuxem from the Holston, at the time he gave them to Mr. Say for description. They certainly do not appear to me to be the same, although in many characters they agree. The animal of *occidentalis* I have not seen; the operculum is spiral; at present I prefer to place it among the *Melania*. Some of the varieties before me are very beautifully furnished with raised revolving striæ. When there is a single one, it gives the shade the appearance of being carinate, as it appears near the centre of the whorl. In some specimens these striæ are more numerous; in a single one I have counted fifteen. There appear to be no bands on the outside, but sometimes purple lines on the inside mark the places of the exterior striæ. There is generally more or less color in the interior and about the columella the base of which is disposed to be angular. The aperture is nearly three-fourths the length of the shell.*—Lea.

The nomenclature of this species is singularly confused. Mr. Lea described the quite young shell of *A. prærosa*, which is then carinate, as *Melania Cincinnatiensis*, and he has considered *costatus* to be the mature form and a synonyme, and distributed shells so labelled. Prof. Haldeman, in his monograph of Leptoxis, declares *costatus*, Anthony, and *occidentalis*, Lea, to be synonymes of *trilineatus*, Say ; and succeeding authors have acquiesced in these views. *Costatus* is, however, a young shell of which *occidentalis* is the mature form. That it is perfectly mature is shown by the deposit of enamel upon the columella of some of the specimens before me. The striæ still appear on the old shell, when the surface is not too much worn. *A. trilineatus* is *never* costate and has three broad, brown bands, and Mr. Anthony informs me that it has never been found in the upper Ohio River, while *costatus* is plentiful at Cincinnati. The figures of *costatus* are from specimens fur-

---

* Since the above was written I have seen in the " Boston Journal of Science " the description and figure by Mr. Anthony of *Anculotus costatus* which in some respects answers to this shell. Mr. A. says that his shell has " about five costæ revolving around it."

nished by Mr. Anthony. The largest one is from one of his types.

6. A. rubiginosa, Lea.

Anculosa rubiginosa, Lea, ii, p. 83, Oct., 1841. Philos. Trans., ix, p. 20. Obs., iv, p. 20. Brot, Mal. Blatt, ii, p. 111, July, 1840.
Anculotus rubiginosus, Lea, Jay, Cat., 4th edit., p. 276. Reeve, Monog. Anc., t. 2, f. 12; t. 6, f. 47.
Leptoxis rubiginosa, Lea, Haldeman, Monog. Lept., f. 59–70. Binney, Check List, No. 383. Chenu, Manuel, i, f. 2035, 2036. Adams, Genera, i, p. 307.
Anculosa Griffithiana, Lea, Philos. Proc., ii, p. 83, Oct., 1841. Philos. Trans., ix, p. 20. Obs., iv, p. 20. Wheatley, Cat. Shells U. S., p. 28.
Anculotus Griffithsianus, Lea, Reeve, Monog. Anculotus, t. 1, f. 8.
Leptoxis Griffithiana, Lea, Binney, Check List, No. 362. Adams, Genera, i, p. 307.

Description.— Shell ovately gibbous, thick, smooth, rusty color; spire rather elevated; sutures impressed; whorls flattened; aperture irregularly ovate, within whitish; columella thick, dark purple.

Habitat.— Warrior River, Alabama.

Diameter, ·40 ; length, ·60 of an inch.

Observations.— A single, and not a very perfect, specimen is before me. The middle of the whorl is flattened, in-

Fig. 727. Fig. 728.

deed a little impressed, and this causes a curve in the outer lip. It is obscurely banded, and the whole of the columella is purple. The aperture is nearly two-thirds the length of the shell. The spire is more exserted than usual in the

Fig. 729. Fig. 730.

 Fig. 731. Fig. 732. *Anculosæ*, but not perfect in this specimen. Four whorls are perceptible.— *Lea.*

Fig. 733. Fig. 734. Fig. 735.   The following description of *A. Griffithiana* by Mr. Lea will better exhibit the usual state of the species.

Anculosa Griffithiana. — Shell ovately gibbous, thick, closely and transversely striate, banded; sutures impressed; whorls four, flattened; aperture ovate, within banded; columella thick, dark purple.

Habitat.— Coosa River, Alabama.

Diameter, ·50; length, ·60 of an inch.

Observations.— The distinctive characters of this species are the transverse striæ and the flattened side. This flatness causes a nob tuse angle below, and one above. One of the two specimens, under examination, is more banded, and has a less number of striæ than the other. The aperture is nearly three-fourths ,the length of the shell.— *Lea.*

Prof. Haldeman figured this last shell in his Monograph as a variëty, but an examination of thousands of specimens from Coosa River, Alabama, proves the entire identity of the two forms by intermediate ones. Perhaps not one specimen in one hundred is entirely smooth, and some are almost costate. The columella is always tinged with purple, and the substance of the shell generally slightly so. It appears to be a very abundant and very distinct species. Among the Coosa River specimens several occurred with the top of the body-whorl plicate.

Mr. Reeve is in error when he says at sp. 47 that sp. 12 does not represent this shell, they equally represent it; also in quoting *A. ampla,* Anthony, as a synonyme of *rubiginosa,* and *Melania compacta,* Anthony, as a synonyme of *A. Griffithiana.*

4.   *Angulated species.*

7.   A. carinata, BRUGUIERE.

Bulimus carinatus, BRUG., Ency. Meth., vers, i, p. 301, 1792.
Paludina dissimilis, SAY, Nicholson's Encyc., 3d Am. edit., 1819.
Anculotus dissimilis, Say, RAVENEL, Cat., p. 11. JAY, Cat., 4th edit., p. 276. REEVE, Monog. Ancul. t. 4, f. 27.
Anculosa dissimilis, Say, WHEATLEY, Cat. Shells U. S., p. 28. HALDEMAN, in Ruppell's Lancaster County, p. 479.
Nitocris dissimilis, Say, ADAMS, Genera, i, p. 308.
Leptoxis dissimilis, Say, HALDEMAN, Monog. Lept., p. 4, t. 4, f. 85-100. BROT, List, p. 24. BINNEY, Check List, No. 355. CHENU, Manuel, i, f. 2049-54.
Helix subcarinata, WOOD, Index, Test. Suppl., t. 7, f. 13. Lister, t. 111, f. 5.
Anculotus carinatus, DEKAY, Moll. N. Y., p. 101, 1843. JAY, Cat., 4th edit., p. 276.
Anculosa carinata, DeKay, WHEATLEY, Cat. Shells U. S., p. 28.
Leptoxis carinata, DeKay, BINNEY, Check List, No. 343. BROT, List, p. 24.

Variety a.

Anculosa carinata, LEA, Proc. Philos., ii, p. 34, April, 1841. Philos. Trans., ix, p. 15. Obs., iv, p. 15.
Leptoxis carinata, Lea, BINNEY, Check List, No. 344.
Nitocris carinata, Lea, ADAMS, Genera, i, p. 308.
Anculosa variabilis, LEA, Philos. Proc., ii, p. 34, April, 1841. Philos. Trans., ix, p. 15. Obs., iv, p. 15. WHEATLEY, Cat. Shells U. S., p. 28.
Leptoxis variabilis, Lea, CHENU, Manuel, f. 2037-39. BINNEY, Check List, No. 394.

BROT, List, p. 26. HALDEMAN, Mo ιog. Le tox ›, p. 4, t. 4, f. 102–9. ADAMS, Genera, i, p. 307.

Variety b.

Anculotus nigrescens, CONRAD, New Fresh-Water Shells, p. 64, t. 8, f. 17, 1834.
DEKAY, Moll. N. Y., p. 102. WHEATLEY, Cat. Shells U. S., p. 28. JAY, Cat.
4th edit., p. 276. MÜLLER, Synopsis, p. 36, 1836.
Leptoxis nigrescens, Conrad, BINNEY, Check List, No. 372. ADAMS, Genera, i,
p. 307.
Anculotus trivittatus, DEKAY, Moll., N. Y., p. 102, t. 7, f. 137, 1843.
Leptoxis trivittata, DeKay, BINNEY, Check List, No. 390. ADAMS, Genera, i, p. 307.

Variety c.

Anculotus monodontoides, CONRAD, New Fresh-Water Shells, p. 61, t. 8, f. 16, 1834.
DEKAY, Moll. N. Y., p. 102. JAY, Cat., 4th edit., p. 276. WHEATLEY, Cat. Shells
U. S., p. 28. REEVE, Monog. Anc., t. 5, f. 37. MÜLLER, Synopsis, p. 41, 1836.
Mudalia monodontoides, Conrad, CHENU, Manuel, i, f. 2046-8.
Leptoxis monodontoides, Conrad, HALDEMAN, Monog. Leptoxis, p. 5, t. 4, 5, f. 124–
133. BINNEY, Check List, No. 370.
Nitocris monodontoides Conrad, ADAMS, Genera, i, p. 308.
Anculotus dentatus, COUTHUOY, Am. Journ. Sci., xxxvi, p. 390, July, 1839. Bost.
Journ. Nat. Hist. ii, p. 185, t. 4, f. 7, Feb., 1839. REEVE, Monog. Anc. t. 5, f. 36.
DEKAY, Moll. N. Y., p. 102. JAY, Cat., 3d edit., p. 63.
Anculosa dentata, Couthuoy, WHEATLEY, Cat. Shells U. S., p. 28.
Leptoxis dentata, Couthuoy, BINNEY, Check List, No. 352.
Nitocris dentata, Couth., ADAMS, Genera, i, p. 308.
Anculosa dentata, LEA, Philos. Proc. iI, p. 34, Apr. 1841.
Leptoxis dentata, Lea, BINNEY, Check List, No. 353.
Anculosa (Mudalia) affinis, HALDEMAN, Monog. Limniades, Cover of No. 3, March
13, 1841.
Anculotus affinis, Haldeman, REEVE, Monog. Anculotus, t. 6, f. 53.
Leptoxis affinis, Haldeman, BINNEY, Check List, No. 337. BROT, List, p. 23.
Nitocris carinata, Lea, ADAMS, Genera, i, p. 308.

Description. — Shell conic, dark horn-color or blackish; whorls about three, with obsolete, distant wrinkles, and an abrupt, acute prominent, carinated line, which revolves on the middle of the body-whorl, and is concealed on the spire by the suture; suture not indented; aperture oval, half as long as the shell, within sanguineous beneath the carina,

Fig. 736. Fig. 737. Fig. 738.

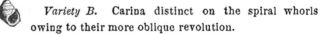

and at base and apex; columella emarginated, a little flattened at the base.

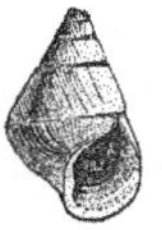

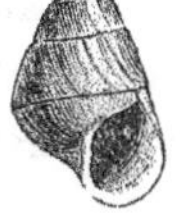

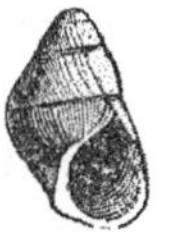

Fig. 739. Fig. 740.

Length, about two-fifths of an inch.

Variety A. Carina obsolete on the ventral portion of the body-whorl.

Fig. 741. Fig. 742.

Variety B. Carina distinct on the spiral whorls owing to their more oblique revolution.

Observations. — The surface of the whorls of this species is generally covered with unequal calcareous matter, resembling a fortuitous accumulation of mud or earth on that part, but which appears to be superposed by the animal, probably with the intention of re-

Fig. 743. Fig. 744. Fig. 745. taining a proper specific gravity. The apex

is often truncated. This species was found by Mr. Thomas Nuttall, during a journey to Pittsburg.— *Say.*

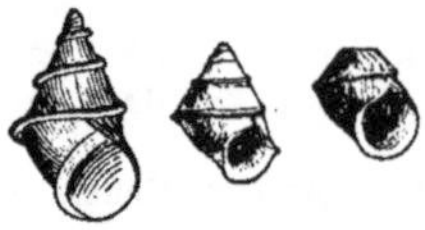

Figure 737 represents a typical shell and figure 743 Mr. Say's variety B.

The following is Mr. DeKay's description of

Anculosa carinatus. — Shell short, pyramidal, thin and fragile; whorls with a distinct, elevated carina, rather suddenly attenuated to the apex, which is frequently eroded; the whorls are polished with incremental striæ ascending to the edge of the carina, where they become multiplied, especially on its lower aspect; suture canaliculate, by the elevated carinæ; aperture subrhomboidal; outer lip simple, angular reflected at the base; pillar lip concave, with a broad callus; outer lip above contiguous to the carina of the preceding whorl; color amber, darker towards the lip.

Length of shell, ·45; extreme width of shell, ·4 of an inch. Length of aperture, ·45 of an inch.

Observations.— This very remarkable species, which may probably form the type of a new genus, is from Lake Champlain. My thanks are due to Dr. B. W. Budd, for an opportunity of adding this to the state collection. I have since obtained others from Cranesport, Broome County, in one of the tributaries of the Susquehanna. These are dark olive-green and

Fig. 746.

many of them ·5–·6 of an inch long. An eminent conchologist pronounces it identical with *A. dissimilis*, but I have not found the description of this species.— *DeKay.*

The figure is copied from that of Mr. DeKay. This species is of protean form and substance, being either thin or ponderous, large or small, carinate or smooth, with or without a tooth on the columella. It is not without much study of numerous individuals from many localities, that I propose to unite forms which eminent conchologists have always considered very distinct, but I find no characters in any of the so-called species here included, which do not become lost in transition forms. In certain parts of Eastern Virginia and Maryland the shell attains but a small growth, becomes stunted and develops a fold on the columella. In this state it becomes *dentata*, Couthuoy, or *monodontoides*, Conrad; while in parts of the Potomac and Susquehanna it becomes large, heavy and inflated.

I have selected a number of figures to show the transition from one form to another. The shells Fig. 748. Fig. 747.   Fig. 749. represented by figures 747, 748, 749 (collected by me, cohabiting with the typical species at Harper's Ferry, Virginia, and at Washington, D.C.),

merge into Mr. Lea's *A. carinata* and *variabilis*. The descriptions of these species here follow accompanied by illustrations of the types.

Anculosa carinata.—Shell ovately conical, carinate, dark olive; spire rather short; sutures small; whorls six; aperture small, round, with-
Fig. 750.   Fig. 751. in whitish, sulcate; columella rather thick, purple.

Habitat.—Roanoke River, Lafayette, Virginia.

Diameter, ·38; length, ·52 of an inch.

Observations.—A single specimen only of this interesting species was sent to me by Dr. Warder. It has some resemblance to *Anculosa dissimilis*, Say. It differs in having a smaller and rounder aperture and in having Fig. 752. Fig. 753. three carinæ, the middle one being the largest. The aperture is rather more than one-third the length of the shell. The carinæ are acute.— *Lea.*

Anculosa variabilis.—Shell obtusely conical, thick, either banded or horn-colored, carinate or smooth; sutures linear; whorls six, flat-
Fig. 754.   Fig. 755. tened; aperture large, nearly round; columella thick white or purple.

Habitat.— Roanoke River, Lafayette, and near Shenandoah Spring Brook.

Observations.—Three specimens are before me, all of which differ more or less. Two of Fig. 758. Fig. 757. Fig. 756. them are rather acutely carinate, with a dark epidermis and three rather large bands, the other is of a rather light horn-color with indistinct bands in the interior, and having no carina

on the lower whorl. On the columella there is a slight swelling. The aperture is about one-half the length of the shell.— *Lea.*

Anculotus nigrescens.—Shell subconical, truncated or much eroded at the apex; superior whorl hardly convex; body-whorl elongated, contracted above on the labrum; columella flattened, obtusely rounded at the base; aperture obovate, rather more than half the length of

Fig. 759.   Fig. 760.   Fig. 763.   Fig. 761.   Fig. 762.   Fig. 764.   Fig.765.

the shell; epidermis blackish; within dark purple. I am indebted to Mr. Hyde for this shell; he informs me it inhabits rivers in Maryland. —*Conrad.*

The cut (fig. 762) is from a type specimen.

Anculosa trivittata.—Shell elliptical; whorls about five, convex; suture impressed; spire short, often eroded, and about the length of the aperture; inner lip arcuated, with a callus; aperture oval, rounded beneath, acute above; color dark olive, with three dark purple, revolving lines on the carina, the central band very narrow.

Length of shell, ·5 of an inch. Length of aperture, ·25 of an inch.

Observations.—These species were obtained from Cranes-
port, in company with the preceding. In some, the bands are obscure or wanting. It appears to be closely allied to *A. melanoides* of Conrad, but is distinguished by the greater number of its volutions.— *DeKay.*

Fig. 766.

The above figures will suffice to show the mutation of form from the carinate varieties, through *trivittata*, DeKay, and *nigrescens*, Conrad, into the small shells with a toothed columella.

The following is Mr. Conrad's description of

Anculotus monodontoides.— Shell subglobose; body-whorl ventricose, not abruptly rounded above; apex eroded; columella with a
Fig. 768. Fig. 767. large, pyramidal tooth at the base; epidermis horn-
colored, with obscure bands; aperture effuse.

Habitat.—Inhabits streams in Virginia; Mr. Hyde.

Observations.—I received a specimen of this curious species from Prof. Green of Jefferson college.—*Conrad.* Fig.770. Fig. 769

Figure 770 is from a type specimen; figure 768, light green in color and a much thinner shell, was
collected by me at Richmond, Virginia.   Entirely identical

with *monodontoides* is *A. dentatus*, Couthuoy, a description of which follows :—

Anculotus dentatus.—Animal much like that of *Melania;* foot broad, short, rounded and thick; body and head black, the latter suborbicular, terminating in a short, proboscidiform mouth, and furnished with two short, rather stout and pointed tentacula, black posteriorly and with faint, grayish, transverse bands on their anterior side; eyes minute, situated on a slight enlargement of the tentacula near their external base.

Operculum elongated, unguiform, thick, corneous, blackish or brown, opaque; spire terminal, increment, coarse and apparent.

Shell rounded or obtusely conical, subdiaphanous, very irregular in its conformation, frequently gibbous and distorted; the color varies from light olive-green to black, according to the age of the specimens; whorls five or six in number, the last constituting the greater portion of the shell, very much inflated and ventricose, and sometimes ornamented with two or three dark brown, transverse bands; spire obtuse, always considerably eroded, unless in very young shells; incremental striæ oblique, in some Fig. 772. Fig. 771.
individuals barely apparent, and in others forming strong ridges on the last whorl; aperture rounded, effuse at the base; right lip thin, sharp and broadly everted; columella dark brown or purple, flattened,
strongly arcuated, with a dentiform projection near the base, which forms a subangular sinus or indentation below it. ᐧ Adjoining the columella is a strongly marked lacuna or fossa, most conspicuous in very old shells, but apparent in every stage of growth, and extending from the base of the shell to the centre of the lower whorl. There is no umbilicus, properly speaking, that region being consolidated by the columella. The internal color is chiefly greenish or brownish, with occasional shades of yellowish-white in old shells.

Habitat.— Inhabits the rapids of the Potomac River, Virginia.

Height, ten-fortieths; diameter of last whorl, eleven-fortieths inch.

Observations.— This shell at first sight might be taken for *Anculotus monodontoides*, Conrad, of Alabama, but may be distinguished from it by the peculiar flattening of the columella, which is deep purple or brown instead of white, and the remarkable fossa in the umbilical region. In that species, moreover, the tooth is situated on the middle of the columella and resembles a plait or fold at that part whereas in ours it is formed by an oblique, inward projection of the columella near the base. The external conformation is exceedingly irregular, varying from subconical to globose, sometimes compressed on the back, at others strongly gibbous. The aperture is also frequently distorted. Young specimens are of a light olive-green color, while older ones are nearly black, and usually covered with an earthy coating. The lower whorl is invariably marked at its base by a broad, dark brown band, and has frequently one on the middle and one on the superior portion. Some of the varieties of this shell, when undistorted, have so great an external resemblance to some of the varieties of *Turbo palliatus*, Say, that a figure of one might answer very well for both. It was found in abundance on the rocks, at the rapids, about a mile above the falls of the river Potomac, apparently delighting in situations where one would imagine it difficult for it to adhere. The only shells found in company with it were *Melania Virginica*, Say, and *Anculotus nigrescens*, Conrad, which latter was in great abundance and variety of form. Some of its less angular varieties closely approached *A. dentatus* in their general appearance, but were easily distinguished by the form of the aperture and the absence of the columellar lacuna.—*Couthuoy.*

No. 772 is a copy of the original figure, and No. 771 is one selected from a number of Maryland specimens kindly loaned to me for examination by Mr. Anthony.

The following description of *Anculosa dentata* was published by Mr. Lea in·the Phil. Proc., but suppressed in Philos. Trans., probably because it was discovered to be a synonyme.

Anculosa dentata.— Shell subglobose, thick, blackish; spire short, obtuse; sutures impressed; whorls convex; aperture large, subrotund; columella thickened, dentate.

Habitat.—Vicinity of Richmond, Virginia; J. A. Warder, M.D.

The following is the only description of *Anculosa affinis*,
Fig. 773. Haldeman. Its claim to specific rank was yielded by

that gentleman, probably, for otherwise he would have published a diagnosis for it.

Anculosa (Mudalia) affinis.— I propose this name for a shell allied to *Paludina dissimilis*, Say; but which differs from it in having a slight tooth upon the columella.

Habitat.—Ohio; Mrs. Say.

The following opinions have been advanced concerning the synonymy of this species :—

Professor Haldeman, Mr. Reeve and Dr. Brot concur in considering *nigrescens* a synonyme of *dissimilis*. The first and last named gentlemen write *carinata*, Lea, and *Nickliniána*, Lea?, synonymes of *A. variabilis*, Lea (*Nickliniana* is a true *Goniobasis*, G. W. T., Jr.). Messrs. Jay, Haldeman and Brot make *dentatus*, Couthuoy, a synonyme of *monodontoides*, Con. Professor Haldeman makes *dentata*, Lea, to be the same as Couthuoy's species.

8. A. dilatata, CONRAD.

Melania dilatata, CONRAD, New Fresh-Water Shells, Appendix, p. 6, t. 9, f. 5, 1834.
Anculotus dilatatus, Conrad, REEVE, Monog. Anculotus, t. 5, f. 38.
Anculosa dilatata, CONRAD, Am. Jour. Sci., n. s., i, p. 407. HANLEY, Conch. Misc., t. 5, f. 38.
Mudalia dilatata, Conrad, CHENU, Manuel de Conchyl., i, f. 2043-5.
Nitocris dilatata, Conrad, ADAMS, Genera, i, p. 308.
Leptoxis dilatata, Conrad, HALDEMAN, Monog. Leptoxis, p. 4, t. 4, f. 111-120. BINNEY, Check List, No. 351. BROT, List, p. 24. CHENU, Manuel. i, f. 2043-5.
Melania Rogersii, CONRAD, New Fresh-Water Shells, Appendix, p. 7, t. 9. f. 7, 1834. JAY, Cat., 4th edit., p. 274.
Anculotus Rogersii, Conrad, REEVE, Monog. Anculotus, t. 4, f. 28.
Leptoxis Rogersii, Conrad, BINNEY, Check List, No. 382.
Nitocris Rogersii, Conrad, ADAMS, Genera, i, p. 308.
Anculotus carinatus, ANTHONY, Bost. Jour. Nat. Hist., iii, pt. 3, p. 394, t. 3, f. 5, July, 1840. REEVE, Monog. Anculotus, t. 5, f. 42.
Leptoxis carinata, Anthony, BINNEY, Check List, No. 342.
Anculotus Kirtlandianus, ANTHONY, Bost. Jour. Nat. Hist., iii, pt. 3, p. 295, t. 3, f. 4, July, 1840. JAY, Cat., 4th edit., p. 276. REEVE, Monog. Anculotus, t. 4, f. 29.
Anculosa Kirtlandiana, Haldeman, WHEATLEY, Cat. Shells U. S., p. 28.
Nitocris Kirtlandianus, Anthony, ADAMS, Genera, i. p. 308.
Melania inflata, LEA, Philos. Trans., vi, p. 17, t. 23, f. 98. Obs., ii, p. 17. WHEATLEY, Cat. Shells U. S., p. 25. BINNEY, Check List, No. 147. TROSCHEL, Archiv fur Naturgesch., ii, p. 226.
Nitocris inflatus, Lea, ADAMS, Genera, i, p. 308.
Leptoxis rapæformis, HALDEMAN, Monog. Leptoxis, p. 4, t. 4, f. 123. BROT, List, p. 25.

Fig. 776.　Fig. 775.　Fig. 774.

Melania dilatata.—Shell subovate, ventricose; spire conical; whorls convex; body-whorl angular in the middle; aperture subovate, half the length of the shell.

Habitat.—Inhabits rivers in Munroe County, Virginia; Mr. William B. Rogers.—*Conrad.*

A. Rogersii, Conrad, universally considered to be a young variety of the preceding is thus described :—

Melania Rogersii.— Shell subovate, with rather distant, prominent spiral lines; whorls convex; body-whorl ventricose; aperture subovate, half the length of the shell; columella obtusely angular at the base.

Fig. 777.

  Variety A. Destitute of revolving lines; whorls gibbous.

  Observations.— Inhabits with the preceding species. It was given me by Professor William B. Rogers, to whom I have dedicated the species. — *Conrad.*

A. carinatus, Anthony, *A. Kirtlandianus,* Anthony, and *A. inflata,* Lea, are all variations of this protean species. Their descriptions follow :—

Anculotus carinatus.— Shell oblong; spire as long as the aperture; volutions four, convex; suture not remarkable; body-whorl angularly ventricose; color olivaceous; from 4–5 elevated, black carinæ, commencing at the upper part of the aperture, traverse the body-whorl; aperture within bluish-white and translucent, the carinæ being very apparent through it.

Fig. 777*a.*

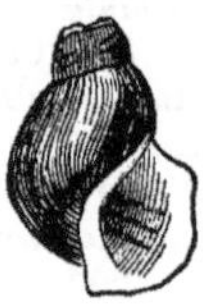

  Extreme length of shell, ¾; breadth, ½ of an inch.

  Observations.— For this beautiful species of *Anculotus* I am indebted to Mrs. Say, who found it at the Falls of the Kanawha, a few weeks since, and kindly presented me with specimens of it for description.— *Anthony.*

Anculotus Kirtlandianus.— Shell turreted, with four convex whorls; spire truncated, the truncation generally destroying one of the volutions; the body-whorl slightly ventricose; color dark olive; aperture subovate; base attenuated, within clouded purple and banded.

Fig. 777*b.* Fig. 777*c.*

  Length of shell, ¾; breadth, ½ of an inch.

  Observations.— Another species which I owe to the kindness of Mrs. Say. It is found in the same situations as *A. carinatus* (Falls of the Kanawha); it resembles very much a *Melania,* the spire being quite as much elevated as in most of the species of that genus; the young are very beautifully banded.— *Anthony.*

Melania inflata. — Shell conical, inflated, dark horn-color; apex obtuse; whorls five, rather convex; columella marked; outer lip spread out.

Habitat.—Indian Creek, Virginia, west of Alleghany Mountains.

Diameter, ·4; length, ·6 of an inch.

Observations.—I am indebted to Mr. Nicklin for this new species, it having been found by him in Indian Creek, between the  Fig. 778. Salt and Red Sulphur Springs. The sinus is so small that at first view it may easily escape observation. The aperture is large, and in this it has some resemblance to a *Paludina.* Near the base of the columella a purple spot may be usually observed. It resembles most in outline the *M. tuberculata* (nobis), but differs in not being angulated, and in being entirely without tubercles. In color it differs entirely. Some individuals have three colored purple bands in the interior, while others are devoid of them.— *Lea.*

Figure 778 is a copy of that of Mr. Lea's. No. 779 is called by Professor Haldeman variety *striata;* No. 780 he names variety *sinuata;* No. 781 variety *iostoma;* No. 782 variety *glauca;* No. 783 variety *solidula* and No. 784 variety *rapæformis.* The last two Prof. Haldeman considers with doubt as new species, and they have been so quoted since by other authors. Messrs. Haldeman and Brot quote the following as synonymes of *dilatata,* Conrad:— *Rogersii,* Conrad; *inflata,* Lea; *Kirtlandianus,* Anthony; *carinatus,* Anthony.

Fig. 781. Fig. 780. Fig. 779.

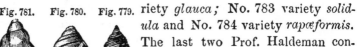

Fig. 783. Fig. 782.

Mr. Lea writes as follows in the Philosophical Transactions, viii, p. 171:—

Fig. 785. Fig. 784.

"Within a few days I have observed in the Boston Journal of Natural History, vol. 3, No. 3, descriptions of two new species of *Anculosa* by Mr. Anthony, *Anculotus carinatus* and *Anculotus Kirtlandianus,* both from the Falls of the Kanawha. Judging from the description and figures, I am led to the conclusion, that both these are identical with *M. inflata,* and from the great variety of this protean species, I am not surprised at its being mistaken. The peculiar character, however, of the angle and channel of the base in this species, is evident throughout. I am not

aware of the animal having been yet observed; when examined it
may prove to be a true *Anculosa*. If so, the synonymy will stand
thus :—

> *Anculosa inflata*, Lea.
> *Melania dilatata*, Conrad.*
> *Melania Rogersii*, Conrad.
> *Anculotus carinatus*, Anthony.
> *Anculotus Kirtlandianus*, Anthony.

The following is from Proc. Bost. Nat. Hist. Soc., i, p. 5,
Feb. 3, 1841.

"The president read a letter from S. J. Whittemore, in which was
an extract from a letter from J. G. Anthony, Esq., of Cincinnati, stat-

Fig. 786. . Fig. 787.

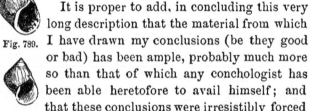

ing that the *Anculotus Kirtlandianus* of Anthony
was identical with the *Melania Rogersii* of Conrad."

It is proper to add, in concluding this very
long description that the material from which

Fig. 788.   Fig. 789.

I have drawn my conclusions (be they good
or bad) has been ample, probably much more
so than that of which any conchologist has
been able heretofore to avail himself; and
that these conclusions were irresistibly forced
upon me against my preconceived convictions.   Should any
conchologist differ from me, the value of this article will still
be scarcely impaired, for I have been careful, particularly with
that object in view, so to arrange the

Fig. 790. Fig. 791. Fig.792.

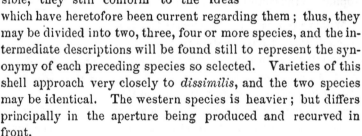

order of the descriptions that, whilst
they exhibit the natural sequence of the
species and its varieties as far as possible, they still conform to the ideas
which have heretofore been current regarding them ; thus, they
may be divided into two, three, four or more species, and the intermediate descriptions will be found still to represent the synonymy of each preceding species so selected.   Varieties of this
shell approach very closely to *dissimilis*, and the two species
may be identical.   The western species is heavier; but differs
principally in the aperture being produced and recurved in
front.

* This description was published prior to Mr. Lea's, and should therefore head the list,
unless it should be degraded to a synonyme, because published as a *Melania* instead of an
Anculosa.

9. A. corpulenta, ANTHONY.

Anculosa corpulenta, ANTHONY, Proc. Acad. Nat. Sci., p. 68. Feb., 1860.
Anculotus corpulentus, Anthony, REEVE, Monog. Anculotus, t. 1, f. 9.
Leptoxis corpulenta, Anthony, BINNEY, Check List, No. 348. BROT, List, p. 24.

Description.—Shell ovate, or broad ovate, smooth, thick; spire rather elevated; composed of 4-6 subconvex whorls; suture decidedly impressed; aperture very broad, ovate, ample, banded inside; columella well rounded, slightly covered with white callus, and with a slight indication of sinus at base.

Habitat.— North Carolina.

Observations.— Cannot well be confounded with any of its congeners; it is unusually elevated for an *Anculosa*, resembling more a *Paludina* in that respect; the whorls are regularly but not abruptly shouldered, and are often excavated with a narrow channel in the middle; striæ and even indistinct carinæ are often visible, but are not a constant character; the bands within the aperture are not always well defined and are sometimes wanting altogether; when present they are generally five in number, and are arrested by a narrow white space at the outer lip; body-whorl often subangulated. Occurs in Dan River, North Carolina, in company with *Anculosa canalifera* (nobis), and appears to be common. Several hundred specimens of various ages are now before me.—*Anthony.*

Fig. 793.

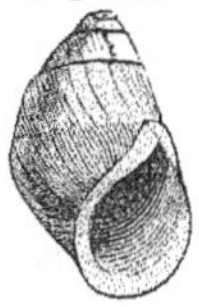

A very distinct and beautiful species most nearly allied to heavy, obsoletely angulated forms of *dissimilis*. The figure is from the type specimen. Other specimens before me are of somewhat larger size and more distinctly angulated.

10. A. melanoides, CONRAD.

Anculotus melanoides, CONRAD, New Fresh-Water Shells, p. 64, t. 8, f. 10, 1834.
DEKAY, Moll. N. Y., p. 102. WHEATLEY, Cat. Shells U.S., p. 26. REEVE, Monog. Anculotus, t. 6, f. 48. MÜLLER, Synopsis, p. 42, 1836.
Leptoxis melanoides, Conrad, HALDEMAN, Monog. Leptoxis, p. 5, t. 5, f. 145. 146. BINNEY, Check List, No. 369.
Nitocris melanoides, Conrad, ADAMS, Genera, i, p. 308.
Anculosa (Mudalia) turgida, HALDEMAN, Supplement to No. 1, Monog. Limniades, Oct., 1840. WHEATLEY, Cat. Shells U. S., p. 28.
Leptoxis turgida, HALDEMAN, Monog. Leptoxis, p. 5, t. 5, f. 151. BINNEY, Check List, No. 393. BROT, List, p. 26.
Leptoxis turgida, Hald., ADAMS, Genera, i, p. 307.

Description.— Shell conical, with three entire volutions; apex eroded; whorls flattened, rounded only at the sutures; lines of growth prominent; body-whorl abruptly rounded; epidermis black-

Fig. 794.   Fig. 795.   ish, obscurely banded; aperture elliptical, about half the length of the shell.

Habitat.— Inhabits rivers in North Alabama.

Length, ½ of an inch.—*Conrad.*

Figure 794 is from a type specimen in the possession of Mr. Anthony. The shell has been cleaned, exhibiting a light green epidermis. *Leptoxis turgida* of Haldeman is identical with this species. The following is the description :—

Leptoxis turgida.— Shell composed of four flat turns; spire and aperture of equal length; posterior (upper) end of the   Fig. 796. labrum advanced upon the body-whorl which swells into the aperture at this point; color light green, translucent.

Habitat.— Alabama.

Length, ½ of an inch.

Observations.— Resembles somewhat the *Paludina (Mudalia) dissimilis,* Say.— *Haldeman.*

5. *Shell smooth, globose, or flattened above.*

1₁. A. trilineata, SAY.

Melania trilineata, SAY, New Harmony Dissem., No. 18, p. 227, Sept. 9, 1829. SAY'S
    Reprint, p. 19, 1840. BINNEY'S edition, p. 144. CATLOW, Conch. Nomenc., p. 189.
Anculosa trilineata, Say, DEKAY, Moll. N. Y., p. 100. WHEATLEY, Cat. Shells U. S.,
    p. 27. JAY, Cat. Shells, 3rd edit., p. 62.
Anculotus trilineatus, Say, JAY, Cat., 4th edit., p. 276. REEVE, Monog. Anculotus,
    t. 5, f. 41b.
Leptoxis trilineata, Say, HALDEMAN, Monog. Leptoxis, p. 5, t. 5, f. 134-144. BINNEY,
    Check List, No. 389. BROT, List, p. 26.

Variety.

Melania viridis, LEA, Philos. Proc. ii, p. 12, Feb. 1841. Philos. Trans., viii, p. 172.
    t. 5, f. 19. Obs., ii, p. 12. DEKAY, Moll. N. Y., p. 95. WHEATLEY, Cat. Shells
    U. S., p. 27. BINNEY, Check List, No. 292. CATLOW, Conch. Nomenc., p. 189.

Description.— Shell subglobose oval, yellowish, more or less tinged with brown; volutions about four, rounded, somewhat wrinkled; spire short, rather more than half the length of the aperture; suture not very deeply impressed; body-whorl with three brownish-black revolving lines, of which the two inferior ones are nearest together, the middle one widest, and the superior one placed nearest the suture and revolving on the spire; the middle one is concealed on the spire by the suture; aperture much dilated, ovate, acute above; labium a little flattened; labrum widely and regularly rounded, without any protrusion near the base; base slightly angulated, without any sinus or undulations; umbilicus none.

Habitat.— Inhabits Falls of the Ohio.

Length, less than ½ of an inch.

Variety A. Inferior band obsolete.

Variety B. Bands obsolete.

Observations.— This species is allied to the preceding (*M. isogona*), but is obviously distinct in its general appearance; the volu- Fig. 797 tions are destitute of a shoulder, and the aperture is ovate, acute above. It is a pretty shell, the bands being very conspicuous, strongly contrasting with the yellow general color, particularly in the young and half grown shell. I obtained about a dozen specimens on the rocky flats of the falls of the Ohio at the lower end of the island which is nearest to Louisville.—*Say.*

Melania viridis described by Mr. Lea is the same as Mr. Say's variety B of *trilineata* and does not exhibit distinctive characters amounting to specific weight. The following is the description :—

Melania viridis.— Shell smooth, subfusiform, rather thick, green; spire short, obtusely conical; sutures linear; whorls five, somewhat convex; aperture ovate, rather large, white.

Fig 798.

Habitat.— Vicinity of Cincinnati, Ohio.

Diameter, ·27; length, ·32 of an inch.

Observations.— Inhabits with the *M. occidentalis*, herein described, and resembles it. It is a smaller species, has one more whorl, has a higher spire, and among nine individuals before me I see no indications of transverse striæ. The aperture is rather more than half the length of the shell.—*Lea.*

In treating *viridis* as a synonyme of *trilineata* I follow the opinions expressed with reference to it by Messrs. Haldeman, Brot, Binney and Anthony. The two former gentlemen together with Dr. Jay, unite in considering *costatus*, Anthony, and *occidentalis*, Lea, as synonymes also. In this opinion I cannot coincide; the two species appear to me to be well separated by the costate surface of Mr. Anthony's species and the uniformly smooth surface of *trilineatus*. Mr. Reeve's figure of *trilineatus* is very poor; the bands are so represented as to appear like ribs. It is by no means certain that *trilineata* is an *Anculosa*. Its small size and smooth surface and general outline suggest its pertinence to the *Amnicolidœ*, to which family several small species, hitherto considered to be *Anculosœ* have been recently removed. It differs from all the *Amnicolidœ*, however, in possessing colored bands. The figure of *trilineata* is from Mr. Say's type in possession of Mr. Anthony. *Viridis* is a copy of Mr. Lea's excellent figure.

12. A. subglobosa, Say.

Melania subglobosa, SAY, Journ. Acad. Nat. Sci., v, p. 128, Sept., 1825. BINNEY'S edit., p. 116. BINNEY, Check List, No. 254. CATLOW, Conch. Nomenc., p. 188. JAY, Cat., 3rd edit., p. 62.
Anculotus subglobosus, Say, CONRAD, New Fresh-Water Shells, p. 60, t. 8, f. 14. DEKAY, Moll. N. Y., p. 103. REEVE, Monog. Anculotus, t. 1, f. 10. JAY, Cat., 4th edit., p. 276.
Anculosa subglobosa, Say, WHEATLEY, Cat. Shells U. S., p. 28.
Leptoxis subglobosa, Say, HALDEMAN, Monog., p. 3, t. 2, f. 40-58. CHENU, Manuel de Conchyl., i, f. 2040-42. BINNEY, Check List, No. 287. BROT, List, p. 25. ADAMS, Genera, i, p. 307.
Melania su globosa, Lea, TROOST, Cat. Shells Tenn., p. 42.
Anculosa gibbosa, LEA, Philos. Proc., ii, p. 34, April, 1841. Philos. Trans., ix, p. 15. Obs., iv, p. 15. WHEATLEY, Cat. Shells U. S., p. 28.
Anculotus gibbosus, Lea, REEVE, Monog. Anculotus, t. 1, f. 3.
Leptoxis gibbosa, Lea, BINNEY, Check List, No. 361. BROT, List, p. 25. ADAMS, Genera, i, p. 307.
Melania globula, LEA, Philos. Proc., ii, p. 12, Feb., 1841. Philos. Trans., viii, p. 174, t. 5, f. 22. Obs., iii, p. 12. DEKAY, Moll. N. Y., p. 95. TROOST, Cat. Shells Tennessee. WHEATLEY, Cat. Shells U. S., p. 25. BINNEY, Check List, No. 126. CATLOW, Conch. Nomenc., p. 187.
Leptoxis globula, Lea, ADAMS, Genera, i, p. 307.

Variety.

Anculosa tintinnabulum, LEA, Philos. Proc., iv, p. 167, Aug., 1845. Philos. Trans., x, p. 67, t. 9, f. 51. Obs., iv, p. 67.
Anculotus tintinnabulum, Lea, REEVE, Monog. Anculotus, t. 2, f. 13.
Leptoxis tintinnabulum, Lea, ADAMS, Genera, i, p. 307.
Melania virgata, LEA, Philos. Proc., ii, p. 13, Feb., 1841. Philos. Trans., viii, p. 175, t. 5, f. 25. Obs., iii, p. 13. DEKAY, Moll. N. Y., p. 95. TROOST, Cat. Shells Tennessee. BINNEY, Check List, No. 290. CATLOW, Conch. Nomenc., p. 189. WHEATLEY, Cat. Shells U. S., p. 27.
Leptoxis virgata, Lea, ADAMS, Genera, i, p. 307.

Description. — Shell subglobose, brownish horn-color; spire but little elevated, not half the length of the aperture; volutions about four; aperture rounded, nearly as broad as long; within more or less tinged with dull red; labium a little flattened.

Length, three-fifths; greatest breadth, eleven-twentieths of an inch.

Observations.— Professor Vanuxem found this curious shell in the north fork of the Holston River, Virginia, where they are extremely abundant. In the old shells the surface, and particularly that of the spire, is considerably corroded, presenting the appearance of having received a fortuitous deposition of calcareous matter. This corrosion, however, does not extend to the destruction of any of the whorls, as is the case with many shells, but its effects seem to be confined to the exterior. It is a second species of my proposed genus *Anculotus*. All the striæ of the operculum are concentric to the superior angle.—*Say.*

Fig. 799. Fig. 800.

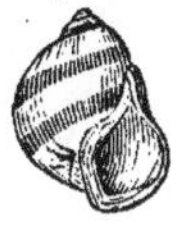

This species, which inhabits an extensive range in Virginia, Tennessee, Alabama and north Georgia, is somewhat variable in outline and ornamentation. The southwest Virginia specimens, which are unicolored, may retain the name of *subglobosa*, as the typical shells, and the young of these = *globula*, Lea, a description of which species follows:—

Melania globula.— Shell smooth, subglobose, dark brown, banded; spire short; sutures impressed; whorls four, rather convex; aperture large, nearly round, within bluish.

Habitat.— Tennessee; Dr. Troost.

Diameter, ·22; length, ·25 of an inch.

Observations.— This is a small, globose species, with two very broad bands, one immediately over and the other below the middle Fig. 801. of the body-whorl. The columella is white, inclined to a rusty hue. The interior of the base is reddish. Some of the specimens are small, and present a variety in which the columella is redder, and the epidermis more yellow, with the same distinctive bands. The aperture is nearly two-thirds the length of the shell.—*Lea.*

The following is the description of

Anculosa gibbosa.— Shell subglobose, gibbous, thick, nearly black, thickly striate; spire short; sutures impressed; whorls rather flattened; aperture subquadrangular, flesh-colored or whitish.

Habitat.—Tennessee.

Diameter, ·50; length, ·68 of an inch.

Observations.— This species is about the size of *Anculosa subglobosa*, Say. It is not so regularly rounded, being flattened on the

Fig. 804.   Fig. 803.   Fig. 802.

upper part of the whorl. The striæ are minute, and seem to be formed by the lines of growth. There is quite a callus on the superior part of the columella, the middle part being deeply impressed. The number of whorls could not be ascertained from my specimens, all of them being more or less eroded.—*Lea.*

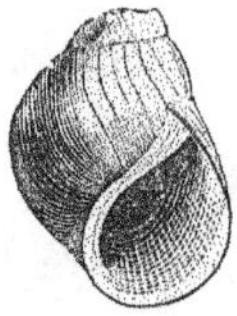

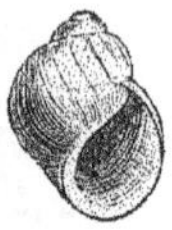

Anculosa tintinnabulum, Lea, is a much stronger variety than the last and may for convenience retain its name, under that of *subglobosa*. It is characterized by the whorls becoming wider, heavier, flattened above the middle and having two broad, dark bands, or maculate with brown. The description of this shell is appended, and also that of its young state, called by Mr. Lea *Melania virgata*.

Anculosa tintinnabulum.— Shell smooth, obtusely conical, bell-shaped, banded, very thick, yellow; spire short; sutures impressed; whorls five, impressed; aperture rather large, round; columella very thick, callous above.

Fig. 805. Fig. 806.  Fig. 807.

Habitat.— Tennessee: Tuscaloosa, Ala.

Diameter, ·48; length, ·70 of an inch.

Observations.— The peculiar, constricted lower whorl, giving a campanulate form to this shell, will distinguish it at once from other species. Six specimens before me are all yellow, with broad, brown bands. A single specimen is perfect enough in the spire to make out five whorls. Two of the specimens are white on the columella, and four are tinted with brown. The outline is very remarkable, in its campanulate form. The mouth, in the perfect specimen, is about two-thirds the length of the shell.—*Lea.*

Melania virgata.— Shell smooth, rounded, rather thin, yellow, double-banded, shining; spire short; sutures linear; whorls convex; aperture large, elliptical, whitish.

Fig. 808.

Habitat.— Tennessee.

Diameter, ·20; length, ·30 of an inch.

Observations.— A single specimen of this small species was sent to me by Dr. Troost. It seems to be mature, and is remarkable for the two broad bands which nearly cover the whorls. The aperture is about half the length of the shell.—*Lea*.

Mr. Reeve's figure of *subglobosa* represents a shell very closely approaching the variety *tintinnabulum*, while his figures of *tintinnabulum* represent respectively, fig. 13*a*, variety *gibbosa*, banded ; fig. 13*b*, probably a young *Leptoxis crassus* of Haldeman.

Professor Haldeman was the first naturalist who detected the specific identity of the shells I have grouped together above, and other gentlemen have since adopted his opinion regarding them.

13. A. prærosa, SAY.

Melania prærosa, SAY, Jour. Acad. Nat. Sci., ii, p. 177, Jan., 1824. BINNEY'S edit. p. 70. CATLOW, Conch. Nomenc., p. 188. SOWERBY, Conch. Man., f. 314.
Anculotus prærosus, Say, CONRAD, New Fresh-Water Shells, p. 59, t. 8, f. 13. JAY, Cat., 4th edit., p. 276. REEVE, Monog. Anculotus, t. 2 f. 15, 16.
Anculotus præmorsa, Say, WOODWARD, Manuel, t. 8, f. 28.
Anculosa prærosa, Say, RAVENEL, Cat., p. 11. WHEATLEY, Cat. Shells U. S., p. 28. ANTHONY, List, 1st and 2d edits. KIRTLAND, Rep. Zool. Ohio, p. 174. DEKAY, Moll. N. Y., p. 103.
Leptoxis prærosa, Say, HALDEMAN, Monog. Lept., p. 2, t. 1, f. 1–18. CHENU, Manuel, i, f. 2030–34. BINNEY, Check List, No. 380. BROT, List, p. 25. ADAMS, Genera, i, p. 307. Morch, YOLDI, Cat., p. 56.
Melania angulosa, MENKE, Syn. Meth., 1st edit., p. 81, 1828. 2d edit., p. 135, 1830. BINNEY, Check List, No. 15.
Melania cruentata, MENKE, Syn. Meth., 1st edit., p. 80, 1828. 2d edit., p. 134, 1830.
Melania ovularis, MENKE, Syn. Meth., 1st edit., p. 80. 2d edit., p. 134. BINNEY, Check List, No. 194.
Melanopsis neritiformis, DESHAYES, Encyc. Meth. Vers., ii, p. 438, No. 14. Anim. Sans Vert., 2d edit., viii, p. 492, 1838.
Lithasia neritiformis, Deshayes, ADAMS, Genera, i, p. 308.
Anculotus augulatus, CONRAD, New Fresh-Water Shells, p. 60, t. 8, f. 15, 1834. DEKAY, Moll. N. Y., p. 102. WHEATLEY, Cat. Shells U. S., p. 27. REEVE, Monog. Anculotus, t. 6, f. 51. JAY, Cat. Shells, 4th edit., p. 276. MÜLLER, Synopsis, p. 40, 1836.
Leptoxis angulata, Conrad, BINNEY, Check List, No. 340. ADAMS, Genera, i, p. 307.
Melania Cincinnatiensis, LEA, Philos. Proc., i, p. 66, Dec., 1838. Philos. Trans., viii, p. 190, t. 6, f 58. Obs., iii, p. 28. JAY, Cat., 4th edit., p. 273. CATLOW, Conch. Nomenc., p. 186.
Anculotus Cincinnatiensis, Lea, DEKAY, Moll. N. Y., p. 95. TROOST, Cat. Shells Tennessee.
Leptoxis Cincinnatiensis, Lea, BINNEY, Check List, No. 346.

Description.— Shell subglobular, oval, horn-color ; volutions three or four, wrinkled across ; spire very short, much eroded in the old shell, so much so as to be sometimes not prominent above the body-whorl; body-whorl large, ventricose, with a very obtuse, slightly impressed, revolving band; aperture suboval, above acute and effuse; within on the side of the exterior lip about four revolving, purplish lines, sometimes dotted, sometimes obsolete or wanting; labium thickened, particularly at the superior termination near the angle, and tinged with purplish; base of the columella somewhat elongated and incurved, meeting the exterior lip at an angle.

Habitat.— Inhabits Ohio River.

Length, about four-fifths of an inch.

Observations.—Found in plenty at the falls of the Ohio. The spire is remarkably curious in the older shells, and the penultimate whorl, between the aperture and the spire, is also remarkably eroded in many older shells. The spire in the young shell is entire, and but

Fig. 809.   Fig. 810.   Fig. 811.

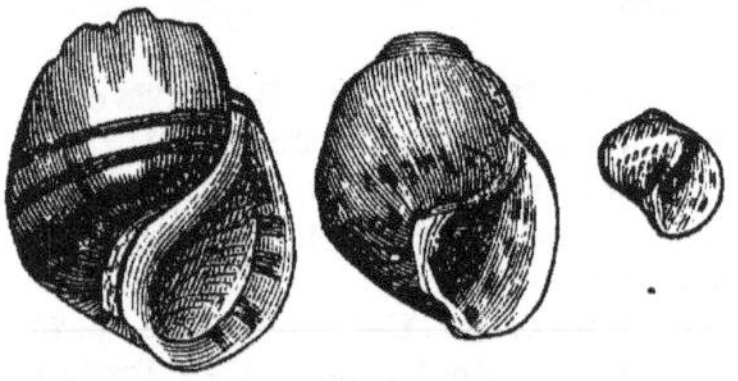

little prominent, though acute, and the bands are distinct on the exterior of the shell. This shell does not seem to correspond with the genus to which I have for the present referred it, and owing to the configuration of the base of the columella, if it is not a *Melanopsis*, it is probable its station will be between the genera *Melania* and *Acathina*. I propose for it the generic name of *Anculosa.—Say*.

The various species described by Menke and Deshayes all appear to be synonymes of *prærosa* judging from the descriptions, translations of which are here given. Prof. Haldeman and Mr. Anthony both agree with me in this opinion. *Melania Cincinnatiensis*, Lea, is only a quite young *prærosa*, as is proved by the extensive suite of specimens before me, for which I am indebted to Mr. Anthony. *Angulatus*, Conrad, represents, as Professor Haldeman truly remarks, a half grown shell in which the carina still lingers. This variety is found only in Alabama. The species is very common, and ranges through Ohio, Indiana, Kentucky, Tennessee, northern Georgia and Alabama.

Melania angulosa.— Shell ovate, truncated, perforated, variable, striate, greenish-brown; whorls five, the last obsoletely angulated above; columella callous, violaceous; lip acute, produced against the columella above.

Habitat.— Ohio River near Cincinnati; Bescke.

Longitude, 8; latitude, 6½ lin.

Melania cruentata.— Shell subglobose, acute at the apex, variable, striated, green, maculate seriately, conspicuously at the ovate oblique aperture, banded with blackish-purple; columella with a reddish callus; lip simple, produced above.

Habitat.— Ohio River near Cincinnati; Bescke.

Longitude, 5; latitude, 4½ lin.—*Menke.*

Melania ovularis. — Shell ovately conoidal, variable, substriate, rather shining greenish, becoming brownish-red, with apex truncate with age; aperture ovate; columella subcallous above; lip rounded above.

Habitat.— Ohio River near Cincinnati.

Longitude, 1 poll.; latitude, 7 lin.

Melanopsis neritiformis.— Shell globose, neritiform; apex very obtuse, reddish-black, smooth; aperture ovately semi-lunar; base scarcely emarginate; columella contorted, callous above, depressed in the middle; outer lip doubly sinuated.

Habitat.— The Ohio and Wabash.—*Deshayes.*

Anculotus angulatus.— Shell subglobose; body-whorl ventricose, contracted above, biangulated; spire very short; volutions carinated at the suture; color olivaceous, with about four series of dark, quadrangular spots on the body-whorl.

Fig. 812.

Observations.— Inhabits Flint River, Morgan County, Alabama, adhering to stones and is common.—*Conrad.*

Melania Cincinnatiensis. — Shell carinate, much depressed, below compressed, brown, three-banded, with two carinæ, Fig. 812*a.* pointed at the apex; whorls four; aperture rounded,

Habitat.— Near Cincinnati, Ohio.

Diameter, ·14; length, ·16 of an inch.

Observations.— This is a very minute species recently taken in the vicinity of Cincinnati, by my brother T. G. Lea. It is very remarkable for its roof-shaped spire, and two carinæ, which are colored. More recently found by Dr. Troost in the Holston, Tennessee.—*Lea.*

Leptoxis retusa, Rafinesque, has been doubtfully referred to this species by Prof. Haldeman.

14. A. tæniata, Conrad.

Anculotus tœniatus, Conrad. New Fresh-Water Shells, p. 63, 1834. DeKay, Moll.
N. Y., p. 103. Jay, Cat., 4th edit., p. 276. Reeve, Monog. Anculotus, t. 6, f. 50,
non t. 2, f. 15. Müller, Synopsis, p. 41, 1833.
Anculosa tœniata, Conrad, Wheatley, Cat. Shells U. S., p. 28.
Leptoxis tœniata, Conrad, Haldeman, Monog. Leptoxis, t. 3, f. 71-73. Binney,
Check List, No. 388. Brot, List, p. 26. Adams, Genera, i, p. 307.
Anculosa Coosaensis, Lea. Proc. Acad. Nat. Sci., p. 54, 1861. Jour. Acad. Nat. Sci.,
v, pt. 3, p. 257, t. 30, f. 65, March, 1863. Obs., ix, p. 76.

Description.— Shell oval, or oblong, olivaceous, with dark green

Fig. 813.   Fig. 814.   spiral bands, four on the body-whorl; one whorl
of the spire not eroded, often longitudinally
produced.

Length, ¾ of an inch.

Observations.— Inhabits friable calcareous
banks of the Alabama River, at Claiborne. It
is a pretty species, remarkable for its dark bands, which resemble
those of *Melania olivula* (nobis) of the same locality.—*Conrad.*

This shell resembles *rubiginosa* and *prœrosa* and appears
to occupy a somewhat intermediate position between the two.
A number of specimens before me, from the Alabama and
Coosa Rivers, including author's examples from the former
stream, indicate the changes which age produces in this species. When half grown it appears to be identical with *Coosaensis,* Lea, whose description and figure are copied below.

Anculosa Coosaensis.—Shell smooth, obtusely conical, thick, dark
horn-color, very much banded; spire elevated, obtuse at the apex;
sutures very much impressed; whorls four, very much constricted
below the sutures, the last large; aperture rounded, white, much
banded within; columella thickened, incurved, dark purple;   Fig. 815.
outer lip acute and expanded.

Operculum rather large, elliptical, dark brown, with the
polar point close to the left edge towards the base.

Diameter, ·34; length, ·55 of an inch.

Observations.—This species is more nearly allied to *tintinnabulum*
(nobis) than any other. It differs in being more elongate, having a
higher spire, having a less dilate aperture, and in usually having four
bands, the *tintinnabulum* usually having two bands, or being without
any. In two of the *Coosaensis,* out of six specimens before me, the

bands are interrupted, changing them to rows of square maculations. Some of the specimens are slightly umbilicate. The aperture is rather more than half the length of the shell.—*Lea.*

15. A. Troostiana, LEA.

Anculosa Troostiana, LEA, Philos. Proc., ii, p. 34. Philos. Trans., ix, p. 15. Obs., iv, p. 15. WHEATLEY, Cat. Shells U. S., p. 28.
Anculotus Troostianus, Lea, REEVE, Monog. Anculotus, t. 4, f. 30.
Leptoxis Troostiana, Lea, HALDEMAN, Monog. Leptoxis, p. 4, t. 3, f. 81. BINNEY, Check List, No. 391. BROT, List, p. 26. ADAMS, Genera, i, p. 307.

Description.— Shell ovately conical, thick, minutely rugose, dark brown; spire somewhat elevated; sutures rather impressed; whorls flattened; aperture rounded, within bluish; columella thick, white or flesh-colored.

Habitat.— Tennessee.

Diameter, ·50; length, ·60 of an inch.

Observations.— There are many specimens before me, all of which in form are unusually alike, for a species of *Anculosa.* It differs from other species which have come under my notice in its spire, which is quite elevated, giving it the aspect of the genus *Melania.* Small, irregular wrinkles, or granulations, may be observed over the whole surface in most specimens, and I believe this will generally be found to be more or less the case with most of the species.—*Lea.*

Fig. 816.

This is a small, ponderous, compact species, with a peculiarly dark epidermis, and is not likely to be confounded with any other. The figure is from one of Mr. Lea's types, which he kindly presented to me. *Anculosa Melanoides,* Conrad (*turgida,* Haldeman), differs from this in being narrower, and in the aperture being produced instead of rounded at the base.

16. A. pinguis, LEA.

Melania pinguis, LEA, Philos. Trans., x, p. 301, t. 30, f. 11 Obs., v, p. 57. BINNEY, Check List, No. 206. BROT, List, p. 40. REEVE, Monog. Melania, sp. 355.

Description.— Shell smooth, inflated, almost round, very thick, dark brown; spire very obtusely conical; sutures impressed; whorls a little convex; aperture very large and rounded, within either white or purple; columella incurved and thickened.

Habitat.— Lebanon, Wilson County, Tennessee.

Diameter, ·34; length, ·53 of an inch.

Observations.— I have three specimens before me from Mr. Safford; two of them are purple within and one white. None of them are

Fig. 817.

perfect on the apex, but I presume that the number of whorls must be five. One of the specimens has four. In outline it is very much like *M. inflata* (nobis), but it differs totally in the form of the columella. In that species the columella is twisted backwards, and makes an angular, oblique channel; in the *pinguis* it is regularly curved, with scarcely a perceptible indentation in place of a channel. The aperture is fully one-half the length of the shell.—*Lea.*

This shell is certainly an *Anculosa*, and is intermediate in its characters between *viridula*, Anthony, and *Kirtlandiana*, Lea. When well cleaned it frequently exhibits a greenish hue. It is rather a common species, and somewhat variable in its proportions, being sometimes prolate and in other specimens from the same locality oblate. Mr. Lea's figure is copied.

17. A. contorta, Lea.

Anculosa contorta, Lea, Proc. Acad. Nat. Sci., p. 187, 1860. Jour. Acad. Nat. Sci., v, pt. 3, p. 253, t. 35, f. 66, March, 1853. Obs., ix, p. 80.
Leptoxis contorta, Lea, Binney, Check List, No. 347. Brot, List, p. 24.

Description.— Shell smooth, ovately rounded, thick, yellowish horn-color; spire raised; sutures deeply impressed; whorls inflated, obscurely and transversely striate; aperture small, nearly round, constricted, yellowish-white within; columella thickened; outer lip acute and expanded.

Habitat.— Coosa River, at Wetumpka, Alabama.

Diameter, ·36; length, ·50 of an inch.

Observations.— A single specimen only was received from Dr. Showalter, which, being much eroded at the apex, prevents Fig. 818. a perfect description being made. But the number of whorls appears to be about four. The form is remarkable for an *Anculosa*, the outline presenting the appearance of a *Paludina;* but the callus on the columella and its whole massiveness forbid its being placed in that genus, while the regular rotundity of the whorls is similar in some measure to it. The aperture is about half the length of the shell.—*Lea.*

18. A. vittata, LEA.

Anculosa vittata, LEA, Proc. Acad. Nat. Sci., p. 188, 1860. Jour. Acad. Nat. Sci., v, pt. 3, p. 256, t. 35, f. 63, March, 1863. Obs., ix, p. 78.
Leptoxis vittata, Lea, BINNEY, Check List, No. 397. BROT, List, p. 26.

Description.— Shell smooth, subglobose, thick, yellowish, very much banded; spire obtuse; sutures impressed; whorls four, inflated, the last large and very much inflated, aperture round, very much contracted in the throat, banded within; columella very much thickened, flattened and purplish; outer lip sharp and expanded.

Habitat.— Coosa River, at Wetumpka, Alabama; E. R. Showalter, M.D.

Diameter, ·30; length, ·33 of an inch.

Observations.— This is a very remarkable species, perhaps more like a much-banded *prærosa*, Say, than any other. It entirely differs from that species in the columella being very thick and flattened, and which nearly fills up half the aperture. The banded varieties of *prærosa* differ very much from each other, while this seems Fig. 819. to be exceedingly regular. The five specimens before me have each four dark brown bands nearly covering up the yellow ground. The upper one is placed immediately under the suture, and is broader than the next two, which are approximate, revolving on the middle of the whorl. The fourth is larger again and revolves near to the base. I have no doubt, judging from the five individuals before me, that the characters of this little species will not be changeable, for they present no difference in phase whatever, although they are of several ages. The aperture is about two-thirds the length of the shell.—*Lea.*

19. A. planospira, ANTHONY.

Melania planospira, ANTHONY, Ann. Lyc. Nat. Hist. N. Y., vi, p. 123, t. 3, f. 24, March, 1854. BINNEY, Check List, No. 208. BROT, List, p. 40. HANLEY, Conch. Misc. Melania, t. 8, f. 67.
Anculotus planospira, Anthony, REEVE, Monog. Anculotus, t. 2 f. 11.

Description.— Shell short-ovate, smooth, rather thick, light horn-colored; body-whorl large, occupying nearly the entire volume of the shell; spire nearly flat, consisting of 4–5 perfectly plane whorls, scarcely elevated above the body-whorl; aperture long narrow ovate; columella rounded, ending in a slight sinus.

Habitat.—Tennessee.

Diameter, ·32 (8 millim.); length, ·50 of an inch (13 millim.). Length of aperture, ·36 (9 millim); breadth of aperture, ·18 ol an inch (4½ millim.).

Observations.— Cannot be confounded with any other species; its Fig. 820. remarkably flat whorls rising like steps, but little above each other, with a distinct and slightly raised rim around the periphery, will alone be sufficient to characterize the species. It seems more like an *Anculosa* in form, but is nevertheless a true *Melania.*   Two bands are visible on the body-whorl and also within the aperture.—*Anthony.*

There are, in Collection Smithsonian, several hundred specimens of this shell from Green River, Kentucky. It is allied to *prœrosa,* but appears distinct in the plane spire.

20. A. ampla, ANTHONY.

Anculosa ampla, ANTHONY. Ann. N. Y. Lyc. Nat. Hist., vi. p. 159, t. 5, f. 22, 23.
Leptoxis ampla, Anthony, BINNEY, Check List, No. 339.   BROT, List, p. 23.

Variety a.

Anculosa elegans, ANTHONY, Proc. Acad. Nat. Sci.. p. 69, Feb., 1860.
Anculotus elegans, Anthony, REEVE, Monog. Anculotus, t. 6, f. 49.
Leptoxis elegans. Anthony, BINNEY, Check List, No. 356.   BROT, List, p. 24.

Variety b.

Anculosa formosa, LEA, Proc. Acad. Nat. Sci., p. 187, 1860.   Jour. Acad. Nat. Sci.
   v. pt. 3. p. 254, March. 1863.   Obs., ix. p. 76.
Leptoxis formosa, Lea, BINNEY, Check List, No. 358.   BROT, List, p. 24.

Description.—Shell ovate-globose, olive-green, with four dark colored bands; spire very short, eroded; whorls 2-3, the last one shouldered, and peculiarly flattened just before Fig. 823. Fig. 822. Fig. 821 completion, and having the shoulder raised into a few very slightly defined tubercles, which in some individuals are hardly perceptible; suture deeply excavated; aperture ovate, showing the dark bands of the exterior; columella brown, excavated and flattened, without basal sinus, giving that portion of the shell much resemblance to a *Littorina.*

Habitat.—Alabama.

Diameter, ·42 (11 millim.); length, ·62 of an inch (16 millim.). Length of aperture, ·42 (11 millim.); breadth of aperture, 35 of an inch (9 millim.).—*Anthony.*

This very beautiful and rather abundant species, although differing very much from all others in its broad, flattened columella, covering the umbilicus completely, in the mouth being broadly inflated and rounded below, and in the whorls being rounded instead of slanting, varies much in itself; so much so in fact as to have caused marked specimens to be described as new species. Among these, the first is *A. elegans*, Anthony, of which the following is the description :—

Anculosa elegans.— Shell subglobose, smooth, thick; spire depressed, consisting of 3-4 flat whorls; color fine, glossy, dark yellow, ornamented with darker bands, of which five are on the body-whorl; aperture obliquely ovate and banded within; columella deeply curved, with a very callous deposit; sinus very small.

Fig. 823.   Fig. 824.

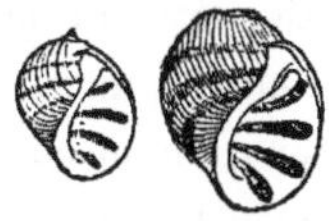

Habitat.— Alabama.

Observations.— A highly ornamental species, which cannot be compared with any other; its bands on a yellow ground render it very lively; it is heavier and smoother than *A. ampla* (nobis), not so broad in the aperture, and far more beautiful; neither is it so much shouldered as that species.—*Anthony.*

An examination of numerous specimens convinces me that *ampla* and *elegans* are only variations of one species. The figures given are all drawn from type specimens. The figure published by Mr. Lea of his *A. formosa*, which is herein copied, is a young *ampla* in form, only differing from the type specimens of that shell in the maculations, but I figure one of the adult shells mentioned by Mr. Lea in his description, which, on account of the very light color, impressed lines and maculations, may remain under the name of *formosa* as a variety.

The following is Mr. Lea's description :—

Anculosa formosa.— Shell smooth, globose, rather thin, semi-transparent, yellowish or saffron color, very much banded and maculate; spire depressed, scarcely conspicuous; sutures depressed; whorls three, the last large and very ventricose; aperture large, rounded; within pale saffron, with dark bands; columella thickened below and above and pale purple; outer lip sharp and very much expanded.

Operculum small, thin, with the polar point below the centre towards the inner edge.

Habitat.— Coosa River, Shelby County, Alabama.

Diameter. ·38: length, ·44 of an inch.

Observations.— I have three specimens before me of this very beautiful species. While it has much resemblance to the rounded varieties
of that protean species, *prærosa*, Say, it may be distinguished by its
being still more globose than its most globose varieties, by its del-
icacy, smoothness and brilliancy. Dr. S. says in his letter that he
thinks it decidedly distinct from all others he has out of many thousands, and that "it is more rotund than any other." The largest
specimen is four-fifths of an inch long, has four well-marked, continuous bands, with rows of maculation between them. The middle-
aged specimen is quite saffron, has the same number of bands with
the rows of maculation, but these bands are somewhat broken up,
and the maculations are not so regular. In the third, the youngest
one, the maculations are almost entirely absent. The largest specimen has a number of impressed, revolving lines, stronger towards

Fig. 826.   Fig. 827.

the base. The description of the operculum is
made from the middle-aged, the only one which
accompanied the three, and in the older ones this
may differ much. In all the specimens before me,
the upper whorls are almost entirely covered by
the last one. In the full grown one, the deep
color of the upper band on the inside continues over on to the callus
of the columella. Two other specimens accompanying these are
considered by Dr. S. to be the same. They are apparently about
half-grown. They differ slightly in form, and totally in the colored
bands, which in these specimens are replaced over the whole surface
with oblong maculations which, at the upper portion of the whorl,
run together, and form an irregular, longitudinal band between low
plications. I have been disposed to think that these two specimens
may prove to be varieties of *picta*, Conrad, which puts on so many
various kinds of bands, but the form is more globose than any *picta*
I have seen. The aperture is nearly the whole of the length of the
shell. Two adult specimens, received since the above was written,
have coarse, transverse striæ and one is without any colored bands,
the whole surface being a yellowish horn-color. The aperture is
about five-sixths the length of the shell.—*Lea.*

21. A. zebra, ANTHONY.

Anculosa zebra, ANTHONY, Proc. Acad. Nat. Sci.. p. 69, Feb., 1860.
Anculotus zebra, Anthony, REEVE, Monog., t. 6, f. 52.
Leptoxis zebra, Anthony, BINNEY, Check List, No. 398. BROT, List, p. 26.

Description.— Shell subglobose, smooth, moderately thick; spire obtusely elevated, but slightly decorticated, and composed of four convex whorls; sutures distinctly impressed; aperture broad, ovate, within bluish, with the epidermal colors seen faintly through; columella rounded, covered with callus, which is thickened at the upper part.

Habitat.— Alabama.

Observations.— This species presents an appearance not often seen in the genus, by its mottled, variegated epidermis; the general ground color is gamboge yellow, but it is varied by Fig. 828. blotches of very dark brown or reddish, often running into diagonal lines, which gives the shell a very lively and pleasant look. Only one other species is described as being similarly marked, viz.: *A. flammata,* Lea; that species I have never seen, but the description does not warrant me in considering the two identical. In old specimens the spire is often produced and somewhat nodulous, while the longitudinal bands become broken into irregular lines, so interrupted as to become scarcely more than quadrangular spots; it is one of our most beautiful species. About a dozen specimens are before me.— *Anthony.*

This species resembles *A. picta,* Conrad, particularly that variety described by Mr. Lea as *flammata,* so much, that its specific distinction may be considered doubtful.

22. A. picta, CONRAD.

Anculosa picta, CONRAD, Am. Jour. Sci., 1st ser., xxx, p. 342. t. 1, f. 15, Jan., 1834.
 WHEATLEY, Cat. Shells U. S., p. 28. HANLEY, Conch. Misc. Melania, t. 5, f. 39.
 MÜLLER, Synopsis, p. 39, 1836.
Anculotus pictus, CONRAD, New Fresh-Water Shells, p. 62. 1834. REEVE, Monog.
 Anculotus, t. 3, f. 26. JAY, Cat., 4th edit , p. 276. DEKAY. Moll. N. Y , p. 103.
Leptoxis picta, Conrad, HALDEMAN, Monog. Lept , t. 3, f. 74-80. BINNEY, Check
 List, No. 377. BROT, List, p. 25. ADAMS, Genera, i, p 307.
Anculosa Foremani, LEA, Philos. Proc., ii. p. 243, Dec., 1842. Philos. Trans., ix,
 p. 29. Obs., iv, p. 29. WHEATLEY, Cat. Shells U. S., p. 28.
Leptoxis Foremani, Lea, BINNEY, Check List, No. 359.
Anculosa flammata, LEA, Philos. Proc., ii. p. 243. Philos. Trans., ix, p. 20. Obs.,
 iv, p. 30.

Anculotus flammatus, Lea, REEVE, Monog. Anculotus, t. 3, f. 18.
Leptoxis flammata, Lea, BINNEY, Check List, No. 357. Conrad, ADAMS, Genera, i, p. 307.

Description.— Shell oval; spire short, convex; apex eroded; body-whorl slightly compressed in the middle; epidermis horn-colored,

Fig. 829.   Fig. 830.   with numerous series of small, angular spots; spots distinct within the labrum; aperture obovate; base regularly rounded.

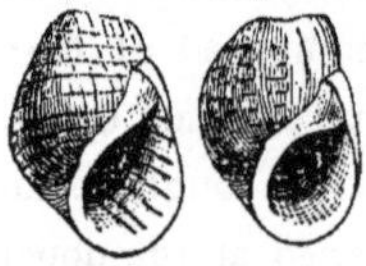

Habitat.—Inhabits pebbles on the bars in the Alabama River, near Claiborne.

Length, five-eighths of an inch.— *Conrad.*

Mr. Conrad's description applies only to a stunted or immature form of this species, which grows considerably larger and assumes some variety in marking. Mr. Lea's descriptions of *A. Foremani* and *A. flammata* are subjoined.

Anculosa Foremani.—Shell smooth, ovately gibbous, thick, yellow, transversely lined; spire very short; sutures impressed; whorls somewhat flattened; columella very thick; aperture rather large, elliptical, whitish.

Habitat.— Alabama.

Diameter, ·40; length, ·50 of an inch.

Observations. — Two of the three specimens under examination have very distinct, capillary, revolving, deep brown lines between the top of the aperture and the base. Above that the   Fig. 831.
space is nearly filled up with two indistinct, interrupted lines which give a clouded appearance to that portion of the shell. The third specimen is of a brighter yellow, with all the lines nearly obliterated. In form this species very closely resembles *A. flammata* herein described, but the capillary lines distinguish it at once, and the columella is thicker at the base. In all the three specimens a slight tinge of brown may be distinguished on the middle of the columella. I dedicate the species to Dr. Foreman, who kindly placed a specimen in my cabinet.—*Lea.*

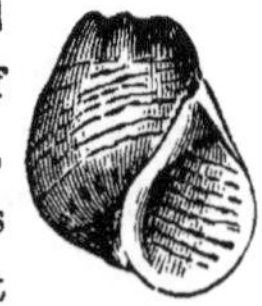

Anculosa flammata.—Shell smooth, ovately gibbous, thick, yellowish, obliquely flammulate; spire very short; sutures impressed; whorls somewhat flattened; columella very thick above; aperture rather large, elliptical, whitish.

Habitat.— Alabama.

Diameter, ·38; length, ·49 of an inch.

Observations.— A single specimen, broken on the outer lip, is before me. The middle of the whorl is slightly flattened. The spire is eroded, and little more than one whorl is presented. The epidermis on this part is nearly perfect, and exhibits a fine, yellow ground with thickly set, oblique, flammulate, brown bands. This species is very distinct from any I know, not being aware that flammulate bands have been before observed in any of this genus. In a single species of *Melania*, somewhat similar bands exist, the *M. breviformis* (Pareyss) from New Holland.— *Lea.*

Fig. 832.

A. picta attains a larger size than the specimens figured. The figure of *A. Foremani* is from a very good specimen named by Mr. Lea; *A. flammata* is drawn from Mr. Reeve's illustration of that shell. I have been doubtful whether or not to include *A. zebra*, Anthony, in the synonymy of this species, but as the shell is much more globose in form than *picta* with a shorter spire and larger aperture proportionally, I conclude to separate it, with, however, a doubt of its specific distinction.

23. A. ornata, ANTHONY.

Anculosa ornata, ANTHONY, Proc. Acad. Nat. Sci., p. 67, Feb., 1860.
Anculotus ornatus, Anthony, REEVE. Monog. Anculotus, t. 3, f. 24.
Leptoxis ornata, Anthony, BINNEY, Check List, No. 375.

Description.— Shell conic, rather thick, smooth; spire elevated, composed of about five convex whorls; suture distinct; color dark yellow, polished, with dark brown bands revolving around the shell; three bands visible on the body-whorl and only one upon the volutions of the spire; aperture ovate, livid and banded within; columella furnished with a callus, often tinted with rose color; sinus very small.

Fig. 833.

Habitat.— North Carolina.

Observations.— A fine species, so much elevated as readily to be taken for a *Melania;* the dark bands on a yellow ground give it a lively appearance; about one hundred specimens are before me, and present very little variation; the dark bands within the aperture are very conspicuous, one being near the upper angle, two others near each other, but widely separated from the first, and a fourth near the base of the shell; the middle bands are often confluent, and all of them are arrested by a broad area before they reach the outer edge.—*Anthony.*

The figure is from a type specimen. The body-whorl is slightly angulated in most of the specimens before me.

24. A. Lewisii, Lea.

Anculosa Lewisii, Lea, Proc. Acad. Nat. Sci., p. 51, 1861. Jour. Acad. Nat. Sci., v, pt. 3, p. 257, t. 35, f. 64, March, 1863. Obs., ix, p. 79.

Description.— Shell smooth, elliptical, rather thick, somewhat inflated, yellowish horn-color; spire obtuse, scarcely exserted, acuminate; sutures scarcely impressed; whorls five, the last very large; aperture large, regularly ovate, whitish within; columella incurved, a little thickened above and below; outer lip acute, somewhat expanded and slightly sinuous.

Operculum rather large, very dark brown, ovate, with the polar point very near the base on the left.

Habitat.— Tennessee; James Lewis, M.D.

Diameter, ·30; length, ·58 of an inch.

Observations.—Dr. Lewis sent me three specimens for examination;

Fig 834.

I presume all he had received from Tennessee. It is quite distinct from any *Anculosa* I have seen. It verges toward the genus *Lithasia* in some of its characters. It reminds one of *Melania obovata*, Say, which probably should be removed from that genus to this. The aperture is more rounded at the base than in that shell, and the spire is much more obtuse, giving the outline of the two shells a very different appearance. It reminds one of the genus *Chilina*, Gray, but cannot be mistaken for that genus. The last whorl is so large that it nearly covers up the spire and leaves only a small portion extruded. Two of the specimens exhibit near the apex quite a disposition in the young to be carinate. In an immature state, therefore, they would present quite a different appearance, as the shoulder would be quite square.— *Lea.*

25. A. squalida, Lea.

Anculosa squalida, Lea, Philos. Proc., iv, p. 167, Aug., 1845. Philos. Trans., x, p. 66, t. 9, f. 50. Obs , iv. p. 66.
Leptoxis squalida. Lea, Binney, Check List, No. 386. Brot, List, p. 25. Adams, Genera, i, p. 307.

Description.— Shell smooth, rounded or elliptical, very thick, dark horn-color; spire obtuse; sutures scarcely impressed; aperture small, nearly round, within white; columella very thick.

Habitat.— Tuscaloosa, Alabama.

Diameter, ·45; length, ·77 of an inch.

Observations.— Dr. Budd submitted five specimens to me, and, as is frequently the case, in this genus, I do not find any Fig. 835. two of the five exactly of the same outline. One is nearly round and presents but a single whorl. Another, a younger and more perfect specimen, is somewhat elliptical, and presents five whorls and a mammilate form. A third specimen is quite elliptical, the spire being obtusely conical. It is a very solid species, with a broad, thick columella, and a considerable callus above. All the five are obscurely banded. This species is allied to *A. prærosa,* Say, but differs somewhat in form, and has bands, not spotted lines. In some of the specimens the aperture is nearly the whole length of the shell.— *Lea.*

26. A. patula, ANTHONY.

Anculosa patula, ANTHONY, Proc. Acad. Nat. Sci., p. 68, Feb., 1860.
Anculotus patulus. Anthony, REEVE, Monog. Anculotus, t. 4, f. 32.
Leptoxis patula, Anthony, BINNEY, Check List, No. 376. BROT, List, p. 25.

Description.—Shell ovate, of a uniform, dark horn color, rather thin; whorls 4–5, convex; sutures very distinct; aperture semicircular, within whitish; columella only slightly rounded, somewhat flattened by a callous deposit, more or less tinged with dirty red.

Habitat.— Tennessee.

Observations. — Resembles none other of the genus; its color, Fig. 836. which is of a dull, dark brown, and its semicircular mouth, remarkable for its length and breadth, are prominent marks of distinction; the body-whorl is very much inflated and angulated or subangulated; the interior aperture is often blotched with regular, dirty brown spots; spire elevated and acute, rapidly diminishing to the apex; the lines of growth are strong, and on some specimens a single prominent varix may be noticed.—*Anthony.*

27. A. viridula, ANTHONY.

Anculosa viridula, ANTHONY, Proc. Acad. Nat. Sci., p. 68, Feb., 1860.
Anculotus viridulus, Anthony, REEVE, Monog. Anculotus, t. 4, f. 34.
Leptoxis viridula, Anthony, BINNEY, Check List, No. 396.

Description.— Shell ovate, of a uniform, dark green color, rather thin; spire much elevated, composed of 4-5 convex whorls; sutures very distinct; aperture ovate, large, about half the length of the

Fig. 837.

shell, livid inside; columella well rounded; has a broad, but not well defined sinus.

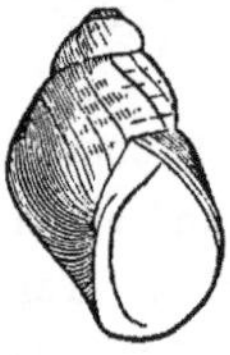

Habitat.— Tennessee.

Observations.—In form and coloring this species resembles *Paludina decisa*, Say, when that is about half grown, and but for its operculum one would hardly deem it an *Anculosa;* it is a plain, unadorned species, not liable to be confounded with any other; its body-whorl is large and subangulated; lines of growth well defined and close; it has a slight disposition to shouldering at the suture; it is not an abundant species so far as at present known.— *Anthony.*

This shell is figured like all the rest of Mr. Anthony's species, from the original type, for the use of which I am indebted to him.   Mr. Reeve thinks this species is identical with *Rogersii*, Conrad; and Dr. Brot believes it to be the same as *dilatata*.   It is a distinct species, but approaches closely to *Kirtlandiana*.   It is found also in North Carolina.

28. A. ligata, ANTHONY.

Anculosa ligata, ANTHONY, Proc. Acad. Nat. Sci., p. 67, Feb., 1860.
Anculotus ligatus, Anthony, REEVE, Monog. Anculotus, t. 3, f. 19.
Leptoxis ligata, Anthony, BINNEY, Check List, No. 367. BROT, List, p. 24.

Description.— Shell ovate, smooth, of a dark green color, rather thick; spire obtusely elevated, composed of about four whorls; suture very distinct; upper whorls flattened; body-whorl con- Fig. 838.
stricted at the middle, banded; aperture ovate, banded within; columella deeply indented, callous; no sinus at base.

Habitat.— Alabama.

Observations.— This species, of which I have some twenty or thirty individuals before me, seems remarkably constant in character for an *Anculosa;* and not readily mistaken for any other; its color, which is a dirty dark green, is but poorly relieved by the faint bands on the whorl; nevertheless it is an interesting species, and one which will always attract attention; its most prominent character is the constriction on the body-whorl, which gives the appearance of a cord being drawn tightly around it while in a yielding state.— *Anthony.*

This species does not resemble very closely the shell described by Mr. Lea as *Anculosa Coosaensis*, although that species possesses (in a less marked degree) the peculiar stricture of the body-whorl. *Ligata* differs in texture and color, and generally possesses three bands only, and none of the numerous specimens I have seen are maculate. *Coosaensis* appears to grow larger and heavier, and is more slender in its proportions, although swelling out more towards the periphery.

Doubtful and Spurious Species.

A. (Paludina) nuclea, LEA, = *Amnicola.*
A. (Paludina) virens, LEA, = *Amnicola.*
A. Spixiana, LEA, REEVE and BROT, = *Angitrema.*
A. incisa, Lea, HALDEMAN, Monog., = *Schizostoma.*
A. cingenda, ANTHONY, MSS., = young of *carinata*, LEA, a variety of *dissimilis.*
A. planulata, Lea, WHEATLEY, Cat. Shells, p. 28, Alabama (desc. not published), = *ampla*, ANTHONY?
?*Mel. carinata*, RAVENEL, Cat., p. 11, Yadkin River, N. C.
?*Mel. costata*, RAVENEL, Cat , p. 11, Dan River, Va., = *dissimilis?*
?*A. subcarinata*, RAVENEL, Cat., p. 11, Susquehanna, = *dissimilis?*
A. integra, SAY, = *Somatogyra.*
A. subglobosa, SAY, = *Somatogyra.*
A. (Paludina) altilis, LEA, = *Somatogyra.*
Paludina altilis, RAVENEL, = *Somatogyra.*
Paludina humerosa, ANTHONY, Proc. Acad. Nat. Sci., p. 71, 1860.

APPENDIX.

The following extracts from a letter recently received from my esteemed correspondent, Dr. James Lewis, who has devoted much time to the study of the Melanians, possess great interest in connection with the uncertainty which pervades the synonymy of the family. Dr. Lewis is well known to conchologists as an acute observer and philosophical naturalist, and his opinions and suggestions are correspondingly valuable.

G. W. T., Jr.

Mohawk, N. Y., Aug. 15, 1873.

Mr. Tryon,
> Dear Sir:

* * * * * * * * * * * *

I do not consider *Goniobasis castanea* to be the same as *G. simplex*. It is more likely (if possible) that *simplex* covers shells that have been named to me by correspondents *G. aterina*, Lea, *G. graminea*, Hald., etc. Probably Haldeman was right when he thought *G. acuto-carinata*, Lea, was a variety of *simplex*. I suspect that it is so for the reason that in the two (contiguous) localities from which I have *acuto-carinata* it occurs associated with species which, *in nearly every other station*, are living with mollusks that have been variously referred to *aterina*, Lea, and *graminea*, Hald. And as this association of similar types with a certain group of species extends over a large area each side of the Holston River, from Jefferson County southwest to Roane County, it seems to me to indicate that the varying forms, of which *aterina* and *acuto-carinata* are types, are simply *one species*, varied somewhat conspicuously in size and perfection of development, and still more varied in degree of carination of the upper whorls, while the *texture* and *color* of the *epidermis* and of the *shells* are less varied than might be expected.

The same mode of reasoning that would fit *aterina, graminea, acutocarinata*, etc., and refer them to *simplex*, would make a unit of all the various shells I refer to *castanea*, including a large mass of unreduced synonymy in which, perhaps, *G. glabra*, Lea, may be a leading term. Of this last, however, I have yet to assure myself. You will observe,

(423)

in passing over some of the earlier descriptions of shells of this group, that many are referred, locally, to the Holston River, or some other river. I have failed to verify these references thus far, and get *Goniobasis* only from creeks, springs, etc. This discrepancy, as referring to *G. glabra*, Lea, renders my endeavors to identify that species just enough uncertain to be always a matter of doubt. Many local references to other early described species are vague and do not define the *station* at all. Now, so far as this element goes, it is apparently an important one in the identification or rediscovery of a species or a type. As regards the group of forms to which Anthony's *G. arachnoidea* belongs, it is spread out over a vast territory. Assuming that Mr. Lea's *Trypanostoma Sycamorense* belongs to this type, we shall find the shells ranging from the northern limits of East Tennessee, along the streams that flow into the main channels of drainage down to Loudon, perhaps farther. The type is pretty constant in two remarkable features combined (*striate-undulate* upper whorls), though sometimes the undulations become obsolete. The synonymy of this type is greater than at present I dare presume to assert.

G. porrecta, Lea, has a pretty suggestive synonymy. Mr. Lea described a small shell from Claiborne, Sycamore County, Tennessee, that was *associated* with *T. Sycamorense*, just as we find *porrecta* with *arachnoidea* in half a dozen places (to be *within* limits). The *association* of species is here suggestive, as in a former case.

As to the Trypanostomas of the creeks of East Tennessee, they are a *perfect series* of differentiations of carinated apices. *One cannot tell where to assign limits.* Limits are apparently obliterated and species have no existence. They are a confused mass and must be referred to one type. It begins with shells that are carinate, doubly, triply carinate down nearly to the last whorl, and ends with shells that have a faint carina *sketched* on the first three or four whorls. I have not the facilities for determining who is to be regarded as the *patron* of this group.

* * * * * * * * * * * *

You remember, perhaps, my unfortunate treatment of *Trypanostoma curtum*, Hald. You also remember that you considered the paper in which it occurred of sufficient importance to honor it with a *critique*. Interested by your suggestions, I again went over the ground covered by the synonymy I suggested, only to flounder in more deeply, and finally to ascertain that one of Say's species (hitherto treated as superfluous) was really entitled to take precedence of *curtum*. * * * * * * * * I am aware that where so much is *uncertain* scarcely any one can make announcements that will be received absolutely. We are very largely at the mercy of opinions, some of which, no doubt, are but the reflex of the idiosyncrasies of the persons with whom they originate.

In regard to Io, I might make a few suggestions, which, when carried to the extent of my investigations, would, perhaps, offer original views. Here again I am restrained as before, and shall not enter into full details. I am of the opinion that Say's *Melania armigera* is an Io. Beyond this, I am unprepared to admit more than one species, though I am aware that others claim more on apparently good grounds. The genus Io, as heretofore limited by yourself, is spread out over the *Upper Tennessee drainage*. It occurs in the principal confluents that unite, forming the Tennessee River, above Chattanooga, and a few specimens have also been found in that portion of the Tennessee River that flows through Jackson County, Alabama. In Clinch River, I have, by Miss Law's aid, obtained perhaps three well marked varieties, one of which, certainly, most naturalists would call a good species. In the Holston and Tennessee I also find varieties one of which seems to have been derived from French Broad River, where only a single form appears. You are aware that a smooth variety (which I have not yet obtained) occurs in the Upper Holston, and varies so much as to be regarded as two species. Following Say's *Melania armigera* through its somewhat extensive distribution, I find that it begins to appear where the conventional Io disappears, and takes the place of "Io" in the Lower Tennessee River, Cumberland River, Wabash River, etc., etc. In the different stations where found, it varies pretty nearly as the typical Io does. In some instances it has varied so much as to have been redescribed as a distinct species, and in one instance (one of my correspondents suggests) a young shell was the occasion of the erection of a new genus. Now taking the parallel between the typical Io and Say's *armigera*, what shall we do? Shall we admit all the species and genera proposed, or will it suffice to write all there is of Io under two species, *fluvialis* and *armigera?* And while we have before us this question of the variability of species, let us inquire how many *species* are there of Say's *Melania nupera?* This species varies in different stations quite as much as *fluvialis* and *armigera*. Specimens entirely smooth are not rare. Others that are *undulate* contrast with the more numerous nodulous specimens. Colors and bands offer contrasts as in *fluvialis* and *armigera*. Now does not *analogy* have *some* weight with us sometimes? But, if it does, can we say that we treat these things *consistently?*

Let us consider the univalves of the Alabama drainage, say of the Black Warrior, Alabama, Coosa and Cahawba Rivers. I have *tried* to identify these, or *some* of these univalves, with those of the Tennessee drainage that circles through northern Alabama, and with the one exception of a MELANTHO, which I believe you separate as a distinct species, I find nothing identical. Perhaps there may be something identical in SOMATOGYRUS, but I have not had opportunity to make satisfactory comparisons. This leads me to question your identification of *Strephobasis Clarkii* (of the Tennessee drainage) with *S. bitæniata* (of the Alabama drainage = Black Warrior River).

I find evidence that leads me to unite *T. annuliferum* and *prasinatum*.

In following out this particular type I am led to infer a considerable number of other synonymes which do not appear in your "Synonymy" published some years ago.

In the Coosa River, abundant studies of synonymy await the patient student who may be favored with unprejudiced duplicates [without labels]. One species of Goniobasis *promises* nearly a dozen synonymes, and if we do not forget the lessons taught us in analogies elsewhere, we shall reduce Schizostoma to within a fifth part of its present limits.

And now let us inquire into the "origin of species," not in the Darwinian sense, but with a view of finding an explanation of the huge synonymy that I plainly see is dawning upon us.

During the last twenty years I have collected many shells and have also received many from correspondents. It has sometimes been my duty to assist my correspondents to identify their species. In many cases in which I have been called on to name species, my correspondents have assorted their shells down to the last variety, and believing each variety to be a species, have *insisted* to have each named separately. This is the key to the origin of many of our species. In other instances, perhaps, parties whose interests increase with the number of species they have at their disposal submit their isolated varieties *separately* for identification. What wonder, then, that the descriptive naturalist should unwittingly fall into a very natural mistake and describe these shells as *new species?*

Very truly yours,

JAMES LEWIS.

ERRATA.

P. MODESTUM, Lea, p. 130. This species must bear the name of its synonyme P. KNOXENSE, Lea, because IO MODESTA, Lea, previously described, is also a *Pleurocera.*

P. TORTUM, Lea, p. 84. This species may be called P. PARKERI, nob., after Mr. Charles F. Parker, a conchologist of Camden, New Jersey. P. TORTUM, Lea, p. 117, will stand as a species.

G. INOSCULATA, Lea, p. 302, read G. OSCULATA, Lea.

G. NIGRINA, Lea, p. 280, is made a synonyme of G. DRAYTONII, but should be cancelled. This species I now consider distinct and I have so treated it, *vide* p. 214.

PAL. NUMEROSA, Anth., p. 421, read HUMEROSA.

INDEX.

ABREVIATIONS. Anc., Anculosa. Ang., Angitrema. E., Eur., Eurycælon. G., Goniobasis. L., Lithasia. M., Meseschiza. P.. Pleurocera. Sch.,Schizostoma. St., Strephobasis.

Abbevillensis, (G.) Lea, 186
abbreviata, (G.) Anth., 159
abjecta, (G.) Hald., 287
abruptum, (P.) Lea, 117
abscida, (G.) Anth., 311
acuta, (G.) Lea, 205
acuta, (G.) Lea, Bell., 282
acutocarinata, (G.) Lea, 228
adusta, (G.) Anth., 272
æqua, (G.) Lea, 307
æqualis, (G.) Hald., 163
affine, (P.) Lea, 69
affinis, (Anc.) Hald., 389
Alabamensis, (G.) Lea, 303
Alabamense, (P.) Lea, 77
Alabamense, (Sch.) Lea. 367
Alabamense, (Seh.) Lea, Reeve, 357
Alexandrensis, (G.) Lea, 284
altipetum, (P.) Anth., 122
alveare, (P.) Conr. 50
ambusta, (G.) Anth., 317
amœna, (G.) Lea, 169
ampla, (Anc.) Anth., 412
ampla, (G.) Anth., 315
amplum, (Sch.) Anth., 360
angulata, (G.) Anth., 221
angulatus, (Anc.) Conr., 405
angulosa, (Anc.) Menke, 405
angusta, (G.) Anth., Reeve, 278
angustispira, (G.) Anth., 278
annuliferum, (P.) Conr., 91
Anthonyi, (Anc.) Budd, 347
Anthonyi, (E.) Budd., 347
Anthonyi, (G.) Lea, 267
Anthonyi, (P.) Lea, 80
Anthonyi, (Sch.) Reeve, 368
approximata, (G.) Hald., 287
arachnoidea, (G.) Anth., 155
aratum, (P.) Lea. 99
arctata, (G.) Lea, 314
armigera, (Ang.) Say, 19
assimilis, (G.) Anth., 232

assimilis, (G.) Lea, Binney, 232
aterina, (G.) Lea, 241
athleta, (G.) Anth., 200
attenuatum, (P.) Lea, 128
auriculæformis, (G.) Lea, 240
auriscalpium, (G.) Menke, 291
auriscalpium, (P.) Menke, 62
auricoma, (G.) Lea, 243

Babylonica, (G.) Lea, 278
Babylonicum, (Sch.) Lea, 369
baculum, (G.) Anth,. 206
Bairdiana, (G.) Lea, 215
Barrattii, (G.) Lea, 232
basalis, (G.) Lea, 331
bella, (Anc.) Lea, 381
bella, (G.) Conr., 157
bella-crenata, (G.) Hald., 157
bellula, (G.) Lea, 329
Bentoniensis, (G.) Lea, 151
bicincta, (G.) Anth., 232
bicolorata, (G.) Anth., 278
bicostatum, (P.) Anth., 85
Binneyana, (G.) Lea, 241
bitæniata, (St.) Con., 47
bivittatum, (P.) Lea, 82
bizonalis, (G.) DeKay, 291
blanda, (G.) Lea, 163, 182
Boykiniana, (G.) Lea, 150
brevis, (Io) Anth., 9
brevis, (L.) Lea, 37
brevispira, (G.) Anth., 277
Bridgesiana, (G.) Lea. 222
Brumbyi, (G.) Lea, 264
Brumbyi, (P.) Lea, 82
brunnea, (G.) Anth., 269
Buddii, (G.) DeKay, 291
Buddii, (G.) Lea, 295
Buddii, (Sch.) Lea, 370
Buddii, (G.) Say, Adams, 295
Buddii, (G.) Lea, Reeve, 318
bulbosa, (G.) Gould, 255

bulbosum, (Sch.) Anth., 361
bullula, (G.) Lea, 333

cadus, (G.) Lea, 168
Cahawbensis, (G.) Lea, 268
cælatura, (G.) Conr., 143
calculoides, (G.) Lea, 330
caliginosa, (G.) Lea, 196
canaliculatum, (P.) Say, 62
canalifera, (Anc.) Anth., 384
canalitium, (P.) Lea, 79
Canbyi, (G.) Lea, 148
Canbyi, (G.) Tryon, 260
cancellata, (G.) Say, 199
capillaris, (G.) Lea, 339
carinata, (Anc.) Anth., 395
carinata, (Anc.) Brug., 388
carinata, (Anc.) DeKay, 388
carinata, (Anc.) Lea, 388
carinata, (Anc.) Rav., 421
carinata, (G) Rav., 287
carinata, (St.) Lea, 43
carinatum, (P.) Lea, 100
carinifera, (G.) Lam , 157
cariniferum, (Sch.) Anth., 353
carinocostata, (G.) Lea, 164
Carolinense, (P.) Lea, 131
casta, (G.) Anth., 219
castanea, (G.) Lea, 271
castaneum, (Sch.) Lea, 355
Catawbæa, (G.) Hald., 232
catenaria, (G.) Lea, Adams, 146, 148
catenaria, (G.) Say, 146
catenoides, (G.) Lea, 148
cerea, (G.) Lea, 197
Chakasahaense, (P.) Lea, 121
chalybæa, (G.) Anth., Brot., 159
Christyi, (G.) Lea, 171
Christyi, (P.) Lea, 110
Cincinnatiensis, (Anc.) Lea, 405
cinctum, (P.) Lea, 99
cinerea, (G.) Lea, 258
cinerelia, (G.) Lea, 195
cingenda, (Anc.) Anth., 421
cinnamomea, (G.) Anth., 221
circincta, (G.) Lea, 200
clara, (G.) Anth., 328
Clarkii, (G.) Lea, 210
Clarkii, (P.) Lea, 79
Clarkii, (St.) Lea, 47
clausa, (G.) Lea, 310
clavaeformis, (G.) Lea, 272
cognata, (G.) Anth., 245
columella, (G.) Lea, 181
comma, (G.) Conr., 205
compacta, (L.) Anth., 36
concinna, (G.) Lea, 207
congesta, (G.) Conr., 239
conica, (P.) Say, 62

conica, (Sch.) Shutt., 380
Conradi, (G.) Brot., 156
Conradi, (P.) Tryon, 106
consanguinea, (L.) Anth., 33
constrictum, (Sch.) Lea, 373
continens, (G.) Lea, 188
contorta, (Anc.) Lea, 410
Coosaensis, (Anc.) Lea, 408
Coosaensis, (G.) Lea, 312
copiosa, (G.) Lea, 325
coracina, (G.) Anth., 180
corneum, (P.) Lea, 122
cornea, (St.) Lea, 45
corneola, (G.) Anth., 192
coronilla, (G.) Anth., 159
corpulenta, (Anc.) Anth., 399
corpulenta, (St.) Anth., 47
correcta, (G.) Brot, 248
corrugata, (G.) Lea, 194
costata, (Anc.) Anth., 385
costata, (Anc.) Rav., 421
costifera, (G.) Hald., 185
costulata, (G.) Lea, 195
Couperii, (G.) Lea, 153
crassa, (Eur.) Hald., 348
crebricostata, (G.) Lea, 204
crebristriata, (G.) Lea, 337
crenatella, (G.) Lea, 298
crepera, (G.) Lea, 311
crispa, (G.) Lea, 173
cristata, (G.) Anth., 217
cruda, (G.) Lea, 218
cruentata, (Anc.) Menke, 405
cubicoides, (G.) Anth., 223
culta, (G.) Lea, 321
Cumberlandiensis, (G.) Lea, 272
Curreyana, (G.) Lea, 185
Currierianum, (P.) Lea, 93
curta, (G.) Menke, 291
curta, (St.) Hald., 40
curtatum, (P.) Lea, 96
curtum, (Sch.) Mighels, 363
curvatum, (P.) Lea, 132
curvicostata, (G.) Anth., 202
curvilabris, (G.) Anth., 159
cuspidata, (G.) Anth., 248
cylindracea, (G.) Con., 290
cylindraceum, (P.) Lea, 108
cylindraceum, (Sch.) Mighels, 361
cylindrica, (G.) Con., Wheatley, 299

Decampii, (G.) Lea, 210
decora, (G.) Lea, 204
decorata, (G.) Anth., 141
demissum, (Sch.) Anth., 372
densa, (G.) Anth., 250
densecostata, (G.) Reeve, 202
dentata, (Anc.) Couth., 389
dentata, (Anc.) Lea, 389

depygis, (G.) Say, 247
Deshayesiana, (G.) Lea, 186
Deshayesiana, (G.) Reeve, 183
difficilis, (G.) Lea, 197
dignum, (P.) Lea, 85
dilatata, (Anc.) Conr., 395
dilatata, (L.) Lea, 29
dislocata, (G) Rav., 191
dissimilis, (Anc) Say, 388
Doolyensis, (G.) Lea, 187
Downieana, (G.) Lea, 151
Downiei, (L.) Lea, 39
Draytonii, (G.) Lea, 286
dubia, (G.) Lea, 274
dubiosa, (G.) Lea, 274
Duttoniana, (Ang.) Lea, 20
Duttonii, (G.) Lea, 175
dux, (P.) Lea, 66

ebenum, (G.) Lea, 242
Edgariana, (G.) Lea, 193
elata, (G.) Anth., 278
elegans, (Anc.) Anth., 412
elegantula, (G.) Anth., 159
elevatum, (P.) Lea, Adams, 95
elevatum, (P.) Say, 95
eliminata, (G.) Anth.. 207
Elliottii, (G.) Lea, 265
ellipsoides, (G.) Lea, 332
elliptica, (G.) Lea, 333
ellipticum, (Sch.) Lea, 359
elongatum, (P.) Lea, 95
Estabrookii, (G.) Lea, 274
Estabrookii, (P.) Lea, 129
Etowahensis, (G.) Lea, 260
Etowahensis,(G.) Lea, Reeve, 148
exaratum, (P.) Menke, 62
exaratum (P.) Lea, 99
excavata, (G.) Anth., 334
excisum, (Sch.) Lea, 356
excuratum, (P.) Conr., 55
excurvatum, (P.) Conr., 55.
exilis, (G.) Hald., 278
exilis. (G.) Hald., Adams, 282
eximium, (P.) Anth , 96
expansa, (G.) Lea, 219

fabalis, (G.) Lea. 305
fallax, (G.) Lea, 302
fasciata, (G.) Menke. 291
fascinans, (G.) Lea. 308
fasciolata, (Ang.) Reeve, 20
fastigiatum, (P.) Anth., 74
filum, (P.) Lea, 65
flammata, (Anc.) Lea, 415
flava, (G.) Lea. 263
flavescens, (G.) Lea, 145
Florencense, (P.) Lea. 77
Florentianus,(Ang.)Lea, Reeve, 16

Florentianus, (L.) Lea, 28
Floridensis, (G.) Reeve, 147
fluvialis, (Io) Say, 4
Foremanii, (Anc.) Lea, 415
Foremanii, (P.) Lea, 52
formosa, (Anc.) Lea, 412
formosa, (G.) Anth., Reeve, 174
formosa. (G.) Conr., 174
fuliginosa, (L.) Lea, 27
fumea, (G.) Lea, 306
funebralis, (G.) Anth., 272
funiculatum, (Sch.) Lea, 370
furva, (G.) Lea, 273
fuscocincta, (G.) Anth., 239
fusiformis, (G.) Lea, 329
fusiformis, (Io) Lea, 4
fusiformis, (L.) Lea, 38

Gabbiana, (G.) Lea, 268
gemma, (G.) DeKay, 291
genicula, (Ang.) Lea, 13
geniculata, (Ang.) Hald., 13
Georgiana, (G.) Lea, 246
Gerhardtii, (G.) Lea, 226
germana, (G.) Anth., 324
gibberosa, (E.) Lea, 345
gibbosa, (Anc.) Lea, 402
gibbosa, (G.) Lea, 244
gibbosa, (Io) Anth., 7
glabra, (G.) Lea, 244
glandaria, (G.) Lea, 327
glandulum, (P.) Anth., 109
glandulum, (Sch.) Lea, 377
glans, (P.) Anth., 109
glans, (Sch.) Lea, 363
glauca, (G.) Anth., 200
globosum, (Sch.) Lea, 357
globula, (Anc.) Lea. 402
gracile, (P.) Lea, 104
gracilior, (G.) Anth., 258
gracilior, (G.) Lea, 332
gracilis, (G.) Lea, Reeve, 258
gracilis, (Io) Lea, 12
gracilis, (G.) Anth., 258
gracilis, (G.) Lea, 178
gracillima, (G.) Anth., 209
gradatum, (P.) Anth., 96
graminea (G.) Hald , 245
granata, (G.) Lea, 141
grata, (G) Anth., 261
gratiosa, (E.) Lea, 342
gravida, (G.) Anth., 323
Griffithiana, (Anc.) Lea, 387
grisea, (G.) Anth., 321
grossum, (P.) Anth., 50
Grosvenorii, (G.) Lea. 278
Grosvenorii, (Mes.) Lea, 350

Haldemani, (G.) Tryon, 282

Haldiana, (G.) Lea, 284
Hallenbeckii, (G.) Lea, 149
harpa, (G.) Lea, 319
Hartmaniana, (G.) Lea, 315
Hartmanii, (G.) Lea, 315
Hartmanii, (P.) Lea, 82
Hartmanii, (Sch.) Lea, 372
hastatum, (P.) Anth., 124
Haysiana, (G.) Lea, 313
Henryanum, (P.) Lea, 90
Hildrethiana, (L.) Lea, 33
Holstonia, (Ang.) Lea, 25
humerosa, (Anc.) Anth., 421
hybrida, (G.) Anth., 238
Hydei, (G.) Con., 140

imbricata, (G.) Anth., 232
imperialis, (L.) Lea, 30
impressa, (G.) Lea, 337
incisum, (Sch.) Lea, 378
inclinans, (G.) Lea, 153
inconstans, (G.) Lea, 188
incrassatum, (P.) Anth., 82
incurta, (G.) Anth., Reeve, 269
incurvum, (P.) Lea, 76
induta, (G.) Lea, 187
inempta, (G.) Anth., 291
inermis, (Io) Anth., 6
infantula, (G.) Lea, 256
inflata, (Anc.) Lea, 395
inflata, (G.) Hald., 328
informis, (G.) Lea, 283
infrafasciatum, (P.) Anth., 74
infuscata, (G.) Lea, 226 -
inornata, (G.) Anth., 278
inosculata, (G) Lea, 266, 302
instabilis, (G.) Lea, 171
intensum, (P.) Anth., 88
intercedens, (G.) Lea, 221
interlineata, (G.) Anth., 275
interrupta, (G.) Hald., 171
intersita, (G.) Hald., 181
intertexta, (G.) Anth., 155
interveniens, (G.) Lea, 170
iostoma, (P.) Anth., 135
iota, (G.) Anth., 236

Jayana, (Ang.) Lea, 17
Jayi, (P.) Lea, 83

Kirtlandia, (G.) Lea, Philippi, 278
Kirtlandiana, (Ang.) Anth., 395
Kirtlandiana, (G.) Lea, 278
Knoxense, (P.) Lea, 130
Knoxvillense, (P.) Lea, 127

labiatum, (P.) Lea, 111
lachryma, (E.) Anth., 343
laciniatum, (Sch.) Lea, 359

laeta, (G.) Jay, 318
lævigata, (G.) Lea, 276
lævis, (G.) Lea, 276
laqueata, (G.) Say, 176
larvaeformis, (G.) Lea, 243
latitans, (G.) Anth., 296
lativittatum (P.) Lea, 100
Leaii, (G.) Brot, 276
Leaii, (G.) Tryon, 163
Leaii, (P.) Tryon, 102
Leaii, (E.) Tryon, 342
Lecontiana, (G.) Lea, 167
Leidyana, (G.) Lea, 164
lepida, (G.) Lea, 344
Lesleyi, (P.) Lea, 53
Lewisii, (Anc.) Lea. 418
Lewisii, (G.) Lea, 331
Lewisii, (P.) Lea, 90
ligata, (Anc.) Anth., 420
ligatum, (P.) Menke, 62
ligatum, (P.) Lea, 67
lima, (Ang.) Conr., 23
Lindsieyi, (G.) Lea, 191
lita, (G.) Lea, 301
Lithasioides, (G.) Lea, 255
littorina, (Anc.) Hald., 384
livescens, (G.) Menke, 248
livida, (G.) Reeve, 278
Louisvillensis, (G.) Lea, 257
lugubre, (P.) Lea, 115
lurida, (Io) Anth., 6
luteola, (G.) Lea, 322
luteum, (P.) Lea, 131
Lyonii, (G.) Lea, 200
Lyonii, (P.) Lea, 124
Lyonii, (St.) Lea, 46

macella, (G.) Lea, 237
mediocris, (G.) Lea, 175
melanoides, (Anc.) Conr., 399
mellea, (G.) Lea, 316
Midas, (E.) Lea, 341
Milesii, (G.) Lea, 250
minor, (P.) Lea, 134
modesta, (Io) Lea, 12
modestum, (P.) Lea, 101, 115. 130
monilifera, (G.) Anth., Jay, 157
moniliferum, (P.) Lea, 57
monodontoides. (Anc.) Conr., 389
monozonalis, (G.) Lea, 176
moriforme, (P.) Lea, 70
mucronatum, (P.) Lea, 114
multilineata, (G.) Say, 291
mutabilis, (G.) Lea, 183
mutata, (G.) Brot, 183

napella, (G.) Anth., 248
napoideum, (P.) Lea, 137
nassula, (G.) Conr., 193

nebulosa, (G.) Conr., 157
negata, (G.) Lea, 336
neglectum, (P.) Anth., 113
neritiformis, (Anc.) Desh., 405
Newberryi, (G.) Lea, 254
Niagarensis, (G.) Lea, 248
Nickliniana, (G.) Lea, 240
nigrescens, (Anc.) Conr., 389
nigrina, (G.) Lea, 214, 286
nigrocincta, (G.) Auth., 236
nigrostoma, (P.) Anth., 135
nitens, (G.) Lea, 182
nitida, (G.) Lea, 182
nobilis, (Io) Lea, 12
nobile, (P.) Lea, 60
nodulosa, (G.) Lea, 196
nodata, (Ang.) Reeve, 22
nodosa, (Io) Lea, 12, 57
nubila, (G.) Lea, 346
nuclea, (L.) Lea, 36
nucleola, (L.) Anth., 32
nuculum, (Sch.) Anth., 361
nupera, (Ang.) Say, 24
nupera, (P.) Say, 50

obesa, (G.) Anth., 162
obliqua, (Sch.) Anth., 378
oblita, (P.) Lea, 85
obovata, (L.) Say, 33
obtusa, (G.) Lea, 168
occata, (G.) Hinds, 146
occidentalis, (Anc.) Lea, 385
occulta, (G.) Anth., 247
Ocoëensis, (P.) Lea, 123
Ohioensis, (G.) Lea, 276
oliva, (G.) Lea, 320
olivaceum, (P.) Lea, 78
olivaria, (St.) Lea, 43
olivata, (G.) Conr., Jay, 308
olivella, (G.) Lea, 171
olivula, (G.) Conr., 308
opaca, (P.) Anth., 135
oppugnata, (G.) Lea, 299
orbicula, (G.) Lea, 325
Ordianum, (P.) Lea, 91
ornata, (Anc.) Anth., 417
ornata, (Sch.) Anth., 366
ornatella, (G.) Lea, 174
ovalis, (G.) Lea, 325
ovalis, (Sch.) Anth., 361
ovoidea, (G.) Lea, 260
ovoideum, (Sch.) Shutt., 356
ovularis, (Anc.) Menke, 405

pagoda, (Sch.) Lea, 365
pagoda, (Sch.) Lea, Reeve, 366, 370
pagodiformis, (G.) Anth., 228
pallescens, (G.) Lea, 266
pallidula, (G.) Anth., 224

papillosa, (G.) Anth., 151
parva, (G.) Lea, 266
parvum, (P.) Lea, 120
patula, (Anc.) Anth., 419
paucicosta, (G.) Anth., 179
paula, (G.) Lea. 232
paula, (G.) Anth., 269
paupercula, (G.) Lea, 192
perangulata, (G.) Conr., 157
percarinata, (G.) Conr., 157
perfusca, (G.) Lea, 269
pergrata, (G.) Lea, 338
pernodosa, (P.) Lea, 50
perstriata, (G.) Lea, 166
picta, (Anc.) Conr., 415
pictum, (P.) Lea, 119
picturata, (P.) Reeve, 119
pilula, (Anc.) Lea, 384
pinguis, (Anc.) Lea, 409
pisum, (Eur.) Hald., 348
planogyrum, (P.) Anth., 105
planospira, (Anc.) Anth., 411
planulata, (Anc.) Lea, 421
plebeius, (G.) Anth., 269
plena, (St.) Anth., 44
plicata, (Anc.) Conr., 381
plicatula, (G.) Lea, 186
plicatum, (P.) Tryon, 94
plicifera, (G.) Lea, 211
ponderosum, (P.) Anth., 66
porrecta, (G.) Lea, 297
Portellii, (G.) Lea, Reeve, 154
Postellii, (G.) Lea, 154
Postellii, (P.) Lea, 75
Potosiensis, (G.) Lea, 253
praemorsa, (Anc.) Say, Woodward, 405
praerosa, (Anc.) Say, 405
Prairiensis, (G.) Lea, 261
prasinatum, (P.) Conr., 81
Prestoniana, (G.) Lea, 250
procissa, (G.) Anth., 139
producta, (P.) Lea, 50
proletaria, (G.) Lea, 188
propinqua, (G.) Lea, 300
propria, (G.) Lea. 336
proteus, (G.) Lea, 344
proxima, (G.) Say, 287
pudica, (G.) Lea, 305
pulchella, (G.) Anth., 257
pulcherrima, (G.) Anth., 230
pumilum, (P.) Lea, 134
pumilum, (Sch.) Lea, 357
pumilum, (St.) Lea. 42
punicea, (G.) Lea, 304
pupoidea, (G.) Anth., 300
purpurea, (G.) Lea, 335
purpurella, (G.) Lea. 190
Pybasii, (G.) Lea, 177

Pybasii, (P.) Lea, 71
pyramidatum, (Sch.) Shutt., 365
pyrenellum, (P.) Conr., 106

quadratum, (Sch.) Anth., 378
quadricincta, (G.) Lea, 262
quadrivittata, (G.) Lea, 335

rapæformis, (Anc.) Hald., 395
rara, (G.) Lea, 304
rarinodosa, (L.) Anth., Reeve, 33
recta, (Io) Anth., 7
rectum, (Sch.) Anth., 373
rectum, (Sch.) Anth., Reeve, 376
regulare, (P.) Lea, 106
rhombica, (G.) Lea, 220
rhombica, (Io) Anth., 7
rigidum, (P.) Anth., 85
Roanense, (P.) Lea, 108
robulina, (Ang.) Anth., 17
robusta, (Io) Lea, 12. 61
robustum, (P.) Lea, 61
robustum, (Sch.) Anth., 376
robusta, (G.) Lea, 318
Rogersii, (Anc.) Conr., 395
roratum, (P.) Reeve, 55
rostellatum, (P.) Lea, 126
rota, (Ang.) Reeve, 19
rubella, (G.) Lea, 285
rubicunda, (G.) Lea, 313
rubiginosa, (Anc.) Lea. 387
rubiginosa, (G.) Lea, 215
rubricata, (G.) Lea, 287
rudens, (G.) Reeve, 212
rufa, (G.) Lea, 290
rufescens, (G.) Lea, 290
rufula, (G.) Hald., 278
rugosa, (G.) Lea, 194

Saffordi, (G.) Lea, 253
salebrosa, (Ang.) Conr., 14
salebrosa, (Sch.) Anth., 376
Sayi, (P.) Desh., 62
Sayi, (P.) Ward, 62
Sayi, (Strombus) Wood, 62
scabrella, (G.) Anth., Reeve, 164
scabriuscula, (G.) Brot, 164
sculptilis, (G.) Lea, 297
Sellersiana, (G.) Say, 180
semicarinata, (G.) Say, 278
semicostata, (G.) Conr., 191
semigradata, (G.) Reeve. 164
Shastaensis, (G.) Lea, 212
Shastaensis, (G.) Lea Reeve, 146
Shelbyensis, (G.) Lea, 306
Showalteriana, (Sch.) Lea. 375
Showalterii, (Anc.) Lea, 382
Showalterii, (G.) Lea, 300
Showalterii, (L.) Lea, 31

Showalterii, (P.) Lea, 71
Showalterii, (Sch.) Lea, 353
silicula, (G.) Gould, 212
simplex, (G.) Say, 250
simplex, (P.) Lea, 132
smaragdinus, (Anc.) Reeve, 381
solida, (L.) Lea, 37
solida, (St.) Lea, 40
solidula, (G.) Lea, 307
sordida, (G.) Lea, 269
Spartenburgensis, (G.) Lea, 225
sparus, (G.) Lea, 197
sphæricum, (Sch.) Anth., 364
Spillmanii, (G.) Lea, 223
Spillmanii, (Io) Lea, 12, 55
Spillmanii, (P.) Lea, 104
Spillmanii, (Sch.) Lea. 369
Spillmanii, (St.) Lea, 44
spinalis, (P.) Lea, 119
spinella, (G.) Lea, 285
spinosa, (Io) Lea, 7
spirostoma, (Io) Anth., 9
Spixiana, (Ang.) Lea, 22
spurca, (P.) Lea, 115
squalida, (Anc.) Lea, 418
Stewardsoniana, (G.) Lea, 144
straminea, (G.) Lea, 322
strenua, (G.) Lea, 164
striata, (G), Lea 203
striatula, (G.) Lea, 203
striatum, (P.) Lea, 126
strictum, (P.) Lea, 101
strigillata, (G.) Muhlfeldt, 291
strigosum, (P.) Lea, 118
Stygia, (Ang.) Say, 22
suavis, (G.) Lea, 321
subangulata, (G.) Anth., 231
subcarinata, (Anc.) Rav., 421
subcarinata, (Anc.) Wood. 388
subcarinata, (G.) Anth., Reeve, 238
subcylindracea, (G.) Lea, 206
subglobosa, (Anc.) Say, 402
subglobosa, (Ang.) Lea, 15
sublirata, (G.) Conr., 146
subrobustum, (P.) Lea. 110.
subsolida, (G.) Lea, 250
substricta, (G.) Hald., 108
substrictum, (P.) Hald., 62
subulæforme, (P.) Lea, 89
subulare, (P.) Lea, 88
succinulata, (G.) Anth., 278
sugillatum, (P.) Reeve. 85
sulcosa, (Anc.) Anth., 382
sulcosa, (G.) Lea, 294
suturalis, (G.) Hald., 183
Sycamorense, (P.) Lea, 94
symmetrica, (G.) Conr., 156
symmetrica, (G.) Hald., 232

tabulata, G. Anth., 230
tæniata, Anc. Cour., 408
tæniolata, (G.) Anth., 318
Taitiana, (G.) Lea, 287
tecta, (G.) Auth., 237
tenebrociuctum, (P.) Anth., 120
tenebrosa, (G.) Lea, 179
tenebrosa, (Io) Lea, 4
tenebrovittata, (G.) Lea, 203
tenera (G.) Anth., 264
Tennesséne, (P.) Lea, 135
terebralis, (G.) Lea, 208
teres,(G.) Lea, 208
textilosa, (G.) Anth., 319
Thorntonii, (G.) Lea, 199
Thorntonii, (P.) Lea, 72
tintinnabulum, (Anc.) Lea, 402
torquata, (P.) Lea, 50
torta, (P.) Lea, 117
tortum, (P.) Lea, 84
torulosa, (G.) Anth., 228
tractum, (P.) Anth., 95
translucens, (G.) Anth., 201
trilineata, (Anc.) Say, 400
tripartita, (G.) Reeve. 203
trivittata, (Anc.) DeKay, 389
trivittata, (L) Lea, 37
trivittatum. (P.) Lea, 73
trochiformis. (G.) Conr., 216
trochulus, (P.) Lea, 137
Troostiana, (Anc.) Lea, 409
Troostiana, (G.) Lea, 296
Troostiana, (G.) Lea, Chenu, 150
Troostii, (P.) Lea, 67
Tryoniana, (G.) Lea, 141, 143
tuberculata, (Anc.) Lea, 381
tuberculata, (Ang.) Lea, 22
Tuomeyi, (Ang.) Lea, 16
Tuomeyi, (G.) Lea, 169
Tuomeyi, (P.) Lea, 103
turbinata, (Eur.) Lea, 348
turgida, (Anc.) Hald., 399
turgidum, (P.) Lea, 133
turrita, (Io) Auth., 11

Ucheensis, (G.) Lea, 232
umbonata. (G.) Lea, 346
unciale, (P.) Hald., 85
undosa, (L.) Anth., 33

undulatum, (P.) Say, 54
univittatum, (P.) Lea, 112

validum, (P.) Anth., 107
Vanuxemensis, (G.) Lea, Wheatley, 312
Vanuxemiana, (G.) Lea, 312
Vanuxemii, (G.) Lea, 250
Vanuxemii, (P.) Lea, 121
variabilis, (Anc) Lea, 388
variabilis, (Io) Lea, 12, 57
varians, (G.) Lea, 140
variata, (G.) Lea, 324
varicosa, (G.) Ward, 278
Vauxiana, (G.) Lea, 242
venusta, (L.) Lea, 28
verrucosa, (Ang.) Raf., 24
verrucosa, (Io) Reeve, 4
versipellis, (G.) Anth., 178
vesicula, (G.) Lea, 162
vestitum, (P.) Conr., 114
vicina, (G.) Anth., 225
Viennaensis, (G.) Lea, 184
virens, (G.) Anth., 253
virens, (Sch.) Anth., 378
virgata, (Anc.) Lea, 402
Virginica, (G.) Gmel., 290
virgulata, (G.) Lea, 327
viride, (P.) Lea, 67
viridicata, (G.) Lea, 189
viridis, (Anc.) Lea. 400
viridula, (Anc.) Anth., 420
viridula, (Io) Lea, 12, 102
viridulum, (P.) Lea, 102
viridulum, (P.) Anth.,125
vittata, (Anc.) Lea, 411
vittata, (G.) Anth., 159
vittata, (L.) Lea, 30
vittatella, (G.) Lea, 283

Warderiana, (G.) Lea, 250
Wardiana,(G.) Lea, Wheatley, 250
Wetumpkaense, (Sch.) Lea, 366
Wheatleyi. (Ang.) Tryon, 21
Whitei, (G.) Lea, 218
Whitei, (P.) Lea, 128

zebra, (Anc.) Anth., 415

SMITHSONIAN MISCELLANEOUS COLLECTIONS
270

CATALOGUE

OF THE DESCRIBED

DIPTERA

OF

NORTH AMERICA.

BY

C. R. OSTEN SACKEN.

[SECOND EDITION.]

WASHINGTON:
SMITHSONIAN INSTITUTION.
1878.

ADVERTISEMENT.

THE present work was undertaken by Baron C. R. Osten Sacken, of Russia, as a revision and extension of a Catalogue of Diptera prepared by him twenty years ago, and published by the Smithsonian Institution in Volume III. of its Miscellaneous Collections. It is, however, not merely a new edition of the volume in question, but an entirely new work, constituting a valuable contribution to our knowledge of the entomology of North America.

SPENCER F. BAIRD,
Secretary Smithsonian Institution.

WASHINGTON, October, 1878.

PREFACE.

The aim of this work requires no explanation. A complete inventory of a branch of entomological science, at a given moment of its existence, is the best means for promoting its advancement. Nor does the imperfection of a publication of this kind require an apology; any fair-minded reader is aware that the chief merit to be expected is completeness, and that whenever this is fairly attained, the usefulness of the work will far surpass its shortcomings. It remains for me therefore, only to explain the rules that I have followed in preparing this Catalogue.

RELATION OF THE PRESENT CATALOGUE TO THAT OF 1858. The first Catalogue of North American Diptera, published by me twenty years ago, was, and was meant to be, merely a compilation of the existing literature on the subject. It brought together a mass of references to the descriptions of about 1800 species, scattered in more than one hundred different works and scientific papers. Although such a publication was an indispensable preliminary step before any study of the North American diptera could be attempted, it conveyed but a very vague idea of the actual composition of the North American fauna of diptera. It was impossible to ascertain, at that time, how many of the specific names, enumerated in the Catalogue, actually represented different species, and how many were mere synonyms; neither was it possible to know, whether the species were placed in the right genera, and even in the right families. In order to give an idea of the extent to which this statement is true, I will quote the genus *Trypeta*, which (excluding the three species named, but not described by T. W. Harris), contains forty-two specific names in the old, and sixty-six in the new Catalogue. But, in comparing these two lists, we find that they have only *eleven* names in common. In other words, of the forty-two so-called species of *Trypeta* of the old Catalogue, only eleven are adopted now as specific names in that genus; the other thirty-one names proved, upon investigation, to be either synonyms, or to represent species which had been erroneously placed in the genus *Trypeta*, or else to be unavailable names, on account of the insufficiency of the descriptions. The difference between *eleven* and *sixty-six* (the number of species in the new Catalogue), represents therefore the addition made to the knowledge of the genus *Trypeta* in North America during the interval between the two catalogues. Other genera give similar results. Thirty-two species of *Dolichopus* were described previous to 1858; the present list contains fifty-nine; but both lists have only *two* specific names in common. Thirty of the earlier descriptions are unrecognizable and therefore useless. The old Catalogue contained 32 names of species of *Eristalis*, occurring in North America, north of Mexico; of these names only *nine* figure as species of Eristalis in the present Catalogue, although the definition of the genus has not been changed since then. The other names of the old Catalogue are either synonyms *(E. dimidiatus,* for instance, has been described under *six* different names), or they belong to other genera, as *Helophilus, Milesia,* even *Xylota.* The genus *Tabanus*, in the old Catalogue, contains one hundred and two names of species, from North America, north of Mexico; among these names only 36 could be adopted; the remainder are either synonyms, or absolutely unavailable, on account of the insufficiency of the descriptions. — These instances will suffice to show that the new Catalogue is, not merely a new edition of the old one, only supplemented by the new species, published between 1858 and 1878; it is a new work, prepared on a different plan.

The process gone through between two editions of a catalogue, (the compilatory and the critical edition), consists in forming collections, in determining them from existing descriptions, and thus making out the synonymies, and then working up each family in monographs. It will be a long time of, course, before this last stage is reached in all the families of North American diptera, and for this reason, this new Catalogue, which represents the *actual* state of our knowledge of these diptera, is not entirely homogeneous; a portion of it only is synonymical and critical, and the rest is still a mere list of names, a compilation. The Catalogue may, in this respect, be divided into three groups of families, representing three stages of our knowledge of the species enumerated:

1. The families of the first group have been worked out in monographs, containing comparative descriptions of all ·the species (as far of course, as represented in the collections), with analytical tables, or else with figures, to facilitate identification. Such families are the *Dolichopodidae*, *Ortalidae* and *Trypetidae* (monographed by Dr. Loew); the *Tipulidae brevipalpi* and *Tabanidae* (monographed by myself). The beginning of a similar work was made by Mr. Loew for the *Ephydrinidae* and *Sciomyzidae* and by me for the genus *Syrphus*.

2. In the families of the second group, collections have been formed, a certain number of earlier descriptions have been identified and synonymies made out; many new species were described; but a monographic treatment is still wanting. Such families are the *Asilidae* (with the exception of the section *Asilina*), the *Bombylidae*, *Syrphidae*, *Tipulidae longipalpi*; also the *Empidae*, *Midaidae*, *Cyrtidae*, *Bibionidae*, *Mycetophilidae* and a number of the smaller families among the *Muscidae acalypterae*.

3. In the families of the third group, collections have been formed, but they are, for the most part, not named. The Catalogue, in such families, is a mere compilation of references to descriptions by earlier writers. Such families are: the *Culicidae*, *Chironomidae*, *Conopidae*, the whole group of *Muscidae calypterae* and the section *Asilina*.

COLLECTION OF TYPE-SPECIMENS. A difference between the old and the new Catalogue, perhaps more important than that already explained, consists in the fact, that the majority of the species

enumerated in the new Catalogue, *are represented in a collection.*
The collection of diptera of the Museum of Comparative Zoölogy
in Cambridge, Mass., contains what may be called the typical
specimens of this Catalogue, that is the types of the descriptions
published by Mr. Loew and by myself, as well as the species
identified by him or by me, from earlier descriptions. That
collection thus contains a little over 2000 named and described
species of diptera from North America *), north of Mexico, besides
a considerable accumulation of unnamed and undescribed materials.
In that collection the american dipterologist now possesses an
advantage not shared by his European colleagues, and that is, of
having very nearly all the typical specimens, necessary for his
work, collected in the same spot. It is highly desirable that this
advantage should, as far as possible, be maintained, and that
describers of new species should deposit their types in the same
Museum, which offers the best guarantees of their permanent
preservation. Sixty years ago, Wiedemann (in the first chapter
of his Magazin für Zoologie), foreseeing the future difficulties of
dipterology, suggested the formation of a central, or as he called
it, *normal* Museum, in some European city, to contain types of
all the described species; no new species were to be published,
without previous comparison in that Museum. May the Museum
in Cambridge realize that idea for America!

LITERATURE. The literary references, which I give in the
notes, are not meant to be a complete index of dipterological
literature, but merely a guide to beginners, who might be easily
deterred by the preliminary work to be gone through, before
attempting the study of any family. Those who intend to go
deeper into the subject will have to form a more complete index
for themselves, by looking over the yearly entomological
Records **), as well as the works in the libraries. That the
majority of the papers quoted by me are those of Dr. Loew,
arises from the fact that for the last 30 years he was the prin-

---

*) These species are marked with a star in the Catalogue.

**) A yearly Record on the progress of entomology is published
in Germany since 1838, in Wiegmann's Archiv für Zoologie. This

cipal dipterological writer in Europe and that the study of his papers cannot enough ·be recommended.

SYSTEM. The systematic distribution of the diptera and the natural affinities of some of the larger and smaller groups, are still matters of uncertainty. I have preserved, with slight modifications, the arrangement adopted by the most recent writers. It has the advantage of adapting, as much as possible, the division in *Orthorhapha* and *Cyclorhapha,* to the sequence of the families, as found in Meigen and other early writers. The Xylophagidae, Stratiomyidae, Coenomyidae, Acanthomeridae, Tabanidae and Leptidae seem to form a natural group, within which it is impossible to bring about a satisfactory linear arrangement. I placed the Asilidae between this group and the Bombylidae, in order to bring together the families provided with a posterior intercalary vein. But I am not at all sure whether this is not a character of secondary importance, and whether Dr. Schiner was not right in placing the Bombylidae nearer to the Tabanidae. The relationship of the Blepharoceridae, Psychodidae, Cyrtidae and Therevidae is likewise uncertain. Orphnephila and Dixa are altogether *incertae sedis.*

Although I consider the *Aphaniptera* as directly related to the *Mycetophilidae,* I have omitted them from my list, because they have hitherto formed a separate object of study.

· GEOGRAPHICAL RANGE. The region, embraced in the present Catalogue is the same as that of its predecessor: all North America, north of the Isthmus of Panama, including the West-Indies. But, instead of enumerating the species promiscuously, as it was done in the earlier Catalogue, I have, within each genus, separated the species occurring north of the Mexican

---

Record was prepared by Erichson from 1838 to 1847; by Schaum from 1848 to 1852; by Gerstaecker from 1853 to 1866; by Brauer from 1867 to 1870; and by Bertkau since that year. In England, the *Zoological Record,* published yearly since 1864, also contains an admirably prepared review of entomological publications. The frequent perusal of these Records cannot enough be recommended to those who wish to become thoroughly acquainted with the literature of any branch of entomology.

boundary, from those which are known to belong south of that line. A species, belonging to two groups simultaneously is placed in the earlier group; within each group the species are arranged alphabetically. — This change was rendered necessary by several considerations of expediency. In the first place, the work of criticism is much more advanced for the diptera of the United States and especially of the northern and middle States, than for those of Mexico, Central-America and the West-Indies: the reason is, that the bulk of the available collections came from the former regions. It was found expedient, therefore, to separate the uncritical and merely compilatory portions of the lists from those, that are more carefully sifted. At the same time, this arrangement offers another advantage in the better survey it affords of the geographical distribution of the diptera. Any one, running over the Catalogue, will now be able at a glance to form an idea of the character of the fauna of the temperate regions of North America, as distinguished from the tropical and subtropical faunae. Finally, this arrangement will be found very convenient in putting the Catalogue to the principal use for which it was intended, that of identifying species of diptera with the existing descriptions. As the Western, and especially the Californian fauna, is very different from the fauna of the Atlantic States, I have formed a third, intermediate group of those species in each genus, that are peculiar to that fauna. Whether this distribution in two or three groups should be maintained in the future editions of the Catalogue, is a question which will have to be decided then, as it has been decided now, on considerations of practical expediency.

Many species living in the lower and warmer regions of Mexico, also occur in Texas, and in the southern States in general. On the other hand mexican species from the higher altitudes, (from Mexico, Puebla etc.) extend quite far north, along the high plateau of North America and in the Rocky Mountains. Thus *Dejeania corpulenta* Wied. and *Dejeania rutilioides* Jaennicke, both first described from Mexico, were found by me in the Rocky Mountains. It is only recently, since I examined the mexican species in the collections in Darmstadt and in Turin, that I was struck with the relationship of the Western and of the Mexican fauna and have been able to identify several species, published by me as new, in my *Western Diptera*. California partakes of this relationship, and shows, at the same time, singular and unexpected coincidences with Europe, not shared by the eastern United States. Future describers of western and southern species will have to bear these facts in mind.

LOCALITIES. The scope of this work did not allow much detail in the matter of localities. Still, as much as the given space allowed, I have inserted the data which I possessed on the subject. Describers of insects, and especially of exotic forms, are often very careless about statements of that kind. It is very probable, for instance that many species, described by Macquart as ·coming from Philadelphia or Baltimore, were merely *sent* from those cities, but collected somewhere else; some of these species have since been received from Texas only. It is to be hoped that future describers will be more accurate about localities *and their altitudes.* California and Mexico, in different altitudes, contain several different faunas and the study of the geographical distribution of insects would reach very erroneous conclusions, if it did not discriminate between these faunas.

SPECIES COMMON TO EUROPE AND TO NORTH AMERICA. A very considerable number of European species is also found in North America, without belonging in the number of imported insects. Some of the species, common to both continents, do not show any perceptible differences; in others, a difference exists, but not such as could be considered a specific character. And thus, by gradations, a point is reached, where the specific difference becomes evident*). A careful study of almost any species, considered as identical, may unexpectedly disclose a minute, but sometimes important distinctive character. Hence all the species of the class in question must be considered as open to challenge.

---

*) About the species common to both continents, and the gradations occurring in the specific differences, compare Loew, in Silliman's Journ., Vol. XXXV,I, p. 3:7.

Authors differ in their mode of treatment of species, the identity of which is doubtful; some prefer at once to describe them as new, others assume the identity, until the difference is proved. For several reasons of a purely practical kind, I prefer the latter method, thus following the principle, laid down by Fabricius (Philos. entomologica): *Locus natalis speciem nunquam distinguit.* Once described as a new species, without indication of its distinctive characters, the species escapes attention; on the contrary, it invites one's notice and challenges criticism, as long as it is quoted as common to both continents. A time will come when it will be possible to subject that whole class of species to a thorough comparative study.

SYNONYMY. It has been my effort throughout to make sure, as much as possible, that every name, which figures in the list, should actually represent a different species. This is reached, in a certain measure, for the fauna north of Mexico (with the exception, of course, of those families, which have not been worked at all: the Muscidae calypterae etc.). To attain this result, I have *first,* made out a number of synonymies by means of an attentive reading of the descriptions; and, *secondly,* I have visited the Museums in London, Paris, Lille, Berlin, Frankfort, Darmstadt, Turin and Vienna, and have seen the types of descriptions, which they contain. Any one, who has visited public Museums for the purpose of examining types of descriptions, knows, that even under the most favorable circumstances, that kind of work is not like work done at home (especially in the difficult families). Moreover, the study of types of descriptions must be based upon a previous knowledge, and a thorough one, of the corresponding species. As I had no collection with me for comparison, and had to rely on my memory, and as my knowledge in the different families of diptera is very unequal, and, in some of them very small, I am far from having exhausted the study of the North American types, contained in those Museums. I am also far from believing, that what I made out is always free from error. Those who in future will take up single families for monographic work, are therefore strongly recommended not to take for granted the synonymies which I give, but to form an opinion for themselves. For synonymies, which are borrowed from other authors, the authority is always quoted in brackets []; synonymies without such a quotation, are my own.

NOMENCLATURE AND PRIORITIES. Readers of the Catalogue will often find, among the synonyms, names which, according to rule, should have the priority, being of earlier date than the adopted specific name. In such cases, I have discriminated between my *rôles* of a monographer and of a catalogue-maker. In those families, which I have described monographically *(Tipulidae brevipalpi, Tabanidae,* the genus *Syrphus),* I have settled the synonymy in a way that, as far as my knowledge goes, I consider as final. In other portions of the Catalogue, the question constantly arose, whether to substitute uncertainty, for certainty, that is, whether specific names by Loew, the types of which exist in the collection of the Museum of Comparative Zoölogy, had to be replaced by their *more or less probable* synonyms from Messrs. Macquart's and Walker's writings? In such cases I have generally given Dr. Loew's names the first place, leaving the question of priority open for the monographer of the future. In the few cases, where I have acted differently, I have given my reasons in a note. Likewise, as a catalogue-maker, I have not replaced current names by some older ones, which I happened to have discovered; the latter will be found in the synonymy. Thus, in looking over the Banksian collection in London, I found that the undoubted type of Fabricius's *Laphria grossa,* is nothing but the common *L. tergissa* Say. In the same way, *Chrysops variegatus* Degeer, is the older name for the wellknown *Chrysops costatus* from Cuba, and *Milesia virginiensis* Drury, the earlier name for *M. ornata* Fab. All these names, not being current, will be found in the synonymy.

Considerations of the same kind have influenced me in the matter of generic names. The name *Anastrepha* Schiner, although earlier than *Acrotoxa* Loew, will be found among the synonyms, because it belongs to the future monographer of the *Trypetidae* to make changes in an existing monograph. I have but sparingly

given synonymies of generic names, and only as far as I have
been able to verify them; merely copying previous authors I
have avoided, as much as possible. These synonymies will be
found very well worked out in Schiner's: Fauna Austriaca, Diptera.

In looking over Agassiz's *Index* and Marschall's *Nomenclator*
many generic names were found to have been preoccupied in
other departments of Zoölogy. Messrs. Harold and Gemminger
(in their Catalogue of Coleoptera) thought that such names
could, without inconvenience, be maintained, provided they did
not occur in the same order of insects. In order to obviate the
possible drawbacks of such a course, without losing its advantages,
I hit upon the expedient of modifying such names by the
addition of the syllable *Neo*. Nine generic names have been
modified in that way. I do not pretend to impose the names
thus formed on dipterology for ever, and look upon them in
the light of a postponement of a change. A satisfactory and
tolerably permanent settlement of many generic groups among
the diptera still belongs to a distant future. It does no good
therefore, to add scores of new generic names to the large
number of useless ones already in existence.

Such generic names, that are not absolutely identical, but
merely resembling, I did not alter. I share the belief of the above
quoted authors that such names can, without any inconvenience,
remain in use simultaneously, not only in different classes of
animals, but even in different orders of insects. For this reason,
I have not altered *Lasiosoma* Winnertz, 1863 (Lasiosomus,
Hemipt. 1861), *Euparyphus* Gerstaecker, 1857 (Euparypha
Mollusca, 1844), *Phortica* Schiner, 1862 (Phorticus, Hemipt.,
1860), *Euxesta* Loew, 1867 (Euxestus Coleopt., 1858), *Brachydeutera* Loew, 1862 (Brachydeuterus, Fishes, 1862), *Euolena*
Loew, 1873 (Evolenes, Coleopt., 1853), *Peronyma* Loew, 1873
(Peronymus, Volitantia, 1868), *Sympycnus* Loew, 1857 (Sympycna.
Neuropt., 1840), *Eurosta* Loew, 1873 (Eurostus Hemipt. 1863),
and some others.

I have not changed any names on philological grounds, but
have adopted some few changes proposed by others, and which
I considered reasonable.

North American types of Fabricius, which must be preserved in his collection in Kiel, I have not seen. Most of them have been redescribed by Wiedemann. A few of the types of Fabricius in the Banksian collection, in the British Museum, also in the Museum of the Jardin des Plantes in Paris, I have been able to identify.

The majority of Wiedemann's North American types are preserved in the Zoölogical Museum in Vienna; but there are some few in the Museum in Berlin; and also in Westermann's collection in Copenhagen.

The types of the Museum in Vienna are contained in three different collections: the general collection, the so-called collection of Wiedemann, and the collection of Winthem. This is in accordance with the statements of Wiedemann at the end of his descriptions („im Wiener Museum", „in meiner Sammlung" and „in v. Winthem's Sammlung"). The original distribution of the types between these three collections, has not, however, been preserved intact; a large number of types from Wiedemann's collection is now found in v. Winthem's, and in some cases even the type, taken from Wiedemann's collection, · has been replaced by another, wrongly named specimen *).

There is no doubt that this transfer of specimens took place at the time, when both collections were owned by v. Winthem. He must have begun the work of incorporating Wiedemann's

---

*) Thus the type of *Tabanus Reinwardtii* is not in Wiedemann's collection, where it should have been, but in v. Winthem's; the T. Reinwardtii at present found in Wiedemann's collection is an entirely different species. Exactly the same is the case with the type of *Asilus aestuans*, and a wrongly named specimen in Wiedemann's collection has led Dr. Schiner to an erroneous conclusion about the identity of that species. Dr. Schiner's paper: Die Wiedemann'schen Asiliden (Verh. Zool Bot. Ges. 1866), was written under the impression that the so-called collection of Wiedemann still contained all the types referred to it in the Auss. Zweifl., and the readers of that paper must not lose sight of that fact in making use of Dr. Schiner's statements. Nearly all the types of *Tabanus* are in v. Winthem's collection, but in other genera, for instance in *Volucella* most of the types are still found in Wiedemann's collection.

collection into his own, without quite finishing this operation. Dr. Hagen, who saw both collections at that time (in 1839), speaks of them as being united; („einverleibt"; see Stett. Ent. Zeitschr. 1844, p. 131). Under such circumstances, the study of these types requires some critical acumen, and a constant reference to both collections; but when attention is paid to Wiedemann's handwriting, to his statements about the number, the sex and the condition of the described specimens, and finally to the square, red labels, with which the types, thus transferred to v. Winthem's collection are marked, but little difficulty will be experienced in finding out the true typical specimens.

Mr. Macquart's types are chiefly preserved in the Museum in Lille, in that of the Jardin des Plantes in Paris, and in the collection of Mr. Bigot, in the same city; the latter collection also contains the diptera which Macquart had described from Mr. Serville's collection. Many types, principally those of the descriptions in the *Histoire Naturelle des Diptères*, I did not find in the above-named collections; they are very probably lost. And as many of the descriptions in that work are too short to be intelligible, they will have to be canceled. I even suspect that several of the species, described there as North American, and which it has not been possible to identify since, belong to other countries. One instance of that kind, *(Ptilogyna fuliginosa,* an australian species), I have traced with certainty.

The types of Mr. Walker's descriptions (including those in in the *Diptera Saundersiana)* are preserved in the British Museum.

Mr. Walker's writings on the order of Diptera are not better than his publications on Lepidoptera, Hemiptera and Orthoptera, as characterized by other authors. The same species are often found described under several different specific names and placed in different genera; well characterized species of a certain genus are placed in the wrong, sometimes in very distant, genera, or even in the wrong family. In the great majority of cases the descriptions of new species were drawn from a single, often hardly recognizable specimen; and when new species happen to be represented by more than one type-specimen, these are almost sure to belong to different species. A few instances will suffice to illustrate the quality of the work of this author. Of the two North American *Eumerus*, described by Walker, the one proves, upon examination, to be a *Helophilus*, the other, the common *Mesograpta geminata*. A North American *Plecia*, described in the *Diptera Saundersiana*, puzzled me for a long time, until I saw the specimen, which proved to be a common female *Dilophus*, with a red thorax. Mr. Walker's *Thereva plagiata* is the well-known *Stichopogon trifasciatus;* his *Asyndulum tenuipes* is *Blepharocera capitata* Loew; and the common *Cordylura bimaculata* is described as *Lissa varipes*. When such blunders are committed with as striking and easily recognizable forms, as Dilophus or Blepharocera, what can be expected from Mr. Walker in the discrimination of species in such genera as *Culex*, *Bibio*, *Chrysops;* *Tabanus*, *Anthomyia* and the smaller acalypterous Muscidae! These doings were not confined to the North American portion of the collections, which Mr. Walker had under his care. To quote a single instance, the *Musca Aluta* n. sp. List etc. IV, p. 911; (the *patria* is given as „Lapland?“, „France?“), is represented in the British Museum by *seven* specimens, which are nothing but our old friend *Stomoxys calcitrans;* an eighth specimen is an *Anthomyia*. The passage at the end of the description: „In one wing of an insect of this species, the lower cross-vein sends forth a stump into the disc“, refers to this latter specimen, and this passage proves that Mr. Walker looked with some attention at it, without perceiving that it belonged to a different, and very easily distinguishable genus, and even to a different family!

Mr. Walker's identifications of the species of former authors are often, I may say in most cases, incorrect. Thus, when in his description of *Tabanus imitans* Walker, he compares it to *T. abdominalis*, Fabr., he means *T. fuscopunctatus*, Macq. which he took for *abdominalis*.

These facts are given as a warning for entomologists not to trouble themselves too much about the interpretation of Mr. Walker's descriptions, because in most cases, they will find themselves misled *by the very data* furnished by him. And it

2

is for this reason, that in several genera, in the choice of which I have been governed by considerations of expediency, I have enumerated Mr. Walker's species separately, at the end of those genera.

What prevented me from carrying out a more complete revision of Mr. Walker's types, was my want of knowledge in many of the families. As I said above, a great deal remains to be done by others. The question has sometimes been raised whether Mr. Walker's descriptions have any claim to priority at all? In my opinion they have, whenever they are recognizable; but they have none, whenever their title to priority can be established only by reference to the type of the description. The characters of some species are so well marked, that a superficial description of a single specimen is sufficient for the recognition of the species; on the contrary, in other species, sometimes in whole genera and families, the specific characters do not lie on the surface, but must be known beforehand or found out. Is such cases Mr. Walker, or any other describer of his type, merely describe *the specimen,* not the species; they do not know the species again, when they see it; consequently, the name they give to that specimen has no scientific meaning at all, and, it seems to me, no claim to priority. A case in point are the North American *Dolichopodidae,* described by Mr. Walker. The elaborate and painstaking criticisms of these descriptions by Dr. Loew (Monogr. etc. Vol. II), prove, that Mr. Walker, either from want of knowledge or from carelessness, did not pay the slightest attention to those characters which serve to distinguish the species of Dolichopus from each other, so that of the twenty‑six so‑called species, described by him, *not a single one* could be recognized. Now I ask whether it would be expedient, with Mr. Loew's monograph in hand, to determine Mr. Walker's type-specimens and then to grant to the names, attached to those types, the priority over Mr. Loew's names? I do not think so, and, for this reason, I would not undertake that task, even if it were possible*). The same reasoning

---

*) Many of Dr. Walker's species of Dolichopus are represented by female specimens, which it would be impossible to determine.

applies to Mr. Walker's descriptions in the genera *Chrysops* and *Tabanus*. A careful study of these descriptions convinced me of their uselessness; the examination of Mr. Walker's types showed, that in most cases, he did not know his own species again, that he described the same species several times in succession (the descriptions being sometimes by the side of each other in his works), that the confused specimens of different species in the same description. Under such circumstances, I did not feel justified in upsetting the nomenclature introduced by me in my monograph.

The authorities of the British Museum, in a most praiseworthy, and truly scientific spirit, have bestowed a great deal of labor upon preserving and labelling Mr. Walker's types. But the task of singling out the original type of the description, from among the specimens added afterwards, is by no means an easy one, often hardly possible. Furthermore, it is a well-known fact that authors are apt not to be very careful with their own types; to remove and diplace them, when made aware of an error; and Mr. Walker, in this respect, was not an exception. Neither his, nor any other types can, therefore be implicitly relied upon, and we have, ultimately, to fall back on the descriptions. — In rescuing those of Mr. Walker's descriptions, which are available and in rejecting the remainder, as useless, we pursue, I think, a course consistent both with justice and scientific expediency.

THE NUMBER OF DESCRIBED NORTH AMERICAN DIPTERA. The number of described Coleoptera from North America, north of Mexico, in Mr. Crotch's Check List is 7450. It is impossible to make a similar statement for the diptera, because, as experience has shown, most of the earlier descriptions are entirely unavailable and represent species which exist merely on paper. The number of described diptera from North America, north of Mexico, contained in the Museum of Comparative Zoölogy in Cambridge, Mass., is a little over 2000. The number of available, but not yet identified, descriptions of earlier authors is not large; and thus we may safely assume that, excluding the unavailable descriptions, the number of described diptera of North America, north of Mexico, will hardly reach 2500. But the undescribed materials, accumulated in the collections, if worked up, would largely increase, perhaps double, that number. Considering the little attention hitherto paid to the order of diptera, these figures seem to prove that the number of existing species of diptera in North America will easily reach and perhaps exceed the number of Coleoptera.

THE FUTURE OF AMERICAN DIPTEROLOGY. Of all orders of insects, the diptera offer probably the most difficulties to the describer. The reason lies in the minuteness of the characters, on which generic and specifics distinction are based. In consequence of this difficulty, there is and was more blundering in this order of insects than in any other, and the mischief done by the incompetent is greater here, than in any other order. By *incompetent* I do not merely mean those, who know little or nothing about diptera; I mean even dipterologists who attempt to write about a family of diptera before having made a special study of it. And in this respect, every one of us, in the course of his career, is often tempted to do some work, which he is incompetent to perform, and every one of us has, some time or other, actually done such work.

In order to preserve, as much as possible, American dipterology against the evils of incompetence, I attempted, several years ago, to draw up some recommendations as to the best course to pursue in that study (in A. S. Packard's Record of American entomology for 1868). As these recommendations have lost nothing of their appropriateness, I may be allowed to reproduce them here.

„If I am asked now what the *desiderata* for the future of this branch of science in America are, I would answer:

„1. Continue the publication of North American diptera in monographs."

„2. Avoid as much as possible the publication of detached species, either singly, or in numbers."

„The cases when the publication of dètached species of Diptera can be really useful in the present state of american dipterology are rare, and will easily suggest themselves to the good sense of the unprejudiced."

„Consciencious monographs are always useful."

„Let monographs be prepared of the families of diptera on the same plan as the monograph of the *Dolichopodidae* by Dr. Loew, or of the *Tipulidae* by me. Let the series of these monographs begin with the larger forms and the more numerous families, as the Tabanidae*), the Asilidae, the Stratiomyidae, the Bombylidae, the Empidae etc. Such a basis being laid with those families, the study of which is comparatively easy, the difficult ones, as the Chironomidae, the Culicidae and the numerous groups of the Muscidae, will follow. The study of these difficult families must be the work of specialists. Mr. Winnertz, of Crefeld, Prussia, devoted more than twenty-five years to the study of the genus *Ceratopogon*, the genus *Cecidomyia* and the family *Mycetophilidae*. During that long period of patient collecting, drawing and describing, he published only four monographs of moderate size. And it is certain that, without such patient collecting, drawing and describing for a number of years, any monograph of such genera as *Ceratopogon* or *Sciara* would have been worthless. Diptera are not like the other orders of insects, where a superficial comparison of two specimens enables one, in most cases, to decide, whether they belong to the same species or not. Each family of diptera requires a special study and a dipterologist may be very well versed in some families, without being able to express any opinion with regard to questions, concerning others."

„*Specialization* is therefore the motto of dipterology. Amateurs may collect and name diptera, but do not let them publish anything, until they have chosen some single family and nearly exhausted it by study and collecting. If they try such a course, they will find that the exhaustive study of a single

---

*) Since writing the above I have published a monographic essay on the Tabanidae.

family is far more remunerative, both in pleasure and in usefulness, than the random description of numerous new species."

But little reflection is necessary to prove that monographic
work is the most advantageous form of work in descriptive
entomology. It implies the greatest concentration of one's
working power, and for this very reason, its greatest economy;
its products are the most lasting, because a good monograph
is not easily supplanted; they are the most useful, because they
facilitate and encourage the study, instead of obstructing it, as
some other kind of work is liable to do.

The productions of unconsciencious and incapable writers
ought not to obstruct better workers and thus to impede the
progress of science. Let no one, attempting a monograph, be
deterred by the number of earlier descriptions in the same family.
The principal effort should be, to collect an abundant material,
representing as nearly as possible the fauna of a given region
in the family selected for work. With such a material the
identification of previous descriptions becomes comparatively easy.
With some perseverance and attention, the available descriptions
will soon be identified and the residue may be neglected, as
useless. It may happen that the whole, or nearly the whole
of the previous descriptions proves to be unavailable; let not
the work be prevented by it. Of the thirty-two earlier descriptions of North-American *Dolichopus*, all but *two*, were unrecognizable; this did not prevent Dr. Loew from writing a
standard monograph of the genus. The next step for the
monographer should be, to prepare descriptions of *all* the species,
because it is a bad plan, in a monograph, simply to refer to the
descriptions of previous authors *). By means of analytical tables,
or of figures, the descriptions should be rendered accessible,
enabling every one, with the monograph in hand, to get at the
name of a given species.

---

*) Erichson expressed the same views in the Preface to his *Ento-
mographieen,* and the passage deserves to be reproduced here:
„Beschreibungen neuer Arten scheinen mir in den meisten Fällen nur dann ein n
wesentlichen Fortschritt der Wissenschaft zu bedingen, wenn eine Uebersicht über die Ab-
theilung, der sie angehören, damit verbunden, und diese als ein Ganzes betrachtet wird.
Es kann in solchen Arbeiten oft hinreichend sein, bei bekannten Arten auf schon vorhan-

It is greatly to be desired that the fauna of the Northern and Middle States should be worked up soon, in order that it may serve as a foundation for the study of the other faunas of the continent. The species, occurring around the centres of civilization should be described first, so as to have the species from the more distant regions *compared with them*. As matters stand now, the opposite state of things is very likely to happen; numerous Western species, brought by explorers, will be described, leaving the Eastern entomologist in doubt, whether the forms which he finds at his door, are the same species, or not.

I tender my sincere thanks to the authorities of the public Museums and owners of private collections, who have kindly assisted me in my work; the authorities of the British Museum, of the Museum du Jardin des Plantes, the Museum of the University in Berlin, the Imperial Zoological Museum in Vienna, and the public Museums in Lille, Frankfort, Darmstadt and Turin. Among the owners of private collections of exotic diptera, I owe a special tribute of gratitude to Mr. Bigot in Paris, Professor Bellardi in Turin and Mr. v. Roeder in Hoym (Anhalt).

Dr. Loew in Guben, my correspondent and collaborator for many years, was unfortunately prevented, by a sudden failure of his health, from assisting me during the preparation of this volume. I have nevertheless used many data, found in his letters, or taken down in looking over his collection of North-American Diptera, (the same, which now is in the Museum of Comparative Zoölogy in Cambridge, Mass.). The large share he has taken in the advancement of North American dipterology speaks for itself.

The greatest share of recognition however, belongs to the Institution under whose auspices, and at whose expense, the

----

dene Beschreibungen zu verweisen, im Allgemeinen habe ich aber gefunden, dass bei diesem Verfahren oft selbst die ausführlich beschriebenen Arten zweifelhaft bleiben, besonders wenn es darauf ankommt sie von nahe verwandten zu unterscheiden, welche als bekannt vorausgesetzt, und nicht näher charakterisirt sind. Versucht man aber, diesen ihre wesentlichen Merkmale beizufügen, findet sich bald, dass man weit sicherer, und ohne merklich grossen Aufwand an Raum, zum Ziele gelangt, wenn man die sämmtlichen Arten gleichmassig beschreibt.'

principal works on North American Diptera, beginning with the *Catalogue* etc. of 1858, have been published. There is not the slightest exaggeration in saying that, without the encouragement and the support, received from the Smithsonian Institution for the last 20 years, the study of North American diptera would have remained far behind the stage which it has reached at present.

The inherent limitation of a Catalogue like the present consists in the fact, that although it is more than a mere compilation, it is less than a monograph. In many respects, the task of the monographer had to be encroached upon: synonymies established, species transferred to the proper genera, European species, occurring in North America, recognized and introduced in the lists etc. The amount of latent labor of this kind, accomplished in this Volume, will reveal itself to those, who will take the trouble to compare it with my earlier Catalogue (for instance in the *Asilidae* or *Syrphidae*). There is some danger in carrying this kind of anticipatory epuration too far, because in performing it, we cannot expect to attain the thoroughness of a monograph. And it is in the belief, that I have reached the point, where it is time for me to stop, that I hand over my work to the public, with a full sense of its imperfections.

C. R. OSTEN SACKEN.

HEIDELBERG, Germany
June 1878.

TABLE OF CONTENTS.

	Page.
Preface	V
Table of Contents	XXV
Authorities	XXVII
List of the new genera and the new species, published in the notes to this volume	XLVII
Explanations, necessary for the use of the Catalogue	XLVIII
Catalogue of North American Diptera	3
Notes	214
Index	265

AUTHORITIES.*)

Amiot. — In the Annales de la Soc. Entom. de France, 1855, Bulletin, p. CIV; remarks upon *Cecid. tritici* Kirby and the identity of the european and american insect, known under that name.

Bellardi, Luigi. — Saggio di ditterologia messicana. — Two parts and Supplement; five plates. — In the Mem. della Reale Accad. delle Scienze di Torino, Ser. II, Vol XIX, 1859, Vol. XXI, 1861—62; also published separately, in 4°, Part I, 77 pages, 2 plates; Part II, 99 pages, 2 plates; Supplement, 28 pages, 1 plate.

Contains the descriptions of about 170 new species of mexican diptera orthorhapha.

Bergenstamm und Loew (P.). Synopsis Cecidomyidarum. — In the Verh. Zool. Bot. Ges. 1876.

A Synopsis of all the litterature on the subject, including the N. A. species; very thorough and complete

Bilimek, Dominik. — Fauna der Grotte Cacahuamilpa in Mexico. — In the Verh. Zool. Bot. Ges. in Wien, 1867, p. 901.

Pholeomyia leucozona n. gen. et sp.

Bigot, Jacques. — Worked up the diptera for Ramon de la Sagra's: Histoire physique, politique et naturelle de l'île de Cuba. Paris 1857 (with a plate). Published in french and in spanish; the french edition is quoted in the catalogue, the spanish has a different pagination.

Twenty five new species.

„ Dipterorum aliquot nova genera. — In the Revue et Magazin de Zool. 1859, p. 305—315; Tab. XI.

Hystrisyphona niger n. gen. et sp., Mexico.

Cryptineura hieroglyphica, n. gen. et sp., United States (=*Chrysogaster nitidus* Wied.)

„ Diptères nouveaux ou peu connus.

V. Asilides exotiques nouveaux (Ann. Soc. Ent. Fr. 1875, p. 237—248).

---

*) It was not intended to give here the full titles of *all* the works and papers quoted in the present volume, but merely of such as contain descriptions of north american diptera.

VI. Espèces exotiques nouv. des genres *Sphixea* et *Volucella* (l. c. p. 469—482).

VII. Espèces nouv. du Genre *Cyphomyia* (l. c. p. 483—488).

VIII. Curie des *Phasides* (l. c. 1876, p. 389—400).

IX. X. Genre *Somomyia* Rondani (l. c. 1877, p. 35—48; 243—259). These papers contain 23 new spec. from Mexico, *two* from the United States, *two* from Haiti, *one* from Jamaica.

Bigot, Jacques. — (without title) in the Bullet. de la Soc. Ent. de Fr. pag. CLXXIV, 1875.

    Thevenemyia californica, n. sp. California.

   " (without title), l. c. pag. XXVI, 1877.

    Carlottaemyia moerens, nov. gen. and spec. from Mexico (=Diacrita costalis Gerst.)

   " (without title), l. c. pag. LXXIII, 1877.

    Macroceromys nov. gen. (Xylophagidae), Mexico.

Bosc. — *Ceroplatus carbonarius*, from Carolina, described in the Dict. d'Hist. Nat., Paris 1802—1804, in 24 Vol., 8vo, chez Déterville et Roret; also in Nouveau Dict. d'Hist. Nat. Paris 1816—19, in 36 Vol.

Brauer, Friedrich. — Monographie der Oestriden, Wien 1863. — With ten plates. The most complete monograph in existence on the subject; it contains the descriptions of all the known american species.

    Cuterebra scutellaris, n. sp. United States.

   " Beschreibung neuer und ungenügend bekannter Phryganiden und Oestriden. — In the Verh. Zool. Bot. Ges. in Wien 1875.

    Hypoderma bonassi, n. sp. (larva), occurring on the american buffalo.

Burgess, E. Two interesting american diptera. — In the Proceed. Boston Soc. N. H. 1878, p. 320—324, with figures.

    Glutops singularis, nov. gen. et sp.; *Epibates Osten Sackenii* n. sp.

Clark, Bracy. — Observations on the genus *Oestrus.* — In the Trans. of the Linn. Soc. Vol III, 1797.

    Oestrus cuniculi, n. sp., Georgia.

   " An essay on the Bots of Horses and other Animals. London 1815, 4°; with two plates.

    Cuterebra horripilum and *Cephenomyia phobifer,* n. sp.

   " Addenda, 1848, 4°, with one plate.

    Cuterebra atrox, n. sp., Mexico.

   " Of the insect called *Oistros* by the ancients and of the true species intended by them under this appellation etc. To which is added a description of a new species of *Cuterebra.* In the Trans. Lin. Soc. Vol XV, p. 402, 1826.

    Cuterebra fontinella, n. sp. Illinois.

Coquebert, A. J. — Illustratio iconographica insectorum quae in museis parisinis observavit et in lucem edidit F. C. Fabricius, praemissis ejusdem descriptionibus. Paris, 1799—1804. In fol. min. With

30 plates. Several American species are figured in this work, but no new ones described.

CURTIS, J. — Description of the Insects brought home by Commander J. Clark Ross. (In his Voyage to the Arctic Regions, 1831.)
Chironomus borealis, *Tipula arctica*, *Helophilus bilineatus*, *Tachina hirta, Anthomyia dubia,* and *Scatophaga apicalis,* are new.

DEGEER, Baron Charles. — Mémoires pour servir à l'Histoire des Insectes. Stockholm, 1752—78. 7 vols. Several American species are described in the 6th vol.

DESVOIDY, Robineau. — Essai sur la tribu des Culicides. In the Mémoires de la Société d'Hist. Nat. de Paris, vol. iii, p. 390—413. 1827.
Five new species from N. America and the West Indies.

„ Essai sur les Myodaires. In the Mémoires des savants étrangers de l'Academie des Sciences de Paris. Vol ii. (1830).
This Essai is a 4to. volume of more than 800 pages, containing a new systematical arrangement of the whole group, and numerous descriptions (among which some eighty new North American species).

„ Histoire naturelle des diptères des environs de Paris. Paris 1863. (Vol. I, XVI and 1143 pages; Vol II, 920 pages).
Posthumous work; contains short descriptions of a few N. A. diptera; no new ones.

DRURY, Drew. — Illustrations of Natural History, wherein are exhibited upwards of two hundred and forty figures of exotic insects. London, 1770—82. 3 vols. (A new edition of this work has been published in 1837, by Westwood, under the title of Illustrations of Foreign Entomology.) Eight N. American and West Indian species are figured.

DUFOUR, Léon. — Révision et Monographie du Genre *Ceroplatus.* In the Annales des Sciences Naturelles, 2e serie, vol. xi, p. 193 (1839), with figures. Contains the description of *Ceroplatus carbonarius Bosc,* from Carolina. Conf. Bosc.

DUMÉRIL, A. M. C. — Considérations Générales sur la classe des Insectes, etc. Strasbourg et Paris, 1823. With plates. No new species.

ERICHSON, F. W. — Die Henopier. Eine Familie aus der Ordnung der Diptern. (In Erichson's Entomographien, Berlin, 1840.) *Ocnaea micans,* new species from Mexico.

ESCHSCHOLZ, Dr. J. F. — Entomographien, in 8vo. Berlin, 1823.
Empis laniventris, and *Musca obscoena,* new species from Unalaschka.

FABRICIUS, J. C. — Systema Entomologiae. Flensburgi, 1775.
„ Mantissa Insectorum. 2 vols. Hafniae, 1787.
„ Entomologia Systematica. 4 vols. Hafniae, 1772—94; Suppl. 1798.
„ Systema Antliatorum. Brunsvigae, 1805

FABRICIUS. O. — Fauna Groenlandica. Hafniae et Lipsiae, 1780. 8vo.
Eighteen diptera are described. A useful commentary to this

book, containing the true interpretation of several of the species, may be found in Schiodte's article on the Arthropods of Greenland. See Schiödte.

Fitch, Dr. Asa. — An Essay upon the wheat-fly and some species allied to it. Albany, 1845.

This is the first edition, which was published in the American Quarterly Journal of Agriculture and Science, vol. ii, No. 2. It contains the descriptions of *Cecidomyia tritici*, Kirby; *Cec. caliptera*, n. sp.; *Cec. thoracica*, n. sp.; *Cec. tergata*, n. sp. A second edition appeared in 1846, in the Transactions of the N. Y. State Agricultural Society, vol v. A new species, *Cec. cerealis*, is separated in this edition from *C. caliptera*, and full descriptions with figures of both are given.

" The Hessian Fly. Albany, 1846. (2d edit. 1847.) With a plate. Published originally in the American Journal of Agriculture and Science, vols iv, v. (1846). Reprinted with some additions in the Transactions of the N. Y. State Agricultural Society, vol. vi, p. 316—376 (1846; in pamphlet-form it bears the date of 1847).

" *Cecidomyia salicis*, n. sp., described in the American Quarterly Journal of Agriculture and Science, vol. i, p. 263.

" Winter Insects of Eastern New York. — In the American Journal of Agric. and Sci., vol. v, pp. 274—284.

N sp. *Culex hiemalis*, *Chironomus nivoriundus*, and *Trichocera brumalis*.

" Survey of Washington County, New-York. — In the 9th. vol. of the Transactions of the N. Y. State Agricultural Society.

Several species occurring in that locality, are mentioned in a popular way.

" First and Second Report on the Noxious, Beneficial, and other Insects of the State of New York. Made to the State Agricultural Society pursuant to an appropriation for this purpose from the Legislature of the State. Albany, 1856. (With four plates.)

Before the publication of the Second Report, the first had been distributed under the title of First Report, etc. 1855 This work contains 21 new American diptera.

" Third, fourth and fifth Reports on the Noxious, beneficial and other insects of the State of New York, made to the State Agricultural Society, pursuant to an annual appropriation for this purpose from the legislature of the State. Albany, 1859. With four plates and many woodcuts.

Cuterebra emasculator n. sp. and several *Cecidomyiae*.

" Sixth, seventh, eighth and ninth Reports etc. etc. Albany, 1865. — With four plates and several woodcuts.

Contains a new edition of the papers on *Cecid. tritici* and *destructor*.

All these reports appeared successively in the Trans. of the N. Y. State Agric. Society and were collected and issued afterwards as separate volumes: Volume I, containing Reports 1 and 2; Vol. II, Reports 3—5; Vol. III, Reports 6—9. Each volume has a title-page, as given above, and a complete index of the contents. In the *first* and *third* volumes the pagination runs through the whole volume; in the *second* volume, a new pagination begins with every report, but, at the same time, the species successively discussed are numbered and these numbers run through the whole volume. For this reason, in quoting this *second* volume, I had to give the *number* of the species referred to, while in quoting the other two volumes, I give the page. — Dr. Fitch's following Reports, which I have seen up to the 12th (1867), do not contain any new species·of N. A. diptera.

FORSTER, J. R. — Novae Species Insectorum. Centuria I. London, 1771.
Tabanus americanus, n. sp. (*T. ruficornis*, Fab,.

GRAY, G. R. — In E. Griffith's Animal Kingdom. (London, 1824—33. 16 vols. With engravings.)
Several N. American species are figured in the 15th vol. The descriptions by Gray are very incomplete.

GREEN, Dr. — Natural History of the Horse Bee. (In Adam's Medical and Agricultural Register, vol. i, p. 53; and in New England Farmer, vol. iv, p. 345.)
Gastrus veterinus, Fab.

GUÉRIN et PERCHERON. — Genera des Insectes. Paris, 1831—35. (With plates.)
Culex mosquito, R. Desv., from Cuba; and *Tabanus flavus*, Macq. (Syn. of *T. mexicanus* Lin.) from the U. States, are figured.

GUÉRIN-MÉNEVILLE, F. E. — Note sur deux Insectes Parasites de la cochenille qui font un grand tort à cette culture en Amérique. (Read in the Academy of Sciences in Paris on the 13th of Nov. 1848. Conf. Guérin's Revue Zoologique, 1848, p. 350.)
Baccha cochenillivora, n. sp. from Guatemala.

,, Iconographie du Règne Animal de G. Cuvier etc. Paris, 1829—44. The insects are in the last (7th) volume.
Leptis Servillei, n. sp. — United States.
Calobata ruficeps, n. sp. — Cuba.
Toxophora americana, n. sp. (figured, not described).
Cuterebra apicalis, n sp. America.

GERSTAECKER, Dr. A. — Beitrag zur Kenntniss der Henopier. — In the Stett. Ent. Zeitschr. 1856, p. 360.
Eulonchus smaragdinus, n sp. California.

,, Beitrag zur Kenntniss exotischer Stratiomyiden. — In the Linn. Entom. Vol. XI, 1857, p. 261; Tab. III
· N. Sp. *Cyphomyia* 3 spec., *Stratiomys* 2 spec., from Mexico; *Chauna ferruginea* from Cuba.

,, Beschreibung einiger ausgezeichneten neuen Diptera aus der Familie Muscariae. — In the Stett. Entom. Zeitschr. 1860, p. 163; with a plate

N. sp. *Pyrgota vespertilio, pterophorina, Toxotrypana curvicauda, Diacrita costalis* from North-America.

GERSTAECKER, Dr. A — Systematische Uebersicht der bis jetzt bekannt gewordenen Mydaiden. — In the Stett. Entom. Zeit. 1868, p. 65, with a plate.

    Leptomydas pantherinus, Mydas lavatus, annularis nov. sp. from N. America.

„ Die zweite deutsche Nordpolfahrt in den Jahren 1869 — 70. Leipzig, 1874. Hymenoptera and Diptera by Gerstaecker; the latter are represented by four species, collected in East Greenland, lat. 73° — 75°: *Tipula truncorum* Melg., *Echinomyia aenea* Stäger, *Cynomyia alpina* Zett., *Calliphora groenlandica* Zett.

GROTE, Aug R. — Description of two new species of North American Brachycerous Diptera. — In the Proc. of the Entom. Soc. Phil. Vol. VI, p. 445, 1866 — 67.

    Sparnopolius coloradensis and *cumatilis*, n. sp. Colorado.

HALDEMAN, Prof. S. S. — Description of several new and interesting Animals. — In the American Journal of Agriculture and Science, vol. vi, p. 193. With figures. 1847. (Reprinted in the Proceedings Boston Soc. N. H. January 1859.)

    Cecidomyia robiniae, n. sp.

HARRIS, Dr. Thaddeus William. — Catalogue of the Insects of Massachusetts. In Prof. Hitchcock's Report on the Geology, Botany, and Zoology of Massachusetts.

    Prof. Hitchcock's Report had two editions; in the first (1833), Dr. Harris mentioned only the generic names of the insects, adding the number of species belonging to each genus. In the second edition (1835), the specific names are also given; many of them are mere collection names, never having been published.

„ A Treatise on some of the Insects of New England, which are injurious to Vegetation. Second edition Boston, 1852.

    The first edition of this work was published in 1841, under the title of A Report on the Insects of Massachusetts, injurious to Vegetation. The second edition contains many additions.

„ A Treatise on some of the Insects injurious to vegetation. Third edition. Boston, 1862. With 8 plates and 278 woodcuts.

    Was published at the expense of the Commonwealth of Massachusetts and is provided with notes by different authors; those on the Diptera are by C. R. Osten Sacken. The quotations in the present volume are from this edition.

„ Entomological correspondence. Edited by Samuel H. Scudder. — Boston, 1869.

    Contains on p. 335—336 descriptions of *Musca harpyia* Harris (= Musca domestica Lin.) and *Musca familiaris* Harris (apparently the same as the european *Pollenia rudis*).

HAUSMANN. — Entomologische Bemerkungen. Braunschweig, 1799.

    Syrphus trifasciatus, n. sp. = Milesia ornata Fab.).

HOLMGREN, A. E. — Insecter fran Nordgroenland samlade af Prof. A. E. Nordenskjöld ar 1870. — In the Ofvers. Kongl. Vetensk. Ak. Förhandl. 1872, p. 100—105.

Contains thirty-nine diptera, among which *six* Ariciae, *one* Scatomyza, *one* Boletina, *one* Sciara are new.

ILLIGER. — Neue Insecten. — In the Magazin fur Insectenkunde, Vol. I, p. 206.

Midas fulvifrons, n. sp. — Georgia.

JAENNICKE, F. — Neue exotische Diptern aus den Museen zu Frankfurt und Darmstadt. — In the Abhandl. d. Senckenb. Ges. Vol. VI; with 2 plates; also separately, in one volume, in 4°, 100 pages; Frankfurt, 1867. Thirty-four new species from Mexico and N. America.

KIRBY, Will. — Fauna Boreali-Americana; or the Zoology of the northern parts of British North America, by J. Richardson, assisted by W. Swainson and Will. Kirby. London, 1829—37. 4 vols.

The fourth volume, containing the entomological part, is by W. Kirby; nine new diptera are described. *(Culex punctor, Tipula pratorum, Arthria analis, Empis luctuosa, geniculata, Tabanus affinis, zonalis, Musca cadaverum, mortisequa.)*

„ A Supplement to the Appendix of Capt. Parry's Voyage in 1819, 1820, containing Mammalia, Birds, Fish, and Marine Invertebrate Animals, by Edw. Sabine; Land Invertebrate Animals, by W. Kirby, etc., in 4to. London, 1824.

Ctenophora Parrii, Chironomus polaris, n. sp.

KIRKPATRICK, J. — The army worm. — Article in the Ohio agricultural Report for 1861.

Exorista leucaniae and *E. Osten Sackenii,* parasites of the army-worm.

LAMARCK, J. B. — Histoire Naturelle des Animaux sans Vertèbres, etc. 1ère édit. 7 vols. Paris, 1815—22. 2e édit. 11 vols. Ibid. 1835—45.

The insects form the third volume of the first, and the fourth of the second edition. I have quoted the first edition. Some typical forms only of American insects are mentioned in this work, and no new species described.

LATREILLE, P. A. — Histoire Naturelle, générale et particulière des Crustacés et des Insectes. 14 vols. Paris, 1792—1805. (This work forms a part of Sonnini's Suites à Buffon.)

„ Genera Crustaceorum et Insectorum, etc. 4 vols. Paris 1806, 7 et 9.

„ The articles on Entomology in the Nouveau Dictionnaire d'Histoire Naturelle, etc. Comp. above Bosc.

All these works contain the mention or description of some typical forms from N. America, but no new species.

LEACH, W. E. — On the genera and species of Eproboscideous Insects. — In the Wernerian Transactions, vol. ii. Edinburgh, 1817.

Olfersia Americana and *Ornithomyia erythrocephala,* n. sp. from N. America.

3

LE BARON, William, M. D — (State Entomologist for Illinois). — Second annual Report on the noxious insects of the State of Illinois, 1872.
Tachina (Exorista) *phycitae*, n. sp.

LINNÉ, Carol. a. — Systema Naturae, etc. Editio XII. Second vol. 1767.

„ Amoenitates Academicae s. Dissertationes variae Phys. Med. Botanicae, ante hac seorsim editae, nunc collectae et auctae. 7 Voll. cum tab. aen. 1749—69.
Asilus aestuans from Pennsylvania, n. sp.

LOEW, Dr. H. — Beschreibung einiger neuen *Tipularia terricola*. In the 5th vol. of the Linnaea Entomologica. Stettin, 1851.
General observations on the genera: *Ptilogyna, Aporosa* and *Toxorhina*, and the descriptions of three new species, *Ap. rufescens, virescens,* and *Tox. fragilis,* from the West Indies.

„ Bemerkungen üb. die Gattung Beris. — In the 7th vol. of the Entomologische Zeitung. Stettin, 1846.
Several American species mentioned; no new ones described.

„ Helophilus. — In the 7th vol. of the Entomologische Zeitung. Stettin, 1846.
Monograph of the genus, mentioning some American species; *H. glacialis,* n. sp. from Labrador.

„ Chauna, genus novum. — l. c. 8th vol. p. 370. Stettin, 1847.
Chauna variabilis, n. sp. from Cuba.

„ Ueber *Tetanocera stictica,* Fab., und ihre nächsten Verwandten, etc. — l. c. 8th vol. p. 114. Stettin, 1847.
Tet. flavescens, n. sp. from Carolina.

„ Ueber *Tetanocera ferruginea,* Meig. und die ihr verwandten Arten. — l. c. 8th vol. p. 194.
Tet. plumosa, n. sp. from Sitka.

„ Bemerkungen über die Familie der Asiliden, etc. in 4to. Berlin, 1850.
Dasypogon anthracinus, n. sp. from Mexico.

„ Ceria. — In the Neue Beiträge zur Kenntniss der Dipteren, by Dr. Loew. Erster Beitrag. Berlin, 1853.
Monograph of the genus; *Ceria pictula* from the U. St.; *C. arietis* and *signifera,* from Mexico, are new.

„ Conops. (l. c.)
Monographical Essai. *Conops genualis, bulbirostris,* and *castanoptera,* n. sp. from the U. States.

„ Neue Diptern (l. c. Zweiter Beitrag. Berlin, 1854).
Pyrgota millepunctata, n. sp. from North America (=*P. valida* Harris).

„ Bombylius. (l. c. Dritter Beitrag. Berlin, 1855.)
Monograph of the genus, containing important synonymical remarks upon several American species; no new ones described.

„ Dipterologische Notizen. Neue Americanische Dolichopoden. — In the Wiener Entomologische Monatsschrift, vol. i, p. 37. Vienna, 1857.

Lyroneurus caerulescens from Mexico, and *Plagioneurus univittatus* from Cuba, new genera and species.

LOEW, Dr. H. — Excursion nach dem Neusiedler See. — In the Neue Beitr. etc. Vierter Beitrag, 1856.

On p. 18 several european species, also occurring in N. A. are mentioned, but a part of these statements is based on erroneous data about the locality. *Helophilus pendulus*, *versicolor*, *floreus*, and *Chrysotoxum bicinctum* have never, as yet, been found in N. America.

„ Ueber die Fliegengattungen *Microdon* und *Chrysotoxum.* — In the Verh. Zool. Bot. Ver. 1856.

Mentions, on p. 614, the occurrence of *Chrysotoxum bicinctum* Lin. in N. America (see the remark to the previous title).

„ Zur Kenntniss der europ. Tabanus-Arten. — In the Verh. d. Zool. Bot. Gesellsch. Wien 1858, p. 573—612.

N. sp. *Tabanus septentrionalis;* Labrador.

„ Ueber einige neue Fliegengattungen. — In the Berl. Entom. Zeitschr. 1858, vol. II, p. 101—122, with a plate.

Plecia longipes n. sp., from New Orleans.

„ Ueber die europ. Helomyzidae und die in Schlesien vorkommenden Arten derselben. — In the Schles. Zeitschr. f. Entom. 1859.

Quoted for the full descriptions of some european species, which also occur in North-America.

„ Die N.-Americanischen Arten d. Gattungen *Tetanocera* und *Sepedon.* — In the Wiener Entom. Monatschr. III, p. 289—300; 1859.

The species here described were later embodied in the paper on Sciomyzidae in the Monographs etc. Vol. I.

„ Diptera americana ab Osten-Sackenio collecta, decas prima. — In the Wiener Entom. Monatschr. IV, p. 79—84; 1860.

Ten new species from the United States; the descriptions were all reproduced in the authors later publications, with the exception of two: *Clinocera maculata* and *C. conjuncta.*

„ Diptera aliquot in insula Cuba collecta. — In the Wiener Entom. Mon. V, p. 33—43; 1861.

Twenty new species.

„ Die Nord-Americanischen Dolichopoden. — In the Neue dipterol. Beiträge, fascicle 8th. 1861.

This paper is superseded by the later Monograph of the N. A. Dolichopodidae in the Monographs, etc. Vol. II.

„ Die americanischen Ulidina. — In the Berl. Entom. Zeitschr. XI, 1867, p. 283—326, with one plate.

Several new N. A. Genera and species. They are all contained in the third volume of the Monographs of the N. A. Diptera.

„ Monographs of the Diptera of North-America, Vol. I—III, with eleven plates. Washington, Smithsonian Institution, 1862—1872.

Vol. I, 1862. — General introduction, Trypetidae, Sciomyzidae, Ephydrinidae and Cecidomyidae (the latter by C. R. Osten-Sacken) [Smithsonian Miscell. Collections, Volume VI].

Vol. II, 1864. — Dolichopodidae [Smiths. Misc. Coll. Vol. VII.
Vol. III, 1872 — Ortalidae and additions to Trypetidae [Smiths.
Miscell. Coll. Vol. XI]*) (For the 4th Volume, see C. R. Osten-
Sacken.)

Loew, Dr. H. — Diptera Americae Septentrionalis indigena. — In the
Berliner Entomol. Zeitschr. Century I, 1861; II, 1862; III and IV,
in 1863; V in 1864; VI in 1865; VII in 1866; VIII and IX in
1869; X in 1872. — Also published separately, in two volumes.
In the present Catalogue, this publication is quoted thus: *Loew*,
Centuriae.

" On the diptera of the Amber-Fauna. — A lecture, delivered at
the meeting of the German association of naturalists and physicians
in Königsberg, translated from the german by C. R. Osten-Sacken,
and published in the Amer. Journ. of Science and Arts, Vol. XXXVII,
May 1864. — The translation contains, on p 317, in a note, a
list of species of diptera which are common to Europe and to
North-America; (this note does not exist in the original german
edition of the lecture).

" Bemerkungen über die von Herrn v. d. Wulp in der Zeitschrift
der niederländischen Entomol. Gesellschaft für 1867 publicirten
N. A. Dipteren. — In the Zeitschr. f. die gesammten Naturw. 1870,
Bd, XXXVI, p. 113—120.
Remarks about the synonymy and the systematic location of
the species in Mr. v. d. Wulp's paper.

" Ueber die Arten d. Gattung Sphyracephala Say. — In the Zeitschr.
f. die Gesammten Naturwissenschaften 1873, Bd. XLII, p. 101.
Remarks on *S. brevicornis* Say; *S. subbifasciata* Fitch declared
its synonym.

" Neue nordamerikanische Dasypogonina. — In the Berl. Entomol.
Zeitschr. 1874, Vol. XVIII, p. 353—377. —
Fourteen new species.

" Neue nordamerikanische Diptera. — In the Berl. Entom. Zeitschr.
1874, p. 378—384.
Six new species.

" Beschreibungen neuer amerikanischen Dipteren. — In the Zeitschr.
f. Gesammte Naturw. 1876; Bd. XLVIII, p. 317—340.
Seventeen new species from North-America.

" Revision der Blepharoceridae. — In the Schles. Zeitschr. f. En-
tomologie, Neue Folge, Heft VI; Breslau 1877. —

---

*) The octavo publications of the Smithsonian Institution are issued in two forms:
separately, or collected in a series of volumes under the general heading of:
Smithsonian Miscellaneous Collections.
Most of the public libraries in North-America and in Europe possess this series,
which is recorded *as such* in their Catalogue. But the separate works which it contains, are,
in most cases, *not* recorded in the Catalogues, unless they have been received separately.
Persons who are not aware of this circumstance have often searched Catalogues in vain
for Dr. Loew's or my publications, while they would have found them under the head of
the Smithsonian Miscellaneous Collections.

The description of *Bibiocephala grandis* O. S. is reproduced here, in german translation.

Loew, Dr. H. — Neue nordamerikanische Ephydrinen. — In the Zeitschrift für die Gesammten Naturwissenschaften, Halle 1878, March—April, p. 192—203.

Fourteen new species.

Macquart, J. — Histoire naturelle des Diptères. — Paris 1834 35; 2 vols, with plates. — Forms a part of the Suites à Buffon, published by Roret.

„ Diptères Exotiques nouveaux ou peu connus. — Two volumes in five parts, and with five Supplements; numerous plates. Paris 1838—1855. — Published originally in the Mémoires de la Société des Sciences et des Arts de Lille; Vol. I, 1838; Vol II, part 1, 1840; part 2, 1841; part 3, 1842; Supplement 1, 1844; Suppl. 2, 1846; Suppl. 3, 1847; Suppl 4, 1849; Suppl. 5, 1855. (The volumes of the separate edition bear somewhat later dates.)

„ Notice sur une nouvelle espèce d'Aricie. — In the Ann. Soc. Entom. de France 1853, p. 675, Tab. XX, No. 2.

Aricia pici, n. sp. San Domingo.

„ Notice sur un nouveau genre de la famille des Pupipares, tribu des Phthiromydes, sous le nom de Megistopoda. — In the Ann. Soc. Entom. de France 1852, p. 331—333, Tab. IV, No. 4.

Megistopoda Pilatei, n. sp. Mexico, Cuba.

Meade, R. H. — Notes on the Anthomyidae of North-America. (In the Entomologists Monthly Magazine, London, April 1878.) No new species; interesting comparison of the european and North-American *Anthomyiae;* list of european species occurring in North-America.

Meigen, F. W. — Systematische Beschreibung der bekannten europäischen zweiflügeligen Insecten. 7 vols. Aachen and Hamm, 1818—1838.

Although this work contains only European species, many of them are common to both continents.

Morris, Miss. — In the Proceedings of the Academy of Natural Sciences of Philadelphia, vol. iv, p. 194 (1849), some remarks have been published by her on the habits of *Cecidomyia culmicola,* n. sp.

Newman, Edw. — Entomological Notes. (In the Entomological Magazine, V, p. 373, 1838.)

Dimeraspis podagra, n. sp. (Microdon globosus Fab.)

Olivier, G. A. — A portion of the entomological volumes of the Encyclopédie Méthodique is by him. In Vol. VIII (1811), under the titles: *Odontomyia, Ocyptera, Ornithomyia,* I found descriptions of several new north-american species, which had been overlooked by previous authors.

Osten-Sacken, C. R. — Catalogue of the described diptera of North-America. Washington, Smithsonian Institution, January 1858 [Smithsonian Misceli. Collections, Vol. III].

Osten-Sacken, C. R. — Appendix to the Smithsonian Catalogue of the described diptera of North-America. October 1859; three pages.

„ New genera and species of north-american Tipulidae with short palpi, with an attempt at a new classification of the tribe: With two plates. — In the Proc. Acad. Nat. Sc. Philad. 1859, p. 197—256.

This paper, as well as the two following, have been entirely superseded by the Monograph of the Tipulidae in the 4th Volume of the Monographs of N. A. Diptera.

„ Appendix to the paper, entitled „New genera and species etc.". — In the Proc. Ac. Nat. Sc. Philad. 1860, p. 15.

„ Description of nine new North-American Limnobiaceae. — In the Proc. Acad. Nat. Sc. Phil. 1861, p. 287—292.

„ On the North-American Cecidomyidae. — In the Monogr. N. A. Diptera, Vol. I, p. 173—205. Washington, April 1862; with a plate and several woodcuts.

Four new species.

„ Characters of the larvae of Mycetophilidae. — In the Proc. Ent. Soc. Phil. I, 1862, p. 151—172, with a plate.

Sciara toxoneura n. sp. (on p. 165).

„ Lasioptera, reared from the gall of a goldenrod. — In the Proc. Entom. Soc. Phil. I, 1863, p. 368—370; also II, p. 77.

Lasioptera solidaginis, n. sp.

„ Description of several new North-American Ctenophorae. — In the Proc. Entom. Soc. Philad. III, 1864, p. 45—49.

Five new species.

„ Description of some new genera and species of N. A. Limnobina. — In the Proc. Entom. Soc. Philad. IV, 1865, p. 224—242.

Six new species.

„ Two new North-American Cecidomyiae. — In the Proc. Ent. Soc. Philad. VI, 1866, p. 219—220.

„ Description of a new species of Culicidae. — In the Trans. Am. Entom. Soc. II, 1868, p. 47—48.

Aëdes sapphirinus, n. sp.

„ On the North-American Tipulidae; part first (Tip. brevipalpi; Cylindrotomina and Ptychopterina). — In the Monographs of the N. A. Diptera, Volume IV, Washington, Smithsonian Institution, January 1869, pages I—XI, and 1—345, with four plates and several woodcuts (Smithsonian Miscellaneous Collections, Volume VIII).*)

Additions and corrections to this volume, will be found at the end of Monographs etc. Vol. III, published in December 1873.

„ Biological notes on Diptera; article first: Galls on Solidago. In the Trans. Am. Entom. Soc. Vol. II, p. 299—303; 1869.

N. sp. *Asphondylia monacha; Cecidomyia anthophila.*

---

*) See the foot-note on page 10.

OSTEN-SACKEN, C. R. — Biol. notes on Diptera, article second: 1. A new american *Asphondylia;* 2. On some undescribed galls of *Cecidomyia.* — In the Trans. Am. Entom. Soc. Vol. III, p. 51—54; 1870—71

„ Biol. notes etc., article third: 1. *Cecidomyia,* living in pine-resin (*Diplosis resinicola* n. sp.). 2. A gall of Cecidomyia on a wild cherry-tree. 3. Additions and corrections. — In the Trans. Am. Entom. Soc. Vol. III, p. 345—347; 1870—71.

„ A list of the Leptidae, Midaidae and Dasypogonina of North-America. — In the Bulletin Buffalo Soc. Nat. Sc. October 1874.
Three new species of Midas.
Additions and corrections to this list are given in the same Bulletin, November 1875, p. 71. (This List is of course entirely superseded by the present publication).

„ Prodrome of a Monograph of the North-American Tabanidae. — In the Memoirs of the Boston Society of Natural History, Vol. II, 1875—78.
Part I. The genera *Pangonia, Chrysops, Silvius, Haematopota, Diabasis* (l. c. p. 365—397).
Part II. The genus *Tabanus,* with an Appendix and Index (l. c. p. 421—479).
Supplement (l. c. p. 555—560).

„ Report on the Diptera, collected by Lieut. W. L. Carpenter in Colorado during the summer 1873. — In Dr. Hayden's U. S. Geological and Geographical Survey of Colorado for 1873. — Washington, 1874 (p. 561—566).
Bibiocephala grandis, n. gen. and sp.

„ Three new galls of Cecidomyiae. — In the Canadian Entomologist, November 1875.
Cecid. verrucicola (on Tilia americana); *Cecid. urnicola* (on Urtica); *Asphondylia recondita* (on Aster patens), *nov. sp.*

„ Note on some Diptera from the Island Guadalupe, Pacific Ocean, collected by Mr. Palmer. — In the Proceed. Boston Soc. of Natural History, October 1875. — No new species.

„ On the North-American species of the gen is Syrphus (in the narrowest sense). — In Proc. Boston Soc. Nat. Hist. October 1875, p. 135—153.
N. sp. *Syrphus amalopis, contumax, torvus* (= topiarius Zett.), *rectus* (= ribesii Lin.)

„ A list of North-American Syrphidae. — In the Bulletin Buffalo Soc. Nat. Sc. November 1875, p. 38—71. —
In the Appendix, descriptions of *nine* new species. Additions and corrections to this list are given in the same Bulletin, May 1876, p. 130. (This List is entirely superseded by the present Catalogue; even the notes, added to it, are reproduced here).

„ Report on the collection of Diptera made in portions of Colorado and Arizona during the year 1873. — In Lieut. Geo. M. Wheeler's Report upon the Explorations and Surveys West of the one hundredth Meridian; Vol. V, Zoology, p. 804—807. — Washington 1875.
N. sp. *Lasia Klottii.*

Osten Sacken, C. R. — Blepharoptera defessa, n. sp. — In Mr. Packard's article: On a new cave-fauna in Utah, in the Bulletin of the U. S Geol. and Geogr. Survey of the Territories, Vol. III, No. 1, p. 168; 1877. (The very bad figure of this *Helomyza* appended to this description, was published without my knowledge.)

" Report on the Diptera collected by Dr. E. Bessels during the Arctic expedition of the *Polaris* in 1872. — In the Proceed. Boston Soc. N. Hist. December 6, 1876.

N. sp. *Tipula Besselsi.*

" Western Diptera, descriptions of new genera and species of Diptera from the region West of the Mississippi, and especially from California. — In the Bulletin of the U. S. Geological and Geographical Survey of the Territories, Vol. III, No. 2, April 30, 1877, page 189 — 354. (A table of contents was printed separately by the author and distributed with his copies.)

One hundred and thirty six new species, and several new genera, principally from California: some few from the Atlantic States.

Palisot de Beauvois, A. M. F. J. — Insectes recueillis en Afrique et en Amérique, etc. in fol. Paris, 1805 – 21. With plates.

Several *Tabani,* one *Chrysops,* and one *Syrphideous* insect from N. America, are described and figured.

Pallas. — Reisen durch verschiedene Provinzen des Russischen Reichs. 1st vol. St. Petersburg, 1771.

On page 475 a *Culex caspius* is described, which Curtis (Ins. of Capt. Ross's voy. identifies with an American species (according to Schiódte Curtis's species is *C. nigripes* Zett.).

Packard, A. S. — Guide to the study of insects, etc. 8ᵛᵒ., with 15 plates and 372 woodcuts. Salem, Mass. First edit. 1868—69; third 1872.

N. sp. *Chironomus oceanicus* Pack., *Ephydra halophila* Pack., *Hippobosca bubonis* Pack. The first two, are described in the following paper; the third is *Olfersia americana* Leach.

" On insects inhabiting salt water. — In the Proc. Essex Instit. Vol. VI, p 41, March 1869.

Ephydra halophila n. sp. and *Chironomus halophilus,* n. sp.

" On insects inhabiting salt water, No. 2. — In the Amer. Journ. of Arts and Sc. 3d. series, Vol. I, p. 100, 1872.

Specific names are given to several larvae, the imagos of which are undescribed (*Ephydra gracilis, californica*).

" In the Report upon the invertebrate animals of Vineyard Sound etc. Washington, D. C. 1874, Mr. Packard mentions several larvae of Diptera, obtained in dredging salt and brackish waters.

Chironomus halophilus, n sp., larva, imago unknown; *Chiron. oceanicus* Packard; *Culex,* larva in brackish waters (no description); *Muscidae* (undetermined larvae described ; *Eristalis* (larva among algae) *Ephydra* (undetermined larva, no description).

PERTY, Maximilian. — Delectus animalium articulatorum quae in itinere per Brasiliam annis 1817 — 20 etc. collegerunt Dr. Spix et Dr. Martius. Monachii, 1830 — 34. 4", with 40 plates.

Several species, described here, occur in Cuba and Mexico.

POEY, Felipe. — Memorias sobre la Historia Natural de la Isla de Cuba; Tomo I⁰, Habana 1851 — 54.

Oecacta furens, nov. gen. et sp.

REICHE, L. — Description de cinq espèces nouvelles d'insectes, provenant de l'expédition aux mers arctiques. — In the Annales de la Soc. Entom. de France, Série 3e, 1857, Bulletin, p. IX.

Anthomyia impudica, n. sp. is a Cordylura.

RILEY, C. V. (State Entomologist of Missouri and Editor of the American Entomologist.) — First annual Report on the noxious, beneficial and other insects of the State of Missouri etc. Jefferson City, 1869.

N. sp. *Lydella doryphorae*, *Anthomyia Zeas*, *Pipiza radicum*.

  „ Second Report etc. 1870.

N. sp. *Asilus missuriensis*, *Exorista flavicauda*.

  „ Third Report etc. 1871.

N. sp. *Masicera archippivora*.

  „ Fourth Report etc. 1872.

N. sp. *Exorista cecropiae*.

  „ Fifth Report etc. 1873.

Galls of *Cecidomyiae* on grape-vine, figured.

  „ Descriptions and natural history of two insects which brave the dangers of *Sarracenia variolaris*. — In the Transact. of the Acad. Nat. Sc. of St. Louis, Vol. III, p. 235—240; 1875.

Sarcophaga sarraceniae, n. sp., larva, pupa, imago described and figured.

  „ Seventh Report etc. 1875.

Biological observations on *Tachina anonyma* and a species of *Sarcophaga*.

  „ Articles in the American Entomologist.

RONDANI, Camillo. — Diptera exotica, revisa et annotata, novis nonnullis descriptis. — Modena 1863 (appeared originally in Archivio Canestrini, III).

N. sp. *Scatina estotilandica*, Labrador.

  „ Osservazioni sopra alquante specie di esapodi ditteri del museo torinense. — In the Nuovi Annali di Bologna, Ser. 3, Vol. II; Sept. — Oct. 1850, p. 165 — 197, with plates.

Tabanus cheliopterus, n. sp. from Carolina.

SAINT-FARGEAU et SERVILLE, authors of a part of the Vol. X of the Encyclopédie Méthodique; north american diptera are mentioned; no new ones.

SAY, Th. — Description of Dipterous Insects of the United States. — In the Journal of the Academy of Natural Sciences in Philadelphia, vol. iii, p. 9 - 54 and 73—104. 1823.

Say, Th. — Description of North-American Dipterous Insects. — l. c. vol. vi, p. 149—178 and 183—188. 1829—30.

„ Keating's Narrative of an Expedition to the Source of St. Peter's River, under the command of S. H. Long. 2 vols. Philadelphia, 1824. — Insects described by Say in the Appendix to the 2d vol.; diptera from p. 357 to p. 378.

„ New Species of N. American Insects, found by Joseph Barabino in Louisiana. Indiana, 1832.

   Sciara dimidiata, Dilophus stygius, n. sp.

„ American Entomology. 3 vols. With plates. Philadelphia, 1824, 25, 28.

   Nineteen diptera are described and figured in this work, seven of which for the first time.

„ *Diopsis brevicornis,* n. sp. — In the Journal of the Academy of Natural Sciences of Philadelphia, vol. i. p. 23.

„ Some account of the insect known by the name of Hessian Fly, etc. — In the Journ. A. N. Sci., Phil., vol. i, 1817.

   Cecidomyia destructor, Say was described for the first time in this paper.

„ The complete writings of Thomas Say on the entomology of North-America, with a memoir of the author by George Ord (edited by John L. Leconte). New-York 1859; Two volumes 8vo.

   In the present volume, the pagination of Say's original papers, as well as that of this new edition of them, are quoted. Some notes are added on the Diptera, by C. R. Osten Sacken.

Shimer, Henry M. D. — Description of a new species of Cecidomyia. — — In the Trans. Amer. Entom. Society, I, p. 281.

   Cecidomyia aceris, n sp.

„ A summers study of Hickory-galls, with descriptions of supposed new insects, bred from them. — In the Traus. Amer. Entom. Soc. II, p. 386, 1869·   •

   On p. 395 there is an imperfect description of an inquilinous Cecidomyia, *C. cossae,* n. sp.

„ Additional notes on the striped squash-beetle *(Diabrotica vittata* Fab.). — In the American Naturalist, V, p. 217.

   Tachina (Melanosphora) *diabroticae,* n. sp. (with figure).

Schiner, Dr. J. R. — Neue oder wenig bekannte Asiliden des K. zoolo-gischen Hofcabinets in Wien. — In the Verh. Zool. Bot. Gesellsch. XVII, p. 355, 1867.

   Five new species from North-America and useful remarks on species, described by other authors.

„ Die Wiedemann'schen Asiliden, interpretirt und in die seither errichteten neuen Gattungen eingereiht. — In the Verh. Zool. Bot. Gesellsch. XVI, p. 649; 1866. —

   Although this paper does not contain any new north-american species, it is important for the classification, and as such, has been quoted; (however, compare about it my Preface).

SCHINER, Dr. J. R. — Reise der Oesterr. Fregatte Novara um die Erde in den Jahren 1857—59; Zoologischer Theil; Diptera; Wien 1868. 1 vol. in 4º, with 4 plates.

Many north-american species, which also occur in South-America, are mentioned in this volume; also many genera are established, which occur in North-America.

SCHIÖDTE, J. G. — Review of the Arthropods of Greenland. Published originally in danish, in Rink's work on Greenland. A german translation, by Mr. Etzel, appeared in the Berl. Entomol. Zeitschrift 1859, p. 134—157. The diptera contain a list of the species hitherto recorded from that country, with a few remarks, but no new species.

STAEGER, C. — Groenland's Antliater. — In Kröjer's Nat. Tidsskrift, new Series, Vol. I, p. 346 369; 1845.

Fifty five diptera are mentioned, *eight* of which, are new.

SWEDERUS, Samuel. Et nytt Genus och femtio nya species af insecter. — In the Vetensk. Acad. Nya Handl. 1787, p. 181 and 276.

Two north-american species: *Musca tomentosa*, which is probably Brachypalpus verbosus, and *Musca* (Syrphus) *monoculus*, I cannot make out the synonymy of the latter.

THUNBERG. — In Act. Soc. Gothoburg. 1819. Pars III, 7, Tab. 7, Fig. 2. — So quoted by Wiedemann, Auss. Zw. I, 110, 4, who reprints Thunberg's description of *Pantophthalmus tabaninus* from the West-Indies.

THOMSON, C. G. — Described the diptera in the volume: Kongliga Svenska fregatten Eugenies Resa etc. Zoologi. Insecta. Diptera, p. 443—614; Tab. IX. 1868.*)

Forty nine new species from California and Panama.

VAN DER WULP, F. M. — Eenige noord-americaansche diptera. — In the Tijdschrift voor Entomol. Nederl. Entomol. Vereeniging, 1867, 2e Ser., II, p. 125—164, Tab. III - V.

Thirty new north-american diptera are described and many of them figured.

„ Nog iets over noord-am ricaansche Diptera. — In the same serial, Vol. IV, p. 80—86, 1869.

Five new species from North-America.

„ Opmerkingen omtrent uitlandsche Asiliden. — In the same serial, Vol. V, 1870.

Stenopogon ochraceus, n. sp.

WALKER, F. — Description of diptera collected by Capt. King in the survey of the Straits of Magellan. Trans. Linn. Soc. London, 1837, T. XVII, p. 331—359.

---

*) Brauer, Bericht über die wissenschaftlichen Leistungen etc. für 1868, contends, that although the title-page bears the year 1868, the volume was actually issued only in 1869; this, in order to secure the priority of the volumes of the Novara Expedition, which appeared in 1868.

Eristalis lateralis n. sp. from Chili, afterwards obtained from Mexico and Jamaica (Walker, List, etc. III, 622).

WALKER, F. — List of the Specimens of Dipterous Insects in the Collection of the British Museum. Four Parts and three Supplements. London 1848—55.

Numerous new species from N. America. The supplements contain a synopsis of the described species of *Tabanidae, Asilidae, Acrocerideae,* and *Stratiomyidae,* from all parts of the world.

„ Insecta Saundersiana, or characters of undescribed Insects in the collection of W. W. Saunders, Esq. *Diptera.* Five parts, with eight plates by Westwood; London 1850 — 56. (Part. I in 1850, Part. II in 1851, Part. III and IV in 1852, Part V in 1856.)

Numerous new north-american species.

„ Characters of undescribed diptera in the collection of Wm. Saunders.   In the Trans. Entom. Soc. N. Ser. IV. 1857, p. 119—158 and 190—235; V, p. 268—334.

About *one hundred* new species from North-America, mostly from Mexico.

„ On some insects of Nova Scotia and Canada. — In the Canadian Entomologist, III, p. 141, October 1871.

A short list of diptera, occurring in Nova Scotia; no new species are described. The species marked with a star also occur in Europe; but some of these data are doubtful. *Bombylius major* Lin. is probably Bombyl. fratellus Wied.; *Helophilus pendulus* Lin. may be H. similis Macq., or some allied species.

„ In the Appendix to „The Naturalist in Vancouver Island and British Columbia", by J. K. Lord, London 1866, 2 Vol., Mr. Walker describes four new species from those regions (l. c. Vol. II, p. 337—339).

Culex pinguis, Laphria columbica, Cuterebra approximata, Eurygaster septentrionalis.

WALSH, Benj. D., M A. — First annual report on the Noxious Insects of the State of Illinois. — In the Appendix to the Transactions of the Illinois State Horticultural Society; Chicago 1868.

Trypeta pomonella n. sp.

„ Insects injurious to vegetation in Illinois; Rock-Island 1861 (Pamphlet).

Exorista (Senometopia) *militaris,* n. sp.

„ On certain remarkable or exceptional larvae, coleopterous, lepidopterous and dipterous. — In the Proc. Boston Soc. Nat. Hist. IX, 1864, p. 286—308.

Midas fulvipes, n. sp.

„ On the insects, coleopterous, hymenopterous and dipterous, inhabiting galls of certain species of willow. — In the Proc. Entom. Soc. Philad. Vol. III, p. 543 - 644 (1864); Vol. VI, p 223 - 288 (1866).

Numerous *Cecidomyiae,* n. sp. and their galls.

„ Larvae in the human body. — In the American Entomologist II, p. 137.

Contains the descriptions of three larvae of *Homalomyia*, designated as *H. Wilsoni*, *Leydii* and *prunivora*. Perfect insect not described.

WALSH, Benj. D. — Mr. Couper's thorn-leaf-gall. In the Canadian Entomologist, I, p. 79. — Short article, referring to the gall of a Cecidomyia, *C. crataegi Bedeguar* Walsh.

WESTWOOD, J. O. — On *Diopsis*, a genus of dipterous insects etc. — In the Trans. Linn. Soc. Vol. XVII, p. 283, 1833—34.
Diopsis (Sphyracephala) *brevicornis* Say; description and figure reproduced from Say.

" Insectorum novorum exoticorum ex ordine dipterorum descriptiones. — In the London and Edinburgh Philosophical Magazine, 1835.
Bittacomorpha, nov. gen.; *Lepidophora aegeriiformis*, Gray; *Pangonia macroglossa*; *Gynoplistia annulata*; all north-american.

" Insectorum nonnullorum novorum (ex ordine dipterorum) descriptiones. — In the Annales de la Société Entomologique de France, 1835, p. 681—685.
Limnobiorhynchus canadensis, nov. gen. et. sp.

" Description of some new exotic *Acroceridae*. — In the Transactions of the Entomological Society, vol V, p. 91. 1848.
Six new species from N. America.

" Synopsis of the dipterous family *Midasiidae*, with descriptions of numerous species. — In Westwood's Arcana Entomologica, vol. I. Plates XIII and XIV. 1841—43.
Five new species from N. America.

" Generis dipterorum monographia *Systropi*. — In Guérin's Magazin de Zoologie 1842.
Systropus foenoides, n. sp. from Mexico.

" Diptera nonnulla exotica descripta. — In the Transactions of the Entomological Society, vol. V, p. 231. 1850.
Ceria daphnaeus, Walk.; from Jamaica, described and figured.

" — Observations on the destructive species of dipterous insects known in Africa under the names of the Tsetse, Zimb and Tsaltsalya. — In the Proceedings of the Zool. Soc. of London, 1850, p. 259—270; with a plate.
Stylomyia confusa Westwood, without locality, is *Stylogaster stylatus* Fabr. from North-America.

" Notae dipterologicae. Monograph of the genus *Systropus*, with notes on the economy of a new species of that genus. — In the Trans. Entom. Soc. London, 1876.
Systropus foenoides Westw. from Mexico; description reproduced from Magaz. de Zool. 1842.

" Notae dipterologicae. Description of new genera and species of the family *Acroceridae*. — In the Trans. Entom. Soc. London 1876.
Pialoidea nov. gen. for *Cyrtus magnus* from Georgia.

WIEDEMANN. C. R. W. — Aussereuropäische Zweiflügelige Insecten. 2
    vols. Hamm 1828—30. With plates.
  „ Diptera exotica. Kiliae 1821.
  „ Analecta entomologica. Kiliae 1824.
  „ Achias, dipterorum genus a Fabricio conditum. Kiliae 1830.
     Sphyracephala (Achias) brevicornis Say; described and figured.
  „ Monographia generis *Midarum.* (In the Nova acta Academiae
    Naturae Curiosorum, vol XV. Bonn 1831. 4to. With three
    plates.)
    Four new species from N. America.
ZETTERSTEDT, J. W. — Insecta lapponica, descripta. 1 vol. in 4to.
    Lipsiae 1838—40.
  „ Diptera Scandinaviae disposita et descripta. 14 vols. Lundae
    1842—1860.
  .   Both of these works contain many diptera common to Lapland
    and the northern parts of the American continent.

———

The Practical Entomologist,

published by the Entomol. Soc. of Philadelphia. Vol I, 1865 — 66,
Vol. II, 1866 — 67.

The American Entomologist,

an illustrated Magazine of popular and practical Entomology, edited
by Benj. D. Walsh and Ch. V. Riley. St. Louis, Mo, Vol. I, 1868;
Vol. II (title changed to Amer. Entom. and Botanist) 1870.

The Canadian Entomologist,

Volume I—VII; 1869—1875. (Voll. I and II published in Toronto;
Voll. III — VII in London, Ont.)

The American Naturalist

a monthly magazine of Natural History, published (until 1877) in
Salem, Mass.

———

    These periodicals have been quoted in the present volume for
the various notices and illustrations of N.-A. Diptera, which they
contain.

LIST

OF THE NEW GENERA AND THE NEW SPECIES
PUBLISHED IN THE NOTES TO THIS VOLUME.

———

I. New genera: **Crioprora** (Syrphidae);
   Diotrepha (Tipulidae).

II. New species:

    Diotrepha mirabilis (Tipulidae). — Southern States.

    Cyrtopogon lyratus (Asilidae). — New-York and New-England.

    Porphyrops signifer (Dolichopodidae). — Northern States.

    Borborus venalicius (Borboridae). — Cuba.

    Arthropeas leptis (Coenomyidae). — Northern States.

III. Changed or modified generic names (the reason for the change is, in every case, explained in the notes):

    Protoplasta in *Idioplasta*.

    Empheria in *Neoempheria* (Mycetophilidae).

    Glaphyroptera in *Neoglaphyroptera* (Mycetophilidae).

    Aspilota in *Neaspilota* (Trypetidae).

    Eristicus in *Neoeristicus* (Asilidae).

    Mochtherus in *Neomochtherus* (Asilidae).

    Itamus in *Neoitamus* (Asilidae).

    Idiotypa in *Neoidiotypa* (Trypetidae).

    Rondania in *Neorondania* (Stratiomyidae).

    Exaireta in *Neoexaireta* (Stratiomyidae).

EXPLANATIONS
NECESSARY FOR THE USE OF THE CATALOGUE.

—

A Star (*) before a specific name means that the species is to be found in the collection of the Museum of Comparative Zoology, in Cambridge, Mass. These stars are omitted only in the family Cecidomyidae.

An interrogation (?) before a specific name means that its position in the genus is doubtful; an interrogation before a synonym, means that the synonymy is uncertain.

An exclamation after a synonymy, means that I have seen the type of the description. I have used this sign whenever I deemed it necessary to inform the reader of that fact; *but the absence of that sign does **not** necessarily mean that I have not seen the type.*

Synonymies. The authority for each synonymy is given after it, in brackets; where no authority is mentioned, my own is assumed.

Genera. Species which I do not know, may sometimes not be placed in the right genera; this applies especially to the species from Mexico and the West Indies.

Loew, in litt. All the data, which I obtained from Mr. Loew, either by letter, or in looking over his North American collection (during my visit in Guben, in September 1877), are quoted in that way.

M. C. Z. Museum of Comparative Zoölogy in Cambridge, Mass. —

Localities. It will be noticed that, in some cases, the localities marked in the catalogue, differ from those which are found in Dr. Loew's Centuries of North-American Diptera. These discrepancies are not errors, or omissions, *but corrections.*

In this Catalogue (as well as in my earlier list), I have not included those species of earlier authors, which were marked simply „America".

New-York is always meant for the State of that name, not for the city.

CATALOGUE
OF NORTH AMERICAN DIPTERA.

I. DIPTERA ORTHORHAPHA.

FAMILY CECIDOMYIDAE.

Cecidomyia.

Meigen, Illiger's Magaz. 1803. ([1])

aceris Shimer, Trans. Amer. Entom. Soc. I, 281. — Illinois; the larva lives on the surface of leaves of *Acer dasycarpum*.

albovittata Walsh, Proc. Entom. Soc. Phil. III, 620; VI, 227. — Illinois; inquilinous on willow-galls.

Amyotii Fitch, Reports Vol. III, 31 (♀). — New-York.

anthophila O Sacken, Trans. Amer. Entom. Soc. II, 302. — New-York; on *Solidago*.

chrysopsidis Loew, Monogr. etc. I, 203; Tab. I, f. 1 (gall.) — Washington, D. C. On *Chrysopsis mariana*.

cornuta Walsh, Proc. Ent. Soc. Phil. III, 625. — On *Salix*.

cossae Shimer, Trans. Amer. Ent. Soc. II, 395. — Illinois; on *Carya*.

culmicola Morris (Miss), Proc. Acad. Nat. Sc. Phil. IV, 194 (1849); No description given; only remarks upon habits etc. Harris, Ins. Injur. Veget. 582. — Pennsylvania.

cupressi-ananassa Riley, Amer. Entom. II, 244 and 273; fig. 153 (gall). — Tennessee, on *Taxodium distichum*.

destructor Say, Journ. Acad. Phil. I, 45, Tab. III, f 1—3; Compl. Wr. I, p. 4 (no figures); Wiedemann, Auss. Zw. I, 21, 1. Other references to the numerous papers concerning this insect may be found in Harris's Ins. Injur. to vegetation, and in Dr. A. Fitch's articles „the Hessian fly" in the Amer. Journ. of Agric. and Science (1846), reprinted, with some additions, in the Trans N. Y. State Agric. Soc. Vol. VI, 1846, p. 316—376; a shorter article, with some new facts, in Dr. Fitch's Reports, Vol. III, p 133—144, Tab. III, f. 2—3, and Appendix, p. 203. According to Loew, in Silliman's Journal, N. Ser. XXXVII, p. 317. this species is the same as the european *Cecid. funesta* Motchulski = *Cecid. secalina* Loew. For the litterature on this subject, see also: Bergenstamm und Loew, Synopsis Cecidomyidarum, 39 (in the Verh. Zool. Bot. Ges. 1876).

gleditchiae O. Sacken, Proc. Ent. Soc. Phil. VI, 219. — Newport, R. J., on *Gleditchia triacanthos*.

grossnlariae Fitch, Reports, Vol. I, 176; Vol. II, No. 150. — On the gooseberry (*Ribes*). ([2]).

hirtipes O Sacken. Monogr. etc I, 195. — Distr. Columbia, on *Solidago*.

orbitalis Walsh, Proc. Ent. Soc. Phil. III, 623; VI, 227. — Inquilinous on willow-galls.

ornata Say, Long's Exped. App. 357; Compl. Wr. I, 242; Wiedemann, Auss. Zw. I, 22, 2. — Pennsylvania.

pseudoacaciae Fitch, Rep. Vol. II, No. 331. — On *Robinia pseudoacacia*.

salicis-batatas Walsh, Proc. Ent. Soc. Phil. III, 601; VI, 225. — On *Salix cordata, discolor, humilis*. ([3]).

salicis-brassicoides Walsh, l. c III, 577; American. Entomol. 105, fig. 84; Packard's Guide 377, f. 282. — On *Salix longifolia*.

salicis-cornu Walsh, l. c. III, 590; VI, 224. — On *Salix humilis*.

salicis-gnaphaloides Walsh, l. c. III, 583; VI, 223. — On *Salix candida, discolor, humilis*. ([4]).

salicis-rhodoides Walsh, l. c. III, 586; VI, 224. — On *Salix humilis*. ([4]).

salicis-strobiloides Walsh, l. c. III, 580. — On *Salix cordata*. ([4] and [5].) Compare also O. Sacken, Monogr. I, 203, where the gall is described for the first time; also Amer. Entom. I, 105, f. 82; Packard's Guide, 377, fig. 280—281.

salicis-strobiliscus Walsh, l. c. III, 582; VI, 223. — On *Salix discolor* and *rostrata*. ([4]).

serrulatae O. Sacken, Monogr. etc. I, 198. — Distr. Columbia, on *Alnus serrulata*.

siliqua Walsh, Proc. Entom. Soc. Philad. III, 591; VI, 224. — On *Salix humilis, cordata? discolor?* According to the author, perhaps the same as *Cec.' salicis* Fitch.

solidaginis Loew, Monogr. etc. I, 194, Tab. I, f. 8. — On *Solidago*.

spongivora Walker, List. etc. I, 30. — Huds. Bay Territ.

Diplosis.

Loew, Dipterol. Beitr. IV, 20; 1850.

atrocularis Walsh, Proc. Ent. Soc. Phil. III, 626; VI, 227. — Rock-Island, Illin., inquilinous on willow-galls.

atricornis Walsh, l. c. III, 628 — Same habits.

annulipes Walsh, l. c. III, 629. -- Same habits.

caliptera Fitch, Essay upon the wheat-fly etc. (*first edition* in the Amer. Quart. Journ. of Agric. and Science, 1845, Vol. II, No. 2, Tab. V, f. 2; *second edition*, Trans. N. Y. State Agricultural Society 1846, Vol. V; *Cecid. cerealis* Fitch is separated from *C. caliptera* in the second edition only). See also Fitch, Reports etc. Vol. III, 90, Tab. II, f. 18 (*Cecidomyia*). — New York, occurs with *Dipl. tritici*.

caryae O. Sacken, Monographs etc. I, 191. — Distr. Columbia; forms galls on the leaves of the hickory.

decemmaculata Walsh, Proc. Entom. Soc Phil III, 631. — Inquilinous on willow-galls.

graminis Fitch, Reports, Vol. III, 90, Tab. II, f. 2, 5 (*Cecidomyia*). — Occurs on wheat, with *D. caliptera*. Synon. *Cecid. cerealis* Fitch, Essay on the wheat-fly, 2ᵈ edition, in the Trans. N. Y. State Agric. Soc. V, 1847 [change of name by Dr. Fitch].

helianthi-bulla Walsh, Proc. Entom. Soc. Phil. VI, 228. — On *Helianthus*.

inimica Fitch, Reports, Vol. III, 88 (*Cecidomyia*). Larva in wheat-heads, in company w. Diplosis *tritici* (although the female alone is described, it is probably a *Diplosis*).

maccus Loew, Monogr. etc. I, 187, Tab. I, f. 11, 12. — Distr. Columbia; habits unknown.

resinicola O. Sacken, Trans. Amer. Ent. Soc. III, 345 (1870 — 71). — Tarrytown, N. Y.; in the resin of *Pinus inops*. The occurrence of the larvae had already been noticed by Mr. Sanborn, in the Proc. Boston Soc. N. H. XII, 93 (1868 — 69).

robiniae Haldeman, Amer. Journ. Agric and Sc. VI, 193, 1847 (with figures); reprinted in Proc. Boston Soc. N. H. VI, 401, 1859 (*Cecidomyia*); Harris, Ins. Injurious to Vegetation, 567 (*id.*); Fitch, Reports, Vol. II, No. 332 (*id.*). — On leaves of *Robinia pseudo-acacia;* Atlantic States.

septemmaculata Walsh, Proc. Ent. Soc. Phil. III, 630; VI, 228. — Inquilinous in willow-galls.

tergata Fitch, Essay on the wheat-fly etc. l. c. f. 3 and 4 (*Cecidomyia*).

thoracica Fitch, Essay on the wheat-fly etc. l. c. f. 5 and 6 (*Cecidomyia*.) (As Dr. Fitch mentions both this and the preceding species as being related to *Dipl. tritici* in size, in the number and form of the joints of the antennae, they must necessarily belong to the genus *Diplosis*).

tritici Kirby, Curtis etc. (*Cecidomyia*); Harris, Ins. Injurious to Veget etc. 592; Fitch, Essay on the wheat-fly etc.; Fitch, Reports, Vol. III, 1 — 88, Tab. II, f. 1, 4 (*id.*); Amyot, Annales de la Soc Entom. de France 1855, Bullet. CIV. — Injurious to wheat in Europe and N. America.

Asphondylia.
Loew, Dipterol. Beitr. IV, 20; 1850.

helianthi-globulus Walsh (*in litt.*), O. Sacken, Trans. Am. Ent. Soc., II, 301. — Rock-Island, Illin., on *Helianthus*.

monacha O. Sacken, Trans. Am. Ent. Soc. II, 300, and III, 347. — New York; on *Solidago*.

recondita O. Sacken, Canadian Entomologist, Nov. 1875. — On *Aster patens,* Long Island, N. Y.

rudbeckiae-conspicua O. Sacken, Trans. Am. Ent. Soc. III, 51. — Pennsylvania; on *Rudbeckia triloba*.

Lasioptera.
Meigen, System. Beschr. I, 88; 1818.

parva Walker, List etc. I, 29. — Huds. B. Terr.

solidaginis O. Sacken, Proc. Entom. Soc. Phil. I, 370. — Larva probably inquilinous in galls on *Solidago*.

ventralis Say, Long's Exped. App. 357; Compl. Wr. I, 242; Wiedemann, Auss. Zw. I, 21, 1. — Pennsylvania.
vitis O. Sacken, Monographs etc. I, 201; gall figured by C. V. Riley,
5th Report, 117; also in Amer. Entomologist, I, 247. — District
Columbia and elsewhere on *Vitis.*

Cecidomyiae known by their galls and larvae only.

agrostis O. Sacken, Monographs etc. I, 204; originally mentioned in
A. Fitch, The Hessian fly, 2d edition, in pamphlet form, p. 38
(„imbricated galls on *Agrostis lateriflora*“).
brachynteroides O. Sacken, Monographs etc. I. 198. — On *Pinus
inops,* producing a swelling at the basis of the leaves.
carbonifera O. Sacken, Monogr. etc. I, 195. — On leaves of *Solidago.*
caryaecola O. Sacken, Monogr. etc. I, 192. — On *Carya;* Distr. Columbia. ([6]).
citrina O. Sacken, Trans. Amer. Ent. Soc. III, 53. — On the terminal
buds of young shoots of *Tilia americana;* New York.
crataegi-bedeguar Walsh, Canad. Ent. I, 79; Proc. Ent. Soc. Phil. VI,
266. — On *Crataegus tomentosa.* (In the same paper Mr. Walsh-
mentions galls on *Crataegus.* which he calls *crataegi-plica, limbus*
and *globulus,* without giving any further description.)
cynipsea O. Sacken, Monogr. etc. 193. — On *Carya.*
erubescens O. Sacken, Monogr. etc. I, 200. — On *Quercus.*
farinosa O. Sacken, Monogr. etc. I, 204. — On leaves of the black-
berry, *Rubus.*
glutinosa O. Sacken, Monogr. etc. I, 193. — On *Carya.*
holotricha O. Sacken, Monogr. etc. I, 193. — On *Carya.* ([6]).
impatientis O. Sacken, Monogr. etc. I. 204; Amer. Entomol. II, 63
(figure of gall). — Deforms flowers of *Impatiens;* Distr. Columbia.
liriodendri O. Sacken, Monogr. etc. I. 204. — On the leaves of *Liriodendron.*
majalis O Sacken, Monogr. etc. I, 204. — On the leaf-ribs of *Quercus
palustris.*
nodulus Walsh, Proc. Ent. Soc. Phil. III, 599. — On *Salix longifolia.*
nucicola O. Sacken, Trans Amer. Ent. Soc. III, 53. — In the husks
of the nuts of *Carya;* New York.
niveipila O. Sacken, Monogr. etc. I, 199. — On Oak-leaves.
ocellaris O. Sacken, Monogr. etc. I, 199. — Produces ocellate spots
on the leaves of *Acer rubrum.*
persicoides O. Sacken, Monogr. etc. I, 193. — On *Carya.* ([6]).
poculum O. Sacken, Monogr. etc I, 201. — On *Quercus.* ([7]).
pini-inopis O. Sacken, Monogr. etc. I, 196. -- Forms a resinous cocoon
on the leaves of *Pinus inops.* Distr. Columbia.
pellex O. Sacken, Monogr. etc. I, 199. — Galls on leaves of *Fraxinus
americana.* Distr. Columbia.
pudibunda O. Sacken, Monogr. etc. I, 202. — On the leaves of *Car-
pinus americana.* Distr Columbia.
racemicola O. Sacken. Monogr. I, 196. — On *Solidago,* among the
racemes. Distr. Columbia.

salicifoliae O. Sacken, Proc. Ent. Soc. Phil. VI, 220. — On *Spirea salicifolia*. Canada.

salicis-aenigma Walsh, Proc. Ent. Soc. Phil. III 608; VI, 227.

salicis-coryloides Walsh, l. c. III, 588; VI, 224. (¹).

salicis-nodulus Walsh, l. c. III, 599.

salicis-semeu Walsh, l. c. III, 607; VI, 226.

salicis-verruca Walsh, l. c. III, 606; VI, 226.

salicis-triticoides Walsh, l. c. III, 598; VI, 225.

salicis-hordoides Walsh, l. c. III, 599.

> N.B. All these are willow-galls, produced by Cecidomyiae; the galls *semen* and *aenigma* Mr. Walsh acknowledges later l. c. VI, 226 to be produced by *Acarus*.

sanguinolenta O. Sacken, Monogr. etc. I, 192. — On *Carya*.

serotinae O. Sacken, Trans. Amer. Entom. Soc. III: 346. — On *Cerasus serotina*; New York.

symmetrica O. Sacken, Monogr. etc. I, 200. — On *Quercus*.

tubicola O. Sacken, Monogr. etc. I, 192. — On *Carya*. (⁶).

tulipiferae O. Sacken, Monogr. etc. I, 202. — On *Liriodendron*.

umbellicola O. Sacken, Trans. Amer. Ent. Soc III, 52 and 347. Among the umbels of *Sambucus racemosa* in New York and New-Jersey.

urnicola O. Sacken, Canadian Entomol. Nov. 1875. — On *Urtica gracilis*; Trenton Falls N. Y.

vaccinii O. Sacken, Monogr. I, 196. — On *Vaccinium*; Distr. Columbia.

verrucicola O. Sacken, Canadian Entomol. Nov. 1875. — On *Tilia americana*, New England.

vitis-coryloides Walsh, Proc. Entom. Soc. Phil. III, 588; l. c. VI, 224; Amer. Entomol. I, 107, figure 86 (figure of the gall); Riley, 5th Report, 116; Packard's Guide, 376, fig. 284. — On *Vitis cordifolia* and *riparia*.

vitis-pomum Walsh and Riley, Amer. Entomol. I, 106; fig. 85; Riley 5th Report, 114, with figure; the latter is reproduced in Packard's Guide, 378, f. 283. — On *Vitis cordifolia*.

viticola O Sacken, Monogr. I, 202. — On *Vitis*. The gall *Vitis-lituus* Riley, Amer. Ent. II, 28, t. 27; also l. c. 113; also 5th Report, 118, is the same as *viticola*.

Observation. In the Western Diptera, 192, I described galls of Cecidomyiae which I observed on the following plants in California.

Juniperus californicus.

Lupinus albifrons.

Audibertia sp.

Garrya fremontii.

Artemisia californica.

Baccharis pilularis.

Tritozyga.

Loew, Monographs etc. I, 178; 1862, Tab. I, f. 13. (Wing.)

The species is not described; it was from Distr. Columbia.

Campylomyza.

Meigen, Syst. Beschr. I, 101; 1818.

scutellata Say, J. Acad. Phil. III, p. 17, 1; Compl. Wr. II, 44; Wiedemann, Auss. Zw. I, 22, 1. — Missouri.

FAMILY MYCETOPHILIDAE. [8]

Mycetobia.

Meigen, System. Beschr. I, 229; 1818.

divergens Walker, Dipt. Saund. 418. — Atlantic States. (I did not
succeed in finding it in the Brit. Mus.)

Ditomyia.

Winnertz, Stett. Ent. Z. VII, 15; 1846.

*euzona Loew, Centur. IX, 1. — New York.

Plesiastina.

Winnertz, Stett. Ent. Z. XIII, 55; 1852.

*lauta Loew, Centur. IX, 3. —. New York.
*tristis Loew, Centur. IX, 2. — Distr. Columbia.

Bolitophila.

Meigen, System. Beschr. I, 220; 1818.

*cinerea Meigen etc., Winnertz, Pilzm. 674. — Europe and North-America.
[Loew in litt.]
disjuncta Loew (undescribed) is likewise common to both continents.
[White Mts., N. H.]

Macrocera.

Meigen, Illiger's Magaz. II, 261; 1803.

*clara Loew, Centur IX, 6. — Distr. Columbia.
*formosa Loew, Centur. VII, 8. — New York.
*hirsuta Loew, Centur. IX, 5. — Distr. Columbia.
*inconcinna Loew, Centur. IX, 7. — Distr. Columbia.

Platyura.

Meigen, Illiger's Magaz. II, 264; 1803.

*diluta Loew, Centur. IX, 9. — Distr. Columbia.
*divaricata Loew, Centur. IX, 8. — Georgia.
fascipennis Say, Long's Exp'd. Append. 360; Compl. Wr. I, 244;
Wied. Auss. Zw. I, 61, 2. — N. W. Territory Say).
*melasoma Loew, Cent. IX, 12. — Distr. Columbia.
*mendica Loew, Centur. IX, 10. — New York.
*mendosa Loew, Centur. IX, 11. — Distr. Columbia.
*subterminalis Say, J. Acad. Phil. VI, 152; Compl. Wr. II, 350. —
Indiana.

Ceroplatus.

Bosc, Actes de la Soc. d'Hist. Nat. de Paris I, 1, 42; 1792.

*carbonarius Bosc, Nouv. Dict. d'Hist. Nat. Ière édit. IV, 543; 2e édit.
  T. V, 585, tab. B, 21, figs. 4, 4; Fabricius, Syst. Antl. 16, 2; Wiede-
  mann, Auss. Zw. I, 61, 3; Dufour, Ann. des Sci. Nat. 2e ser. T.
  XI (1839), 202; Macquart, Dipt. Exot. I, 1, 77, tab. XI, fig. 1. —
  Carolina.

Asyndulum.

Latreille, Hist. Nat. des Crust. et des Ins. XIV. 290; 1804.

*coxale Loew, Centur. IX, 4. — Huds. B. Territ.

Observation. For *Asyndulum tenuipes* Walker, List etc.
I, 86, see *Blepharocera capitata* Loew.

Diomonus.

Walker, List, etc. I, 87; 1848.

*nebulosus Walker, List, etc. I, 87. — Huds. B. Territ.

Neoempheria.

Empheria, Winnertz, Pilzm. 1863. [9].

*balioptera Loew, Centur. IX, 13. — Illinois.
*didyma Loew, Centur. IX, 14. — English River.
  Sciophila bimaculata Loew, Centur. VII, 9 (change of name by
  Loew).
*nepticula Loew, Centur. IX, 15. — Georgia.

Polylepta.

Winnertz, Pilzm. 1863.

*fragilis Loew, Centur. IX, 16. — Massachusetts.

Sciophila.

Meigen, System. Beschr. I. 245; 1818.

*appendiculata Loew, Centur. IX, 19. — New York.
*biseriata Loew, Centur. IX, 20. — Red River of the North.
bifasciata Say, Long's Exped. App. 363; Compl. Wr. I, 246; Wiedemann, Auss. Zw. I, 62, 1. — N. W. Territory (Say). [perhaps
  an *Empheria*. — Loew in litt.]
grisea Walker, List, etc. I, 92. — Huds. B. Territ.
hirticollis Say, Long's Exped. App. 362; Compl. Wr. I, 246; Wiedemann, Auss. Zw. I, 64, 6. — N. W. Territ. (Say).
littoralis Say, Long's Exped. App. 361; Compl. Wr. I, 245; Wiedemann, Auss. Zw. I, 64, 5. — Lake Superior.
obliqua Say, Long's Exped. App. 363; Compl. Wr. I, 247; Wiedemann, Auss. Zw. I, 63, 3. — N. W. Territory (Say).
*obtruncata Loew, Centur. IX, 18. — Distr. Columbia.

*onusta Loew, Centur. IX, 17. — Distr. Columbia.
*tantilla Loew, Centur. IX, 21. — Distr. Columbia.

popocatepetli Bellardi, Saggio etc. I, 11. — Mexico.

Observation. For *Sc. bimaculata* Loew, Centur. VII, 9, see *Neompheria didyma*.

Lasiosoma.
Winnertz, Pilzm. 1863.

fasciata Say, Journ. Ac. Phil. III, 26, 1; Compl. Wr. II, 50 *(Sciophila)*; Wiedemann, Auss. Zw. I, 62, 2 *(id.)*. — Pennsylvania; Maryland.
*quadratula Loew, Centur. IX, 22. — Maine.
*pallipes Say, Long's Exp. App. 361; Compl. Wr. I, 245 *(Sciophila)*; Wiedemann, Auss. Zw. I, 63, 4 (id.). — N. W. Territory (Say).

Tetragoneura.
Winnertz, Stett. Ent. Z. 1846, 18.

This genus occurs in the U. States according to Loew, Monographs etc. I, 14, although no species has, as yet, been described.

Eudicrana.
Loew, Centur. IX, 23; 1869.

*obumbrata Loew, Centur IX, 23. — New York.

Syntemna.
Winnertz, Pilzm. 1863.

*polyzona Loew, Centur. IX, 24. — Middle States.

Phthinia.
Winnertz, Pilzm. 1863.

*tanypus Loew, Centur. IX, 26. — New York.

Boletina.
Staeger, Kröjer's Tidskr. III, 234, 1840.

*tricincta Loew, Centur. IX, 25. — Maryland, Wisconsin.
groenlandica Staeger, Groenl. Antliater 17, 18; Holmgren, Ins. Nord-groenl. — Greenland.
arctica Holmgren, Ins. Nordgroenl. Oefv. Kongl. Vetensk. Acad. Förh 1872, No 6. — Northern Greenland.

Gnoriste.
Meigen, System. Beschr. I, 1818; Winnertz, Pilzm. 778.

*megarrhina O Sacken, Western Diptera, 193. — Yosemite Valley, Cal.

Neoglaphyroptera.
Glaphyroptera Winnertz, Pilzm. 1863. ([9]).

*bivittata Say, J. Acad. Phil. VI, 152 *(Leja)*; Compl. Wr. II, 351. — Indiana (Say); Atlantic States.

Glaphyroptera lateralis v. d. Wulp, Tijdschr. v. Entom. 2 Ser. II, 131, Tab. III, f. 3. 4. [Loew, Zeitschrift für Ges. Naturw. Vol. XXXVI, 113.]

*decora Loew, Centur. IX, 28. — Georgia.

*melaena Loew, Centur. IX, 27. — New-York.

*oblectabilis Loew, Centur. IX, 31. — Middle States.

*opima Loew, Centur. IX, 29. — Connecticut.

*sublunata Loew, Centur. IX, 30 — New York.

*ventralis Say, Long's Exped. App. 364; Wiedem., Auss. Zw. I, 65, 2 (*Leja*). — N. W. Territ. (Say).

*Winthemii Lehmann, Insect. spec. nonnullae etc. Winnertz, Pilzm., 789. — Europe and North-America.

   Mycetophila maculipennis Say, Long's Exp. App. 365; Compl. Wr. I, 248; Wied. Auss. Zw. I, 66, 2. [Loew in litt.]

   Leja trifasciata Walker, List, etc. I, 93. — Huds. B. Territ. [Loew in litt.]

*varia Walker, List, etc., I, 93 (*Leja*). — Huds. B. Terr. (Wk.).

Leja.

Meigen, System. Beschr. I, 253; 1818.

*abbreviata Loew, Cent. IX, 33. — Middle States.

*sororcula Loew, Centur. IX, 32. — New York.

unicolor Walker, List, etc. I, 93. — Huds. B. Terr.

punctata Bellardi, Saggio etc. App. 5, f. 3. — Mexico.

Acnemia.

Winnertz, Pilzm. 1863.

*psylla Loew, Centur. IX, 34. — Maryland.

Docosia.

Winnertz, l. c. 1863.

*dichroa Loew, Centur. IX, 35. — Distr. Columbia.

Rhymosia.

Winnertz, l. c. 1863.

*filipes Loew, Centur. IX, 36. — Connecticut.

Allodia.

Winnertz, l. c. 1863.

*crassicornis Stannius, Obs. de Mycet. 1831, 22, 20; Winnertz, l. c. 828. — Europe and North-America; Pennsylvania, Maryland. [Loew in litt.]

Trichonta.

Winnertz, l. c. 1863.

*foeda Loew, Centur. IX, 38. — Middle States.

*vulgaris Loew, Centur. IX, 37. — Distr. Columbia.

Zygomyia.
Winnertz, l. c. 1863.
* ignobilis Loew, Centur. IX, 39. — Middle States.
* ornata Loew, Centur. IX, 40. — Pennsylvania.

Epicypta.
Winnertz, l. c. 1863.
* pulicaria Loew, Cent. IX, 41. — Pennsylvania.

Mycothera.
Winnertz, l. c. 1863.
* paula Loew, Centur. IX, 42. — Middle States.

Mycetophila.
Meigen, Illiger's Magaz. II, 263. 1803.
* bipunctata Loew, Centur. IX, 44. — Wisconsin.
* discoidea Say, J. Acad. Phil. VI, 153; Compl. Wr. II, 351. — Indiana.
* extincta Loew, Centur. IX, 43. — Middle States.
* fallax Loew, Centur. IX, 50. — Middle States.
  ichneumonea Say, J. Acad. Phil. III, 16, 1; Compl. Wr. II, 43; Wiede-
    mann, Auss. Zw. I, 67, 3. — Pennsylvania.
* inculta Loew, Centur. IX, 46. — Middle States.
* monochaeta Loew, Centur. IX, 54. — Distr. Columbia.
* mutica Loew, Centur. IX, 45. — Middle States.
  nubila Say, J. Acad. Phil. VI, 6, 153; Compl. Wr. II, 352. — Indiana.
* pinguis Loew, Centur. IX, 47. — Maine; English River.
* polita Loew, Centur. IX, 53. — New York.
* procera Loew, Centur. IX, 55. — New York.
* punctata Meigen etc.; Winnertz, l. c. 916. — Europe and North-
    America (Pennsylvania; Loew, _in litt._).
* quatuornotata Loew, Centur. IX, 52. — Maryland.
* scalaris Loew, Centur. IX, 48. — Middle States.
  sericea Say, Long's Exped. App. 365; Compl. Wr. I, 248; Wiede-
    mann, Auss. Zw. I, 66, 1. — N. W. Territ.
* sigmoides Loew, Centur. IX, 51. — Middle States.
* trichonota Loew, Centur. X, 49. — Distr. Columbia.

    Observation. Mr. Walker's species:
        bifasciata, Walker, List, etc. I, 96. — Huds. B. Terr.
        contigua Walker, List, etc. I, 96. — Nova Scotia.
        despecta Walker, List, etc. I, 101. — Huds. B. Terr.
        laeta Walker, List, etc. I, 97. — Nova Scotia.
        obscura Walker, List, etc. I, 101. — Huds. B. Terr.
        parva Walker, List, etc. I, 97. — Huds. B. Terr.
        plebeja Walker, List, etc. I, 100. — Huds. B. Territ.
        propinqua Walker, List, etc. I, 96. — Nova Scotia.

Sciara.
Meigen, Illiger's Magaz. II, 263; 1803; _Molobrus_ Latr. [10].

  abbreviata Walker, List, etc. I, 109. — Huds. B. Terr.

atrata Say, Long's Exp. App. 366, 1. Compl. Wr. I, 249; Wied. Auss.
  Zw. I, 70, 9. — N. W. Terr. (Say).
dimidiata Say, Spec. of Amer. Ins. found by Jos. Barabino 15. Compl.
  W. I, 308. — Louisiana.
exigua Say, Long's Exp. App. 367, 4; Compl. Wr. I, 249; Wied. Auss.
  Zw. I, 69, 7. — N. W. Terr. (Say).
exilis Say, J. Acad. Phil. VI, 154; Compl. Wr. II, 352. — Indiana.
femorata Say, J. Acad. Phil. III, 78, 1; Compl. Wr. II, 70; Wied. Auss.
  Zw. I, 70, 8. — Pennsylvania.
flavipes Meigen, etc. Staeger, Groenl. Antliater. — Europe, Greenland.
fraterna Say, Long's Exped. App. 367, 3; Compl. Wr. I, 249; Wied.
  Auss. Zw. I, 69, 6. — N. W. Terr. (Say).
fuliginosa Fitch, First and Second Report, etc. 255 *(Molobrus)*. —
  New York.
groenlandica Holmgren, Ins. Nordgroenl. Oefv. Kongl. Vet. Acad.
  Förh. 1872, No. 6 — North-Greenland.
inconstans Fitch, l. c. 255 *(Molobrus)*. — New York.
iridipennis Zetterstedt, Ins. Lapponica; Staeger, Groenl. Antliater. —
  Greenland.
lurida Walker, List, etc. I, 106. Dipt. Saunders, 418. — Trenton Falls.
mali Fitch, First and Second Report etc. 254 *(Molobrus)*. — New York.
nigra Wiedemann, Dipt. Exot. I, 44, 7. Auss. Zw. I, 68, 3. — Savannah.
*__ochrolabis__ Loew, Centur. IX, 57. — New York.
perpusilla Walker, List, etc. I, 106. — Huds. B. Terr.
polita Say, Long's Exp. App. 366, 2; Compl. Wr. I, 249; Wied. Auss.
  Zw. I, 70, 10. — N. W. Terr.
punctata Walker, List, etc. I, 106. — N. America.
robusta Walker, List etc. I, 105. — Huds. B. Terr.
rotundipennis Macquart, Dipt. Exot. I, 2, 178; Bellardi, Saggio etc.
  I, 13. — Carolina (Macq.); Mexico (Bellardi).
*__sciophila__ Loew, Centur IX, 56. — Distr. Columbia.
vulgaris Fitch, First and Second Report etc. 255 *(Molobrus)*. —New York.

atra Macquart, Dipt. Exot. I, 1, 78; Bellardi, Saggio etc. I, 12. —
  Brazil (Macq.); Mexico (Bellardi); Schiner (Novara, 11) thinks this
  is *Sciara americana* Wiedem.
gigantea Macquart, Dipt. Exot. 1er Suppl. 19; Bellardi, Saggio etc.
  I, 13 — New Granada (Macq.); Mexico (Bellardi).
unicolor Say, J. Acad. Phil. VI, 153; Compl. Wr. II, 351. — Mexico.

Trichosia.
Winnertz, Beitr. z. Monogr. d. Sciarinen, 1867. ([10])
*__hebes__ Loew, Centur. IX, 58. — New York.

Zygoneura.
Meigen, System. Beschr. Vol. VI, 1830;
Winnertz, Beitr. z. Monogr. d. Sciarinen.
*__toxoneura__ O. Sacken, Proc. Ent. Soc. Phil. 1862, 165 *(Sciara)*. —
  Distr. Columbia.

FAMILY SIMULIDAE.

Simulium.

Latreille, Hist. Nat. Crust. et Ins. XIV, 294; 1804. ([11]).

decorum Walker, List etc. I, 112. — Huds. B. Terr.
*invenustum Walker, List, etc. I, 112. — Huds. B. Terr.
*piscicidium Riley, Amer. Ent. II, 367 (♀). — Mumford, N. Y.
*venustum Say, J Acad. Phil. III, 28; Compl. Wr. II, 51; Wied., Auss.
    Zw. I, 71, 1. — Ohio; Distr. Columbia.
*vittatum Zetterstedt, Ins. Lapp. 803; Dipt. Scand. X, 3423; Staeger,
    Groenl. Antliater; Holmgren, Ins. Nordgroenl. p. 104. — Greenland.
    Culex reptans O. Fabricius (non Linné) Fauna Groenl. 211, 173.
    [Staeger and Schiödte, Berlin. Ent. Z. 1859, 112.]

cinereum Bellardi, Saggio etc. I, 13. — Mexico.
metallicum Bellardi, Saggio etc. I, 14. — Mexico.
mexicanum Bellardi, Saggio etc. App. 6. — 'Mexico.
ochraceum Walker, Trans. Ent. Soc. N. Ser. V, 332. — Mexico.
quadrivittatum Loew, Centur. II, 2. — Cuba.

> **Observation.** *Simulium molestum* Harris, Ins. Inj. to Veget.
> 3d edit. 601 has never been described; *Simulium nocivum* Harris,
> I. c. 602 is a *Ceratopogon.*

FAMILY BIBIONIDAE.

Bibio.

Geoffroy, Hist. Nat. des Ins. II, 571, 3; 1764; *Hirtea* Fabricius,
    Zetterstedt etc. ([12]).

*albipennis Say, J. Acad. Phil. III, 77, 3; Compl. Wr. II, 69; Wiede-
    mann, Auss. Zw. I, 80, 7; Macquart, Dipt. Exot. I, 1, 88, 5,
    tab. XIII, f. 2. — Atlantic States.
articulatus Say, J. Acad. Phil. III, 77, 4; Compl. Wr. II, 69; Wied.
    Auss. Zw. I, 81, 8. — Pennsylvania. ([13]).
*abbreviatus Loew, Centur. V, 9. — Distr. Columbia.
*basalis Loew, Centur. V, 11. — New Hampshire.
baltimoricus Macquart, Dipt. Exot. 5e Suppl., 17, 12. — Baltimore.
brunnipes Fabricius, Ent. Syst. IV, 250, 80 *(Tipula)*; Syst. Antl. 54,
    15 *(Hirtea)*; Wiedemann, Auss. Zw. I, 81, 10. — Newfoundland (Fab.)
    Tipula rufipes Fabricius, Mant. Ins. II, 327, 69 [Wied.].
canadensis Macquart, Dipt. Exot. I, 2, 179; (?) Bellardi, Saggio etc.
    I, 18. — Canada, Mexico (Bellardi).
castanipes Jaennicke, Neue Exot. Dipt. 10. — Illinois.
*femoratus Wiedemann, Dipt. Exot. I, 35, 2; Auss. Zw. I, 79, 4. —
    Atlantic States.
    Bibio fuscipennis Macquart, Dipt. Exot. I, 1, 87, 3. (Loew in litt.)
*fraternus Loew, Centur. V, 8. — Distr. Columbia.

*gracilis Walker, List, etc. I, 123. — Nova Scotia.
*inaequalis Loew, Centur. V, 3. — Sitka.
*longipes Loew, Centur. V, 12. — Distr. Columbia.
*lugens Loew, Centur. V, 6. — Winnipeg.
*nigripilus Loew, Centur. V, 10. — Winnipeg.
*obscurus Loew, Centur. V, 5. — Huds. B. Terr.
   pallipes Say, J. Acad. Phil. III, 76, 1; Compl. Wr. II, 68; Wiede-
      mann, Auss. Zw. I, 81, 9; — Pennsylvania. (Compare also: Van
      der Wulp, Tijdschr. etc. 2d Ser. IV, 81.)
*rufithorax Wiedemann, Auss. Zw. I, 78. 2. — Pennsylvania, Florida.
   senilis v. d. Wulp, Tijdschr. Ent. 2d Ser. IV, 81. — Wisconsin.
   thoracica Say, Long's Exp. App. 368; Compl. Wr. I, 250; Wiede-
      mann, Auss. Zw. I, 78, 1. — Florida.
* variabilis Loew, Centur. V, 7. — New Hampshire, Sitka.
* xanthopus Wiedemann, Auss. Zw. I, 80; Macquart, Dipt. Exot. I,
   1, 88, 4. — Atlant. States.

* hirtus Loew, Cent. V, 2; O. Sacken, Western Diptera, 211. — California.
* nervosus Loew, Centur. V, 4. — California.

   criorrhinus Bellardi, Saggio etc. I, 17; Walker, Trans. Ent. Soc. N.
      S. V, 331. — Mexico.
   dubius Bellardi, Saggio etc. I, 18 — Mexico.
   fuligineus Bellardi, Saggio etc. I, 19. — Mexico.
   piceus Bellardi, Saggio etc. I, 17. — Mexico.

        Observation. Mr. Walker's species;
        fumipennis Walker, List, etc. I, 122. — Huds. B. Terr.
        humeralis Walker, l. c. 121. — Nova Scotia.
        scita Walker, l. c. 122. — Nova Scotia.
        striatipes Walker, l. c. — Nova Scotia.
        vestita Walker, l. c. — Nova Scotia.

Dilophus.

    Meigen, Illiger's Magaz. II, 264; 1803.

*breviceps Loew, Centur. IX, 59. — New Hampshire.
*dimidiatus Loew, Centur. VIII, 3. — New York.
*longiceps Loew, Centur. I, 14. — Illinois.
*orbatus Say, J. Acad. Phil. III. 77, 5 (*Bibio*); Compl. Wr, II, 69;
      Wiedemann, Auss. Zw. I, 77, 6. — Pennsylvania; Mexico (Bel-
      lardi, Saggio etc. I, 19).
*obesulus Loew, Centur. IX, 60. — Distr. Columbia.
*serotinus Loew, Centur. I, 15. — Illinois.
   spinipes Say, J. Acad. Phil. III, 79, 2; Wiedemann, Auss. Zw. I,
      75, 1. — Missouri.
   stigmaterus Say, J. Acad. Phil. III, 78, 1; Wiedemann, Auss. Zw. I,
      76, 4. — Missouri.
   stygius Say, Ins. of Louisiana, coll. by J. Barabino; Compl. Wr. I,
      309. — Louisiana (there is an earlier *D. stygius* Say, from
      Mexico).

thoracicus Say, J. Acad. Phil. III, 80, 3; Wiedemann, Auss. Zw. I,
77, 5. — Pennsylvania, Maryland.
*****tibialis** Loew, Centur. IX, 61. — Sitka.

maculatus Bellardi, Saggio etc. I, 19; tab. I, f. 5. — Mexico.
minutus Bellardi, Saggio etc. App. 7. -- Mexico.
stygius Say, J. Acad. Phil. VI, 155; Compl. Wr. II, 352. — Mexico.

> **Observation.** Mr. Walker's species.
> **fulvicoxa** Walker, List, etc. I, 117. — Huds. B. Terr.
> **serraticollis** Walker, List, etc. 1, c. — Huds. B. Terr.

Hesperinus.

Walker, List, etc. I, 81, 1848; *Spodius* Loew, Berl. Ent. Z. II, 101;
Tab. I, f. 1–15; 1858. ([12]).
*****brevifrons** Walker, List, etc. I, 81. — British Possessions; White
Mts., N. H. and Colorado Mts., in the alpine region.

Plecia.

Wiedemann, Auss. Zw. I, 72; 1828. ([14]).

ruficollis Fabricius, Wiedemann, Auss. Zw. I, 72; Macquart, Hist. Nat.
Dipt., Atlas, Tab. IV f. 17; Bellardi, Saggio etc, I, 15. — South
America; Mexico; Florida. (Lake Harney, by Messrs. Hubbard
and Schwarz.)
*****heteroptera** Say, J. Acad. Phil. III, 77, 2; Compl. Wr. II, 69 *(Bibio);*
Wiedemann, Auss. Zw. I, 80, 6 *(id.)* — Atlantic States.
Penthetria atra Macquart, Hist. Nat. Dipt. I, 175, 2. Compare also
Van der Wulp, Tijdschr. etc. 2[d] Ser. IV, 81.
Eupeitenus ater Macquart, Dipt. Exot. I, 1, 85; Tab. XII, f. 3. —
Philadelphia.
Plecia longipes Loew, Berl. Ent. Z. II, 109. — New Orleans.

bicolor Bellardi, Saggio etc. I, 16. — Mexico.
heros Say, J Acad. Phil. VI, 154 *(Penthetria);* Compl. Wr. II, 352
(id.) — Mexico.
nigerrima Bellardi, Saggio etc. I, 14. — Mexico.
rostrata Bellardi, Saggio etc. I, 15. — Mexico.
rufithorax Walker, List, etc. I, 116. — Jamaica.
vittata Bellardi, Saggio etc. App. 7, f. 4. — Mexico.

> Observation. *Plecia bimaculata* Walker, Dipt. Saund. 422,
> United States, is the female of one of the common North-American
> *Dilophus.*

Scatopse.

Geoffroy, Hist. Nat. d. Ins. II, 545; 1764. ([15]).

*****atrata** Say, Long's Exp. App. 367; Compl Wr. I, 250; Wiedemann,
Auss. Zw. I, 71, 1. — Philadelphia.
Scatopse recurva Loew, Linn. Entom. I, 330, Tab. III, f. 4. —
Europe. (Loew. Sillim. Journ. N. Ser. Vol. XXXVII, 317.)

*notata Linn., Meigen etc. — This common european species, also
occurs in N. Am.

pulicaria Loew, Linn. Entom. I, 338, Tab. III, f. 10. — Europe, and also
in Wisconsin, according to v. d. Wulp, Tijdschr. etc. 2d Ser. IV, 80.

*pygmaea Loew, Centur. V, 13. — Distr. Columbia.

> **Observation.** The following three species of Mr. Walker's are mentioned
> separately, as their very short descriptions do not show any tangible differences
> and the identification would be, I should say, impossible.
> nitens Walker, List, etc. I, 114. — Huds. B. Terr.
> obscura Walker, List, etc. 114. — Huds. B. Terr.
> pusilla Walker, List, etc. I, 114, — Huds. B. Terr.

Aspistes.

Meigen, Syst. Beschr. I, 319, 1818; *Arthria* Kirby, Fauna Bor. Am.
311; 1837. ([16].

*analis Kirby, Fauna Bor. Am. Ins. 311, 1; Tab. V, f. 8. *(Arthria).* —
Arctic America.

> *Aspistes borealis* Loew, Stett. Ent. Z. 1847, 69. — North of Europe
> and North-America (About the occurrence in N. A. see Loew in
> Sillim. Journ. I. c. 317).

FAMILY BLEPHAROCERIDAE. [17].

Blepharocera.

Blepharicera Macquart, Ann. Soc. Ent. de Fr. II, 1, 61; 1843;
Asthenia Westwood 1842; preocc.

*capitata Loew, Centur. IV, 43. — Distr. Columbia; White Mts., N. H.
Asyndulum tenuipes Walker, List etc. I, 86. — Huds. Bay Territ. (!)

*yosemite O. Sacken, Western Diptera, 195. — Yosemite Valley, Calif.

Bibiocephala.

O. Sacken, in Hayden's Report on Geol. Survey Color. Territ. 1873;
translated by Loew in Zeitschr. für Entomol. Neue Folge, Heft 6,
Breslau 1877, p. 95.

*grandis O. Sacken, Hayden's Report 1873, 564; translated by Loew,
I. c. 98. — Rocky Mountains, Colorado.

> Observation. For *Asthenia americana* Walker, List etc. I, 28,
> see the note ([19]).

Paltostoma.

Schiner, Verh. Zool. Bot. Ges. 1866, p. 931; Novara etc. p. 27.

superbiens Schiner, Novara etc. p. 28, Tab. II, f. 4, — South-America.
(I quote this species, because I have seen specimens from Mexico,
in Mr. Bellardi's collection, which may perhaps belong to it. ([19]).

5

FAMILY CULICIDAE.

Megarrhina.

R. Desvoidy, Essai etc. in the Mém. de la Soc. d'hist. nat. de Paris III, 412; 1827.

* haemorrhoidalis Fabricius, Ent. Syst. IV, 401, 5 *(Culex)*; Syst. Antl. 35, 8. *(id.)*; Wiedemann, Dipt. Exot. I, 6, 1 *(id.)*; Auss. Zw. I, 2 *(id.)* — Cayenne; Cuba.

> Observation. *Megarrhina ferox* Wied. (Brazil), mentioned in my first Catalogue, is omitted here, as its occurrence in Georgia (Walker, List, etc. I, 1) is exceedingly doubtful.

Culex.

Linné, Fauna Suecica, 1761.

* annulatus Meigen etc. — Europe and the North West of North-America (brought by R. Kennicott from Mackenzie River).

Boscii R. Desvoidy, Culicides etc. *(Psorophora)*. — Carolina.

* ciliatus Fabricius, Entom. Syst. IV, 401, 6; Syst. Antl. 35, 10; Coquebert, Ill. Icon. Ins. Tab. XVII, f. 7; St. Fargeau et Serville, Encycl. Méthod. X, 658; Wiedemann, Auss. Zw. I, 3, 5; Macquart, Hist. Nat. Dipt. I, 36, 15; Dipt. Exot. 4e Suppl. 11, Tab. I, f. 1. — Atlantic States.

Culex molestus Wiedemann, Dipt. Exot. I, 7, 4 [Wied.].

Culex conterrens Walker, Dipt. Saunders, 427 [!]. — U. S.

consobrinus Rob. Desvoidy, Culicides, 408, 27. — Pennsylvania.

musicus Say, J. Acad. Phil. VI, 149; Compl. Wr. II, 348. — Indiana.

nigripes Zetterstedt, Insecta Lapponica; Dipt. Scand. IX, 3458, 5; Staeger, Groenl. Antliater; Holmgren, Ins. Spetsb; Ins. Nordgroenl. 104. — Spitzbergen, Greenland.

Culex pipiens O. Fabricius, Fauna Groenl. 209, 171 [Schiödte].

Culex caspius Pallas in Curtis, Ins. Capt. Ross's Voyage, LXXVI [Schiödte].

punctor Kirby, Fauna Bor. Amer., Insects 308, 1. — Arctic America.

pungens Wiedemann, Auss. Zw. I, 9, 16. — New Orleans.

rubidus R. Desvoidy, Culicides etc. — Carolina.

taeniatus Wiedemann, Auss. Zw. I, 10, 18. — Georgia.

* taeniorhynchus Wiedemann, Dipt. Exot. I, 43, 1; Auss. Zw. I, 8, 13. — Atlantic St.; Mexico (Wied.); S. America (Schiner, Novara, 31).

Culex damnosus Say, Journ. Acad. Phil. III, 11, 3; Compl. Wr. II, 40. (Change of name by Wied.)

Culex sollicitans Walker, Dipt. Saund. 427. [!] — U. S.

testaceus v. d. Wulp, Tijdschr. v. Eutom. 2d Ser. II, 128, Tab. III, f. 1. — Wisconsin.

* triseriatus Say, Journ. Acad. Phil. III, 12, 4; Compl. Wr. II, 40; Wiedemann, Auss. Zw. I, 11, 19. — Pennsylvania (Say).

incidens Thomson, Eugenie's Resa etc. 443. — California.

pinguis Walker, in Lord's Naturalist etc. II, 337. — Vancouver.

Bigoti Bellardi, Saggio etc. App. 3, fig. 1. — Mexico.
cubensis Bigot, R. de la Sagra's Hist. etc. 786. — Cuba.
*****fasciatus** Fab. Syst. Antl. 36, 13; Wiedemann, Auss. Zw. I, 8, 13. — Jamaica.
> *Culex mosquito* R. Desv. Culicides etc. 390; Guérin et Percheron, Genera etc. (figured carefully) Dipt. tab. ii, fig. 1. Macq. Hist. Nat. Dipt. I, 35, 8. — Cuba.

frater R. Desvoidy, Culicides etc. (he quotes *C. fasciatus* Wied. as synonym, but distinguishes it from *C. fasciatus* Fab.) — West Indies.
posticatus Wiedemann, Dipt. Exot. I, 43, 2; Auss. Zw. I, 9, 15. — Mexico.
mexicanus Bellardi, Saggio etc. I, 5. — Mexico.

> **Observation.** Mr. Walker's species of Culex, omitted in the preceding list, are given here:
>> **excitans** List, etc. I, 4. — Georgia.
>> **excrucians** Dipt. Saund. 429. — Nova Scotia.
>> **impatiens** List etc. I, 5. — Huds. B. Terr.
>> **impiger** List etc. I, 6 — Huds. B. Terr.
>> **implacabilis** List etc. I, 7. — Huds. B. Terr.
>> **perturbans** Dipt. Saund. 428. — United States.
>> **provocans** List etc. I, 7. — Nova Scotia.
>> **stimulans** List etc. I, 4. — Nova Scotia.
>> **territans** Dipt. Saund. 428. — United States.
>
> About the typical specimens of these species in the Brit. Mus. see the note. [20].

Anopheles.

Meigen, Syst. Beschr. I, 10, 1818. [21].

annulimanus v. d. Wulp, Tijdschr. v. Ent. 2ᵈ Ser. II, 129, Tab. III, f. 2. — Wisconsin.
*****crucians** Wiedemann, Auss. Zw. 12, 1. — Maryland (Say).
> *Culex punctipennis* Say, Journ. Ac. Phil. III, 9, 1; Compl. Wr. II, 39 [Wied].

ferruginosus Wiedemann, Auss. Zw. I, 12, 2. — New Orleans (Wied); On the Mississippi (Say).
> *Culex quinquefasciatus* Say, Journ. Ac. Phil. III, 10, 2; Compl. Wr. II, 39. [Change of name by Wied.]

maculipennis Meigen ⎰ European species, which also occur in N. A.
nigripes Staeger ⎱ according to Loew, Sillim. Journ. N. Ser. Vol. XXXVII, 317.
*****quadrimaculatus** Say, Long's Exp. App. 356; Compl. Wr. I, 241; Wiedemann, Auss. Zw. I, 13, 4. — Atlantic States and Canada, also in the South of Europe.
> *Culex hiemalis* Fitch, Winter Insects etc.
> *Anopheles pictus* Loew, Dipt. Beitr. I, 4. — South of Europe. [Loew, Sillim. Journ. N. Ser. Vol. XXXVII, 317.]

albimanus Wiedemann, Auss. Zw. I, 13, 3. — San Domingo.

Aëdes.

Meigen, Syst. Beschr. I, 13; 1818.
*****fuscus** O. Sacken, Western Diptera, 191. — Cambridge, Mass.
*****sapphirinus** O. Sacken, Trans. Amer. Ent. Soc. II, 47. — New York, Distr. Columbia.

Corethra.

Meigen, Illiger's Magaz. II, 260; 1803.

*punctipennis Say, Journ. Acad. Phil. III, 16; Compl. Wr. II, 43;
Wiedemann, Auss. Zw. I, 14, 1. — Pennsylvania (Say).
*trivittata Loew, Centur. II, 1. — Maine; Yukon River, Alaska.

FAMILY CHIRONOMIDAE. [22].

Diamesa.

Meigen, Syst. Beschr. VII, 72; 1830.

Waltlii Meigen etc. Staeger, Groenl. Antliater. — Europe, Greenland.

Chironomus.

Meigen, Illig. Mag. II, 260; 1803.

albistria Walker, List, etc. I, 17. — Huds. Bay.
anticus Walker, List, etc. I, 21. — Georgia.
aterrimus Meigen, etc., Staeger, Groenl. Antliater. — Europe, Greenland.
attenuatus Walker, List, etc. I, 20. — Huds. Bay.
basalis Stacger, Groenl. Antliater. 351, 6; Holmgren, Ins. Nordgroenl.
    105. — Greenland.
bimacula Walker, List, etc. I, 15. — Huds. Bay.
borealis Curtis, Ins. of Ross's Voy. LXXVII. — Arctic America.
brunneus Walker, List, etc. I, 21. — Huds Bay..
byssinus Meigen, etc. Staeger, Groenl. Antliater. — Europe, Greenland.
confinis Walker, List, etc. I, 15. — Huds. Bay.
crassicollis Walker, l. c. 18. — Huds. Bay.
cristatus Fabr. Syst. Antl. 39, 4. Wiedemann, Auss. Zweifl. I, 14, 1.
    Macquart, Hist. Nat. Dipt. I, 50, 10. — New York (Fab.)
devinctus Say, Journ. Acad. Phil. VI, 150; Compl. Wr. II, 349. —
    Indiana.
festivus Say, Journ. Acad. Phil. III, 13, 2; Compl. Wr. II, 41; Wied.
    Anal. Entom. 10.; Auss. Zw. I, 16, 5. — Illinois (Say).
fimbriatus Walker, l. c. 20. — Huds. Bay.
flavicingula Walker, l. c. I, 20. — Huds. Bay.
frigidus Zetterstedt, Insecta Lapponica; Dipt. Scand. IX, 3516, 33;
    Staeger, Groenl. Antliater; Holmgren, Ins. Nordgroenl. 105. —
    Greenland; also Northern Europe
geminatus Say, J. Acad. Phil. III. 14, 4; Compl. Wr. II, 42. —
    Pennsylvania.
glaucurus Wiedemann, Auss. Zweifl. I, 15, 3. — Atlantic States.
    Chironomus stigmaterus Say, Journ. Acad. Phil. III, 15, 6; Compl.
    Wr. II, 42. [Change of name by Wied.]
hyperboreus Staeger, Groenlands Antliater; Zetterstedt, Dipt. Scand.
    IX, 3487. — Greenland; also Northern Europe.
*intermedius Staeger, Kröjer's Tidskr. II, 559. — Europe and N. W.
    of North-America (brought together with plumosus, of which it
    may be only a smaller variety).

jucundus Walker, List, etc. I, 16. — Georgia.
lasiomerus Walker, l. c. I, 19. — Huds. Bay.
lasiopus Walker, l. c. I, 19. — Huds. Bay.
lineola Wiedemann Auss. Zw. I, 17, 6. — Pennsylvania.
    Chironomus lineatus Say, J. Acad Phil. III, 14, 5; Compl. Wr. II,
    42. [Wied.].
lobifer Say, J. Acad. Phil. III, 12, 1; Compl. Wr. II, 41. *(C. lobiferus)*;
    Wiedemann, Auss. Zweifl. I, 16, 4; Macquart, Hist. Nat. Dipt. I,
    50, 12. — Pennsylvania.
modestus Say, J. Acad. Phil. III, 13, 3; Compl. Wr. II, 41; Wiede-
    mann, Auss. Zw. I, 18, 8. — Pennsylvania.
nigritibia Walker, List, etc. I, 16. — Huds. Bay.
*****nivoriundus** A. Fitch, Winter Insects, I. — New-York.
pellucidus Walker, l. c. 21. — Huds. Bay.
oceanicus Packard, Proc. Essex Instit. VI, 42 (figure of larva on p.
    43, of imago on p. 45). — Salem, Mass.
picipes Meigen etc., Staeger, Groenlands Antliater. — Europe, Greenland.
*****plumosus** Linné, Meigen etc. — Europe and N. W. of North-America
    (brought by R. Kennicott from Mackenzie River).
polaris Kirby, Suppl. to App. to Parry's First Voyage; Curtis, Ins.
    of Ross's Voyage, LXXVII tab. A, figs. 14 and 2. — Arctic
    America; Greenland.
pumilio Holmgren, Ins. Spetsb. 41; Ins. Nordgroenl. 105. — Spitz-
    bergen and Greenland.
redeuns Walker. Dipt. Saund. 422. — U. States.
stercorarius Zetterstedt, Dipt. Scand. IX, 3571, 97; Holmgren, Ins.
    Nordgroenl. 105. — Greenland; also in Europe.
taenionotus Say, J. Acad. Phil. VI, 149; C. Wr. II, 349. — Indiana.
trichomerus Walker, List, etc. I, 21. — Huds. Bay.
tricinctus Meigen, I, 41, 49. — Europe and N. America (Loew *in litt.*).
unicolor Walker, List, etc. I, 19. — Nova Scotia.
variabilis Staeger, Groenl. Antliater; Zetterstedt, Dipt. Scand. IX,
    3519; — Greenland; also in the North of Europe.

*****octopunctatus** Loew, Wien. Entom. Monatschr. V, 33. — Cuba.

    Observation: *Chiron. riparius* Meig., *Chloris* M., *pedellus*
    Lin., *viridis* Macq. are european species, also occurring in North-
    Am., according to van der Wulp, Tijdschr. voor Entom. 2ᵈ Ser. II, 126.

Tanypus. (²³).

Meigen, Illiger's Magaz. II, 261; 1803.

*****annulatus** Say, J. Acad. Phil. III, 15, 1; Compl. Wr. II, 43; Wiede-
    mann, Auss. Zw. I, 19, 3. — Pennsylvania.
baltimoreus Macquart, Dipt. Exot. 5ᵉ Suppl. 15, 1. — Baltimore.
*****bellus** Loew Centur. VII, 4. — Distr. Columbia.
choreus Meigen etc. — Europe and North-America (Loew in Sillim.
    Journ. XXXVII, 317; Walker, Dipt. Saund. 422).

crassinervis Zetterstedt, Ins. Lapponica; Dipt. Scandin. IX, 3599, 5;
    Staeger, Groenl. Antliater, p.354, 11. — Greenland; also in Lapland.
*decedens Walker, List, etc. I, 22. — Huds. B. Terr.
*flavicinctus Loew, Centur. I, 2. — Pennsylvania.
futilis v. d. Wulp, Tijdschr. voor Entom. 2ᵈ Ser. II, 130. — Wisconsin.
*hirtipennis Loew, Centur VII, 6. — Maine.
pictipennis Zetterstedt, Ins. Lapponica 818, 5; Staeger, Groenl Ant-
    liater. — Greenland.
*pilosellus Loew, Centur. VII, 7. — Dist. Columbia.
*pinguis Loew, Centur. I, 1. — New York.
*pusillus Loew, Centur. VII, 5. — Distr. Columbia.
*scapularis Loew, Centur. VII, 1. — Distr. Columbia.
*thoracicus Loew, Centur. VII, 3. — Distr. Columbia.
tibialis Staeger, Groenl. Antliater. -- Greenland.
tibialis Say. J. Acad. Phil. III, 15, 2; Compl. Wr. II, 43; Wiedemann,
    Auss. Zw. I, 20, 4. — Pennsylvania.
*tricolor Loew, Centur. I, 3. — New York.
turpis Zetterstedt, Ins. Lapp. 811, 8 (Chironomus); Staeger, Groenl.
    Antl. 350, 3 (?? query by Zetterstedt, Dipt. Scand. IX, 3596). —
    Lapland; also Greenland?
*humeralis Loew, Centur. VII, 2. — Cuba.

    Observation: *Tanypus monilis Lin.*, a european species,
occurs in North-America (Wisconsin) according to Van der Wulp,
Tijdschr. v. Entom. 2ᵈ Ser. II, 126. *T. annulatus* Say looks very
much like *T. monilis Lin.*, and if Mr. Van der Wulp's identification is correct, I should have taken both for the same species.

Chasmatonotus.
Loew, Centur. V, I; 1864.

*unimaculatus Loew, Centur. V, 1. — White Mts, N. H.
*bimaculatus O. Sacken, Western Diptera, 191. — Catskill, Mountain
    House, N. Y.; Quebec (Can.).

Ceratopogon. [24]
Meigen, Illig. Magaz. II; 1803.

*albiventris Loew, Centur. I, 7. — Georgia.
*argentatus Loew, Centur. I, 5. — Distr Columbia.
basalis Walker, List, etc. I, 27. — Trenton Falls.
*bimaculatus Loew, Centur. I, 6. — Distr. Columbia.
*festivus Loew, Centur. I, 13. — Pennsylvania.
*longipennis Loew, Centur. I, 10. — Pennsylvania.
*lineatus Meigen, Syst. Beschr. etc. I, 80. — Europe and North-America
    [the latter according to Loew, in Sillim. Journ. N. Ser. XXXVII, 317].
obscurus Walker, List etc. I, 26. - Huds. B. Terr.
*opacus Loew, Centur. I, 9. — Distr. Columbia.
parvus Walker, List, etc. I, 26. — Huds. B. Terr.
*plebejus Loew, Centur. I, 11. — Pennsylvania.
*rufus Loew, Centur. I, 12. — Pennsylvania.

scutellatus Say, J. Acad Phil. VI, 150; Compl. Wr. II, 349. — Indiana.
*setulosus Loew, Centur. I, 8. — Distr. Columbia.
sordidellus Zetterstedt, Ins. Lapp. 820, 6; Dipt. Scand. IX, 3640;
Staeger, Groenl. Antliater. — Greenland.
Culex pulicans (misprint for *pulicaris*) O. Fabricius, Fauna Groenlandica [Schiódte].
transiens Walker, List, etc. I, 25. — Huds. B. Terr.
*trivialis Loew, Centur. I, 4. — Distr. Columbia.

*genualis Loew, Centur. VI, 1. — Cuba.

Oecacta.

Poey, Memorias etc. Vol. I; 1851.
furens Poey, Memorias etc. I, 236, Tab. XXVII. — Cuba.

Heteromyia.

Say, Americ. Entom. Vol. II; 1825.
*fasciata Say, N. Am. Entom. Vol. II. Tab. XXXV; Compl. Wr. I, 79. —
Atlantic States.

Observation: If this genus be adopted, it will have to include
several other species, now placed in the genus Ceratopogon;
Cerat. argentatus Loew among them.

FAMILY ORPHNEPHILIDAE.

Orphnephila.

Haliday, Zool. Journ. V, 350; Tab. XV, f. 1—9; 1831; *Thaumalea*
Ruthe 1831; *Chenesia* Macquart 1834.
*testacea Ruthe, Isis 1831, 1211 (1831); Haliday, l. c. (*O. devia*). —
Europe and North-America; New York. [About the identity see
Loew, Monogr. etc. I, 6.]

Observation. *Orphnephila* is a very heterogeneous form,
which cannot well be referred to any of the existing families.

FAMILY PSYCHODIDAE.

Psychoda.

Latreille, Précis etc.; 1796.
alternata Say, Long's Exped. App. 358; Compl. Wr. I, 242; Wiedemann, Auss. Zw. I, 23. — Pennsylvania.
degenera Walker, List etc. I, 33. — Huds. Bay Territ.

FAMILY TIPULIDAE.[25].

SECTION I. LIMNOBINA.

Dicranomyia.

Stephens, Catal. Brit. Ins. 1829.
O. Sacken, Proc. Ac. Nat. Sc. Phil. 1859 and Monogr. IV, 53.

*badia Walker, List etc. I, 46. *(Limnobia)*; O. Sacken, Mon. etc. IV, 72. Tab. III, f. 2, forceps. — United States and British Possessions (Quebec); also in California.
Dicranomyia humidicola, O. Sacken, Proc. Ac. Nat. Soc. Phil. 1859, 210.

*brevivena O. Sacken, Mon. etc. IV, 66. — New York, Distr. Columbia.

*distans O. Sacken, Proc. Ac. Nat. Sc. Phil. 1859, 211; Mon. etc. IV, 67. — Florida.

*diversa O. Sacken, Proc. Ac. Nat. Sc. Phil. 1859, 212; Mon. etc. IV, 64. — Distr. Columbia.

*defuncta O. Sacken, Proc. Ac. Nat. Sc. Phil. 1859, 213; Monogr. etc. IV, 76. — Distr. Columbia; New York; Maine; Canada; California.
Limnobia simulans Walker, List, etc. I, 45. [26]).

*floridana O. Sacken, Mon. etc. IV, 67. — Florida.

*gladiator O. Sacken, Proc. Ac. Nat. Sc. Phil. 1859, 212; Mon. etc. IV, 63; Tab. III, f. 4, forceps. — Distr. Columbia.

*globithorax O. Sacken, Mon. IV, 74. — New Hampshire; Distr. Columbia.

*haeretica O. Sacken, Mon. etc. IV, 70; Tab. I, f. 3, wing. — New York; Fort Resolution, Huds. B. Terr.

*halterata O. Sacken, Mon. etc. IV, 71. — Labrador.

*immodesta O. Sacken, Proc. Ac. Nat. Sc. Phil. 1859, 211; Mon. etc. IV, 62. — Distr. Columbia; New York; Maine.

*liberta O. Sacken, Proc. Ac. Nat. Sc. Phil. 1859, 209; Mon. etc. IV, 69; Tab. III, f. 3, forceps. — Atlantic States and Canada; a similar species occurs in Europe.

*longipennis Schummel, Beitr. etc. 104, 2 *(Limnobia)*. — O. Sacken, Mon. etc. IV, 61; Tab. I, f. 1, wing. — New York; Massachusetts; Quebec, Can.; also in Europe.
Dicranomyia immemor O. Sacken, Proc. Ac. N. Sc. Phil. 1861, 287.

*morioides O. Sacken, Mon. etc. IV, 73. — New York.
Dicranomyia morio O. Sacken (nec Fabr.), Proc. Ac. N. Sc. Phil. 1859, 212.

*pubipennis O Sacken, Proc. Ac. Nat. Sc. Phil. 1859, 211; Mon. etc. IV, 73; Tab. I, f 2, wing. — Distr. Columbia; New York.

*pudica O Sacken, Proc. Ac. Nat. Sc. Phil. 1859, 212; Mon. etc. IV, 64. — Illinois.

*rara O Sacken, Mon. etc. IV, 75. — New York.

*rostrifera O Sacken. Mon. etc. IV, 65. — New York.

*stulta O. Sacken, Proc. Acad. N. Sc. Phil. 1859, 210; Mon. etc. IV, 68. — New York, Canada.

*marmorata O. Sacken, Proc. Acad: N. Sc. Phil. 1861, 288; Mon. etc.
IV, 77. Compare also Western Diptera 197. — California.

Geranomyia.

Haliday, Ent. Mag. I, 154; 1833; *Aporosa* Macquart, 1838; *Plettusa*
Philippi 1865. Compare O. Sacken, Monogr. etc. IV, 78.

*canadensis Westwood, Ann. Soc. Ent. France 1835, 683 (*Limnobio-*
rhynchus). — O. Sacken, Mon. etc. IV, 80. — North America, from
Canada to Florida; also in California.
Geranomyia communis O. Sacken, Proc. Acad. Nat. Sc. Phil. 1859, 207.
*diversa O. Sacken, Proc. Acad. Nat. Sc. Phil. 1859, 207; — Mon. etc.
IV, 80. — New York.
*rostrata Say, Journ. Acad. Nat. Sc. Phil. III, 22, 6 (*Limnobia*); Compl.
Wr. II, 47; Wiedemann, Auss. Zw. I, 35, 20. *(id.).* — O. Sacken,
Proc. Acad. Nat. Sc. Phil. 1859, 207; Mon. etc. IV, 79. — Atlantic
States, Canada and Cuba (apparently the same species).

intermedia Walker, List, etc. I, 47 (*Limnobia*). — Jamaica.
mexicana Bellardi, Saggio etc. App. 4 (*Aporosa*). — Mexico.
*rufescens Loew, Linn. Ent. V, 396, Tab. II, f. 9—12 (*Aporosa*). —
Portorico.
*virescens Loew, Linn. Ent. V, 396 (*Aporosa*). — St. Thomas.

Rhipidia.

Meigen, Syst. Beschr. I, 1818; O. Sacken Mon. etc. IV, 81 and III,
in Add. and Corr.

*maculata Meigen, Syst. Beschr. etc. I, 153, Tab. V, f. 9—11. —
O. Sacken, Proc. Acad. Nat. Sc. Phil. 1859, 208; Monogr. etc.
IV, 82. — Europe and Atlantic States of North America.
*fidelis O. Sacken, Proc. Acad. Nat. Sc. Phil. 1859, 209; Mon. etc.
IV, 83. — New York; Illinois; Canada.
*domestica O. Sacken, Proc. Acad. Nat. Sc. Phil., 1859, 208; Mon. etc.
IV, 84; Tab. III, f. 5, forceps. — Atlantic States and apparently
the same species in Brazil.

Limnobia.

Meigen, Syst. Beschr. I, 1818; O. Sacken, Mon. etc. IV, 84.

*cinctipes Say, Journ. Acad. Nat. Sc. Phil. III, 21, 4; Compl. Wr. II,
47; Wiedemann, Auss. Zw. I, 32, 15. O. Sacken, Proc. Acad.
Nat. Sc. Phil. 1859, 214; Mon. etc. IV, 88. — Atlantic States.
*hudsonica O. Sacken, Proc. Acad. Nat. Sc. Phil. 1861, 289; Mon. etc.
IV, 91. — Slave Lake, Huds. B. Terr.
*immatura O. Sacken, Proc. Acad. Nat. Sc. Phil. 1859, 214; Mon. etc.
IV, 89. — Distr. Columbia; Wisconsin; Maine.
*indigena O. Sacken, Proc. Acad. Nat. Sc. Phil. 1859, 215; Mon. etc.
94; Tab. III, f. 7, forceps. Atlantic States and Colorado; Canada.

*parietina O. Sacken, Proc. Acad. Nat. Sc. Phil. 1861, 289; Mon. etc.
    IV. 93. — Trenton Falls, N. Y.; White Mts. N. H.
*sociabilis O. Sacken, Mon. etc. IV, 95. — Illinois.
*solitaria O. Sacken, Proc. Acad. Nat. Sc. Phil. 1859, 215; Mon. etc.
    IV, 90; Tab. III, f. 6, forceps. — New York, New Hampshire,
    Maine and far north in British America.
*triocellata O. Sacken, Proc. Acad. Nat. Sc. Phil. 1859, 216; Mon. etc.
    IV, 92. — Distr. Columbia, New York, Wisconsin.
*tristigma O. Sacken, Proc. Acad. Nat. Sc. Phil. 1859, 216; Mon. etc.
    IV, 95. — Illinois.

*californica O. Sacken, Proc. Acad. Nat. Sc. Phil. 1861, 288; Mon. etc.
    IV. 96. — California.
*sciophila O. Sacken, Western Diptera, 197. — Marin and Sonoma
    Co, Cal.

livida Say, Journ. Acad. Phil. VI, 151; Compl. Wr. II, 349. — Mexico.

Trochobola.

O. Sacken, Mon. etc. IV, 97; 1868; *Discobola* O. Sacken, 1865.

*argus Say, Long's Exp. App. 358; Compl. Wr. I, 243 (*Limnobia*);
    Wiedemann, Auss. Zw. Ins. I, 33, 17 (*id.*); O. Sacken, Proc. Acad.
    Nat. Sc. Phil. 1859, 217 (*id.*); Mon. etc. IV, 98. Tab. I, f. 4,
    wing. — Massachusetts; Maine; New York; New Yersey; Nova
    Scotia, Canada. ([27]).

SECTION II. LIMNOBINA ANOMALA.
Rhamphidia.

Meigen, Syst. Beschr. VI; 1830; *Megarhina* and *Ilelius* St. Fargeau, 1825;
    O. Sacken, Mon. IV, 103.

*flavipes Macquart, Dipt. Exot. 5e Suppl. 17. Tab. I, f. 4 (wing). —
    O. Sacken, Monogr. etc. IV, 105. — Atlantic States.
    Rhamphidia prominens Walker, Dipt. Saund. 435.
    Rhamphidia brevirostris O. Sacken, Proc. Acad. Nat. Sc. Phil. 1859, 222.

Elephantomyia.

O. Sacken, Proc. Acad. Nat. Sc. Phil. 1859; Monogr. etc. IV, 106,

*Westwoodi O. Sacken, Mon. etc. IV, 109, Tab. I, f. 5, wing; Tab. III,
    f. 8, forceps. — N. America, from Quebec to Florida.
    Elephantomyia canadensis O. Sacken (nec Westwood), Proc. Acad.
    Nat. Sc. Phil. 1859, 221.

Toxorrhina.

Loew, Linn. Entom. V, 400; 1851; O. Sacken, Mon. etc. IV, 109.

*magna O. Sacken, Proc. Phil. Ent. Soc. 1865, 232; Mon. etc. IV, 114. —
    New Jersey.

*muliebris O. Sacken, Proc. Phil. Ent. Soc. 1865, 233; Mon. IV, 115;
see also Additions to Vol. IV at the end of Mon. Vol. III. —
Princeton, Mass.; Tarrytown, N. Y.

fragilis Loew, Linn. Ent. V, 401, Tab. II, f. 16—18. — Portorico.

Dicranoptycha.

O. Sacken, Proc. Acad. Nat. Sc. Phil. 1859. Mon. etc. IV, 116.

*germana O. Sacken, Proc. Acad. Nat. Sc. Phil. 1859, 217; Mon. etc.
IV, 117. — Trenton Falls, N. Y.
*sobrina O. Sacken, Proc. Acad. Nat. Sc. Phil. 1859, 218; Mon. IV, 118;
Tab. I, f. 8, wing; Tab. III, f. 12, forceps. — Distr. Columbia;
a similar species in California.
Dicranoptycha sororcula O. Sacken, Proc. Ac. Nat. Sc. Phil. 1859, 218.
*nigripes O. Sacken, Proc. Acad. Nat. Sc. Phil. 1859, 218; Mon. etc.
IV, 119; Tab. III, f. 11, forceps. — Dalton, Georgia.

Elliptera.

Schiner, Wiener Entom. Monatsschr. VII, 222, 1863.
O. Sacken, Monogr. IV, 122

*clausa O. Sacken, Western Diptera, 197. — Yosemite Valley, Cal.

Antocha.

O. Sacken, Proc. Acad. Nat. Sc. Phil. 1859; Mon. etc. IV, 125.

*opalizans O. Sacken, Proc. Acad. Nat. Sc. Phil. 1859, 220; Mon. etc.
IV, 126, Tab. III, f. 10. — Europe and N. America (from Distr.
Columbia to Fort Resolution, Huds. B. Terr.)
Antocha saxicola O. Sacken, Proc. Acad. Nat. Sc. Phil. 1859, 220.

Atarba.

O. Sacken, Monogr. etc. IV, 127: 1868.

*picticornis O. Sacken, Mon. etc. IV, 128, Tab. I, f. 13, wing. —
Delaware; Distr. Columbia, Trenton Falls, N. Y.

Teucholabis.

O. Sacken, Proc. Acad. Nat. Sc. Phil. 1859, Monogr. etc. IV, 129.

*complexa O. Sacken, Proc. Acad. Nat. Sc. Phil. 1859, 223; Monogr.
etc. IV, 132. — Distr. Columbia, New York, Illinois.
*chalybeiventris Loew, Wiener Monatschr. 1861, 33. (*Rhamphidia*).
(About the location in this genus compare O. Sacken, Monogr. IV,
132.) — Cuba.

Diotrepha.

nov. gen.

*mirabilis, n. sp. see the note.[28]. — Georgia; Texas; Cuba (?)

SECTION III. ERIOPTERINA.

Rhypholophus.

Kolenati Wiener Ent. Monatschr. IV, 1860.
O. Sacken, Monogr. etc IV, 141; *Dasyptera* Schiner 1863.

fascipennis Zetterst. Dipt. Scand. X, 3777 *(Erioptera;* description reproduced in Monogr. etc. IV, App. I, 328). — Greenland (according to Staeger's *Groenl. Antliater* in Krojer's Tidskrift, etc. 1845, 355, 16) ([29]).

* holotrichus O. S-cken, Proc. Acad. Nat Sc. Phil. 1859, 227 *(Erioptera);* Monogr. etc. IV, 141. — Distr. Columbia; New York.
* innocens O. Sacken, Mon. etc. IV, 142. — Distr. Columbia; N. Jersey.
* meigenii O. Sacken, Proc. Acad. Nat. Sc. Phil. 1859, 226 *(Erioptera);* Monogr. etc. IV, 144. — United States and Canada.
* monticola O. Sacken, Monogr. etc. IV, 145. — White Mts., N. H.
* nigripilus O. Sacken, Monogr. etc. IV, 142. — Distr. Columbia.
* nubilus O. Sacken. Proc. Acad. Nat. Sc. Phil. 1859, 227 *(Erioptera);* Monogr. etc. IV, 141, Tab. I, f. 14, wing. — Distr. Columbia; New York.
* rubellus O. Sacken, Monogr. etc. IV, 144, Tab. I, f. 15, wing. — New York; Delaware.

Erioptera.

Meigen, Illig. Magaz. II; 1803.
O. Sacken, Monogr. IV, 146.([30]).

Subgenus *Erioptera* (O. Sacken, Monogr. IV, 151); *Trichosticha* Schiner 1863 (ex parte).

* chrysocoma O. Sacken, Proc. Acad. Nat. Sc. Phil. 1859, 226: Monogr. etc. IV, 156. — Atlantic States and Canada (Quebec) etc.
* chlorophylla O. Sacken, Proc. Acad. Nat. Sc. Phil. 1859, 226; Monogr. etc. IV, 157, Tab. I, f. 16, wing. — Atlantic States and Canada (Quebec).
* septemtrionis O. Sacken, Proc. Acad. Nat. Sc. Phil. 1859, 226; Monogr. etc. IV, 155. — Northern States; also Distr. Columbia.
* straminea O. Sacken, Monogr. etc. IV, 157. — Middle States.
* villosa O. Sacken, Proc. Acad. Nat. Sc. Phil. 1859, 229; Monogr. etc. IV, 155. — Middle States.
* vespertina O. Sacken, Proc. Acad. Nat. Sc. Phil. 1859, 226, Monogr. etc. IV, 157, Tab. IV, f. 20, forceps. — Distr. Columbia; Wisconsin; Florida; Canada (Quebec).

Subgenus Acyphona (O. Sacken, Monogr. etc. IV, 151).

* armillaris O. Sacken, Monogr. etc. IV, 158. — Distr. Columbia; New York; Canada (Quebec).
* graphica O. Sacken, Proc. Acad. Nat. Sc. Phil. 1859, 227; Monogr. etc. IV, 159. — Distr. Columbia.

*venusta O. Sacken, Proc. Acad. Nat. Sc. Phil. 1859, 227; Monogr. etc. IV, 158; Tab. I, f. 17, wing; Tab. IV, f. 16, forceps. — Atlantic States.

Subgenus Hoplolabis (O. Sacken, Monogr. etc. IV, 152).

*armata O. Sacken, Proc. Acad. Nat. Sc. Phil. 1859, 227; Monogr. etc. IV, 160; Tab. I, f. 18, wing; Tab. IV, f. 14, forceps. — Atlantic States and Canada (Quebec).
*bipartita O. Sacken, Western Diptera 199. — Environs of San Francisco, Cal.

Subgenus Mesocyphona (O. Sacken, Monogr. etc. IV, 152).

*caloptera Say, Journ. Acad. Nat. Sc. Phil. III, 17, 1; Compl. Wr. II, 44 (*E. caliptera*); Wiedemann, Auss. Zw. I, 23, 1; O. Sacken, Proc. Acad. Nat. Sc. Phil. 1859, 226; Monogr. etc. IV, 161; Tab. IV, f. 15, forceps. — Atlantic States, as far West as Colorado, north to Quebec, Canada; also in Cuba.
*parva O. Sacken, Proc. Acad. Nat. Sc. Phil. 1859, 227; Monogr. etc. IV, 162. — Distr. Columbia; New Jersey.
*dulcis O. Sacken, Western Diptera, I, 198. — Lake Tahoe, Sierra Nevada, Cal.

Subgenus Molophilus (Curtis, Brit. Entomol. 1833; O. Sacken, Monogr. etc. IV, 153; *Erioptera* Schiner 1863).

*forcipula O. Sacken, Monogr. etc. IV, 163. — New Jersey (a similar species in California see Western Diptera, 200).
*hirtipennis O. Sacken, Proc. Acad Nat. Sc. Phil. 1859, 228; Monogr. etc. IV, 163. — Distr. Columbia; New Jersey.
*pubipennis O. Sacken, Proc. Acad. Nat. Sc. Phil. 1859, 228; Monogr. etc. IV, 162. — Distr. Columbia.
*ursina O. Sacken, Proc. Acad. Nat. Sc. Phil. 1859, 228; Monogr. etc. IV, 164. — Distr. Columbia; Maryland; (a similar species in California, see Western Dipt. 200.)

Trimicra.

O. Sacken, Proc. Acad. Nat. Sc. Phil. 1861; Monogr. etc IV, 165.

*anomala O. Sacken, Proc. Acad. Nat. Sc. Phil. 1861, 290; Monogr. etc. IV, 167; Tab. II, 1, wing — Distr. Columbia; New York; Rhode Island; also in California, see Western Diptera, 200; Oaxaca, Mexico (Coll. Bellardi).

Chionea.

Dalman, K. Vetensk, Acad. Handl. 1816; O. Sacken, Monogr. IV, 168.

scita Walker, List etc. I, 82. — North America.
*valga Harris, Ins. Inj. to Veget. etc. 3d ed. 601 fig. 260. — Massachusetts.
Chionea aspera Walker, List, etc. I, 82. — Huds. B. Terr.

Symplecta.

Meigen, Syst. Beschr. VI, 1830; O. Sacken, Monogr. IV, 170.

*punctipennis Meigen, Eur. Zw. Ins. I, 147. Tab. V, f. 7. *(Limnobia)*; id. l. c. VI, 283 *(Symplecta)*; — O Sacken, Proc. Acad. Nat. Sc. Phil. 1859, 228; Monogr. etc. IV, 171; Tab. I, f. 20, wing; Tab. IV, f. 21, forceps. — Atlantic States, incluing Colorado; Canada (Quebec); also in California and Chili; see Western Diptera 200. ([31] .

Symplecta cana Walker, List etc. I, 48.

Gnophomyia.

O. Sacken, Proc. Acad. Nat. Sc. Phil. 1859; Monogr. etc. IV, 172.

*luctuosa O. Sacken, Proc. Acad. Nat. Sc. Phil. 1859, 224; Monogr. etc. IV, 174. — Florida.

Limnobia nigricola Walker, Trans. Entom. Soc. Lond. V, N. S., Pt. VII, 66.

*tristissima O. Sacken, Proc. Acad. Nat. Sc. Phil. 1859, 224; Monogr. etc. IV, 175; Tab, II, f. 5, wing; Tab. IV, f. 19, forceps and ovipositor. — Atlantic States and Canada.

Goniomyia.

Gonomyia Megerle, in Meigen's Syst. Beschr. I, 1818; O. Sacken, Proc. Acad. Nat. Sc. Phil. 1859; Monographs, etc. IV, 176, name amended in *Goniomyia;* compare also Additions, at the end of Monographs, Vol. III. ([32]).

*blanda O. Sacken, Proc. Acad. Nat. Sc. Phil. 1859, 231; Mon. etc. IV, 182; Tab. IV, f. 17, forceps. — Distr. Columbia; New York; South Carolina.

*cognatella O. Sacken, Proc. Acad. Nat. Sc. Phil. 1859, 230; Mon. etc. IV, 181; Tab. IV, f. 18, forceps. — Distr. Columbia.

*manca O. Sacken, Mon. etc. IV, 178. — N. Jersey.

*subcinerea O. Sacken, Proc. Acad. Nat. Sc. Phil. 1859, 231; Mon. etc. IV, 181; Tab. II, f. 4, wing. — Distr. Columbia; New York; Canada (Quebec.

*sulphurella O. Sacken, Proc. Ac. N. Sc. Phil 1859, 230: Mon. etc. IV, 180; Tab. II, f. 2, wing. — Distr. Columbia; New York; Canada (Quebec .

[About the occurrence of this genus in California, see my Western Diptera.]

Empeda.

O. Sacken, Mon..etc. IV, 183, 1868.

*stigmatica O. Sacken, Mon. etc. IV, 184. — New York.

Cryptolabis.

O. Sacken, Proc. Acad. N. Sc. Phil.; 1859; Mon.. etc. IV, 185.

*paradoxa O. Sacken. Proc. Acad Nat. Sc. Phil. 1859, 225; Mon. etc. IV, 186, Tab. II, f. 11, wing; Tab. III, f. 3, forceps and ovipositor. — Virginia.

Cladura.

O. Sacken, Proc. Acad. Nat. Sc. Phil. 1859; Mon. etc. IV, 187.

*flavoferruginea O .Sacken, Proc. Acad. Nat. Sc. Phil. 1859, 229; Mon.
etc. IV, 188; Tab. IV, f. 22, forceps. — Distr. Columbia.
*indivisa O. Sacken, Proc. Ac. N. Sc. Phil. 1861, 291; Mon. etc.
IV, 189 (Wing figured on p. 34). — New York; Massachusetts;
Canada (Quebec).

Sigmatomera.

O. Sacken, Mon. etc. IV, 137; 1868.

flavipennis O. Sacken. Monogr. etc. Vol. III (in the Additions and
Corrections). — Mexico.

SECTION IV. LIMNOPHILINA.

Epiphragma.

O. Sacken, Proc. Acad. Nat. Sc. Phil. 1859; Mon. etc, IV, 193.

*fascipennis Say, Journ. Acad. Nat. Sc. Phil. III, 19, 1; Compl. Wr. II,
45 (*Limnobia*); Wiedemann, Auss. Zw. I, 31, 14 (*id*) — O. Sacken,
Mon. etc. IV, 194. — Atlantic States; Canada (Quebec).
Epiphragma paronina O. Sacken, Proc. Acad. Nat. Sc. Phil.
1859, 239.
*solatrix O. Sacken, Proc. Acad. Nat. Sc. Phil. 1859, 238; Mon. etc. IV,
195; Tab. II, f. 8, wing. — Distr. Columbía.

Limnophila.

Macquart, Hist. Nat. Dipt. I; 1834.
O. Sacken, Monogr. IV, 196.

*adusta O. Sacken, Proc. Acad. Nat. Sc. Phil. 1859, 235; Mon. etc. IV,
215. — Atlantic States and Canada (Quebec).
*aprilina O. Sacken, Proc. Acad. Nat. Sc. Phil. 1859, 235; Mon. etc. IV,
223; Tab. IV, f. 23, forceps. — Distr. Columbia; White Mts., N. H.
*areolata O. Sacken, Proc. Acad. Nat Sc. Phil. 1859, 237; Mon. etc.
IV, 214. — New York; Maryland; Distr. Columbia.
*brevifurca O. Sacken, Proc. Acad Nat. Sc. Phil. 1859, 237; Mon. etc.
IV, 221. — Distr. Columbia; Quebec (Canada).
*contempta O. Sacken, Mon. etc IV, 218. — Middle States.
carbonaria Macquart, Dipt. Exot. I, 1, 66 (Description reproduced
in Mon. IV. Appendix.) — Carolina.
*cubitalis O. Sacken, Mon. etc. IV, 229 — Virginia; Ohio.
*fasciolata O. Sacken, Mon. etc. IV, 206. — Massachusetts.
Limnophila fasciata O. Sacken (nec Schummel), Proc. Acad. Nat.
Sc. Phil. 1859, 234.
*fratria O. Sacken, Mon. etc. IV, 220. — Northern States.
*fuscovaria O. Sacken, Proc. Acad. Nat Sc. Phil 1859, 240; Mon. etc.
IV, 225. — Atlantic States and Canada (Quebec).

gracilis Wiedemann, Auss. Zw. I, 28, 8 *(Limnobia;* description reproduced in Monogr. etc. IV, Appendix). — Pennsylvania.

*imbecilla O. Sacken, Proc. Acad. Nat. Sc. Phil. 1859, 237; Mon. etc. IV, 213. — Maryland; New York.

***inornata** O. Sacken, Mon. etc. IV, 20; see also Additions at the end of Mon. Vol. III. — Massachusetts; Tarrytown, N. York.

*lenta O. Sacken, Proc. Acad. Nat. Sc. Phil. 1859, 241; Mon. etc. IV, 231. — Middle States.

*luteipennis O. Sacken, Proc. Acad. Nat Sc. Phil. 1859, 236; Mon. etc. 217; Tab. II, f. 10, wing; Tab. IV, f. 25, forceps. — United States and Canada (Quebec); California.

(?) *Limnobia biterminata* Walker, Dipt. Saund. 437.

*macrocera Say, Journ. Acad. Nat. Sc. Phil. III, 20, 2. *(Limnobia);* Compl. Wr. II, 46; Wiedemann, Auss. Zw. 1, 34, 19. *(id);* — Macquart, Hist. Nat. Dipt. I, 108, 2 *(Cylindrotoma);* — O. Sacken, Proc. Acad. Nat. Sc. Phil. 1859, 234 *(Lasiomastix);* Mon. etc. IV, 204. — United States and Canada (Quebec).

*montana O. Sacken, Proc. Acad. Nat. Sc Phil. 1859, 240 *(Dactylolabis);* Mon. etc. IV, 227; Tab. II, f. 7, wing; Tab. IV, fig 26, forceps. — United States and Canada (Quebec); California.

*munda O. Sacken, Monogr. etc. IV, 226. — White Mts. N. H.; Canada (Quebec).

*niveitarsis O. Sacken, Mon. etc. IV, 209. — Delaware; Maryland.

*poetica O. Sacken, Mon. etc. IV, 207. — Massachusetts.

*quadrata O. Sacken, Proc. Acad. Nat. Sc. Phil. 1859, 241; Mon. etc. 230, Tab II, f. 9, wing. — United States and Canada (Quebec.)

*recondita O. Sacken, Mon. etc. IV, 212. — New York; Pennsylvania; Georgia.

*rufibasis O. Sacken, Proc. Ac. N. Sc. Phil. 1859, 239, *(Prionolabis);* Mon. etc. IV, 225; Tab. II, f. 3, wing; Tab. IV, f. 27, forceps. — Distr. Columbia; New York; Mass.

*tenuicornis O. Sacken, Mon. etc. IV, 208. — White Mts., N. H.

*tenuipes Say, Journ. Acad. Nat. Sc. Phil. III, 21, 3; Compl. Wr. II, 46 *(Limnobia);* O. Sacken, Proc. Acad. Nat. Sc. Phil. 1859, 235; Mon. etc. IV, 210. — U. States; Canada (Quebec).

Limnobia humeralis Wiedemann (non Say), Auss. Zw. I, 34. ([33]).

*toxoneura O. Sacken, Proc. Acad. Nat. Sc. Phil. 1859, 236; Mon. etc. IV, 213. — N. York.

*ultima O. Sacken, Proc. Acad. Nat. Sc. Phil. 1859, 238, Mon. etc. IV, 222; Tab. IV, f. 24, forceps. — Distr. Columbia; Maine; Canada (Quebec) and farther North, as far as Alaska.

*unica O. Sacken, Mon. IV. 205. — White Mts., N. H.

*damula O. Sacken, Western Diptera, 201. — San Bernardino, Cal.

nebulosa Bellardi, Saggio etc. I, 6; Tab. I, f. 4, wing. *(Tipula).* — Mexico.

undulata Bellardi, Saggio etc. Append., 3, Tab. I, f. 2. — Mexico.

Observation. *L. tenuipes, luteipennis, aprilina, montana, munda (?), adusta (?),* or species exceedingly like them, also occur in California, see my Western Dipt. 201.

Phyllolabis.
O. Sacken, Western Dipt. 202; 1877.

*claviger O. Sacken, Western Dipt. 203. — California.
*encausta O. Sacken, Western Dipt. 204. — California.

Ulomorpha.
O. Sacken, Mon. etc. IV, 232; 1868.

*pilosella O. Sacken, Proc. Acad. Nat. Sc. Phil. 1859, 342; Mon. etc. IV, 233 *(Limnophila)*. — Trenton Falls, N. York.

Trichocera.
Meigen, in Illiger's Magaz., 1803; O. Sacken, Monogr. IV, 233.

*bimacula Walker, List, etc. I, 84. — Nova Scotia.
brumalis Fitch, Winter Insects, etc. (1848). — New York.
gracilis Walker, List, etc. I, 84. — New York Factory.
hiemalis (De Geer) Zetterstedt, Dipt. Scand. X, 4041; Holmgren, Ins. Nordgroenl. — Northern Greenland.
maculipennis Meigen, etc., Staeger, Groenl. Antliater. — Europe, Greenland.
*regelationis Lin., O. Fabricius, Fauna Groenl. 202, 157 *(Tipula)*. — Europe and North America.
scutellata Say, Long's Exp. App. 360; Compl. Wr. I, 244; Wiedemann, Auss. Zw. I, 60, 1. — Falls of Kakabikha, beyond Lake Superior (Say).
*trichoptera O. Sacken, Western Dipt. 204. — Marin Co., Calif.

Observation. *Gynoplistia annulata* Westwood, Lond. and Edinb. Philos. Mag. 1835, from Newfoundland, has never been found in North America since. Compare about it, Mon. IV, 42. Its description is reproduced in the Appendix to the same volume.
Limnobia ignobilis and *turpis* Walker, Dipt. Saund. are not recognizable in the descriptions; I did not see them in the Brit. Mus. Compare about them my remarks in Monogr. etc. IV, 40, 41.
Limnobia stupens Walker, Trans. Ent. Soc. N. Ser. V, 333 (from Mexico), seems to belong either to the *Limnophilina* or the *Amalopina.*

SECTION V. ANISOMERINA.
Anisomera. ([34]).
Meigen, Syst. Beschr. I; 1818; O. Sacken, Mon. etc. IV, 242.
Hexatoma Latreille; 1809. *Nematocera,* Meigen; 1818.

*megacera O. Sacken, Proc. Acad. Nat Sc. Phil. 1859, 242; Mon. etc. IV, 243; Tab. II, f. 12, wing. — Distr. Columbia; Maryland.

6

Eriocera.

Macquart, Dipt Exot. I, 1, 74; 1838; O Sacken, Monogr. etc. IV, 244.

*brachycera O. Sacken, Western Dipt 204. — White Mts. N. H.

*fuliginosa O. Sacken, Proc. Acad. Nat. Sc. Phil. 1859, 243; Monogr. etc. IV, 255; Tab. IV, f. 28, forceps. — Virginia; Distr. Columbia.

*longicornis Walker, List, etc I, 82 (Limnobia); O. Sacken, Proc. Acad. Nat. Sc. Phil. 1859, 245 (Arrhenica); Monogr. etc. IV, 253. — New York; Maine; Massachusetts; Illinois; Canada.

*spinosa O. Sacken, Proc. Acad. Nat. Sc. Phil. 1859, 244 (Arrhenica); Monogr. etc. IV, 252; Tab. IV, f. 29, forceps. — New York; Massachusetts. NB. The description of the female, given l. c. belonges to E. brachycera; see O. Sacken, Western Dipt. 205.

*Wilsonii O. Sacken, Monogr. etc. IV, 255. — Delaware.

*californica O. Sacken, Western Diptera, 204. — California.([35]).

Observation. In Mr. Bellardi's mexican collection, I saw four species of Eriocera, all with four posterior cells and short antennae in both sexes.

Penthoptera.

Schiner, Wiener Ent. Mon. VI; 1863. O. Sacken, Monogr. IV, 256.

*albitarsis O. Sacken, Monogr. etc. IV, 257. — Pennsylvania; Connecticut.

SECTION VI. AMALOPINA.

Amalopis.

Haliday, in Walker's Ins. Brit Dipt. III, XV; 1856; O Sacken, Monogr. etc. IV, 260; 1868; Crunobia Kolenati 1860.

*auripennis O. Sacken, Proc. Acad. Nat. Sc. Phil. 1859, 247; Monogr. etc. IV, 268. — Massachusetts.

*calcar O. Sacken, Proc. Acad. Nat. Sc Phil. 1859, 247; Monogr. etc. IV, 268; Tab. II, f. 14, wing. — Wisconsin; White Mts., N. H.; Canada (Quebec).

*hyperborea O. Sacken, Proc. Acad. Nat. Sc. Phil. 1861, 292; Monogr. etc. IV, 269. — Labrador.

*inconstans O. Sacken, Proc. Acad. Nat. Sc Phil. 1859, 247; Monogr. etc. IV, 266; Tab. II, f. 15, wing; Tab. IV, f. 30, forceps. — Atlantic States and Canada (Quebec).

*vernalis O. Sacken, Proc. Acad. Nat. Sc. Phil. 1861, 291; Monogr. etc. IV, 270. — White Mts., N. H.; Distr. Columbia.

[Amalopis calcar, or a closely resembling species, and Amalopis nov. sp. occur in California; see O. Sacken, Western Dipt. 205.]

Pedicia.

Latreille, Genera etc. Vol. IV; 1809; O. Sacken, Monogr. IV, 273.

*albivitta Walker, List., etc. I, 37; O. Sacken, Proc. Acad. Nat. Sc. Phil. 1859, 248; Monogr etc. IV, 273 — New York; Connecticut; Massachusetts (a chiefly northern species).

Pedicia contermina Walker, List, etc. I, 38. — Nova Scotia. (I believe
this to be a mere variety)
(The *Tipula rivosa* of O. Fabricius, Fauna Groenl 200, 156 is not
Pedicia rivosa Lin., but, according to Schiödte, in Berl. Ent.
Zeitschr. 1859, 152, *Tipula nodulicornis* Zetterstedt.)
*obtu₃a O. Sacken, Western Dipt. 205. — Marin Co., Cal.

Ula.

Haliday, Entom. Magaz. I; 1833; O. Sacken, Monogr. etc. IV, 274.
*elegans O. Sacken, Monogr. etc. IV, 276. — White Mts., N. H.
*pauper O. Sacken, Monogr. etc. IV, 277. — Distr. Columbia.
   Ula pilosa O Sacken (non *Schummel*) Proc. Acad. Nat. Sc. Phil.
   1859, 251.

Dicranota.

Zetterstedt, Ins. Lappon. 1840; O. Sacken, Monogr. etc. IV, 278.
*rivularis O. Sacken, Proc. Acad. Nat. Sc Phil. 1859, 249; Monogr. etc.
   IV, 281; Tab. II, f. 16, wing — Distr. Columbia.
*eucera O. Sacken, Monogr. etc. IV, 281. — Distr. Columbia.

Plectromyia. *

O. Sacken, Monogr. etc. IV, 282; 1868.
*modesta O. Sacken, Monogr. etc. 284; Tab. II, f. 18, wing; — White
Mts., N. H.

Rhaphidolabis.

O. Sacken, Monogr. etc. IV, 284; 1868.
*tenuipes O. Sacken, Monogr. etc. IV, 287; Tab. II, f. 17, wing. —
Maryland; New York.
*flaveola O. Sacken, Monogr. etc. IV, 288. — Maryland; Massachusetts.
   [A Rhaphidolabis, resembling R. tenuipes occurs in California;
   see my Western Dipt]

SECTION VII. CYLINDROTOMINA.
Cylindrotoma.

Macquart, H. N. Dipt. I; 1834.
O. Sacken, Monogr. etc. IV, 296.
*americana O. Sacken, Proc. Ent. Soc. Phil. 1865, 236; Monogr. IV,
   299. — White Mts, N. H.
*nodicornis O. Sacken, Proc. Ent. Soc. Phil. 1865, 239 *(Triogma);*
   Mon. etc. IV, 301; Tab. II, f. 7, wing. (*Liogma*, nov. gen. is
   proposed for it in Monogr. IV, 298.) — Northern States, not
   rare, Canada (Quebec).

Triogma.

Schiner, Wien. Ent. Mon. VII; 1863; O Sacken, Monogr. etc, IV, 303.
*exsculpta O. Sacken, Proc. Ent. Soc Phil. 1865, 239; Monogr. etc.
   IV, 304. — Pennsylvania.

Phalacrocera.

Schiner, Wiener Ent. Mon VII; 1863; O. Sacken, Monogr. etc. IV, 305.

*tipulina O. Sacken, Proc. Ent. Soc. Phil. 1865, 241; Monogr. etc. IV, 308. — White Mts., N. H.

SECTION VIII. PTYCHOPTERINA.

Ptychoptera.

Meigen, Illiger's Magaz., 1803; O. Sacken, Monogr. IV, 309. ([36]).

quadrifasciata Say, Long's Exp. App. 359; Compl. Wr. I, 244; Wiedemann, Auss. Zw. I, 60, 2. (Description reproduced in Monogr. Vol. IV. Appendix.) — Pennsylvania

*rufocincta O. Sacken, Proc. Acad. Nat. Sc. Phil. 1859, 252; Monogr. etc. IV, 313; Tab. II, f. 19, wing. — Atlantic States and Canada (Quebec).

*lenis O. Sacken, Western Dipt. 206. — Yosemite, Cal.; Georgetown, Colorado.

(?) Ptychoptera metallica Walker, List, etc. I, 80; description reproduced in Monogr. IV, Appendix. — Hudson Bay's Territory.([37]).

Bittacomorpha.

Westwood, Lond. and Edinb. Philos. Magaz. VI, 281; 1835.
O. Sacken, Monogr. etc. IV; 313.

*clavipes Fabricius, Spec. Ins. II, 404, 19; Mant. Ins. II, 323, 21; Ent. Syst. IV, 239, 25 *(Tipula)*; Syst. Antliat. 22, 4 *(Ptychoptera)*; Wiedemann, Auss. Zw. I, 59 *(id.)*; Westwood, Lond. and Edinb. Phil. Magaz 1835, 281; O. Sacken, Monogr etc. IV, 315; Tab. II, f. 20, wing; Tab. IV, f. 31, forceps. — From Newfoundland to Florida and Texas. — I have also seen specimens from Oregon (Collection of Mr. Henry Edwards in San Francisco); from Clear Creek Cañon, Colorado (Coll. of. J. D. Putnam, Davenport, Jowa), but I have not compared them with specimens from the Atlantic States. Specimens from California in Mr. Verrall's collection in London have a shining thorax and a shorter submarginal cell; they may belong to a different species. Two specimens from Brazil, in the Vienna Museum, do not differ from the typical ones. Still, the occurrence in Brazil of this insect, as well as of *Pyrgota undata,* requires confirmation.

Idioplasta.

Protoplasa O. Sacken, Proc. Acad Nat. Sc. Phil. 1859; Monogr. etc. IV, 316. ([38]).

*Fitchii O. Sacken, Proc. Acad. Nat. Sc. Phil. 1859, 252 *(Protoplasa)*; Monogr. etc. IV, 319 *(id.)*; figure of wing, on p. 317. — New York; Georgia.

*vipio O. Sacken, Western Diptera, 208 *(Protoplasta)*. — San Mateo, Cal.

SECTION IX. TIPULINA.

Longurio.

Loew, Centur. VIII. 2; 1869.

*testaceus Loew, Centur. VIII, 2. — Massachusetts.
(?)longipennis Macquart, Dipt. Exot. I, 57, 9 (Tipula); Bigot, R. de
la Sagra, etc. 786 (id.). — Cuba.

Holorusia.

Loew, Centur. IV, 2; 1863.

*rubiginosa Loew, Centur. IV, 1. — California (not rare about
S. Francisco).

Tipula.

Linné, Anim. per Sueciam observata; 1736. (³⁹).

*abdominalis Say, J. Acad. Phil. III, 18; Compl. Wr. II, 45 (Cteno-
phora); Wiedemann, Auss. Zw. I, 37 (id.). — Northern Atlantic
States and Canada (seems common about Quebec; also received
from Kansas, Wisconsin and Kentucky).
Tipula albilatus Walker, List, I, 65 (!).
*angulata Loew, Centur. V, 22. — Massachusetts.
*angustipennis Loew, Centur. IV, 19. — Massachusetts, Huds. B. Terr.
(?) Tipula glomerata Walker, List, etc. I, 70. — North America. (⁴²).
annulicornis Say, J. Acad. Phil. VI, 151; Compl. Wr. II, 850. —
Indiana.
*apicalis Loew, Centur. IV, 2. — Maine; Dobb's Ferry, N. Y.
*appendiculata Loew, Centur. IV, 20. — Saskatchewan.
*arctica Curtis, Ross's Exp. LXXVII, Tab. A, f. 15; Holmgren, Ins.
Nordgroenl. 105. — Greenland, Arctic America.
Tipula rivosa, O. Fabr. (non Linné), Fauna Groenl., 156 (Syno-
nymy by Schiödte, Berl. Ent. Z. 1859, 152).
Tipula nodulicornis Zetterstedt, Ins. Lapp. 841, 8; Staeger, Groenl.
Antliater 355; Zetterstedt, Dipt. Scand. X, 3934 [Schiodte]. (⁴⁰). —
*balioptera Loew, Centur. IV, 15. — English River, H. B. T.
*bella Loew, Centur. IV, 29. — Connecticut, Massachusetts, New York,
Canada.
(?) Tipula furca Walker, List, etc. I, 70. — North America. (⁴²).
*Besselsi O. Sacken, Proc. Boston Soc Nat. Hist. Decemb. 6, 1876. —
Polaris Bay, Lat. 82.
borealis Walker, List, etc. I, 66. — Nova Scotia.
*caloptera Loew, Centur. IV, 30. — Red River of the North; Massa-
chusetts.
*canadensis Loew, Centur. V, 19. — Huds. B T.
*centralis Loew, Centur. V, 21, — Huds. B. T.
*cincta Loew, Centur. IV, 24. — Distr. Columbia; White Mts., N. H.
*costalis Say, J. Acad. Phil., III. 23, 2; Compl. Wr. II, 48; Wiede-
mann, Auss. Zw. I, 51, 17. — Middle and Northern States; Canada.

*cunctans Say, J. Acad. Phil. III, 23, 1; Compl. Wr. II, 48; Wiede-
   mann, Auss. Zw. I. 45, 8. — Pennsylvania.
   Tipula casta Loew, Centur. IV, 25. ([41]).
*discolor Loew, Centur. IV, 12. — Massachusetts.
 disjuncta Walker, Dipt. Saunders. 442. — United States.
*dejecta Walker, Dipt Saunders. 442. — Atlantic States.
 dorsimacula Walker, List, etc. I, 69. — Nova Scotia.
 duplex Walker, List, etc I, 66. — Nova Scotia.
* eluta Loew, Centur. IV, 27. — Distr. Columbia.
*fasciata Loew, Centur. IV, 6. — Sharon Springs, N. Y; Pallissa-
   des, N. J.
 filipes Walker, List, etc. I, 65. — Florida.
*flavicans Fabricius, Syst Antl. 24, 5 (*flavescens*, in erratis *flavicans*);
   Wiedemann, Dipt. Exot. I, 25, 5; Auss. Zw. I, 48, 13. — United
   States and Canada.
*fragilis Loew, Centur. IV, 7. — Maine.
*fraterna Loew, Centur. V, 14. — Distr. Columbia.
 frigida Walker, List, etc. I, 68. — Nova Scotia.
*fuliginosa Say, J. Acad. Phil. III, 18, 1: Compl. Wr. II, 44 (*Ctenophora*); Wiedemann, Auss. Zw. I, 40, 5. (*id.*). — Middle and
   Northern States. ([43]).
*grata Loew, Centur. IV, 11, — Distr. Columbia, New York.
*hebes Loew, IV, 18. — Connecticut, Illinois, Maine.
*infuscata Loew, Centur. IV, 26. — New York; Distr. Columbia.
*ignobilis Loew, Centur. IV, 9. — Distr. Columbia; White Mts., N. H
*latipennis Loew, Centur. V, 20. — White Mts., N. H.; Canada.
*longiventris Loew, Centur. IV, 5. — Illinois; Maine; Lake Winnipeg.
*macrolabis Loew, Centur. V, 17. — Huds. B. Terr.
 maculipennis Wiedemann, Auss. Zw. I, 46, 9; — Northern States;
   Nova Scotia (Walker, List, etc. I, 67).
   Tipula maculatipennis, Say, Long's Exp. App., 359; Compl. Wr. I,
   243 (name modified by Wiedemann).
*pallida Loew, Centur. IV, 16. — Massachusetts.
 platymera Walker, Dipt. Saund. 441. — Canada.
 pratorum Kirby, Fauna Bor. Amer. Ins 310. — Arctic America.
 puncticornis Macquart, Dipt. Exot. 4e Suppl. 15, 22; Tab. I, f. 6. —
   North-America.
 resurgens Walker, List, etc. I, 67. — Newfoundland.
 simulata Walker, Dipt. Saund 441. — Canada.
*septentrionalis Loew, Centur. IV, 4. — Labrador.
*serrulata Loew, Centur. V, 18. — Fort Resolution, Huds. B. Terr.
*serta Loew, Centur IV, 14. — Lake Winnipeg, Huds. B. Terr.; Massa-
   chusetts; Canada.
*speciosa Loew, Centur. IV, 22. — Illinois, Distr. Columbia.
*strepens Loew, Centur. IV, 28. — New York; White Mts., N. H.
*subfasciata Loew, Centur. IV, 13. — English River, Huds. B. Terr.
*submaculata Loew, Centur IV, 23. — Massachusetts; Western N. York.
*suspecta Loew, Centur. IV, 8. — Distr. Columbia.

*tephrocephala Loew, Centur. V, 23. — White Mts., N. Hampshire; New Jersey.
*ternaria Loew, Centur. V, 15. — Huds. B. Terr.
*tessellata Loew, Centur IV, 3. — Labrador.
*tricolor Fabricius, Ent. Syst. IV, 235, 9; Syst Antl. 26, 13; Wiedemann, Dipt. Exot. I, 22, 1; Auss. Zw. I, 44, 6. — Atlantic States.
triplex Walker, List, etc. I, 66. — Nova Scotia.
*trivittata Say, J. Acad. Phil. III, 26, 6; Compl. Wr. II, 50; Wiedemann, Auss. Zw. I, 42, 4. — Atlantic States
truncorum Meigen etc.; Gerstaecker, Die 2te deutsche Nordpolfahrt etc. — Europe and East Greenland.
*umbrosa Loew, Centur. IV, 31. — Louisiana.
*valida Loew, Centur. IV, 21. — Massachusetts, Illinois, New York.
*versicolor Loew, Centur. IV, 17. — Illinois.

*beatula O. Sacken, Western Diptera, 209. — California (Marin Co., not rare).
*fallax Loew, Centur. IV, 10. — California.
*pubera Loew, Centur. V, 16. — California (Marin and Sonoma Co.).
*praecisa Loew, Centur. X, 2; O. Sacken, Western Diptera, 209. — California, common.
*spernax O. Sacken, Western Diptera, 210. — Sierra Nevada, Calif.

associans Walker, Trans. Ent. Soc. Nat. Sc. V, 333. — Mexico.
Craverii Bellardi, Saggio, etc. I, 7; Tab. I, f. 1 (wing). — Mexico. (Schiner, Novara etc. 35, considers this species a synonym of *Tip. obliquefasciata* Macquart, Dipt. Exot. Suppl I, 15, 15, Tab. I, 'f. 10.); it is also very like *T. pubera* Loew, from California.
dispellens Walker, Trans. Ent. Soc. N. S. V, 333. — Mexico.
Edwardsii Bellardi, Saggio, etc. I, 8; Tab. I, f. 2 (wing). — Mexico.
quadrimaculata Bellardi, Saggio, etc. I, 9; Tab. I, f. 3 (wing). — Mexico.

Observation. *Tipula atra* Linné, in O. Fabricius, Fauna Groenl. is an *Empis*.
Tip. pennicornis Linné, ibid. perhaps *Ctenophora?*
Tipula monoptera Linné, ibid. perhaps *Sciara?*

Pachyrrhina.

Pachyrhina, Macquart, Hist. Nat. Dipt. I, 88, 1834.
*abbreviata Loew, Centur. IV, 36. — Mississippi.
*altissima O. Sacken, Western Diptera, 210. — Pike's Peak, Col.; Taos Peak, N. M, above tree-line.
*collaris Say, J. Acad. Phil. III, 23, 2; Wiedemann, Auss. Zw. I, 51, 17. — Massachusetts; Pennsylvania; Distr. Columbia.
*eucera Loew, Centur. IV, 39. — Distr. Columbia.
*ferruginea Fabricius, Syst. Antl 28, 19 (*Tipula*); Wiedemann, Dipt. Exot. I, 28, 9; Auss. Zw. I, 53, 21 (id.); Macquart, Dipt Exot. 4e Suppl, 13; Tab. I, f. 3. — United States and British Possessions, common; California, see O. Sacken, Western Dipt., 211.

*gracilicornis Loew, Centur. V, 32. — Western New York.
*incurva Loew, Centur. IV, 32 — Atlantic States.
   (?) *Tipula alterna* Walker, List, etc. I, 72. — Nova Scotia. ([42]).
*lugens Loew, Centur. V, 26.   White Mts., N. H.; Canada.
*macrocera Say, J. Acad. Phil. III, 24, 3; Compl. Wr. II, 48; Wiede-
   mann, Auss. Zw. I, 52, 18; Macquart, Hist. Nat. Dipt. I, 108, 2. —
   Atlantic States.
*nobilis Loew, Centur. V, 24. — White Mts., N. H.
*occipitalis Loew, Centur. V, 30. — Huds. B. Terr. (Yukon River.)
*pedunculata Loew, Centur. IV, 33. — Saskatchewan; Illinois; Cats-
   kill, N. Y.
*polymera Loew, Centur. IV, 40. — Illinois; Ohio.
*punctum Loew, Centur. IV, 34. — Illinois; Maine.
*sodalis Loew, Centur. V, 29. — Connecticut
*suturalis Loew, Centur. IV, 37. — Georgia, Florida.
*tenuis Loew, Centur. IV, 41. — Sharon Springs, N. Y.; Virginia.
*unifasciata Loew, Centur. IV, 35. — Middle States
*unimaculata Loew, Centur. V, 23. — New York; Illinois.
*virescens Loew, Centur V, 25. — Distr. Columbia (Lw.); New Jersey.
*vittula Loew, Centur. V, 27. — Huds. B. Terr.
*xanthostigma Loew, Centur. V, 31. — Illinois.

affinis Bellardi, Saggio, etc. I, 10. *(Tipula).* — Mexico.
*circumscripta Loew, Centur. IV, 38. — Cuba.
mexicana Macquart, Dipt. Exot. Suppl. 1, 12, 8. — Mexico.
nigrolutea Bellardi, Saggio, etc. I, 11 *(Tipula);* Walker, Trans. Ent.
   Soc. Nat. Sc. V, 343. — Mexico.
proxima Bellardi, Saggio, etc. I, 9 *(Tipula).* — Mexico.
quadrilineata Macquart, Dipt. Exot. I, 1, 50. — Mexico.

Stygeropis.

Loew, Centur. IV, 42, 1863; *Prionocera* Loew, Stett. Ent. Z. 170; 1844.

*dimidiata Loew, Centur. VI, 2. — Huds. B. Terr.
*fuscipennis Loew, Centur. VI, 3. — Illinois.
*sordida Loew, Centur. IV, 42. — Lake Winnipeg.
Parrii Kirby, Suppl. to App. to Capt. Parry's first Voy. 1824
   (Ctenophora). — Arctic America.

Dolichopeza.

Curtis, British Entomology, 62, 1825. Meigen, System. Beschr. VI, 1830,
   p. 283, Tab. 65, f. 10, 11 (on the plate, it is called *Leptina).*

*annulata Say, Journ. Acad. Phil. VI, 151 *(Tipula);* Compl. Wr. II,
   350; Wiedemann, Auss. Zw. I, 54, 22 *(id.).* — Pennsylvania (Say);
   Middle States.

     Observation. I place *Tip. annulata* Say provisionally in this
genus, to which it is closely allied,-although, in some respects,
it is different. It has a discal cell; the forceps of the male has

a different structure etc. One or two other species, as yet un-described, occur in the United States, which are still more like the European *D. sylvicola*, although they also have a discal cell.

Ctenophora. ([44]).
Meigen, Illiger's Magaz. II, 263; 1803.

*apicata O. Sacken, Proc. Ent. Soc. Phil. 1864, 46. — New Hampshire.
dorsalis Walker, List, etc. I, 76. — Newfoundland.
*frontalis O. Sacken, Proc. Ent. Soc. Phil. 1864, 48. — Massachusetts.
  (?) *Ctenophora succedens* Walker, Dipt. Saund., 448. — Canada.
*fumipennis O. Sacken, Proc. Ent. Soc. Phil. 1864, 47. — Virginia.
*nubecula O. Sacken, Proc. Ent. Soc. Phil. 1864, 45. — Illinois.
*topazina O. Sacken, Proc. Ent. Soc. Phil. 1864, 47. — Virginia.
*angustipennis Loew, Centur. X, 3; O Sacken, Western Diptera, 211.—
  California (among the redwoods in the Coast-Range, not rare).

Observation. For *Ptilogyna fuliginosa* Macquart, see the note. ([45]).

FAMILY DIXIDAE.
Dixa.
Meigen, Syst. Beschr. I, 216; 1818.

*centralis Loew, Centur. III, 3. — New York.
  (?) *Dixa nova* Walker, List, etc. I, 85. — New York Factory.
*clavata Loew, Centur. VIII, 1. — Massachusetts.
*fusca Loew, Centur. III, 5. — New York.
*marginata Loew, Centur. III, 1. — Distr. Columbia.
*notata Loew, Centur. III. 4. — Maryland.
*terna Loew, Centur. III, 2. — New York.
  (?) *Dixa recens* Walker, List, etc. I, 85. — New York Factory.
*venosa Loew, Centur. X, 1. — Texas.

Observation. About an undescribed *Dixa* from California, compare O. Sacken, Western Diptera, 196.

FAMILY RHYPHIDAE.
Rhyphus.
Latreille, Hist. Nat. etc. XIV, 291; 1804.

*alternatus Say, J. Acad. Phil. III, 27, 2; Compl. Wr. II, 51; Wiede-mann, Auss. Zw. I, 82, 1. — Atlantic States.
*fenestralis Scopoli; Meigen, Syst. Beschr. I, 323. — Europe and North America (Loew, Sillim. Journ. l. c.).
*punctatus Meigen, etc. — Europe and North America (Loew, Sillim. Journ. l. c.).
  Rhyphus marginatus Say, J. Acad. Phil. III, 27, 1; Compl. Wr. II, 50; Wiedem. Auss. Zw. I, 82, 2 (Loew, Sillim. Journ. N. Ser. XXXVII, 317).

scalaris Wiedemann, Auss. Zw. II, 618, 8. — Georgia.

taeniatus Bellardi, Saggio, etc. App. 5, f. 15. — Mexico.

FAMILY XYLOPHAGIDAE.

Rhachicerus.

Rachicerus, Haliday, in Walker, List, etc V, 103; 1854.

*fulvicollis Haliday, Walker, List, etc. I, 124; V, 104. — Georgia.
*obscuripennis Loew, Centur. III, 6. — Illinois; Detroit, Mich.

*honestus O. Sacken, Western Diptera, 211. — California.
*nigripalpus Loew, Berl. Ent. Z. 1874, 379. — Mexico.
*varipes Loew, Centur. III, 7. — Cuba.

Xylophagus.

Meigen, in Illiger's Magaz. II, 266; 1803.

*abdominalis Loew, Centur. IX, 64. — Texas
*fasciatus Walker, List, etc. I, 128. — Huds. B. Terr.
*longicornis Loew, Centur. IX, 62. — Massachusetts.
*lugens Loew, Centur. III, 8. — Illinois; Pennsylvania; White Mts., N. H.
  persequus Walker, Dipt. Saund., 1. — North America.
  reflectens Walker, List, etc. I, 12. — New York.
*rufipes Loew, Centur. IX, 63. — Massachusetts; Canada.
  triangularis Say, Journ. Acad. Phil. III, 30; Compl. Wr. II, 52;
    Wiedemann, Auss. Zw. I, 85, 2. — Missouri. (Macquart, Dipt.
    Exot. I, 1, 171, suspects that this is a *Subula.*)

Subula.

Meigen, Syst. Beschr. II, 15; 1820; Macquart, H. N. Dipt.

*americana Wiedemann, Dipt. Exot. I, 51, 1; Auss. Zw. I, 84, 1
  (*Xylophagus*). — Distr. Columbia; Illinois.
  Subula tenthredinoides v. d. Wulp, Tijdschr. voor Entom. 2d Ser. II,
    132; Tab. III, f. 5—7. — Wisconsin [„Is but a dark variety of
    S. americana"; Loew, Zeitschr. f. Ges. Naturw. XXXVI, 114].
  fasciata Say, Journ. Acad. Phil. VI, 155; Compl. Wr. II, 353 (*Xylo-
    phagus*). — Indiana (may this not be the same as *Arthropeas
    americana* Loew?).
*pallipes Loew, Centur. III, 9. — Atlantic States.

Bolbomyia.

Loew, Bernstein u. Bernsteinfauna, 39, 1850.([46]).

*nana Loew, Centur. II, 5. — District Columbia (compare about this
  species Dr. Loew's article, On the Diptera of the amber fauna,
  translated in Sillim. Journ Vol. XXXVII, 313).

Dialysis.

dissimilis Walker, Dipt. Saund., 4; List, etc. I, 128 (*X. Americanus* Wied. ?). — Locality not given, but probably North America, from the comparison to X. americanus.

NB. According to Loew, Monogr. etc. I, 16 the bristle-like fourth antennal joint ascribed by Mr. Walker to this species renders it very doubtful, whether it is properly referred to the Xylophagidae. I do not remember having seen it in the Brit. Mus.

Macroceromys.

Bigot, Ann. Soc. Ent. de Fr. 1877, Bulletin LXXIII.

fulviventris Bigot ♀ (not described). — Mexico. (The genus is referred by the author to the Xylophagidae.)

FAMILY COENOMYIDAE. [47]

Coenomyia.

Latreille, Précis des Caract. génér. etc. 1797; *Sicus* Fabr. [47a].

*pallida Say, Long's Exped. Append. 369; Amer. Ent. II, plate XX; Compl. Wr. I, 42 and 251; Wiedemann, Auss. Zw. I, 86, 1; Harris, Ins. New Engl., 407; Macquart, Dipt. Exot. 5e Suppl. 38, 1. — Atlantic States.

' Observation. Mr. Loew (Sillim. Journ N. Ser. XXXVII, 317) states that this species is the same as the european *C. ferruginea*. About *Sicus crucis* Fabr. Ent Syst. IV, 264, 7, and Syst. Antl. 76, 5, from the West Indies, Wiedemann (Auss. Zw. I, 86) says, that it is in no way different „from *errans* and hence, the same as *Coen. ferruginea* Meig.“.

Arthropeas.

Loew, Stett. Ent. Z. 1850, 302—308.

*americana Loew, Centur. I, 16. — Northern Wisconsin; Massachusetts.
*leptis nov. spec. See the note [48]. — White Mts, N. H.

FAMILY STRATIOMYIDAE.

SECTION I. BERIDINA (Loew, Mon. I, 17).

Metoponia.

Macquart, Dipt. Exot. 2e Suppl. 28; 1847.

*fuscitarsis Say, J. Acad. Phil III, 29, and VI, 155; Compl. Wr II, 52, and 353 (*Beris*). — Atlantic States and Canada.
Sargus dorsalis Say, Long's Exped. App. 377; Compl. Wr. I, 257; Wiedemann, Auss. Zw. I, 540, 3 (*Beris*).
Sargus pallipes Wiedemann, Auss. Zw. II, 41.
Beris lata Walker, List, etc. I, 127.
Beris brevis Walker, List, etc. I, 127.

*obscuriventris Loew, Centur. IV, 45. — Distr. Columbia; Connecticut [Loew, Beschr. Europ. Dipt. III, 72, mentions a species from Siberia which he thinks may be identical with this].

*similis Loew, Centur. IV, 44. — New York.

Beris.

Latreille, Hist. Nat. des Crust. et des Ins. XIV, 340; 1804.([49]).

*viridis Say, Long's Exped. App. 368, I; Compl. Wr. I, 251; Wiedemann, Auss. Zw. I, 83, 2. — Atlantic States and Brit. Possessions. *Beris quadridentata* Walker, List, etc. I, 127.

mexicana Bellardi, Saggio, etc. I, 20, Tab. I, 6. — Mexico.

Neoexaireta.

Exaireta Schiner, Verh. Zool. Bot Ges. 1867, 309; Novara etc. p. 71, 1868; *Diphysa* Macquart, Dipt. Exot. I, 1, 172 (ex parte). [50]).

rufipalpis Wiedemann, Auss. Zw. II, 619, 10 *(Xylophagus)*; Macquart, Dipt. Exot. I, 1, 172 *(Diphysa)*. — Mexico.

SECTION II. SARGINA (Loew, Monogr. etc. I, 17).

Sargus.

Fabricius, Ent. Syst. Suppl. 566; 1798. ([51]).

debilis Walker, Dipt. Saund. 83. — United States.

*decorus Say, Long's Exp. App. 376; Compl. Wr. I, 257; Wiedemann, Auss. Zw. II, 38, 19. — North America, common. *Sargus marginatus* v. d. Wulp, Tijdschr. v. Ent. 2d Ser. II, 134 [Loew, Zeitschr. für Ges. Naturw. XXXVI, 119].

*elegans Loew, Centur. VII, 10. — New York; Kentucky; Florida; Massachusetts.

trivittatus Say, J. Acad. Phil. VI, 159; Compl. Wr. II, 355. — Indiana.

xanthopus Wiedemann, Auss. Zw. II, 40. — Pennsylvania.

Alchidas Walker, List, etc. III, 517. — Jamaica.

aureus Bellardi, Saggio, etc. I, 42, Tab. I 20. — Mexico.

Bagosas Walker, List, etc. III, 518. — Jamaica.

bicolor Wiedemann, Auss. Zw. II, 41. — Porto Rico.

caesius Bellardi, Saggio, etc. I, 40; Tab. I, 18. — Mexico.

clavatus Walker, List, etc. V, 93. Bellardi, Saggio, etc. I, 41. (The identification of W.'s description is given as doubtful.) — Brazil (Walker); Mexico (Bell.).

*lateralis Macquart, H. N. Dipt. I, 262; Bigot, in R. de la Sagra etc., 800. — Cuba.

latus Bellardi, Saggio, etc. I, 41; Tab. I, f. 19. — Mexico.

linearis Walker, List, etc. V, 328. — Mexico.

*lucens Loew, Centur. VII, 11. — Cuba.

nigrifemoratus Macquart, Dipt. Exot. 2e Suppl. 31, 10; Bellardi, Saggio, etc. I, 42. — Mexico.

*pleuriticus Loew, Centur. VII, 13. — Cuba.
Sallei Bellardi, Saggio, etc. I. 43; Tab. I, f. 21. — Mexico.
speciosus Macquart, Dipt. Exot. 1e Suppl., 56, 9; Bellardi, Saggio, etc. I, 40. — Yucatan (Macquart); Mexico (Bellardi).
stramineus Fabricius, Syst. Antl. 253 *(Scaeva)*; Wiedemann, Auss. Zw. II, 39; Bellardi, Saggio, etc. I, 44. — Mexico.
subinterruptus Bellardi, Saggio, etc. I, 44; Tab. I, f. 22; Walker, Trans. Ent. Soc. N. Ser. V, 271. — Mexico.
*tricolor Loew, Centur. VII, 12. — Cuba.
versicolor Bellardi, Saggio, etc. App. 13, f. 8. — Mexico.

Chloromyia.
Duncan, Magaz. Zool. and Bot. 1837; *Chrysomyia* Macquart, Hist. Nat. Dipt. I, 262; 1834. ([52]).
*viridis Say, J. Acad. Phil. III, 87; Compl. Wr. II, 77 *(Sargus)*; Wiedemann, Auss. Zw. 39 *(id.)*. — United States and Canada (Quebec); also in California.

Microchrysa.
Loew, Verh. Zool. Bot. Ver. 1855.
*polita Linné, Meigen, etc. *(Sargus)*. — Europe and North America.

Ptecticus.
Loew, Verh. Zool. Bot. Ver. 1855.([53]).
*testaceus Fabricius, Syst. Antl. 257, 6 *(Sargus)*; Wiedemann, Auss. Zw. II, 35, 15 *(id.)*; Macquart, Dipt. Exot. I, 1, 203, 5, and Suppl. I, 57 *(id.)*; Bellardi, Saggio, etc. I, 45 *(id.)*; Rondani, Studi Ent. I, 103 *(id.)*. — South America (Fabr., Wied.); Yucatan (Macq.); Mexico (Bellardi). —
NB. Specimens occurring in the United States (I have a pair from Genesseo, Western New York) are somewhat different; the four last joints of the tarsi are white, instead of joints 2 and 3 only, as stated in the descriptions.

Chrysochlora.
Latreille, Familles Natur. 1825; Macquart, Dipt. Exot. I, 1, 198; Loew, Verh. Zool. Bot. Ver. 1855.
purpurea Walker, Trans. Ent. Soc. Nat. Ser. V, 271. — Mexico.

Chrysonotus.
Loew, Verh. Zool. Bot. Ver. 1855.
*nigricornis Loew, Centur. VII, 14. — Distr. Columbia; Western New York.

Nothomyia.
Loew, Centur. VIII, 4; 1869.
*calopus Loew, Centur. VIII, 5. — Cuba.
*scutellata Loew, Centur. VIII, 4. — Cuba.

SECTION III. HERMETINA (Loew, Monogr. etc. I, 18).
Hermetia.
Latreille Hist. Nat. des Crust. etc. XIV, 338; 1804.

*chrysopila Loew, Centur. X, 11. — Texas.
*illucens Linné, Syst. Nat. II, 979, 2 *(Musca illucens* and *leucopa);*
  Degeer, Ins. VI, 205, 3, pl. XXIX, fig. 8 *(Nemotelus);* Fabricius
  Mant. II, 327, 2 *(Bibio);* Entom. Syst. IV, 253, 2 *(Mydas);* Syst.
  Antl. 62, 1 *(Hermetia);* Latreille, Dict. d'Hist. Nat. XXIV, 194,
  54; Hist Crust. et Ins. XIV, 338; Gen. Crust. et Ins. IV, 271;
  Lamarck, Hist. Anim sans Vert. III, 355, 2 *(Xylophagus);* Wiede-
  mann, Auss. Zw. II, 22, 1, pl. VII, f. 3; Guérin et Percheron,
  Genera, etc. Dipt. Tab. IV; Macquart, Hist. Nat. Dipt. I, 228, I;
  Dipt. Exot. I, 1, 177, 1, pl. XXI, f. 2; Bigot, in R. de la Sagra
  etc. 799; Bellardi, Saggio, etc I, 26. - South America, West Indies,
  Mexico; also in the United States. (I received a specimen labelled
  New York)
*sexmaculata Macquart, Hist. Nat. Dipt. I, 229, 4. — Porto Rico (Macq.);
  Cuba; Florida (M. C. Z.)
aurata Bellardi, Saggio, etc. I, 27, Tab. I, f. 8. — Mexico.
coarctata Macquart, Dipt. Exot Suppl. I, 50, 2, pl V, fig. 4; Bellardi,
  Saggio, etc. I, 24. — Merida de Yucatan (Macq.); Mexico (Bellardi):
  South America (Schiner, Novara, 70).
*lativentris Bellardi, Saggio, etc. I, 27; Tab. I, f. 9; also App. 8. —
  Mexico.
planifrons Macquart, Dipt. Exot Suppl. I, 50, 3 — Yucatan.

SECTION IV. ODONTOMYINA.
Oxycera.
Meigen, Illiger's Magaz. I4, 265; 1803. ([54]).

*centralis Loew, Centur. III, 14. — Red River of the North.
*maculata Olivier, Encycl. Méthod. VIII, 600, 4; Macquart, Dipt. Exot.
  I, 2, 190. — Carolina; Distr. Columbia. Massachusetts.
picta v. d. Wulp, Tijdschr., v. Ent. 2ᵈ Ser. II, 133; compare also Loew,
  Zeitschr. f. ges. Naturw. XXXVI, 117. — Wisconsin.
*nnifasciata Loew, Centur. III, 15. — Pennsylvania.
variegata Olivier, Encycl. Méthod. VIII, 600, Macquart, Dipt. Exo⁴.
  I, 2, 191. — North Carolina.
*Crotchi O. Sacken, Western Dipt, 212. — California.
Liburna Walker, List, etc III, 528. — Jamaica.
metallica Wiedemann, Auss Zw. II, 60. — St. Thomas [Loew *in 'it*
  suspects this species to be a *Nothomyia].*

Euparyphus.
Gerstaecker, Linn. Entom. XI, 1857. ([55]).

*bellus Loew, Centur. VII 18. — Massachusetts.
*brevicornis Loew, Centur. VII, 16. — Distr. Columbia.

*stigmaticalis Loew, Centur. VII, 17. — Distr. Columbia; Western N Y.
*tetraspilus Loew, Centur. VII, 15. — New York; Quebec, Can.
elegans Wiedemann, Auss. Zw. II, 58, 8 (Cyphomyia); Gerstäcker,
    Linn. Entom. XI, 316. — Mexico.

Odontomyia.
Meigen, Klassific. etc. I, 128; 1804. ([56]).

*binotata Loew, Centur. VI, 22. — Illinois; Texas.
brevipennis Olivier, Encycl. Méthod. VIII, 434, 13. — Carolina.
canadensis Walker, List, etc. V, 310 (Stratiomys). — Canada.
*cincta Olivier, Encycl. Méthod. VIII, 432, 3; Macquart, Dipt. Exot I,
    2, 189. — Carolina; Illinois.
*flavicornis Olivier, Encycl. Méthod. VIII, 433, 9; Macquart, Hist. Nat.
    Dipt. I, 248, 4. — North America
*hieroglyphica Olivier, Encycl. Méth. VIII, 434. — Carolina; Distr.
    Columbia.
*inaequalis Loew, Centur. VI, 24. — Fort Resolution, Huds. B. Terr.
intermedia Wiedemann, Auss. Zw. II, 64, 5. — North America.
interrupta Olivier, Encycl. Méthod. VIII, 433, 8. — Carolina.
*lasiophthalma Loew, Centur. VI, 23. — New York; New Jersey.
limbipennis Macquart, Dipt. Exot. Suppl. 2, 30, 24. — America? ([57]).
*microstoma Loew, Centur. VI, 28. — Massachusetts; New York.
*nigerrima Loew, Centur. X, 6. — Middle States.
*nigrirostris Loew, Centur. VI, 19. — Northern Wisconsin.
obscura Olivier, Encycl. Méthod. VIII, 433, 7; Macquart, Dipt. Exot. I,
    2, 189. - Carolina.
Paron Walker, List, etc. III, 536. — Trenton Falls, New York.
*pilimana Loew, Centur. VI, 27. — Illinois.
*plebeja Loew, Centur. X, 5. — Connecticut.
*varipes Loew, Centur. VI, 21. — Carolina.
vertebrata Say, Long's Exped. App. 369; Compl. Wr. I, 251; Wiede-
    mann, Auss. Zw. II, 73, 20; Bellardi, Saggio, etc. I, 38. — N. W.
    Territory (Say).
*virgo Wiedemann, Auss. Zw. II, 69, 13. — Georgia.

*arcuata Loew, Centur. X, 4. — California.
*megacephala Loew, Centur. VI, 20. — California.

affinis Bellardi, Saggio, etc. I, 35, Tab. I, 12. — Mexico.
albomaculata Macquart, Dipt. Exot. I, 1, 189, 12. — San Domingo.
dorsalis Fabricius, Syst. Antl. 82, 20; Wiedemann, Auss. Zw. II, 66
    (Stratiomys). — South America (Wied ; but Fabricius has „in
    Americae insulis").
dissimilis Bellardi, Saggio, etc. I, 35, Tab. I, f. 13, 14. — Mexico.
emarginata Macquart, Dipt. Exot I, 1, 190, 14. — Mexico.
femorata Bellardi, Saggio, etc. I, 37. — Mexico.
flavifasciata Macquart, Dipt. Exot. 4e Suppl. 53, 36. — Mexico.
Lefebvrei Macquart, Dipt. Exot. I, 1, 189, 13; comp. also Walker,
    List, etc. V, 311; Bellardi, Saggio, etc. I. 33. — Mexico

maculifrons Walker, List, etc. III, 536. — Honduras.
prasina Jaennicke, Neue Exot. Dipt. 16. — Mexico.
quadrimaculata Bellardi, Saggio, etc. I, 37; Tab. I, f. 15. — Mexico.
rubricornis Macquart, Dipt. Exot. Suppl. I, 53, 21. — Yucatan.
* rufipes Loew, Centur. VI, 25. — Cuba.
* scalaris Loew, Centur. VI, 26. — Cuba.
tritaeniata Bellardi, Saggio, etc. I, 38; Tab. I, f. 17. — Mexico.
Truquii Bellardi, Saggio, etc. I, f. 11. — Mexico.
vicina Macquart, Dipt. Exot. I, 188, 11. — Cuba.
viridis Bellardi, Saggio, etc. I, 36; Tab. I, f. 16. — Mexico.

Stratiomyia.

Stratiomys, Geoffroy, Hist. Nat. d. In. II, 475; 1764. ([58]).
Stratiomyia, as amended by Loew, Centur. VII, 4.*)
* angularis Loew, Centur. VI, 16. — Philadelphia.
* apicula Loew, Centur. VI, 13. — Illinois.
* constans Loew, Centur. X, 8. — Texas.
* discalis Loew, Centur. VI, 14. — Illinois.
flaviceps Macquart, Hist. Nat. Dipt. I, 245. — Philadelphia.
  Statiomys coronata, Guérin, Iconogr. Texte, 544; Tab. 98, f. 6.
* laticeps Loew, Centur. VII, 20. — Huds. B. Terr.
* lativentris Loew, Centur. VI, 8. — Lake Superior.
lineolata Macquart, Dipt. Exot. 4e Suppl. 48, 6; Tab. III, f. 5. —
  Virginia.
* marginalis Loew, Centur. VI, 17. — Philadelphia.
* Meigenii Wiedemann, Auss. Zw. II, 61, 2 (Tab. VIII, f. 7). — Savannah.
nigrifrons Walker, List, etc. III, 531. — Huds. B. Terr.
* norma Wiedemann, Auss. Zw. II, 62, 3. — North America.
* nigriventris Loew, Centur. VI, 15. — Nebraska.
* normula Loew, Centur. VI, 5. — New York.
* notata Loew, Centur. VI, 18. — Nebraska. -
nymphis Walker, List, etc. III, 530. — Huds. B. Terr.
* obesa Loew, Centur. VI, 11. ' — Illinois.
* picipes Loew, Centur. VII, 21. — Massachusetts; Canada; Lake
  Winnipeg.
  Stratiomys ischiaca (Harris) Walker, List, etc. III, 529. — Massa-
    chusetts.
  Stratiomys badius Walker, List, etc. III, 529. — New Hampshire [See
    Walker, l. c. 1157, where the habitat originally stated „New
    Holland", is declared erroneous, and the synonymy with *S. ischiaca*
    is acknowledged].
pulchella Macquart, Dipt. Exot. I. 1, 180, 3; Tab. XXII, f. 2. —
  Georgia.
* quadrigemina Loew, Centur. VI, 4. — Connecticut.
* quaternaria Loew, Centur. VI, 12. — Illinois.

---

*) Geoffroy, in translating Réaumur's *mouche armée*, evidently meant to make the
name *Stratiomyia* and not *Stratiomys*, which is nonsensical.

robusta Walker, List, etc. V, 37. — North America.
*senaria Loew, Centur. VI, 7. — Florida.
*unilimbata Loew, Centur. VI, 6. — Wisconsin.
vicina Macquart, Dipt. Exot. I, 1, 181, 4 („the male of *S. flaviceps?"*
    Macq.). — Philadelphia.

*barbata Loew, Centur. VI, 9. — California.
*insignis Loew, Centur. X, 7.— California.
*maculosa Loew, Centur. VII, 19. — California.
*melastoma Loew, Centur. VI, 10. — California.

bimaculata Bellardi, Saggio, etc. App. 10; fig. 7. — Mexico.
constricta Walker, Trans. Ent. N. Ser. V, 268. — Mexico.
euchlora Gerstaecker, Linn. Ent. XI, 328. — Mexico.
fenestrata Gerstaecker, Linn. Ent. XI, 327. — Mexico.
Gerstaeckeri Bellardi, Saggio, etc. I, 31; Tab. I, f. 10. — Mexico.
goniphora Say, J. Acad. Phil. VI, 161; Compl. Wr. II, 356. — Mexico.
*mutabilis Fabricius, Ent. Syst. IV, 266; Syst. Antl., 81; Wiedemann,
    Auss. Zw. II, 63, Tab. VIII, f. a — d; Perty, Del. Anim. etc.
    Tab. 38, 14; Bellardi, Saggio, etc. I, 30; compare also Schiner,
    Novara etc. 61. — Mexico, Brazil.
    Stratiomys fasciata Fabricius, Ent. Syst. IV, 266; Syst Antl., 81
    [Bellardi].
pinguis Walker, Trans. Ent. Soc. N Ser. V, 270. — Mexico.
subalba Walker, List, etc. V, 45; Bellardi, Saggio, etc. I, 31. — Mexico,
    Brazil.
trivittata Say, J. Acad. Phil. VI, 160; Compl. Wr. II, 356. -- Mexico.

Cyphomyia.
Wiedemann, Zool. Mag. 1, 3, 55, 1819; Analecta etc. 1824;
    Gerstaecker, Linn. Ent. XI, 263. (59).

albitarsis Fabricius, Syst. Antl. 80, 12 ♀ *(Stratiomys);* Gerstaecker, Linn.
    Ent. XI, 300; Bellardi, Saggio, etc. I, 22. — Guyana, Columbia
    (Gerst.); Yucatan (Macq.); Mexico (Bell.); South America (Schin.,
    Novara, 53).
    Cyphomyia fenestrata Macq. Dipt. Exot. Suppl. I, 48, ♂ ♀ [Gerst.].
auriflamma Wiedemann, Zool. Mag. I, 3, 54; Auss. Zw. Vol. II, 54;
    Tab. VIII, f. 1; Macquart, Hist. Nat. Dipt. I, 242; Guérin, Iconogr.
    Tab. XCVIII, f. 5; Gerstaecker, Linn. Ent. XI, 276; Bellardi,
    Saggio, etc. I, 21. — Mexico, Brazil, Guyana.
    Cyphomyia chrysodota Perty, Del. An. Art. 184; Tab. XXXVI,
    f. 14, ♂ [Gerst.].
    Cyphomyia cyanea Macquart, Hist. Nat Dipt. I, 242, ♀ [Gerst.].
*marginata Loew, Centur. VI, 31. — Cuba.
pilosissima Gerstaecker, Linn. Ent. XI, 292. — Mexico.
*rubra Loew, Cent. VI, 30. — Cuba.
similis Bellardi, Saggio, etc. I, 23; Tab. I, f. 7. — Mexico.
scalaris Bigot, Ann. Soc. Ent. 1875, 487. — Mexico.
simplex Walker, Trans. Ent. Soc. N. Ser. V, 268. — Mexico.

tomentosa Gerstaecker, Linn. Ent. XI, 294; Bellardi, Saggio, etc. I, 22. — Mexico.

varipes Gerstaecker, Linn. Ent. XI, 283; compare also Schiner, Novara, 52. — Mexico; Columbia (Schiner).

Acanthina.

Wiedemann, Auss. Zw. II, 50: 1830; compare also Gerstaecker, Linn. Ent. XI, 335.

nana Bellardi, Saggio, etc. App., 9. — Mexico.

*ornata Macquart, Dipt. Exot. Suppl. I, 51; Tab. V, f. 5; Bellardi, Saggio, etc. I, 28. — Brazil, Mexico.

Chordonota.

Gerstaecker, Linn. Ent XI, 311; 1857.

carbonaria Bellardi, Saggio, etc. App. 11. — Mexico.

fuscipennis Bellardi, Saggio, etc. App. 11, f. 6. — Mexico.

Clitellaria.

Meigen, Illiger's Magaz. II, 265; 1803. ([60]).

*subulata Loew, Centur. VI, 29. — Virginia.

*lata Loew, Centur. X, 9. — California.

*rustica O. Sacken, Western Diptera, 213. — California (Marin and Sonoma Co.).

Anchialus Walker, List, etc. III, 522 („var. chalybeae Wied.?" according to Walker l. c. IV, 1157). — Jamaica.

fenestrata Macq., Dipt Exot. 1e Suppl. 54, 3 (Ephippium). — Yucatan.

Halala Walker, List, etc. III, 523. — Honduras.

obesa Walker, Trans. Ent. Soc. N. Ser. V, 270. — Mexico.

Euryneura.

Schiner, Verh. Zool. Bot. Ges. 1867, 308; Novara etc. p. 56, 1868.

pygmaea Bellardi, Saggio, etc. App. 12, fig. 5 (Clitellaria); considered an Euryneura by Schiner, l. c. — Mexico.

Neorondania.

Rondania, Jaennicke, Neue Exot. Dipt. 1867. ([61]).

obscura Jaennicke, Neue Exot. Dipt., 17. — Mexico.

chalybea Wiedemann, Anal. Ent. 30, 36; Auss Zw. II, 49, 4 (Clitellaria); — Jaennicke, Tab. I, f. 4. — St. Thomas.

Nemotelus.*)

Geoffroy, Hist. Nat. d. Ins. II, 542; 1764. ([62]).

albirostris Macquart, Dipt. Exot. 4e Suppl. 55, 3; Tab. III, f. 8. — Virginia.

---

*) Nematotelus would be more correct, but the name, one of the oldest in dipterology, is too venerable for a change.

carneus Walker, List, etc. III, 521. — Huds. B. Terr.
*canadensis Loew, Centur. III, 12. — Fort Resolution, Huds. B. Terr.
*carbonarius Loew, Centur. VIII, 6. — Massachusetts.
*crassus Loew, Centur. III, 10. — Rhode Island.
*glaber Loew, Centur. X, 10. — Texas.
pallipes Say, J. Acad Phil. III, 29; Compl. Wr. II, 52; Wiedemann, Auss. Zw. II, 45, 2. - Pennsylvania.
*unicolor Loew, Centur. III, 11. — Illinois.

*acutirostris Loew, Centur. III, 13. — Cuba.
polyposus Say, J. Acad. Phil. VI, 160; Compl. Wr. II, 356. — Mexico.

> Observation. *N. nigrinus* Fall from Europe, according to v. d. Wulp, l. c. 126 also occurs in the United States.

SECTION V. PACHYGASTRINA.
Pachygaster.
Meigen, Illiger's Magaz. II, 266; 1803. [63].
*pulcher Loew, Centur. III, 16. — Distr. Columbia.

Chauna.
Loew, Stett. Ent. Zw. VIII, 370; 1847.

Gerstaecker, Linn. Ent. XI, 338.

*variabilis Loew, Stett. Ent. Zw. VIII, 370, Tab. I, f. 11—15. — Cuba.
Chauna ferruginea Gerstaecker, Linn. Ent. XI, 340; Tab. III, f. 7. [Synonymy according to Loew, Berl. Ent. Z. Vol. II, 349; who acknowledges that Gerstaecker's error was due to the imperfect description and figure of the antennae of *Ch. variabilis.*]

FAMILY ACANTHOMERIDAE.
Acanthomera.
Wiedemann, Dipt. Exot. 60; 1821.

Bellardii Bigot; Bellardi, Saggio, etc. App. 16, f. 11. — Mexico.
Bigotii Bellardi, Saggio, etc. App. 16, f. 10. — Mexico.
crassipalpis Macquart, Dipt. Exot. 2e Suppl. 27, 5; Tab. I, f. 3 (*female*). — Guatemala.
picta Wiedemann, Dipt. Exot., 61, Tab. II, f. 2; Auss. Zw. I, 108, Macquart, Dipt. Exot. I, 167; Guérin, Iconogr., Tab. XCVIII, f. 3. Bellardi, Saggio, etc. I, 76. — Brazil, Mexico.
seticornis Wiedemann, Auss. Zw. I, 108, 1; Macquart Dipt. Exot. I, 1, 168, 3; Tab. XX, f. 1 and Suppl. 2ᵃ, 27. — Brazil (Wied.); Guatemala (Macquart). Macquart suspects that this is the male of his *A. crassipalpis*.
tabanina Thunberg, Act. Soc. Gothob. 1819, 111, 7; Tab. VII, f. 2 (*Pantophthalmus*); Wiedemann, Auss. Zw. I, 110, 4. — West Indies.

FAMILY TABANIDAE. [64].

Pangonia.

Latreille, Hist. Nat. des Crust. et des Ins. III, 437; 1802. [65].

*chrysocoma O. Sacken, Prodrome etc. I, 368. — Trenton Falls, New York; Delaware.

fusiformis Walker, Dipt. Saund. 19. — North America.

isabellina Wiedemann, Auss. Zw. I, 112, 3 *(Silvius)*. — North América. [66].

macroglossa Westwood, London and Edinburgh Philos. Magaz. 1835; reproduced in O. Sacken, Prodrome, 368. — Georgia.

*pigra O Sacken, Prodrome etc. I, 367. — New York, Kentucky.

*rasa Loew, Centur. VIII, 7; O. Sacken, Prodrome etc. I, 366. — Illinois; Wisconsin; New York.

*tranquilla O. Sacken, Prodrome etc. I, 367. — Pennsylvania; Massachusetts; White Mts., N. H.; Quebec, Can.

*hera O. Sacken, Western Diptera, 214. — San Francisco, Cal.

*incisa Wiedemann, Auss. Zw. I, 90, 6. — Arkansaw (Say); Colorado Springs, Col.

Pangonia incisuralis Say, J. Acad. Phil. III, 31; Amer. Entom. pl. XXXIV; Compl. Wr. I, 75 [change of name by Wiedemann].

aurulans Wiedemann, Auss. Zw. II, 620, 12. — Mexico.

atrifera Walker, Trans. Ent. Soc. New series V, 272. — Mexico.

flavohirta Bellardi, Saggio, etc. I, 49. — Mexico.

fulvithorax Wiedemann, Auss. Zw. I, 89; Bigot, R. de la Sagra etc., 797. — Brazil (Wied.); Cuba (Bigot).

incerta Bellardi, Saggio, etc. I, 52. — Mexico.

nigronotata Macquart, Dipt. Exot. 4e Suppl. 27, 56; Tab. II, f. 5; Bellardi, Saggio, etc. I, 51. — Mexico.

planiventris Macquart, Dipt. Exot. 4e Suppl. 26, 55. — Mexico.

rhinophora Bellardi, Saggio, etc. I, 46; Tab. II, f. 1. — Mexico.

rostrifera Bellardi, Saggio, etc. I, 47. — Mexico.

Sallei Bellardi, Saggio, etc. I, 50. — Mexico.

Saussurei Bellardi, Saggio, etc. I, 49; Tab. II, f. 4. — Mexico.

semiflava Wiedemann, Auss. Zw. II, 622, 16; Bellardi, Saggio, etc. I, 51; Tab. II, f. 2. — Mexico.

Pangonia bicolor Macquart, Dipt. Exot. 4e Suppl. 27, 57 (Bellardi).

tenuirostris Walker, Trans. Ent. Soc. N. Ser. V, 272. — Mexico.

Wiedemauni Bellardi, Saggio, etc. I, 48; Tab. II, f. 3. — Mexico.

Pangonia basilaris Wiedemann, Auss. Zw. II, 621. [The name was changed by Bellardi.]

Chrysops.

Meigen, in Illiger's Magaz., 1803. [67].

*aestuans van der Wulp, Tijdschr. v. Ent. 2. Ser. II, 135; Tab. III, f. 8, 9; O. Sacken, Prodrome etc. I, 378, — North Western States.

(?) *Chrysops moerens* Walker, List, etc. I, 201. — Nova Scotia.

*atropos O. Sacken, Prodrome etc. I, 372. — Florida.

    Chrysops divisus Walker, List, etc. I, 204.

*callidus O. Sacken, Prodrome etc. I, 379. — Middle States.

*celer O. Sacken, Prodome etc. I, 376. — Middle States; Massachusetts.

*delicatulus O. Sacken, Prodrome etc., I, 380. — North Conway, N. H.

*excitans Walker, Dipt. Saund, 72; O. Sacken, Prodrome etc. I, 373. — Northern United States and British Possessions.

*fallax O. Sacken, Prodrome etc. I, 392. — Middle and Northern States.

*flavidus Wiedemann, Dipt. Exot. I, 105, 5; Auss. Zw. I, 199, 7; O. Sacken, Prodrome etc. I, 385. — Atlantic States; British Possessions.

    Chrysops pallidus Bellardi, Saggio, etc. I, 73; Tab. II, f. 16. — Mexico.

    Chrysops canifrons Walker, List, etc. I, 197. — Florida.

*frigidus O. Sacken, Prodrome etc. I, 384; also II, 474. — Northern States and British Possessions.

*fugax O. Sacken, Prodrome etc. I, 375. — Northern States and British Possessions.

    (?) *Chrysops carbonarius* Walker, List, etc. I, 203 (*ex parte*).

    (?) *Chrysops ater* Macquart, Dipt. Exot. 4e Suppl. 40, 18. — Newfoundland.

*hilaris O. Sacken, Prodrome etc. I, 391. — Middle and Northern States; Canada.

*indus O. Sacken, Prodrome etc. I, 383. — Western New York, Canada.

Iugens Wiedemann, Dipt. Exot. I. 109, 12; Auss. Zw. I, 212, 26. — Georgia (Wied.).

*morosus O. Sacken, Prodrome etc. I, 389; also II, 474. — Maryland; Florida; Texas.

    (?) *Chrysops trinotatus* Macquart, Dipt. Exot. I, I, 161, 9.

*mitis O. Sacken, Prodrome etc. I, 374. — British Possessions; Lake Superior.

    (?) *Chrysops provocans* Walker, Dipt. Saund, 73.

*moechus O. Sacken, Prodrome etc. I, 387. — Middle and Southern States.

*montanus O. Sacken, Prodrome etc. I, 382. — Catskill Mountain House, New York.

*niger Macquart, Dipt. Exot. I, 1, 161, 10; O. Sacken, Prodrome etc. I, 377. — Atlantic States and British Possessions.

    (?) *Chrysops carbonarius* Walker, List, etc. I, 203 (Var. *β*.).

nigripes (Zetterstedt) Loew, Verh. Zool. Bot. Ges. 1858, 623. — Lapland; Sitka.

*obsoletus Wiedemann, Dipt. Exot. I, 108, 10; Auss. Zw. I, 211, 25; O. Sacken, Prodrome etc. I, 393. — Middle and Northern States.([68]).

*plangens Wiedemann, Auss. Zw. I, 210, 22 (♀); O. Sacken, Prodrome etc. I, 393. — Atlantic States.

    Chrysops fuliginosus Wiedemann, Dipt. Exot. I, 109, 11; Auss. Zw. I, 210, 23 (♂).

*pudicus O. Sacken, Prodrome etc. I, 381 and II, 474. — Massachusetts, Florida; Long Island, New York.

*sordidus O. Sacken, Prodrome etc. I, 376. — White Mts., N. H.;
  British Possessions.
*striatus O. Sacken, Prodrome etc. I, 391. — District Columbia; Illinois.
  Chrysops furcatus Walker, List, etc. I, 199.
  Chrysops vittatus Bellardi (non Wiedemann), Saggio, etc. I, 74. —
  Mexico.
*univittatus Macquart, Dipt. Exot. 5e Suppl. 36, 21; O. Sacken, Pro-
  drome etc. I, 387. — Middle States.
  (?) Chrysops fascipennis Macquart, Hist. Nat. Dipt. I, 216.
*vittatus Wiedemann, Dipt. Exot. I, 106, 7; Auss. Zw. I, 200, 8;
  Macquart Dipt. Exot. 5e Suppl. 37, 22; O. Sacken, Prodrome etc.
  I, 390. — Middle and Northern States.
  Chrysops areolatus Walker, List., etc. I, 197.
  Chrysops lineatus Jaennicke, Neue Exot. Dipt. 26.

*fulvaster O. Sacken, Western Diptera, 221. — Colorado; Utah.
*noctifer O. Sacken, Western Diptera, 220. — Sierra Nevada, Calif.
*proclivis O. Sacken, Western Diptera, 222. — Marin Co. Calif.
*surdus O. Sacken, Western Diptera, 223. — Sierra Nevada, Cal.
*quadrivittatus Say, Journ. Acad Phil. III. 33, 1; Compl. Wr. II, 54;
  Wiedemann, Auss. Zw. I, 200, 9. — Near the Rocky Mts. (Say);
  Nebraska. ([69]).

affinis Bellardi, Saggio, etc. I, 70; Tab. II, f. 14. — Mexico.
apicalis Bellardi, Saggio, etc. I, 73. — Mexico.
crucians Wiedemann, Auss. Zw. I, 211. — Brazil (Wied.), Cuba
  (Jaennicke, Neue Exot. Dipt., 41).
*costatus Fabricius, Ent. Syst. IV, 373, 45 (Tabanus); Syst. Antl.
  112, 8; Palisot, Ins. Dipt., 223; Tab. III, f. 7; Wiedemann, Dipt.
  Exot. I, 104, 4; Auss. Zw. I, 198, 5; Macquart, Dipt. Exot. I,
  1, 160, 8; Bigot, in R. de la Sagra, etc. 798; Guérin, Iconogr.
  Texte, III, 542; Tab. XCVII, f. 3. (Called Chr. molestus on the
  plate.) — S. America (Fab.); Cuba (Macq.); Jamaica (Wk.).
  Tabanus variegatus Degeer, VI, Tab. XXX, f. 7 (Synon. very probable).
frontalis Macquart, Dipt. Exot. I, 1, 160, 7. Walker, List, etc. V,
  284. — West Indies.
geminatus Wiedemann, Auss. Zw. I, 205, 16; Macquart, Dipt. Exot.
  4e Suppl. 39. — Patria ignota (Wied.); Mexico (Macq.).
inornatus Walker, List, etc. I, 198. — West Indies; Brazil.
lateralis Wiedemann, Auss. Zw. I, 209, 21; Walker, List, etc. I, 200;
  V, 288. — Patria ignota (Wied.); Honduras (Walk.).
latifasciatus Bellardi, Saggio, etc. I, 71; Tab. II, f. 15. — Mexico.
megaceras Bellardi, Saggio, etc. I, 74; Tab. II, f. 18. — Mexico.
scalaratus Bellardi, Saggio, etc. I 72; Tab. II, f. 19. — Mexico.
subcaecutiens Bellardi, Saggio, etc. I, 69; Tab. II, f. 13. — Mexico.
virgulatus Bellardi, Saggio, etc. I, 71; Tab. II, f. 17. — Mexico.

  Observaticn. Chrysops sepulchralis (Fabricius?) Kirby, Fauna
Bor. Am. Ins. 314, 1, is omitted in the above list, because it is

very probably identical with one of the species enumerated therein; but the description is too vague for identification; moreover the identity of the species with the true *C. sepulchralis* Fabricius seems very doubtful.

Silvius.

Meigen, System. Beschr. III, 27, 1820. ([70]).

*gigantulus Loew, Centur. X, 12 *(Chrysops)*; O. Sacken, Western Diptera, 215. — California; Washington Territory; Vancouver Island; Colorado.

Silvius trifolium O. Sacken, Prodrome etc. I, 395. ([71]).

Observation. For *Silvius isabellinus* Wied., see *Pangonia*.

Lepidoselaga.

Lepiselaga Macquart, Dipt Exot. I, 1, 153, 1838; about its relation to *Hadrus* Perty, compare Loew, Dipt. Sudafrica's I, 31.

*lepidota Wiedemann, Auss. Zw. I, 193 *(Tabanus)*; Perty, Delectus etc. 183, Tab. XXXVI, f. 9 *(Hadrus)*; Macquart, Dipt. Exot. I, 1, 154; Tab. XVIII, f. 3; Bellardi, Saggio, etc. I, 75 *(Hadrus)*; according to Loew, Century VIII, 8, only the female, described by Bellardi, ,belongs here. — Guyana, Brazil (Perty, Macquart); Mexico (Bellardi).

Haematopota crassipes Fabricius, Syst. Antl. 108, 4 [Loew, Centur. VIII, 8].

*recta Loew, Centur. VIII, 8. — New Granada, Mexico.

Hadrus lepidotus Bellardi, Saggio, etc. I, 75, male. [Loew, Centur. VIII, 8.]

Haematopota.

Meigen, in Illiger's Magaz. 1803.

punctulata Macquart, Dipt. Exot. I, 1, 163, 2. — Carolina.

*americana O. Sacken, Prodrome etc. I, 395. — North West of the United States and the British Possessions.

Dichelacera.

Macquart, Dipt. Exot. I, 1, 112, 1838.

abiens Walker, List, etc. I, 191. — West Indies.

scapularis Macquart, Dipt. Exot. 2e Suppl. 15, 9; Bellardi, Saggio, etc. I, 53; Tab. II, f. 12. — Mexico.

Observation. *Dichelacera fasciata* Walker, Dipt. Saund. 68 is erroneously stated to be from North America. The typical specimen in the Brit. Mus. is south american.

Diachlorus.

O. Sacken, Prodrome etc. II, 475, 1876; *Diabasis*, Macquart, Hist. Nat. Dipt. I. 207, Dipt. Exot. I, 1, 150, was preoccupied by a genus of Coleoptera.

*ferrugatus Fabricius, Syst. Antl. 111, 2 *(Chrysops)*; Wiedemann, Dipt. Exot. I, 94, 56 *(Tabanus)*; Auss. Zw. I, 186, 113 *(id.)*; Osten

Sacken, Prodrome etc. I, 396 *(Diabasis)*; id. II, 475. — Southern
States; Mexico; Brazil; West Indies; Honduras.
Diabasis ataenia Macquart, Dipt. Exot. I, I, 152, 3.
Chrysops approximans Walker, List, etc. I, 198 (!).
Chrysops convergens Walker, List, etc. I, 198 (!).
Tabanus Rondanii Bellardi, Saggio, etc. I, 68; Tab. II, f. 11.
Tabanus americanus Palisot de Beauvais, Dipt. Tab. III, f. 6.

Therioplectes.

Zeller, Isis 1842 (ex parte); O. Sacken, Prodrome etc. II, 425; 1876.

**affinis Kirby, Fauna Bor. Amer. IV, 313, 1 *(Tabanus)*; O. Sacken,
Prodrome etc. II, 466. -- Northern United States and British
Possessions.
Tabanus triligatus Walker, List, etc. V, 183 (!). — Arctic America.
**astutus O. Sacken, Prodrome etc. II, 471 *(Tabanus)*. — White Mts.,
N. H.; Manlius, N. Y.; Southington, Conn.
carolinensis Macquart, Dipt. Exot. I, 1, 145, 47 *(Tabanus)*. —
Carolina. ([72]).
**cinctus Fabricius, Ent. Syst. IV, 366, 18 *(Tabanus)*; Syst. Antl. 97,
20 *(id.)*; Meigen, Syst. Beschr. etc. II, 42, 16 *(id.)*; Wiedemann,
Dipt. Exot. I, 67, 10 *(id.)*; Auss. Zw. I, 119, 12 *(id.)*; Harris, N.
Engl. Ins. 3d edit. 602, f. 261 *(id.)*; O. Sacken, Prodrome etc. II,
464. — Atlantic States; Mexico (?Walker, List, etc. I, 153).
**epistates O. Sacken, Prodrome etc. Supplem. 555. — |Huds. B. Terr.
Tabanus socius O. Sacken, Prodrome etc. II, 467 (name changed
because there is an earlier *T. socius* Walker).
**flavipes Wiedemann, Auss. Zw. I, 137, 41 *(Tabanus)*; O. Sacken,
Prodrome etc. II, 462. — Labrador.
**illotus O. Sacken, Prodrome etc. II, 469. — British Possessions in
North America.
**lasiophthalmus Macquart, Dipt. Exot. I, 1, 143, 45 *(Tabanus)*;
O. Sacken, Prodrome etc. II, 465. — Atlantic States and British
Possessions.
Tabanus notabilis Walker, List, etc. I, 166 (!).
Tabanus punctipennis Macquart, Dipt. Exot. 2e Suppl. 23, 108;
compare also O. Sacken, Prodrome etc. II, 473. - Philadelphia (!).
**microcephalus O. Sacken, Prodrome etc. II, 470. — White Mts., N. H.;
Trenton Falls, N. Y.; Massachusetts.
**septentrionalis Loew, Verh. Zool. Bot. Ges. 1858, 593 *(Tabanus)*;
O. Sacken, Prodrome etc, II, 467. — Labrador.
**trispilus Wiedemann, Auss. Zw. I, 150 *(Tabanus)* (!); O. Sacken, Pro-
drome etc. II, 464. — Northern and Middle States; Illinois.
vicinus Macquart, Dipt. Exot. I, 1, 143, 44 *(Tabanus)*. — Carolina.
**zonalis Kirby, Fauna Boreali-Americana, IV, 314, 2 *(Tabanus)*;
O. Sacken, Prodrome etc. II, 463. — Northern States, **as far**
West as Oregon, British Possessions.
Tabanus tarandi Walker, List, etc. I, 156 (!).
Tabanus terrae novae Macquart, Dipt. Exot. 4e Suppl. 35, 109 !).

Tabanus flavocinctus Bellardi, Saggio, etc. I, 61 (!). ([73]).

* **phaenops** O. Sacken, Western Diptera, 217. — Sierra Nevada, Cal.
* **procyon** O. Sacken, Western Diptera, 216. — Marine Co., Sonoma Co., Cal.
* **rhombicus** O. Sacken, Prodrome etc. II, 472; Western Diptera, 218. — Rocky Mountains, Colorado.
* **sonomensis** O. Sacken, Western Diptera, 216.'— Marin and Sonoma Co., California.

* **quadripunctatus** Fabricius, Syst. Antl. 99, 29 *(Tabanus);* Wiedemann, Dipt. Exot. I, 77, 30 *(id.);* Auss. Zw. I, 151, 63 *(id.).* — Brazil (Wied.); Mexico (Bellardi); Central America (M. C. Z.).
Tabanus nigropunctatus Bellardi, Saggio, etc. I, 67. ([74]).

Tabanus.
Linné, Fauna Suecica; 1761. ([75]).

* **abdominalis** Fabricius, Syst. Antl. 96, 15 (Museum Bosc.) (!); O. Sacken, Prodrome etc. II, 434 and Supplement. - Kentucky, Georgia. ([76]).
(?) *Tabanus abdominalis* Palisot Beauvois, Ins. 101, Tab. II, f. 4.
* **Actaeon** O. Sacken, Prodrome etc. II, 443. — Massachusetts; Connecticut; Minnesota; Wisconsin; Canada.
* **americanus** Forster, Nov. Spec. Centur. I, 100; O. Sacken, Prodrome, etc. II, 457. — Middle and Southern Atlantic States.
Tabanus plumbeus Drury, Ins I, Tab. 44, 2.
Tabanus ruficornis Fabricius, Syst. Ent. 789, 8; Ent. Syst. IV, 365, 14; Syst. Antl. 96, 14; Wiedemann, Dipt. Exot. I, 62; Auss. Zw. I, 112, 1.
Tabanus limbatus Palisot-Beauvois, Ins. Dipt. Tab. I, f. 2. .
* **annulatus** Say, Journ. Acad. Phil. III, 32, 2; Compl. Wr. II, 53; Wiedemann, Auss. Zw. I, 185; O. Sacken, Prodrome etc.; Suppl. 555. — Missouri; Cumberland Gap, Ky.; Georgia; Kansas.
* **atratus** Fabricius, System. Ent. 789, 9; Ent. System. IV, 366, 16; System. Antl. 96, 16; Wiedemann, Dipt. Exot. I, 63, 2; Auss. Zw. I, 114, 3; Macquart, Dipt. Exot. I, 1, 142, 41; Bellardi, Saggio, etc. I, 58; Harris, Ins. N. Engl., 3d edit. 602; O. Sacken, Prodrome etc. II, 454. — Atlantic States; Mexico (coll. Bellardi!).
Tabanus niger Palisot-Beauvois, Ins. Dipt. Tab. I, f. 1.
Tabanus americanus Drury, Ins. I, Tab. 44, f. 3.
Tabanus validus Wiedemann, Auss. Zw. I, 113, 2 (!).
* **catenatus** O. Sacken (non Walker', Prodrome etc. II, 433. — Atlantic States.
Tabanus recedens Walker, List, etc. I, 147 (!). ([77]).
* **cerastes** O. Sacken, Prodrome etc. II, 462. — Kentucky; Wisconsin.
Tabanus hirtioculatus Macquart, Dipt. Exot. 5e Suppl. 33, 128; compare also O. Sacken, Prodrome II, 473. ([78]).
cingulatus Macquart, Dipt. Exot I, 1, 144, 46. — Philadelphia.
* **coffeatus** Macquart, Dipt. Exot. 2e Suppl, 23, 109 (♂!); O Sacken, Prodrome etc. II, 441. — Distr. Columbia; Delaware; New York, Florida, Massachusetts.

(?) *Tabanus nigripes* Wiedemann, Auss. Zw. I, 142, 50 (♂).

*costalis Wiedemann, Auss. Zw. I, 173, 94; O. Sacken, Prodrome etc. II, 450. — Atlantic States.

(?) *Tabanus costalis* Bellardi, Saggio, etc. 63. — Mexico.

Tabanus vicarius Walker, List, etc. I, 137 (!).

Tabanus baltimorensis Macquart, Dipt. Exot. 5e Suppl. 34, 129 (!).

*cymatophorus O. Sacken, Prodrome etc. II, 441. — Kentucky.

*Endymion O. Sacken, Prodrome etc., Supplement, 556. — Georgia.

*exul O. Sacken, Prodrome etc. Supplement, 557. — District Columbia; Maryland; Pennsylvania; New Jersey.

Tabanus abdominalis Wiedemann (non Fabricius), Dipt. Exot. I, 65, 6; Auss. Zw. I, 116, 7 (!).

*fronto O. Sacken, Prodrome etc. II, 431. — Georgia.

(?) *Tabanus cheliopterus* Rondani, Nuovi Annali d. Sc. N. di Bologna; descr. reproduced in O. Sacken, Prodr. II, 473. — Carolina. ([79]).

*fulvulus Wiedemann, Auss. Zw. I, 153, 66; O. Sacken, Prodrome etc. II, 451. — Middle States; Kentucky.

*fuscopunctatus Macquart, Dipt. Exot. 4e Suppl. 34, 108 (!); O. Sacken, Prodrome etc. II, 432; the male in the Supplement, 559. — South Carolina; Georgia; Florida.

Tabanus imitans Walker, List, etc. I, 147. — Georgia. ([80]) (!)

*giganteus Degeer, Ins. VI, 226, 1; Tab. XXX, f. 1; O. Sacken, Prodrome etc. II, 458. — Middle and Southern Atlantic States; Kansas.

Tabanus lineatus Fabricius, Spec. Ins. II, 455, 4; Ent. Syst. IV, 363, 5; Syst. Antl. 94, 3; Wiedemann, Dipt. Exot. I, 63, 3; Auss. Zw. I, 115, 4.

Tabanus bicolor Macquart, Dipt. Exot. 2e Suppl. 21, 105, *female* (!).

Tabanus caesiofasciatus Macquart, Dipt Exot. 5e Suppl. 32, 126; *male* (!).

gracilis Wiedemann, Auss. Zw. I, 156, 71. — Georgia. ([81]).

*lineola Fabricius, Ent. Syst. IV, 369, 33; Syst. Antl. 102, 41; Coquebert, Illustr. Iconogr. 112, Tab. XXV, f. 6.; Wiedemann, Dipt. Exot. I, 81, 36; Auss. Zw. I, 170, 89; Harris, Ins. N. Engl. 3d edit. 602, f. 262; Palisot-Beauvois, Dipt. Tab. II, fig. 6 (doubtful); O. Sacken, Prodrome etc. II, 448. — Atlantic States; Mexico. ([82]).

Tabanus simulans Walker, List, etc. I, 182.

(?) *Tabanus scutellaris* Walker, Dipt. Saunders. 27.

*longus O. Sacken, Prodrome etc. II, 447; also in the Supplement, 559. — Middle Atlantic States.

*lugubris Macquart, Dipt. Exot. I, 1, 145, 48; O. Sacken, Prodrome etc. II, 456. — South Carolina.

Tabanus ater Palisot-Beauvois, Ins.; Dipt. II, f. 5.; Wiedemann, Dipt. Exot. I, 74, 23; Auss. Zw. I, 136, 39 (the latter only *ex parte*).

*Megerlei Wiedemann, Auss. Zw. I, 132, 32 (!); O. Sacken, Prodrome etc. II, 457. — Florida.

* **melanocerus** Wiedemann, Auss. Zw. I, 122, 16 (!); O. Sacken, Prodrome etc. II, 440. — Middle and Southern Atlantic States.

(?) *Tabanus exaestuans* Linné, System. Nat. II, 1000, 8; Degeer VI,
229, 8; Tab. XXX, f. 5; Fabricius, Ent. System. IV, 365, 13;
System. Antl. 96, 12; Compare also O. Sacken, Prodrome etc. II,
441. — Surinam.

* **mexicanus** Linné, System. Nat. II, 1000, 10; Fabricius, Spec. Ins. II,
457, 16; Ent. System. IV, 367, 22; Syst. Antl. 98, 25; Wiedemann, Dipt. Exot. I, 76, 29; Auss. Zw. I, 147, 58; Macquart Dipt.
Exot. I, 1, 143, 43; O. Sacken, Prodrome etc. II, 459. — South
Carolina; Florida; Missouri; New Jersey; Mexico; South America.

Tabanus punctatus Fabr., Ent. System. IV, 368, 25.

Tabanus inanis Fabr., l. c. 26.

Tabanus ochroleucus Meigen, System. Beschr. II, 62, 41.

Tabanus olivaceus Degeer, VI, 230. 6; Tab. XXX, f. 6.

Tabanus sulphureus Palisot-Beauvois, Ins. 222, Dipt. Tab. III, f. 3.

Tabanus flavus Macquart, Hist. Nat. Dipt. I, 200, 13; Guérin et
Percheron, Genera etc. Dipt. II.

Tabanus viridiflavus Walker, Newman's Zool. VIII, App. LXVI
(„fide Walker", thus quoted by Bellardi, Saggio I, 59).

* **molestus** Say, Journ. Acad. Phil. III, 31, 1; Compl. Wr. II, 53;
Wiedemann, Auss. Zw. I. 125, 21 (!); O. Sacken, Prodrome etc.
II, 438. — Distr. Columbia; Kentucky; Georgia; Missouri.

* **nigrescens** Palisot-Beauvois, Dipt. Tab. II, f. 2; Wiedemann, Auss.
Zw. I, 116, 6 (translation from Palisot); O. Sacken, Prodrome
etc. II, 453. — New York; Massachusetts; New Jersey, Pennsylvania; Maryland; Tennessee; Canada.

* **nigrovittatus** Macquart, Dipt. Exot. 2e. Suppl., 24, 111; O. Sacken,
Prodrome etc. II, 449. — Massachusetts; Rhode Island; New
York; New Jersey.

* **nivosus** O. Sacken, Prodrome etc. II, 445. — New Jersey.

* **Orion** O. Sacken, Prodrome etc. II, 442. — Canada; Massachusetts;
Connecticut.

* **psammophilus** O. Sacken, Prodrome etc. II, 445. — Florida.

* **pumilus** Macquart, Dipt. Exot. I, 1, 146, 51; O. Sacken, Prodrome
etc. II, 448. — Middle and Southern Atlantic States.

* **Reinwardtii** Wiedemann, Auss. Zw. I, 130 (!); O. Sacken, Prodrome
etc. II, 461. — Northern and Middle Atlantic States; Canada.

Tabanus erythrotelus Walker, Ins. Saund. 25; Tab. II, f. 1.

* **rufus** Palisot-Beauvois, Dipt. Tab. II, f. I; p. 100; Wiedemann, Auss.
Zw. I, 117, 8 (translation of Palisot's description); O. Sacken
Prodrome etc. II, 456 *female;* the *male* is described in the
Supplement, 559. — South Carolina; Georgia; Florida.

Tabanus fumipennis Wiedemann, Auss. Zw. I, 119, 11 (!) *Male.*

* **sagax** O. Sacken, Prodrome etc. II, 452. — Illinois, Minnesota.

* **stygius** Say, Journ. Acad. Phil. III, 33, 3; Compl. Wr. 54; Wiedemann,
Auss. Zw. I, 131, 31 (!); O. Sacken, Prodrome etc. II, 454. —
Middle and Southern States.

*sulcifrons Macquart, Dipt. Exot. 5e Suppl. 33, 127 (!) — Baltimore (Macq.). ([83]).

   Tabanus tectus O. Sacken, Prodrome etc. II, 436. — Pennsylvania.
*tener O. Sacken, Prodrome etc. II, 440. — Georgia, Florida.
   (?) *Tabanus unicolor* Macquart, Dipt. Exot. 2e Suppl. 22, 107. — Carolina. ([85]).
*trijunctus Walker, List, etc. V, 182; O. Sacken, Prodrome etc. II, 432. — Florida.
*trimaculatus Palisot-Beauvois, Dipt. Tab. I, f. 5; Wiedemann, Auss. Zw. I, 137, 40 (transl. of Palisot's description); ibid. 132, 33, (Wiedemann's own description, doubtfully identified with Palisot's); Macquart, Dipt. Exot. I, 1, 142; O. Sacken, Prodrome etc. 439. — Middle and Southern States ; Illinois, Kansas.
   Tabanus quinquelineatus Macquart, Hist. Nat. Dipt. I, 200, 11.
*turbidus Wiedemann, Auss. Zw. I, 124, 20 (!); O. Sacken, Prodrome etc. II, 430. — Georgia, Kentucky. ([84]).
   (?) *Tabanus fusconervosus* Macquart, Dipt. Exot. I, 1, 147, 52 (no locality).
*variegatus Fabricius, Syst. Antl. 95, 10; Wiedemann, Dipt. Exot. I, 67, 11; Auss. Zw. I, 120, 13; O. Sacken, Prodrome etc. II, 437. — Middle States. ([86]).
*venustus O. Sacken, Prodrome etc. II, 444. — Northern Texas ; Kansas.
*vivax O. Sacken, Prodrome etc. II, 446. — Trenton Falls, New York; Maine.
   (?) *Tabanus marginalis* Wiedemann, Auss. Zw. I, 166, 84. ([87]).
*Wiedemanni O. Sacken, Prodrome etc. II, 455; Supplem. 559. — Florida; Georgia; Cumberland Gap, Ky.
   Tabanus ater Wiedemann (non Palisot-Beauvois), Auss. Zw. I, 136, 39 (ex parte; non Dipt. Exot.).

*punctifer O. Sacken, Prodrome etc. II, 453; Western Diptera, 220. — Colorado Mts.; Yellowstone; Utah; Sonora; California.
*aegrotus O. Sacken, Western Diptera etc., 219. — California (Marin Co.).

*albiscutellatus Macquart, Dipt. Exot. 4e Suppl. 34, 107, Tab. II, f. 9. — Mexico.
*albonotatus Bellardi, Saggio, etc. I, 56; Tab. II, f. 5. — Mexico; Tampico.
alteripennis Walker, Trans. Ent. Soc. N. Ser. V, 274. — Mexico.
aurantiacus Bellardi, Saggio, etc. I, 67; Tab. II, f. 9. — Mexico.
Bigoti Bellardi, Saggio, etc. I, 59. — Mexico.
   Tabanus apicalis Macquart, Dipt. Exot. 2e Suppl. 20. [Bellardi].
bipartitus Walker, List, etc. I, 158. — Honduras.
caliginosus Bellardi, Saggio, etc. I, 68, Tab. II, f. 10. — Mexico.
carneus Bellardi, Saggio, etc. I, 62. — Mexico.
circumfusus Wiedemann, Auss. Zw. II, 624, 21. — Mexico.
commixtus Walker, Trans. Ent. Soc N. Ser. V. 273. — Mexico.
completus Walker, List, etc. I, 185. — St. Thomas.

De filippii Bellardi, Saggio, etc. I, 57. -- Mexico.
dorsifer Walker, Trans. Ent. Soc. N. Ser. V, 273. -- Mexico.
ferrifer Walker, Dipt. Saund. I, 30. — West Indies.
lucidulus Walker, List, etc. I, 188. — Jamaica.
luteo-flavus Bellardi, Saggio, etc. I, 60. — Mexico.
longiappendiculatus Macquart, Dipt. Exot. 5e Suppl. 32, 125, —
    Honduras.
obliquus Walker, Dipt. Saund. I, 28. — West Indies.
propinquus Bellardi, Saggio, etc. I, 65. — Mexico.
purus Walker, Trans. Ent. Soc. N. Ser. V, 274. — Mexico.
quinquevittatus Wiedemann, Dipt. Exot. I, 84, 39; Auss. Zw. I, 173,
    93; Bellardi, Saggio, etc. I, 65. — Mexico. [88].
oculus Walker, List, etc. I, 157. — Honduras, Columbia.
parallelus Walker, List, etc. I, 187. — West Indies.
parvidentatus Macquart, Dipt. Exot. I, 1, 142, 40; Walker, List, etc.
    V, 189. — West Indies.
rubescens Bellardi, Saggio, etc. App. 15. — Mexico.
rufiventris Macquart, Dipt. Exot. I, 1, 141, 39; Walker, List, etc. I,
    180; Bigot, R. de la Sagra, 798. — Cuba, Jamaica.
Sallei Bellardi, Saggio, etc. I, 61; Tab. II, f. 7. — Mexico.
stigma Fabricius, Syst. Antl. 104, 50; Wiedemann, Dipt. Exot. I, 92,
    53; Auss. Zw. I, 180, 104. — South America and St. Thomas
    (Wied.).
subsimilis Bellardi, Saggio, etc. 66. — Mexico.
subtilis Bellardi, Saggio, etc. App. 14; f. 9. — Mexico.
subruber Bellardi, Saggio, etc. I, 55. — Mexico.
    Tabanus ruber Macquart, Dipt. Exot. 1er Suppl. 42, 87 (change of
    name by Bellardi).
Sumichrasti Bellardi, Saggio, etc. I, 56. — Mexico.
tinctus Walker, Dipt. Saund. 29. — West Indies.
trilineatus Latreille, Humb. et Bompl. Rec. d'Obs. de Zool. fasc. X,
    116—117; Tab. XI, f. 6; Wiedemann, Dipt. Exot. I, 84; Auss.
    Zw. I, 168; Bellardi, Saggio, etc. I, 63. — Mexico.
Truquii Bellardi, Saggio, etc. 64; Tab. II, f. 6. — Mexico.

Observation. The following species, the descriptions of which are unrecognizable, have not been included in the above lists:

Linné: *Tabanus calens*, System. Nat. II, 1000, 6.
Palisot-Beauvois: *T. ferrugineus, nebulosus, pallidus, palpinus.*
Macquart: *Tabanus nanus* Dipt. Exot. Suppl. 1, 42, 88. — Texas. The
    name is preoccupied by Wiedemann for an african species. About the
    possible synonymy compare my Prodrome II, 445.
Tabanus dorsonotatus Dipt. Exot. 2e Suppl. 22, 106. — Carolina. In Mr.
    Bigot's collection I found a *Tab. dorsomaculatus* from Carolina, with a
    label in Macquart's handwriting, which I take to be this species. It is
    an unrecognizable specimen, which has evidently been mouldy and washed
    with some liquid afterwards. The name be better dropped.
Tabanus Novae Scotiae Dipt. Exot. 2e Suppl. 24, 110. In Mr. Bigot's collec-
    tion; the type is a female, not unlike a small *T. Actaeon*, the abdominal
    triangles however have a golden-yellow pubescence.
Walker: *T. comes* List, etc. V, 173. (Synon. *Tab. inscitus* List, etc. I, 172.) —
    British Possessions.

T. confusus, List, etc. I, 147. — Georgia.
T. conterminus Dipt. Saund. 24. — United States.
T. derivatus List, etc. I, 151. — North America.
T. duplex List, etc. V, 173 (Synom. T. imitans, List, etc. I, 173. — Huds. B. Terr.
T. frontalis List, etc. I, 172. — Nova Scotia.
T. fulvofrater List, etc. I, 181. — Illinois.
T. incisus Dipt. Saund. 26. — Cap. Breton.
T. intermedius List, etc. I, 173 — Huds. B. Terr.
T. leucomelas List, etc., I, 175. — Georgia.
T. mutatus Dipt. Saund. I, 23. — United Staates.
T. patulus List, etc. I, 175. — Georgia.
T. proximus List, etc. I, 147. — Florida.
T. rufofrater Dipt. Saund. I, 26. — Georgia.
T. scitus List, etc. I, 181. —

Some remarks about these species will be found in O. Sacken, Prodrome etc. II, 472—474. In the notes, which I took at the Brit. Mus. I remarked that *Tab. patulus* and *derivatus* are unknown to me.

Mr. Walker's identifications of the species of other authors are very often incorrect, and hence the comparisons to such species, occurring in his descriptions, are not to be relied on. Thus *T. b. melanocerus* Wied., *bicolor* Wied., *abdominalis* Fabr. etc. were incorrectly identified by him in the Brit. Museum.

Atylotus.

O. Sacken, Prodrome etc. II, 426, 1876; definition amended in the Western Diptera, 215.

*bicolor Wiedemann, Dipt. Exot. I, 96, 58 *(Tabanus)*; Auss. Zw. I, 118, 115 (♂) *id.*; O. Sacken, Prodrome etc. II, 460. — New York; Pennsylvania; Illinois; Canada.
Tabanus ruficeps Macquart, Dipt. Exot. 5e Suppl. 35, 130 *male* [!]
Tabanus fulvescens Walker, List, etc. I, 171; O. Sacken, Prodrome etc. 460. — Massachusetts; Canada. [89]).

*insuetus O. Sacken, Western Diptera etc., 219. — Webber Lake; Sierra Nevada; Cal.

(?) **Craverii** Bellardi, Saggio, etc. I, 60 *(Tabanus)*. — Mexico. [90]).

FAMILY LEPTIDAE.

SECTION I. PSAMMORYCTERINA. [91].

Triptotricha.

Loew, Centur. X, 15; id. Berl. Ent. Z. 1874, 381, note.

*fasciventris Loew, Berl. Ent. Z., 1874, 380. — Pennsylvania.
*rufithorax Say, J. Acad. Phil. III, 36, 5; Compl. Wr. II, 56 *(Leptis)*; Wiedemann, Auss. Zw. I, 223 *(id.)*. — Pennsylvania; New York; Kentucky.

*discolor Loew, Berl. Ent. Z. 1874, 379. — San Francisco.
*lauta Loew, Centur. X, 15; compare also Berl. Ent. Z. 1874, 382. — California.

Pheneus.

Walker, Dipt. Saund. 155; 1856.

tibialis Walker, Dipt. Saund. 156, Tab. IV, f. 3. — Jamaica.

Observation. Mr. Walker refers this genus to the Asilidae. I place it here on the authority of Mr. Loew (in litt.).

SECTION II. LEPTINA.

Chrysopila.

Macquart, Dipt. du nord de la France; 1827.

*basilaris Say, J. Acad. Phil. III, 36, 4; Compl. Wr. II, 55 (Leptis); Wiedemann, Auss. Zw. I, 228, 16 (id.). — Pennsylvania.

*fasciata Say, J. Acad. Phil. III, 37, 7; Amer. Entom. Tab. XIII (Leptis); Compl. Wr. I, 28; Wiedemann, Auss. Zw. I, 225, 9 (id.) — Middle and Northern States.
Leptis par Walker, List, etc. I, 215.

*foeda Loew, Centur. I, 18. — Illinois.

*modesta Loew, Centur. X, 14. — Texas.

*ornata Say, J. Acad. Phil. III, 34, 1; Compl. Wr. II, 54; Amer. Ent. Tab. XIII (Leptis); Wiedemann, Auss. Zw. I, 221, 1 (id.); Walker, List, etc. I, 213 (re-described, the identification being doubtful). — Atlantic States (common).

propinqua Walker, List, etc. I, 215. — Trenton Falls.
Leptis simillima Walker, List, etc. I, 215. — Trenton Falls (♂; synonymy by Walker with a doubt).

*proxima Walker, List, etc. I, 214. — Northern States and British Possessions.

*quadrata Say, J. Acad. Phil. III, 35, 3; Compl. Wr. II, 55 (Leptis); Wiedemann, Auss. Zw. I, 226, 11 (id.). — North America (common).
Leptis fumipennis Say, J. Acad. Phil. III, 37, 6; Compl. Wr. II, 56; Wiedemann, Auss. Zw. I, 227, 12 (id.); Walker, List, etc. I, 217 (♂).
Leptis reflexa Walker, List, etc. I, 216 (♀).
Chrysopila dispar v. d. Wulp, Tijdschr. v. Ent. 2ᵈ Ser. II, 143; Tab. IV, f. 6—11.

*rotundipennis Loew, Centur. I, 19. — Georgia.

Servillei Guérin, Iconogr. etc., Texte III, 541; Tab. XCVI, f. 3 (Leptis). — North America. [92]

*thoracica Fabricius, System. Antl. 70, 4 (Leptis); Wiedemann, Auss. Zw. I, 222, 2 (id.); Macquart, Dipt. Exot. II, 1, 32; Tab. III, bis, f. 3. — Eastern North America common.

*velutina Loew, Centur. I, 17. — Illinois, Kentucky.

*humilis Loew, Berl. Ent. Z. 1874, 379; O. Sacken, Western Diptera, 223 (translation of Dr. Loew's description). — San Francisco.

basalis Walker, Trans. Ent. Soc. N. Ser. V, 285. — Mexico.

*ludens Loew, Wien. Entom. Mon. V, 34. — Cuba.
mexicana Bellardi, Saggio, etc. II, 96. — Mexico.
nigra Bellardi, Saggio, etc. App. 27. — Mexico.
trifasciata Walker, Trans. Ent. Soc. N. Ser. V, 284. — Mexico.

Leptis.
Fabricius, System. Antl. 69, 1805.

*albicornis Say, J. Acad. Phil. III, 38, 9; Compl. Wr. II, 56; Amer.
  Entom. Tab. XIII; Compl. Wr. I, 27; Wiedemann, Auss. Zw. I,
  223. — Pennsylvania; South Carolina (M. C. Z.).
Boscii Macquart, Dipt. Exot. II, 1, 30, 2. — Carolina.
*dimidiata Loew, Centur. III, 17. — Sitka.
*hirta Loew, Centur. I, 21. — Illinois.
intermedia Walker, List, etc. I, 212 (Rhagio). — Huds. B. Terr.
*mystacea Macquart, Dipt. Exot. II, 1, 30, 1; Tab. III, bis, f. 2;
  Walker, List, etc. I, 212 and IV, 1153 (Rhagio), re-described, the
  identification being doubtful. — Eastern North America (not rare).
*ochracea Loew, Centur. II, 3. — New York.
*punctipennis Say, J. Acad. Phil. III, 34, 2; Compl. Wr. II, 55;
  Wiedemann, Auss. Zw. I, 227. — Middle and Northern States
  (common.).
  Atherix filia Walker, List, etc. I, 219. (⁹³).
*plumbea Say, J. Acad. Phil. III, 39, 10; Compl. Wr. II, 56; Wiede-
  mann, Auss. Zw. I, 228. — Middle States.
  Leptis griseola v. d. Wulp, Tijdschr. v. Ent. 2d Ser. II, 142, Tab. IV,
  f. 5. [Loew, Zeitschr. f. Ges. Naturw. 1870, 115].
*terminalis Loew, Centur. 1, 20. — New York.
*scapularis Loew, Centur. I, 22. — Illinois, New York, Distr. Columbia.
vertebrata Say, J. Acad. Phil. III, 38, 8, Amer. Ent. Tab. XIII;
  Compl. Wr. I, 27. — Florida.

*costata Loew, Centur II, 4; O. Sacken, Western Dipt., 223. — California.
*incisa Loew, Centur. X, 16; O. Sacken, Western Dipt., 223. — California.

bitaeniata Bellardi, Saggio, etc. App. 26, f. 14. — Mexico.
cinerea Bellardi, Saggio etc. II, 95. — Mexico. (⁹⁴).
polytaeniata Bellardi, Saggio, etc., App. 27, 13. — Mexico.

Ptiolina.
Zetterstedt, Dipt. Scand. I, 226; 1843; compare also Frauenfeld, Verh.
Z. B. Ges. 1867, 495.

fasciata Loew, Centur. IX, 65. — British North America.
*majuscula Loew, Centur. IX, 66. — British North America.

Atherix.
Meigen, Illiger's Magaz. II, 271; 1803.

*variegata Walker, List, etc. I, 128. — Northern States and British
  Possessions.

* **vidua** Walker, List, etc. IV, 1153. — Huds. B. Terr.
* **varicornis** Loew, Centur. X, 13. — California.

latipennis Bellardi, Saggio, etc. II, 93. — Mexico.
longipes Bellardi, Saggio, etc. II, 94; Tab. II, f. 17. — Mexico.

Spania.

Meigen, System. Beschr. VI, 335; 1830.
edeta Walker, List, etc. III, 489. — Huds. B. Terr. ([95]).

Glutops.

E. Burgess, Proc. Boston Soc. N. Sc. 1878, 320, with figures. ([96]).
singularis Burgess, l. c. — Springfield, Mass.

FAMILY ASILIDAE. ([97])

SECTION I. DASYPOGONINA.

DIVISION A. — FRONT TIBIAE WITHOUT SPURS.

Leptogaster.

Meigen, Illiger's Magaz. 1803; *Gonypes* Latr. 1804.

* **badius** Loew, Centur. II, 6. — Illinois.
* **brevicornis** Loew, Centur. X, 23. — Texas.
  carolinensis Schiner, Verh. Z. B. Ges. 1866, 696. — Carolina.
  Gonypes nitidus Macquart, Dipt. Exot. I, 2, 155. ([98]).
* **eudicranus** Loew, Berl. Ent. Z. 1874, 353. — Texas.
* **favillaceus** Loew, Centur. II, 12. — Connecticut.
* **flavipes** Loew, Centur. II, 15. — Atlantic States (not rare).
    (?) *Leptogaster flavicornis* v. d. Wulp, Tijdschr. v. Ent. 2d Ser II, 136; Wisconsin. [Loew in Zeitschr. für ges. Naturw. XXXVI, 120.]
* **incisularis** Loew, Centur. II, 11. — Illinois.
* **histrio** Wiedemann, Auss. Zw. I, 535, 5. — Pennsylvania.
    Leptogaster annulatus Say, J. Acad. Phil. III, 75, 1; Compl. Wr. II, 68. [Change of name by Wiedemann.]
* **murinus** Loew, Centur. II, 9. — Nebraska.
  ochraceus Schiner, Verh. Z. B. Ges. 1867, 359. — Pennsylvania.
* **pictipes** Loew, Centur. II, 7. — Illinois.
* **tenuipes** Loew, Centur. II, 14. — District Columbia.
* **testaceus** Loew, Centur. II, 10. — New York.
* **varipes** Loew, Centur. II, 8. — Distr. Columbia.

cubensis Bigot, R. de la Sagra's Hist. etc. 792 *(Gonypes)*. — Cuba.
fervens Wiedemann, Auss. Zw. II, 646. — Mexico.
* **obscuripes** Loew, Centur. II, 13. — Cuba.
    Leptogaster Ramoni Jaennicke, Neue Exot. Dipt. 46. [Loew].
Truquii Bellardi, Saggio, etc. II, 87; Tab. II, f 18. — Mexico.

8

Ceraturgus.

Wiedemann, Analecta, 12, 1824; Auss. Zw. I, 414; 1828.

aurulentus Fabricius, System. Antl. 166, 11 *(Dasypogon)*; Wiedemann, Dipt. Exot. I, 228, 26 *(id.)*; Analecta etc. 12; Auss Zw. I,
414, 1, Tab. V, f. 5; Macquart, Hist. Nat. Dipt. I, 239, 1;
Tab. VII, f. 4 *(head)*. — New York (Fab.).

*cruciatus Say, J. Acad. Phil. III, 52, 6; Compl. Wr. II, 66 *(Dasy-
pogon)*, *female;* Wiedemann, Auss. Zw. I, 381, 24 *(id.)*. — Arkansas
(Say); New York.
   Ceraturgus fasciatus Walker, List, etc. II, 367, *male* [Loew Beschr.
   Eur. Dipt. III, 124].
   Dasypogon cornutus Wiedemann, Auss. Zw. I, 382 (Without loca-
   lity); I saw the type in Vienna.

*lobicornis O. Sacken, Western Diptera, 237. — Idaho, California.

dimidiatus Macquart, Dipt. Exot. 2e Suppl. 35, 56 *(Dasypogon)*;
Walker, List, etc. VI, 428; Bellardi, Saggio, etc. II, 61 *(Ceraturgus)*. — Mexico.
ruflpennis Macquart, Dipt. Exot. 2e Suppl. 32, 2; Bellardi, Saggio, etc.
II, 59. — Mexico.
vitripennis Bellardi, Saggio, etc. II, 60. — Mexico (can hardly be a
Ceraturgus).

Observation. For *Cerat. niger* Macquart see *Taracticus.* ([99]).

Dioctria.

Meigen, Illiger's Magaz.; 1803.

*Albius Walker, List, etc. II, 301. — New York, Massachusetts, etc.;
California (? see O. Sacken, Western Diptera, 287).

*resplendens Loew, Centur. X, 21. — California.
*pusio O. Sacken, Western Diptera, 238. — California.

Echthodopa.

Loew, Centur. VII, 27, 1866; Compare also
Loew's Beschr. Eur. Dipt. II, 78, *observ.*

*formosa Loew, Centur. X, 22. — Pennsylvania.
*pubera Loew, Centur. VII, 27 — Nebraska.

Plesiomma.

Macquart, Dipt. Exot. I, 2, 54; 1838.

*unicolor Loew, Centur. VII, 35. — Pecos River, Western Texas and
New Mexico.
*funesta Loew, Wien. Ent. Mon. V, 35; Centur. VII, 31. — Cuba.
   Dioctria lugubris Jaennicke, Neue Exot. Dipt. 48. — Cuba (Loew
   in litt.).
*indecora Loew, Centur. VII, 33. — Cuba.

*leptogastra Loew, Centur. VII, 32. — Cuba.
*lineata Fabricius, Spec. Ins. II, 465, 28; En+om. System. 386, 47
    (*Asilus*)! System. Antl. 167, 13; Wiedemann, Dipt. Exot. I, 221,
    12 (*Dasypogon*); Auss. Zw. I, 385, 29 (*id.*); (?) Schiner, Verh.
    Zool. Bot. Ges. 1867, 374. — West Indies (St. Thomas; Loew
    in litt.).
longiventris Schiner, Verh. Zool. Bot. Ges. 1867, 375. — Cuba.
macra Loew, Wien. Ent. Monatschr. V, 35; Centur. VII, 34. – Cuba.

Microstylum.
Macquart, Dipt. Exot. I, 2, 26; 1838.

*galactodes Loew, Centur. VII, 44. — Pecos River, Western Texas;
    Kansas.
*morosum Loew, Centur. X, 27. — Dallas, Texas. ([100]).

Ospriocerus.
Loew, Centur. VII, 51, 1866.

*Aeacus Wiedemann, Auss. Zw. I, 390 (*Dasypogon*); O. Sacken, Western
    Diptera, 290. — Nebraska; Colorado.
    Dasypogon abdominalis Say, Long's Exped. App. 375; Compl.
    Wr. I, 255 [Change of name by Wied.].
    (?) *Dasypogon spathulatus* Bellardi, Saggio, etc. II, 82; Tab. I,
    f. 9; [Loew, Centur. VII, 51]. — Mexico.
*eutrophus Loew, Berl. Ent. Z. 1874, 355. — Texas; Kansas.
*Rhadamanthus Loew, Centur. VII, 52. — Pecos River, Western Texas.
*Minos O. Sacken, Western Diptera, 291. — Colorado.
*Aeacides Loew, Centur. VII, 51. — California.

Ablautatus.
Loew, Berl. Ent. Z. 1874, 377; O. Sacken, Western Diptera, 289.
Ablautus, Loew, Centur. VII, 63, 1866.

*trifarius Loew, Centur. VII, 63. — California.
*mimus O. Sacken, Western Diptera, 289. — San Bernardino, Cal.

Stenopogon.
Loew, Linn. Entom. II, 453; 1847.

*consanguineus Loew, Centur. VII, 48. — Nebraska.
*inquinatus Loew, Centur. VII, 47. — Nebraska.
*latipennis Loew, Centur. VII, 49. — Pecos River, Western Texas
    („May 28").
*longulus Loew, Centur. VII, 50. — Pecos River, Texas.
*modestus Loew, Centur. VII, 46. — Red River of the North.
    subulatus Wiedemann, Auss. Zw. I, 375, 14 (*Dasypogon*); Walker,
    List, etc. I, 311 and VI, 422 (*id.*). — Georgia.

*breviusculus Loew, Centur. X, 28 — California.
*gratus Loew, Centur. X, 31. — California.

Stenopogon univittatus Loew, Centur. X, 29, ♀ [Synonymy suggested by Mr. Loew himself in Berl. Ent Z. 1874, 358].
*obscuriventris Loew, Centur. X, 30. — California.
*morosus Loew, Berl. Ent. Z. 1874, 356. — Sierra Nevada, Cal.
*californiae Walker, List, etc. II, 322 *(Dasypogon)*. — California.

Scleropogon.
Loew, Centur. VII, 45; 1866.

ochraceus v. d. Wulp, Tijdschr. Ent. 2ᵈ Ser. V, 212; Tab. IX, f. 6 *(Stenopogon)*. — North America. [101].
*picticornis Loew, Centur. VII, 45. — California.
*helvolus Loew, Berl. Ent. Z. 1874, 355. — Texas.

Truquii Bellardi, Saggio, etc. II, 76; Tab. I, f. 10 *(Stenopogon ?)*. — Mexico.

Sphageus.
Loew, Centur. VII, 55; 1866.

*chalcoproctus Loew, Centur. VII, 55. — Cuba.

Dicolonus.
Loew, Centur. VII, 56; 1866.

*simplex Loew, Centur. VII, 56. — California.

Archilestris.
Loew, Berl. Ent. Z. 1874, 377; *Archilestes,* Schiner, Verh. Z. B. Ges. 1866, 672; id. Novara, 168. [102].
*magnificus Walker, List, etc. VI, 427 *(Dasypogon);* Bellardi, Saggio, etc. II, 79; Tab. I, f. 11 *(Microstylum)*. — Mexico.

Dizonias.
Loew, Centur. VII, 53; 1866.

*bicinctus Loew, Centur. VII, 54. — Pecos River, Western Texas; Dallas, Texas; Florida.
Dasypogon tristis Walker, Dipt. Saund. 93. [103]. — United States.
Dasypogon quadrimaculatus Bellardi, Saggio, etc. II, 80; Tab. I. f. 8. — Mexico.
*phoenicurus Loew, Centur. VII, 53. — Tamaulipas, Mexico.
Lucasi Bellardi, Saggio, etc. II, 81; Tab I, f. 7 *(Dasypogon)*. — Mexico.

Callinicus.
Loew, Centur. X, 32; 1872.

*calcaneus Loew, Centur. X, 32. — Marin and Sonoma Co., California.

Anisopogon.
Loew, Berl. Ent. Z. 1874, 377; *Heteropogon* Loew, Linn. Ent. II, 488, 1847.
*gibbus Loew, Centur. VII, 58 *(Heteropogon)*. — Pennsylvania.

Dasypogon macerinus Walker, List, etc. II, 356. — Trenton Falls.
*lautus Loew, Centur. X, 34 (*Heteropogon*). — Texas.
*phoenicurus Loew, Centur. X, 33 (*Heteropogon*). — Texas.

humilis Bellardi, Saggio, etc. II, 77 (*Heteropogon*). — Mexico.

Cyrtopogon.
Loew, Linn. Ent. II, 516; 1847. ([104]).

*bimacula Walker, Dipt. Saund., 102, Tab 4, f. 1; (*Luarmostus* n. gen.).
Male. — Huds. B. Terr.; White Mts., N. H.
Dasypogon melanopleurus Loew, Centur. VII, 61 [Loew, Berl. Ent.
Z. 1874, 365, Note 2d.]. *Female.*
*chrysopogon Loew, Centur. VII, 59. — New England and Canada.
(?) *Dasypogon Falto* Walker, List, etc. II, 355. — Nova Scotia.
*Lutatius Walker, List, etc. II, 357 (*Dasypogon*). — Nova Scotia
(Walk.); Western New York; Massachusetts. ([105]).
*lyratus nov. sp., see the note ([105]). — Catskill Mts., New York;
White Mts., N. H.
*marginalis Loew, Centur. VII, 60; compare also Berl. Ent. Z. 1874,
365, Note 2d. — Massachusetts, Canada.

*aurifex O. Sacken, Western Dipt., 300. — Sierra Nevada, Cal.
*callipedilus Loew, Berl. Ent. Z. 1874, 358; O. Sacken, Western
Diptera, 296. — Sierra Nevada, Cal.
*cerussatus O. Sacken, Western Diptera, 308. — Sonoma Co., Cal.
*cretaceus O. Sacken, Western Diptera, 302. — Sierra Nevada, Cal.
*cymbalista O. Sacken, Western Diptera, 297. Sierra Nevada, Cal.
*evidens O. Sacken, Western Diptera, 306. — Sierra Nevada, Cal.
*leucozonus Loew, Berl. Ent. Z. 1874, 364; O. Sacken, Western
Diptera, 299. — Sierra Nevada, Cal.
*longimanus Loew, Berl. Ent. Z. 1874, 360; O. Sacken, Western
Diptera, 303. — Marin Co., Cal.
*montanus Loew, Berl. Ent. Z. 1874, 362; O. Sacken, Western Diptera,
298. — Sierra Nevada, Cal.
*nugator O. Sacken, Western Diptera, 307. — Sierra Nevada, Cal.
(?)*nebulo O. Sacken, Western Diptera, 309. — Sierra Nevada, Cal.
*plausor O. Sacken, Western Diptera, 297. — New Mexico; Utah;
Idaho.
*profusus O. Sacken, Western Diptera, 305. — Northern New Mexico
*princeps O. Sacken, Western Diptera, 302. — Sierra Nevada, Cal.
*positivus O. Sacken, Western Diptera, 307. — Sierra Nevada, Cal.
*rattus O. Sacken, Western Diptera, 308. — Sierra Nevada, Cal.
*rejectus O. Sacken, Western Diptera, 307. — Sierra Nevada, Cal.
*sudator O. Sacken, Western Diptera, 307. — Sierra Nevada, Cal.

Pycnopogon.
Loew, Linn. Ent. II, 526; 1847.

*cirrhatus O. Sacken, Western Diptera, 293. — Mariposa Co., Cal.

Holopogon.

Loew, Linn. Ent. II, 473; 1847.

*guttula Wiedemann, Dipt. Exot. I, 228, 27 *(Dasypogon);* Auss. Zw.
I, 411, 74 *(id.);* Walker, List, etc. II, 355 (description given, the
identification having appeared doubtful). — Atlantic States.

philadelphicus Schiner, Verh. Zool. Bot. Ges. 1867, 360; compare
also Loew, Berl. Ent. Z. 1874, 367, note. — Philadelphia.

*phaeonotus Loew, Berl. Ent. Z. 1874, 366. — Texas.

*seniculus Loew, Centur. VII, 62. — Nebraska.

Daulopogon.

Loew, Berl. Ent. Z. 1874, 377; *Lasiopogon* Loew, Linn. Ent. II, 508; 1847.

*opaculus Loew, Berl. Ent. Z. 1874, 367. — Illinois. ·

*tetragrammus Loew, Berl. Ent. Z. 1874, 368. — Canada.

*arenicola O. Sacken, Western Diptera, 310. — San Francisco, Cal.

*bivittatus Loew, Centur. VII, 57 *(Lasiopogon;* compare also Loew,
Berl. Ent. Z. 1874, 370, note). — California.

Psilocurus.

Loew, Berl. Ent. Z. 1874, 373, note.

*nudiusculus Loew, Berl. Ent. Z. 1874, 370. — Texas.

Stichopogon.

Loew, Linn. Ent. II, 500; 1847.

*argenteus Say, J. Acad. Phil. III, 51, 4; Compl. Wr. II, 65 *(Dasy-*
pogon); Wiedemann Auss. Zw. I, 409, 69 *(id.).* — Atlantic States
(not rare on sea-beaches).

*trifasciatus Say, J. Acad. Phil. III, 51, 3; Compl. Wr. II, 64 *(Dasy-*
pogon). — Atlantic States; common.

    Thereva plagiata Harris, Cat. Ins. Mass. Walker, List, etc. I, 223
(description given). (!)

candidus Macquart, Dipt. Exot. Suppl. I, 67, 48 *(Dasypogon);* Bellardi,
Saggio, etc. II, 78. — Vera Cruz, Mexico.

    Dasypogon gelascens Walker, Trans. Ent. Soc. N. Ser. V, 277
[Bellardi].

    Dasypogon fasciventris Macquart, Dipt. Exot. 4e Suppl. 69, 75;
Tab. VI, f. 13. [Bellardi, l. c. 79, states on Bigot's authority
that this is only a variety of *S. candidus* Macq. The original
specimen is in M. Bigot's collection.]

Holcocephala.

Jaennicke, Neue Exot. Dipt. 51, 1867 (instead of *Discocephala* Macquart Dipt. Exot. I, 2, 50, 1838, which is preoccupied. Loew adopts
this change in Berl. Ent. Z. 1874, 377).

*abdominalis Say, J. Acad. Phil. III, 50, 2; Compl. Wr. II, 64, *(Dasy-*

pogon); Wiedemann Auss. Zw. I, 412, 75 *(id.)*. — Atlantic States (not rare in damp situations).
Discocephala rufiventris Macquart, Dipt. Exot. I, 2, 50, 1; Tab. IV, f. 2. — Carolina; Brazil.
Dasypogon Aeta Walker, List, etc. II, 362.
Dasypogon laticeps v. d. Wulp, Tijlschr. v. Ent. 2d Ser. II, 137; Tab. III, 10—16. [Loew, Z. f. Ges. Naturw. Vol. XXXVI, 115.]
*calva Loew, Centur. X, 35 *(Discocephala)*. — Texas (Loew); Western New York (M. C. Z.).

affinis Bellardi, Saggio, etc. II, 86, Tab. I, 13 *(Discocephala)*. — Mexico.
deltoidèa Bellardi, Saggio, etc. II, 85; Tab. I, f. 12 *(Discocephala)*. — Mexico.
divisa Walker, Trans. Ent. Soc. N. Ser. V, 279 *(Discocephala)*. — Mexico.
interlineata Walker l. c. 279 *(Discocephala)*. — Mexico.
longipennis Bellardi, Saggio, etc. II, 86; Tab. I, f. 14 *(Discocephala)*. — Mexico.
minuta Bellardi l. c. 83 *(Discocephala)*. — Mexico.
nitida Wiedemann, Auss. Zw. II, 603 *(Dasypogon)*; Walker, List, etc. VI, 503 *(Dasypogon)*; Bellardi l. c. 84 *(Discocephala)*. — Mexico.

DIVISION B. FRONT TIBIAE WITH A SPUR AT THE TIP.
Nicocles.
Jaennicke, Neue Exot. Dipt. 47, 1867; *Pygostolus* Loew, Centur. VII, 28; this name as preoccupied, is given up by Loew, Centur. X, 24, Nota.
*pictus Loew, Centur. VII, 30 *(Pygostolus)*. — Distr. Columbia.
Discocephala Amastris Walker, List, etc II, 362. — Georgia.
*politus Say, J. Acad. Phil. III, 52, 5; Compl. Wr. II, 65 *(Dasypogon)* *female*; Wiedemann, Auss. Zw. 1, 405, 63 *(id.)*; — Walker. List, etc. VI, 421 *(id.)*. — Pennsylvania, Maryland (Say); Massachusetts (O. S.).
Pygostolus argentifer Loew, Centur. VII, 28; *male*. [Loew *in litt.*]

*aemulator Loew, Centur. X, 25 *(Pygostolus)*. — California.
*dives Loew, Centur. VII, 29 *(Pygostolus)*. — California (Sonoma Co.).

analis Jaennicke, Neue Exot. Dipt. 47; Tab. I, 13. — Mexico.

Clavator.
Philippi, Verh. Zool. Bot. Ges. 1865, 699; Tab 26, f. 31.
O. Sacken, Western Dipt, 291.
*sabulonum O. Sacken, Western Dipt., 292. — San Bernardino, Cal

Blacodes.
Loew, Berl. Ent. Z. 1874, 377; *Blax*, Centur. X, 24; 1872.
*bellus Loew, Centur. X, 24 *(Blax)*. — Texas.

Taracticus.

Loew, Centur. Vol. II. 240, Nota; 1872.

*octopunctatus Say, J. Acad. Phil. III, 49; Compl. Wr. II, 63 *(Dioctria)*; Wiedemann, Auss. Zw. I, 365 *(id).*; Walker, List, etc. VI, 387 *(id.)*. — Atlantic States.
niger Macquart, Dipt. Exot. I, 2, 25; Tab. II, f. 1 *(Ceraturgus)*. — North America (Macq.); Mexico (Walker, List, etc. VI, 378).

Diogmites.

Loew, Centur. VII, 36, 1866; *Deromyia* Philippi 1865 (?).([106]).

angustipennis Loew, Centur. VII, 41. — Kansas; Matamoras; Mexico.
*discolor Loew, Centur. VII, 37. — Pennsylvania.
  (?) *Dasypogon rufescens* Macquart, Hist Nat. Dipt. I, 295, 8; Walker, List, etc. VI, 426. — Philadelphia. ([107]).
*hypomelas Loew, Centur. VII, 42. — Pecos River, New Mexico.
*misellus Loew, Centur. VII, 39. — Distr. Columbia.
*platypterus Loew, Centur. VII, 36. — Illinois.
*symmachus Loew, Centur. X, 26. — Texas.
*umbrinus Loew, Centur. VII, 43. — New York, Massachusetts, Illinois.
  Dasypogon basalis Walker, Dipt. Saund., 95. — Atlantic States. ([108]).
  Dasypogon Herennius Walker, List, etc. II, 339. — Cincinnati.

*annulatus Bigot, R. de la Sagra, etc. 789; Tab. XX, f. 3 *(Senobasis)*. — Cuba. ([109]).
  Dasypogon secabilis Walker, Trans. Ent. Soc. N. Ser. V, 276; Bellardi, Saggio, etc. II, 63; Tab. I, f. 4 *(Saropogon?)*. — Mexico [Loew *in litt.*].
  Senobasis auricinctus Schiner, Verh. Zool. Bot. Ges. 1867, 371. — Surinam [Loew *in litt.*].
affinis Bellardi, Saggio, etc. II, 73 *(Saropogon)*. — Mexico.
bicolor Jaennicke, Neue Exot. Dipt. 49 *(Saropogon)*. — Panama,
Bigotii Bellardi, Saggio, etc. II, 70 *(Saropogon)*. — Mexico.
*bilineatus Loew, Centur. VII, 40. — Cuba.
brunneus Fabricius, Mant. Ins. II, 359, 20 *(Asilus)*; Entomol. System. IV, 382, 28 *(id.)*; System. Antl. 165, 9 *(Dasypogon)*; Wiedemann, Dipt. Exot. I, 219, 9 *(id.)*; Auss. Zw. I, 382 *(id.)*. Macquart, Dipt. Exot. I, 2, 34, 4 *(id)* ([110]). Bellardi, Saggio, etc. II, 67 *(Saropogon)*. — Cayenne (Fab); Mexico (Bellardi); Philadelphia (Macq.).
Craverii Bellardi, Saggio, etc. II, 68 *(Saropogon)*. — Mexico.
Cuantlensis Bellardi, Saggio, etc. II, 68 *(Saropogon)*. — Mexico.
dubius Bellardi l. c. 74 *(Saropogon)*. — Mexico.
goniostigma Bellardi, Saggio, etc. II, 65; Tab. I, f. 6 *(Saropogon)*. — Mexico.
Jalapensis Bellardi, Saggio, etc. II, 65; Tab. I, f. 5 *(Saropogon)*. — Mexico.
nigripes Bellardi, Saggio, etc. II, 75 *(Saropogon)*. — Mexico.
nigripennis Macquart, Dipt. Exot. 2e Suppl. 34, 55; Tab. I, f. 6 *(Dasypogon)*; Bellardi, Saggio, etc. II, 75 *(Saropogon)*. — Mexico.

pseudojalapensis Bellardi, Saggio, etc., App. 25 *(Dasypogon)*. — Mexico.

rubescens Bellardi, Saggio, etc. II, 71 *(Saropogon)*. — Mexico.

Sallei Bellardi, Saggio, etc. II, 70 *(Saropogon)*. — Mexico.

*ternatus Loew, Centur. VII, 38. — Cuba.

tricolor Bellardi, Saggio, 72 *(Saropogon)*. — Mexico. [Probably *Diogmites*, but not certain. Loew, *in litt.]*

virescens Bellardi, Saggio, 72 *(Saropogon)*. — Mexico.

*   *   *

Duillius Walker, List, etc. II, 340 *(Dasypogon)*. — Honduras. ([111]).

Saropogon.
Loew, Linn. Ent. II, 439; 1847.

*adustus Loew, Berl. Ent. Z. 1874, 375. — Texas.

*combustus Loew, l. c. 374. — Texas.

Lastaurus.
Loew, Bem. über die Fam. der Asiliden, Berlin 1851, 11.

anthracinus Loew, Bem. über die Fam. der Asiliden, 12. — Mexico. [Schiner (Verh. Z. B. Ges. 1867, 373) identifies this species with *Dasypogon lugubris* Macq. Dipt. Exot Suppl. 1, 64, from Surinam; whether correctly or not, the insufficiency of my materials does not enable me to decide. — Loew, *in litt.*]

Observation. For *Dasypogon sexfasciatus* Say and *Dasypogon albiceps* Macq. see the genus *Laphystia* (Laphrina).

The following species I do not know and cannot refer them to the new genera formed at the expense of Dasypogon in Meigen's and Wiedemann's sense:

Dasypogon angustus Macquart, Dipt. Exot. 3e Suppl. 20, 59; Tab. I, f. 11. — Haiti.

Dasypogon cepphicus Say, Journ. Acad. Phil. VI, 158; Compl. Wr. II, 354. — Mexico.

Dasypogon mexicanus Macquart, Dipt. Exot. 1er Suppl. 68, 49; Tab. VI, f. 10. — Mexico.

Dasypogon nigritarsis Macquart, Dipt. Exot. 1er Suppl. 68, 50. — Mexico.

Dasypogon parvus Bigot, R. de la Sagra, etc. 789; Tab. 20, f. 2. — Cuba. [Mr. Bigot told me that the original type has been accidentally destroyed in his collection.]

The occurrence of *Dasypogon teutonus* Linn. in North America seems very improbable, although Macquart, Dipt. Exot. 4e Suppl. pages 8 and 64, mentions it as received from Florida. Hitherto not a single Asilida, common to Europe and North America, has been recorded with certainty.

SECTION II.   LAPHRINA. (*).

Megapoda.
Macquart, Hist. Nat. Dipt. I, 228, 1834; Dipt. Exot. I, 2, 59.

cyaneiventris Macquart, Dipt. Exot. 1er Suppl. 71, 3; Tab. VII, f. 12. — Mexico.

---

*) In this and in the following Section (Asilina), I followed Schiner's views (in „die Wiedemann'schen Asiliden", Verh. Z. B. Ges. 1866, 649), whenever I had no opinion of my own. Schiner, Verh. Z. B. Ges. 1866, 652 gives an analytical table for determining the genera.

Atomosia.

Macquart, Dipt. Exot. I, 2, 73; 1838.

glabrata Say. J. Acad. Phil. III, 53, 2; Compl. Wr. II, 66 (Laphria). — Atlantic States.
*puella Wiedemann, Auss. Zw. I, 531 (Laphria). — Locality unknown to Wied. (North America, according to Schiner, Verh. Z. B. Ver. 1866, 706, top of second column). — Atlantic States.
Laphria pygmaea Macquart, Hist. Nat. Dipt. I, 287, 30. — Georgia.
(?) Laphria Echèmon Walker, List, etc. II, 386. — Ohio.
pusilla Macquart, Dipt. Exot. I, 2, 76, 6. — North America.
*rufipes Macquart, Dipt. Exot. 2e Suppl. 39, 9. — Philadelphia (Macq.).

Beckeri Jaennicke, Neue Exot. Dipt., 51. — Mexico.
(?) Bigoti Bellardi, Saggio, etc, II, 20. — Mexico (the query is Bellardi's).
*incisuralis Macquart, Dipt. Exot. I, 2, 76, 4; Tab. VII, f. 1; Bigot, in R. de la Sagra etc. 788. — Cuba.
Macquartii Bellardi, Saggio, etc. II, 20. — Mexico.
sericans Walker, Trans. Ent. Soc. N. Ser. V, 282. — Mexico.
similis Bigot, in R. de la Sagra etc., 788; Tab. XX, f. 4. — Cuba.
tibialis Macquart, Dipt. Exot. 1er Suppl. 76, 8. — Yucatan.

Cerotainia.

Schiner, Verh. Zool. Bot. Ges. 1866, 673; id. Novara, 170.

*macrocera Say, J. Acad. Phil. III, 73, 3; Compl. Wr. II, 67 (Laphria); Wiedemann, Auss. Zw. I, 531, 57 (id.). — Pennsylvania.

nigripennis Bellardi, Saggio, etc. II, 19. (Atomosia). — Mexico (placed in this genus by Schiner, Verh. Z. B. Ges. 1866, 706).

Dasyllis.

Loew, Bem. über die Fam. der Asiliden, 20; 1851.

*flavicollis Say, Long's Exped. App. 374, 2; Compl. Wr. I, 255 (Laphria); Wiedemann, Auss. Zw. I, 519, 34 (id.). — N. W. Territory (Say); Massachusetts (Harris, Catal.); Atlantic States.
Laphria melanopogon Wiedemann, Auss. Zw. I, 520, 36 ♀ [Synonymy suggested by Wiedemann and borne out by the type in Vienna].
*lata Macquart, Dipt Exot. 4e Suppl. 75 (Laphria). — Texas (Macq.); Louisiana. ([112]).
Mallophora analis Macquart, Dipt. Exot. 1er Suppl. 78, 20 (Synonymy and change of name by Macquart).
*posticata Say, Long's Exped. App. 374, 1; Compl. Wr. I, 255 (Laphria); Wiedemann, Auss. Zw. I, 518, 32 (id.); Macquart, Dipt. Exot. I, 2, 69, 17 (id.) — N. W. Territory (Say); Massachusetts (Harris Cat.). — Atlantic States.
*sacrator Walker, List, etc. II, 382 (Laphria). — Nova Scotia (Walk.); Quebec; White Mts., N H ; Catskill, New York

* **saffrana** Fabricius, System. Antl. 160, 18; *(Laphria)*; Wiedemann, Dipt. Exot. I, 234, 4 *(id.)*; Auss. Zw. I, 504, 9 *(id.)*. — Carolina (Fab.); Georgia (Wied.).
* **thoracica** Fabricius, Syst. Antl. 158, 10 *(Laphria;* in the *erratum* the name is changed for *L. fulvithorax)*; Wiedemann, Dipt. Exot. I, 236, 8 *(Laphria)*; Auss. Zw. I, 511, 21 *(id.;* Wiedemann does not adopt the change of name, proposed by Fabricius in erratis and l. c. states the reason); Macquart, Dipt. Exot. I, 2, 68, 14 *(Laphria)*. — North America (Fab.); also in the West Indies (Macq.).
  Laphria Alcanor Walker, List, etc. II, 383 (!). ([113]).
  Laphria affinis Macquart, Dipt. Exot. 5e Suppl., 54, 45. — Baltimore. ([114]).
* **tergissa** Say, J. Acad. Phil. III, 74, 5; Compl. Wr. II, 67 *(Laphria)*; Wiedemann, Auss. Zw. I, 502 5 *(id.)*. — Pennsylvania (Say).
  Laphria grossa Fabricius, Spec. Ins. II, 460, 1; System. Antl. 153, l. ([115]).
  Laphria analis Macquart, Dipt. Exot. I, 2, 68, 15. ([116]).
  Laphria flavibarbis Harris, Ins. N. Engl. 3d edit. 604. ([117]).

* **astur** O. Sacken, Western Dipt. 285. — California, common.
* **columbica** Walker, in Lord's Naturalist etc. II, 388! *(Laphria)*; description reproduced in O. Sacken, Western Diptera, 285. — Vancouver's Island.

* **fascipennis** Macquart, Hist. Nat. Dipt. I, 284, 20 *(Laphria)*. — Cayenne (Macq.); Central America (Loew). .
  Laphria praepotens Macquart, Dipt. Exot. Suppl. 1, 74. — (Loew *in litt*). According to Schiner, Novara etc. 172, this species is a *Dasyllis*.

   Observation. The *Laphria flavipila* Macquart, Hist. Nat. Dipt. I, 282, 8, United States, is omitted in the above list, as it is impossible to make out, what it is.

Pogonosoma.
Rondani, Dipt. it. Prodr. I, 160; 1856.

* **dorsata** Say, Amer. Entom. I, Tab. VI *(Laphria)*; Wiedemann, Auss. Zw. I, 506, 12 *(id.)*. — Pennsylvania (Say).
  melanoptera Wiedemann, Auss. Zw. I, 514, 26 *(Laphria)*. — Patria unknown (Wied.); South Carolina (Schiner, Verh. Z. B. Ges. 1866, 707; it is not explained however on what authority this statement is made, which is the more singular, as l. c. 691, Dr. Schiner states that the species is unknown to him).

Laphria. (*)
Meigen, in Illiger's Magaz. II, 1803.

* **Aeatus** Walker, List, etc. II, 381. — Nova Scotia; Huds. B. Terr (Walk.); White Mts., N. H.

---

*) Several of the species mentioned here as *Laphriae,* probably belong to *Dasyllis.*

*bilineata Walker, List, etc. IV, 1156. — Huds. B. Terr. (Walker); Canada; Colorado (M. C. Z.).

carolinensis Schiner, Verh. Zool. Bot. Ges. 1867, 380. — Carolina.

flavescens Macquart, Dipt. Exot. I, 2, 69, 16. — Pyrenees in Europe and Carolina in North America (Macquart's statement).

georgina Wiedemann, Dipt. Exot. I, 235, 6; Auss. Zw. I, 506. — Savannah.

lasipus Wiedemann, Auss. Zw. I, 502, 6 (lasipes, in erratis lasipus). — Kentucky.

melanogaster Wiedemann, Dipt. Exot. I, 236, 7; Auss. Zw. I, 507, 14; Macquart, Dipt. Exot. 1er Suppl., 75, 30. — Savannah and Mexico (Wied.); Texas (Macq.).

*Sadales Walker, List, etc. II, 378. — New York (Walk.); White Mts., N. H.

*sericea Say, J. Acad. Phil. III, 74, 4; Amer. Entom. I, Tab. VI; Wiedemann, Auss. Zw. I, 508, 16. — United States (Say).

terrae novae Macquart, Dipt. Exot. I, 2, 69, 18. — Newfoundland.

*rapax O. Sacken, Western Diptera, 286. — Sierra Nevada, Cal.

*vultur O. Sacken, Western Diptera, 286. — California.

Amandus Walker, List, etc. II, 373. — Guatemala.

componens Walker, Trans. Entom. Soc. N. Ser. V, 281. — Mexico.

homopoda Bellardi, Saggio, etc. App. 20, f. 16. — Mexico.

triligata Walker, Trans. Ent. Soc. N. Ser. V, 281. — Mexico.

Olbus Walker, List, etc. II, 375; Macquart, Dipt. Exot. 5e Suppl. 53; Tab. II, f. 3. — Guatemala (Walk.); Honduras (Macq.).

Pseudorus.

Walker, Dipt. Saund. 103; 1850—56.

bicolor Bellardi, Saggio, etc. II, 11; Tab. I, f. 20. — Mexico.

Lampria.

Macquart, Dipt. Exot. I, 2, 60; 1838.

*bicolor Wiedemann, Auss. Zw. I. 522, 40 (Laphria). — Patria unknown (Wied). — Middle and Southern States.

Laphria saniosa Say, J. Acad. Phil. VI, 158; Compl. Wr. II, 355.

Laphria Antaea Walker, List, etc. II, 379 and VII, 527 (= „saniosa Say?" Walk.).

Laphria megacera Macquart, Hist. Nat. Dipt. I, 284, 18 (!).

*rubriventris Macquart, Hist. Nat. Dipt. I, 284, 19 (Laphria). — Philadelphia (Macq.); Texas. ([118]).

*felis O. Sacken, Western Diptera, 286. — Sierra Nevada, Cal.

circumdata Bellardi, Saggio, etc. II, 15; Tab. I, f. 17. — Mexico.

clavipes Fabricius, Syst. Antl. 162, 27 (Laphria); Wiedemann, Dipt. Exot. I, 237, 9 (id.); Auss. Zw. II, 513, 23 (id.); Macquart, Dipt. Exot.

I, 2, 61; Bellardi, Saggio, etc. II, 13; Tab. I, f. 15. — Brazil (Fabr.); Mexico (Bell.).

mexicana Macquart, Dipt. Exot. 2e Suppl., 37, 3; Bellardi, Saggio, etc. II, 13. — Mexico.

Laphystia.

Loew, Linn. Ent. II, 538; 1847.

*sexfasciata Say, J. Acad. Phil. III, 50, 1; Compl. Wr. II, 64 *(Dasypogon)*; Wiedemann, Auss. Zw. I, 408, 68 *(id.)*. — Missouri (Say); New Jersey, Florida (M. C. Z.).

(?) albiceps Macquart, Dipt. Exot. 1er Suppl. 69, 51 *(Dasypogon)*. — Texas.

Observation. Dr. Schiner (Verh. Zool. Bot. Ges. 1866, 698) places *L. sexfasciata* Say, in the genus *Laphyctis*; Loew objects to it in Berl. Ent. Z. 1874, p. 373.

Andrenosoma.

Rondani, Dipt. it. Prodr. I, 160; 1856.

*pyrrhacra Wiedemann, Auss. Zw. I, 517, 31 *(Laphria)*. — Savannah, Missouri; Brazil (the latter locality also in Schiner, Novara etc., 175). *Laphria fulvicauda* Say, J. Acad. Phil. III, 53; Amer. Ent. I, Tab. VI *(id.)*. [Name changed by Wied.]

cinerea Bellardi, Saggio, etc. II, 16; Tab. I, f. 16 *(Lampria)*. — Mexico.
cincta Bellardi, Saggio, etc. II, 18; Tab. I, f. 19 *(Laphria)*. — Mexico.
formidolosa Walker, Trans. Ent. Soc. N. Ser. V, 280; Bellardi, Saggio, etc. II, 17; Tab. I, f. 18 *(Laphria)*. — Mexico. [118]
xanthocnema Wiedemann, Auss. Zw. I, 509, 18; Macquart, Dipt. Exot. I, 2, 67, 12. — West Indies (Macq.); Brazil (Wied.). [118]

SECTION III. ASILINA. [119]

Mallophora.

Macquart, Hist. Nat. Dipt. I, 300; 1834.

ardens Macquart, Hist. Nat. Dipt. I, 302, 4; Dipt. Exot. I, 2, 89, 12; Tab. VIII, f. 2. — North America (Macq.).
*bomboides Wiedemann, Dipt. Exot. I, 203, 37 *(Asilus)*; Auss. Zw. I, 476, 77 *(id.)*; Macquart, Hist. Nat. Dipt. I, 302, 2; Dipt. Exot. I, 2, 89, 11. — Georgia (Wied.).
clausicella Macquart, Dipt. Exot. 4e Suppl. 79, 27; Tab. VII, f. 8. — Virginia („perhaps a variety of *M. heteroptera?*" Macq.).
fulviventris Macquart, Dipt. Exot. 4e Suppl. 77, 24. -- Mexico; Texas? (Macq.)
*laphroides Wiedemann, Auss. Zw. I, 483 *(Asilus)*. — Kentucky. *Mallophora heteroptera* Macquart, Dipt. Exot. I, 2, 90, 13; Tab. VIII, f. 3. — Philadelphia.
(?) *Mallophora minuta* Macquart, Hist. Nat. Dipt. I, 302, 5.

*orcina Wiedemann, Auss. Zw. I, 477, 79 *(Asilus)*. — Georgia (Wied.); Distr. Columbia.

Amphinome Walker, List, etc. II, 387 *(Asilus)*. — Honduras. [Loew *in litt.*; supposes this to be a Proctacanthus; I could not find the specimen in the Br. Mus.]
Craverii Bellardi, Saggio, etc. II, 22. — Mexico.
fulvianalis Macquart, Dipt. Exot. 4e Suppl. 78, 25 („perhaps ♀ of *fulviventris"* Macq.). — Mexico.
infernalis Wiedemann, Dipt. Exot. I, 202 *(Asilus)*; Auss. Zw. I, 475 *(id.)*; Macquart, Hist. Nat. Dipt. I, 301; Perty, Delectus etc. 181, Tab. XXXVI, f. 5 *(Asilus)*. — Brazil; Mexico.
*Macquartii (Loew *in litt.*), Macquart, Dipt. Exot. I, 2, 89, 10; Bigot in R. de la Sagra etc. 790 (described by both as *M. scopifera* Wied.). — Cuba. [120].
pica Macquart, Dipt. Exot. 4e Suppl. 78, 26. — Mexico.
robusta Wiedemann, Auss. Zw. I, 478, 81 *(Asilus)*; Macquart, Dipt. Exot. 1er Suppl. 78. — No locality in Wiedem.; Yucatan (Macq.).
scopifer Wiedemann, Auss. Zw. I, 478, 83 *(Asilus)*. — Brazil (Wied.); Columbia, S. A. (Schiner, Novara). [120].

Observation. *Trupanea perpusilla* Walker, Dipt. Saund., 123. — United States; I saw the specimen in the Brit. Mus., it appeared to me like a small *Mallophora*.

Promachus.

Loew, Linn. Ent. III, 390; 1848; *Trupanea* Macquart (preocc.).

*apivorus Fitch, Reports, Vol. III, 251 – 256; Tab. IV, f. 7 *(Trupanea)*; Riley, 1st. Report, 168 *(id.)*. — Nebraska; North Missouri. [121].
*Bastardii Macquart, Dipt. Exot. I, 2, 104, 30 *(Trupanea)*. — North America.
Asilus Laevinus Walker, List, etc. II, 392 (!). — Massachusetts.
Promachus philadelphicus Schiner Verh. Z. B. Ges. 1867, 389. — Pennsylvania (!).
Trupanea rubiginis Walker, Dipt. Saund., 123 — North America (!).
quadratus Wiedemann, Dipt. Exot I 201, 34; Auss. Zw. I, 485, 90 *(Asilus)*. — Georgia. [122].
*rufipes Wiedemann, Auss. Zw. I, 487, 93 *(Asilus)*. — America (Wied.); Georgia (M. C Z.)
* vertebratus Say, J. Acad. Phil. III, 47; Compl. Wr. II, 62 *(Asilus)*; Wiedemann, Auss. Zw. I, 485, 91 *(id.)*; Macquart, Dipt. Exot. I, 2, 103, 27 *(Trupanea)*. — Missouri (Say); Illinois (M. C. Z.).

cinctus Bellardi, Saggio, etc. II, 25; Tab. II, f. 2. — Mexico.
fuscipennis Macquart, Dipt. Exot. 1er Suppl. 81, 44; Tab. VIII, f. 4 *(Trupanea)*; Bellardi, Saggio, etc. II. 24; Tab. II, f. 1., Var A; Schiner, Novara etc. p. 177. — New Granada (Macq.); Mexico (Bell.). [123]

magnus Bellardi, Saggio, etc. II, 26. — Mexico.
pulchellus Bellardi, Saggio, etc. II, 29; Tab. II, f. 5. — Mexico.
quadratus Bellardi, Saggio, etc. II, 27; Tab. II, f. 3. — Mexico. ([124]).
trapezoidalis Bellardi, Saggio, etc. II, 28, Tab. II, f. 4. — Mexico.
Truquii Bellardi, Saggio, etc. II, 30; Tab. II, f. 6. — Mexico.

Observation. *Asilus ultimus* Walker, Dipt. Saund., 136, United States, is a *Promachus,* and if I recollect right, *P. Bastardii.*

Erax.
Macquart, Dipt. Exot. I, 2, 107; 1838.

*aestuans Linné, System. Nat. II, 1007, 5; Amoen. Acad. VI, 413, 95 *(Asilus)*; Fabricius, System. Ent. IV, 379, 8 *(Asilus)*; System. Antl. 164, 2 *(Dasypogon)*; Olivier, Encyclop. Méth, I, 264; Wiedemann, Dipt. Exot. I, 200, 32; Auss. Zw. I, 467, 63 *(Asilus)*; Macquart, Hist. Nat. Dipt. I, 312, 36 *(Asilus)*; Dipt. Exot. I, 2, 115, 19; Bigot, in R. de la Sagra, etc. 791. — North America; Cuba (according to Macquart also in Brazil). ([125]).
albibarbis Macquart, Dipt. Exot. I, 2, 118, 26; Comp. Schiner, Verh. Z. B. Ges. 1867, 395. — North America.
*ambiguus Macquart, Dipt. Exot. 1er Suppl. 84, 34 — Galveston, Texas; Merida, Yucatan (Macq.); Georgia (M. C. Z.).
Asilus interruptus Macquart, H. N. Dipt. I, 310, 29. — Georgia. ([126]).
apicalis Wiedemann, Dipt. Exot. I, 191, 16 *(Asilus)*; Auss. Zw. I, 443, 28 *(id.)*. — North America. ([127]).
*Bastardi Macquart, Dipt. Exot. I, 117, 25; Tab. 9, f. 7; Riley, 2d Report, 124 (figure of larva, pupa, imago). — North America.
completus Macquart, Dipt. Exot. I, 2, 117, 23; Tab. IX, f. 9. — North America.
femoratus Macquart, Dipt. Exot. I, 2, 115, 20. — Carolina.
incisuralis Macquart, Dipt. Exot. I, 2, 117, 24. — Philadelphia.
lateralis Macquart, Dipt. Exot. I, 2, 116, 21. — Philadelphia.
macrolabis Wiedemann, Auss. Zw. I, 458, 51 *(Asilus)*. — Kentucky.
niger Wiedemann, Dipt. Exot. I, 196, 26; Auss. Zw. I, 460, 53 *(Asilus)*. — Georgia.
notabilis Macquart, Dipt. Exot. I, 2, 110, 6; Tab. IX, f. 8. —'America
pogonias Wiedemann, Dipt. Exot. I, 198, 29; Auss. Zw. 460, 54 *(Asilus)*. — North America.
Asilus barbatus Fabricius, System. Antl. 169, 22 name changed by Wied.).
rufibarbis Macquart, Dipt. Exot. I, 2, 116, 22. — North America.
tibialis Macquart, Dipt. Exot. I, 118, 27. — Philadelphia; Cayenne, Guyana (Macq.).
vicinus Macquart, Dipt. Exot. 1er Suppl. 85, 36. — Galveston, Texas

affinis Bellardi, Saggio, etc. II, 41. - Mexico.
aper Walker, List, etc. VII, 621. — Mexico
anomalus Bellardi, Saggio, etc II, 32; Tab II, f. 7. — Mexico.
argyrogaster Macquart, Dipt. Exot. 1er Suppl. 84, 35. — Yucatan. ([126])

bicolor Bellardi, Saggio, etc. II, 47. — Mexico.
bimaculatus Bellardi, Saggio, etc. II, 45; Tab. II, f. 11. — Mexico
(Bellardi); Columbia, S. A. (Schiner, Novara, 182).
carinatus Bellardi, Saggio, etc. II, 36; Tab. II, f. 9. — Mexico.
caudex Walker, List, etc. II, 404. — West Indies.
cinerascens Bellardi, Saggio, etc. II, 39; Tab. II, f. 10; Compare
also Schiner. Verh. Z. B. Ges. 1867, 394. — Mexico.
cingulatus Bellardi, Saggio, etc. II, 42. — Mexico.
comatus Bellardi, Saggio, etc. II, 34. — Mexico.
eximius Bellardi, Saggio, etc. II, 38. — Mexico.
flavofasciatus Wiedemann, Auss. Zw. I, 470, 68. — Brazil (Wied.);
Honduras (Walker, List, etc. II, 400).
fortis Walker, List, etc. VII, 623. — San Domingo.
fulvibarbis Macquart, Dipt. Exot. 3e Suppl. 28, 44; Tab. II, f. 13. —·
Haiti.
Haitensis Macquart, Dipt. Exot. 3e Suppl. 28, 45; Tab. II, f. 10. — Haiti.
Haloesus Walker, List, etc. II, 405. — Jamaica.
invarius Walker, Dipt. Saund. 131. — Jamaica.
lascivus Wiedemann, Auss. Zw. I, 474, 75. — Brazil (Wied.); Honduras (Walker, List, etc. II, 400). ([128]).
   Asilus Amarynceus Walker, List, etc. II, 400 (no locality). —
   [Synonymy according to Walker, List, etc. VII, 637.]
maculatus Macquart, Dipt. Exot. I, 2, 111, 9; Tab. IX, f. 6. —
Guyana; Columbia (S. A.); Guadeloupe. ([126]).
Loewii Bellardi, Saggio, etc. App. 21, f. 17. — Mexico.
marginatus Bellardi, Saggio, etc. II, 46. — Mexico.
nigrimystaceus Macquart, Dipt. Exot. 2e Suppl. 41, 40. — Guadeloupe.
parvulus Bellardi, Saggio, etc. II, 35; Tab. II, f. 8. — Mexico.
pumilus Walker, List, etc. VII, 640. — Vera Cruz.
quadrimaculatus Bellardi, Saggio, etc. II, 44; Tab. II, f. 13. —
Mexico.
rufitibia Macquart, Dipt. Exot. 3e Suppl. 27, 42; Tab. II, f. 11. —
Haiti; Rio Negro (S. Amer.).
stylatus Fabricius, System. Ent. IV, 795, 17; Ent. System. IV, 384,
38 *(Asilus)*; System. Antl. 171, 31 *(Dasypogon)*; Wiedemann,
Dipt. Exot. I, 198, 30 *(Asilus)*; Auss. Zw. I, 462, 57 *(id.)*; Tab.
VI, f. 6. — West Indies.
tricolor Bellardi, Saggio, etc. II, 40; Tab. II, 12. — Mexico.
unicolor Bellardi, Saggio, etc. II, 37. — Mexico.

Observation. *Erax Dascyllus* Walker, List, etc. II, 401,
Massachusetts; the fragment in the Brit. Mus. is not recognizable.
Erax Antiphon Walker, List, etc. VII, 618. Short diagnosis only;
at the same time the author quotes in the synonymy:
Asilus Antiphon List, etc. II, 397, with the remark: „the previous
description of this species is erroneous". This previous description refers evidently to some other species and gives no habitat.
I do not find anything about this species in my notes taken in
the Brit. Mus.

Neoeristicus.

Eristicus Loew, Linn. Ent. III, 396; 1848. ([129]).

Bellardii Schiner, Novara etc. 182 *(Erax)* — Columbia, S. A. (Schiner); Mexico (Bell.).

Erax nigripes Bellardi, Saggio, etc. II, 48 *(Eristicus)*, change of name by Schiner.

villosus Bellardi, Saggio, etc. II, 49 *(Eristicus)*. — Mexico.

Proctacanthus.

Macquart, Dipt. Exot. I, 2, 120; 1838.

*brevipennis** Wiedemann, Auss. Zw. I, 431, 10 *(Asilus)*. — Kentucky (Wied.); Florida (O. S.).

fulviventris Macquart, Dipt. Exot. 4e Suppl. 88, 12. — Florida. ([130]).

*heros** Wiedemann, Auss. Zw. I, 427, 4 *(Asilus)*. — Kentucky (Wied.); South Carolina (M. C. Z.).

longus Wiedemann, Dipt. Exot. I, 183, 1; Auss. Zw. I, 426, 3 *(Asilus)*; Macquart, Hist. Nat. Dipt. I, 307, 18 *(Asilus)*; Dipt. Exot. I, 2, 123, 6. (compare also Schiner, Verh. Zool. Bot. Ges. 1866, 682, 3). — Georgia.

micans Schiner, Verh. Zool. Bot. Ges. 1867, 397. — North America.

*Milbertii** Macquart, Dipt. Exot. I, 2, 124, 8. — North America.

(?) *Asilus Agrion* Jaennicke, Neue Exot Dipt. 57. — Illinois. ([131]).
Asilus missuriensis Riley, 2d Report 122, fig. 89. — Missouri.

nigriventris Macquart, Dipt. Exot. I, 2, 124, 9. — Philadelphia; Carolina (Macq).

*philadelphicus** Macquart, Dipt. Exot. I, 2, 123, 7; — Philadelphia (Macq.).

Craverii Bellardi, Saggio, etc. II, 50. — Mexico.

rufiventris Macquart, Dipt. Exot. I, 2, 123, 5; Tab. X, f. 2. — San Domingo, Honduras.

Eccritosia.

Schiner, Verh. Zool. Bot. Ges. 1866, 674.

plinthopyga Wiedemann, Dipt. Exot. I, 184, 4 *(Asilus)*; Auss. Zw. I, 432, 11 *(id.)*; Bigot, in R. de la Sagra etc. 791 *(id.)*. — Cuba.

Asilus.

Linné, Fauna Suecica; 1761. ([132]).

femoralis Macquart, Dipt. Exot. 2e Suppl. 45, 61. — Philadelphia.

longicella Macquart, Dipt. Exot. 4e Suppl. 95, 77; Tab. IX, f. 5. — North America (with a doubt).

*Novae Scotiae** Macquart, Dipt. Exot. 2e Suppl. 46, 62. — Nova Scotia.

*sericeus** Say, J. Acad. Phil. III, 48, 2; Compl. Wr. II, 63; Wiedemann, Auss. Zw. I, 429, 8. — United States.

Asilus Herminius Walker, List, etc. II, 410 (!).

tibialis Macquart, Hist Nat. Dipt. I, 313, 33. — Philadelphia.

9

apicalis Bellardi, Saggio, etc. II, 57. — Mexico. ([133]).
atripes Fabricius, System. Antl. 170, 29 (*Dasypogon*); Wiedemann,
    Dipt. Exot. I, 195, 24; Auss. Zw. I, 155, 46. — West Indies.
inamatus Walker, Trans. Ent. Soc. N. Ser. V, 283. — Mexico.
infuscatus Bellardi, Saggio, etc. II, 56; Tab II, f. 15. — Mexico.
megacephalus Bellardi, Saggio, etc. II, 58; Tab. II, f. 14. — Mexico.
mexicanus Macquart, Dipt. Exot. 1er Suppl. 94, 55. — Mexico.
perrumpens Walker, Trans. Ent. Soc. N. Ser. V, 283. — Mexico.
vittatus Olivier, Encycl. Méth. I, 263, 4. — San Domingo.

Observation.

Asilus Alethes Walker, List, etc, II, 454. — New York.
Asilus Antimachus Walker, List, etc. II, 454. — Trenton Falls N. Y.
Asilus Lecythus Walker, List, etc. II, 451. — Nova Scotia.
Asilus Orphne Walker, List, etc. II, 456. — New York.
Asilus Paropus Walker, List, etc. II, 455. — New York.
Asilus Sadyates Walker, List, etc. II, 453. — Ohio.
The specimens exist in the Brit. Mus. and belong to the different genera, in
which Asilus has been subdivided; most of them, if not all, will coincide with
previously described species.
Asilus ultimus Walker Dipt. Saund. 136, is a Promachus.
For *Asilus Agrion* Jaennicke, see *Proctacanthus Milbertii.*

Philonicus.
Loew, Linn. Ent. IV, 144; 1849.

taeniatus Bellardi, Saggio, etc. II, 55. — Mexico.
Tuxpanganus Bellardi, Saggio, etc. App. 22. — Mexico.

Lophonotus.
Macquart, Dipt. Exot. I, 2, 125; 1838: Loew, Linn. Ent. III, 423, 1848,
    modifies the limits of the genus.

humilis Bellardi, Saggio, etc. II, 51. — Mexico.

Neomochtherus.
Mochtherus Loew, Linn. Ent. IV, 58; 1849. ([134]).

gracilis Wiedemann, Auss. Zw. I, 445, 31 (*Asilus*). — Savannah. ([135]).

Truquii Bellardi, Saggio, etc. II, 52. — Mexico.
fuliginosus Bellardi, Saggio, etc. II, 52. — Mexico.

Neoitamus.
Itamus Loew, Linn Ent. IV, 84; 1849. ([134]).
*aenobarbus Loew *in litt.* — Northern and Middle.States.

Epitriptus.
Loew, Linn. Ent. IV, 108; 1849.

(?) albispinosus Bellardi, Saggio, etc. II, 54 (the query is Bellardi's). —
Mexico.
niveibarbus Bellardi, Saggio, etc. II, 53. — Mexico.

Machimus.
Loew, Linn. Ent. IV, 1; 1849.

avidus v. d. Wulp, Tijdschr. v. Ent. 2d Ser., IV, 82. — Wisconsin.

Stilpnogaster.
Loew, Linn. Ent. IV, 82; 1849.

anceps v. d. Wulp, Tijdschr. v. Ent. 2d Ser. IV, 84. — Wisconsin.

Tolmerus.
Loew, Linn. Ent. IV, 94; 1849.

*annulipes Macquart, Dipt Exot. I, 2, 149, 36 *(Asilus)*. — Carolina (Macq.); Atlantic States and Canada.
notatus Wiedemann, Auss. Zw. I, 451, 40 *(Asilus)*. — Savannah.

Ommatius. (135a).
Illiger; Wiedemann, Auss. Zw. I, 418; 1828.

tibialis Say, J. Acad. Phil. III, 49; Compl. Wr. II, 63; Wiedemann, Auss. Zw. I, 422, 6. — Pennsylvania.

fuscipennis Bellardi, Saggio, etc. App. 23. — Mexico.
marginellus Fabricius, Spec. Ins. II, 464, 22 *(Asilus)*; Ent. System. 384, 36 *(id.)*; System. Antl. 170, 28 *(Dasypogon)*; Wiedemann, Dipt. Exot. I, 213, 1; Auss. Zw. I. 421, 5; Tab. VI, f. 5. — West Indies; Macquart, Dipt. Exot. I, 2, 134, 4 has it trom Brazil. (136).
parvus Bigot, Ann. Soc. Entom. 1875, 247. — Mexico.
pumilus Macquart, Dipt. Exot. 2e Suppl., 42, 6; Tab. I, f. 10; Bellardi, Saggio, etc. II, 59. — Mexico.
Saccas Walker, List, etc. II, 474. — Jamaica.
vitreus Bigot, Ann. Soc. Ent. 1875, 245. — Haity.

Emphysomera.
Schiner, Verh. Zool. Bot. Ges. 1866, 665; id. Novara, p. 195.

pilosula Bigot, Ann. Soc Ent. 1875, 243. — Mexico.
bicolor Bigot, Ann. Soc. Ent. 1875, 244. — Mexico.

FAMILY MIDAIDAE. (137).
Leptomidas.
Leptomydas, Gerstaecker, Stett. Ent. Z. 1868, 81.

*venosus Loew, Centur VII, 26. — Pecos River, Western Texas.

*pantherinus Gerstaecker, Stett. Ent. Z. 1868, 85; O. Sacken, Western Diptera, 280. — California (Lone Mt. San Francisco, O. Sacken).
*tenuipes Loew, Centur. X, 20 *(Midas)*. — California.

Midas. (138).
Mydas Fabricius, Entom. System. IV, 252; 1794.

*audax O. Sacken, Bul. Buff. S. N. H. 1874, 186 (the descriptions of this and of the two following species, are reproduced in the note. — Kentucky. (177).

* **carbonifer** O. Sacken, l. c. 186. — Western New York.

* **chrysostomus** O. Sacken, I. c. 187. — Texas.

* **clavatus** Drury, Illustr. of Nat. Hist. I, 103; Tab. 44, f. 1 and Vol. II, App. *(Musca);* Westwood, Arc. Ent., 51, 14. — Atlantic States (rare in Massachusetts).

> *Nemotelus asiloides* Degeer, VI, Tab. XXIX, f. 6.

> *Bibio illucens* Fabricius, System. Ent. 756, 1. ([140]).

> *Bibio filata* Fabricius, Spec. Ins. II, 412; Mantissa, 328, 1; Ent. System. IV, 252 *(Mydas);* System. Antl. 60, 1 *(id.);* Olivier, Encycl. Méth. VIII, 83, 1; Wiedemann, Dipt. Exot. 116, 2; Auss. Zw. I, 240, 3; Monogr. Midar. Tab. 53, f. 8 (for the quotations from Latreille and Dumeril, see Wiedemann).

crassipes Westwood, Arcan. Ent. I, 51; Tab. XIII, f. 3. — North America (?).

fulvipes Walsh, Proc. Bost. Soc. N. H. IX, 306. — Illinois.'

fulvifrons Illiger, Magaz. I, 206; Wiedemann, Monogr. Mid. 47; Tab. LIII, f. 13. — Georgia.

incisus Macquart, Dipt. Exot. I, 2, 11; Tab. I, f. 1. — Carolina. (Mexico, according to Jaennicke, l. c. p. 46.)

* **luteipennis** Loew, Centur. VII, 23. — Pecos River, Western Texas.

maculiventris Westwood, Lond. and Edinb. Phil. Mag. 1835, Arc. Ent. I, 53; Tab. XIII, f. 5. — Georgia.

pachygaster Westwood, Arc. Ent. I, 53; Tab. XIII, f. 4. — Georgia.

parvulus Westwood, Arc. Ent. I, 53; Tab. XIII, f. 6. — Georgia (Westw.); Florida (Walk.).

* **simplex** Loew, Centur. VII, 25. — Pecos River, Western Texas.

* **tibialis** Wiedemann, Monogr. Mid. 42; Tab. LIII, f. 6; Bellardi, Saggio, etc. II, 6. — Maryland; Michigan; Mexico (Bellardi).

* **xanthopterus** Loew, Centur. VII, 24. — Pecos River, Western Texas.

> *Mydas lavatus* Gerstaecker, Stett. Ent. Z. 1868, 96. — Mexico.

* **ventralis** Gerstaecker, Stett. Ent. Z 1868, 102. — California.

> *Midas rufiventris* Loew, Centur. VII, 22 (change of name by Gerst.).

annularis Gerstaecker, Stett. Ent. Z. 1868, 100. — Mexico.

basalis Westwood, Arc. Ent. I, 53, Bellardi, Saggio, etc. II, 10. — Mexico.

bitaeniatus Bellardi, Saggio, etc. II, 7; Tab. I, f. 1. — Mexico.

* **gracilis** Macquart, Hist. N. Dipt. I, 274; Tab. VII, f. 1. — South America (Macq.); Cuba (Loew *in litt.*).

interruptus Wiedemann, Monogr. Mid. 46; Tab. LIII, f. 12. — Mexico.

> *Midas tricinctus* Bellardi, Saggio, etc. II, 8; Tab. I, f. 2 [Gerst.].

militaris Gerstaecker, Stett. Ent. Z. 1868, 99. — Mexico.

> *Midas vittatus* Macquart, Dipt. Exot. 4e Suppl. 60; Tab. IV, f. 6; Bellardi, Saggio, etc. II, 7 [change of name by Gerst.].

rubidapex Wiedemann, Monogr. Mid. 40; Tab. 52, f 2 ($\mathring{\jmath}$); Auss. Zw. II, 626; Bellardi, Saggio, etc. II, 5. — Mexico.

senilis Westwood, Arc. Ent. I, 52. — Mexico.

subinterruptus Bellardi, Saggio, etc. II, 10; Tab. I, f. 3. — Mexico.

*tricolor Wiedemann, Monogr. Mid. 42; Tab. 53, f. 5; Bigot, R. de la Sagra, etc. 799. — Cuba.

Observation. According to Mr. Walker, List, etc. I, 228, *Dolichogaster* (Midas) *brevicornis* Wied. (variet. *iopterus* Wied.) from Brazil, also occurs in Florida and Massachusetts.

Raphiomidas.
O. Sacken, Western Diptera, 281; 1877.

*episcopus O. Sacken, l. c. 282. — Southern California.

Apiocera.
Westwood, London and Edinburgh Phil. Magaz. 1835; the same, Arcana etc.; *Pomacera* Macquart Suppl. 2, p. 47, 1847; *Anypenus* Philippi, Verh. Zool. Bot. Ges. 1865, 702; Tab. 25, f. 26.

*haruspex O. Sacken, Western Diptera, 283. — Yosemite Valley, Cal.

FAMILY NEMESTRINIDAE. [141].
Hirmoneura.
Meigen, System. Beschr. II, 132; 1820.

*clausa O. Sacken, Western Diptera, 225. — Dallas, Texas. [142].

brevirostris Macquart, Dipt. Exot. Suppl. I, 101, 8; Tab. 20, f. 1. — Yucatan.

FAMILY BOMBYLIDAE. [143].
Exoprosopa.
Macquart, Dipt. Exot. II, 1, 35; 1840.

*caliptera Say, J. Acad. Phil. III, 46, 7; Compl. Wr. II. 62 *(Anthrax)*; O. Sacken, Western Dipt., 233 — Arkansas (Say); Cheyenne, Wyo.; Tehuacan, Mexico (Coll. Bellardi).
*decora Loew, Centur. VIII, 19.   Wisconsin (Loew); Georgia, Texas, Illinois, Iowa, Red River of the North.
*dodrans O. Sacken, Western Dipt., 234. — Colorado Springs, Col.
*dorcadion O. Sacken, Western Dipt., 231. — White Mts., N. H.; Maine; Rocky Mts., Col.; Sierra Nevada, Cal.; Washington Terr. *Anthrax capucina* Fabricius, Ent. System. IV, 259, 12; System. Antl. 123; Wiedemann and later authors have erroneously referred these quotations to a european species.
(? *Anthrax californiae* Walker, Dipt. Saund., 172. [144].
*doris O. Sacken, Western Dipt., 235. — Humboldt Desert, Nevada.
*emarginata Macquart, Dipt. Exot. II, 1, 51, 40. — Philadelphia (Macq.); Virginia, Missouri.
*fascipennis Say, Long's Exped. App. 373, 4; Compl. Wr. I, 254 *(Anthrax)*; Wiedemann, Auss. Zw. I, 284, 39 *(id.)*. — Atlantic States (especially Middle States); Cuba.

Anthrax noctula Wiedemann, Auss. Zw. II, 635, 45 (!).
Exoprosopa coniceps Macquart, Dipt. Exot. 4e Suppl. 108, 63;
   Tab. X, f. 9; Bigot, R. de la Sagra etc. 793 (!). — Virginia (Macq.);
   Cuba (Bigot).
Exoprosopa philadelphica Macquart, Dipt. Exot. II, 1, 52, 41;
   Tab. XVIII, f. 1.([145]).
*fasciata Macquart, Dipt. Exot. II, 1, 51, 38; Tab. XVII, f. 6; O. Sacken,
   Western Dipt., 231. — Atlantic States.
Exoprosopa longirostris Macquart, Dipt. Exot. 4e Suppl. 108, 62;
   Tab. X, f. 8 (!. — Virginia.
Exoprosopa rubiginosa Macquart, Dipt. Exot. II, 1, 51, 39; *ibid.*
   Suppl. I, 111. — Philadelphia; Columbia (South America). ([14C]).
Mulio americana v. d. Wulp, Tijdschr. etc. 2d Ser., 141; Tab. IV,
   f. 1 — 4.
pueblensis Jaennicke, Neue Exot. Dipt. 34; Tab. II, f. 21. — Mexico
   (Jaenn.); Texas (Coll. v. Roeder).
*sima O. Sacken, Western Dipt., 231. — Humboldt Desert, Nevada.
*titubans O. Sacken, Western Dipt., 233. — Denver, Col.

*Agassizii Loew, Centur. VIII, 24. — California.
*bifurca Loew, Centur. VIII, 23. — California.
*eremita O. Sacken, Western Dipt., 236. — Northern California. ([147]).
*gazophylax Loew, Centur. VIII, 18. — California.

anthracoidea Jaennicke, Neue Exot. Dipt., 32; Tab. II, f. 18. —
   Mexico.
blanchardiana Jaennicke, l. c. 33; Tab. II, f. 20. — Mexico.
*cerberus Fabricius, Ent. System. IV, 256, 1 (*Anthrax*); System. Antl.
   118, 1 (*id.*); Wiedemann, Dipt. Exot. I, 118, 1 (*id.*); Auss. Zw. I,
   253, 2; Tab. III, f. 1 (*id.*); Macquart, Hist. Nat. Dipt. I, 400, 1
   (*id.*); Dipt. Exot. II, 1, 38, 6; Tab. XVI, f. 5. -- South America
   (Wied. Macq.); Jamaica (Walker, List, etc. II, 238); Cuba (M. C. Z.).
clotho Wiedemann, Auss. Zw. II, 635 (*Anthrax*). — Mexico.
*cubana Loew, Centur. VIII, 22. — Cuba.
ignifer Walker, List, etc. II, 243 (*Anthrax*). — Jamaica. ([148]).
Kaupii Jaennicke, Neue Exot. Dipt. 32; Tab. II, f. 17 (wing). — Mexico.
lacera Wiedemann, Auss. Zw. II, 633, 44 (*Anthrax*). — Mexico.
Latreillii Wiedemann, Auss. Zw. II, 633, 43 (*Anthrax*). — Mexico.
limbipennis Macquart, Dipt. Exot. Suppl. I, 110, 50; Tab. XX, f. 3. —
   Yucatan.
*nubifera Loew, Centur. VIII, 25. — Cuba.
Orcus Walker, List, etc. II, 237 (*Anthrax*). — Mexico.
parva Loew, Centur. VIII, 26. — Cuba.
Pilatei Macquart, Dipt. Exot. Suppl. I, 110, 49; Tab. XX, f, 2. —
   Yucatan.
Proserpina Wiedemann, Auss. Zw. I, 257, 6 (*Anthrax*); Macquart,
   Dipt. Exot. II, 1, 38, 7; Bigot, in R. de la Sagra etc. 793. —
   No locality (Wied.); San Domingo (Macq.); Cuba (Bigot.).

rostrifera Jaennicke l. c. 33; Tab. II, f. 19. — Mexico.
subfascia Walker, List, etc. II, 249 (Anthrax). — Jamaica.
*sordida Loew, Centur. VIII, 21. — Matamoras.
Thomae Fabricius, System. Antl. 135, 32 (Anthrax); Wiedemann, Dipt.
    Exot. I, 129, 13 (id.); Auss. Zw. I, 271, 22 (id.). — St. Thomas.
*trabalis Loew, Centur. VIII, 20. — Mexico.
trimacula Walker, List, etc. II, 250 (Anthrax). — Jamaica. ([149]).

> **NB.** *Anthrax Satyrus* Fabr. from Australia, or China (compare Wiedemann, Auss. Zw. I, 322, 95) is referred by Mr. Walker, List, etc. II, 243 to a species from Georgia. The ground is not stated.

Dipalta.

O. Sacken, Western Dipt., 236, 1877.

*serpentina O. Sacken, Western Dipt., 237. — Georgia; Colorado; California; Mexico (Coll. Bellardi).

Anthrax.

Scopoli, Ent. Carniol.; 1763. ([150]).

albipectus Macquart, Dipt. Exot. 3e Suppl. 34, 80; Tab. III, f. 12. — North America.
albovittata Macquart, Dipt. Exot. 4e Suppl. 113, 90; Tab. X, f. 15. — North America (?).
*alternata Say, J. Acad. Phil. III, 45, 5: Compl. Wr. II, 61; Wiedemann, Auss. Zw. I, 303, 66. — Middle States.
    Anthrax consanguinea Macquart, Dipt. Exot. II, I, 69, 42; Tab. XXI. f. 1. — Philadelphia.
cedens Walker, Dipt. Saund., 190. — United States.
*celer Wiedemann, Auss. Zw. I, 310, 77; Macquart, Dipt. Exot. II, 1, 69, 43. — Kentucky; Georgia (Philadelphia in Macquart).
*Ceyx Loew, Centur. VIII, 30. — Virginia; Georgia.
    (?) Anthrax demogorgon Walker, List, etc. II, 265. — Florida.
(?) connexa Macquart, Dipt. Exot. 5e Suppl. 76, 96; Bigot, in R. de la Sagra etc. 794. — Baltimore (Macq.); Cuba (Bigot).
costatus Say, Long's Exped. App. 373, 5; Compl. Wr. I, 254; Wiedemann, Auss. Zw. I, 314, 82. — N. W. Territory (Say).
edititia Say, J. Acad. Phil. VI, 157; Compl. Wr. II, 353. — No locality.
*flaviceps Loew, Centur. VIII, 29. — Tamaulipas.
floridana Macquart, Dipt. Exot. 4e Suppl. 112, 89; Tab. X, f. 14. — Florida.
*fulviana Say, Long's Exped. App. 372, 3; Compl. Wr. I, 253; Wiedemann, Auss. Zw. I, 290, 47. — North Western States and British Possessions; Georgetown, Colo.
*fulvohirta Wiedemann, Dipt. Exot. I, 149, 46; Auss. Zw. I, 308, 73; Macq. Dipt. Exot. II, 1, 69, 41; Meigen, Syst. Beschr. II, 158, 26; Tab. XVII, f. 11 (A. cypris, erroneously described as European). — Middle States.
    Anthrax conifacies Macquart, Dipt. Exot. 4e Suppl, 112, 88; Tab. X, f. 13. — Virginia.

Anthrax separata Walker, Dipt. Saund., 177.
fuscipennis Macquart, Hist. Nat. Dipt. I, 410, 33. — North America.
gracilis Macquart, Dipt. Exot. II, 1, 76, 64; Tab. XXI, f. 1. — Philadelphia.
*halcyon Say, Long's Exp. App. 371 *(Alcyon);* Compl. Wr. I, 252; Wiedemann, Auss. Zw. I, 288, 44; Tab. III, f. 6; Macquart, Dipt. Exot. II, 1, 68; Tab. XIX, f. 6. — North Western States and British Possessions; Colorado. ([151]).
*hypomelas Macquart, Dipt. Exot. II, 1, 76, 63; Tab. XXI, f. 1. — North America (Macq.); Pennsylvania, Wisconsin. ([150]).
*lateralis Say, J. Acad. Phil. III, 42, 2; Compl. Wr. II, 59; Wiedemann, Auss. Zw. I, 318, 89. — Atlantic States; Colorado.
Anthrax Bastardi Macquart, Dipt. Exot. II, 1, 60, 13. ([150]).
*lucifer Fabricius, System. Ent. 759, 13; Mant. Ins. II, 329, 21 *(Bibio);* Ent. System. IV, 262, 21; System. Antl. 126, 40; Wiedemann, Dipt. Exot. I, 142, 36; Auss. Zw. I, 294, 53; Bigot, in R. de la Sagra etc. 794. — West Indies; Georgia; Texas (see O. Sacken, Western Diptera 240).
Anthrax fumiflamma Walker, Dipt. Saund., 184.
*mucorea Loew, Centur. VIII, 43. — Nebraska.
*nigricauda Loew, Centur. VIII, 38. — Massachusetts (Lw.); Canada.
*palliata Loew, Centur. VIII, 32. — Illinois.
(?) *Anthrax incisa* Walker, Dipt. Saund., 187. — North America.
*parvicornis Loew, Centur. VIII, 36. — Illinois.
*pertusa Loew, Centur. VIII, 28. — Western Texas.
*scrobiculata Loew, Centur. VIII, 39. — Illinois.
*sinuosa Wiedemann, Dipt. Exot. I, 147, 42; Auss. Zw. I. 301, 64; O. Sacken, Western Dipt., 239. — Georgia (Wied.); Southern and Middle States; California.
Anthrax concisa Macquart, Dipt. Exot. II, 1, 68, 37. — Carolina (!).
Anthrax nycthemera Macquart (nec Hoffmannsegg), Dipt. Exot. II, 1, 67, 33 (!).
Anthrax assimilis Macquart, Dipt. Exot. Suppl. I, 114, 73. — Galveston, Texas.
*stenozona Loew, Centur. VIII, 40. — Illinois.
*tegminipennis Say, Long's Exped. App. 371, 2; Compl. Wr. I, 253; Wied. Auss. Zw. I, 289, 46. — N. W. Territory (Say); Iowa; Brit. N. America; Maine.
vestita Walker, List, etc. II, 258. — Nova Scotia.

*alpha O. Sacken, Western Dipt., 239. — Sierra Nevada, Cal.; Cheyenne, Wyo.
*curta Loew, Centur. VIII, 35. — California.
*diagonalis Loew, Centur. VIII, 33. — California.
*fuliginosa Loew, Centur. VIII, 31. — California.
*molitor Loew, Centur. VIII, 42. — California.

(?) àbbreviata Wiedemann, Auss. Zw. II, 637, 49. — Mexico.
*adusta Loew, Centur. VIII, 41. — Cuba.

Astarte Wiedemann, Auss. Zw. II, 637, 48. — Mexico.
*bigradata Loew, Centur. VIII, 37. — Cuba.
castanea Jaennicke, Neue Exot. Dipt. 30; Tab. II, f. 15 (wing). — Mexico.
cyanoptera Wiedemann, Auss. Zw. II, 638, 51. — Mexico.
delicatula Walker, List, etc. II, 266. — Jamaica.
faunus Fabricius, System. Antl. 126, 38; Dipt. Exot. I, 139, 30; Auss. Zw. I, 292, 50; Macquart, Dipt. Exot. II, 1, 75, 61; Tab. XXI, f. 1. — West Indies.
funebris Macquart, Dipt. Exot. II, 1, 66, 30; Tab. 21, f. 10. — San Domingo.
gorgon Fabricius, System. Antl. 126, 41; Wiedemann, Auss. Zw. I, 303, 67. — West Indies.
Nero Fabricius, System. Antl. 127, 45; Wiedemann, Dipt. Exot. I, 149, 47; Auss. Zw. 316, 85. — West Indies.
nudiuscula Thomson, Eug. Resa, etc., 482. — Panama.
paradoxa Jaennicke, Neue Exot. Dipt. 31; Tab. II, f. 16 (wing). — Mexico.
*proboscidea Loew, Centur. VIII, 27. — Sonora.
pusio Macquart, Dipt. Exot. II, 1, 76, 62; Tab. XXI, f. 1; Bigot, R. de la Sagra etc. 794. — Cuba.
quinquepunctata Thomson, Eug. Resa, etc. 484. — Panama.
*sagata Loew, Centur. VIII, 34. — Matamoras.
translata Walker, Dipt. Saund., 182. — West Indies.
trifigurata Walker, Trans. Ent. Soc. N. Ser. V, 285. — Haity.

Hemipenthes.

Loew, Centur. VIII, 44; 1869.

*morioides Say, J. Acad. Phil. III, 42, 1; Compl. Wr. II, 58 (Anthrax); Wiedemann, Auss. Zw. I, 309, 75 (id.). — Missouri (Say).
seminigra Loew, Centur. VIII, 44. ([152]). — Saskatchewan; Canada.

Argyramoeba.

Argyromoeba Schiner, Wien. Ent. Monatschr. 1860; amended by Loew, in Centur. II, 290.

*albofasciata Macquart, Dipt. Exot. II, 1, 67, 34; Tab. XXI, f. 12 (Anthrax). — Georgia (Macq.)
Anthrax analis Macquart, Hist. Nat. Dipt. I, 407, 25 (change of name by Macq.).
*analis Say, J. Acad. Phil. III, 45, 4; Compl. Wr. II, 60 (Anthrax); Wiedemann, Auss. Zw. I, 313, 80 (id.). — Atlantic States and Canada; Georgia (Say); Massachusetts, Illinois, Maryland etc.
Anthrax georgica Macquart, Hist. Nat. Dipt. I, 406, 19; Dipt. Exot. II, I, 68, 38; Tab. 21, f. 11 (!). ([153]).
*antecedens Walker, Dipt. Saund. 193 (Anthrax). — United States (Walk.).
*argyropyga Wiedemann, Auss. Zw. I, 313 (Anthrax) male. — (No habitat in Wied.); Virginia; Georgia.
Argyramoeba contigua Loew, Centur. VIII, 50 (female).

*Cephus Fabricius, System. Antl. 124, 25 *(Anthrax)*; Wiedemann, Auss.
 Zw. I, 297, 58 *(id.)*; Macquart, Dipt. Exot. II, 59, 12 *(id.)*. —
 South America (Fab., Wied.); Georgia; Virginia.
*fur O. Sacken, Western Dipt., 244. — Texas. ([154]).
*limatulus Say, J. Acad. Phil. VI, 157; Compl. Wr. II, 354 *(Anthrax)*. —
 Indiana (Say); Colorado (?); California (?); compare O. Sacken,
 Western Dipt., 243.
*Oedipus Fabricius, System. Antl. 123, 22 *(Anthrax)*; Wiedemann, Dipt.
 Exot. I, 124, 8 *(id.)*; Auss. Zw. I, 262, 12 *(id.)*. — United States
 (reaches quite far in the N. W. of the Brit. Possessions; according
 to Schiner, occurs also in South America); Mexico (Coll. Bellardi).
 Anthrax irrorata Say, J. Acad. Phil. III, 46, 6; Compl. Wr. II, 61.
 Anthrax irrorata Macquart, Dipt. Exot. II, 1, 60; Tab. XX, f. 6.
*obsoleta Loew, Centur. VIII, 47. — Missouri.
*pauper Loew, Centur. VIII, 48. — Illinois.
*Pluto Wiedemann, Auss. Zw. I, 261, 11 *(Anthrax)*; O. Sacken, Western
 Dipt., 244. — Kentucky (Wied.); occurs from Texas to Canada.
*Simson Fabricius, System. Antl. 119, 5 *(Anthrax)*; Wiedemann, Dipt.
 Exot. I, 122, 6 *(id)*; Auss. Zw. I, 259, 9; Tab. III, f. 2 *(id.)*;
 Macquart, Dipt. Exot. II, 1, 59, 11 *(id.)*. — Atlantic States; also
 in Columbia, South America (Schiner, Novara, 120).
 Anthrax scripta Say, J. Acad. Phil. III, 43, 3; Compl. Wr. II, 59.
 Nemotelus tigrinus Degeer, VI, Tab. 29, f. 11 [Wied.].
*stellans Loew, Centur. VIII, 46. — Oregon.

*Delila Loew, Centur. VIII, 45. — California.

*euplanes Loew, Centur. VIII, 49. — Cuba.
(?) disjuncta Wiedemann, Auss. Zw. II, 639, 53 *(Anthrax)*. — Mexico.
 Gideon Fabricius, System. Antl. 125, 27 *(Anthrax)*; Wiedemann, Auss.
 Zw. I, 311, 79 *(id.)*. — South America (Fabr., Wied.); Jamaica
 (Walker).
 Leucothoa Wiedemann, Auss. Zw. II, 638, 50 *(Anthrax)*. — Mexico.

Triodites.
O. Sacken, Western Dipt., 245; 1877.
*mus O. Sacken, Western Dipt., 246. — California, Utah.

Lomatia.
Meigen, System. Beschr. VI, 324; 1830; *Stygia* Meig. (preocc.); *Stygides*
 Latreille, Fam. Natur. 1825, 491.

 elongata Say, J. Acad. Phil. III, 41, 1; Compl. Wr. II, 58 *(Stygia)*;
 Wiedemann, Auss. Zw. I, 315 and 561; Tab. II, f. 6. — Pennsylvania. ([155]).

Oncodocera.
Macquart, Dipt. Exot. II, 1, 83; 1840.
*leucoprocta Wiedemann, Auss. Zw. I, 330 *(Mulio) male*. — No locality.
 (Wied.); Georgia; Virginia, Illinois, Wisconsin, Kentucky, Mexico.

Oncodocera dimidiata Macquart, Dipt. Exot. II, 1, 84 *(female);* Tab. 15, f. 1.

Anthrax terminalis Wiedemann, Auss. Zw. II, 639. — Mexico (!).

*valida Wiedemann, Auss. Zw. II, 636, 47 *(Anthrax).* — Mexico.

Anisotamia eximia Macquart, Dipt. Exot. 4e Suppl. 115; Tab. XI, f. 2 [1]. ([156]).

Leptochilus.

Loew, Centur. X, 40; 1872.

*modestus Loew, Centur. X, 40. — Texas.

Aphoebantus.

Loew, Centur. X. 39; 1872.

*cervinus Loew, Centur. X, 39. — Texas.

Bombylius. ([157]).

Linné, Fauna Suecica; 1761.

*atriceps Loew, Centur. IV, 49. — Florida, Virginia (Loew); New York; Connecticut (M. C. Z.).

*fratellus Wiedemann, Auss. Zw. I, 583, 17. — Georgia (Wied.); Northern States and Brit. Possessions (M. C. Z.).

Bombylius vicinus Macquart, Dipt. Exot. II, 1, 98, 30 [Loew, Neue Beiträge etc. III, 14].

Bombylius albipectus Macquart, Dipt. Exot. 5e Suppl. 82, 71; Tab. IV, f. 10. — Baltimore.

Bombylius aequalis Harris (nec Fab.), Ins. Injur. to Veget. 3d edit. 606 f. 263. ([158]).

Bombylius major Kirby (nec Linné), Fauna Bor. Amer. Ins. 312, 1.

*mexicanus Wiedemann, Dipt. Exot. I, 166, 10; Auss. Zw. I, 338, 11; Loew, Neue Beiträge etc. III, 24. — Middle and Southern States; Mexico.

(?) *Bombylius fulvibasis* Macquart, Dipt. Exot. 5e Suppl. 82, 72 [Loew *in litt.*]. ([159]).

Bombylius philadelphicus Macquart, Dipt. Exot. II, 1, 99, 33; Tab. VI, f. 3 and Tab. VII, f. 3 [Loew *in litt.*].

*pulchellus Loew, Centur. IV, 47. — Illinois.

*pygmaeus Fabricius, Mant. Ins. II, 367, 13; Ent. System. IV, 411, 19; System. Antl. 135, 32; Olivier, Encycl. Méth. I, 328, 22; Wiedemann, Auss. Zw. I. 351, 34; Lamarck, Anim. sans vert. III, 407, 4; Kirby, Fauna boreali-americana, Ins, 312, 2. — Atlantic States and Brit. Possessions (M. C. Z. has a specimen from Virginia).

*validus Loew, Centur. IV, 48 — Illinois; Virginia (Lw.); New York, Georgia.

*varius Fabricius, System. Antl. 132, 17; Wiedemann, Dipt. Exot. I, 163, 6; Auss. Zw. I, 335, 7; Loew, Neue Beitr. etc. III, 29. — Middle States.

*albicapillus Loew, Centur. X, 42; O. Sacken, Western Dipt., 249. —
Marin and Sonoma Co.; Cal.
*aurifer O. Sacken, Western Dipt., 249. — Sierra Nevada, Cal.
*cachinnans O. Sacken, Western Dipt., 250. — Sonoma Co., Cal.
*lancifer O. Sacken, Western Dipt. 251., — San Francisco; Yosemite
Valley.
*metopium O. Sacken, Western Dipt., 249. — Marin Co., Cal.
*major Linné, Fabricius, Meigen, etc.; O. Sacken, Western Dipt., 248. —
Europe and California.

bicolor Loew, Wien. Ent. Monatschr., V, 34. — Cuba.
*haemorrhoicus Loew, Centur. IV, 46. — Cuba.
helvus Wiedemann, Dipt. Exot. I, 164, 6 b; Auss. Zw. I, 336, 8. —
Mexico.
plumipes Drury, Illustr. etc. II; Tab. XXXIX, f. 3; Wiedemann, Auss.
Zw. I, 351, 50. — Jamaica.
*ravus Loew, Centur. IV, 50. — Matamoras.
*semirufus Loew, Centur. X, 41. — San Domingo.

Comastes.

O. Sacken, Western Dipt., 256; 1877.([160]).

*robustus O. Sacken, Western Dipt., 257. — Waco, Texas.
rufus Olivier, Encýcl. Méth. I, 327, 8 (Bombylius). — West Indics
Bombylius basilaris Wiedemann, Zool. Magaz. III, 46, 7 b:
Dipt. Exot. I, 164, 7; Auss. Zw. I, 335 [Loew, Neue Beitr
etc., III, 29, 51].

Systoechus.

Loew, Neue Beitr. etc, III, 34; 1855 (ex parte); O. Sacken, Western
Dipt, 250—253.

*candidulus Loew, Centur. IV, 51; O. Sacken, Western Dipt., 253. —
Wisconsin (Lw.); Illinois, Kansas.
*solitus Walker, List, etc. II, 288 (Bombylius); O. Sacken, Western
Dipt., 253. — Georgia, Florida.
*vulgaris Loew, Centur. IV, 52; O. Sacken, Western Dipt., 253. —
Nebraska (Lw.); Iowa; Colorado; Illinois.

*oreas O. Sacken, Western Dipt., 254. — Sierra Nevada, Cal.

Anastoechus.

O. Sacken, Western Dipt., 251; 1877.

*barbatus O. Sacken, Western Dipt., 252. — Cheyenne, Wyoming; the
same, or a similar species, all over the United States.

Pantarbes.

O. Sacken, Western Dipt., 254; 1877.

*capito O. Sacken, Western Dipt., 256. — Sonoma Co., Cal.

Sparnopolius.

Loew, Neue Beitr. etc., III, 43; 1855.

*brevicornis Loew, Centur. X, 43; O. Sacken, Western Dipt., 259. — Texas.
*coloradensis Grote, Proc. Ent. Soc. Phil. VI, 445; O. Sacken, Western Dipt., 259; — Colorado.
cumatilis Grote, Proc. Ent. Soc. Phil. VI, 445. — Colorado.
*fulvus Wiedemann, Dipt. Exot. I, 172, 22 (Bombylius); Auss. Zw. I, 347, 27 (id.); Loew, Neue Beitr. etc, III, 43. — Atlantic States.
Bombylius L'herminieri Macquart, Dipt. Exot. II, 1, 103, 44 [!]; Tab. VII, f. 7.
Bombylius brevirostris Macquart, Dipt. Exot. II, 1, 103, 43 [!]. ([101]).

apertus Macquart, Dipt. Exot. 2e Suppl. 54, 50. (Bombylius). — Guadeloupe [Loew in litt. supposes this to belong to Dischistus].

Lordotus.

Loew, Centur. IV, 53; 1863.

*gibbus Loew, Centur. IV, 53; O. Sacken, Western Dipt., 258. — Matamoras (Lw.); Colorado; California.
Adelidea flava Jaennicke, Neue Exot. Dipt. 39. — Mexico. ([162]).
*planus O. Sacken, Western Dipt., 258. — California.

Ploas.

Latreille, Dict. d'hist. nat. Vol. XXIV; 1804.
Meigen, System. Beschr. II, Tab. 19, f. 6.

pictipennis Macquart, Dipt. Exot. II, 1, 107, 2; Tab. IX, f. 3. — Carolina.

*amabilis O. Sacken, Western Dipt., 261. — Yosemite Valley, Cal.
*atratula Loew, Centur. X, 44. — California.
*fenestrata O. Sacken, Western Dipt., 260. — California.
*nigripennis Loew, Centur. X, 45. — California.
*obesula Loew, Centur. X, 46. — California.
*rufula O. Sacken, Western Dipt., 261. — California.
*limbata Loew, Centur. VIII, 51. — New Mexico.

Paracosmus.

O. Sacken, Western Dipt., 262; 1877; Allocotus Loew, Centur. X, 48; 1872. ([163]).

*Edwardsii Loew, Centur. X, 48 (Allocotus). — San Francisco, Cal.

Phthiria.

Meigen in Illig. Mag. II, 268; 1803; Poecilognathus Jaennicke, Neue Exot. Dipt., 43.

punctipennis Walker, List, etc II, 294. -- Georgia.

*sulphurea Loew, Centur. III, 18; O. Sacken, Western Dipt., 262. —
  New Jersey (Lw.); Texas and Colorado.
*scolopax O. Sacken, Western Dipt., 263. — Manitou, Colorado.

*egerminans Loew, Centur. X, 47. — California.
*humilis O. Sacken, Western Dipt., 264. — Sonoma Co., California.
*notata Loew, Centur. III, 19. — California.

thlipsomyzoides Jaennicke, Neue Exot. Dipt. 43; Tab. I, f. 11
  (*Poecilognathus* nov. gen.). — Mexico. ([104]).

, Geron.
Meigen, System. Beschr. II, 223; 1820.

*calvus Loew, Centur. IV, 54. — New York.
holosericeus Walker, List, etc. II, 295. — Georgia.
*macropterus Loew, Centur. IX, 76. — New York.
*senilis Fabricius, Ent. System. IV, 411, 17; System Antl. 135, 31
  (*Bombylius*); Wiedemann, Auss. Zw. I, 357, 1; Macquart, Dipt.
  Exot. Suppl. I, 119. — West Indies (Wied.); Galveston, Texas (Macq.).
*subauratus Loew, Centur. IV, 55; compare also IX, 77, Nota. —
  Pennsylvania.
*vitripennis Loew, Centur. IX, 77. — Middle States.

*albidipennis Loew, Centur. IX, 78. — California.

insularis Bigot, in R. de la Sagra etc. 792. (*Bombylius*). — Cuba.
rufipes Macquart, Dipt. Exot. Suppl. I, 119. — Yucatan.

Systropus.
Wiedemann, Nova Dipt. Genera, 1820; *Cephenus* Latreille, Fam.
Natur. 1825, 496.

*macer Loew, Centur. IV, 56; about the larva see O. Sacken, Western
  Dipt, 265. — Atlantic States (I have seen it from Kansas as the
  most western locality).

*foenoides Westwood, Magazin de Zoologie 1842. Ins. Tab. 90. —
  The same in Trans. Ent. Soc. London 1876, 578. — Mexico.

Lepidophora.
Westwood, Lond. and Edinb. Phil. Mag. 1835.

*aegeriiformis Westwood, Lond. and Edinb. Phil. Mag. 1835; VI,
  447; Macquart, Dipt. Exot. Suppl. I, 115, 1; Tab. X, f. 1;.Gray,
  in Griffith's Anim. Kingd. XV, Ins. 2, 779; Tab. 126, f. 6 (*Ploas*). —
  Georgia; Illinois; Kansas.
appendiculata Macquart, Dipt. Exot. Suppl. I, 118, 2; Tab. XX, f.
  4 (*Toxophora*). — Galveston, Texas.
ledipocera Wiedemann, Auss. Zw. I, 360, 1; Tab. V, f. 4 (*Toxophora*);
  Macquart, Dipt. Exot. II, 1, 119; ibid. Suppl. I, 119. — No patria
  (Wied); North America? (Macq.).

Toxophora.

Meigen, in Illig. Mag. II. 270; 1803.

*Amphitea Walker, List, etc. II, 298; O. Sacken, Western Dipt. 267. — Florida (Walk.); Middle and Southern States.

americana Guérin, Iconogr. etc. Insectes, Tab. 95, f. 1 (No description). — North America.

leucopyga Wiedemann, Auss. Zw. I, 361, 2; Macquart, Dipt. Exot. II, 1, 117; Tab. XIII, f. 1. — No locality in Wiedemann; Carolina (Macq.); Georgia (Walker, List, etc. II, 298 „Synon. of *T. fulva?*" ([165]).

Toxophora fulva Gray, Griffith's Anim. Kingd. XV, Ins. 2, 779; Tab. 126, f. 5.

*fulva O. Sacken (non Gray), Western Dipt., 267. — Georgia.

*virgata O. Sacken, Western Dipt., 266. — Texas, Georgia.

Epibates.

O. Sacken, Western Dipt. 268; 1877. ([166]).

funestus O. Sacken, Western Dipt., 271. — White Mts., N. H.

Harrisii O. Sacken, Western Dipt., 273. — Atlantic States (?).

*niger Macquart, Hist. Nat. Dipt. I, 390 (*Apatomyza*); Dipt. Exot. II, 1, 111, 1; Tab. IV, f. 1 (*id.*); O. Sacken, Western Dipt., 273. — Georgia. ([167]).

Cyllenia aegiale Walker, List, etc. II, 296 and ibid. IV, 1154.

*luctifer O. Sacken, Western Dipt., 271. — Vancouver Isl.

*magnus O. Sacken, Western Dipt., 272. — Vancouver Isl.

*marginatus O. Sacken, Western Dipt., 272. — San Francisco, Cal.

*muricatus O. Sacken, Western Dipt., 272. — Sierra Nevada, Cal.; Colorado Mts. (9000 feet altitude; Morrison).

Osten Sackenii Burgess, Proc. Boston Soc. N. H., 1858, 323; Tab. IX, f. 1. — Southern Colorado; Upper Leavenworth Valley, Kansas.

Thevenemyia.

Bigot, Bullet. Soc. Ent. de France 1875, CLXXIV. ([168]).

californica Bigot, l. c. — California.

FAMILY THEREVIDAE.

Psilocephala.

Zetterstedt, Ins. Lapp. 525, Nota; 1840; Dipt. Scand. I, 211.

*erythrura Loew, Centur. IX, 75. — Middle States.

*melampodia Loew, Centur. VIII, 12. — Illinois.

*munda Loew, Centur. VIII, 13. — Wisconsin.

*melanoprocta Loew, Centur. VIII, 15. — Northern United States.

*nigra Say, J. Acad. Phil. III, 40, 2; Compl. Wr. II, 57 (*Thereva*); Wiedemann, Auss. Zw. I, 235, 12 (*id.*). — United States.

Thereva haemorrhoidalis Macquart; Dipt. Exot II, 1, 26, 9 (♂).
*notata Wiedemann, Dipt. Exot. I, 114, 8; Auss. Zw. I, 236, 14
(*Thereva*). — Georgia.
*pictipennis Wiedemann, Dipt. Exot. 113, 6 (*Thereva*); Auss. Zw. I,
235, 11 (*id.*). — Georgia.
*platancala Loew, Zeitschr. für Ges. Naturw. Dec. 1876., 321. — Texas.
*rufiventris Loew, Centur. VIII. 17. — Nebraska
*scutellaris Loew, Centur. IX, 74. — Distr. Columbia.
*variegata Loew, Centur. IX, 73. — Canada.

*costalis Loew, Centur. VIII, 16. — California.
*laevigata Loew, Zeitschr. für Ges. Naturw. Dec. 1876, 319. — San
Francisco.
*longipes Loew, Centur. VIII, 11. — Cuba.
nigra Bellardi, Saggio, etc. II, 92 (Dr. Schiner, in Novara etc. 146,
identifies this species with one from Chile, but changes the name
for *P. penthoptera* on account of *P. nigra* Say). — Mexico.
*platycera Loew, Centur. II, 290, line 3 from bottom.
Thereva laticornis Loew, Centur. VIII, 14. — Cuba [change of
name by the author].
univittata Bellardi, Saggio, etc. II, 90. — Mexico.
Sumichrasti Bellardi, Saggio, etc. II, 91. — Mexico.

Thereva.

Latreille, Précis etc. 1796; *Thereua* (Loew). ([168]).

*albiceps Loew, Centur. IX, 69. — Red River of the North; Lake
. Winnipeg.
albifrons Say, J. Acad. Phil. VI, 156; Compl. Wr. II, 353. —
Indiana.
*candidata Loew, Centur. VIII, 10. — Northern United States;
Canada. ([169]).
corusca Wiedemann, Auss. Zw. I, 232, 7. — East Florida,
Thereva tergissa Say, J. Acad. Phil. III, 39, 1 (Compl. Wr. II, 57).
*flavicincta Loew, Centur. IX, 70. — Northern Wisconsin River;
White Mts., N. H.
frontalis Say, Long's Exped. App. 370; Compl. Wr. I, 252; Wiedemann, Auss. Zw. I, 230, 2. — N. W. Territory (Say).
*gilvipes Loew, Centur. IX, 71. — Massachusetts.
*strigipes Loew, Centur. IX, 72. — Lake Winnipsg.
ruficornis Macquart, Dipt. Exot. II, 1, 25, 8. — Carolina.

*comata Loew, Centur VIII, 9. — California.
*fucata Loew, Centur. X, 37. — California.
*hirticeps Loew, Berl. Ent. Zool. 1874, 382. — San Francisco.
*melanoneura Loew, Centur. X, 36. — California.
*melanophleba Loew, Zeitschr. f. Ges. Naturw. 1876, 317. — San
Francisco.
*vialis O. Sacken, Western Dipt. 274. — Yosemite Valley, Calif.

crassicornis Bellardi, Saggio, etc II, 88; Tab. II, f. 16. — Mexico.
argentata Bellardi, Saggio, etc. II, 89. — Mexico.

> **Observation.** Mr. Walker's Therevae:
> **conspicua** Walker, List, etc. I, 223. — Nova Scotia.
> **germana** Walker, List, etc. I, 222. — Florida.
> **nervosa** Walker, List, etc. I, 223. — Georgia.([170]).
> **senex** Walker, List, etc. I, 224 — Nova Scotia.
> **varia** Walker, List, etc. I, 221. — Florida.
> **vicina** Walker, List, etc. I, 222. — Nova Scotia.
> *Thereva plagiata* (Harris) Walker is *Stichopogon trifasciatus* (Say).
> These species are represented in the Brit. Mus. by a single specimen each,
> except T. germana, of which there are two. Most of them will coincide I think
> with Say's and Loew's species; the others will hardly be recognizable from Mr.
> Walker's descriptions.

Xestomyza.

Wiedemann, Nova Dipt. Genera, 1820.

*****planiceps** Loew, Centur. X, 38. — California.

> Observation. The genera *Baryphora* Loew, Stett. Ent. Z.
> 1844 p. 123; Tab. II, f. 1—5, and *Cionophora* Egger, Verh. Zool.
> Bot. Ver. 1854; Tab. I, f. 1, 2 are evidently related to *Xesto-
> myza*, although Schiner has, perhaps prematurely, united them with
> it. The antennae of *Baryphora*, as figured by Loew, are remar-
> kably like those of *Tabuda*, but look very different from the
> figure of the antennae of *Cionophora*.

Tabuda.

Walker, Dipt. Saund., 197; 1850—56.

*****fulvipes** Walker, Dipt. Saund., 197; Tab. VI, f. 4. — New Jersey
(Evett, Proc. Ent. Soc. Phil. I, 217); (Walker gives no locality);
Georgia (coll. v. Roeder).

FAMILY SCENOPINIDAE.

Scenopinus.

Latreille, Hist. Nat. des Cr. et des Ins. XIV; 1804. ([171]).

*****bulbosus** O. Sacken, Western Dipt, 275. — Missouri.
*****fenestralis** Linné, Meigen, etc. — Europe and North America.
> *Scenopinus pallipes* Say, J. Acad. Phil. III, 100; Compl. Wr. II,
> ·86: Wiedemann, Auss. Zw. II, 233 [Loew, in Sillim. Journ N.
> S. XXXVII, 318].

*****laevifrons** Meigen, Loew, Verh. Zool. Bot. Ver. 1857. — Europe and
North America. [The american specimens were identified by Loew;
compare Sillim. Journ. l. c.]

*****nubilipes** Say, J. Acad. Phil. VI, 170; Compl. Wr. II, 362. — Indiana
(Say); Cuba; Florida [Loew, *in litt.*].

*****albidipennis** Loew, Centur. VIII, 53. — Cuba.

10

Pseudatrichia.

O. Sacken, Western Dipt., 275; nomen novum vice *Atrichia*, Loew, Centur. VII, 76; 1866.

longurio Loew, Centur. VII, 76 (*Atrichia*). — Mexico.

FAMILY CYRTIDAE. [172]

Acrocera.
Meigen in Illiger's Magaz.; 1803.

*bimaculata Loew, Centur. VI, 53. — Distr. Columbia.
bulla Westwood, Trans. Ent. Soc. V, 98. — New York.
fasciata Wiedemann, Auss. Zw. II, 16, 2; Erichson, Ent. I, 166, 4. — Georgia.
fumipennis Westwood, Trans. Ent. Soc. V, 98. — Georgia.
nigrina Westwood, Trans. Ent. Soc. V, 98. — Georgia.
obsoleta v. d. Wulp, Tijdschr. v. Ent. 2e Ser. II, 139; Tab. III, f. 17. — Wisconsin.
subfasciata Westwood, Trans. etc. V, 98. — New York.
unguiculata Westwood, Trans. etc. V, 98. — Georgia.

Opsebius.
Costa, Rendic. di Soc. R. Borbon. Acad. d. Sc. V. 20; 1856.

Pithogaster Loew, Wien. Ent. Monatschr. I, 33, 1857. [173].
*gagatinus Loew, Centur. VI, 34. — Pennsylvania.
*sulphuripes Loew, Centur. IX, 68. — Sharon Springs, N. Y.

*diligens O. Sacken, Western Dipt., 278. — Vancouver's Isl.
*paucus O. Sacken, Western Dipt., 279. — California.

Pialoidea.
Westwood, Trans. Ent. Soc. Lond. 1876, 514.

magna Walker, List, etc. III, 511 (*Cyrtus*). — Georgia.

Ocnaea.
Erichson, Entomogr.; 1840.

micans Erichson, Entomogr. I, 155, 1. — Mexico.
*helluo O. Sacken, Western Dipt., 278. — Dallas, Texas.

Apellcia.
Bellardi, Saggio, etc. Append. 19, 1862.

vittata Bellardi, Saggio, etc. App. p. 19, fig, 12. — Mexico.

Pterodontia.
Gray, in Griffith's Anim. Kingd. 1832; see also Westwood, Tr. Ent Soc V.

analis Westwood, Trans. Ent. Soc. V, 97. — Georgia.
NB. There is another *Pt. analis* Macq. from New Granada.

flavipes Gray, in Griffith's Anim. Kingd. CXXVIII, f. 3; Westwood, Trans. Ert. Soc. V, 96. — Georgia.
*misella O. Sacken, Western Dipt., 277. — Oregon.

Eulonchus.
Gerstaecker, Stett. Ent. Zeit.; 1856.
*marginatus O. Sacken, Western Dipt., 277. — Napa Valley, California.
*sapphirinus O. Sacken, Western Dipt., 276. — Sierra Nevada, California.
*smaragdinus Gerstaecker, Stett. Ent. Z. 1856, 360; O. Sacken, Western Dipt , 276. — San Francisco, California.
*tristis Loew, Centur X, 19. — Coast Range Mts., California.

Lasia.
Wiedemann, Analecta etc.; 1824.
*Kletti O. Sacken, in Lieut. Wheeler's Report Expl. and Surveys etc. Vol. V, Zool. 804; with woodcuts. — Camp Apache, Arizona.

Oncodes.
Latreille, Précis etc. 154; 1796.
*costatus Loew, Centur. IX, 67. — Massachusetts.
*dispar Macquart, Dipt. Exot. 5e Suppl. 67, 1; Tab. II, f. 12 *(Henops).*— Baltimore.
*eugonatus Loew, Centur. X, 18. — Texas.
*incultus O. Sacken, Western Dipt., 279. — White Mts., N. H.
*pallidipennis Loew, Centur. VI, 32. — Pennsylvania.

*melampus Loew, Centur. X, 17. — California.

Philopota.
Wiedemann, Auss. Zw. II, 17; Tab. 9, f. 1; 1830.
Truqnii Bellardi, Saggio, etc. I, 77; Tab. II. f. 20. — Mexico.

FAMILY EMPIDAE.

SECTION HYBOTINA.

Hybos.
Meigen, in Illiger's Magaz II; 1803.
purpureus Walker, List, etc. III, 486 — Georgia.
reversus Walker, l. c. 487. — Trenton Falls.
subjectus Walker, l. c. 487. — Huds. B. Terr.
*triplex Walker, List, etc. III, 486. — Trenton Falls. [174]
  Hybos .duplex Walker, List, etc. III, 486.

dimidiata Loew, Wien. Ent. Monatschr. V, 36. — Cuba.
dimidiata Bellardi, Saggio, etc. II, 97. — Mexico.
(This and the preceding species where published in the same year, 1861.)

Syneches.

Walker, Dipt. Saund., 165; 1850—56; Loew, Dipternfauna Südafrika's, 259;
Pterospilus Rondani. ([175]).

*albonotatus Loew, Centur. II, 18. — Distr. Columbia.
*pusillus Loew, Centur. I, 25. — New York; Chicago.
*rufus Loew, Centur. I, 24. — New York; Chicago.
*simplex Walker, Dipt. Saund., 165; Tab. V, f. 7 *(Syneches)*. — Atlantic
    States.
    Syneches punctipennis v. d. Wulp, Tijdschr. v. Ent. 2d Ser. II, 139;
    Tab. III, f. 18—21 [Loew, Zeitschr. f. Ges. Naturw. Vol XXXVII, 115].
*thoracicus Say, J. Acad. Phil. III, 76, 1; Compl. Wr. II, 68 *(Hybos)*;
    Wiedemann, Auss. Zw. I, 538, 3 *(id.)*; Macquart, Dipt. Exot. I,
    2, 156, 1; Tab. XIII, f. 1 *(id.)*. — Atlantic States.

Syndyas.

Loew, Dipternfauna Südafrika's, 260; 1860. ([175]).

*dorsalis Loew, Centur. I, 26. — New York.
*polita Loew, Centur. I, 27. — Carolina.

Brachystoma.

Meigen, System. Beschr. III, 12; 1822.

*binummus Loew, Centur. II, 16. — Distr. Columbia.
*nigrimana Loew, Centur. II, 17. — Illinois.
*serrulata Loew, Centur. I, 23. — Georgia; Ohio.

    Observation. In a note to Centur. II, 17 Loew proposes for
these three species the formation of a new genus, *Blepharoprocta*,
distinguished by the first submarginal cell being closed.

Ocydromia.

Meigen, System. Beschr. II, 311; 1820.

peregrinata Walker, List, etc. III, 488. — Trenton Falls.
glabricula Fallen, Meigen, etc. — Europe and Sitka (Loew, *in litt.*).

SECTION EMPINA. ([176]).

Empis.

Linné, Fauna Suecica; 1763; Meigen, System. Beschr. III, 15.

Abcirus Walker, List, etc. III, 494. — Georgia.
Aghastus Walker, List, etc. III, 496. — Huds. B. Terr.
Amytis Walker, List, etc. III, 493. — New York.
*armipes Loew, Centur. I, 32. — New York.
Colonica Walker, List, etc. III, 498. — Nova Scotia.
Cormus Walker, List, etc. III, 496. — Huds. B. Terr.
distans Loew, Centur. VIII, 54. — Georgia.
Eudamides Walker, List, etc. III, 493. — North America.

geniculata Kirby, N. Am. Zool. Ins. 311, 2. — British America.
laniventris Eschscholz, Ent. I, 113, 83; Wiedemann, Auss. Zw. II, 6,
    12; Macquart, Dipt. Exot. I, 2, 162 (*Eriogaster* n. gen.). — Unalaschka.
*labiata Loew, Centur. I, 33. — Distr. Columbia.
*laevigata Loew, Centur. V, 49. — White Mts., N. H.
*leptogastra Loew, Centur. III, 30. — Distr. Columbia.'
*longipes Loew, Centur. V, 51. — Lake George, N. Y.; New Jersey.
luctuosa Kirby, N. Am. Zool. Ins. 311, 1. — British America.
*nuda Loew, Centur. II, 20. — Illinois.
Ollius Walker, List, etc. III, 493. — Nova Scotia
*obesa Loew, Centur. I, 28. — Massachusetts.
*pallida Loew, Centur. I, 30. — New York.
*poeciloptera Loew, Centur. I, 31. — New York.
*poplitea Loew, Centur. III, 29. — Sitka.
reciproca Walker, Trans. Ent. Soc. N S. IV, 147. -- United States.
*rufescens Loew, Centur. V, 52. — White Mts, N. H.
*sordida Loew, Centur. I, 29. — Distr. Columbia.
*spectabilis Loew, Centur. II, 21. — Maryland.
*stenoptera Loew, Centur. V, 50. — White Mts., N. H.
*varipes Loew, Centur. I, 34. — Pennsylvania.

*barbata Loew, Centur. II, 19. — California.

atra Wiedemann, Auss. Zw. II, 1, 1. — St. Croix.
bicolor Bellardi, Saggio, etc. II, 98. — Mexico.
cyanea Bellardi, Saggio, etc. II, 98. — Mexico.
*spiloptera Wiedemann, Auss. Zw. II, 5, 10. - Mexico.
    Empis picta Loew, Centur. III, 28 and Vol. I, 261, where the
    synonymy is acknowledged.
suavis Loew, Centur. VIII, 56 — Mexico (type in Berl. Mus).
superba Loew, Wien. Ent Mon. V, 36; Centur. VIII, 57. — Cuba
totipennis Bellardi, Saggio, etc. II, 99. — Mexico.
violacea Loew, Centur. VIII, 55. — Mexico (type in Berl. Mus.).

Pachymeria.

Stephens, System. Catal. 1829; Macquart, H. N. Dipt. I, 333, *Pachymerina;*
    but in Vol. II, 657 he adopts Stephens's earlier name. ([177]).

*brevis Loew, Centur. II, 22. — Distr. Columbia.
*pudica Loew, Centur. I, 35; Wien. Ent. Monatschr. VIII, 12, 5 (the
    Pachymeria tumida quoted there as a synonym of *P. pudica,*
    does not exist). — Distr. Columbia.

Iteaphila.

Zetterstedt, Ins. Lapponica 541; 1840. ([178]).

*Macquartii Zetterstedt, Ins. Lapponica 541. — Northern Sweden;
    also in North America (White Mts.; Quebec).

Microphorus.

Macquart, Dipt. du Nord etc. 140; 1827; *Trichina* Meigen. ([178]).
drapetoides Walker, List, etc. III, 489. — Huds. B. Terr.

Rhamphomyia.

Meigen, System. Beschr. III; 1822.

Agasicles Walker, List, etc. III, 499. — Huds. B. Terr.
americana Wiedemann, Auss. Zw. II, 8, 3. — North America.
Anaxo Walker, List, etc III, 500. — Huds. B. Terr.
* angustipennis Loew, Centur. I, 55. — New York.
* aperta Loew, Centur. II, 27. — Illinois.
* basalis Loew, Centur. V, 54. — White Mts., N. H.
* brevis Loew, Centur. I, 52. — Distr. Columbia.
* candicans Loew, Centur. V, 61. — White Mts., N. H.
* clavigera Loew, Centur. I, 53. — New York.
cilipes Say, J. Acad. Phil. III, 95, 2; Compl. Wr. II, 83 *(Empis)*;
     Wiedemann, Auss. Zw. II, 7, 2. — Ohio.
* conjuncta Loew, Centur. I, 56. — Distr. Columbia.
Cophas Walker, List, etc. III, 499. — New York.
* corvina Loew, Centur. I, 51. — New York.
* crassinervis Loew, Centur. I, 59. — New York.
Dana Walker, List, etc. III, 502. — Huds. B Terr.
Daria Walker, List, etc. III, 503 — New York.
* debilis Loew, Centur I, 45. — Saskatchewan.
* dimidiata Loew, Centur. I, 36. — Maryland; Massachusetts.
Ecetra Walker, List, etc III, 500. — Georgia.
* exigua Loew, Centur. II, 32. — Illinois; Distr. Columbia.
•   expulsa Walker, Trans. Ent. Soc. N. S. IV, 148. — United States.
Ficana Walker, List, etc III, 501. — Huds. B. Terr.
flavirostris Walker, List, etc. III, 501. — Huds. B. Terr.
* frontalis Loew, Centur. II, 28. — Illinois.
* fumosa Loew, Centur. I, 39. — New York; Distr Columbia.
* gilvipes Loew, Centur. I, 48. — New York; Illinois.
* glabra Loew, Centur. I, 41. — Virginia; Illinois; Distr. Columbia.
* gracilis Loew, Centur. I, 43. — Pennsylvania.
* hirtipes Loew, Centur. V, 59. — White Mts., N. H.
* impedita Loew, Centur. II, 31. — Illinois; Distr. Columbia.
* incompleta Loew, Centur. III, 31. — Distr. Columbia.
* irregularis Loew, Centur V, 60. — White Mts., N. H.
laevigata Loew, Centur. I, 37. — Nebraska.
* leucoptera Loew, Centur. I, 62. — Distr. Columbia.
* limbata Loew, Centur. I, 60. — Distr. Columbia.
* liturata Loew, Centur. I, 61. — Distr. Columbia
* longicauda Loew, Centur. I, 38. — Distr. Columbia.
* longicornis Loew, Centur. I, 47. — Distr. Columbia.
* longipennis Loew, Centur I, 46. — Distr. Columbia.
* luctifera Loew, Centur. I, 50. — New York.
* luteiventris Loew, Centur. V, 57. — White Mts, N. H.

*macilenta Loew, Centur. V, 55. — White Mts., N. H.
Mallos Walker, List, etc. III, 502. — Huds. B. Terr.
Minytus Walker, List, etc. III, 502. — Huds. B. Terr.
*mutabilis Loew, Centur. II, 26. — Illinois.
*nana Loew, Centur. I, 64. — Maryland.
*nigricans Loew, Centur. V, 58. — White Mts, N. H.
nigrita Zetterstedt, Ins. Lapp. 567; Stager, Groenl. Antl. 357, 22;
    Holmgren, Ins. Nordgroenl, 100. — Greenland.
    Empis borealis Fabricius, Fauna Groenl 211, 174 [Schiödte].
nitidivittata Macquart, Dipt. Exot., 1er Suppl. 97, 2. — Galveston,
    Texas.
Phemius Walker, List, etc. III, 500. — Huds. B. Terr.
*pectinata Loew, Centur. I, 49. — Distr. Columbia.
*polita Loew, Centur II, 29. — Illinois; Distr. Columbia.
*priapulus Loew, Centur. I, 54. — Maryland.
pulchra Loew, Centur. I, 40. — New York.
*pulla Loew, Centur. I, 44. — Connecticut.
*pusio Loew, Centur. I, 63. — Maryland.
quinquelineata Say, J. Acad. Phil. III, 95; Compl. Wr. II, 82 *(Empis)*;
    Wiedemann, Auss. Zw. II, 7, 1. — Missouri.
rufirostris Say, J. Acad. Phil. III, 159; Compl. Wr. II, 355. — Indiana.
*rava Loew, Centur. II, 25. — Illinois.
*rustica Loew, Centur. V, 56. — White Mts., N. H.
*scolopacea Say, J. Acad. Phil. III, 96, 3; Compl. Wr. II, 83 *(Empis)*;
    Wiedemann, Auss Zw. II, 8, 4. — Pennsylvania.
*sellata Loew, Centur. I, 42. — Distr. Columbia.
*soccata Loew, Centur. I, 67. — Mississippi.
*sordida Loew, Centur. I, 58. — Distr. Columbia.
*testacea Loew, Centur. II, 24. — Illinois; Maryland; Distr. Columbia.
*tristis Walker, Trans. Ent. Soc. N. S. IV, 148. — United States.
*umbilicata Loew, Centur. I, 65. — Pennsylvania; Maine ("Mexico'
    in the Centuries is an error).
*umbrosa Loew, Centur. V, 53. — White Mts., N. H.
*ungulata Loew, Centur. I, 66. — Maine ("Mexico" in the Centuries
    is erroneous).
*unimaculata Loew, Centur. II, 33. — Illinois; Distr. Columbia.
*vara Loew, Centur. I, 57. — Nebraska.
*vittata Loew, Centur, II, 23. — Illinois.

*luctuosa Loew, Centur. Vol. II, 290, line 2 from bottom. (Change
    of name.)
    Rhamphomyia lugens, Loew, Centur. II, 30. — California.

Hilara.

Meigen, System. Beschr. III; 1822.

*atra Loew, Centur. II, 42. — Illinois.
*basalis Loew, Centur. II, 45. — Illinois.
*brevipila Loew, Centur. II, 41. — Illinois.

*femorata Loew, Centur. II, 35. — Maryland.
*gracilis Loew, Centur. II, 44. — Pennsylvania.
*leucoptera Loew, Centur. II, 43 — Florida.
*lutea Loew, Centur. III, 33. — Distr. Columbia.
*macroptera Loew, Centur. III, 32. — Distr. Columbia.
  migrata Walker, List, etc. III, 491. — Huds. B. Terr.
*mutabilis Loew, Centur. II, 40. — Illinois.
*nigriventris Loew, Centur. II, 38. — Pennsylvania.
  plebeja Walker, Trans. Ent. Soc. N. S. IV, 148. — United States.
*seriata Loew, Centur. V, 63. — White Mts., N. H
*testacea Loew, Centur. V, 64. — White Mts., N. H. (the typical spe-
  cimens are from New Rochelle, N. Y.).
  transfuga Walker, List, etc. III, 492. — Huds. B. Terr.
*tristis Loew, Centur. V, 62. — White Mts., N. H.
*trivittata Loew, Centur. II, 39. — Illinois.
*umbrosa Loew, Centur. II, 34. — Illinois.
*unicolor Loew, Centur. II, 37. — Maryland.
*velutina Loew, Centur. II, 36. — Distr. Columbia.

Hormopeza.
Zetterstedt, Ins. Lapp. 540; 1840.

*brevicornis Loew, Centur. V, 65. — Yukon River, Alaska.
*nigricans Loew, Centur. V, 66. — Yukon River, Alaska.

Gloma.
Meigen, System. Beschr. III, 14; 1822.

Phthia Walker, List, etc. III, 492. — Trenton Falls, N. Y. [„Is not
  a Gloma"; Loew in litt.]
*obscura Loew, Centur. V, 68. — White Mts., N. H.
*rufa Loew, Centur. V, 67. — White Mts., N. H.

Cyrtoma.
Meigen, System. Beschr. IV, 1; 1824.

*femorata Loew, Centur. V, 69. — White Mts., N. H.
*halteralis Loew, Centur. II, 46. — Distr. Columbia.
*longipes Loew, Centur. II, 47. — Illinois; Pennsylvania.
*pilipes Loew, Centur. II, 48. — Illinois. Vid. Nr. 411.
*procera Loew, Centur. V, 70. — Sitka.

Leptopeza.
Macquart, Dipt. du Nord etc.; 1827.

*flavipes Meigen, System. Beschr. II, 353. — Europe and North America
  (Saskatchewan Riv.).

SECTION TACHYDROMINA.
Stilpon.
Loew, Neue Beiträge VI, 34, line 21 from top; also p. 43; 1859. [179]
*varipes Loew, Centur. II, 58. — Pennsylvania.

Drapetis.
Meigen, System. Beschr. III; 1822. [179].

*divergens Loew, Centur. X, 62 — Texas.
*gilvipes Loew, Centur. X. 61. — Texas.
nigra Meigen, Macquart, etc. — Europe and North America (according to Walker, List, etc. III, 511).
*pubescens Loew, Centur. II, 57. — New York.
*unipila Loew, Centur. X, 60. — Texas.

Tachydromia.
Meigen, Illiger's Magaz. 1803; System. Beschr. III, 67, Divis. B (on the plate, the genus is called *Sicus*); Loew, Schles Z. für Entom. 1863. *Platypalpus* Macquart, Dipt. du Nord etc.; Schiner, Fauna Austriaca. Compare note [180].

All the species enumerated below where described by Dr. Loew as *Platypalpus*; but in the Centuries, Vol. II, page 289 he recommends to change the name for *Tachydromia*.

*aequalis Loew, Centur. V, 75. — Illinois.
Alexippus Walker, List, etc. III, 510. — Huds. B. Terr.
*apicalis Loew, Centur. V, 79. — Pennsylvania.
*debilis Loew, Centur. III, 37. — Distr. Columbia.
*discifer Loew, Centur. III, 36. — Distr. Columbia.
*flavirostris Loew, Centur. V, 80 — White Mts., N. H.
*laeta Loew, Centur. V, 81. — White Mts., N. H.
*lateralis Loew, Centur. V, 78. — White Mts., N. H.
*mesogramma Loew, Centur. III, 38. — Distr. Columbia; New York.
*pachycnema Loew, Centur. V, 77. — Distr. Columbia; Tarrytown, New York.
*trivialis Loew, Centur. V, 76. — Maine; Distr. Columbia.
vicarius Walker, Trans. Ent Soc. N. Ser. IV, 149. — United States.

Bacis Walker, List, etc. III, 510. — Jamaica.

Phoneutisca.
Loew, Centur. III, 35; 1863.
*bimaculata Loew, Centur. III, 35. — Sitka.

Tachypeza.
Meigen, System. Beschr. VI, p. 341, 1830; and VII, p. 94, 1838 (*Tachydromia* Meig. Div. A.; *Tachydromia* Macquart, Schiner). [180].
*clavipes Loew, Centur. V, 73. — Illinois.
fenestrata Say, J. Acad. Phil. III, 95; Compl. Wr. II, 82 (*Sicus*); Wiedemann, Auss. Zw. II, 12, 1 (*Tachydromia*). — Middle States.
maculipennis Walker, List, etc. III, 507 (*Tachydromia*). — Huds. B. Terr.
portaecola Walker, l. c. III, 506 (*Tachydromia*). — Huds. B. Terr.
postica Walker, Trans. Ent. Soc. N. S. IV, 149 (*Tachydromia.* — United States.

*pusilla Loew, Centur. V, 74. — Illinois.
*rapax Loew, Centur. V, 71. — Illinois.
*rostrata Loew, Centur. V, 72. — White Mts., N. H.; New York.
similis Walker, List, etc. III, 506 *(Tachydromia)*. — Huds. B. Terr.
vittipennis Walker, Trans. Ent. Soc. N. Ser. IV, 149 *(Tachydromia)*. —
    United States.
*Winthemi Zetterstedt, Insecta Lapp. 548; Dipt. Scand. I, 321. —
    Northern Europe; White Mts., N. H. (Found by me on the walls
    of the Half-Way House on Mount Washington).

Ardoptera.
Macquart, Dipt. du Nord etc.; 1827. ([181]).

*irrorata Fallen, Meigen, etc ; Walker, Ins. Brit. I, 103, 1; Tab. III,
    f. 5. — Europe and North America. [Loew *in litt.*]

Synamphotera.
Loew, Zeitschr. für Ges. Naturw. Vol. XI, 453; 1858; compare also
    the same, Beschr. Eur. Dipt. II, 255. ([182]).

*bicolor Loew, Centur. III, 34. — Sitka.

Hemerodromia.
Meigen, System. Beschr. III, 1822. ([183]).

albipes Walker, List, etc. III, 505. — Huds. B. Terr.
*defecta Loew, Centur. II, 55. — Distr. Columbia.
*notata Loew, Centur. II, 53. — Illinois; Pennsylvania.·
*obsoleta Loew, Centur. II, 52. — Illinois; Maryland.
precatoria Meigen, etc. — Europe and North America ͵Huds. B. Terr.
    according to Walker, List, etc. III, 505).
*scapularis Loew, Centur. II, 54. — Maryland.
superstitiosa Say, Long's Exped. App. 376; Compl. Wr. I, 258;
    Wiedemann, Auss. Zw. II, 11, 1. — N. W. Territory (Say). ([184]).
*valida Loew, Centur. II, 51. — Huds. B. Terr.
*vittata Loew, Centur. II, 56. — Distr. Columbia (Loew); Goat Isl.,
    Niagara Falls.
    (?) *Ochthera empiformis* Say, J. Acad. Phil. III, 99; Compl. Wr. II,
    85; compare Loew, Monogr. I, 159.

Clinocera.
Meigen, Illiger's Magaz. II, 271; 1803 ([185]).

*binotata Loew, Zeitschr. für ges. Naturw. 1876, 325. — New York.
*fuscipennis Loew, Zeitschr. für ges. Naturw. 1876, 324. — White
    Mts., N. H.
*lineata Loew, Centur. II, 50. — Pennsylvania.
*simplex Loew, Centur. II, 49. — Huds B. Terr.
    (?. *Heliodromia longipes* Walker, List, etc. III, 504. — Huds. B. Terr.
*conjuncta Loew, Wiener Ent. Monatschr. IV, 79. — Middle States.
*maculata Loew, Wiener Ent. Monatschr. IV, 79. — Middle States.

FAMILY DOLICHOPODIDAE. (¹⁸ᶜ).

Hygroceleuthus.

Loew, Neue Beitr. V, 1857; Monogr. II, 16.

*latipes Loew, Neue Beitr. VIII, 5; Monogr. II, 17. — Red River of the North; Illinois.

*afflictus O. Sacken, Western Dipt., 313. — Marine Co., California.
*crenatus O. Sacken, Western Dipt., 312. — Sonoma Co., California.
lamellicornis Thomson, Eugenies Resa, 511 (*Dolichopus*); compare also O, Sacken, Western Diptera, 313. — California.

Dolichopus.

Latreille, Précis etc.; 1797.
Loew, Monogr. II, 18.

*acuminatus Loew, Neue Beitr. VIII, 12, 4; Monogr. II, 34. — Illinois.
*albiciliatus Loew, Centur. II, 59; Monogr. II, 31. — Illinois, Western New York.
*batillifer Loew, Neue Beitr. VIII, 15, 10; Monogr. II, 45. — Atlantic States.
*bifractus Loew, Neue Beitr. VIII, 19, 17; Monogr. II, 53. — Northern United States.
*brevimanus Loew, Neue Beitr. VIII, 14, 8; Monogr. II, 39. — Distr. Columbia.
*brevipennis Meigen; Loew, Monogr. II, 37. — Europe; British North America (Fort Resolution)
*chrysostomus Loew, Neue Beitr. VIII, 23, 24; Monogr. II, 67. — Distr. Columbia.
*comatus Loew, Neue Beitr. VIII, 23, 25; Mónogr. II. 69. — Middle States.
*cuprinus Wiedemann, Auss. Zw. II, 230; Loew Neue Beitr. VIII, 20, 19; Monogr. II, 55. — Atlantic States.
Dolichopus cupreus Say, J. Acad Phil. III, 86, 9; Compl. Wr. II, 76. [Change of name by Wied.].
*detersus Loew, Centur. VII, 79. — Western New York.
*dorycerus Loew, Centur. V, 85; Monogr. II, 326 - White Mts., N. H.
*discifer Stannius; Loew, Monogr. II, 71. — Europe; British North America; New York; White Mts., N. H.; Sitka.
Dolichopus tanypus, Loew, Neue Beitr. VIII, 24, 26 [Loew].
*eudactylus Loew, Neue Beitr. VIII, 16, 11; Monogr. II, 46. — Massachusetts; New York.
*funditor Loew, Neue Beitr. VIiI, 22, 23; Monogr. II, 66. — Middle States.
*fulvipes Loew, Centur. II, 61; Monogr. II, 61. — Illinois; White Mts., N. H.; New York.
*gratus Loew, Neue Beitr. VIII, 11, 1; Monogr. II, 29. — New York; New Jersey.

groenlandicus Zetterstedt, Dipt. Scand. II, 528; Staeger, Groenl. Antl. 358, 23; Holmgren, Ins. Nordgroenl, 100. — Greenland.

*hastatus Loew, Monogr. II, 59. — Sitka.

*incisuralis Loew, Neue Beitr. VIII, 25, 28; Monogr. II, 74. — New York.

*laticornis Loew, Neue Beitr. VIII, 12, 2; Monogr. II, 29. — Connecticut.

*lobatus Loew, Neue Beitr. VIII, 24, 27; Monogr. II, 72. — Illinois; British North America.

*longimanus Loew, Neue Beitr. VIII, 14, 7; Monogr. II, 39. — British North America and Northern United States.

*longipennis Loew, Neue Beitr. VIII, 21, 20; Monogr. II, 57. — Middle States.

*luteipennis Loew, Neue Beitr. VIII, 18, 15; Monogr. II, 51. — Distr. Columbia; Illinois.

*melanocerus Loew, Centur. V, 86; Monogr. II, 330. — Canada.

*nudus Loew, Monogr. II, 41. — Brit. North America (Fort Resolution).

*ovatus Loew, Neue Beitr. VIII, 13, 5; Monogr. II, 35. — Middle States; Illinois.

*pachycnemus Loew, Neue Beitr. VIII, 13, 6; Monogr. II, 36. — Middle States.

*palaestricus Loew, Centur. V, 84; Monogr. II, 328. — White Mts., N. H.

*platyprosopus Loew, Centur. VII, 80. — British North America.

*plumipes Scopoli, Loew, Monogr. II, 60. — Europe; Sitka; Quebec. Dolichopus pennitarsis Fallen (Loew, l. c.).

*praeustus Loew, Centur. II, 62; Monogr. II, 68. — Illinois.

*pugil Loew, Centur. VII, 77. — Canada; Massachusetts.

*quadrilamellatus Loew, Centur. V, 83, Monogr. II, 331. — New Jersey.

*ramifer Loew, Neue Beitr. VIII, 19, 16; Monogr. II, 52. — Northern United States, Nebraska, Lake Winnipeg.

*ruficornis Loew, Neue Beitr. VIII, 21, 21; Monogr. II, 63. — Middle States.

*sarotes Loew, Centur. VII, 81. — Illinois.

*scapularis Loew, Neue Beitr. VIII, 22, 22; Monogr. II, 64. — Middle States.

*scoparius Loew, Monogr. II, 70, — Northern Atlantic States.

*setifer Loew, Neue Beitr. VIII. 12, 3; Monogr. II, 31. — Distr. Columbia; New York; Newport, R. J.

*setosus Loew, Centur. II, 63; Monogr. II, 73. — Massachusetts.

*sexarticulatus Loew, Monogr. II, 62. — Distr. Columbia.

*socius Loew, Centur. II, 60; Monogr. II, 40. — Illinois; Western New York.

*splendidus Loew, Neue Beitr. VIII, 14, 9; Monogr II, 44. — Illinois.

*splendidulus Loew, Centur. V, 82; Monogr. II, 327. — White Mts., N. H.

*Stenhammari Zetterstedt, Dipt. Scand. II, 521. — Northern Sweden and Lapland; Sloop Harbor, Labrador, July 19. (A. S. Packard)

*subciliatus Loew, Monogr. II, 43. — Brit. North America (Fort Resolution).

*tener Loew, Neue Beitr. VIII, 17, 13; Monogr. II, 49. — Chicago.

*terminalis Loew, Centur. VII, 78. — Western New York (Genesseo'.
* tetricus Loew, Monogr. II, 33 — Brit. North America (Fort Resolution).
*tonsus Loew, Neue Beitr. VIII, 16, 12; Monogr. II, 47. — Distr.
    Columbia.
*variabilis Loew, Neue Beitr VIII, 17, 14; Monogr. II, 50. — New York.
*vittatus Loew, Neue Beitr. VIII, 20, 18; Monogr. II, 55. — Illinois;
    New York.
* xanthocnemus Loew, Monogr. II, 21. — Sitka.

aurifer Thomson, Eug. Resa etc. 512.
* canaliculatus Thomson, Eugenies Resa, 512;   O. Sacken, Western
    Dipt. 315. — California (Marin Co.).
*corax O. Sacken, Western Dipt. 314. — Sierra Nevada, Cal.
metatarsalis Thomson, Eugenies Resa 512 — California.
*pollex O. Sacken. Western Dipt. 314. — Sierra Nevada, Cal.

[The following species of *Dolichopus*, published by previous authors have not
been identified by Mr. Loew, and most of them never will be, on acount of their
incomplete descriptions. These descriptions are reproduced in the Appendix to
Monogr Vol. II, page 289—320. A critical examination, by Mr. Loew, of these
species is given in the same volume page 20—24.]

abdominalis Say, J. Acad. Phil. VI, 170; Compl. Wr. II, 362. — Indiana.
adjacens Walker, List, etc. III, 661. — Huds. B.
affinis Walker, List, etc. III, 659. — Nova Scotia.
bifrons Walker, Dipt. Saund. III. 212 [perhaps *Pelastoneurus* Lw. 1. c.].
    — United States.
ciliatus Walker, List, etc, III, 661. — Huds. B.
coercens Walker, List, etc. III, 661. — New York.
confinis Walker, l. c. 664. — Huds. B.
consors Walker, Dipt. Saund. III, 213. — United States.
conterminus Walker, List, etc. III, 664. — New York.
contingens Walker, Dipt. Saund. III, 213. — United States.
contiguus Walker, List, etc. III, 663. — New York.
discessus Walker, List, etc. III, 662. — Massachusets.
distractus Walker, l. c. III, 662. — New York.
exclusus Walker, l. c. III, 663. — Huds. B. Terr.
finitus Walker, l. c. III, 662. -- New York.
hebes Walker, Dipt. Saund. III, 213. — United States.
heteroneurus Macquart, Dipt. Exot. 4e Suppl. 128, 5; Tab. XII,´f. 10.
    [*Pelastoneurus* or*Paraclius?* — Lw. l. c.]. — North America.
ineptus Walker, Dipt. Saund. III, 214. — United States.
irrasus Walker, List, etc. III, 767. — Florida.
lamellipes Walker, List, etc. III, 660. — Huds. B. Terr.
maculipes Walker, Dipt. Saund. III, 214 [perhaps *Pelastoneurus* — Lw.
    l. c.]. — United States.
obscurus Say, J. Acad. Phil. III, 85, 4; Wiedemann, Auss. Zw. II, 232, 6·
    [evidently a *Gymnopternus* — Lw. l. c.]. — Pennsylvania.
pulcher Walker, Dipt. Saund. III, 215 [perhaps *Gymnopternus* — Lw.
    l. c.]. — United States.
remotus Walker, List, etc. III, 666. — North America.
separatus Walker, 1 c. 665. — Huds. B. Terr.
sequax Walker, l. c. III, 666. — Huds. B. Terr.
soccatus Walker. List, etc. III, 666. — Huds. B. Terr.
terminatus Walker, List, etc. III, 665. — North America.
varius Walker, Dipt. Saund. III, 21 . — United States.

Gymnopternus.

Loew, Neue Beitr. V; 1857; Monogr. II, 75.

*albiceps Loew, Neue Beitr. VIII, 30, 7; Monogr. II, 85. — Middle States.
*barbatulus Loew, Neue Beitr. VIII, 29, 2; Monogr. II, 82. — Middle States.
* chalcochrus Loew, Monogr. II, 335. — New York; Distr. Columbia.
* coxalis Loew, Centur. V, 87; Monogr. II, 335. — New York.
* crassicauda Loew, Neue Beitr. V.II, 35, 20; Monogr. II, 96 — New York.
* debilis Loew, Neue Beitr. VIII, 35, 19; Monogr. II, 95. — Pennsylvania.
* despicatus Loew, Neue Beitr. VIII, 33, 13; Monogr. II, 90. — Middle States.
* difficilis Loew, Neue Beitr. VIII, 33, 14; Monogr. II, 91. — New York.
* exiguus Loew, Monogr. II, 337. — Illinois.
* exilis Loew, Neue Beitr. VIII, 30, 5; Monogr. II, 84. — Pennsylvania.
*fimbriatus Loew, Neue Beitr. V.II, 32, 12; Monogr. II, 89. — Maryland.
* flavus Loew, Neue Beitr. VIII, 28, 1; Monogr. II, 80. — Pennsylvania.
* frequens Loew, Neue Beitr. VIII, 32, 10; Monogr. II, 88. — Middle States.
* humilis Loew, Monogr II, 336. — New York; Illinois.
* laevigatus Loew, Neue Beitr. VIII, 31, 9; Monogr. II, 87. — Middle States.
* lunifer Loew, Neue Beitr. VIII, 32, 11; Monogr. II, 89. — New York.
* meniscus Loew, Centur. V, 88; Monogr. II, 336. — Distr. Columbia.
* minutus Loew, Neue Beitr. VIII, 35, 21; Monogr. II, 96. — Middle States.
* nigribarbus Loew, Neue Beitr. VIII, 33, 15; Monogr. II, 91. — Pennsylvania
* opacus Loew, Neue Beitr. VIII, 34, 17; Monogr. II, 93. — New York.
* parvicornis Loew, Neue Beitr. VIII, 34, 16; Monogr. II, 92. — Middle States.
* phyllophorus Loew, Centur. VII, 82. — Lake George, N. Y.
* politus Loew, Neue Beitr. VIII, 34, 18; Monogr. II, 94 and 334. — New York.
* pusillus Loew, Monogr. II, 334. — Illinois.
* scotias Loew, Neue Beitr. VIII, 29, 3; Monogr. II, 81. — British North America (Lake Winnipeg).
* spectabilis Loew, Neue Beitr. VIII, 80, 5; Monogr. II, 85. — New York.
* subdilatatus Loew, Neue Beitr. VIII, 31, 8; Monogr. II, 86. — Middle States.
* subulatus Loew, Neue Beitr. VIII, 29, 2; Monogr. II, 80. — New York.
* tristis Loew, Monogr. II, 83. — Sitka.
* ventralis Loew, Neue Beitr. VIII, 36, 22; Monogr. II, 97. — New York; Distr. Columbia.

Observation. *Dol. obscurus* Say, is probably a *Gymnopternus;* compare Loew, Monogr. II, 20.

Paraclius.

Paracleius, Bigot, Ann. Soc. Ent. 1859, 215; amended in Loew, Monogr. II, 97; 1864.

albonotatus Loew, Monogr. II, 102. — New Orleans.
*claviculatus Loew, Centur VII, 83. — New Rochelle, New York.
*pumilio Loew, Centur. X, 63. — Texas.

*arcuatus Loew, Neue Beitr. VIII, 39, 4; Monogr. II, 101. — Cuba.

Pelastoneurus.

Loew, Neue Beitr. VIII; 1861; Monogr. II, 103.

*abbreviatus Loew, Centur. V, 89; Monogr. II, 338. — New Rochelle, New York.
*alternans Loew, Centur. V, 91; Monogr. II, 339. — New Rochelle, New York.
*cognatus Loew, Monogr. II, 109. — Middle States; Texas.
*furcifer Loew, Centur. X, 64. — Texas.
*laetus Loew, Neue Beitr. VIII, 38, 3; Monogr. II, 106. — Georgia; Distr. Columbia.
*lamellatus Loew, Centur. V, 90; Monogr II, 338. — New York.
*longicauda Loew, Neue Beitr. VIII, 37, 1; Monogr. II, 104. — New York.
*lugubris Loew, Neue Beitr. VIII, 38, 2; Monogr. II, 105. — Trenton Falls, New York.
*vagans Loew, Neue Beitr. VIII, 39, 5; Monogr. II, 108. — Middle States.

Polymedon.

O. Sacken, Western Dipt, 317; 1877.

*flabellifer O. Sacken, Western Dipt., 317. — Sonoma Co., California.

Tachytrechus.

Stannius, Isis 1831; Loew, Neue Beitr. V, 1857; Monogr. II, 109.

*angustipennis Loew, Centur. II, 64; Monogr. II, 113. — Distr. Columbia; also in California, see O. Sacken, Western Dipt., 315.
*binodatus Loew, Centur. VII, 84. — Saratoga, New York.
*moechus Loew, Neue Beitr. VIII, 40, 1; Monogr. II, 110. — Trenton Falls, New York.
*vorax Loew, Neue Beitr. VIII, 41, 2; Monogr. II, 112. — Distr. Columbia.

*sanus C. Sacken, Western Dipt., 316. — Sierra Nevada, California.

Observation. *Tachytrechus moechus* and *sanus* belong to the new genus *Macellocerus* Mik, Schulprogr. d. Acad. Gymn. in Wien, 1878. —

About *Orthochile derempta* Walker, List, etc. see the note ([187]).

Hercostomus.

Loew, Neue Beitr. V, 1857; Monogr. II, 116. ([188]).

*unicolor Loew, Monogr. II, 117. — Fort Resolution, Huds. B. Terr.

Diostracus.

Loew, Neue Beitr. VIII; 1861; Monogr. II, 120.

* prasinus Loew, Neue Beitr. VIII, 44, 1; Monogr. II, 121. — New York.

Argyra.

Macquart, Hist. Nat. Dipt. I, 456; 1834; Loew, Monogr. II, 123.

* albicans Loew, Neue Beitr. VIII, 45, 1; Monogr. II, 125. — Distr. Columbia.
* albiventris Loew, Monogr. II, 128 — Sitká.
* calceata Loew, Neue Beitr. VIII, 47, 4; Monogr. II, 131. — Middle States.   •
* calcitrans Loew, Neue Beitr. VIII, 46, 3; Monogr. II, 130. — New York.
* cylindrica Loew, Monogr. II, 132. — Sitka.
* minuta Loew, Neue Beitr. VIII, 46, 2; Monogr. II, 129. — Distr. Columbia.
* nigripes Loew, Monogr. II, 127. — Sitka.

Synarthrus.

Loew, Neue Beitr. V; 1857; Monogr. II, 134.

barbatus Loew, Neue Beitr. VIII, 48, 2; Monogr. II, 138. — Middle States.
* cinereiventris Loew, Neue Beitr. VIII, 48, 1; Monogr. II, 137. — Middle States; Texas.
* palmaris Loew, Monogr. II, 135. — Sitka.

Rhaphium.

Meigen, Illiger's Magaz. II; 1803;
Loew, Neue Beitr. V; Monogr. II, 140.

* lugubre Loew, Neue Beitr. VIII, 49, 1; Monogr. II, 141. — Carolina.

Porphyrops.

Meigen, System. Beschr. IV, 45; 1824; Monogr. II, 142.

* fumipennis Loew, Neue Beitr. VIII, 51, 3; Monogr. II, 146. — Middle States.
* longipes Loew, Centur. V, 92; Monogr. II, 340. — White Mts., N. H., Canada.

*melampus Loew, Neue Beitr. VIII, 50, 1; Monogr. II, 144. — Atlantic States.
*nigricoxa Loew. Neue Beitr. VIII, 51, 2; Monogr. II, 145. — Maryland.
pilosicornis Walker, List, etc. III, 653. — Huds. B. Terr.
*rotundiceps Loew, Neue Beitr. VIII, 51, 4; Monogr. II, 146. — Distr. Columbia.
*signifer, n. sp. see the note ([189]). — New York.

Leucostola.

Loew, Neue Beitr. V; 1857; Monogr. II, 151.

* cingulata Loew, Neue Beitr. VIII, 53, 1; Monogr. II, 152. — Distr. Columbia.
(*Eutarsus eques*, Loew, Monogr. II, 154, is from Venezuela.)

Diaphorus.

Meigen, System. Beschr. IV; 1824; Loew, Monogr. II, 156.

*lamellatus Loew, Monogr. II, 165. — Middle States.
*leucostomus Loew, Neue Beitr. VIII, 58, 5; Monogr. II, 166. — Distr. Columbia; Maryland.
*mundus Loew, Neue Beitr. VIII, 57. 2; Monogr. II, 161. — Pennsylvania.
*opacus Loew, Neue Beitr. VIII, 56, 1; Monogr. II, 160. — New York.
*sodalis Loew, Neue Beitr. VIII, 58, 4; Monogr. II, 163. — New York.
*spectabilis Loew, Neue Beitr. VIII, 57, 3; Monogr. II, 162. — Distr. Columbia.

*subsejunctus Loew, Centur. VI, 83. — Cuba.
*interruptus Loew, Wien. Ent. Monatschr. V, 37; Neue Beitr. VIII, 59; Monogr. II, 168. — Cuba.

Asyndetus.

Loew, Centur. VIII, 58; 1869; compare also Loew, Beschr. Eur. Dipt. II, 296.

* ammophilus Loew, Centur. VIII, 58. — Newport, R. I.
* appendiculatus Loew, Centur. VIII, 59. — Newport, R. I.

Lyroneurus.

Loew, Wien. Ent. Monatschr. I, 37; 1857; Monogr. II, 169.

* caerulescens Loew, Wien. Ent. Mon. I, 39; Neue Beitr. VIII, 60, 1; Monogr. II, 170. — Mexico.

Chrysotus.

Meigen, System. Beschr. IV, 1824; Loew, Monogr. II, 171. ([190]).

* affinis Loew, Neue Beitr. VIII, 64; Monogr. II, 178. — Middle States.
* auratus Loew, Neue Beitr. VIII, 65; Monogr. II, 183. — New York.

11

* cornutus Loew, Monogr. II, 174. — Distr. Columbia.
* costalis Loew, Neue Beitr. VIII, 64; Monogr. II, 179. — Florida; Maryland.
* discolor Loew, Neue Beitr. VIII, 65; Monogr. II, 182. — Middle States.
* longimanus Loew, Neue Beitr. VIII, 62; Monogr. II, 175. — Middle States.
* obliquus Loew, Neue Beitr. VIII, 63; Monogr. II, 176. — New York. ([101]).
* pallipes Loew, Neue Beitr. VIII, 66; Monogr. II, 183. — Middle States. ([110]).
* picticornis Loew, Monogr. II, 184. — Distr. Columbia; Texas.
* subcostatus Loew, Monogr. II, 181. — Illinois.
* validus Loew, Neue Beitr. VIII, 63, 2; Monogr. II, 175. — Middle States.
* vividus Loew, Monogr. II, 178. — Distr. Columbia.

> The following species, described by previous authors as *Chrysotus*, either do not belong to this genus, or can not be recognized, on account of the insufficiency of the descriptions. Mr Loew discusses them in Monogr. II, 172, and the descriptions are reproduced in the Appendix to the same volume.

abdominalis Say, J. Acad. Phil. VI, 169, 3; Compl. Wr. II, 362. — Indiana.
concinnarius Say, J. Acad. Phil. VI, 168; 2; Compl. Wr. II, 361. — Mexico.
incertus Walker, List, etc. III, 651. — United States.
nubilus Say, J. Acad. Phil. VI, 168, 1; Compl. Wr. II, 361. — Indiana.
viridifemora Macquart, Dipt. Exot. 4e Suppl. 124, 2; Tab. XII, f. 3. — North America.

Sympycnus.

Loew, Neue Beitr. V, 1857; Monogr. II, 185. ([102]).

* frontalis Loew, Neue Beitr. VIII, 67; Monogr. II, 188. — Pennsylvania.
* lineatus Loew, Neue Beitr. VIII, 67; Monogr. II, 189. — Virginia; New York.
* nodatus Loew, Centur. II, 68; Monogr. II, 191. — Illinois.
* tertianus Loew, Monogr. II, 187. — Sitka.

Campsicnemus.

Haliday, in Walker's Ins. Brit. Dipt. I, 187; 1851; Loew, Monogr. II, 193.

* claudicans Loew, Monogr. II, 194 — Sitka.
* hirtipes Loew, Neue Beitr. VIII, 68; Monogr. II, 193. — Pennsylvania; New York.

Plagioneurus.

Loew, Wien. Ent. Monatschr. I, 43; 1857; Monogr. II, 196.

* univittatus Loew, Wien. Ent. Mon. I, 43; Neue Beitr. VIII, 69; Monogr. II, 196. — Cuba; Brazil.

Liancalus.

Loew, Ncue Beitr. V, 1857; Monogr. II, 198.

*genualis Loew, Neue Beitr. VIII, 70; Monogr. II, 199. — Middle
States.
* querulus O. Sacken, Western Dipt., 318. — Sonoma Co., California.

Scellus.

Loew, Neue Beitr. V, 1857; Monogr. II, 200.

* avidus Loew, Monogr. II, 207. — Fort Resolution, Huds. B. Terr.
* exustus Walker, Dipt. Saund. 211 (Medeterus); Loew, Neue Beitr.
VIII, 71; Monogr. II, 203. — Middle States; Illinois.
*filifer Loew, Monogr. II, 209. — Fort Resolution; Huds. B. Terr.
* spinimanus Zetterstedt, Dipt. Scand. II, 445 (Hydrophorus); Loew,
Monogr. II, 205. — Fort Resolution, Huds. B. Terr.
Hydrophorus notatus Zetterstedt, Ins. Lapp. 701 [Lw.].

* monstrosus O. Sacken, Western Dipt., 319. — British Columbia.
* vigil O. Sacken, Western Dipt., 318. — Sierra Nevada, California.

Hydrophorus.

Fallen, Dolichopod. 1825; Wahlberg, Oefv. of k. vet. akad. forh. 1844;
Loew, Monogr. II, 211.

* aestuum Loew, Centur. VIII, 60. — Newport, R. I.
* cerutias Loew, Centur. X, 65. — Texas.
* innotatus Loew, Monogr. II, 212. — Sitka.
* parvus Loew, Centur. II, 67; Monogr. II, 216. — Pennsylvania.
* pirata Loew, Neue Beitr. VIII, 71, 1; Monogr. II, 214. — Pennsylvania.
* viridiflos Walker, Dipt. Saund., 212. — North America. (I refer to
this species so·· specimens from Massachusetts.)

> **Observation.** The following species, described as Mediterus, belong, in
> part at least, to Hydrophorus; those of Mr. Walker's are discussed by Mr. Loew
> in Monogr. II, 215. Mr. Say's two species I do not find mentioned in Mr. Loew's
> Monogr. The description of all these species are reproduced in the Appendix to
> Monogr., Vol. II.

alboflorens Walker, List, etc. III, 656. — Nova Scotia.
chrysologus Walker, List, etc. III, 655. — Huds. B. Terr.
exustus Walker, Dipt. Saund., 211. — North America.
glaber Walker, List, etc. III, 655. — Huds. B. Terr.
lateralis Say, J. Acad. Phil. VI, 169, 1; Compl. Wr. II, 362. —
Indiana.
punctipennis Say, J. Acad. Phil. VI, 170, 2; Compl. Wr. II, 362. --
Mexico.

Medeterus.

Medetera Fischer, Notice sur une mouche carnivore, 1819; Loew, Monogr. II, 218 ([190]).

*nigrines Loew, Neue Beitr. VIII, 73; Monogr. II, 218. — Middle States.
*veles Loew, Neue Beitr. VIII, 73; Monogr. II, 219. — Florida.

breviseta Thomson, Eugen. Resa, etc. 510. — California (this species in probably a *Hydrophorus).*

Chrysotimus.

Loew, Neue Beitr. V, 1857; Monogr. II, 20.

*delicatus Loew, Neue Beitr. VIII, 74; Monogr. II, 222. — New York.
*pusio Loew, Neue Beitr. VIII, 74; Monogr. II, 221. — New York.

Xanthochlorus.

Loew, Neue Beitr. V, 1857; Monogr. II, 223.

*helvinus Loew, Neue Beitr. VIII, 75; Monogr. II, 224. — Chicago.

Saucropus.

Loew, Neue Beitr. V, 1857; Monogr. II, 224.

*carbonifer Loew, Centur. IX, 84. — New York. (I found it at Lloyd's Neck, Long Island; also in the Central Park N. York. — O. S.).
*dimidiatus Loew, Neue Beitr. VIII, 75; Monogr. II, 225. — Florida; Distr. Columbia.
*rubellus Loew, Neue Beitr. VIII, 76; Monogr. II, 226. — Berkeley Springs, Virginia.
*superbiens Loew, Neue Beitr. VIII, 76; Monogr. II, 227. — Florida; Distr. Columbia; New York.
*tenuis Loew, Monogr. II, 228. — Middle States.

Psilopus.

Meigen, System. Beschr. VI, 1824; Loew, Monogr. II, 229.

*bicolor Loew, Neue Beitr. VIII, 96; Monogr. II, 280. — Middle States.
*calcaratus Loew, Neue Beitr. VIII, 93; Monogr. II, 272. — Carolina.
*caudatulus Loew, Neue Beitr. VIII, 93; Monogr. II, 271. — Missouri; Illinois.
*ciliatus Loew, Neue Beitr. VIII, 88; Monogr. II, 260. — Florida.
(?) *Psilopus mundus* Wiedemann, Auss. Zw. II, 227.
*comatus Loew, Neue Beitr. VIII, 89; Monogr. II, 262; — Middle States.
*filipes Loew, Neue Beitr. VIII, 99; Monogr. II, 286. — Middle States (South America, in Schiner, Novara, 213).

*inermis Loew, Neue Beitr. VIII, 93; Monogr. II, 272. — Pennsylvania.
*pallens Wiedemann, Auss. Zw. II, 219; Loew, Neue Beitr. VIII, 97; Monogr. II, 275. — New York; Newport, R. I.; Sag Harbour, L. I.([193]).

Psilopus albonotatus, Loew, Neue Beitr. V, 4. — Island Rhodus; Asia minor [Loew].

*patibulatus Say, J. Acad. Phil. III, 87 and VI, 168; Compl. Wr. II, 76 and 361 *(Dolichopus)*; Wiedemann, Auss. Zw. II, 225; Loew, Neue Beitr. VIII, 85; Monogr. II, 251. — Atlantic States.

Psilopus amatus Walker, List, etc. III, 648 [Loew].
Psilopus inficitus Walker, List, etc. III, 649 [Loew].

*psittacinus Loew, Neue Beitr. VIII, 96; Monogr. II, 281 — Florida.
*scaber Loew, Neue Beitr. VIII, 85; Monogr. II, 250. — Pennsylvania.
*scobinator Loew, Neue Beitr. VIII, 91; Monogr. II, 268. — New York; Illinois.
*scintillans Loew, Neue Beitr. VIII, 94; Monogr. II, 273. — Middle States.
*sipho Say, J. Acad. Phil. III, 84; Compl. Wr. II, 75 *(Dolichopus)*; Wiedemann, Auss. Zw. II, 218; Loew, Neue Beitr. VIII, 83; Monogr. II, 248. — Atlantic States.

Psilopus gemmifer Walker, List, etc. III, 646 [Loew].

*tener Loew, Centur. II, 71; Monogr. II, 284. — Pennsylvania.
ungulivena Walker, Trans. Ent. Soc. N. S. IV, 149. — United States.
*variegatus Loew, N. Beitr. VIII, 95; Mon. II, 278. — Florida; Cuba.

castus Loew, Centur. VI, 84. – Cuba.
*chrysoprasius Loew, Neue Beitr. VIII, 90; Monogr. II, 266. — Cuba; (Brazil, Schiner, Novara, 213).

Psilopus chrysoprasi Walker, List, etc. III, 646. [Lw.].

dimidatus Loew, Centur. II, 70; Monogr. II, 246. — Mexico; (South America, Schiner, Novara, 212).
*dorsalis Loew, Centur. VI, 85. — Cuba.
*jucundus Loew, Neue Beitr. VIII, 87; Monogr. II, 258. — Cuba.

Psilopus sipho Macquart, Dipt. Exot. II, 2, 119; Tab. 21, f 1 [Loew].

*melampus Loew, Centur. II, 69; Monogr. II, 253. — Mexico (South America, Schiner, Novara, 212).
*pilosus Loew, Neue Beitr. VIII, 86; Monogr. II, 256. — Cuba.

The following species were not identified by Mr. Loew in preparing his work; they are discussed in Monogr. etc. II, pag. 231—243; the original descriptions are reproduced in the Appendix to the same volume:

albicoxa Walker, List, etc. III, 651. — Ohio; Massachusetts, Nova Scotia.
caudatus Wiedemann, Auss. Zw, II, 224, 23. — Georgia.
delicatus Walker, List, etc. III, 645. — New York.
femoratus Say, J. Acad. Phil. III, 86, 5 *(Dolichopus)* and VI, 168, 11; Compl. Wr. II, 76 and 361; Wiedemann, Auss. Zw. II, 226, 28. — Pennsylvania.
nigrifemoratus Walker, List, etc. III, 650. — Nova Scotia.

Sayi Wiedemann, Auss. Zw. II, 219, 13; Say, J. Acad Phil. III, 85,
2 (*Dolichopus unifasciatus*). — Pennsylvania.
virgo Wiedemann, Auss. Zw II, 224, 24. — New York.
haereticus Walker, Trans. Ent Soc. N. Ser. V, 286. — Mexico.
incisuralis Macquart, Dipt. Exot. Suppl. I, 120, 21; Tab. XX, f. 6. —
Yucatan.
lepidus Walker, Dipt. Saund. 207. — Mexico.
longicornis Fabricius, System. Ent. 783, 52; Ent. System. IV, 341,
124 (*Musca*); System. Antl. 269, 14 (*Dolichopus*); Wiedemann,
Auss. Zw. II, 220, 14. — West Indies.
　(?) *Psilopus radians* Macquart, Hist. Nat. Dipt. I, 450, 6; Dipt.
Exot. II, 2, 121, 18. — Amer. Sept. |Loew, Monogr. II, 240].
macula Wiedemann, Auss. Zw. II, 219, 12. — West Indies.
portoricensis Macquart, Hist Nat. Dipt. I, 450, 7; Dipt. Exot. II, 2,
121, 17 and I[er] Suppl. 120; Tab. XI, f. 7 (*wing*). — Porto Rico;
also in Columbia, South Amer.
peractus Walker, Trans. Ent. Soc. N. Ser. V, 286. — Mexico.
permodicus Walker, Trans. Ent. Soc. N. Ser. V, 287. — Mexico.
solidus Walker, Trans. Ent. Soc. N. Ser. V, 286. — Mexico.
suavium Walker, List, etc. III, 648. — Jamaica.

Observation. *Psilopus diffusus* Wiedemann and *P. guttula* Wiedemann, of
my former Catalogue, are stated by Mr. Loew to be Brazilian species, and not
North American; in Monogr. Vol. II, 235 and 237 he gives full descriptions of them.

FAMILY LONCHOPTERIDAE.

Lonchoptera.

Meigen, in Illiger's Magaz. II, 1803.

*lutea Panzer, Meigen, System. Beschr. IV, 107. — Europe and North
America.
*riparia Meigen, System. Beschr. IV, 108. — Europe and North America.
[The american specimens of these species do not show any
apparent difference from European ones.]

II. DIPTERA CYCLORHAPHA.

FAMILY SYRPHIDAE.

Mixogaster.

Macquart, Dipt. Exot. II, 2, 14, 1842.

mexicanus Macquart, Dipt. Exot. 1er Suppl. 123; Tab. X, fig. 15. — Mexico.

Microdon.

Meigen, Illiger's Magaz. II, 1803; *Aphritis* Latreille, 1804. ([174]).

*   **aurulentus** Fabricius, System. Antl 185, 8 *(Mulio)*; Wiedemann, Auss. Zw. II, 86, 10; Macquart, Dipt. Exot. II, 2, 12, 4; Tab. II, f. 1 *(Aphritis)*. — Carolina (M. C. Z. has a specimen from Illinois, which may belong here).
*   **baliopterus** Loew, Centur. X, 56. — Texas.
*   **coarctatus** Loew, Centur. V, 47. — Distr. Columbia.
*   **fulgens** Wiedemann, Auss. Zw. II, 82, 1; Macquart, Dipt. Exot. 1er Suppl. 122 *(Aphritis)*. -- Georgia (Wied); Florida; Guyana (Macq.). *Microdon euglossoides* Gray, in Griffith's Animal Kingdom; Ins. II; Tab 125, f. 2 [Walker, List, etc. III, p. 538.]
*   **fuscipennis** Macquart, Hist. Nat. Dipt. I, 488, 3 *(Ceratophyia)*. — Philadelphia (Macq.); Texas. ([195]). *Microdon Agapenor* Walker, List, etc. III, 539. — Georgia. [Walker, List, etc. IV, 1157, where a new generic name, *Mesophila*, is proposed.]
*   **globosus** Fabricius, System. Antl. 185, 9 *(Mulio)*; Wiedemann, Auss. Zw. II, 86, 11; Macquart, Dipt. Exot. II, 2, 12, 5; Tab. I, f. 4 *(Aphritis)*. — Carolina (Fab.); Atlantic States. *Dimeraspis podagra* Newman, Ent. Mag. V, 373. [Walker, List, etc., III, p. 540.]
    rufipes Macquart, Dipt. Exot II, 2, 11; Tab. II, f. 3 *(Aphritis)*. — Philadelphia.
*   **tristis** Loew, Centur. V, 45. — Virginia (Lw.); New York and northward, as far as Mackenzie River.

*   **inaequalis** Loew, Centur. VII, 70. — Cuba.
*   **laetus** Loew, Centur. V, 46. — Cuba.
*   **trochilus** Walker, Dipt. Saund. 216. — Mexico (this may be the same as *M. aurifex* Wied. II, 85, from Brazil).

Observation. For *Chymophila splendens* Macquart, Hist. Nat. Dipt. I, 486 etc., see the note ([196]).

Chrysotoxum.
Illiger's Magaz. II, 1803. ([197]).

*derivatum Walker, List, etc. III, 542. — Huds. B. Terr.; Yukon River, Alaska; Colorado Mts.

flavifrons Macquart, Dipt. Exot. II, 2, 17, 2; Tab. III, f. 2. — Newfoundland.

*laterale Loew, Centur. V, 42. — Nebraska.

*pubescens Loew, Wiener Ent. Monatschr. IV, 83, 10; Centur. V, 43. — Distr. Columbia.

*ventricosum Loew, Centur. V, 44. — Distr. Columbia.

nigrita Fabricius, Ent. System. IV, 292, 49 (Syrphus); System. Antl. 183, 1 (Mulio); Wiedemann, Auss. Zw. II, 88, 2. — Jamaica.

Paragus.
Latreille, Hist. Nat. Crust. et Ins. XIV, 358; 1804.

*angustifrons Loew, Centur. IV, 64. — Virginia.

*bicolor Fabricius, Meigen, etc. — Europe and North America.

*dimidiatus Loew, Centur. IV, 63. — Distr. Columbia.

> Observation. Paragus transatlanticus Walker, List, etc. III, 544, Trenton Falls, is represented in the Brit. Mus. by two specimens, both types; only one of them is a Paragus.
> For Paragus aeneus Walker, see Orthoneura. ([198]).

Pipiza.
Fallén, Dipt Suec. Syrphi, 58; 1816.

buccata Macquart, Dipt. Exot. II, 2, 107; Tab. XVIII, f. 2. — Carolina.

*calcarata Loew, Centur. VI, 42. — New York.

*femoralis Loew, Centur. VI, 38. — Illinois.

*festiva Meigen (or a species closely allied to it). — Canada.

*fraudulenta Loew, Centur. VI, 41. — Illinois.

*nigribarba Loew, Centur. VI, 40. — New York.

radicum Riley, 1st Rep. p. 121, f. 66; Amer. Ent. I, p. 83. — Illinois (apparently the same as femoralis Loew).

*salax Loew, Centur. VI, 39. — Pennsylvania.

divisa Walker, Trans. Ent. Soc. N. Ser. IV, 156. — Vera Cruz.

Psilota.
Meigen, System. Beschr. III, 256; 1822.

flavidipennis Macquart, Dipt. Exot. 5e Suppl. 97; Tab. V, f. 5 (compare the remark in Loew, Monogr. I, 27). — Philadelphia.

Triglyphus.
Loew, Oken's Isis. 1840, 512.

*modestus Loew, Centur. IV, 62. — New York.

*pubescens Loew, Centur. IV, 61. — Wisconsin.

Chrysogaster.

Meigen, Illiger's Magaz. II, 1803. ([199])

*latus Loew, Centar. IV, 59. — British North America (English River).
*nigripes Loew, Centur. IV, 60. — New York.

Observation. *Chrysogaster Apisaon* Walker, List III, 572. — New York.
„   *Antitheus* l. c. 572. — New York.
„   *recedens* Walker, Dipt. Saund., 228. — United
States. Mr. Walker's types in the Brit. Mus. are single specimens, in very poor condition. Upon comparison, they will probably prove identical with Mr. Loew's species of *Chrysogaster* and *Orthoneura.*

Orthoneura.

Macquart, Hist. Nat. Dipt. I, 563; 1834.

*nitida Wiedemann, Auss. Zw. II, 116, 1 (*Chrysogaster*). — North America.
Cryptineura hieroglyphica Bigot, Rev. et Magaz. de Zool. 1859.
*pictipennis Loew, Centur. IV, 58. — New York.
*ustulata Loew, Centur. IX, 80. — Orange, N. J.

*nigrovittata Loew, Zeitschr. für Ges. Naturw. December 1876, p. 323. —
San Francisco.

Observation. *Paragus aéneus* Walker, List, etc. III, 545, Ohio, is an
Orthoneura. ([198])

Chilosia.

Cheilosia Meigen, System. Beschr. III, p. 296; 1822. ([200])

*capillata Loew, Centur. IV, 65. — Distr. Columbia.
*comosa Loew, Centur. IV, 66. — British America.
*cyanescens Loew, Centur. IV, 67. — Illinois.
*leucoparea Loew, Centur. IV, 69. — Carolina.
*pallipes Loew, Centur. IV, 70. — Distr. Columbia, White Mts., N. H.;
California.
*plumata Loew, Centur. IV, 68. — Virginia.
*tristis Loew, Centur. IV, 71. — Red River of the North.

Observation. *Syrphus Aesyctes* Walker, List, etc. III, 591, Huds. B. Terr.
Syrphus latrans l. c. 575, Huds. B. Terr. are both *Chilosiae.*

Melanostoma.

Schiner, Wiener Ent. Monatschr. IV, 213; 1860.

ambigua (Fallen?) Zetterstedt, Ins. Lapp. 608, 38(?) (*Syrphus*); Dipt.
Scand. II, 757, 60 (*id.*); variety in Staeger, Groenl. Antl. p. 361,
29(?). [The quotations and queries are Schioedte's, in the Berl.
Ent. Zeitschr. 1859, p. 153.] — Greenland.
*scalaris Fabricius, Panzer, etc. (*Syrphus*). — Europe and North
America (common).
Syrphus mellinus (Linné), Fabricius, Meigen, etc. See description
in Schiner, Fauna Austr. Dipt. I, 291.
*obscura Say, Amer. Ent. I; Tab. XI (*Syrphus*), Compl. Wr. I, 23;
Wiedemann, Auss. Zw. II, 131 (*id.*). — Atlantic S.ates.

trichopus Thomson, Eugen. Resa, 502 *(Syrphus)*. — California.
*tigrina O. Sacken, Western Dipt., 323. — California.

> **Observation.** *M. gracilis* Meig. and *M. maculosus* Meig., both European, are stated to occur in N. America by Mr. Walker, List, etc., III, 588 – 589. Mr. Verrall informs me that „those two species are synonyms of *M. scalaris* Fab. But Mr. Walker's *Syrphus maculosus* has two representatives in the British Museum, both *Platychiri*, one *resembling P. immarginatus* Zett., the other *resembling P. scambus* Staeger."

Platychirus.

Platycheirus St. Fargeau et Serv. Encycl. Méth. T. X, 513; 1825.

*hyperboreus Staeger, Groenl. Antl. 362, 30 *(Syrphus)*; Holmgren, Ins. Nordgroenl. p. 100 *(Scaeva)*. — Greenland (Staeger, Holmgren). Pennsylvania, Virginia, etc. (M. C. Z.).
Naso Walker, List, etc. III, 587 *(Syrphus)*. — Huds. B Terr.
Pacilus Walker, Dipt. Saund. 240 *(Syrphus)*. Compl Wr. II, 79. ([201]).
*quadratus Say, J. Acad. Phil. III, 90, 4 *(Scaeva)*; Wiedemann, Auss. Zw. II, 135, 32 *(Syrphus)*. — Atlantic States.
Syrphus fuscanipennis Macquart, Dipt Exot. 5e Suppl. 95, 58.
*peltatus Meigen, System. Beschr. III, 334 *(Syrphus)*. — Europe; North America (Sitka, according to Loew; Western New York, in M. C. Z.).

Pyrophaena.

Schiner, Wiener Ent. Monatschr. IV, p. 213; 1860.

*ocymi Fabricius, Panzer, Meigen, System. Beschr. III, 337 *(Syrphus)*. — Europe; North America (Massachusetts, White Mts., N. H., Quebec; Athabasca Lake, etc.).
*rosarum Fabricius etc., Meigen, System. Beschr. III, 338 *(Syrphus)*. — Europe; North America (Massachusetts; White Mts., N. H.).

Leucozona.

Schiner, Wiener Ent. Monatschr. IV, 214; 1860. ([202]).

*lucorum Linné, etc., Meigen, System. Beschr. III, 313; Tab. 30, f. 27 *(Syrphus)*; Curtis, Brit Ent. 753 *(id.)*. — Europe; North America (British Possessions, Quebec).

Catabomba.

O. Sacken, Western Dipt. 325; 1877 ([203]).

*pyrastri Linné, Meigen, etc. *(Syrphus)*; O. Sacken, Western Dipt., 325. — Europe; California, Utah, Colorado; also in Chile (according to Macquart).
Syrphus transfugus Fabricius, Ent. System. IV, 306, 104.
Syrphus affinis Say, J. Acad. Phil. III, 93, 9; Compl. Wr. II, 81; Wiedemann, Auss. Zw. II, 117, 2. — Arkansas.

Eupeodes.

O. Sacken, Western Dipt., 328; 1877.

' rolucris O. Sacken, Western Dipt., 329. — California, Utah, Colorado.

Syrphus.

Fabricius, System. Ent. 1775. ([204]).

*abbreviatus (Zetterstedt), Schiner, Fauna Austr. I, 311; O. Sacken, Proc. Bost. Soc. N. H. 1875, 144. — Europe and North America (Massachusetts .

alcidice Walker, List, etc. III, p. 579. -- Huds. B. Terr. ([205]).

*amalopis O. Sacken, Proc. Bost. Soc. N. H. 1875, 148. — White Mts., N. H.

*americanus Wiedemann, Auss. Zw. II, 129; O. Sacken, Proc. Bost. Soc. N. H. 1875, 145. — Atlantic States (Massachusetts; Michigan; Texas); British Possessions; the same or a similar species in California, see O Sacken, Western Dipt., 327.

*contumax O. Sacken, Proc. Bost. Soc. N. H. 1875, 147. — White Mts., N. H.

(?) *Syrphus adolescens* Walker, List, etc. III, 584. — Huds. B. Terr.; Nova Scotia. ([211]).

*diversipes Macquart, Dipt. Exot. 4° Suppl. 155, 54; O. Sacken, Proc. Bost. Soc. N. H. 1875, 149. - White Mts., N. H. (common); Catskill Mt. House, N. Y.; Lake Superior; Newfoundland (Macq.).

(?) *Syrphus cinctellus* Zetterstedt, Schiner, etc. — Europe.

dimidiatus Macquart, Hist. Nat. Dipt. I, 537, 10. — Georgia.

*geniculatus Macquart, Dipt. Exot. II, 2, 101, 24; Tab. XVII, f. 5; O. Sacken, Proc. Bost. Soc. N. H. 1875, 159. — Newfoundland (Macq.); White Mts, N. H ([206]).

*lapponicus Zetterstedt, Dipt. Scand. II, 701, 3; Staeger, Groenl. Antl. 360, 23. — Europe and North America (Greenland; White Mts., N. H.); a similar species in California, see in O. Sacken, Western Dipt., 326. ([207]).

Syrphus Agnon Walker, List, etc. III, 579. — Nova Scotia; Huds. B. Terr.

Syrphus arcucinctus Walker, List, etc. III, 580. — Huds B. Terr. ([208]).

*Lesueurii Macquart, Dipt. Exot. II, 2, 92, 10; Tab. XVI, f. 3 (♀); O. Sacken, Proc. Bost. Soc. N. H. 1875, 143. — Northern and Middle States (probably also in Europe)

Epistrophe conjungens Walker, Dipt. Saunders, 242; Tab. VI, f. 5 (♂).

*ribesii Linné, etc. — Europe and North America.

Syrphus rectus O. Sacken, Proc. Bost. Soc. N. H 1875, 140.

(?) *Syrphus philadelphicus* Macquart, Dipt. Exot. II, 2, 93, 11; Tab. XVI, f. 2. ([203]).

tarsatus Zetterstedt, Ins. Lapp. 601, 2; Dipt. Scand. II, 730, 33; Staeger, Groenl. Antl. 360, 27. — Europe and Greenland.

*torvus O. Sacken, Proc. Bost. Soc. N. H. 1875, 139. — Atlantic States.

Syrphus topiarius Zetterstedt (non Meigen); Staeger Groenl. Antl. 360, 26. — Europe and Greenland.

(?) *Scaeva concava* Say, J. Acad. Phil. III, 89, 3; Compl. Wr. II, 78; Wiedemann, Auss. Zw. II, 130 (*Syrphus*). ([209]).

*umbellatarum O. Sacken, Proc. Bost. Soc. N H. 1875, 151. — White Mts, N. H.

(?) *Syrphus umbellatarum* Schiner, Fauna Austr. I, p. 307. — Europe.
(?) *Syrphus guttatus* in Walker's List, etc. III, p. 536. — Huds. B.
    Terr. ([210]).
Syrphus sexquadratus Walker, List, etc. III, 586. — Huds. B. Terr.;
    Nova Scotia.

fumipennis Thomson, Eugen. Resa, 499. — California.
*intrudens O. Sacken, Western Dipt., 326. — Coast Range, California.
*opinator O. Sacken, Western Dipt., 327. — Marin Co., California.
*protritus O. Sacken, Western Dipt., 328. — Marin Co., California.

Antipathes Walker, List, etc. III, 589. — Jamaica.
colludens Walker, Trans. Ent. Soc. N. Ser. V, 292. — Mexico.
delineatus Macquart, Dipt. Exot. 1er Suppl. 139, 37; Tab. XI, f. 13. —
    Mexico; (perhaps an *Allograpta?*)
*jactator Loew, Wiener Ent. Mon. V, 40; Centur. VI, 46. — Cuba.
limbatus Fabricius, Syst. Antl. 251, 10 *(Scaeva)*; Wiedemann, Auss. Zw.
    II, 133, 30. — West Indies.
mutuus Say, J. Acad Phil VI, 164, 2; Compl. Wr. II, 358. — Mexico.
*nigripes Loew, Centur. VI, 44. — Cuba.
*praeustus Loew, Centur. VI, 45. — Cuba.
quadrifasciatus Bigot, in R. de la Sagra, etc., 804; Tab. 20, f. 5. —
    Cuba.
radiatus Bigot, in R. de la Sagra, etc., 804. — Cuba.
*simplex Loew, Wien Ent. Mon. V, 40; Centur. VI, 43. — Cuba.
stegnus Say, J. Acad. Phil. VI. 163, 1; Compl. Wr. II, 358. — Mexico.

> **Observation.** *Scaeva dryadis* Holmgren, Ins. Spetsb. 26. —
> Spitzbergen and Greenland (Holmgr. Ins. Nordgroenl. 100). Not
> having seen the description of this species, I cannot tell whether
> it is a true *Syrphus*, a *Platychirus*, or a *Melanostoma*.
> *Scaeva arcuata* Fallèn, which Holmgren, Ins. Nordgroenl, has
> from Greenland, belongs to what I call the group of Syrphus
> Lapponicus; for this reason I have not quoted it in the above list.
> *Syrphus sexmaculatus* Palisot-Beauvois, Ins. 224, Dipt. Tab. III,
> f. 8. — Southern States, San Domingo. This species evidently
> belongs to some other genus than Syrphus. The author compares
> it to *Syrphus tympanitis* Fabr. and says that it may be a mere
> variety, or the other sex of that species. *Syrphus tympanitis*
> Fabr. Syst. Antl. 226, 10. is, I think, *a Volucella.*

> For *Syrphus Aesyctes* and *latrans* Wk., see *Chilosia.*
>   "   "   *oestriformis* Wk., see *Eristalis.*
>   "   "   *Naso* and *Pacilus* Wk., see *Platychirus.*
>   "   "   *Corbis , coalescens , Gurges, Quintius, interrogans,* Wk., see
>   *Mesograpta.*
>   "   "   *dimensus* Wk., see *Allograpta.*
>   "   "   *profusus* Wk., see *Milesia.*
>   "   "   *hecticus* Jaennicke, see *Mesograpta polita.*

Didea.

Macquart, Hist. Nat Dipt. I, p. 508, 1834; *Enica*, Meigen, 1838.
*fascipes Loew, Centur. IV, 82. — Pennsylvania. ([212]).

*laxa O. Sacken, Bullet. Buff Soc. Nat. Hist. III, 66; reproduced in the note (^{212}). — White Mts., N. H, Lake Superior.

Mesograpta.

Loew, Centur. Vol. II, p. 210; *Mesogramma* Loew, Centur. VI, 47; 1865.

* Boscii Macquart, Dipt. Exot. II, 100, 23; Tab. XVII, f. 2 (*Syrphus*). — Carolina (Macq.); Alabama, Florida.

Syrphus Gurges Walker, Dipt. Saund., 236. — United States.

*geminata Say, J. Acad. Phil. III, 92, 7; Compl. Wr. II, 80; Wiedemann, Auss. Zw. II, 145, 50 (*Syrphus*). — Atlantic States; California.

Syrphus interrogans Walker, Dipt. Saund., 238. — North America.

Eumerus privernus Walker, Dipt. Saund., 225.

Toxomerus notatus Macquart, Dipt. Exot 5º Suppl., 93.

*marginata Say, J. Acad. Phil. III, 92, 6; Compl. Wr. II, 80 (*Scaeva*); Wiedemann, Auss. Zw. II, 146, 52 (*Syrphus*). — Atlantic States and California.

*polita Say, J. Acad. Phil. III, 88, 1; Compl. Wr. II, 77 (*Scaera*); id. American Ent. I. Tab. XI (*Syrphus*); Compl. Wr. I, 24; Wiedemann, Auss. Zw. II, 132, 28 (*id.*). — Atlantic States; Cuba.

Syrphus cingulatulus Macquart, Dipt. Exot. 4º Suppl. 155, 53 (1).

Syrphus hecticus Jaennicke, Neue Exot Dipt. 90. — Illinois.

*parvula Loew, Centur. VI, 47. — Florida.

*planiventris Loew, Centur. VI, 49. — Florida.

Syrphus Quintius Walker, Dipt. Saund., 239. — United States.

limbiventris Thomson, Eugenies Resa, 495 (*Syrphus*). — California.

anchorata Macquart, Dipt. Exot. II, 2, 97; Tab. 16, f. 8 (*Syrphus*). — Brazil; North America.

*arcifera Loew, Centur. VI, 52. — Cuba.

ectypus Say, J. Acad. Phil. VI, 165, 3 (*Syrphus*); Compl. Wr II, 359. — Cuba.

*laciniosa Loew, Centur. VI, 50. — Cuba.

minuta Wiedemann, Auss. Zw. II, 146 (*Syrphus*); Bigot, in R. de la Sagra, etc, 806. — Brazil (Wied.); Cuba (Bigot).

*poecilogastra Loew, Centur. VI, 51. — Cuba

*pulchella Macquart, Dipt. Exot. 1er Suppl. 138, 36; Tab. XI, f. 12 (*Syrphus*). — San Domingo.

*subannulata Loew, Centur. VI, 48. — Cuba.

Observation. *Syrphus coalescens* Walker,Dipt. Sau d., 237, — North America. *Syrphus corbis* Walker, Dipt. Saund , 237. — North America. Both are *Mesograptae*, each represented by a single specimen in the Brit. Mus. I find in my notes that both produced on me the impression of *M. Boscii*, although the description of *S. coalescens* reads more like that of *M. planiventris* Loew; the female, described by Walker, is probably a different species.

Sphaerophoria.

St. Fargeau et Serville, Encycl Method X, 513, 1825; Macquart, Dipt. du Nord, 1829; *Melithreptus* Loew, Oken's Isis 1840, 573. (^{213}).

*cylindrica Say, Amer. Ent. I; Tab. XI (*Syrphus*); Compl. Wr. I, 22; Wiedemann, Auss. Zw. II, 138, 38 (*id.*). — North America (common,.

Sphaerophoria contigua Macquart, Dipt. Exot. 2e Suppl. 62, 4.
strigata Staeger, Groenl. Antl. 362, 31; Holmgren, Ins. Nordgroenl. 100 („an varietas *S. pictae"?* Holmgren·. — Greenland.
picta Macquart; Zetterstedt, Dipt. Scand. II, 772, 7. — Europe and Greenland (Holmgren, Ius. Nordgroenl. 100).

infumata Thomson, Eugenies Resa, 501 *(Syrphus)*. — California.
*micrura O. Sacken, Western Dipt., 330. — San Francisco.
*sulphuripes Thomson, Eugenies Resa, 501 *(Syrphus)*; O. Sacken, Western Dipt., 330. — California.

> **Observation.** Mr. Walker mentions the European *S. hieroglyphica. mentastri* and *scripta* as occuring in Nova Scotia (Walker, List. etc., III, p. 593).

Allograpta.

O. Sacken, Bulletin Buff. Soc. N. H. III, 49; 1876. ([214]).

?emarginata Say, J. Acad. Phil. III. 91, 5 *(Scaeva);* Compl. Wr. II, 78; Wiedemann, Auss. Zw. II, 119, 4 *(Syrphus)*. — Florida (Say,; Virginia; Delaware (Ent. Soc. Phil.).
*obliqua Say, J. Acad. Phil. III, 89, 2 *(Scaeva);* Compl. Wr. II, 78; Amer. Ent. I; Tab. XI; Compl. Wr. I, 23; Wiedemann, Auss. Zw. II, 138, 39 *(Syrphus)*. — North America; also in South America (Schiner, Dipt. Novara, etc., 353).
Syrphus securiferus Macquart, Dipt. Exot. II, 2, 100, 22 and 1er Suppl. 139 (♀) (!).
Sphaerophoria Bacchides Walker, List, etc III, 594 (!).
Syrphus signatus v. d. Wulp, Tijdschr. v. Ent. 2e Ser. II, 144; Tab. IV, f. 12.
Syrphus dimensus Walker, Dipt. Saund , 235 (!).
*fracta O. Sacken, Western Dipt., 331. — Southern California.

Xanthogramma.

Schiner, Wien. Ent. Monatschr. IV, 215; 1860.

*felix O. Sacken, Bulletin Buff Soc. N. H. III, 67 (reproduced in the note ([215]). — West Point, N. Y.; Pennsylvania; Illinois.

Doros.

Meigen, Illiger's Magaz. II; 1803.

*aequalis Loew, Centur. IV, 84. — Pennsylvania.
*flavipes Loew, Centur. IV, 83. — Pennsylvania (Lw.); New York.

> **Observation.** For *Doros Balyras* Walker, see *Temnostoma.*

Ascia.

Meigen, System. Beschr. III, 193; 1822.

*globosa Walker, List, etc. III, 546. — Trenton Falls, N. Y.

Sphegina.

Meigen, System. Beschr. III, 193; 1822.

*infuscata Loew, Centur. III, 23. — Sitka.

*lobata Loew, Centur III, 21. — Northern and Middle States; Canada.
*rufiventris Loew, Centur. III, 22. — New York; White Mts., N. H.; Canada.

Ocyptamus.

Macquart, Hist. Nat. Dipt. I, 554; Tab. XII, f. 13; 1834; compare also Loew, Dipt. Südafrika's 293.

*Amissas Walker, List, etc. III, 589 (*Syrphus*). — Georgia. [21ᵣ]
*fuscipennis Say, J. Acad. Phil. III, 100 (*Baccha*); Compl. Wr. II, 86. — Atlantic States.
   Ocyptamus fascipennis Macquart, Hist. Nat. Dipt. I, 554, 2; Tab. 12, f. 13.
*longiventris Loew, Centur. VII, 66. — Distr. Columbia.
   Radaca Walker, List, etc. III, 590 (*Syrphus*). — Florida. [216]

*conformis Loew, Centur. VII, 67. — Cuba.
   dimidiatus Fabricius, Ent. System. IV, 310, 118 (*Syrphus*); System. Antl. 254, 25 (*Scaeva*); Wiedemann, Auss. Zw. II, 140, 42 (*Syphus*). — West Indies (Wied.); Brazil (Schiner, Novara).
   funebris Macquart, Dipt. Exot. II, 2, 105: Bigot, in Ramon de la Sagra, etc., 807. — „Teneriffa, but more probably America" (Macq.); Cuba (Bigot); Brazil (Schiner).
*latiusculus Loew, Centur, VII, 68. — Cuba.
*scutellatus Loew, Centur. VII. 69. — Cuba.

Baccha. (*)

Fabricius, System. Antl. 199; 1805.

*aurinota (Harris' Walker, List, etc. III, 548. — Atlantic States (Massachusetts; White Mts.; New York, etc.).
   Baccha fascipennis Wiedemann, Auss. Zw. II, 96. — No locality given.
   Babista Walker, List, etc. III, 549. — Georgia.
*cognata Loew, Centur. III, 27. — New York (erroneously Northern Wisconsin in the Centuries).
   costata Say, J. Acad. Phil. VI, 161; Compl. Wr. II, 357. — Indiana.
*lugens Loew, Centur. III, 24. — Northern Wisconsin.
   lineata Macquart, Dipt. Exot. 1ᵉʳ Suppl. 139, 4; Tab. XX, f. 5. — Texas or Yucatan (Macquart).
*obscuricornis Loew, Centur. III, 26. — Sitka.
*Tarchetius Walker, List, etc. III, 549. — Georgia.

*lemur O. Sacken, Western Dipt., 331. — California; Wyoming Terr.
*angusta O. Sacken, Western Dipt., 332. — California.
   Baccha elongata Fabricius, the common european species, is, I believe, the same as *B. angusta*.

(*) Some of the species placed among the Bacchae, may perhaps belong to *Ocyptamus*.

*capitata Loew, Centur. III, 25. — Cuba.
*clavata Fabricius, Ent. System. IV, 298, 73 *(Syrphus)*; System. Antl.
200, 3 *(id.)*; Wiedemann, Auss. Zw. II, 94, 4. — West Indies
(Wied.'; Brazil (Schiner).
cochenillivora Guérin, Rev. Zool. 1843, 350; Bull. Soc. Ent. 1848,
LXXXI. — Guatemala.
cubensis Macquart, Dipt. Exot. 4e Suppl. 161, 5. — Cuba.
cylindrica Fabricius, Spec. Ins. II, 429, 41 *(Syrphus)*; Ent. System.
IV, 298, 74 *(id.)*; System. Antl. 199, 2; Wiedemann, Auss. Zw.
II, 92. — West Indies.
*notata Loew, Centur. VII, 65. — Cuba.
*parvicornis Loew, Wien. Ent. Mon. V, 41; Centur. VII, 64. — Cuba.

Myiolepta.

Newman, Ent. Magaz. V, 373; 1838.

*aerea Loew, Centur. X, 53. — Illinois.
*nigra Loew, Centur. X, 52. — Pennsylvania.
*strigilata Loew, Centur. X, 54. — Texas.
*varipes Loew, Centur. IX, 79. — Virginia.

Rhingia.

Scopoli, Ent. Carniol. 358; 1763.

*nasica Say, J. Acad. Phil. III, 94; Compl. Wr. II, 81; Wiedemann,
Auss. Zw. II, 115. 1. — Atlantic States.

Brachyopa.

Meigen, System. Beschr. III, 260; 1822.

*notata O. Sacken, Bulletin Buff. Soc. N. H. III, 68 (reproduced in
the note [217]). — White Mts., N. H.
*vacua O. Sacken, l. c. ([217]). — Quebec, Canada.
*ferruginea Fallén, Syrph. 34, 3; Meigen. System. Beschr. III, 263. —
Europe and North America (Saskatchevan). [Loew in litt.]

Volucella.

Geoffroy, Hist. des Ins. II, 1764; *Cenogaster* Duméril, Exposition etc.
1801 and Dict. d'Hist. Natur. (Levrault in Strasburg, publisher) 1817.

*esuriens Fabricius, Ent. System. IV, 281, 10 *(Syrphus)*; System. Antl.
226, 9 *(id.)*; Wiedemann, Auss. Zw. II, 197, 4. — West Indies
(Fabr.); Texas; also in South America (Schiner, Novara).
Volucella mexicana Macquart, Dipt. Exot. II, 2, 25; Tab. V, f. 3. —
Mexico (Macq.); Island Santa Rosa, California (O. Sacken,
Western Dipt., 393).
Volucella dispar Macquart, Dipt. Exot. 4e Suppl. 123, Tab. XI, f.
2. — New Granada. [Schiner, Novara, etc., 356.]
Volucella Maximiliani Jaennicke, Neue Exot. Dipt., 87. — Mexico.
([218]. [Schiner, Novara, 356, from comparison ot typical specimens.]
*evecta Walker, Dipt. Saund., 251 — Atlantic States and British
Possessions (White Mts., N. H.; Massachusetts; Detroit, Michigan).

Volucella plumata Macquart (non Fabr.', Dipt. Exot. 4e Suppl. 131.

*__*fasciata__ Macquart, Dipt. Exot. II, 2, 22, 2; Tab V, f. 2. — Carolina (Macq.); Texas; Colorado (O. Sacken, Western Dipt., 334); Meztitlan ,Mexico, collect. Bellardi!).

*__*pusilla__ Macquart, Dipt. Exot. II. 2, 21, 1; Tab. V, f. 1 („perhaps a variety of *V. fasciata*" Macq.). — Cuba (Macq.); Florida (M. C. Z.). ([210]).

*__*vesiculosa__ Fabricius, System. Antl. 226, 11 *(Syrphus)*; Wiedemann, Auss. Zw. II, 201, 11; Macquart, Dipt. Exot. 3e Suppl 39; Tab. IV, f. 3. — North America (Pennsylvania; Maryland; Kentucky); South America (Wied.).

*__*avida__ O. Sacken, Western Dipt., 333. — California (O. S.); Tehuacan, Mexico (Coll. Bellardi).

*__*satur__ O. Sacken, Western Dipt., 333. — Colorado, Utah.

*__*abdominalis__ Wiedemann, Auss. Zw. II, 196, 2; Macquart, Dipt. Exot. II, 2, 25, 8. — Cuba.

amethystina Bigot, Ann. Soc. Ent. de Fr. 1875, 479. — Mexico.

aperta Walker, Trans. Ent. Soc. N. Ser. V, 292. — Mexico.

*__*apicalis__ Loew, Centur. VI, 36. — Cuba.

castanea Bigot, Ann. Soc. Ent. Fr. 1875, 476. — Mexico.

chalybescens Wiedemann, Auss. Zw. II, 204. — Brazil (Wied.); Cuba (Jaennicke, Neue Exot. Dipt. p. 4).

Haagii Jaennicke, Neue Exot. Dipt., 89. — Mexico.

Iata Wiedemann, Auss. Zw. II, 195. — Mexico.

metallifera Walker, List, etc. III, 636. — Mexico, Venezuela.

mellea Jaennicke, Neue Exot. Dipt., 88. — Mexico.

nigrifacies Bigot, Ann. Soc. Ent. 1875, 479. — Mexico.

*__*obesa__ Fabricius, System. Ent. 763, 5 *(Syrphus)*; Ent. Sytem. IV, 282 *(id.)*; System. Antl. 227 *(id.)*; Wiedemann, Auss. Zw. II,199; Macquart, Hist. Nat. Dipt. I, 494, 5; St. Fargeau et Serville, Encycl. Méth. X, 786 *(Ornidia)*. — In the tropics everywhere; West Indies; South America; Asia; Africa (Mr. Bellardi's collection contains a specimen of from New Orleans).

picta Wiedemann, Auss. Zw. II, 201; Bigot, in R. de la Sagra etc. 802. — Brazil (Wied.); Cuba (Bigot).

pulchripes Bigot, Ann. Soc. Ent. Fr. 1875, 480. — Mexico.|

postica Say, J. Acad. Phil. VI, 166, 2; Compl. Wr. II, 360. — Mexico.

purpurifera Bigot, Ann. Soc. Ent. Fr 1875, 477. — Mexico.

*__*sexpunctata__ Loew, Wien. E t. Monatschr. V, 39; Centur. VI, 37. — Cuba.

tibialis Macquart, Dipt. Exot. 1er Suppl. 123, 14. — Yucatan.

tricincta Bigot, Ann. Soc. Ent. Fr. 1875, 477. — Mexico.

tristis Bigot, Ann. Soc. Ent. Fr. 1875, 482. — Mexico.

varians Bigot, Ann. Soc. Ent. Fr. 1875, 481. — Mexico.

viridula Bigot, Ann. Soc. Ent. Fr. 1875, 481. — Mexico.

violacea Say, J. Acad. Phil. VI, 166, 1; Compl.· Wr. II, 360. — Mexico.

12

variegata Bigot, Ann. Soc. Ent. Fr. 1875, 478. — Mexico.

Observation. *Volucella vacua* Fabricius is quoted by Walker, List, etc. III, 637 from Georgia and Florida.

Temnocera.

St. Fargeau et Serville, Encycl. Méth. X, 786, 1825; Macquart, Dipt. Exot. II, 2, 27. ([220]).

*megacephala Loew, Centur. IV, 57. — California.
*setigera O. Sacken, Western Dipt., 334. — Northern New Mexico (O. S.); Tehuacan, Mexico (Collect. Bellardi).

pubescens Loew, Wien. Ent. Monatschr. V, 38; id. Centur. VI, 35. — Cuba.
*purpurascens Loew, Centur. VIII, 52. — Hayti.
unilecta Walker, Trans. Ent. Soc. N. Ser. V, 292. — Mexico.
viridula Walker, Trans. Ent. Soc. N. Ser. V, 292. — Mexico.

Copestylum.

Macquart, Dipt. Exot. Suppl. 1er, 124; 1846.

*marginatum Say, J. Acad. Phil. VI, 167, 3; Compl. Wr. II, 360 (*Volucella*). — Mexico (Say); Waco, Texas (O. Sacken, Western Dipt., 233).

NB. Is *C. flaviventris* Macq. Suppl. 1, 125; Tab. X, f. 16 from Venezuela, a different species? The descriptions read remarkably alike.

Sericomyia.

Meigen, in Illiger's Magaz. II, 1803.

*chalcopyga Loew, Centur. III, 20. — Sitka.
*limbipennis Macquart, Dipt. Exot. 2e Suppl. 58, 2 (*female*). — Atlantic States and Canada.
    Sericomyia chrysotoxoides, Macquart, Dipt. Exot. II, 2, 19, 1; Tab. III, f. 3 bis. (*male*).
    Sericomyia filia Walker, List, etc. III, 596.
*militaris Walker, List, etc. III, 595. — Huds. B. Terr.; Nova Scotia; White Mts., N. H.; Colorado Mts.; Red River of the North.
*sexfasciata Walker, List, etc., III, 596. — Huds. B. Terr.

Observation. *Volucella lappona* O. Fabricius, Fauna Groenl. 208, 169, must be a *Sericomyia;* whether it is *Seric. lappona* Linn. I do not know; Schiödte omits it in his enumeration.

Arctophila.

Schiner, Wien. Ent. Monatschr. IV, 215; 1860.

*flagrans O. Sacken, Buffalo Bull. Soc. N. Hist. III, 69; Western Dipt. 335. — Rocky Mts., Colorado.

Eristalis.

Latreille, Dict. d'Hist. Nat.; H. N. Crust. et Ins. XIV, 363; 1804.

*aeneus Scopoli, Fabricius, Meigen (System. Beschr. etc. III, 384, 2).
— Europe and North America (common); occurs also in Algiers, the Canary Islands, Malta, Syria (Schiner, die Oesterr. Syrphiden, 120).

Eristalis sincerus Harris, Ins. Injur. to Veget. 3ᵈ edt. 609. [The identity with the European species is acknowledged by Loew, in Sillim. Journ., Vol. XXXVII, 317.]

Eristalis cuprovittatus Wiedemann, Auss. Zw. II, 190, 54.

albiceps Macquart, Dipt. Exot. II, 2, 56, 41. — Carolina. ([221]).

*atriceps Loew, Centur. VI, 64. — White Mountains, N. H.; Canada.

Eristalis compactus Walker, List, etc. III, 619. — Huds. B. Terr. ([222]).

*Androclus O. Sacken (non Walker), Western Dipt., 337. — Quebec; Western New York, White Mts., N. H.; Utah; Yucon River, Alaska. ([223]).

*Bastardi Macquart, Dipt. Exot. II, 2, 35, 7; Tab. IX, f. 1. — North America (common in the Atlantic States and British Possessions).

Eristalis nebulosus Walker, List, etc. III, 616 (!).

(?) *Eristalis semimctallicus* Macquart, Dipt. Exot. 4e Suppl. 140, 65. — Nova Scotia, Canada ([224]).

*dimidiatus Wiedemann, Auss. Zw. II, 180, 41. — Atlantic States. ([225]).

Eristalis inflexus Walker, List, etc. III, 617.

Eristalis niger Macquart, Hist. Nat. Dipt. I, 505, 15.

Eristalis L'Herminieri Macquart, Dipt Exot. II, 2, 55, 38 *(male)*.

Eristalis chalybeus Macquart, Dipt Exot. II, 2, 55, 39 *(male and female)*.

Eristalis incisuralis Macquart, Dipt. Exot. 4e Suppl. 139, 64 *(female)*.

*flavipes Walker, List, etc. III, 633. — British Possessions; White Mountains, N. H.; Massachusetts; Newport, R. I.; Detroit, Mich. ([226]).

Milesia Barda Say, J. Acad. Phil. VI, 163; Compl. Wr. II, 357; *female* (for the *male*, see *Mallota Barda).

*inornatus Loew, Centur. VI, 68. — Red River of the North (Loew).

*latifrons Loew, Centur. VI, 65. — Matamoras (Loew); Texas; Iowa,

*melanostomus Loew, Centur. VI, 69. — British Possessions; Oregon; Minnesota; Massachusetts; Illinois.

Eristalis flavipes Walker, List, etc. III, 633; Var. β [Loew].

*obscurus Loew, Centur. VI, 67. — Red River of the North.

oestriformis Walker, List, etc. III, 573 *(Syrphus)*. — Huds. B. Terr. ([227]).

*pilosus Loew, Centur. VI, 70. — Greenland.

*saxorum Wiedemann, Auss. Zw. II, 158, 9; Macquart, Dipt. Exot II, 2, 33, 5. — Savannah (Wied.); Philadelphia (Macq.); Massachusetts (M. C. Z.).

Eristalis pervagus (Harris) Walker, List, etc. III, 618.

*tenax Linné, etc. Europe and North America ([228]); also Cape of
Good Hope and China (Schiner, Dipt. Austriaca, Syrphidae, 10;
also Siberia and Japan (Loew, Wien. Ent. Monatschr. II, 101).

*transversus Wiedemann, Auss. Zw. II, 188, 51; Macquart, Dipt. Exot.
II, 2, 33, 4; Tab. IX, f. 12. — Atlantic States.

(?) *Eristalis philadelphicus* Macquart, Dipt. Exot. II, 2, 34, 6;
Tab. VIII, f. 4 ([229]).

Eristalis pumilus Macquart, Dipt. Exot. II, 2, 57, 43. — North
America.

Eristalis vittatus Macquart, Hist. Nat. Dipt. I, 507, 19. — North
America.

*vinetorum Fabricius, Ent. System. Suppl. 562; System. Antl. 235, 13
(Syrphus); Wiedemann, Auss. Zw. II, 163, 15; Macquart, Dipt.,
Exot. II, 2, 41, 16. — Cuba (Fab.); Brazil (Schiner, Novara, 361);
Pennsylvania (Carlisle Springs, August 1860); Florida; Matamoras.

Eristalis trifasciatus Say, J. Acad. Phil. VI, 165; Compl. Wr. II,
359. — Indiana (the locality „Mexico" given in the Compl. Wr.
of Say, is erroneous).

Eristalis uvarum Walker, List, etc. III, 623. — Jamaica [Loew
in litt.].

(?) *Eristalis thoracicus* Jaennicke, Neue Exot. Dipt. 91. — Mexico.

*hirtus Loew, Centur. VI, 66; O. Sacken, Western Dipt., 335. —
California, Colorado.

Eristalis temporalis Thomson, Eugenies Resa, 490.

*stipator O. Sacken, Western Dipt., 336. — California, Colorado.

*atrimanus Loew, Centur. VI, 62. — Cuba.

Bellardii Jaennicke, Neue Exot. Dipt. 92. — Mexico.

cubensis Macquart, Dipt. Exot. II, 2, 42, 19 (" ♀ of *albifrons* or
variety of *annulipes* Macq. ?" Macquart). — Cuba.

diminutus Walker, List, etc. III, 622. — Mexico.

expictus Walker, Trans. Ent. Soc. N. Ser. V, 291. — Mexico.

familiaris Walker, Trans. Ent. Soc. N. Ser. V, 290. — Mexico.

femoratus Macquart, Dipt. Exot. II, 2, 40, 15; Tab. IX, f. 6; also
1er Suppl. 130; Tab. IX, f. 6. — Rio Janeiro; Columbia, S. A.;
Yucatan. [Syn. of *E. furcatus* Wiedemann, Auss. Zw. II, 176.
34; Brazil and Montevideo. Verrall *in lit.*].

guadalupensis Macquart, Dipt. Exot. II, 2, 32, 3. — Guadeloupe.

*Gundlachi Loew, Centur. VI, 61. — Cuba.

*hortorum Fabricius, System. Ent. 764, 11; Ent. System. IV, 286, 29
(Syrphus); System. Antl. 236, 16; Wiedemann, Auss. Zw. II,
169, 24. — West Indies.

Musca surinamensis Degeer, VI, 145; Tab. XXIX, f. 1.

impositus Walker, Trans. Ent Soc. N. Ser. V, 289. — Hayti.

lateralis Walker, Linn. Trans. XVII, 347, 42. — Brazil; Chili;
Guyana; Mexico; Jamaica (Walker, List, etc. III, 622).

mexicanus Macquart, Dipt. Exot. 2e Suppl. 59, 54. — Mexico.

semicirculus Walker, Dipt. Saund., 249. — Honduras.
*seniculus Loew, Centur. VI, 63. — Cuba.
testaceicornis Macquart, Dipt. Exot. 4ᵉ Suppl. 138, 62. — Mexico.
tricolor Jaennicke, Neue Exot. Dipt. 92. — Mexico.

Observation.

Eristalis Androclus Walker, List, etc. III, 612. — British Possessions.
Eristalis fratcr Walker, List, etc. III, 614.
Eristalis chalepus Walker, Dipt. Saund., 247; Canada.
All three are Helophili; see the note [230].
Eristalis intersistens Walker, List, etc. III, 615; Trenton Falls, seems to be Xylota badia.
Eristalis decisus Walker, List, etc. III, 604; Trenton Falls, is Helophilus similis.
Eristalis Everes Walker, Dipt. Saund., 246; North America. I could not find it in the British Museum, and have for this reason omitted it as unrecognizable, from the above list.
Two species of Macquart's are also omitted from the List of described species:
Eristalis basilaris Macquart, Hist. Nat. Dipt. I, 502, 4. — North America.
Eristalis inflatus Macquart, l. c. 507, 18. — North America.
I did not find the types of these two species, either in Lille, or in Paris and the descriptions do not apply to any of the known species.

Pteroptila.

Loew, Centur. VI, 59, 1865; Plagiocera Macquart, Dipt. Exot. II, 2, 59. [231].

acuta Fabricius, System. Antl. 189, 7 (Milesia); Wiedemann, Auss. Zw. II, 110, 8 (id.). — Carolina.
*crucigera Wiedemann, Auss. Zw. II, 105, 2 (Milesia); Macquart, Dipt. Exot. II, 2, 60, 1 (Plagiocera), Tab. X, f. 7; also 1ᵉʳ Suppl. 134. — Florida; Georgia; Dallas, Texas; Yucatan (Macq.).
Mallota milesiformis Macquart, Hist. Nat. Dipt. I, 500 [Synonymy by Macquart].

cincta Drury, Ins. I, 109; Tab. XLV, f. 6 (Musca). — Jamaica, San Domingo.
Syrphus pinguis Fabricius, System. Ent. 763, 6; Ent. System, IV, 282, 16; System. Antl. 233, 6 (Eristalis); Wiedemann, Auss. Zw. II, 193, 61 (id.).
Milesia Ania Walker, List, etc. III, 564; Macquart, Dipt. Exot. 5ᵉ Suppl. 94, 9 [I found both of these synonymies in the Berlin Museum].
*decora Loew, Centur. VI, 59. — Cuba.
*pratorum Fabricius, System. Ent. 765, 13; Ent. System. IV, 286, 31 (Syrphus); System. Antl. 236, 18 (Eristalis). — West Indies.
*ruficrus Wiedemann, Auss. Zw. II, 105, 3 (Milesia). — Cuba.
zonata Loew, Centur. VI, 60. — Mexico.

Helophilus.

Meigen, in Illiger's Magaz. II, 1803. [232].

*chrysostomus Wiedemann, Auss. Zw. II, 174 (Eristalis). — Savannah (Wied.); New York; White Mts, N. H.

*borealis Staeger, Groenl. Antl. 359, 25; Loew, Stett. Ent. Zeitschr. VII, 123. — Greenland.
*divisus Loew, Centur. IV, 78. — Distr. Columbia.
*glacialis Loew, Stett. Ent. Zeitschr. VII, 121. — Labrador.
*groenlandicus O. Fabricius, Fauna Groenl. 208, 170 (Tabanus); Loew, Stett. Ent. Zeitschr. VII, 119. — Arctic America; Greenland; Twin Lakes (Colorado); Labrador; also in Europe, Sweden.
   Helophilus arcticus Zetterstedt, Ins. Lapp. 595, 2; Dipt. Scand. II, 678, 2 (ex parte); VIII, 3117, 2; Staeger, Kroejer's Tidskr. N. R. I, 359; Holmgren, Nordgroenl. Ins. 100. [Loew and Schioedte].
   Helophilus bilineatus Curtis, Ins. of Ross's Exp. LXXVIII [Schioedte, Berl. Ent. Zeitschr. 1859, 153).
   (?) Helophilus latro Walker, List, etc. III, 607. — Huds. B. Terr.; Nova Scotia.
*hamatus Loew, Centur. IV, 79. — Fort Resolution, Huds. B. Terr.
*integer Loew, Centur. IV, 76. — New York.
*laetus Loew, Centur. IV, 77. — New York; Northern Wisconsin; Illinois.
*latifrons Loew, Centur. IV, 73. — Northern States; Nebraska; Red of the North; California (O. Sacken, Western Dipt., 333).
*lineatus Fabricius, Meigen, Curtis (Brit. Ent.) etc., Loew, Stett. Ent. Zeitschr. 1846, 167. — Europe; North America (Massachusetts; Illinois; Quebec, Canada.).
   (?) Helophilus stipatus Walker, List, etc. III, 602. — Trenton Falls. ([233]).
   Helophilus Anausis Walker, List, etc. III, 603. — Huds. B. Terr.
Novae Scotiae Macquart, Dipt. Exot., 2e Suppl. 60, 10. — Nova Scotia.
*obscurus Loew, Centur. IV, 74. — Fort Resolution, Huds. B. Terr.; South Park, Colorado ([234]).
*obsoletus Loew, Centur. IV, 75. — Fort Resolution, Huds. B. Terr.
porcus Walker, List, etc. III, 551 (Eumerus). — Huds. B. Terr. ([235]).
*similis Macquart, Dipt. Exot. II, 2, 64, 7. — Georgia (Macq.); United States; Canada.
   Helophilus fasciatus Walker, List, etc. III, 605. — Trenton Falls.
   Eristalis decisus Walker, List, etc. III 604. — Trenton Falls.
   Helophilus susurrans Jaennicke, Neue Exot. Dipt. 94. — Illinois. ([236]).

*polygrammus Loew, Centur. X, 55. — California (Sierra Nevada); Oregon (O. Sacken, Western Dipt., 338; Mexico (? I saw in the Berlin Mus. a specimen very like this species).
femoralis Walker, List, etc. III, 603. — Mexico.
mexicanus Macquart, Dipt. Exot. II, 2, 64, 6; Tab. IX, f. 2. — Mexico.

Observation. Eristalis Androclus and frater (Walker, List, etc.) and E. chalepus (Walker, Dipt. Saund.) are Helophili; see the observation at the end of Eristalis, and the Note ([230]).
   About the occurrence in North America of Heloph. pendulus, versicolor, floreus, see the Note ([197]).
   For Helophilus albiceps Macq. see Polydonta curvipes.

Teuchocnemis.

O. Sacken, Bull. Buff. Soc. N. H, III, 58; 1876. ([237]).

*Bacuntius Walker, List, etc. III, 563 *(Milesia)*. — Georgia; Texas. ([238]).
*lituratus Loew, Centur. IV, 81 *(Pterallastes)*. — Pennsylvania.

Pterallastes.

Loew, Centur. IV, 80; 1863.

*thoracicus Loew, Centur. IV, 80. — Pennsylvania.

Mallota.

Meigen, System. Beschr. III, 377; 1822; *Imatisma* Macquart, Dipt. Exot.
II, 2, 67; 1842.

*posticata Fabricius, System. Antl. 237, 21 *(Eristalis)*; Wiedemann,
Auss. Zw. II, 194, 62 (translation from Fabric); Macquart, Dipt.
Exot. II, 2, 68; Tab. XII, f. 2 *(Imatisma)*. — Atlantic States;
the same, or a similar species in California (O. Sacken, Western
Dipt., 338).
Syrphus cimbiciformis Fallen, *Eristalis cimbiciformis* Meigen. The
north of Europe (the identity of this species with the N. American
one is acknowledged by Mr. Loew in Neue Beitr., IV, 18 and
in Sillim. J. Vol. XXXVII, 317).
*barda Say, J. Acad. Phil. VI, 163; Compl. Wr. II, 357 *(Milesia) male*;
(the female described by Say is that of *Eristalis flavipes* Walker;
compare note ([226]) Catskill, N. Y.; Massachusetts; White Mts , N.H.
Eristalis coactus Wiedemann, Auss. Zw. II, 165 (without locality).
Merodon Balanus Walker, List, etc. III, 599. — New York.
Bautias Walker, List, etc. III, 600 *(Merodon)*. — Georgia. ([239]).
bipartita Walker, List, etc. III, 599 *(Merodon)*. — Georgia.

Merodon.

Meigen, Illiger's Magaz. II; 1803.

No american species are as yet recorded. The european *Merodon
narcissi* has been occasionally introduced to the United States in
dutch bulbs and the fly reared from them by Mr. F. G. Sanborn
(see Packard's Guide, 399).
For *Merodon Bautias, Balanus, bipartitus* Walker, see *Mallota*.

Polydonta.

Macquart, Dipt. Exot. 4e Suppl. 144; 1849.

*curvipes Wiedemann, Auss. Zw. II, 149, 3 *(Merodon)*. — Northern
States, and British Possessions; the same, or a similar species
in California and Colorado; see O. Sacken, Western Dipt., 338.
Polydonta bicolor Macquart, Dipt. Exot. 4e Suppl. 144, 1; Tab.
XIII, f. 6 *(male)*.
Helophilus albiceps Macquart, Dipt. Exot. 1er Suppl. 132, 9; Tab
XI, f. 7 *(female)*.
Merodon morosus Walker, List, etc. III, 599 *(female)*.

Tropidia.

Meigen, System. Beschr. III, 346; 1822.

albistylum Macquart, Dipt. Exot. 2e Suppl. 60, I; Tab. II, f. 10. — North America.

*__mamillata__ Loew, Centur. I, 68. — Illinois.

*__quadrata__ Say, Amer. Ent. I; Tab. VIII; Compl. Wr. I, 14 *(Xyloto)*; Wiedemann, Auss. Zw. II, 101. 6 *(id.)*; Macquart, Dipt Exot. II, 272. — United States (Massachusetts, White Mts., N. H.; New York); California (O. Sacken, Western Dipt., 338).

Criorrhina.

Criorhina Hoffmannsegge *(in litt.)*, was introduced as a subgenus of *Milesia* in Meigen, System. Beschr. III, 236; 1822, appears as such in St. Fargeau et Serville, Encycl. Méth. X, 518, 1825; adopted as a genus in Macquart, Hist. Nat. Dipt. I, 497; 1834.

*__analis__ Macquart, Dipt. Exot. II, 2, 79; Tab. XV, f. 2 *(Milesia)*. — North America (Macq.).

*__armillata__ O. Sacken, Bull. Buff. Soc. N. H. III, 68 (reproduced in the note [240]).

Crioprora.

nov. gen. [241].

*__cyanogaster__ Loew, Centur. X, 51; *(Brachypalpus)*. — Pennsylvania.

*__alopex__ O. Sacken, Western Dipt., 338 *(Pocota)*. — California.

*__cyanella__ O. Sacken, Western Dipt., 339 *(Pocota.)* — California.

Brachypalpus.

Macquart, Hist. Nat. Dipt. I, 523; 1834.

Amithaon Walker, List, etc. III, 567 *(Milesia)*. — North Carolina. [242].

*__frontosus__ Loew, Centur. X, 50. — Distr. Columbia, Texas, Massachusetts.

(?) *Xylota Oarus* Walker, List. etc. III, 558. — Trenton Falls.

*__verbosus__ (Harris) Walker, List, etc. III, 568. — Connecticut, Canada, Virginia.

Musca tomentosa Swederus, Vetensk. Ak. Nya Handl.; 1787.

Xylota.

Meigen, Sytem. Beschr. III, 211; 1822. [243].

Aepalius Walker, List, etc. III, 557. — Georgia. [244].

Anthreas Walker, List, etc. III, 556. — Trenton Falls, New York.

*__angustiventris__ Loew, Centur. VI, 53. — Illinois; Western New York.

Baton Walker, List, etc. III, 554 („perhaps synon. with *ejuncida“* Wk.). — Florida; Nova Scotia.

*__barbata__ Loew, Centur. V, 40. — Sitka.

*__bicolor__ Loew, Centur. V, 39. — Illinois (Lw.); Englewood, N. J. (O. S.).

*__chalybea__ Wiedemann, Auss. Zw. II, 98. — No locality (Wied.) Northern and Middle States (Illinois; Pennsylvania).

communis Walker, List, etc. III, 557. — Huds. B. Terr. (perhaps the same as *obscura* Lw.).

curvipes Loew, Neue Beitr. II, 19, 71. — Europe and North America; White Mts., N. H. (About the identity of the species, see O. Sacken, Bull. Buff. Soc N. H. III, 70, also reproduced in the note (245).

***ejuncida** Say, Amer. Ent. I; Tab. VIII; Compl. Wr. I, 15; Wiedemann, Auss. Zw. II, 100, 5. — Florida; Pennsylvania (Say); New England (common) (246).

flavifrons Walker, List, etc. III, 537. — Huds. B. Terr.

***fraudulosa** Loew, Centur. V, 41. — Illinois, Wisconsin, White Mts., N. H.

***pigra** Fabricius, Meigen, etc. — Europe and North America.
  Xylota haematodes Fabricius, System. Antl., 193, 21 *(Milesia)*; Say, Amer. Ent. I; Tab. VIII; Compl. Wr. I, 16; Wiedemann, Auss. Zw. II, 99, 3; Macquart, Dipt. Exot. II, 2, 73; Tab. XIII, f. 4. — North America. [About the specific identity, see Loew, Sillim. Journ. Vol. XXXVII, 317]

Libo Walker, List, etc. III, 556 — Nova Scotia.

***metallica** Wiedemann, Auss. Zw. II, 102, 8. — Georgia.

***obscura** Loew, Centur. VI, 55. — Red River of the North.

***quadrimaculata** Loew, Centur. VI, 56. — Illinois.

***subfaciata** Loew, Centur. VI, 57. – Red River of the North.

***vecors** O. Sacken, Bull. Buff. Soc. N. H., III, 69 (reproduced in the note (245). — White Mts., N. H.

arcuata Say, J. Acad. Phil. VI, 162; Compl. Wr. II, 357. — Mexico.

***pachymera** Loew, Centur. VI, 54. — Cuba.

***pretiosa** Loew, Wien. Ent. Monatschr. V, 39; Centur. VI, 53. — Cuba.

subcostalis Walker, Trans. Ent. Soc. Phil. N. S. V, 291 — Mexico.

Observation. For *Xylota Oarus* Walker, see *Brachypalpus frontosus.*

Syritta.

St. Fargeau et Serville, Encycl. Méthod. X, 808; 1825.

***pipiens** Linné, Meigen, etc. — Europe and North America (common); also in California, Nevada, Utah.
  Xylota proxima Say, Amer Ent. I; Tab. VIII; Compl. Wr. I, 16; Wiedemann, Auss. Zw. II, 102, 9. (About the identity of the European and North American species, compare Loew, Sillim. Journ. l. c.)

Eumerus.

Meigen, System. Beschr. III, 202; 1822.

No species from North America have been as yet recorded. For *Eumerus porcus* Walker, see *Helophilus porcus;* for *Eumerus privernus* Walker, see *Mesograpta geminata.* (247).

Genus novum? [248].

*badia Walker, List, etc. III, 559 (*Xylota*). — New York (Walker);
  White Mts., N. H ; Maine.
  (?) *Eristalis intersistens* Walker, List, etc. III, 615. — Trenton
  Falls.
notata Wiedemann, Auss. Zw. II, 109, 7 (*Milesia*). — Macquart, Dipt.
  Exot. II, 2, 80, 2; Tab. XV, f. 5 (*id.*). — Georgia; Carolina.
  Syrphus profusus Walker, List, etc. III, 578. — Georgia.

Somula.
Macquart, Dipt. Exot. 2e Suppl. 57; 1847.

*decora Macquart, Dipt. Exot. 2e Suppl. 57, 1; Tab. II, f. 11. — Middle
  States.

Chrysochlamys.
Walker (Rondani), Ins. Brit. I, 279; 1851. [249].

*buccata Loew, Centur. IV, 72; O. Sacken, Western Dipt., 340. —
  Alleghany Mts., Virginia.
.*dives O. Sacken, Western Dipt., 341. — Kentucky.
*nigripes O. Sacken, Western Dipt, 341. — Massachusetts.
*croesus O. Sacken, Western Dipt., 341. — Utah.

Spilomyia.
Meigen, in Illiger's Magaz. II; 1803. [250].

*fusca Loew, Centur. V, 34. — Pennsylvania, Massachusetts, White
  Mts., N. H.
*hamifera Loew, Centur. V, 33. — Pennsylvania; Virginia; Florida;
  Kentucky.
*longicornis Loew, Centur. X, 49. — Massachusetts; Pennsylvania;
  Texas; Kansas.

Temnostoma.
St. Fargeau et Serville, Encycl. Méth. X, 518; 1825.

*aequalis Loew, Centur. V, 36. — British North America; New Eng-
  land (White Mnts., N. H., not rare). [251].
*alternans Loew, Centur. V, 37. — Pennsylvania (Lw.); Quebec, Can.;
  White Mts., N. H.
*Balyras Walker, List, etc. III, 577 (*Doros*). — New York; White
  Mts., N. H.
  Temnostoma obscura Loew, Centur. V, 35. — British America. [252].
* excentrica Harris, Ins. of New England, etc. 3d ed., 609; f. 267
  (*Milesia*). About O. Sacken's description, given in the same
  volume, compare the note [251]. — New England (Harris ;
  Illinois (O. Sacken).

Lepidomyia.
Loew, Centur. V, 38; 1864.

*calopus Loew, Centur. V, 38. — Cuba.

Milesia.

Latreille, Hist. Nat. des Crust. et des Ins. XIV, 361; 1804.

*ornata Fabricius, System. Antl. 188, 5; 1805; Wiedemann, Auss. Zw.
II, 106, 4; Macquart, Dipt. Exot. II, 2, 81, 4; Tab. 15, f. 4. —
United States, from New England to Texas, Florida and Kansas;
Guadeloupe (Macq.).
Musca virginiensis Drury, Illustr. II; Tab. XXXVII, f. 6; 1773.
[Wied.]
Syrphus trifasciatus Hausmann, Ent. Bemerk. II, 67, 10; 1799.
[Wied.]
*limbipennis Macquart, Dipt. Exot. 4e Suppl. 147, 8; Tab. XIV, f. 3.
— North America (Macq.'; Florida. (²⁵³).

Sphecomyia.

Latreille, Fam. Natur. du Règne Anim.; 1825; Dict. Classique d'Hist.
Nat. XV, 545; 1829; *Tyzenhausia* Gorski; 1852. (²⁵⁴).

*vittata Wiedemann, Auss. Zw. II, 87 *(Chrysotoxum)*. — Unknown
locality (Wied.); New York; Virginia; White Mts., N. H.;
Colorado (O. Sacken, Western Dipt., 341).
Psarus ornatus Wiedemann, Auss. Zw. II, 91, 1; Tab. IX, f. 7;
Macquart, Hist. Nat. Dipt. I, 491, 2; Dipt. Exot. II, 2, 18, 1;
Tab; III, f. 3. — Georgia (Wied.).
*brevicornis O. Sacken, Western Dipt., 341. — Sierra Nevada, Cal.

Mixtemyia.

Macquart, Hist. Nat. Dipt. I, 491; 1834.

*quadrifasciata Say, Long's Exped. App. 377 *(Paragus)*; Compl. Wr.
I, 257; Wiedemann, Auss. Zw. II, 91, 2 *(Psarus)*; Macquart,
Hist. Nat. Dipt. I, 491; Tab. XI, f. 8. — Canada (Quebec);
White Mts., N. H.; Cambridge; Mass.; Connecticut.

*ephippium O. Sacken, Bull. Buff. Soc. N. H. III, 70 (reproduced in
the note (²⁵⁵). — Mexico.

Ceria.

Fabricius, System. Ent. IV, 277; 1794. (²⁵⁶).

*abbreviata Loew, Centur. V, 48; compare also X, 57, nota 2. —
Pennsylvania, New York.
pictula Loew, Neue Beitr. I, 17. — Southern States.
*signifera Loew, Neue Beitr. I, 19. — Mexico (Lw.); Texas (M. C. Z.;
determination by Lw.).

*tridens Loew, Centur. X, 57. — Sierra Nevada, Cal.

arietis Loew, Neue Beitr. I, 17. — Mexico.
cacica Walker, Trans. Ent. Soc. N. Ser. V, 2ɔ7. — Mexico.

Daphnaeus Walker, List, etc. III, 537; Westwood, Trans. Ent. Soc. V, 231; Tab. XXXIII, f. 7; Loew, Neue Beitr. etc. I, 18. — Jamaica.
tricolor Loew, Wien. Ent. Monatsch. V, 37. — Cuba.

FAMILY CONOPIDAE.

Conops.

Linné, Fauna Suecica; 1761. ([257]).

aethiops Walker, List, etc. III, 671. — North America.
analis Fabricius, System. Antl. 175, 3; Wiedemann, Auss. Zw. II, 237, 5; Macquart, Dipt. Exot. II, 3, 14, 12; Tab. I, f. 3. — South America (Fabr.); Carolina (Macq.).
brachyrrhynchus Macquart, Dipt. Exot. II, 3, 15, 13; Tab. I, f. 8. — North America.
bulbirostris Loew, Neue Beitr., etc. I, Conops, 30. — North America (Loew *in litt.*).
castanopterus Loew, Neue Beitr., etc. I, Conops, 33. — Savannah.
costatus Fabricius, System. Antl. 175, 4; Wiedemann, Auss. Zw. II, 238, 6; Macquart, Dipt. Exot. II, 3, 14, 11; Tab. I, f. 4. — South America (Fabr.); Carolina (Macq.).
***excisus** Wiedemann, Auss. Zw. II, 234, 1 and 236, 3 *(C. excisa* ♀ and *C. sugens* ♂*);* Loew, Neue Beitr., etc. I, Conops, 28. — Georgia, Florida.
flaviceps Macquart, Dipt. Exot. II, 3, 15, 14. — North America.
fulvipennis Macquart, Dipt. Exot. II, 3, 13, 10; Tab. I, f. 9. — Georgia.
***genualis** Loew, Neue Beitr., etc. I, Conops, 32. — Middle States.
marginatus Say, J. Acad. Phil. III, 82, 1; Compl. Wr. II, 73; Wiedemann, Auss. Zw. II, 240, 9; Loew, Neue Beitr., etc. I, Conops, 34. — Missouri.
***pictus** Fabricius, Ent. System. IV, 391, 3; System. Antl. 176, 5; Macquart, Dipt. Exot. II, 3, 13, 9 *(ex parte).* — West Indies (Fabr.); Carolina (Macq.).
Conops Ramondi Bigot, in Ramon de la Sagra etc. 808; Tab. XX, f. 6. [Loew *in litt.;* see note [258]].
***sagittarius** Say, J. Acad. Phil. III, 83, 2; Loew, Neue Beitr., etc. I, Conops, 31. — Atlantic States.
Conops nigricornis Wiedemann, Auss. Zw. II, 236, 4. [Wied.].
tibialis Say, J. Acad. Phil. VI, 171; Compl. Wr. II, 363. — Indiana.

Stylogaster.

Macquart, Hist. Nat. Dipt. II, 38; 1835; Dipt. Exot. II, 3, 17.
Stylomyia Westwood, Proc. Zool. Soc. of London, 1850, 269.

***stylatus** Fabricius, Syst. Antl. 177, 11 *(Conops);* Wiedemann, Auss. Zw. II, 243, 2 *(Myopa);* Macquart, Dipt. Exot. II, 3, 17; Tab. II, f. 3. — Pennsylvania, Delaware; also in Brazil (Fabr., Wied.).

Myopa biannulata Say, J. Acad. Phil. 81, 3; Compl. Wr. II. 72.
Stylomyia confusa Westwood, Proc. Zool. Soc. London, 1850, 269. —
No locality. ([25]).

Oncomyia.

Loew, Centur. VII, Nr. 73, thus amends the earlier name *Occemyia* Rob.
Desv., Dipt. des Env. de Paris, 50; 1853.

*abbreviata Loew, Centur. VII, 73. — Distr. Columbia.
*loraria Loew, Centur. VII, 74. — White Mts., N. H.

Zodion.

Latreille, Précis etc.; 1796.

abdominale Say, J. Acad. Phil. III, 84. 2; Compl. Wr. II, 74; Wiedemann, Auss. Zw. II, 242, 2. — Rocky Mountains.
*nanellum Loew, Centur. VII, 75. - Distr. Columbia.
occidentis Walker, List, etc. III, 676. — Ohio.

splendens Jaennicke, Neue Exot. Dipt. 97. — Mexico.

Dalmania.

Dalmannia Rob. Desv. Ess. Myod. 248, 1830; *Dalmania (id.)*, Myopaires;
the latter adopted by Loew, Centur. Vol. II, p. 290. *Stachynia* Macquart, Dipt. du Nord, 1833—34. ([26]).

*nigriceps Loew, Centur. VII, 71. — Virginia (Lw.); Massachusetts.

Myopa.

Fabricius, System. Ent. p. 798; 1775.

americana Wiedemann, Auss. Zw. II, 242, 3 *(Zodion)*. — Montevideo
(Wied.); North America (Walker, List, etc III, 678).
apicalis Walker, List, etc. III, 679. — North America.
bistria Walker, List, etc. III, 679. — North America.
*clausa Loew, Centur. VII, 72. — Maine.
fulvifrons Say, J. Acad. N. Sc. Phil. III, 83; Compl. Wr. II, 74
(Zodion); Wiedemann, Auss. Zw. II, 241, 1 *(id.)* — Pennsylvania,
Maryland (Say).
Myopa rubrifrons Rob. Desovidy, Ess. Myod. 247, 17 [Walker,
List, etc. III, 678].
longicornis Say, Journ. Acad. N. Sc. Phil. III, 81, 2; Compl. Wr. II,
72; Wiedemann, Auss. Zw. II, 245, 4. — Missouri.
obliquefasciata Macquart, Dipt. Exot. 1er Suppl. 141, 1. — Texas.
vesiculosa Say, J. Acad. N. Sc. Phil. III, 80, 1; Compl. Wr. II, 72;
Wiedemann, Auss. Zw. II, 245, 3. — Pennsylvania (Say); Massachusetts (Harris, Catal.).
vicaria Walker, List, etc. III, p 679. — Nova Scotia.

conjuncta Thomson, Eugen. Resa, Dipt. 515. — California.

Observation. For *Myopa biannulata* Say, see *Stylogaster
stylatus*. For *Myopa nigripennis* Gray, see *Pyrgota undata*.

FAMILY PIPUNCULIDAE.

Pipunculus.

Latreille, Hist. Nat. des Crust. et des Ins.; 1804. ([261]).

*cingulatus Loew, Centur. VI, 73. — Distr. Columbia.
*fasciatus Loew, Centur. X, 59. — Texas.
*fuscus Loew, Centur. VI, 71. — Maryland.
 lateralis Walker, Dipt. Saund., 216. — North America.
*nigripes Loew, Centur. VI, 75. — Pennsylvania.
*nitidiventris Loew, Centur. VI, 72. - Distr. Columbia.
 reipublicae Walker, List, etc. III, 639. — New York.
*subopacus Loew, Centur. VI, 74. — Distr. Columbia.
*subvirescens Loew, Centur. X, 58. - Texas.
 translatus Walker, Trans. Ent. Soc. N. Ser. IV, 150. — United States.

FAMILY PLATYPEZIDAE.

Callomyia.

Meigen, Klassification etc., I, 2, 311; 1804.

*divergens Loew, Centur. VI, 77. — Pennsylvania.
*notata Loew, Centur. VI, 77. — Pennsylvania.
*talpula Loew, Centur. IX, 81. — New Hampshire.
*tenera Loew, Centur. IX, 82. — New York.

Platypeza.

Meigen, in Illiger's Magaz. II, 272; 1803.

*anthrax Loew, Centur. IX, 83. — New York.
*flavicornis Loew, Centur VI, 79. - Pennsylvania.
*obscura Loew, Centur. VI, 80. — Pennsylvania.
*pallipes Loew, Centur. VI, 81. — Distr. Columbia.
*velutina Loew, Centur. VI, 79. — Pennsylvania.

Platycnema.

Zetterstedt, Dipt. Scand. I, 332; 1842.

*imperfecta Loew, Centur. VI, 82. — Distr. Columbia.

FAMILY OESTRIDAE. ([262])

Gastrophilus.

Leach, on the gen. and sp. of Eprob. ins. etc. 1817; *Gastrus* Meigen.

*equi Fabricius, Meigen, Latreille, B. Clark etc. A. Fitch, Survey of
 Washington Co., N. Y. (in Trans. N. Y. Agric. Soc. Vol. IX, 799;
 Oestrus); Harris, Ins. of N. Engl. 3d edit. 623; Tab. VIII, f. 2;
 Brauer, Oestriden, 68; Tab, I, f. 1; Tab. V, f. 1; Tab. VII, f. 1
 — 3 (larva). — Europe and North America; on horses.

haemorrhoidalis Linné, Fabricius, Meigen, Clark etc. Harris, Ins. of
N. Engl. 623. Brauer, Oestriden, 83; Tab. I, f. 5; Tab. VII, f. 4
(larva). — Europe and North America; on horses.
*****nasalis** Linné, Meigen, etc. Brauer, l. c. 86; Tab. I, f. 7; Tab. VII,
f. 6 (larva). – Europe and North America; on horses (I have
seen specimens from New York, Utah and Kansas).
Gastrus veterinus Clark, Fabricius, Fallen; Green, Natur. Hist. of
the horse-bee in Adams's medical and agricultural register, Vol. I,
53; New England Farmer, Vol. IV, 345; Harris Ins. N. Engl. 3ᵈ
edit. 623.
Oestrus subjacens Walker, List, etc. III, 687. — Nova Scotia
[Brauer suggests this synonymy, which I can confirm, after
having seen the specimens in the Brit. Mus.].
pecorum Fabricius, Fallen, Meigen, etc. Walker, List, etc. III, 686;
Brauer, Oestriden, 75; Tab. I, f. 4; Tab. VII, f. 5 and 7 (larva). —
Europe, and according to Walker, Jamaica.

Hypoderma.
Clark, Essay on bots etc.; 1815.

bonassi Brauer, Verh. Zool. Bot. Ges., 1875, 75 (the larva alone is
described). — On the buffalo.
*****bovis** De Geer, Fabricius, etc, Brauer, Oestriden, 125; Tab. II, f. 2;
Tab. V, f. 4; Tab. VIII, f. 1ᵃ and 7; Fitch, Survey, etc. 799;
Harris, Ins. N. Engl. 3ᵈ edit. 624. — Europe and North America
(on oxen).
*****lineata** Villers, Olivier, etc. Brauer, Oestriden, 122; Tab. II, f. 3;
Tab. V, f. 8 (larva). — Europe and North America (specimens
from Kentucky in the Vienna Museum; from Texas in M. C. Z.).
On sheep or oxen (?).
Oestrus supplens Walker, List, etc. III, 685; Brauer, Oestriden, 129
[merely a translation of Walker's description. Brauer suggests
that this may be *H. lineata*; the specimens I saw in the Brit.
Museum are either *lineata* or *bovis*]. — Nova Scotia.

Oedemagena.
Latreille, Fam. Natur.; 1825.

tarandi Linné, Fabricius, Meigen, etc. — Brauer, Oestriden, 131. —
On the reindeer; Europe and North America (the latter according
to Palisot in Macquart, Dipt. Exot. II, 3, 25; according to Brauer
the Vienna Museum possesses an american specimen).

Oestrus.
Linné, Fauna Suecica. 1761.

*****ovis** Linné, Fabricius, Meigen, etc. Brauer, Oestriden, 151; Tab. III,
f. 1; Tab. VI, f. 1; Tab. VII, f. 10 (larva); A. Fitch, Survey oⱼ
Washington, Co. (l. c. 799). -- Europe and North America; on sheep

Cephenomyia.

Cephenemyia Latreille, Fam. Natur.; 1825; amended by Brauer.

Ulrichii Brauer, Oestriden, 199; Tab. III. f. 8; Tab. IX, f. 7 (larva). — Europe (on *Cervus Alces*); North America (only larvae were seen by Brauer from this part of the world).

phobifer Clark, Essay etc., 69; Tab. II, f. 30 *(Oestrus)*; Brauer, Oestriden, 213 and also 291; Tab. V, f. 11 (Referred to the genus with a doubt, as this author never saw the insect). — Georgia.

Observation. A larva of this genus found in the throat of *Cervus macrotis* Say in the North Western territories, is described by Brauer, l. c. 211 and figured on his Tab. IX, f. 9. The fly from it is not yet known.

Cuterebra.

Clark, Essay on the Bots; 1815; *Trypoderma* Wiedemann, Loew.

americana Fabricius, System. Ent. 774, 6; Ent. System. IV, 315, 14; System. Antl. 288, 21 *(Musca)*; Wiedemann, Auss. Zw. II, 258, 3 *(Trypoderma)*; Macquart, Dipt. Exot. II, 3, 23, 5; Brauer, Oestriden, 242; Tab. IV, f. 2; Tab. VI, f. 7 (head). — United States and Mexico.

Cuterebra cauterium Clark, Essay on Bots 70; Tab. II, f. 3 (Brauer).

approximata Walker, in Lord's Naturalist etc. II, 338. — Vancouver's Isl.

*****buccata** Fabricius, Mant. Ins. 305, 1; Ent. System. IV, 230, 1; System. Antl. 227, 1 *(Oestrus)*; Wiedemann, Auss. Zw. II, 259, 4 *(Trypoderma)*; Olivier, Encycl. Méth. VIII, 464; Macquart, Hist. Nat. Dipt II, 47, 2; Brauer, Oestriden, 429; Tab. IV, f. 4; Tab. VI, f. 9 (head). — Kentucky, Pennsylvania, Carolina (Fabr.); Massachusetts (Harris).

Cuterebra purivora Clark, Essay on Bots, etc. 70, 4; Tab. II, f. 29. [Wied.].

cuniculi Clark, Trans. Lin. Soc. III, 299; Essay on Bots 70, 1; Tab. II, f. 26; Fabr., Syst. Antl. 230, 9 *(Oestrus)*; Wiedemann, Auss. Zw. II, 256, 1 *(Trypoderma)*; Olivier, Encycl. Méth. VIII, 464, 2; Macquart, Hist. Nat. Dipt. II, 47, 1; Tab. XIII, f. 17; Brauer, Oestriden, 240. — Georgia, Massachusetts (Brauer, l. c. doubts the specific distinctness of this species from *C. horripilum*).

emasculator Fitch, Reports, Vol. II, Nr. 210; Brauer, Oestriden, 232 (Translation of Dr. Fitch's account, with remarks). — North America; on *Tamias striatus*.

fontinella Clark, Trans. Lin Soc. XV, 410; Joly, Réch. sur les Oestrides, 289. Brauer, Oestriden, 242 reproduces Clark's description. — Illinois.

*****horripilum** Clark, Essay etc., 70; Tab. II, f. 27; Brauer, Oestriden, 235; Tab. IV, f. 6; Tab. VI. f. 11 (head); Wiedemann, Auss. Zw. II, 237 *(Trypoderma)*. — New York, Georgia, Nova Scotia.

*scutellaris Loew, Brauer, Oestriden, 230; Tab. IV, f. 3: Tab. 6, f. 10 (head). — North America (according to Brauer probably synonymous with *C. emasculator*).

analis Macquart, Dipt. Exot. II, 3, 22; Tab. II, f. 5; Joly, Rech. 278 (Fig.); Brauer, Oestriden, 237; Tab. IV, f. 1, 1ª; Tab. VI, f. 8 (head). — Brazil and Mexico.

apicalis Guérin, Iconogr. etc. 547; Tab. 101, f. 1. — America (according to Brauer l. c. 240, probably the male of the preceding species).

atrox Clark, Essay etc. Addenda; Brauer, Oestriden, 241. — Mexico.

terrisona Walker, List, etc. III, 685. — Brauer, Oestriden, 244. — Guatemala. (Brauer, who merely translates Walker's description, holds this to be the same as *C. americana*)

Dermatobia.

Brauer, Verh. Zool. Bot. Ges.; 1860.

The so — called *Oestrus hominis* of Central and South America belongs here. The description of all the known larvae, as well as of the known imagos are collected in Brauer, Oestriden, 251 — 269; Tab. X. All the references will be found there. Here I will quote only Say, „On the South Amer. species of Oestrus, which inhabits the human body", in the Journ. Acad. N. Sci. Phil. II, 354, 1822; Compl. Wr. II, 32.

FAMILY TACHINIDAE.[263]

SECTION I. PHASINA.

Phasia.

Latreille, Hist. Nat. des Crust. et des Ins. XIV, 379; 1804.

atripennis Say, J. Acad. Phil. VI, 172, 1; Compl. Wr. II, 363. — Indiana.

Hyalomyia.

Rob. Desvoidy, Myod. 298; 1830.

occidentis Walker, Dipt. Saund., 260. — United States.
*triangulifera Loew, Centur. IV, 85. — New York.

Trichopoda.

Latreille, in Cuvier's Règne animal Vol. V; 1829.

ciliata Fabricius, System. Antl. 315, 9 *(Ocyptera)*; Wiedemann, Auss. Zw. II, 273, 8; Macquart, Dipt. Exot. II, 3, 77, 2; Tab. IX, f. 1. — South America (Fabr., Wied.). — Carolina (Macq.).

cilipes Wiedemann, Auss. Zw. II, 276, 11. — Carolina.
Thereva pennipes Fabricius, System. Antl. 219, 8 (change of name by Wiedemann.)

flavicornis R. Desvoidy, Myod. 284. — Carolina.

*formosa Wiedemann, Auss. Zw. II, 268, 1; Macquart, Hist. Nat. Dipt.
II, 194, 1; Tab. XV, f. 8. — Georgia.
hirtipes Fabricius, System. Antl. 219, 9 *(Thereva)*; R. Desvoidy, Myod.
284; Wiedemann, Auss. Zw. II, 276, 12. — Carolina.
*lanipes Fabricius, System. Antl. 220, 10 *(Thereva)*; Wiedemann, Auss.
Zw. II, 270, 4; R. Desvoidy, Myod., 284, 5. — Georgia.
*pennipes Fabricius, Ent. Syst. IV, 348, 149 *(Musca)*; System. Antl.
327, 5 *(Dictya)*; Wiedemann, Auss. Zw. II, 274, 9; R. Desvoidy,
Myod., 283, 1. — Atlantic States.
Phasia jugatoria Say, J. Acad. Phil. VI, 172, 2; Compl. Wr. II, 364.
plumipes Fabricius, System. Antl. 220, 11 *(Thereva)*; Wiedemann,
Auss. Zw. II, 277, 13; R. Desvoidy, Myod. 285, 6. — Carolina.
*pyrrhogaster Wiedemann, Auss. Zw. II, 271. — Cuba; Texas (Loew
in litt.).
*radiata Loew, Centur. IV, 89. — Distr. Columbia.
*trifasciata Loew, Centur. IV, 90. — Connecticut.

haitensis R. Desvoidy, Myod. 285. — San Domingo.
mexicana Macquart, Dipt. Exot. 1er Suppl. 172, 3 — Mexico.
nigricauda Bigot, Ann. Soc. Ent. Fr. 1876, 394. — Mexico.

Himantostoma.
Loew, Centur. IV, 87; 1863.
*sugens Loew, Centur. IV, 87. — Illinois.

Xysta.
Meigen, System. Beschr. IV, 181; 1824.
* didyma Loew, Centur. IV, 86. — Illinois.

SECTION II. GYMNOSOMINA.

Gymnosoma.
Meigen, in Illiger's Magaz. II, 1803.
* filiola Loew, Centur. X, 66. — Texas.
fuliginosa R. Desvoidy, Myod. 237. — Carolina.
occidua Walker, List, etc. IV, 692. — Nova Scotia.
* par Walker, List, etc. IV, 692. — Nova Scotia.

Cistogaster.
Latreille, in Cuvier's Règne animal. Vol. V; 1829.
* divisa Loew, Centur. IV, 88. — Connecticut.
immaculata Macquart, Dipt. Exot. II, 3, 76; Tab. VIII, f. 7. — Carolina.

SECTION III. OCYPTERINA.

Ocyptera. [264].
Latreille, Hist. Nat. des Crust. et des Ins. XIV, 378; 1804.
arcuata Say, J. Acad. Phil. VI, 173; Compl. Wr. II 363. — Indiana.
[Not an *Ocyptera*, Loew *in litt.*].

aurata R. Desvoidy, Myod. 226 *(Hemyda)*. — Philadelphia. [Not an *Ocyptera,* Loew *in litt.*].

carolinae R. Desvoidy, Myod. 232 *(Parthenia);* Macquart, Dipt. Exot. II, 3, 75. — Carolina.

Dosiades Walker, List, etc. IV, 695. — Nova Scotia.

Epytus Walker, List, etc. IV, 694. — Georgia.

Euchenor Walker, List, etc. IV, 696. — Massachusetts; Newfoundland.

liturata Olivier, Encycl. Méthod. VIII, 423, 1. — Carolina.

Dotadas Walker, List, etc IV, 694. — Jamaica.

Ervia.

Rob. Desvoidy, Myod. 225, 1830; Macquart, Dipt. Exot. II, 3, 74.

triquetra Olivier, Encycl. Méthod. VIII, 423, 2 *(Ocyptera)*; Rob. Desvoidy, Myod. 225. — Carolina.

Lophosia.

Meigen, System. Beschr. IV, 216; 1824.

setigera Thomson, Eugen. Resa, etc. 527. — California.

SECTION IV. PHANINA.

Wahlbergia. (*)

Zetterstedt, Dipt Scand. I; 1842.

*****brevipennis** Loew, Centur. IV, 91. — Nebraska.

SECTION V. TACHININA.

Dejeania.

Rob. Desvoidy, Myod. 33, 1830; Macquart, Dipt. Exot. II, 3, 32; 1843.

*****corpulenta** Wiedemann, Auss. Zw. II, 280 *(Tachina);* Schiner, Novara etc. 337 (I suspect that Macquart's *D. corpulenta* in Hist. Nat. Dipt. II, 77, 22; Dipt. Exot. II, 3, 35, 4; 1er Suppl. 143; Tab. XII, f. 2, is some other species). — Mexico (Wied.); South America (Schiner); Rocky Mts., in Colorado (O. Sacken). ([265]).

Dejeania rufipalpis Macquart, Dipt. Exot. II, 3, 35, 5; Tab. III, f. 1. — Mexico.

Dejeania vexatrix O. Sacken, Western Dipt., 343.

*****rutilioïdes** Jaennicke, Neue Exot. Dipt. 137. — Mexico (Jaenn.); San Diego, Cal.; Manitou, Colorado (O. S., Western Dipt.). ([266]).

analis Macquart, Dipt. Exot. II, 3, 34, 3; Tab. III, f. 3; Bigot, in R. de la Sagra etc. 809 *(Echinomyia).* — Mexico (Macq.); Cuba (Bigot).

---

*) Schiner (Fauna Austr. Dipt. I. p. 419) revives the older name *Besseria* R Desvoidy; but as R. Desvoidy himself, in his later work, *Diptères des envir de Paris* ignores *Besseria* and adopts *Wahlbergia,* we may do the same here.

armata Wiedemann, Auss. Zw. II, 287, 11 *(Tachina)*; Macquart, Dipt. Exot. 4e Suppl. 168; Tab. XV, f, 7. — Cuba; Brazil (Macquart and Schiner, Novara etc. 337).

Hystricia.

Macquart, Dipt. Exot. II, 3, 43; 1843; compare also Schiner, Dipt. of the Novara etc. 331, foot-note.

* **vivida** Harris, Ins. New Engl. 3ᵈ Edit., 612; Tab. VIII, f. 1 *(Tachina)*. — United States, common. (²⁶⁷).

Hystricia testacea Macquart, Dipt. Exot. II, 3, 44; Tab. IV, f. 4. — North America and Mexico.

Tachina finitima Walker, List, etc. IV, 70. — Nova Scotia (!).

(?) *Tachina abrupta* Wiedemann, Auss. Zw. II, 293, 22. — North America.

ambigua Macquart, Dipt. Exot. 4e Suppl. 172, 9. — Mexico.
amoena Macquart, Dipt. Exot. II, 3, 44, 2. — Mexico.

Hystrisyphona.

Bigot, Rev. et Mag. de Zool. 1859, 309.

niger Bigot, l. c. — Mexico.

Jurinia.

R. Desvoidy, Myod. 34; 1830.
Macquart, Dipt. Exot. II, 3, 37.

* **algens** Wiedemann, Auss. Zw. II, 285, 8 *(Tachina)*. — North America (Wied.); New England and British possessions, common; also farther South.
amethystina Macquart, Dipt. Exot. II, 3, 42, 9; Tab. III, f. 7, and 1er Suppl. 147. — Georgia, Venezuela.
apicifera Walker, List, etc. IV, 718. — North America.
aterrima R. Desvoidy, Myod. 36. — United States.
Boscii R. Desvoidy, Myod. 36. — United States.
candens Walker, List, etc. IV, 720. — Nova Scotia.
decisa Walker, List, etc. IV, 715. — Huds. B. Terr.; Nova Scotia.
georgica Macquart, Hist. Nat. Dipt. II, 79, 31. — Georgia.
fuscipennis Jaennicke, Neue Exot. Dipt. 83. — North America.
* **hystrix** Fabricius, System. Ent. 777, 21 *(Musca)*; Ent. System. IV, 325, 55 *(id.)*; System. Antl. 310, 8 *(Tachina)*; Olivier, Encycl. Méthod. VIII, 22, 59 *(Musca)*; Wiedemann, Auss. Zw. II, 283, 6; Macquart, Hist. Nat. Dipt. II, 79, 30 *(Echinomyia)*. — America (Fabr.); Kentucky (Wied.).

Jurinia metallica R. Desvoidy, Myod. 35.

Musca pilosa Drury, Ins. I; Tab. XLV, f. 7 [Wied.].

leucostoma R. Desvoidy, Myod. 37. — North America.
virginiensis Macquart, Dipt. Exot. 4e Suppl. 171, 16. — Virginia.

echinata Thomson, Eugen. Resa, 516. — California.

analis Macquart, Dipt. Exot. II. 3, 39, 1; Tab. III, f. 8. — Brazil, Mexico.
apicalis Jaennicke, Neue Exot. Dipt. 82. — Mexico.
basalis Walker, List, etc. IV, 713. — Jamaica.
contraria Walker, List, etc. IV, 716. — Mexico.
debitrix Walker, Trans. Ent. Soc. N. Ser. V, 296. — Mexico.
epileuca Walker, List, etc. IV, 716. — Jamaica.
flavifrons Jaennicke, Neue Exot. Dipt. 82. — Mexico.
innovata Walker, Trans. Ent. Soc. N. Ser. V, 296. — Mexico.
lateralis Macquart, Dipt. Exot. II, 3, 42, 8; Tab. III, f. 10. — Mexico.

Echinomyia.
Dumeril, Exposit. d'une Méthode Natur. etc. 1798.

aenea Zetterstedt, Dipt. Scand. VIII, 3217; Gerstaecker, Die 2te deutsche N rd olfahrt etc — East Greenland.
Anaxias Walker, List, etc. IV, 726. — Nova Scotia.
florum Walker, List, etc. IV, 722 (*Fabricia*). — Huds. B. Terr., Nova Scotia.
haemorrhoa v. d. Wulp, Tijdschr. v. Ent. 2d Ser. II, 145; Tab. IV, f. 13—16. — Wisconsin.
iterans Walker, List, etc. IV, 727. — Nova Scotia.
Leschenaldi R. Desvoidy, Myod. 42 (*Peleteria*). — North America.
Lapilaei R. Desvoidy, Myod. 44; id. Dipt. des env. de Paris I, 642. — Newfoundland.
picea R. Desvoidy, Myod. 44; id. Dipt. des env. de Paris I, 642; Macquart, Dipt. Exot. II, 3, 37, 2; Tab. III, f. 4. — Nova Scotia.
punctifera Walker, List, etc. IV, 728. — Massachusetts.

californiae Walker, Dipt. Saund., 270 (*Fabricia*). — California.
filipalpis Thomson, Eugen. Resa, 517. — California.
basifulva Walker, List, etc. IV, 725. — Jamaica.

Cyphocera.
Cuphocera, Macquart Ann. Soc. Ent. de France II, 3, 267; 1845; amended in *Cyphocera* by Rondani and Loew.

ruficauda v. d. Wulp, Tijdschr. v. Ent. 2d Ser II, 146; Tab. IV, f. 17—20 (*Schineria*); Loew, in Zeitschr. f. Ges. Naturw. XXXVI, 114, refers the species to the present genus. — Wisconsin.

Gymnochaeta.
Rob. Desvoidy, Myod. 371; 1830.

*alcedo Loew, Centur. VIII, 61. — United States.

Micropalpus.
Macquart, Hist. Nat. Dipt. II, 80; 1835.

distinctus R. Desvoidy, Myod. 54 (*Linnemyia*). — Philadelphia.

piceus Macquart, Hist. Nat. Dipt. II, 84, 11. — Carolina.
Marshamia analis Rob. Desvoidy, Myod. 58 [Macq.].

californiensis Macquart, Dipt. Exot. 4e Suppl. 175, 18. — California.

albomaculatus Jaennicke, Neue Exot. Dipt. 80. — Mexico.
flavitarsis Macquart, Dipt. Exot. II, 3, 47, 4; Tab. V, f. 1; 1er Suppl.
152, 11; Tab. XIII, f. 13; 3e Suppl. 45; Schiner, Dipt. of the
Novara etc. 334 *(Saundersia)*. — Mexico (Macq.); South America
(Schiner). ([268]).
ornatus Macquart, Dipt. Exot. II, 3, 47, 5; Tab. IV, f. 6; Schiner,
Dipt. of the Novara etc. 333 *(Saundersia)*. — Mexico; Columbia
(S. America).
rufipes Jaennicke, Neue Exot. Dipt. 79. — Panama.

Gonia.
Meigen, in Illiger's Magaz. II; 1803.

albifrons Walker, List, etc. IV, 798. — Huds. B. Terr.
auriceps Meigen, System. Beschr. V, 5, 7. — Europe and Georgia,
North America (Walker, List, etc. IV, 798).
*frontosa Say, J. Acad. Phil. VI, 175; Compl. Wr. II, 365. — Upper
Missouri (Say).
philadelphica Macquart, Dipt Exot. II, 3, 51, 6. — Philadelphia.
angusta Macquart, Dipt. Exot. II, 3, 51, 7; Tab. V, f. 5. — Locality
unknown (Macq.); Jamaica (Walker, List, etc. IV, 798).
crassicornis Fabricius, System. Antl. 301, 84 *(Musca)*; Wiedemann,
Auss. Zw. II, 345, 4. — West Indies.
chilensis Macquart, Dipt. Exot. II, 3, 50, 5; Tab. V, f. 4; Bigot, in
R. de la Sagra etc. 809. — Cuba; Chili (Macquart says that this
species differs from the european *G. capitata* only in the absence
of black at the end of the abdomen).

Nemoraea.
R. Desvoidy, Myod. 71; 1830.
Schiner, Fauna Austr. I, 447. ([269]).

Clesides Walker, List, etc. IV, 757. — North America.
*leucaniae Kirkpatrick, Ohio Agric. Report for 1860, 358 *(Exorista)*;
Riley, 2d Rep. 51, f. 17 *(id.)*. — Parasite of *Leucania unipuncta*.
Exorista Osten Sackenii, Kirkpatrick, l. c., according to Riley l. c.,
only a variety of the former.
Senometopia militaris Walsh, Insects injurious to Vegetation in
Illinois (Pamphlet containing a detailed description of this fly,
with a figure. It is dated Sept. 1861. The description is repro-
duced by Packard, Entom. Report on the army-worm and grain-
aphis, in the Scientif. Survey of the State of Maine 1861); Amer.
Entom II, 101. Occurs in the West, as well as in the Eastern
States, according to Packard.

Masurius Walker, List, etc. IV, 753 (*Erigone*). — North America.
Pyste Walker, List, etc. IV, 754 (*Erigone*). — Nova Scotia.
trixoides Walker, List, etc. IV, 760. — Georgia.

intrita Walker, Trans. Ent. Soc. N. Ser. V, 297. — Mexico.

Exorista.
Meigen, in Illiger's Magaz. II; 1803. ([270]).

Areos Walker, List, etc. IV, 766 (*Lydella*). — North America.
cecropiae Riley 4th Rep. 108. Also Amer. Ent. II, 101. — On *Attacus Cecropia*.
doryphorae Riley, Amer. Ent. I, 46, f. 35; the same, First Rep. 111, f. 48 (*Lydella*); parasite on *Doryphora decemlineata*.
Epicydes Walker, List, etc. IV, 785 (*Aplomyia*). — Huds. B. Terr.
flavicauda Riley, 2d Rep. 51 (f. 18). — Missouri.
Hybreas Walker, List, etc. IV, 785 (*Aplomyia*). — Huds. B. Terr.
irrequieta Walker, List, etc. IV, 789 (*Aplomyia*). — Nova Scotia.
Mella Walker, List, etc. IV, 767 (*Lydella*). — Nova Scotia.
Panaetius Walker, List, etc. IV, 767 (*Lydella*). — Nova Scotia.
Pansa Walker, List, etc. IV, 787 (*Aplomyia*). — Nova Scotia.
phycitae Le Baron, 2d Rep. 123 (parasite of caterpillar of *Phycita nebulo* in Illinois). — Also Riley, 4th Rep. 40.
violenta Walker, List, etc. IV, 788 (*Aplomyia*). — Nova Scotia.

cessatrix Walker, Trans. Ent. Soc. N. Ser. V, 305 (*Lydella*). — Mexico.
? indita Walker, l. c. 306 (*Lydella*). — Mexico.
lepida R. Desvoidy, Myod. 153 (*Zenillia.*) — Cuba.
rubrella R. Desvoidy, Myod 179 (*Carcellia*). — San Domingo.

Tachina.(*)
Meigen, in Illiger's Magaz. II; 1803.

addita Walker, Dipt. Saund., 290. — United States.
albifrons Walker, Dipt. Saund., 283. — United States.
Ampelus Walker, List, etc. IV, 732. — Nova Scotia.
ancilla Walker, Dipt. Saund., 299. — United States.
antennata Walker, Dipt. Saund., 298. — United States.
atra Walker, Dipt. Saund., 273. — Georgia.
convecta Walker, Dipt. Saund., 277. — United States.
degenera Walker, List, etc. IV, 733. — Huds. B. Terr.
disjuncta Wiedemann, Anal. Ent. 45, 88; Auss Zw. II, 295, 24. — North America.
Dydas Walker, List, etc. IV, 748. — Huds. B. Terr.
exul Walker, Dipt. Saund., 277. — United States.
hirta Curtis, Ins. Ross's Exp. LXXIX. — Arctic America.
insolita Walker, Dipt. Saund., 277; Tab. VII, f. 2. — United States.
interrupta Walker, Dipt. Saund., 295. — Georgia.

*) This is not *Tachina* in Schiner's sense, but a congeries of species published by authors under that head, and which could not be disposed of elsewhere.

Melobosis Walker, List, etc. IV, 743. — Florida.
obconica Walker, Dipt. Saund., 296. — United States.
signifera Walker, List, etc. IV, 708. — Nova Scotia.
speculifera Walker, l. c. 731. — North America.
unifasciata R. Desvoidy, Myod. 105 *(Latreillia)*. — Philadelphia.

albincisa Wiedemann, Auss. Zw. II, 334, 98. — St. Thomas.
breviventris Wiedemann l. c. II, 297, 28. — Brazil (Wied.); Jamaica
  (Walker, List, etc. IV, 712).
crudelis Wiedemann, l. c. II, 300, 35. — West Indies.
cubaecola Jaennicke, Neue Exot. Dipt. 74; Tab. II, f. 6. — Cuba.
distincta Wiedemann, Anal. Ent. 45; Auss. Zw. II, 334, 99. — West
  Indies. [According to Macquart, Dipt. Exot. II, 3, 59, this is a
  Masicera].
elegans Bigot, in R. de la Sagra etc. 810; Tab. 20, f. 7. — Cuba.
hirta Drury, Ins. 109; Tab. XLV, f. 4 *(Musca)*. — Jamaica.
occidentalis Wiedemann, Auss. Zw. II, 335. — St. Thomas.   [Also
  referred to Masicera by Macquart, Dipt. Exot. II, 3, 59.]
potens Wiedemann, Auss. Zw. II, 312; Bigot, in R. de la Sagra etc.
  810. — Brazil (Wied.); Cuba (Bigot) [Macquart, Dipt. Exot. II, 3,
  58, refers this species to *Eurygaster*].
pusilla Wiedemann, Auss. Zw. II, 337, 104. — West Indies.
saltatrix Wiedemann, l. c. 300, 36 — West Indies.
trivittata Wiedemann, Auss. Zw. II, 300, 34. — West Indies.
subvaria Walker, Dipt. Saund., 299. — West Indies.

Observation. *Tachina anonyma* (Masicera?) Riley, 4th Rep.
129, 5th Rep. 133 and 7th Rep. 178 has never been described.
It was bred from different moths, and also from the migratory
Grasshopper *Caloptenus spretus.*

Masicera.
Macquart, Hist. Nat. Dipt. II, 118; 1835.

archippivora Riley, 3d Rep. 150. — Missouri (parasitic on *Danaus
  archippus* and other caterpillars).

cubensis Macquart, Dipt. Exot. 3e Suppl. 46, 13; Tab. V, f. 5;
  Bigot, in R. de la Sagra etc, 813. — Cuba.
expergita Walker, Trans. Ent. Soc. N. Ser. V, 304. — Mexico.
disputans Walker, Trans. Ent. Soc. N. Ser. V, 303. — Mexico.
gentica Walker, Trans. Ent. Soc. N. Ser. V, 302. — Mexico.
necopina Walker, Trans. Ent. Soc. N. Ser. 303. — Mexico.

Observation. Macquart, Dipt. Exot. II, 3, 59 refers *Tachina
distincta* Wied. and *T. occidentalis* Wied., both from the West
Indies, to the genus Masicera; they will be found among the
Tachinae.

Phorocera.
R. Desvoidy, Myod. 131; 1830.
Schiner, Fauna Austr. I, 488.

Demylus Walker, List, etc IV 779. — North America (?).

prisca Walker, List, etc. IV, 780. — Nova Scotia.
Theutis Walker, List, etc. IV, 778. — Nova Scotia.
claripennis Macquart, Dipt. Exot. 3e Suppl. 49, 10; Tab. V, f. 8. —
North America.

botyvora R. Desvoidy, Myod. 138. — Cuba (bred from the chrysalis
of a *Botys*).

Baumhaueria.

Meigen, System. Beschr. VII, 251; 1838.

analis v. d. Wulp, Tijdschr. v. Ent. 2d Ser. II, 148; Tab. IV, f. 21—23. —
Wisconsin.

Belvoisia.

R. Desvoidy, Myod. 103; 1830.

* **bifasciata** Fabricius, System. Ent. 777, 19 *(Musca)*; Ent. System. IV,
325, 53 *(id.)*; System. Antl. 299, 78 *(id.)*; Latreille, Dict. d'Hist.
Nat. XXIV, 195, 373 *(Ocyptera)*; Wiedemann, Auss. Zw. II, 305,
44 *(Tachina)*; R. Desvoidy, Myod. 104 *(Latreillia ♂)*; R. Desvoidy,
Dipt. des environs de Paris I, 563 *(Lalage)*; Macquart, Hist. Nat.
Dipt. II, 104, 19 *(Nemoraea ♂)*; Dipt. Exot. II, 3, 57; Tab. VI,
f. 2; Bigot in R. de la Sagra etc. 813 *(Nemoraea)*; Riley, Fifth
Report 140, with figure). — North and South America. ([271]).
Belvoisia bicincta R. Desvoidy, Myod. 103, ♀.
Senometopia bicincta Macquart, Hist. Nat. Dipt. II, 112.

Metopia.

Meigen, Illiger's Magaz. II; 1803. (*)
Schiner, Fauna Austr. I, 498. ([272]).

grisea R. Desvoidy, Myod. 131 *(Araba)*. — North America.

Xychus Walker, List, etc. 770 IV, *(Ophelia)*. — Jamaica.

Senotainia.

Macquart, Dipt. Exot. 1er Suppl. 167; 1846.

rubriventris Macquart, l. c. 167; Tab. XX, f. 8. — Galveston, Texas.

Miltogramma.

Meigen, Illiger's Magaz. II; 1803.

trifasciata Say, J. Acad. Phil. VI, 174; Compl. Wr. II, 363. — Indiana.

erythrocera Thomson, Eugen. Resa etc. 523. — California.

biseta Thomson, Eugen. Resa etc. 524. — Panama.

---

*) Agassiz, Index universàlis, erroneously has 1828

Blepharopeza.

Blepharipeza Macquart, Dipt. Exot. II, 3, 54, 1843; amended by Loew,
Centur. X, 67.

bicolor Macquart, Dipt. Exot. 1er Suppl. 158, 4; Tab. XX, f. 7. —
Galveston, Texas.

*adusta Loew, Centur. X, 67. — California.

rufipalpis Macquart, Dipt. Exot. II, 3, 55, 1; Tab. VI, f. 1; Bigot,
in R. de la Sagra etc. 815. — Cuba, Mexico.

Eurygaster. (272a).
Macquart, Dipt. Exot. II, 3, 57; 1843.

septentrionalis Walker, Lord's Natur. in Vancouver's Island, II,
339. — Vancouver's Island.

commentans Walker, Trans. Ent. Soc N. S. V, 300. — Mexico.
desita Walker, l. c. 299. — Mexico.
fertoria Walker, l. c. 300. — Mexico.
habilis Walker, l. c. 301. — Mexico.
modestus Bigot, R. de la Sagra etc. 812. — Cuba.
obscurus Bigot, l. c. 812. — Cuba.
postica Walker, Trans. Ent. Soc. N. S. V, 301. — Mexico.
saginata Walker, Trans. Soc. N. Ser. V, 298. — Mexico.

Degeeria.
Meigen, System. Beschr. VII, 249; 1838.

lateralis Macquart, Dipt. Exot. 3e Suppl. 48, 2; Tab. V, f. 6. —
North America.

Clytia.
R. Desvoidy, Myod. 287; 1830.

atra R. Desvoidy, Myod. 288, 2. — Carolina.

Scopolia.
R Desvoidy, Myod. 268; 1830.

lateralis Macquart, Dipt. Exot. II, 3, 71; Tab. VIII, f. 3. — North
America.

nigra Bigot, in R. de la Sagra etc. 814; Tab. XX, f. 8. — Cuba.

Euthera.
Loew, Centur. VII, 85; 1866.

*tentatrix Loew, Centur. VII, 85. — New York, Texas.

Ptilocera.
Macquart, Hist. Nat. Dipt. II, 169; 1835.

americana Macquart, Hist. Nat. Dipt. II, 173. — Philadelphia.

Observation. This genus, now abandoned, seems to have principally contained Tachinina, approaching the *Dexina* in their appearance. Schiner places the european species under the head of *Phyto* Rob. Desvoidy.

FAMILY DEXIDAE.

Proscna.

St. Fargeau et Serville, Encycl. Méthod. X, 500; 1825.

*mexicana Macquart, Dipt. Exot. 4e Suppl. 231; Tab. XXI, f. 12. — Mexico.

Dexia.

Meigen, System. Beschr. V, 33; 1826.

abdominalis R. Desvoidy, Myod. 306, 2 *(Estheria).* — Nova Scotia.
Abzoe Walker, List, etc. IV, 846. — Georgia.
albifrons Walker, Dipt. Saund., 317. — United States.
analis Say, J. Acad. Phil. VI, 177, 2; Compl. Wr. II, 366. — Indiana.
analis R. Desvoidy, Myod. 315, 3 *(Zelia).* — Carolina.
apicalis R. Desvoidy, Myod. 316, 4 *(Zelia).* — Carolina.
canescens Walker, Dipt. Saund., 310. — United States.
cerata Walker. List, etc. IV, 847. — North America.
Cremides Walker, List, etc. IV, 842. — North America.
dives Wiedemann, Auss. Zw. II, 377, 15. — Kentucky.
Halone Walker, List, etc. IV, 837. — Georgia.
Harpasa Walker, List, etc. IV, 840. — North America.
melanocera R. Desvoidy, Myod. 312, 2. — Carolina.
Ogoa Walker, List, etc. IV, 841. — Nova Scotia.
pedestris Walker, Dipt. Saund., 313. — United States.
postica Walker, List, etc. IV, 310. — Georgia.
punctata R. Desvoidy, Myod. 308, 3 *(Dinera).* — Philadelphia.
Prexaspes Walker, List, etc. IV, 837 *(Estheria).* Georgia.
Pristis Walker, List, etc. IV, 841. — Massachusetts.
rostrata R. Desvoidy, Myod. 315, 1 *(Zelia).* — North America.
rufipennis Macquart, Dipt. Exot. II, 3, 87, 3; Tab. X, f. 3. — Nova Scotia.
tibialis R. Desvoidy, Myod. 306, 1 *(Estheria).* — Nova Scotia.
triangularis v. d. Wulp, Tijdschr. v. Ent. 2d Ser. II, 149; Tab. V, f. 1 5. — Wisconsin.
velox R. Desvoidy, Myod. 316, 5 *(Zelia).* — Carolina.
*vertebrata Say, J. Acad. Phil. VI, 176, 1; Compl. Wr. II, 366. — Indiana.

fuscanipennis Macquart, Dipt. Exot. 1er Suppl. 188, 7; Tab. XX, f. 11. — Yucatan.
perfecta Walker, Trans. Ent. Soc. N. S. V, 307. — Mexico.
plumosa Wiedemann, Auss. Zw. II, 370; Bigot, in R. de la Sagra etc. 815. — Brazil (Wied.); Cuba (Bigot).

rubriventris 'Macquart, Dipt. Exot. 1er Suppl. 188, 6; Tab XX, f. 10. — Yucatan.

strenua R. Desvoidy, Myod. 315, 2 *(Zelia)*. — San Domingo.

Thomae Wiedemann, Auss. Zw. II, 379. — St. Thomas (Wied.); Jamaica (Walker, List, IV, 840).

Sericocera.

Macquart, Hist. Nat. Dipt. II, 165; 1835.

pictipennis Macquart, Dipt. Exot. II, 3, 67, 1; Tab. VII, f. 5. — Philadelphia.

Observation. This genus of Macquart's seems to have contained a mixture of heterogeneous forms, which Schiner distributed among the genera *Olivieria, Peteina* (Section Tachinina) and *Mintho, Thelaira* and *Melania* (Section Dexina). *S. pictipennis* Macquart, judging from the figure, belongs to the Dexidae.

Melanophora.

Meigen, in Illiger's Magaz. II; 1803.

? diabroticae Shimer, Amer. Naturalist, V, 219; f. 60 (the author calls it *Melanosphora*, perhaps Melanophora?). — Illinois (parasitic on *Diabrotica vittata)*.

distincta R. Desvoidy, Myod. 273 *(Linnemyia)*. — Europe; Philadelphia.

nigripes R. Desvoidy, Myod. 58 *(Marshamia)*. — North America.

* roralis Linné etc.; Meigen, System. Beschr. IV, 284. — Europe and North America (see Loew, Sillim. Journ. Vol. XXXVII, p. 318).

Illigeria.

R. Desvoidy, Myod. 273; 1830.

Aclops Walker, List, etc. IV, 796. — Georgia.

Corythus Walker, List, etc. IV, 797. — Georgia.

Helymus Walker, List, etc. IV, 795. — Maine.

Observation. Judging from the descriptions, the insects, which Mr. Walker places in this genus, have very little in common, and belong to different genera.

Theresia.

R. Desvoidy, Myod 325; 1830.

tandrec R. Desvoidy, Myod 326. — Carolina.

Microphthalma.

Macquart, Dipt. Exot. II, 3, 84; 1843.

nigra Macquart, Dipt. Exot. II, 3, 85, 1; Tab. X, f. 2. — North America.

Megaprosopus.

Macquart, Dipt. Exot. II, 3, 83; 1843.

rufiventris Macquart, Dipt Exot. II, 3, 84, 1; Tab. X, f. 1. — Mexico.

FAMILY SARCOPHAGIDAE.

Sarcophaga.

Meigen, System. Beschr. V, 14; 1826. ([73]).

acerba Walker, List, etc. IV, 824. — Nova Scotia.
aegra Walker, List, etc. IV, 821. — Massachusetts.
Anaces Walker, List, etc. IV, 833. — North America.
anxia Walker, List, etc. IV, 818. — North America.
argyrocephala Macquart, Dipt. Exot. 1er Suppl. 192, 25. — Galveston, Texas.
aspera Walker, List, etc. IV, 825. — North America (?).
assidua Walker, Dipt. Saund., 328. — United States.
aterrima R. Desvoidy, Myod. 336, 3 *(Peckia)*. — Carolina.
avida Walker, List, etc. IV, 822. — Nova Scotia.
basalis Walker, Dipt. Saund., 323. — United States.
comes Walker, Dipt. Saund., 323. — United States.
consobrina R. Desvoidy, Myod. 344, 24 *(Myophora)*. — Philadelphia.
derelicta Walker, Dipt. Saund., 322. — United States.
fulvipes Walker, Dipt. Saund., 328. — United States.
Georgina Wiedemann, Auss. Zw. II, 357, 4; Harris, Ins. Injur. to Veget. 3d edit. 613. — Georgia (Wied.); British Possess. (Walker, List, etc. IV, 829); Massachusetts (Harris, Catal.).
importuna Walker, List, etc. IV, 819. — North America (?).
L'herminieri R. Desvoidy, Myod. 339, 5 *(Myophora)*. — Carolina.
lanipes R. Desvoidy, Myod. 336, 5. — Carolina.
pallipes Walker, Dipt. Saund., 329. — United States.
querula Walker, List, etc. IV, 821. — North America (?).
rabida Walker, List, etc. IV, 823. — Nova Scotia.
rapax Walker, l. c. IV, 818. — North America (?).
rediviva Walker, l. c. IV, 823. — Huds. B. Terr.
* **sarraceniae** Riley, Trans. St. Louis Acad. of N. Soc. III, 239. — Missouri.
stimulans Walker, List, etc. IV, 817. — North America.
vigil Walker, List, etc. IV, 831. - Nova Scotia.
viridescens R. Desvoidy, Myod. 342, 13 *(Myophora)*. — Nova Scotia.

pallinervis Thomson, Eugen. Resa, etc. 535. — California, Honolulu.

* **chrysostoma** Wiedemann, Auss. Zw. II, 356, 2 (compare also Schiner, Novara 313). — West Indies, Brazil.
conclausa Walker, Trans Ent. Soc. N. S. V, 309. — Mexico.
cubensis R. Desvoidy, Myod. 342, 4 *(Myophora)*. — Cuba.
cubensis Macquart, Dipt. Exot. II, 3, 106, 20; Tab. XII, f. 6; Bigot, in R. de la Sagra etc. 819. — Cuba.
despensa Walker, Trans. Ent Soc. N. Ser. V, 309. — Mexico.
effrenata Walker, Trans. Ent. Soc. N. Ser. V, 309. — Mexico.
fervida R. Desvoidy, Myod. 341, 10 *(Myophora)*. — San Domingo.

fortipes Walker, Trans. Ent. Soc. N. Ser. V, 310. — Haity.
fulvipes Macquart, Dipt. Exot. II, 3, 105, 19; Tab. XII, f. 5. — Cuba.
immanis Walker, List, etc. IV, 815. — Honduras.
innota Walker, Trans. Ent. Soc. N. S. V, 308. — Mexico.
intermutans Walker, Trans. Ent. Soc N. Ser. V, 308. — Mexico.
incerta Bigot, in R. de la Sagra, etc. 818. — Cuba.
incerta Walker, Dipt. Saund., 324. — Jamaica.
lambens Wiedemann, Auss. Zw. II, 365, 23. — West Indies; Brazil.
muscoides Bigot, R. de la Sagra, etc. 816. — Cuba.
obsoleta Wiedemann, Auss. Zw. II, 367, 29. — West Indies.
occidua Fabricius, Ent. System. IV, 315, 12 *(Musca)*; System. Antl.
    288, 19; Wiedemann, Auss. Zw. II, 368, 31. — West Indies.
pusilla Bigot, R. de la Sagra, etc. 817. — Cuba.
perneta Walker, Trans. Ent Soc. N. Ser. V, 303. — Mexico.
plinthopyga Wiedemann, Auss. Zw. II, 360, 10; Walker, Lin. Trans.
    XVII, 352, 57. — St. Thomas (Wied.); Brazil (Walker, Lin.
    Trans.), Jamaica, Demerara, Nova Scotia (Walker, List, etc. IV, 820).
plumipes R. Desvoidy, Myod. 336, 4 *(Peckia)* — San Domingo.
rubella Wiedemann, Auss. Zw. II, 357; 5 — Antigoa.
trigonomaculata Macquart, Dipt. Exot. II, 3, 106, 21; Tab. XIII,
    f. 2. — Mexico.
trivittata Macquart, Dipt. Exot. II, 3, 105, 18; Tab. XII, f. 3;
    Bigot. in R. de la Sagra etc. 816. — Cuba, Mexico

Observation. *S. nudipennis* Loew in litt. is mentioned in
Packard's Guide, etc. 408, as being bred from the nests of
Pelopaeus flavipes. It has never been described and is therefore
omitted. *Sarcophaga carnaria* Linné, quoted in Harris's Catal.
Ins. Mass., in Riley's Seventh Report, 180, and in other writings,
is omitted here for the reason stated in the note ([273]). Macquart,
Dipt. Exot. II, 3, 95, asserts that he had *Sarcophaga carnaria*
from Hayti; this requires confirmation. About a *Sarcophaga*
attacking grasshoppers in Iowa, see Report of the Departt. of
Agriculture, Washington 1867, page 36.

Phrissopoda.

Phrissopodia Macquart, Hist. Nat. Dipt. II, 222; 1835.
Phrissopoda Macquart, Dipt. Exot. II, 3, 96.

praeceps Wiedemann, Auss. Zw. II, 355 *(Sarcophaga;* referred to the
present genus by Macquart, Dipt Exot. II, 3, 96). — Cuba.
Peckia imperialis R. Desvoidy, Myod. 335; Macquart, Hist. Nat.
Dipt. II, 223; Tab. XVI, f. 1 *(Phrissopodia).* — Cuba; also Port
Jackson, Australia, according to Macquart, Dipt. Exot. II, 3, 96.
[Synonymy by Macquart, with a doubt.]

Cynomyia.

R. Desvoidy, Myod., 363; 1830.
Schiner, Fauna Austr. I, 574.

alpina Zetterstedt, Insecta Lapponica 651, 7; Dipt. Scand. IV, 1304;
    Gerstaecker, Die 2te deutsche Nordpolfahrt etc. Lapland; East Greenl.

cadaverina R. Desvoidy, Myod. 365, 3. — Carolina.
flavipalpis Macquart, Dipt. Exot. 4e Suppl. 236, 3. — Newfoundland.
mortuorum Linné, Meigen, etc. *(Sarcophaga)*; — O. Fabricius, Fauna
    Groenl. 206, 166 *(Musca)*; Staeger, Groenl. Antl., 363, 32;
    Holmgren, Ins. Nordgroenl. 101. — Greenland.

FAMILY MUSCIDAE.
Stomoxys.
Geoffroy, Hist. des Ins. I; 1764.

*calcitrans Linné, Meigen, etc.; Harris, Ins. of N. Engl. 3d edit. 614,
    f. 270. — Europe and North America (comp. Loew, Sillim. J. l. c.).
Cybira Walker, List, etc. IV, 1159 (Addenda). — Nova Scotia.
dira R. Desvoidy, Myod. 387, 8. — North America.
inimica R. Desvoidy, Myod, 387, 6. — North America.
parasita Fabricius, Ent. System. IV, 394, 3; System. Antl. 280, 3;
    Wiedemann, Auss. Zw. II, 252, 11 (merely a translation from
    Fabricius). — North America.
? **occidentis** Walker, Dipt. Saund., 332 *(Musca).* — United States.

Idia.
Meigen, System. Beschr. V, 9, 102; 1826. ([273a]).

viridis Wiedemann, Analecta etc. 50; Auss. Zw. II, 354, II. —
    North America.

Mesembrina.
Meigen, System. Beschr. V, 10, 103; 1826.

Latreillii R. Desvoidy, Myod. 401, 2. — Nova Scotia.
pallida Say, J. Acad. Phil. VI, 175; Compl. Wr. II, 366. — Indiana.
*resplendens Wahlberg, K. vet. Ak. Fórh. 1844, 66. — Europe (Lap-
    land) and North America (comp. Loew, Sillim. J. l. c.).

anomala Jaennicke, Neue Exot. Dipt., 69; Tab. II, f. 4. — Cuba.

Calliphora.
R. Desvoidy, Myod. 433; 1830.

aurulans R. Desvoidy, Myod. 437, 11. — Carolina; Nova Scotia.
compressa R. Desvoidy, Myod. 438, 16. — Carolina (Desv.); Huds. B.
    Terr. (Walker, List, etc. IV, 893).
*erythrocephala Meigen, System. Beschr. V, 62; Schiner, Fauna Austr.
    I, 584. — Europe and North America (comp. Staeger, Groenl.
    Antl.).
    Volucella vomitoria Fabricius, Fauna Groenl. 207, 167 (?) [Schiödte].
groenlandica Zetterstedt, Ins. Lapp. 657, 16; Dipt. Scand. IV, 1330
    (Musca); Staeger, Groenl. Antl. 363; Gerstaecker, 2te deutsche
    Nordpolfahrt etc.; Holmgren, Ins. Nordgroenl. 101. — Northern
    Europe and Greenland.

Volucella caesar O. Fabricius, Fauna Groenl. 207, 168 [Schiödte].
Ilerda Walker, List, etc. IV, 908 *(Melinda)*. — Huds. B. Terr.
Lilaea Walker, List, etc. IV, 894. — Huds. B. Terr.
mortisequa Kirby, N. Amer. Zool. Ins. 317. — Arctic America
(Lat. 65). ([274]).
myoidea R. Desvoidy, Myod. 436, 8. — Philadelphia.
obscoena Eschscholz, Entomographieen I, 113, 84 *(Musca)*; Wiedemann,
Auss. Zw. II, 392 *(id.)*. — Island Unalaska ([275]).
splendida Macquart, Dipt. Exot. 1er Suppl. 196, 17. — Texas.
terrae novae Macquart, Dipt. Exot. 4e Suppl. 244, 29. — Newfoundland.
viridescens R. Desvoidy, Myod. 437, 12. — Carolina; Florida (Walker,
List, etc IV, 895).
*****vomitoria** Linné, Fabricius, Meigen etc. *(Musca)*. — Europe and
North America (also in Guyana; Macquart, Dipt. Exot. II, 3, 127).
Calliphora vicina R. Desvoidy, Myod. 435, 5. — Philadelphia (is
either *vomitoria* or *erythrocephala)*.

femorata Walker, Trans. Ent. Soc. N. Ser. V, 310. — Mexico.
(?) **rutilans** Fabricius, Spec. Ins. II, 436, 6 *(Musca)*; Ent. System.
IV, 314, 7 *(id.)*; System. Antl. 287, 13 *(id.)*; Wiedemann, Auss.
Zw. II, 392, 14 *(id.)*. — South America (Wied.); Fabricius has:
„in Americae insulis".
socors Walker, Trans. Ent. Soc. N. Ser. V, 311. — Mexico.
stygia Fabricius, Spec. Ins II, 438 *(Musca)*; Ent. System. IV, 317,
22 *(id.)*; System. Antl. 290, 31 *(id.)*; Olivier, Encycl. Méth. VIII,
14 *(id.)*; Wiedemann, Auss. Zw. II, 398, 15 *(id.)*. — Newfoundland (Fabr., Wied.). ([276]).

Pollenia.
R. Desvoidy, Myod. 412; 1830.

*****rudis** Fabricius *(Musca)*; Meigen, System. Beschr. V, 66 *(id.)*. — Europe
and North America (see Loew, Sillim J. I. c.).
Musca familiaris Harris, Ent. correspondence 336. — New England.
vespillo Fabricius, Meigen, etc. *(Musca)*. — Europe and Nova Scotia
(Walker, List, etc. IV, 907).

Graphomyia.
R. Desvoidy, Myod. 403; 1830.

americana R. Desvoidy, Myod. 404. — North America (Schiner,
Novara 304, described another Gr. americana, from S. America).
? **contigua** Walker, Dipt. Saund., 449 *(Musca)*. — United States.
Idessa Walker, List, etc. IV, 908. — Huds. B. Terr.
serva Walker, Dipt. Saund., 349 *(Musca)*. — United States.

Lucilia.
R. Desvoidy, Myod. 452; 1830.

brunnicosa R. Desvoidy, Myod. 459. — North America.

caesar Linné, Fabricius, Meigen, etc. *(Musca)*. — Europe and North America; Massachusetts and Huds. B. Terr. Walker, List, etc. IV, 879; Philadelphia, R. Desvoidy, Myod. 452.

caeruleiviridis Macquart, Dipt. Exot. 5e Suppl. 113, 62. — Baltimore.

carolinensis R. Desvoidy, Myod. 457. — Carolina.

compar R. Desvoidy, Myod. 457. — Philadelphia.

consobrina Macquart, Dipt. Exot. 3e Suppl. 57, 42 („var. L. fraternae"? Macq). — North America.

cornicina Fabricius, Meigen, System. Beschr. V, 57 *(M. caesarion)*. — Europe and North America (according to v. d. Wulp, Tijdschr. etc. 2d Ser. IV, 80).

fraterna Macquart, Dipt. Exot. 3e Suppl. 57, 41. — North America.

fulvifacies R. Desvoidy, Myod. 467 *(Phormia)*; Dipt. des envir. de Paris II, 848 *(id.)*. — Paris, France; Philadelphia. —

Heraea Walker, List, etc. IV, 881. — North America.

lepida Desvoidy, Myod. 453. — France, Philadelphia, Nova Scotia.

***macellaria** Fabricius, System. Ent. 776, 14 *(Musca)*; Ent. System. IV, 319, 28 *(id.)*; System. Antl. 292, 42 *(id.)*; Olivier, Encycl. Méth. VIII, 14, 14 *(id.)*; Wiedemann, Auss. Zw. II, 405, 36 *(id.)*; Macquart, Dipt. Exot. II, 3, 147, 28; Tab. XVII, f. 9; Bigot in R. de la Sagra etc. 820 — Brazil, Cuba, United States.

Lucilia hominivorax Coquerel, Ann. Soc. Ent. 1858, 173; Tab. IV, f. 2.

mollis Walker, List, etc. IV, 892 *(Phormia)*. — Huds. B. Terr.

muralis Walker, List, etc IV, 888. — Huds. B. Terr.

nigrina Bigot, Ann. Soc. Ent. Fr. 1877, 247. — Illinois.

philadelphica R. Desvoidy, Myod. 466 *(Phormia)*. — Philadelphia.

regina Meigen, System. Beschr. V, 58 *(Musca)*. — Europe and North America (according to Harris, Cat. Ins. Mass.).

rufipalpis Jaennicke, Neue Exot. Dipt. 67. — Illinois.

Sayi Jaennicke, Neue Exot. Dipt. 67. — Illinois.

terrae novae Macquart, Dipt. Exot. 4e Suppl. 251, 57; Tab. XXIII, f. 1. — Newfoundland.

terrae novae R. Desvoidy, Myod. 467 *(Phormia)*. — Newfoundland.

?proxima Walker, Dipt. Saund. 341 *(Musca)*. — California.

stigmaticalis Thomson, Eugen. Resa, 544. — California.

argentifera Bigot, Ann. Soc. Ent. Fr. 1877, 251. — Mexico.

brunnicornis Macquart, Dipt. Exot. II, 3, 142, 15. — Mexico.

Cluvia Walker, List, etc. IV, 885. — West Indies.

callipes Bigot, Ann. Soc. Ent. Fr. 1877, 249. — Mexico.

flavigena Bigot, Ann. Soc. Ent. Fr. 1877, 249. — Mexico.

fulvinota Bigot, Ann. Soc. Ent. Fr. 1877, 251. — Mexico.

insularis Walker, Dipt. Saund. 340 *(Musca)*. — West Indies.

meridensis Macquart, Dipt. Exot. 1er Suppl. 199, 33. — Yucatan.

mexicana Macquart, Dipt. Exot. II, 3, 143, 17; Tab. XVIII, f, 7. — Mexico.

mutabilis Bigot, Ann. Soc. Ent. Fr. 1877, 248. — Mexico.

14

nigriceps Macquart, Dipt. Exot. II, 3, 143, 16. — Mexico.
pallidibasis Bigot, Ann. Soc. Ent. Fr. 1877, 247. — Mexico.
picicrus Thomson, Eugen. Resa, 543. — Panama.
pueblensis Bigot, Ann. Soc. Ent. Fr. 1877, 250. — Mexico.
putrida Fabricius, Ent. System. IV, 316, 16 *(Musca)*; System. Antl.
     288, 24 *(id.)*; Wiedemann, Auss. Zw. II, 404, 35 *(id.)*. — South
     America (Wied.); Cuba (Jaennicke, Neue Exot. Dipt. 4).
ruficornis Macquart, Dipt. Exot. 1er Suppl. 198; compare . also
     Schiner, Novara, 304. — Columbia, S. Amer. (Macq.); Cuba
     (Bigot, in R. de la Sagra 821); Chile (Schiner).
surrepens Walker, Trans. Ent. Soc. N. Ser. V, 312. − Mexico.
violacea Macquart, Dipt. Exot. 2e Suppl. 83, 34. — Mexico.

Chrysomyia.
R. Desvoidy, Myod. 444; 1830.

caerulescens R. Desvoidy, Myod. 447, 8. — Carolina.
certima Walker, List, etc. IV. 873. — Florida.
L'herminieri R. Desvoidy, Myod. 446, 6. — Carolina.
hyacinthina R. Desvoidy, Myod. 450, 16; Macquart, Dipt. Exot. II,
     3, 148, 29 *(Lucilia)*. — South America (R. Desv.); North America
     (Macq.).
turbida Walker, Dipt. Saund., 336 *(Musca)*. — United States.
aztequina Bigot, Ann Soc. Ent. Fr. 1877, 252. — Mexico.
decora R. Desvoidy, Myod. 448, 10. — West Indies.
Plaei R. Desvoidy, Myod. 448, 11. — West Indies.
tibialis R. Desvoidy, Myod. 446, 5. — San Domingo.

Somomyia.
Rondani, Atti del Accad. delle Sci. di Bologna, 1861; Prodromus, IV, 9.

Sylphida Bigot, Ann. Soc. Ent. Fr. 1877, 45, 17. — New Orleans.
semiviolacea Bigot, l. c. 46, 18. — Porto Rico.
soulouquina Bigot, l. c. 47, 20. — Hayti.

Pyrellia.
R. Desvoidy, Myod. 462; 1830.

cadaverina Linné, Meigen, System. Beschr. V, 59, 19 *(Musca)*. —
     Europe and North America (Fitch, Survey etc. 801).
cadaverum Kirby, Fauna Bor. Amer Ins. 316, 1 („very near to *Musca*
     cadaverina", says Kirby). — Arctic America, lat. 65.
occidentis Walker, Dipt. Saund., 347 *(Musca)*. − United States.

     NB. On page 332 of the same volume, Walker described another
     Musca occidentis (see *Stomoxys*).

*__setosa__ Loew, Centur. VIII, 63. — Illinois

frontalis Thomson, Eugen. Resa, etc. 545. — California.

basalis Walker, Dipt. Saund., 347. — West Indies.

centralis Loew, Centur. VIII, 62. — Cuba.

ochricornis Wiedemann, Auss. Zw. II, 408, 41 *(Musca)*; Macquart, Dipt. Exot. II, 3, 149, 3; Tab. XX, f. 5; Bigot, in R. de la Sagra etc. 821. — Brazil (Wied.); Cuba (Macq.; Bigot).

scordalus Walker, Trans. Ent. Soc. N. Ser. V, 313. — Mexico.

specialis Walker, Trans. Ent. Soc. N. Ser. V, 312 — Mexico.

suspicax Walker, l. c. — Mexico.

Ormia.

R. Desvoidy, Myod. 428; 1830; *Ochromyia*, Macquart, Hist. Nat. Dipt. II, 250; Dipt. Exot. II, 3, 132.

punctata R. Desvoidy, Myod. 428, 1; Macquart, Hist. Nat. Dipt. II, 250, 3 *(Ochromyia)*. — West Indies (R. Desv.); Jamaica (Walker, List, etc. IV, 868).

Musca.

Linné, Fauna Suecica; 1763.

corvina Fabricius, Meigen, System. Beschr. V, 69, 32. — Europe and North America (Nova Scotia, Walker, List, etc. IV, 900). Occurs also in the East Indies, Manilla, Taiti, etc. (see Schiner, Novara 307).

*domestica Linné, etc. — Europe and North America (the common house-fly; see Loew, in Sillim. Journ. l. c.; about the occurrence in Cuba, see Bigot in R. de la Sagra, 822).
Musca harpyia Harris, Ent. Correspondence 335.

basilaris Macquart, Dipt. Exot. II, 3, 153, 8. — Brazil (Macq.); Jamaica (Walker, List, etc. IV, 901).

pusilla Macquart, Dipt. Exot. 3e Suppl. 59, 16; Tab. VI, f. 13. — Hayti.

sensifera Walker, Trans Ent. Soc. V, 314. — Mexico.

NB. *Musca cloacaris* O. Fabricius, Fauna Groenl. 204, 163, may be *Scatophaga litorea* Fall., according to Schiódte, Berl. Ent. Zeitschr. 1859, 153.
Musca vivax O. Fabricius, l. c. 206, 165 (I do not know.)

Cyrtoneura.

Curtoneura Macquart, Hist. Nat. Dipt. II, 274; 1835; amended by later authors.

*micans Macquart, Dipt. Exot 5e Suppl. 116, 10. — Baltimore.

*stabulans Fallen, Meigen, System. Beschr. V, 75, etc. *(Musca)*. — Europe and North America (see Loew, in Sillim. Journ. l. c.). Occurs also in New Zealand (Schiner, Novara, 304).

quadrisetosa Thomson, Eugen. Resa. 549. — California.

recurva Thomson, Eugen. Resa, 548. — California.

mexicana Macquart, Dipt. Exot, II, 3, 158, 4; Tab. XXI, f. 9. —
Mexico.

Myospila.

Rondani, Prodrom. Dipt. Ital. I, 91, 9; 1856.
Schiner, Fauna Austr. Dipt. I, 598.

*meditabunda Fabricius; Panzer; Meigen, System. Beschr. V, 79
(*Musca*). — Europe and North America (see Loew, Sillim. Journ.
l. c.; compare however the observation at the end of the genus
Spilogaster).

FAMILY ANTHOMYIDAE. [277]

Aricia.

R. Desvoidy, Myod. 486; 1830.

bispinosa Zetterstedt, Dipt. Scand. IV, 1428; Holmgren, Ins. Nordgroenl. 101. — Northern Sweden; Greenland.
cinerella v. d. Wulp, Tijdschr v. Ent. 2ᵈ Ser. II, 150. — Wisconsin.
deflorata Holmgren, Ins. Nordgroenl. 102. — Greenland.
denudata Holmgren, Ins. Spetsb. 30; Ins. Nordgroenl. 101. — Spitzbergen and Greenland.
dorsata Zetterstedt, Dipt. Scand. IV, 1472, 82; Holmgren, Ins.
Spetsb. 29; Ins. Nordgroenl. 101. — Lapland; Spitzbergen,
Greenland.
frenata Holmgren, Ins. Nordgroenl. 103. — Greenland.
Fabricii Holmgren, Ins. Nordgroenl. 101. — Greenland.
icterica Holmgren, Ins. Nordgroenl. 102. — Greenland.
incerta Walker, Dipt. Saund., 354. — United States.
moesta Holmgren, Ins. Nordgroenl 102. — Greenland
morioides Zetterstedt (perhaps *morio* Zett ? I do not find an *A.
morioides* Zett.). — Europe and North America (see Loew, Sillim.
Journ. l. c.).
pauxilla Holmgren, Ins. Spetsb. 32; Ins. Nordgroenl. 101. — Spitzbergen, Greenland.
proxima v. d. Wulp, Tijdschr. v. Ent. 2ᵈ Ser. IV, 85. — Wisconsin.
pruinosa Macquart, Dipt. Exot. 1ᵉʳ Suppl. 201, 4 — Galveston, Texas.
ranunculi Holmgren, Ins. Spetsb. 34; Ins. Nordgroenl. 101. — Spitzbergen, Greenland.
solita Walker, Dipt. Saund., 354. — United States.
tarsalis Walker, Dipt. Saund, 355. — United States.
tristicula Holmgren, Ins. Nordgroenl. 102. — Greenland.

circulatrix Walker, Trans. Ent. Soc. N. Ser. V, 316. — Mexico.
procedens Walker, Trans. Ent. Soc. N. Ser. V, 315. — Mexico.
rescita Walker, Trans. Ent. Soc. N. Ser. V, 315. — Mexico.

Observation. R. H. Meade Esq , in Bradford, England,
having published a most interesting article: Notes on the An-

thomyidae of North America (Ent. Monthly Magazine, April
1878, p. 250—252), I have reproduced his conclusions below, at
the end of each corresponding genus; compare also the note [277]
for the general conclusions.

About *Aricia* he writes:

The genus *Polietes* (Rond.) of which the well-known (european)
M. lardaria F. is the principal species, is not represented in the
(North American) collection.

„In the genus *Hyetodesia (Aricia* pt. Macq.). I determined
seven distinct (North American) species, several of which closely
resemble european, as *Musca lucorum* Fall., *A. lugubris* Meig.,
and *A. obscurata* Meig., but none of them, I think, are quite
identical."

„In the genus *Mydaea (Aricia* pt. Macq.). I found ten species,
only one of which was similar to any in Europe, viz. the common
M. pagana F., which has a yellow scutellum."

Spilogaster.

Macquart, Hist. Nat. Dipt. II, 293; 1835.

*angelicae Meigen, System. Beschr. V, 117, 59 *(Musca)*. — Europe
and North America (see Loew, Sillim. Journ. I. c. *Hylemyia
angelicae*).

*urbana Meigen, System. Beschr. V, 118, 60 *(Musca)*. — Europe and
North America (see Loew, Sillim. Journ. l. c. *Hylemyia urbana)*;
Lake Winnipeg; Connecticut.

terminalis Walker, Dipt. Saund., 356. — United States.

Observation. Mr. Meade says (l. c.):

„In *Spilogaster* there where eleven (North American) species, one
or two of which closely resembled european species, but were,
however, distinct. One fly in this genus possessed several interesting characters, which deserve especial notice. There was only
one male in the collection and it bore a remarkable resemblance
to *Cyrtoneura* (Myospila) *meditabunda* F. The fifth longitudinal
vein was curved in a similar manner towards the fourth vein,
though in a less degree; the spots upon the abdomen and the
general color, size and appearance, were also very like those of
that fly; but it differed in having the eyes naked and the arista
furnished with much shorter hairs."

Hydrophoria.

Rob. Desvoidy, Myod., 503; 1830.

„The genus was represented by three (N. A) species, all of
small size, one of which was similar to *Musca ambigua* Fallen."
(R. H. Meade, l. c. p. 251.)

Hydrotaea.

R. Desvoidy, Myod. 509; 1830.

*armipes Fallen, Dipt. Succ. Musc. 75, 86; Zetterstedt, Dipt. Scand.
IV, 1434, 44. — Europe and North America (see Loew, Sillim.
Journ. l. c. and Meade, Ent. Monthly Mag. April 1878).

*dentipes Meigen, System. Beschr. V. 144, 105; Staeger, Groenl. Antl.
363, 35. — Europe and North America (see Loew, Sillim. Journ.
l. c. and Meade, Ent. Monthly Mag. April 1878).

ciliata Fabricius; Meigen, System. Beschr. V, 159 *(Musca spinipes*
Fallen); Staeger, Groenl. Antl. — Europe and Greenland.

irritans Fallen, Dipt. Suec. Musc. 62, 58; Zetterstedt, Dipt. Scand.
IV, 1431, 10; Staeger, Groenl. Antl. 363, 35. — Europe and
Greenland.

Observation. „I found only two species belonging to the
genus *Hydrotaea*, both of which seemed identical with the common european *M. dentipes* F. et *M. armipes* Fall." (Meade, I. c.)

Lasiops.

Meigen, System. Beschr. VII, 323; 1838.

„The genus *Lasiops* contained two (N. A.) species, one closely
resembling *L. cunctans* Meig." (R. H. Meade, l. c. p. 251.)

Ophyra.

R. Desvoidy, Myod. 516; 1830.
Schiner, Fauna Austr. I, 619.

aenescens Wiedemann, Auss. Zw. II, 435, 29 *(Anthomyia)*; Macquart,
Dipt Exot. 1er Suppl. 203, 4. — New Orleans (Wied.); Texas
(Macq.).

*leucostoma Wiedemann, Zool. Mag. I, 82 *(Anthomyia)*; Meigen, System.
Beschr. V, 160 *(id.).* — Europe and North America (Loew, in
Sillim. Journ. l. c. and Meade, in Ent. M. Mag. April 1878,
p. 251); Atlantic States, common.

Drymeia.

Meigen, System. Beschr. V, 204; 1826.

„In the genus *Drymeia*, I found, as in Europe, one well marked
species only, which exhibited all the peculiar characters seen in
the *M. hamata* of Fallèn, but was quite distinct from that common fly." (R. H. Meade I. c.).

Limnophora.

R. Desvoidy, Myod 517; 1830.

contractifrons Zetterstedt, Ins. Lapp. 683, 97 *(Anthomyza)*; Dipt.
Scand. IV, 1463 *(Aricia).*
Anthomyza arctica Zetterstedt, Ins. Lapp. 669, 34 *(Varietas)*; Staeger,
Groenl. Antl. — North of Europe and Greenland.

*diaphana Wiedemann, Zool. Mag. I, 81, 31 *(Anthomyia)*; Meigen,
System. Beschr. V, 189, 185 *(id.).* — Europe and North America
(see Loew, Sillim. Journ. l. c.).

*stygia Meigen, System. Beschr. V, 155, 127 *(Anthomyia).* — Europe
and North America (see Loew, Sillim. Journ. I. c. *Anthom. stygia)*;
Sitka.

tiiangulifera Zetterstedt, Ins. Lapp. 680, 83 *(Anthomyza)*; Staeger, Groenl. Antl. 364, 40. — Europe and Greenland.

trigonifera Zetterstedt, Ins. Lapp. 669, 33 *(Anthomyza)*; Dipt. Scand. IV, 1466 *(Aricia)*; Staeger, Groenl. Antl. 364, 38. — Europe and Greenland.

Observation. „The genus *Limnophora* contained eight (N. A.) species, two or three of which closely resembled european ones; but none of them appeared quite identical. In the european species of this family, of which the *A. compuncta* Wied. is the type, the eyes of the males are sometimes separated by a rather wider space than is usual among the Anthomyidae, except in *Coenosia*, *Lispa* etc., and this character was marked in an exaggerated degree in all the american species, so that it was difficult to determine by the eyes alone, whether they should be placed in the genus *Limnophora* or *Coenosia*." R. H. Meade, I. c.

Eriphia.

Meigen, System. Beschr. V, 206; 1838.

? **Acela** Walker, List, etc. IV, 962.
Arelate Walker, List, etc. IV, 961.
biquadrata Walker, 1. c. 963.
ciliata Walker, 1. c. 961.
flavifrons Walker, 1 c. 966.
grisea Walker, 1. c. 962.
Lamnia Walker, 1. c. 964.
Iata Walker, 1 c. 963.
marginata Walker, 1. c. 964.
pretiosa Walker, 1. c. 965.

Huds. B. Terr.

Hylemyia.

Rob. Desvoidy, Myod. 550; 1830.

* **deceptiva** Fitch, Reports, Vol. I, 801; Tab. I, f. 3. — New York.

frontata Zetterstedt, Ins. Lapp. 669, 35; Dipt. Scand. IV, 1453, 64; Staeger, Groenl. Antl. 363, 37. — Europe (Lapland) and Greenland.

* **pici** Macquart, Ann. Soc. Ent. 1853, 657; Tab. XX, Nr. 2 *(Aricia)*. — San Domingo; The larva lives in a swelling on the wing of *Picus striatus.*

Hylemyia angustifrons Loew, Wien. Ent. Monatschr. V, 41. — Cuba [Loew *in litt.*].

. **probata** Walker, Trans. Ent. Soc. N. Ser. V, 318. — Mexico.

Anthomyia. (*)

Meigen, in Illiger's Magaz II; 1803.

brassicae (Bouché?), A. Fitch, Report XI, 40. — Europe and North America (injurious to cabbage). ([278])

---

*) I have prefixed a ? before those species which are Anthomyiae in the wider sense· only, not in that of Schiner.

campestris R. Desvoidy, Myod. 585 *(Egle)*. — Europe and North America (Philadelphia).

ceparum (Meigen, Bouché) A. Fitch, Report. XI, 31; Walsh, Amer. Ent. II, 110, f. 72. (^{279}).

? **communis** Walker, Dipt. Saund., 366. — United States.

Dejeanii R. Desvoidy, Myod. 558, 4 *(Nerina)*. — Philadelphia.

? **dubia** Curtis, Ins. Ross's Exp. LXXIX. — Arctic America.

? **raphani** Harris, Ins. of New Engl. 3^d edit. 617; Fitch, Report XI, 59 (injurious to radish plants). — New England; New York.

ruficeps Meigen, System. Beschr. V, 177, 162; Staeger, 366, 43. — Europe and Greenland.

? **similis** Fitch, Reports I, 301. — New York.

scatophagina Zetterstedt, Ins. Lapp. 677, 69 *(Anthomyza)*; Dipt. Scand. IV, 1510, 120 *(Aricia)*; Staeger, Groenl. Antl. — North of Europe and Greenland.

striolata Fallen; Meigen, System. Beschr. V, 173, 156; Zetterstedt, Ins Lapp. 684, 103; Staeger, Groenl. Antl., 365, 42. — Europe and Greenland.

****tarsata** v. d. Wulp, Tijdschr. v. Ent. 2^d Ser. II, 151; Tab. V, f. 6. — Wisconsin.

? **Zeas** Riley, 1st Report 154; Tab. II, f. 24 (injurious to indian corn). — Missouri.

? **leucoprocta** Wiedemann, Auss. Zw. II, 433. — West Indies.

? **protrita** Walker, Trans. Ent. Soc. N. Ser. V, 317. — Mexico.

micropteryx Thomson, Eugen. Resa 555. — California.

ochripes Thomson, l c. 553. — California.

ochrogaster Thomson, l. c. 557. — California.

Species described in Mr. Walker's List etc. IV. They are left in the subdivisions adopted by him.

A. Feeler-bristle feathered or hairy, Meigen. Dipt. V, Tab. 44, f. 1, 2.

a. Legs black.

* Eyes hairy.

Rugia Walker, l. c. 923. — Huds. B. Ters.

** Eyes non hairy.

palposa Walker, l. c. 926. — Huds B. Terr.

spinosa Walker, l. c. 926. — Huds. B. Terr.

Apina Walker, l. c. 927. — Nova Scotia.

Anane Walker, l. c. 927. — Huds. B. Terr.

Lipsia Walker, l. c. 928. — Huds. B. Terr.

Pylone Walker, l c. 928. — North America.

nigripennis Walker, l. c. 929. — Huds. B. Terr.

Omole Walker, l. c. 930. —　　　　　　　　 "

similis Walker, l. c. 930. —

nigra Walker, l. c. 931. — Huds. B. Terr.
Teate Walker, l. c. 931. —   „
nigrifrons Walker, l. c. 932. —   „
Barpana Walker, l. c. 933. — Nova Scotia.
Narina Walker, l. c. 933. — Nova Scotia.

b. Legs wholly or mostly yellow.

* Eyes hairy.

Luteva Walker, l. c. 934. — Nova Scotia.

** Eyes not hairy.

Bysia Walker, l. c. 936. — Nova Scotia.
Troene Walker, l. c. 936. —   „
Aemene Walker, l. c. 937. —   „
Alcathoe Walker, l. c. 937. —   „
Lysinoe Walker, l. c. 938. —   „
Ausoba Walker, l. c. 938. —   „
Signia Walker, l. c. 939. —   „
Geldria Walker, l. c. 940. —   „
Alone Walker, l. c. 941. — Huds. B. Terr.
soccata Walker, l. c. 941. —   „

B. Feeler-bristle downy or bare; legs black; eyes not hairy.

Narona Walker, l. c. 945. — Florida.
Donuca Walker, l. c. 946. — Nova Scotia.
Brixia Walker, l. c. 946. —   „
Alaba Walker, l. c. 948. — North America.
Idyla Walker, l. c. 948. — Huds. B. Terr.
Uxama Walker, l. c. 948. —   „
Tinia Walker. l. c. 949. —
Badia Walker, l. c. 950. —   „
Perrima Walker, l. c. 950. —   „
Viana Walker, l. c. 951. — Nova Scotia.
Acra Walker, l. c. 951. — Huds. B. Terr.
Isura Walker, l c. 952. — Nova Scotia.
determinata Walker, l. c. 955. — „
Opalia Walker, l. c. 956. —   „

Observation. Mr. Meade (Entom. Monthly Mag April 1878) says about N. A. Anthomyiae: „In this genus, as now restricted, I determined eight species, one of which seemed identical with *Musca radicum*, Lin. and another with *M. pluvialis* Lin.

Chortophila.

Macquart, Hist. Nat. Dipt. II, 323, 1825; Rondani, Dipt. Ital. Prodr.

„A large number of small flies in the (North American) collection could be referred to the genus *Chortophila*. I made out as many as twenty nine distinct species, several of which were similar to european forms, viz.　*C. floccosa* Macq , *A. angustifrons* Meigen, *A. gilva* Zett. , *A. vittigera* Zett. and *A. flavoscutellata* Zett." (R. H. Meade, in Ent. Monthly Magaz., April 1878, p. 252.)

Azelia.

Rob. Desvoidy, Essai sur les Myodaires, 1830; Loew, Die deutschen Arten d. Gatt. Azelia (Ent. Miscellen etc. Breslau 1874).

Mr. Meade says about the North American Azeliae (Ent. Monthly Magaz. April 1878).

„The only species in this genus corresponded with *A. Stregeri* Zett." According to Loew, 1. c. the latter in the same with *A. cilipes* Haliday, Ann. Nat. Hist. II, p. 105, which is the older name.

Atomogaster.

Macquart, Hist. Nat Dipt. II, 329; 1835.

*albicincta Fallen, Meigen, etc. — Europe and North America (Loew· *in litt.*); Nebraska, Texas.

Homalomyia.

Bouché, Naturgesch. d. Ins. I, 88; 1834.

*canicularis Linné, Meigen, System. Beschr. V, 143, 104 *(Anthomyia)*. — Europe and North America (see Loew, Sillim. Journ. 1. c. and Meade, Ent Monthly Mag. 1878, April).

*manicata Meigen, System. Beschr. V, 140, 100 *(Anthomyia)*; Zetterstedt, etc — Europe and North America (see Loew, 1. c.).

prunivora Walsh, Amer. Ent. II, 137 (description of imago and larva). — Illinois. ([280]).

*scalaris Fabricius; Meigen, System. Beschr. V, 141, 102 etc. *(Anthomyia)*. — Europe and North America (see Loew, Sillim. Journ. 1. c. and Meade, Ent. Monthly Mag. 1878, April).

Fannia saltatrix R. Desvoidy, Myod. 567 [Schiner].

*serena Fallen, Musc. 76, 88. — Europe (Sweden) and North America (Loew *in litt.*).

*spathulata Zetterstedt, Dipt. Scand. IV, 1543. — Europe (Lapland) and North America (Loew *in litt.*.

*subpellucens Zetterstedt, Dipt. Scand. IV, 1561, 176. — Europe (Lapland) and North America (Loew, Sillim. Journ. I. c.).

*tetracantha Loew, Centur. X, 69. — Middle States.

femorata Loew, Wiener Ent. Monatschr. V, 42, 18; Centur. X, 68. — Cuba.

Observaticn. „There were five (N. A.) species, belonging to this genus, three of which seemed identical with the common european *M. canicularis* L., *A. scalaris* M., and *A. incisurata* Zett. It is most probable that these common flies, which abound in and about our houses in Europe, have been imported into America, like the house fly, *M. domestica.*" (R. H. Meade, 1. c.)

Dialyta.

Meigen, System. Beschr. V, 203; 1826. ([281]).

? cupreifrons Walker, List, etc. IV, 966. — Huds. B. Terr.

Lispe.

Lispa Latreille, Precis etc.; 1796. ([282]).

* flavicincta Loew, Stett. Ent. Zeit VIII, 27. — Europe and North America, Huds. B. Terr. (Loew *in litt.*).
* consanguinea Loew, Wiener Ent. Monatsch. II, 8. — Europe and North America, Texas (Loew *in litt.*).
hispida Walker, List, etc. IV, 971. — Huds. B. Terr.
* sociabilis Loew, Centur. II, 72. — Distr. Columbia.
simillima Walker, List, etc. IV, 972. — Huds. B. Terr.
* uliginosa Fallen, Dipt. Suec. (*Musca*) 93, 2; Loew, Stett. Ent. Zeitschr. VIII, 24. — Europe and North America (Loew, in Sillim. Journ. l. c. and Meade, in Ent. Monthly Magaz. April 1878, p. 252).

Observation. „The genus *Lispa* contained three (N. A.) species, one similar to *L. tentaculata* Degeer, and another to *L. uliginosa* Fall." (Meade, l. c.)

Caricea.

Rob. Desvoidy, Myod., p. 530; 1830.

„This genus contained but one species, which seems to be very common in America, as there were numerous specimens of it in the collection; it was of considerable size and the females bore a remarkable resemblance to those of *M. impuncta* Fall., but the males were very different and quite characteristic of the genus." (Meade. l. c.)

Coenosia.

Meigen, System. Beschr. V, 210; 1826.

* calopyga Loew, Centur. X, 71. — Pennsylvania.
incisurata v. d. Wulp, Tijdschr. v. Ent. 2d Ser. IV, 84. — Wisconsin.
* modesta Loew, Centur. X, 72. — Distr. Columbia.
* nivea Loew, Centur. X, 70. — Pennsylvania.
(For *Coenosia tricincta* Loew, Centur. IX, 83, see Cordylura, where it has been transferred by Loew *in litt.*).
fuscopunctata Macquart, Dipt. Exot. 4e Suppl. 270, 4. — North America.

Mr. Walker's species:
antica Walker, Dipt. Saund., 367. — United States.
atrata Walker, Dipt. Saund., 369. — United States.
intacta Walker, Dipt Saund., 369. — United States.
intacta Walker (bis!) Trans. Ent. Soc. N. S. V, 318. — North America.

lata Walker, Dipt. Saund., 368. — United States.
sexmaculata Walker, List, etc. IV, 970. — Huds. B. Terr.
solita Walker, Dipt. Saund, 368. — Huds. B. Terr.
spinosa Walker, List, etc. IV, 967. — Huds. B. Terr.
substituta Walker, List, etc IV, 971. — Massachusetts.

> Observation: Mr. Meade (Ent. Monthly Magaz. April 1878)
> made out sixteen north american species of *Coenosia*, many of
> which were very similar in their characters to european ones;
> but he could only identify one, which was apparently identical
> with *A. pygmaea* Zett.

Schoenomyza.

Haliday, Ent. Mag. 1833. ([283]).

* chrysostoma Loew, Centur. IX, 86. - New Hampshire.
* dorsalis Loew, Centur. X, 73. — Distr. Columbia.

FAMILY CORDYLURIDAE.

Cordylura.

Fallen, Spec Ent. etc.; 1810. ([284]).

*acuticornis Loew, Centur. IX, 94. — British North America.
*adusta Loew, Centur. III, 41. — New Jersey; White Mts., N. H.
*albibarba Loew, Centur. IX, 96. — White Mts, N. H.
*angustifrons Loew, Centur. III, 45 — Wisconsin.
*bimaculata Loew, Wiener Ent. Monatschr. IV, 81, 3; Centur. III,
    40. — Atlantic States; Canada.
    Cordylura maculipennis v. d. Wulp, Tijdschr. v. Ent. 2d Ser. II,
    152; Tab. V, f. 7—9. [Loew, Zeitschr. f. Ges. Naturw. XXXVI,
    116, 9.]
    Lissa varipes Walker, List, etc. IV, 1046. — Ohio (!).
*capillata Loew, Centur. X, 77. — White Mts, N. H.
*cincta Loew, Centur. III, 47. — Distr. Columbia.
*confusa Loew, Centur. III, 43. — British N. A.
    Cordylura pubera Linné, in Walker, List, etc. IV, 972. — Huds.
    B. Terr.
*cornuta Loew, Centur. III, 48. — British possessions; White Mts,
    N. H. (the patria „British Columbia in the Centuries, is erroneous).
*flavipes Loew, Centur. III, 46 — Wisconsin.
*fulvibarba Loew, Centur. X, 76. — Fort Resolution, Huds. B. Terr.
*gagatina Loew, Centur. IX, 93. — Canada
*gilvipes Loew, Centur. III, 49. — English River, Lake Winnipeg.
*glabra Loew, Centur. IX, 90. — White Mts., N. H.
*gracilipes Loew, Centur. IX, 87. — White Mts., N. H.
*haemorrhoidalis Meigen, System. Beschr. V, 237; — Staeger, Groenl.
    Antl. 366. - Europe and North America; Greenland (Staeger);
    White Mts., N. H. (Loew *in litt.*).

impudica Reiche, Ann. Soc. Ent. de Fr. 1857, Bullet. p. 77 (*Anthomyia*). — Greenland (is a Cordylura, according to Loew, Berl. Ent. Zeitschr. 1858, 347).
*inermis Loew, Centur. IX, 88. — White Mts., N. H.
*latifrons Loew, Centur. IX, 92. — Middle States.
*lutea Loew, Centur. X, 75. — Sitka.
*megacephala Loew, Centur. IX, 94. — Distr. Columbia.
*munda Loew, Centur. IX, 91. — Fort Resolution, Huds. B. Terr.
*nana Loew, Centur. V, 94. — Canada.
pictipennis Loew, Wiener Ent. Monatschr. VIII, 22. — Siberia and North America.
*pleuritica Loew, Centur. III, 42. — English River, Winnipeg; Massachusetts; Connecticut
*praeusta Loew, Centur. V, 93. — Canada.
qualis Say, J. Acad. Phil. VI, 176; Compl. Wr. II. 366. — Indiana [„eyes approximate above“, cannot be *Cordylura!* Loew, *in litt.*].
*scapularis Loew, Centur. IX, 89. — English River, Winnipeg.
*setosa Loew, Wiener Ent. Monatschr. IV, 81, 4; Centur. III, 44. — Distr. Columbia.
*terminalis Loew, Centur. III, 39. — Pennsylvania.
*tricincta Loew, Centur. IX, 83 (*Coenosia); transferred to Cordylura, by Loew, in litt. — White Mts, N. H.
*variabilis Loew, Zeitschr f. Ges. Naturw. 1876, 326. — Massachusetts.
*vittipes Loew, Centur. X, 74. — Sitka.
*unilineata Zetterstedt, Dipt. Scand. V, 2010. — Sweden, Lapland; also in Sitka (Loew *in litt.*).

> **Observation.** Species from Mr. Walker's, List, etc.
> **Aea,** l. c. IV, 978. — Huds B. Terr.
> **bicolor,** l. c. 974. — Huds. B. Terr.
> **cupricrus,** l. c. 974. — Huds. B. Terr.
> **flavipennis,** l. c. 975. — Huds. B. Terr.
> **imperator,** l. c. 975. — Huds. B. Terr.
> **longa,** l. c. 976. — Huds. B. Terr.
> **tenuior,** l. c. 977. — Huds. B. Terr.
> **volucricaput,** l. c. 977. — Huds. B. Terr.

Hydromyza.

Fallen, Dipt. Suec. Hydromyz.; 1823.

*confluens Loew, Centur. III, 50. — English River, Lake Winnipeg.

Scatophaga.

Meigen, Illiger's Magaz. II; 1803; *Scatomyza* Fallen; *Pyropa* Illiger.

ariciiformis Holmgren, Ins. Nordgroenl. 103. — Greenland.
apicalis Curtis, Ins. Ross's Exp. LXXX. — Arctic. America.
bicolor Walker, List, etc. IV, 982. — Huds. B. Terr.
canadensis Walker, Trans. Ent. Soc. N. Ser. IV, 218. — Canada.

exotica Wiede nann, Auss. Zw. II, 44?, 3. — New Orleans.

fuscinervis Zetterstedt, Dipt. Scand. V, 1974, 11; Holmgren, Ins. Nordgroenl. 107. — Lapland and Greenland

intermedia Walker, List, etc. IV, 980. — Nova Scotia.

litorea Meigen, etc. Staeger's Groenl. Antl. p. 366, 46. — Europe and Greenland.

nigripes Holmgren, Ins. Spetsb. 34; Ins. Nordgroenl. 103. — Spitzbergen and Greenland.

pallida Walker, List, etc. IV, 981. — Huds B. Terr.

pubescens Walker, List, etc. IV, 982. — Huds. B. Terr.

*__squalida__ Meigen, etc ; Staeger, Groenl. Antl. 366, 45. — Europe and and North America (the occurrence in the latter is confirmed by Loew, in Sillim. Journ. XXXVII, p. 318); Nova Scotia (Walker, List, etc. IV, 981).

Pyropa furcata Say, J. Acad. Phil. III, 98; Compl. Wr. II, 85 [Loew, l. c].

Scatophaga furcata Wiedemann, Auss. Zw. II, 449, 5 (merely a translation from Say).

Scatophaga postilena Harris, Catal. Ins. Mass.

*__stercoraria__ Linné, etc. — Europe and North America (Occurrence confirmed by Loew, in Sillim. Journ., XXXVII, 318). ([2t.b]).

thinobia Thomson, Eugen. Resa, 563. — California.

Fucellia.

Rob. Desvoidy, Ann. Soc. Ent. de Fr. 2ª Ser. X, 269—271; 1841; *Halithea* Haliday (preoccupied).

*__fucorum__ Fallen, Zetterstedt, etc *(Scatomyza)*; Curtis's Ins. Ross's Exp. LXXX; Staeger, Groenl. Antl. 366, 47. — Europe and North America.

Scatina.

Rob. Desvoidy, Myod., 629; 1830; compare also Rondani, Prodr. I, 102.

estotilandica Rondani, Archiv. etc. Canestrini III, fasc. 1, p. 35. — Labrador.

Observation. Mr. Rondani, in the same place, mentions *Scatophaga diadema* Wiedemann (Montevideo), as having been received from Labrador.

FAMILY HELOMYZIDAE. [286]

Helomyza.

Fallen, Heteromyz., 3, 1820; Loew, Schl. Z. f. Ent. 1859, 17.

*__apicalis__ Loew, Centur. II, 86. — Distr. Columbia.

*__assimilis__ Loew, Centur. II, 87. -- Huds. B. Terr.

borealis Bohemann, Ins. Spetsb. 573, 15; Holmgren, Ins. Spetsb. 35;
Ins. Nordgroenl. 104. — Spi'zbergen and Greenland.
*lateritia Loew, Centur. II, 89. — Connecticut.
*longipennis Loew, Centur. II, 90. — New York.
*plumata Loew, Centur. II, 88. — New York.
quinquepunctata Say, J. Acad. Phil. III, 101; Compl. Wr. II, 86;
Wiedemann, Auss. Zw. II, 588, 3. — Cow Island, Missouri River.
tibialis Zetterstedt, Ins. Lapp. 767; Staeger, Groenl. Antl., 366, 50;
Holmgren, Ins. Nordgroenl 104. — Lapland and Greenland.
*Zetterstedtii Loew, Schles. Z. f. Ent. 1859, Helomyzidae 63. — North
of Europe and North America (Loew *in litt.*).

*limbata Thomson, Eugen. Resa, etc. 569. — California [There is an
earlier *H. limbata* Walker, Loew *in litt.*].

> **Observation.** Mr. Walker's species of *Helomyza* are:
> fasciata Walker, List, etc. IV, 1094. — Nova Scotia.
> lateralis Walker, l. c. IV, 1095. — North America.
> tincta Walker, List, etc. IV, 1092. — Nova Scotia.

Scoliocentra.

Schles. Zeitschr. f. Ent. 1859, 43.

*fraterna Loew, III, 51. — Sitka.
*helvola Loew, II, 80. — Illinois.
[There are two more species, as yet undescribed, in the collections.]

Anorostoma.

Loew, Schles. Z. f. Ent. 1859, 47.

*marginata Loew, Centur. II, 81. — Brit. North America.

Allophyla.

Loew, Schles. Z. f. Ent. 1859, 43.

*laevis Loew, Centur. II, 85. — Brit. North America. [„hardly differs
from the european *A. nigricornis* Meig., except in the coloring
of the antennae". Loew, l. c.].

Blepharoptera.

Loew, Schles. Z. f. Ent. 1859, 57.
Blephariptera Macquart, Hist. Nat. Dipt. II, 412; 1835.

*biseta Loew, Schl. Z. f. Ent. 1859, 62. — Europe and Sitka (Loew *in litt.*).
carolinensis R. Desvoidy, Myod 629, 11 *(Scatophaga)*; referred here
by R. Desvoidy in Ann. Soc. Ent.; 1841, p. 258, foot-note.
*cineraria Loew, Schl. Z. f. Ent. 1859, 67. — Europe and British N. A.
Blepharoptera armipes Loew, Centur. II, 83 (Loew *in litt.*).
*defessa O. Sacken, in Packard's: Cave fauna in Utah (Bulletin U. S.
Geol. and Geogr. Survey, Vol. III, No. 1). — Kentucky. ([287]).
*discolor Loew, Centur. X, 78. — White Mts., N. H.

geniculata Zetterstedt, Ins. Lapp. 767, 12 *(Helomyza)*; Staeger, Groenl. Antl. 366, 49 *(id.)*; Holmgren, Ins. Nordgroenl. 104. — North of Europe and Greenland.

iners Meigen, System. Beschr. VI, 57, 22 *(Helomyza)*; Loew, Schles. Z. f. Ent., 859, 63. — Europe and North America [see Loew, in Sillim. Journ. XXXVII, 318].

*leucostoma Loew, Centur. III, 54. — Sitka.

*lutea Loew, Centur. III, 52. — Sitka.

*pectinata Loew, Centur. X, 79. — Texas.

*pubescens Loew, Centur II, 82. — Massachusetts.

*tristis Loew, Centur. II, 84. — Lake Winnipeg.

Occothea.

Loew, Schles. Z. f. Ent. 1859, 54.

fenestralis Fallen, etc. compare Loew, l. c. — Europe; Siberia; North America (New York, Loew *in litt.*).

Tephrochlamys.

Loew, Schles. Z. f. Ent. 1859, 72.

*rufiventris Meigen, System. Beschr. VII, 58 *(Helomy-a)*; Loew, Schles. Z. f. Ent. 1859, 77. — Europe and Canada (Loew *in litt.*).

Heteromyza.

Fallen, Heteromyz. 1; 1820; Loew, Schles. Z. f. Ent. 1859, 70.

Observation. Whether the following species belong to Heteromyza in Loew's or even in Fallen's sense, is, of course, doubtful. According to Loew (Schles. Zeitschr. f. Ent. 1859, 9), *H. buccata* is no Heteromyza at all, but is related to the family *Phycodromidae.*

buccata Fallen, Meigen, etc. Walker, List; etc IV, 1088. — Europe and Nova Scotia (according to Walker).

eriphides Walker, l. c. 1088. — Huds. B. Terr.

flavipes Walker, l c. 1089. — Huds. B. Terr.

fusca Macquart, Dipt. Exot. II, 3, 203, 3; Tab. XXV, f 12. — North America.

FAMILY SCIOMYZIDAE. [288]

Sciomyza.

Fallen, Sciomyzidae 11; 1820.

*albocostata Fallen, Sciomyz 12, 3; Zetterstedt, Dipt Scand. V, 2098; Schiner, Fauna Austr. II, 47. — Europe; North America [Loew in Sillim. Journ. XXXVII, 318].

*apicata Loew, Zeitschr. f. Ges. Naturw. 1876, 331. — Fort Resolution, Huds. B. Terr.

*humilis Loew, Zeitschr f. Ges. Naturw. 1876, 330. — Texas.

*longipes Loew, Zeitschr. f. Ges. Naturw. 1876, 328. — White Mts., New Hampshire.

*luctifera Loew, Centur. I, 71; Monogr. I, 107. — Pennsylvania.
*nana Fallen, Loew, Monogr. I, 104. — Europe; United States, Canada.
*obtusa Fallen, Loew, l. c. 105. — Europe, United States.
*pubera Loew, l. c. 106. — Middle States.
*tenuipes Loew, Centur. X, 80. — Middle States.
*trabeculata Loew, Centur. X, 81. — Texas.
vittata Haliday, Ent. Mag. 1833. — Europe and North America (Masschusetts; Loew *in litt.*).

obscuripennis Bigot, R. de la Sagra etc. 826. — Cuba.

Mr. Walker described four Sciomyzae from North America; the three first are discussed by Mr. Loew in Monogr. I, 104:
    antica Walker, Dipt. Saund. 400. — United States.
    nigripalpus Walker, List, etc. IV, 1068. — Huds. B. Terr.
    parallela Walker, Dipt. Saund. 401. — United States.
    transducta Walker, Trans. Ent. Soc. N. Ser. V, 320. — North America.

Tetanocera.

Latreille, Genera Crust. et Ins. IV, 1809; *Tetanocerus* Duméril, 1801.

*ambigua Loew, Centur. V, 95. — Maine.
*arcuata Loew, Wien. Ent. Monatschr. III, 292; Monogr. I, 115. — Middle States.
*clara Loew, Monogr. I, 109. — New York.
*combinata Loew, Wien. Ent. Monatschr. III, 295; Monogr. I, 116. — United States and Canada.
*costalis Loew, Monogr. I, 118. — Illinois.
*flavescens Loew, Stett. Ent. Z. VIII, 123; Wien. Ent. Monatschr. III, 291; Monogr. I, 113. — Carolina (Lw.); Western New York (M. C. Z.; determ. by Loew *in litt.*, who suspects that *T. flavescens* is only a larger form of *arcuata*).
*pallida Loew, Wien. Ent. Monatschr. III, 294; Monogr. I, 113. — Middle States.
*pictipes Loew, Wien. Ent. Monatschr. III, 292; Monogr. I, 111. — Atlantic States and Canada; Bermudas.
*plebeja Loew, Monogr. I, 120. — Atlantic States and Canada.
*plumosa Loew, Stett. Ent. Z. VIII, 201; Wien. Ent. Monatschr. III, III, 296; Monogr. I, 121. — Middle and Northern States; Canada.
Tetanocera vicina Macquart, Dipt. Exot. II, 3, 180; Tab. XXIV, f. 7 [Lw.].
Tetanocera Struthio Walker, List, etc. IV, 1086 [Lw.].
*rotundicornis Loew, Centur. I, 70; Monogr. I, 123. — Brit. North America.
*saratogensis Fitch, Reports I, 68; Wien. Ent. Monatschr. III, 256; Monogr. etc. I, 119. — Atlantic States; Canada.
*sparsa Loew, Monogr. I, 117. — Middle States.
*triangularis Loew, Centur. I, 69; Monogr. I, 122. — Brit. North America.
*valida Loew, Monogr. I, 110. — New York; Quebec, Canada.

15

pectoralis Walker, Trans. Ent. Soc. N. Ser. V, 321. — Mexico.
*spinicornis Loew, Centur. VI, 86. — Cuba.

> **Observation.** The three remaining species, mentioned in my first Cata-
> logue are:
> Boscii R. Desvoidy, Myod. 690, 8 *(Pherbina)*. — Carolina.
> canadensis Macquart, Dipt. Exot. II, 3, 181, 4; Tab. XXIV, f. 5. — Canada.
> guttularis Wiedemann, Auss. Zw. II, 584, 3; Macquart, Dipt. Exot. II, 3,
> 181, 3. — Montevideo (Wied.); Philadelphia (Macq.). The remarks of
> Dr. Loew on these species are reproduced in the note([289]).

Sepedon.

Latreille, Hist. Nat. des Crust. et des Ins. XIV, 305; 1804.

*armipes Loew, Wien. Ent. Monatschr. III, 298; Monogr. I, 126. —
Middle States.
*fuscipennis Loew, Wien. Ent. Monatschr. III, 299; Monogr. I, 124. —
Middle States.
*macropus Walker, List, etc. IV, 1078; Monogr. I, 125. — Jamaica,
Cuba.
*pusillus Loew, Wien. Ent. Monatschr. III, 299; Monogr. I, 127. —
Middle States.

Dryomyza.

Fallen, Sciomyz.; 1820.

*anilis Fallen; Loew, Monogr. I, 128. — Europe and North America
(Middle States).
convergens Walker, List, etc. IV, 983. — Nova Scotia.
*simplex Loew, Monogr. I, 128. — Middle States.

maculiceps Walker, Trans. Ent. Soc. N. Ser. V, 319. — Mexico.

Actora.

Meigen, System. Beschr. V, 403; 1826.

ferruginea Walker, List, etc. IV, 1066. — Nova Scotia.

FAMILY PSILIDAE.

Loxocera.

Meigen, Illiger's Magaz.; 1803. ([290]).

*collaris Loew, Centur. IX, 97. — Distr. Columbia.

*cylindrica Say, J. Acad. Phil. III, 98; Compl. Wr. II, 84; Wiedemann,
Auss. Zw. II, 528. — Atlantic States.
*fallax Loew, Centur. IX, 98. — Canada.
*pectoralis Loew, Centur. VIII, 64. — Distr. Columbia.
*pleuritica Loew, Centur. VIII, 65. — New York; Connecticut.
quadrilinea Walker, Trans. Ent. Soc. N. Ser. V, 329. — United
States.

Psila.

Meigen, Illiger's Magaz. II; 1803.

bicolor Meigen, System. Beschr. V, 358. — Europe and North America.
(Sitka; Lake Winnipeg; Loew, in Sillim. J. XXXVII, 318 asserts
the specific identity.)

*****bivittata** Loew, Centur. VIII, 67. — Connecticut, Quebec, Canada.

*****collaris** Loew, Centur. VIII, 68. — Connecticut.

*****dimidiata** Loew, Centur. VIII, 69. — Red River of the North.

*****lateralis** Loew, Wien. Ent. Monatschr. IV, 81, 5; Centur. VIII, 66. —
Distr. Columbia.

*****levis** Loew, Centur. VIII, 71. — White Mts., N. H.

*****sternalis** Loew, Centur. VIII, 70. — Middle States.

Chyliza.

Fallen, Opomyz. 6; 1820.

*****apicalis** Loew, Wien. Ent. Monatschr. IV, 82, 6; Centur. VIII, 72. —
Distr. Columbia.

metallica Walker, List, etc. IX, 1045. — Huds. B. Terr.

nigroviridis Walker, Trans. Ent. Soc. N. Ser. V, 330. — United
States.

*****notata** Loew, Centur. IX, 99. — Distr. Columbia.

FAMILY MICROPEZIDAE.

Calobata.

Meigen, Illiger's Magaz.; 1803; *Ceyx* Duméril, Exposit. etc.; 1801.

*****Alesia** Walker, List, etc. IV, 1048. — Huds. B. Terr. (Walk.); New
England (M. C. Z.).

*****antennipennis** Say, J. Acad. Phil. III, 97, 1; Compl. Wr. II, 83 (*C.
antennaepes*); Wiedemann, Auss Zw. II, 546, 14. — Pennsylvania
(Say); Maryland, Kentucky (M. C. Z.)

*****geometra** R. Desvoidy, Myod. 736, 1 (*Neria*). — Carolina (R. D.);
Texas, Kentucky (M. C. Z.).

*****lasciva** Fabricius, Suppl. 574, 111 (*Musca*); System. Antl. 262; Wiede-
mann, Auss. Zw. II, 535; Schiner, Dipt. of the Novara etc. 253
(gives a fuller description). — Cayenne (Fabr.); Cuba (Jaennicke,
Neue Exot. Dipt. 4); New York (M. C. Z.).

Calobata albimana Macquart, Dipt. Exot. II, 3, 245; Tab. XXXIII,
f. 3. — Philadelphia; Cuba; Java; Port Jackson, Australia
[Schiner, Novara, etc. 253].

? *Calobata valida* Walker, Dipt. Saund., 390. — United States.

Calobata ruficeps Guérin, Iconogr. etc. III, 553; Tab. 103, f. 7. —
Cuba.

Taenioptera trivittata Macquart, Hist. Nat. Dipt. II, 491, 1; Tab.
XX, f. 9. — North America. ([271])

*****nebulosa** Loew, Centur. VII, 89. — Florida.

*pallipes Say, J. Acad. Phil. III, 97, 2; Compl. Wr. II, 84; Wiede-
mann, Auss. Zw. II, 548, 3 *(Micropeza)*. — Missouri (Say); Huds.
B. Terr. (M. C. Z.).
*univitta Walker, List, etc. IV, 1049. — New York.

Aloa Walker, List, etc. IV, 1053. — Jamaica.
erythrocephala Fabricius, System. Antl. 260, 1; Wiedemann, Auss.
Zw. II, 532, 1. — Brazil (Fabr.); Mexico (Walker, List, etc.
IV, 1055).
fasciata Fabricius, System. Ent. 781, 43 *(Musca)*; Ent. System. IV,
336, 102 *(id.)*; System. Antl. 262, 9; Wiedemann, Auss. Zw. II,
536, 7. — West Indies.
*maculosa Loew, Centur. VII, 88. — Cuba.
*placida Loew, Centur. VII, 90. — Cuba.

NB. *C. angulata* Loew, Centur. VII, 87 and *C. platycnema* Loew,
Centur. VII, 86, are from New Granada.

Observation. Mr. R. Desvoidy, Myod. 736—38 describes four species of ˥
genus *Neria,* which he identifies with *Nerius* Fabricius. One of these species,
which I believe to have recognized, is a *Calobata* (*C. geometra,* see above). It is very
probable, that the other three species likewise are *Calobatae* and have nothing to
do with the genus *Nerius* Fab., as defined by Wiedemann, Auss. Zw. II, 549:

> *Neria atripes* R. Desvoidy, ⎫
> „   *carolinensis* R. Desvoidy, ⎬ all from Carolina.
> „   *longipes* (Fab.), R. Desvoidy, ⎭

The descriptions are very short, and it seems probable, judging from them, that
all three apply to differently colored individuals of the same species.

Micropeza.

Meigen, Illiger's Magaz.; 1803. ([292]).

*producta Walker, List, etc. IV, 1056. — Georgia (Walk.); Cuba
(Loew, Berl. Z. 1868, 167).

divisa Wiedemann, Auss. Zw. II, 540 *(Calobata)*. — Mexico.
pectoralis Wiedemann, Auss. Zw. II, 540 *(Calobata)*. — Mexico.
[These two species are placed here in accordance with Mr. Loew's
statement in the Berl. Ent. Z. 1868, 393, 394.]

Lissa.

Meigen, System. Beschr. V, 370 (1826); this genus is provisionally placed in this family in accordance with Loew, Monogr. I, 39.

Lissa varipes Walker, List, etc. IV, 1046. — Ohio, is *Cordylura bimaculata* Loew. — The two other species, *L. carbonaria*
(New York), and *cornuta* (Huds. B. Terr.), both l. c. 1047, do
not seem to belong to Lissa at all.

FAMILY ORTALIDAE. [293].

SECTION I. PYRGOTINA.

Pyrgota.

Wiedemann, Auss. Zw. II, 581; 1830; Loew, Monogr. III, 72.

*filiola Loew, Zeitschr. f. Ges. Naturw. Dec. 1876, 332. — Texas.
Pyrgota debilis O. Sacken, Western Dipt. 343. — Kentucky.

fenestrata Macquart, Dipt. Exot. Suppl. 4, 281; Tab. XXVI, f. 1
 (*Oxycephala*). — North America [Macquart gives no locality, but
 says: „same locality as *Oxycephala fuscipennis*", which is Pyrgota
 undata]. [294].

pterophorina Gerstaecker, Stett. Ent. Z. XXI, 190; Tab. II, f. 6;
 Loew, Monogr. III, 81. — Carolina.

*undata Wiedemann, Auss. Zw. II, 581; Tab. X, f. 6; Macquart, Hist.
 Nat. Dipt. II, 423; Tab. XVIII, f. 23; Harris, Ins. Injur. to Veget.
 3ᵈ edit. 610 f. 268 (*Sphecomyia*); Gerstaecker, Stett. Ent. Z. XXI,
 188; Tab. II, f. 7 and 7ᵃ; Loew, Monogr. III, 77. — Not rare
 especially in the northern States, from Massachusetts to Kansas.
 (A specimen exactly like *P. undata* is labelled „Brazil" in the Vienna
 Museum. This occurrence requires confirmation, like that of
 Bittacomorpha clavipes, recorded from Brazil in the same Museum.)
 Myopa nigripennis Gray, Griffith's Animal Kingdom, Tab. 125, f. 5.
 Oxycephala fuscipennis Macquart, Dipt. Exot. II, 3, 198; Tab. XXVI,
 f. 6 [!]. — No locality. (Macq. 4ᵉ Suppl. 281, America.)

*valida Harris, Ins. Injur. to veget. 3ᵈ edit. 611 (*Sphecomyia*). — Nor-
 thern and Middle States. [295].
 Pyrgota millepunctata Loew, Neue Beitr. II, 22, 50; Monogr. III, 74.
 ?Oxycephala maculipennis Macquart, Dipt. Exot. Suppl. I, 210;
 Tab. XVIII, f. 12.

vespertilio Gerstaecker, Stett. Ent. Z. XXI, 189; Tab. II, f. 8; Loew,
 Monogr. III, 79. — Carolina.

Toxotrypana.

Gerstaecker, Stett. Ent. Z. XXI, 191; 1860.

curvicauda Gerstaecker, Stett. Ent. Z. XXI, 194; Tab. II, f. 9. —
 West Indies (Island St. Jean, in the small Antilles).

SECTION II. PLATYSTOMINA.

Amphicnephes.

Loew, Monogr. III, 83; 1873.

*pertusus Loew, Monogr. III. 84; Tab. VIII, f. 1. — Distr. Columbia;
 Connecticut; Carolina; Texas.

Himeroëssa.

Loew, Monogr. III, 85; 1873.

*pretiosa Loew, Monogr. III, 85; Tab. VIII, f. 2. — Cuba.

Rivellia.

R. Desvoidy, Myod. 729; 1830; Loew, Monogr. III, 44 and 87.

Boscii R. Desvoidy, Myod. 730, 3. — Carolina [compare Loew, Monogr. III, 93, Obs. 2].

*conjuncta Loew, Monogr. III, 88; Tab. VIII, f. 3. — Maryland.

*flavimana Loew, Monogr. III, 92; Tab. VIII, f. 7. — Nebraska.

(?) *Herina metallica* v. d. Wulp, Tijdschr. v. Ent. 2d Ser. II, 154; Tab. V, f. 10. — Wisconsin [Mr. Loew, in the Zeitschr. f. Ges. Naturw. XXXVI, 116 identified this species with *R. viridulans,* a synonymy, which he gives up in Monogr. Vol. III].

*micaus Loew, Monogr. III, 94. — Texas.

*pallida Loew, Monogr. III, 95; Tab. VIII, f. 8. — Distr. Columbia.

*quadrifasciata Macquart, Hist. Nat. Dipt. II, 433, 8 (*Herina*); Loew, Monogr. III, 90; Tab. VIII, f. 5. — Nebraska.

*variabilis Loew, Monogr. III, 91; Tab. VIII, f. 6. — Distr. Columbia (?).

*viridulans R. Desvoidy, Myod. 729, 2; Loew, Monogr. III, 88; Tab. VIII, f. 4. — New York, Georgia, Distr. Columbia.

Trypeta quadrifasciata (Harris), Walker, List, etc. IV, 993, f. 5 [Lw.].

Herina rufitarsis Macquart, Dipt. Exot. 5e Suppl., 123, 7; Tab. VII, f. 5 [Lw.].

Tephritis melliginis Fitch, First Report 65. — United States [Lw.].

NB. For *Ortalis Ortoeda* Walker, quoted by Mr. Loew among the synonyms, see note ([296]).

Stenopterina.

Loew, Monogr. III, 96; l. c 22; modified from *Senopterina* Macquart, Hist. Nat. Dipt. II, 453; 1835.

*caerulescens Loew, Monogr. III, 97. — Texas.

Herina splendens Macq. Suppl. I, 209. — Columbia. ([297]).

mexicana Macquart, Dipt. Exot. II, 3, 208; Tab. 29, f. 2 (*Herina*); compare also Loew, Monogr. III, 98, *Observation* 2, where this species is, by mistake called *metallica.* — Macquart's description is reproduced in Monogr. III, 199. — Mexico.

Myrmecomyia.

R. Desvoidy, Myod. 721; 1830; Loew, Monogr. III, 99.

*myrmecoides Loew, Wien. Ent. Monatschr. IV, 83 (*Cephalia*); Monogr. III, 100; Tab. VIII, f. 9. — Distr. Columbia.

SECTION III. CEPHALINA.

Tritoxa.

Loew, Monogr. III, 102; 1873.

*cuneata Loew, Monogr. III, 107; Tab. VIII, f. 11. — Nebraska.

*flexa Wiedemann, Auss. Zw. II, 483, 11 (*Trypeta*); Loew, Monogr. III, 102; Tab. VIII, f. 10. — Northern Red River; Illinois.

Trypeta arcuata Walker, Dipt. Saund. 383; Tab. VIII, f. 3 [Loew].
*incurva Loew, Monogr. III, 104; Tab. VIII, f. 12. — Illinois, Kansas,
Distr. Columbia, Texas.

Camptoneura.

Macquart, Dipt. Exot. II, 3, 200; 1843; Loew, Mon. III, 108.
*picta Fabricius Ent. System. IV, 355 (*Musca*); System. Antl. 330
(*Dictya*); Wiedemann, Auss. Zw. II, 489 (*Trypeta*); Macquart, Dipt.
Exot. II, 3, 201; Tab. 27, f. 4; Loew, Monogr. III, 109; Tab.
VIII, f. 13. — United States.
Tephritis conica Fabricius, System. Antl. 318, 10 [Lw.].
Delphinia thoracica R. Desvoidy, Myod. 720, 1 [Lw.].
Urophora nigriventris Macquart, Dipt. Exot. 5e Suppl. 124, 18. (298).

Diacrita.

Gerstaecker, Stett. Ent. Z. XXI, 195; 1860; Loew, Monogr III, 111.
*aemula Loew, Monogr. III, 114; Tab. VIII, f. 15. — California.
*costalis Gerstaecker, Stett. Ent. Z. XXI, 197; Tab. II, f. 10, and 10a;
Loew, Monogr. III, 111; Tab. VIII, f. 14. — Mexico (Oaxaca).
Carlottaemyia moerens Bigot, Bull. Soc. Ent. de France XXVI, 1877
[Synonymy by Mr. Bigot, l. c 1877, CXXXII].

Idana.

Loew, Monogr. III, 115; 1873.
*marginata Say, J. Acad. Phil. VI, 183, 2; Compl. Wr. II, 368 (*Ortalis*);
Loew, Monogr. III, 115; Tab. VIII, f. 16. — Virginia; Pennsylvania.

SECTION IV. ORTALINA.
Tetropismenus.
Loew, Zeitschr. f. Ges. Naturw. Dec. 1876, 333.
*hirtus Loew, l. c. — San Francisco.

Tetanops.
Fallen, Dipt. Suec. Ortalidae; 1820; Loew, Monogr. III, 119.
*integra Loew, Monogr. III, 121; Tab. VIII, f. 18. — Illinois.
*luridipennis Loew, Monogr. III, 119; Tab. VIII, f. 17. — Nebraska.

Tephronota.
Loew, Zeitschr. f. d. Ges. Naturw. 1868, 6; Monogr. III, 122; 1873.
*humilis Loew, Monogr. etc. III, 121; Tab. VIII, f. 24. — New York,
Virginia, Texas; Wisconsin (v. d. Wulp).
Herina ruficeps v. d. Wulp, Tijdschr. v. Ent. IX, 156; Tab. V, f. 11.
[Loew]. (299).
(?) *Trypeta Narytia* Walker, List, etc. IV, 1020 (ex parte). — Florida. (300).

Ceroxys.

Macquart, Hist. Nat. Dipt. II, 437; 1835; Loew, Monogr. III, 125.

* canus Loew, Monogr. III, 129; Tab. VIII, f. 22; Berl. Ent. Z. II,
   374 (Ortalis). — Yukon River, Alaska; Nebraska (the same or a
   very similar species occurs in Europe).
* obscuricornis Loew, Monogr. III, 126; Tab. VIII, f. 20. — Nebraska.
* ochricornis Loew, Monogr. III, 126; Tab. VIII, f. 21. — Northern
   Wisconsin River.
* similis Loew, Monogr. III, 127; Tab. VIII, f. 23. — Connecticut;
   Quebec, Canada (ressembles very much the european C. crassi-
   pennis).

Anacampta.

Loew, Zeitschr. f. d. Ges. Naturw. 1868, 7; Monogr. III, 129; 1873.

* latiuscula Loew. Monogr. III, 130; Tab. VIII, f. 19 — California.
* pyrrhocephala Loew, Zeitschr. f. Ges. Naturw. 1876, 335. — California.

SECTION V. PTEROCALLINA.

Pterocalla.

Rondani, Esame di varie specie d'insetti ditteri Braziliani; Torino, 1848;
        Loew, Monogr. III, 132. ([305]).

   strigula Loew, Monogr. III, 133; Tab. VIII, f. 30. — Georgia (type
   in the Berl. Museum).

Stictocephala.

Loew, Monogr. III, 134; 1873.

* cribellum Loew, Monogr. III, 134; Tab. VIII, f 26 — Nebraska.
* cribrum Loew, Monogr. III, 135; Tab. VIII, f. 25 — Middle States.
* corticalis (Fitch) Loew, Monogr. III, f. 136; Tab. VIII, f. 28. —
   New York.
* vau Say, J. Acad. Phil. VI, 184, 4; Compl. Wr. II, 369 (Ortalis);
   Loew, Monogr. III, 138; Tab. VIII, f. 29. — Atlantic States.

Callopistria.

Loew, Monogr. III, 140; 1873.

* annulipes Macquart, Dipt. Exot. 5e Suppl. 121 (Platystoma); Loew,
   Monogr. III, 141; Tab. VIII, f. 27. — Atlantic States.

Myennis.

R. Desvoidy, Myod. 717, 1830; Loew, Monogr. III, 142.

   scutellaris Wiedemann, Auss. Zw. II, 484 (Trypeta); Loew, Monogr.
   I, 92 Tab. II, f. 26, 27 (Trypeta?); Monogr. III, 143. — Mexico.

SECTION VI. ULIDINA.

Oedopa.

Loew, Berl. Ent. Z. 1867, 287; Monogr. III, 146.

*capito Loew, Berl. Ent. Z XI, 287; Tab. II, f. 2; Monogr. III, 146; Tab. IX, f. 1—3. — Nebraska.

Notogramma.

Loew, Berl. Ent. Z. 1867, 289; Monogr. III, 148.

*stigma Fabricius, Ent. System. Suppl. 563, 72 *(Musca)*; System. Antl. 303, 96 *(id)*; Wiedemann, Auss. Zw. II, 565, 1 *(Ulidia)*; Loew, Monogr. III, 148; Tab. IX, f. 5. — Cuba.
Notogramma cimiciformis Loew, Berl. Ent. Zeitschr. XI, 289; Tab. II, f. 3 [Loew].
Dacus obtusus Fabricius, System. Antl. 278, 30 [Loew].

Seoptera.

Seioptera, Kirby, Introd. to Ent. II, 305; 1817 (Letter XXIII); also Stephens, Catalogue (1829); defined for the first time and modified in *Seoptera* by Loew, Berl. Ent. Z. 1867, 295; also in Monogr. III, 151.
Myodina Rob. Desvoidy, Essai etc. 1830.

*colon (Harris) Loew, Berl. Ent. Z. XI, 296; Tab. II, f. 6; Monogr. III, 152; Tab. IX, f. 6. — Illinois.
*vibrans Linné, Meigen, etc. *(Ortalis)*. — Europe and the Eastern United States and Canada (Quebec). [The differences between the two species are explained by Loew in Monogr. III, 153; the occurrence of *S. vibrans* in N. A. is mentioned by O. Sacken in a note at the end of volume, immediately after the plates].

Acrosticta.

Loew, Berl. Ent. Z. 1867, 293; also Monogr. III, 151.

*dichroa Loew, Berl. Ent. Z. 1874, 384. — San Francisco.

Ulidia.

Meigen, System. Beschr. V, 385; 1826; compare Loew, Monogr. III, 63.

*rubida Loew, Zeitschr. f. Ges. Naturw. 1876, 337. — California.

Euxesta.

Loew, Berl. Ent. Z. 1867, 297; Monogr. III, 153. ([305]).

*nitidiventris Loew, Monogr. III, 157. — Texas.
*notata Wiedemann, Auss. Zw. II, 462, 9 *(Ortalis)*; Loew, Berl. Ent. Z. XI, 300; Tab. II, f. 9; Monogr. III, 156; Tab. IX, f. 9. — Atlantic States (New York, Illinois, etc.).
*scoriacea Loew, Zeitschr. f. Ges. Naturw. 1876, 336. — Texas.

*abdominalis Loew, Berl. Ent. Z. XI, 307; Tab. II, f. 15; Monogr. III, 164; Tab. IX, f. 15. — Cuba.

alternans Loew, Berl. Ent. Z. XI, 307; Tab. II, f. 16; Monogr. III, 165; Tab. IX, f. 16. — Brazil? Cuba?

* **annonae** Fabricius. Ent System. IV, 358, 189 *(Musca)*; System. Antl. 320, 19 *(Tephritis)*; Wiedemann, Auss Zw. II, 4ɛ3 *(Ortalis)*; Loew, Berl. Ent. Z. XI, 305; Tab. II, f. 13; Monogr. III, 162; Tab. IX, f. 13; compare also *Amethysa annonae* in Schiner, Novara, 2ᴧ3. — Cuba (South America, Schiner). (³⁰¹).
  Urophora quadrivittata, Macquart, Hist N. Dipt. II, 456 [Lw.].

* **binotata** Loew, Berl. Ent. Z. XI, 304; Tab. II, f. 12; Monogr. III, 160; Tab. IX, f. 12. — Cuba.

  costalis Fabricius, Ent. System. IV, 360, 196 *(Musca)*; Syst. Antl 278 *(Dacus)*; Wiedemann, Auss. Zw. II, 464 *(Ortalis)*; Loew, Berl. Ent. Z. XI, 301; Tab. II, f. 10; Monogr. III, 158; Tab. IX, f. 10. — West Indies.
  Dacus aculeatus Fabricius, System. Antl. 275 [Lw.].

* **eluta** Loew, Berl. Ent. Z. XI, 312; Tab. II, f. 19; Monogr. III, 168; Tab. IX, f. 18. — Cuba.

* **pusio** Loew, Loew, Berl. Ent. Z. XI, 299; Tab. IX, f. 8; Monogr. III, 155; Tab. IX, f. 8. — Cuba.

* **quaternaria** Loew, Berl. Ent. Z. XI, 302; Tab. II, f. 11; Monogr. III, 159; Tab. IX, f. 11. – Cuba.

* **spoliata** Loew, Berl. Ent. Z. XI, 298; Tab. II, f. 7; Monogr III, 154; Tab. IX f. 7.　　　Cuba.

* **stigmatias** Loew, Berl Ent. Z. XI, 310; Tab. II, f. 18; Monogr. III, 166; Tab. IX, f. 17. — Cuba; Brazil

* **Thomae** Loew, Berl. Ent Z. XI, 306; Tab. II, f. 14; Monogr. III, 163; Tab. VIII, f. 14. — St. Thomas.

Chaetopsis.

Loew, Berl. Ent. Z. XI, 315; 1867; Monogr. III, 169.

* **aenea** Wiedemann, Auss. Zw. II, 462 *(Ortalis)*; Loew, Berl. Ent. Z. XI, 315; Tab II, f. 21; Monogr. III, 170; Tab IX, f. 19. — Atlantic States; Canada; Cuba; the Bermudas.
  Ortalis trifasciata, Say, Journ. Acad. Phil. VI, 184; Compl. Wr. II, 368 [Lw.].
  Urophora fulvifrons Macquart, Dipt. Exot. 5e Suppl., 125; Tab. VII, f. 8 (Lw.)
  Trypeta Narytia Walker, List, etc. IV, 1020; synon. ex parte [!]. — Florida. (³⁰ᵃ).
  Ortalis Massyla Walker, List, etc. IV, 992; reproduced in Monogr. III, 199 [1]. — North America.
  Ortalis Ortoeda Walker, List, etc. IV, 992. — North America. (²⁹⁶).
  Trypeta (Aciura) aenea v. d. Wulp, Tijdschr. v. Ent. 2ᵈ Ser. II, 157; Tab. V, f. 12—14 [Lw].

* **debilis** Loew, Berl. Ent. Z XI, 318; Tab. II, f. 22; Monogr. III, 172; Tab. IX, f. 20. — Cuba.

Stenomyia.

Loew, Berl. Ent. Z. 1867, 320; Monogr. III, 173.

*tenuis Loew, Berl. Ent. Z. XI, 321; Tab. II, f. 24; Monogr. III, 174; Tab. IX, f. 21. — Georgia; Texas.

Eumetopia.

Macquart, Dipt. Exot. 2e Suppl. 87; 1847; Loew, Berl. Ent. Z. 1867, 322; Monogr. III, 175.

*rufipes Macquart, Dipt. Exot. 2e Suppl. 88; Tab. VI, f. 2; Loew, Berl. Ent. Z. XI, 322; Tab. II, f. 25; Monogr. III, 175; Tab. IX, f. 22. — Distr. Columbia; Texas.

*varipes Loew, Centur. VI, 87; Berl. Ent. Z. XI, 323; Tab. II, f. 26; Monogr. III, 176; Tab. IX, f. 23. — Cuba.

SECTION VII. RICHARDINA.

Coniceps.

Loew, Monogr. III, 177; 1873; compare also the same, Beschr. Europ. Dipt. III, 292.

*niger Loew, Monogr. III, 178. — Texas.

Stenomacra.

Loew, Monogr. III, 180; 1873.

*Guerini Bigot, in R. de la Sagra, etc. 322; Tab. XX. f. 9 *(Sepsis)*; Loew, Monogr., etc. III, 180; Tab. IX, f. 25. — Cuba.

Neoidiotypa.

Idiotypa Loew, Monogr. III, 183; 1873. ([302]).

*appendiculata Loew, Monogr. III, 183; Tab. IX, f. 26. — Cuba.

Steneretma.

Loew, Monogr. III, 186; 1873.

*laticauda Loew, Monogr. III, 187. — Texas.

Coelometopia.

Coilometopia Macquart, Dipt. Exot. 2e Suppl. 91, 1847; Loew, Monogr. III, 188.

bimaculata Loew, Monogr. III, 189. — Cuba.

Observation. *Hemixantha spinipes* Loew and *Melanoloma affinis* Loew, described in Monogr. III, 190—193, are from Brazil.

Epiplatea.

Loew, Berl. Ent. Z. 1867, 324; Monogr. III, 194.

*erosa Loew, Berl. Ent. Z. XI, 325; Tab. II, f. 25; Monogr. III, 194; Tab. IX, f. 24. — Cuba.

Ortalide described by brevious writers, but not known to Mr. Loew, when he repared his Monograph. The descriptions are reproduced in the Appendix to Monographs etc. Vol. III, 197 —203, and discussed by Mr. Loew (except *Ortalis platystoma* Thomson, which was added by me). I reproduce Dr. Loew's comments, together with my remarks on the original types seen by me.

Ortalis ligata Say, J. Acad. Phil. VI, 83; Compl. Wr. II, 368. — Mexico. [Probably *Rivellia*. — Lw·].

Meckelia philadelphica R. Desvoidy, Myod. 715. — Philadelphia. [Probably *Ceroxys or Anacampta*. — Lw.].

Ortalis basalis Walker, Dipt. Saund., 373. — United States. [Not *Ceroxys*, as Walker suggests; perhaps an Ulidina. — Lw.] I could not find it in the Brit. Mus.; the *Ortalis basalis* which I saw there, is from Tasmania.

Ortalis Massyla Walker, List, etc. IV, 992. — North America. [Seems to be an Euxesta. — Lw.]. I took it for *Chaetopsis aenea*.

Ortalis (?) diopsides Walker, List, etc. IV, 995. — Huds. B. Terr. [Belongs perhaps to the Ulidina. — Lw.].

Ortalis (?) costalis Walker, List, etc. IV, 995. — Huds. B. Terr. [Probably likewise an Ulidina? — Lw.]. Represented in the Brit. Mus. by a fragment without a head, and with only one wing; looks like *Sepsis*.

Ortalis bipars Walker, Trans. Ent. Soc. N. Ser. V, 326. — United States. (I could not find it in the Brit. Mus.).

Bricinnia flexivitta Walker, Trans. Ent. Soc. N. Ser. V, 324. — Mexico. (I did not find this species in the Brit. Mus.).

Urophora interrupta Macquart, Hist. Nat. Dipt. II, 459. — North America. [Is a *Rivellia* of difficult interpretation. — Lw., Monogr. III, 337, 32.]

Urophora antillarum Macquart, Dipt. Exot. 4e Suppl. Tab. XXVI, f. 17. — West Indies. [Almost undoubtedly an Ulidina. — Lw.]. The typical specimen in Mr. Bigots collection is an exceedingly soiled, hardly recognizable specimen, but looks very much like an *Euxesta*.

Ulidia fulvifrons Bigot in R. de la Sagra, etc. 826. — Cuba. [Not an *Ulidia;* may belong to the Ulidina. — *Ulidia metallica*, described in the same place belongs to the *Agromyzidae*. — Lw.]. I have not seen the specimen in Mr. Bigots collection.

Ortalis platystoma Thomson, Eugen. Resa etc. 572. — Panama.

FAMILY TRYPETIDAE. [303].

Trypeta.
Meigen, in Illiger's Magaz. II, 1803.

Subgenus Hexachaeta.
Loew, Monogr. III, 219; Observ. 2; 1873.

*eximia Wiedemann, Auss. Zw. II, 477; Loew, Monogr. etc III, 216. — Brazil; Mexico.

Tephritis fasciventris Macquart, Dipt. Exot. 4e Suppl. 291; Tab. XXVII, f. 3 [Lw.].

Subgenus Acrotoxa.
Loew, Monogr. III, 227—231; 1873.

Anastrepha, Schiner, Novara etc. 263, 1868. ([304]).
Leptoxyda, Macquart, Hist. Nat. Dipt. II, 452, 1835.
Leptoxys, Macquart, Dipt. Exot. II, 3, 216.
amabilis Loew, Monogr. III, 219. — Mexico.
fraterculus Wiedemann, Auss. Zw. II, 524 *(Dacus);* Loew, Monogr. III, 222; Tab. X, f. 6. — Cuba, Brazil, New Granada, Peru.
Trypeta unicolor Loew, Monogr. I, 70; Tab. II, f. 6 [Lw.].
ludens Loew, Monogr. III, 223; Tab. XI, f. 19. — Mexico.
suspensa Loew, Monogr. I, 69; Tab. II, f. 5; ibid. III, 219; Tab. X, f. 5. — Cuba (Loew); South America, Schiner, Novara etc. 263.
tricincta Loew, Monogr. III, 225. -- Hayti.

Observation. *Trypeta obliqua* Macquart, *Ocresia* Walker, and perhaps *Acidusa* Walker, all from North America, belong to the present subgenus (for the full quotations, see at the end of the genus *Trypeta).*
Five brazilian Acrotoxae are described and figured by Mr. Loew in Monogr. III, 229—230; Tab. XI, f. 20—24: *parallela, hamata, integra, consobrina, pseudo-parallela.*

Subgenus Stenopa.
Loew, Monogr. III, 234; 1873.

vulnerata Loew, Monogr. III, 232. — Massachusetts.

Subgenus Acidia
R. Desvoidy, Myod. 720; 1830; Loew, Europ. Bohrfliegen, 34; 1862.

fratria Loew, Monogr. I, 67; Tab. II, f. 4; Monogr. III, 235; Tab. X, f. 4. — Atlantic States.
(?) *Trypeta liogaster* Thomson, Eugen. Resa, 578, 251. — California [Lw.].
fausta O. Sacken, Western Diptera, 346. — Alpine Region of Mt. Washington, N. H.
suavis Loew, Monogr. I, 75; Tab. II, f. 10; ibid. III, 235; Tab. X, f. 10. — Middle States.

Subgenus Epochra.
Loew, Monogr. III, 238; Observ., 1873.

canadensis Loew, Monogr. III, 235. — Canada, Maine.

ᐧ Subgenus Straussia.
Strauzia Rob. Desvoidy, Myod. 718; 1830; Loew, Monogr. III, 243.

longipennis Wiedemann, Auss. Zw. II, 483; Loew, Monogr. I, 65; ibid. III, 238; Tab. X, f. 2, 3. — Atlantic States; Colorado ,O. Sacken Western Dipt. 345).

Strauzia armata R. Desvoidy, Myod. 719, 2 (♂). [Lw.].
Strauzia inermis R. Desvoidy, Myod. 718, 1 (♀). |Lw].
Tephritis trimaculata Macquart, Dipt. Exot. II, 3, 226, 8; Tab.
　XXXI, f. 3. [Lw.].
Trypeta cornigera Walker, List, etc. IV, 1010. [Lw.].
Trypeta cornifera Walker, List, etc. IV, 1011. [Lw.].

Subgenus Zonosema.
Loew, Europ. Bohrfliegen; 1862.

*basiolum O. Sacken, Western Diptera, 348. — Massachusetts.

Subgenus Spilographa.
Loew, Europ. Bohrfliegen, 39; 1862.

*electa Say, Journ. Acad. Phil. VI, 185, 1; Compl. Wr. II, 369; Loew,
　Monogr. I, 71, 6; Tab. II, f. 7; Monogr. III, 244; Tab. X, f. 7.
　— Florida, Kansas.
*flavonotata Macquart, Dipt. Exot. 5e Suppl. 125; Tab. VII, f. 9
　(*Tephritis*); Loew, Monogr. III, 245. — Baltimore (Macq.);
　Yukon River, Alaska (Lw).

Subgenus Oedicarena.
Loew, Monogr. III, 247; Observ.; 1873.

tetanops Loew, Monogr. III, 245; Tab. XI, f. 15. — Mexico.
*persuasa O. Sacken, Western Diptera, 344. — Colorado.

Subgenus Peronyma.
Loew, Monogr. III, 250; Observ. 2; 1873. (³⁰⁵).

sarcinata Loew, Centur. II, 73; Monogr. III, 247; Tab. XI, f. 16. —
　South Carolina.
　(?) *Tephritis quadrifasciata* Macquart, Dipt. Exot. II, 3, 226; Tab.
　XXX, f. 8. — Georgia [Lw.].

Subgenus Plagiotoma.
Loew, Monogr. III, 252; Observ. 2; 1873.

*obliqua Say, J. Acad. Phil. VI, 186, 3; Compl. Wr. II, 370; Loew,
　Monogr. I, 99 and III, 251; Tab. XI, f. 14. — Pennsylvania;
　Indiana; Texas; Schiner, Novara, etc. 267, has it from Brazil.

*discolor Loew, Monogr. I, 64; Tab. II, f. 1; ibid. III, 250; Tab. X,
　f. 1. — Cuba.

Observation. *Plagiotoma biseriata*, a brazilian species, is
described by Mr. Loew in Monogr. III, 252.

Subgenus Trypeta.
Loew, Europ. Bohrfliegen, 51; 1862.

*palposa Loew, Monogr. I, 74; Tab. II, f. 9; Monogr. III, 253; Tab.

X, f. 9. — Northern Wisconsin River (Lw.); compare O. Sacken, Western Diptera, about the specimens from Colorado.

*florescentiae Linné, Meigen, etc.; Loew, Monogr. III, 254. — Europe and North America (Canada).

Subgenus Oedaspis.

Loew, Europ Bohrfliegen, 46; 1862.

*atra Loew, Centur. II, 74; Monogr. III. 256; Tab. XI, f. 17. — New York; Mexico.
*gibba Loew, Monogr. III, 260. — Texas.
*penelope O. Sacken, Western Diptera, 346. — Western New York.
*polita Loew, Monogr. I, 77; Tab. II, f. 12; ibid. III, 257; Tab. X, f. 12. — Washington; New York; Connecticut; Mississippi.

Observation. *Oedaspis nigerrima* Loew, from Brazil, is described in Monogr. III, 258; Tab. XI, f. 18.

Subgenus Rhagoletis.

Loew, Europ. Bohrfliegen, 44; 1862; compare also Monogr. III, 267.

*cingulata Loew, Monogr. I, 76; Tab II, f. 11; Monogr. III, 263; Tab. X, f. 11. — Middle States; Long Branch, N. J.
*pomonella Walsh, First Rep. Illin. etc. 29—33; fig. 2. (This description is reproduced in the article: The apple-worm and apple-maggot, in the Amer. Journ. of horticulture, Boston, Dec. 1867.) Loew, Monogr. III, 265. — Illinois.
*tabellaria Fitch, First Rep. 66; Loew, Monogr. III, 263. — New York; Canada.

Subgenus Aciura.

Rob. Desvoidy, Myod. 773; 1830; Loew, Europ. Bohrfliegen, 29; 1862.

*insecta Loew, Monogr. I, 72; Tab. II. f. 8; Monogr. III, 268; Tab. X, f. 8. — Cuba; (Florida?); Schiner, Novara etc. 265 has the same species from South America.

Observation. *Aciura phoenicura* Loew, from Brazil is described Monogr. III, 269; Tab. XI, f 12.

Subgenus Blepharoneura.

Loew, Monogr. III, 271; Observ ; 1837.
*poecilogastra Loew, Monogr. III, 270. — Cuba.

Subgenus Acrotaenia.

Loew, Monogr. III, 274; Observ.; 1873.
testudinea Loew, Monogr. III, 272; Tab. XI, f. 13. — Cuba.

Subgenus Eutreta.

Loew, Monogr. etc. III, 275; Observ. 3; 1873. Syn. *Icaria* Schiner, Novara, 267 (1868). ([366]).

* Diana O. Sacken, Western Diptera, 347. — Missouri.

*rotundipennis Loew, Monogr. I, 79; Tab. II, f. 14; ibid. III, 276, Tab. X, f. 14. — Middle States.

*sparsa Wiedemann, Auss. Zw. II, 492; Loew, Monogr. I, 78; Tab. II, f. 13; ibid. III, 274; Tab. X, f. 13. — United States (including Texas, Colorado, California) and Canada.

Trypeta caliptera Say, Journ. Acad. Phil. VI, 187, 3; Compl. Wr. II, 370. [Lw.].

Platystoma latipennis Macquart, Dipt. Exot. II, 3, 200; Tab. XXVI, f. 8. [Lw.]

Acinia novaeboracensis Fitch, First Rep. 67. [Lw.].

Subgenus Carphotricha.

Loew, Europ. Bohrfliegen, 77, 1862; compare also Monogr. III, 279.

*culta Wiedemann, Auss. Zw. II, 486; Loew, Monogr. I, 94; Tab. II, f. 29; ibid. III, 276; Tab. XI, f 3. — Savannah; Carolina, Texas, Kansas.

Acinia fimbriata Macquart, Dipt. Exot. II, 3, 228, 5; Tab. XXXI, f. 5. [Lw.].

Subgenus Eurosta.

Loew, Monogr. III, 280; Observ. 5; 1873. ([305]).

*comma Wiedemann, Auss. Zw. II, 478; Loew, Monogr. I, 93; Tab. II, 28; ibid. III, 280; Tab. XI, f. 2; Macquart, Dipt. Exot. II, 3, 229 (*Acinia*). — Kentucky; Maryland; Massachusetts.

*latifrons Loew, Monogr. I, 89; Tab. II, f. 22; ibid. III, 283; Tab. X, f. 22. — Connecticut, Wisconsin, Carolina, Detroit, Mich., White Mts., N. H.

Trypeta cribrata v. d. Wulp, Tijdschr. v. Ent. 2d Ser. Vol. II, 158; Tab. V, f. 15 [Lw.].

*solidaginis Fitch, First Rep. 66 (*Acinia*); Loew, Monogr. I, 82; Tab. II, 16; ibid. III, 279; Tab. X, f. 16. — Atlantic States and Canada.

Tephritis asteris Harris, Ins. Injur. to veget 3d edit 620. [Lw.].

Subgenus Acidogona.

Loew, Monogr. III, 285; Observ.; 1873.

*melanura Loew, Monogr. III, 283; Tab. XI, f. 6. — Distr. Columbia.

Subgenus Neaspilota.

Aspilota Loew, Monogr. III, 286; Observ.; 1873. ([307]).

*alba Loew, Centur. I, 72; Monogr. I, 100; ibid. III, 285; Tab. XI, f. 11. — Pennsylvania; Missouri; Colorado. ([30r]).

*albidipennis Loew, Centur. I, 73; Monogr. I, 100; ibid. III, 286; Tab. XI, f. 10. — Pennsylvania.

*vernoniae Loew, Centur. I, 74; Monogr. I, 101; ibid. III, 286; Tab. XI, f. 8. — Pennsylvania

Subgenus Icterica.

Loew, Monogr. III, 287; Observ.; 1873.

* **circinata** Loew, Monogr. III, 288. — New York.
* **seriata** Loew, Monogr. I, 84; Tab. II, f. 18; ibid. III, 287, Tab. X, f. 18. — Illinois; Detroit, Michigan; Massachusetts.

Lichtensteinii Wiedemann, Auss Zw. II, 497; Loew, Monogr. etc. I, 92; Tab. II, f. 25; ibid. III, 289; Tab. XI, f. 9. — Mexico.

Subgenus Ensina.

Rob. Desvoidy, Myod. 751; 1830; Loew, Europ. Bohrfliegen, 64; compare also Monogr. III, 291; Observ. 2.

* **humilis** Loew, Monogr. I, 81; Tab. II, f. 17; ibid. III, 291; Tab. X, f. 17. — Cuba; Key-West, Florida; the Bermudas. (I have seen specimens from Colorado, apparently belonging here; Western Diptera, 345.)

Acinia picciola Bigot, R. de la Sagra etc. 824; Tab. XX, f. 10 [Lw.].

> **Observation.** *Ensina peregrina* Loew, from Brazil, is described in Monogr. III, 292, Tab. X, f. 30.
> *Trypeta aurifera* Thomson, California, is an *Ensina;* compare below, at the end of the genus Trypeta.

Subgenus Tephritis.

Latreille, Hist. Nat. des Crust. et des Ins. XIV, 389, 1804; compare also Loew, Europ. Bohrfliegen 96 and Monogr. III, 295.

* **angustipennis** Loew, Germ. Zeitschr. V, 382; Tab. II, f. 4; id. Eur Bohrfl. 113, Nr. 24; Monogr. III, 293 where the rest of the synonymy may be found). — Europe (Scandinavia) and North America (Yukon River, Alaska).
* **albiceps** Loew, Monogr. III, 302; Tab. XI, f. 5. — Canada; Maine.
* **clathrata** Loew, Monogr. I, 80; Tab. II, f. 15; ibid. III, 297; Tab. X, f. 5. — Middle States.
* **euryptera** Loew, Monogr. III, 304. — West Point, N. Y.
* **finalis** Loew, Centur. II, 78; Monogr. III, 296; Tab. XI, f. 4. — Texas; California.

geminata Loew, Centur. II, 75; Monogr. III, 298; Tab. XI, f. 1. — Pennsylvania.
* **platyptera** Loew, Monogr. III, 306. — Connecticut.

fucata Fabricius, Ent. System. IV, 359, 194 *(Musca);* System. Antl. 321, 24 *(Tephritis);* Wiedemann, Auss. Zw. II, 505; Loew, Monogr. III, 301. — West Indies? (Fabr.); South America (Wied.)

> **Observation.** *Trypeta acutangula* and *genalis* Thomson, from California, probably belong to the subgenus *Tephritis;* compare below, the end of the genus Trypeta.

16

Subgenus Euaresta.

Loew, Monogr. III, 295; also 308; Observ.; 1873.

*aequalis Loew, Monogr. I, 86, Tab. II, f. 20; ibid. III, 308; Tab. X,
f. 20. — Illinois, Ohio, Maryland (about the specimens from
Colorado, compare O. Sacken, Western Dipt, 345)
*bella (Fitch) Loew, Monogr. I, 88; Tab. II, f. 23; ibid. III, 311;
Tab. X, f. 23. — Atlantic States.
*festiva Loew, Monogr. I, 86; Tab. II, f. 21; ibid. III, 309; Tab. X,
f. 21. — Pennsylvania; Connecticut; Illinois; Ohio, Quebec,
Canada.
*mexicana Wiedemann, Auss. Zw. II, 551; Loew, Monogr. III, 317;
Tab. X, f. 28. - Texas; Mexico.
*pura Loew, Monogr. III, 320. — Massachusetts.

*melanogastra Loew, Monogr. I, 90; Tab. II, f. 24; ibid. III, 315;
Tab. X, f. 24. — Cuba.
timida Loew, Centur. II, 76; Monogr. III, 312; Tab. X, f. 25. —
Mexico.

Observation. *Euaresta spectabilis, obscuriventris, tenuis* Loew,
from Brazil, are described in Monogr. III, 309, 313, 316; Tab.
X, f. 27, 26, 29.

Subgenus Urellia.

R. Desvoidy, Myod. 774; 1830; Loew, Europ. Bohrfliegen, 117.

*abstersa Loew, Centur. II, 77; Monogr. III, 323; Tab. XI, f. 7. —
North America; Cuba.
*actinobola Loew, Monogr. III, 326. — Texas.
*solaris Loew, Monogr. I, 84; Tab. II, f. 19; ibid. III, 325; Tab. X,
f. 19. — Georgia (about the specimens from California, compare
O. Sacken, Western Dipt., 345).

*polyclona Loew, Monogr. III, 324. — Cuba.

Observation. *Trypeta Mevarna* Walker, Florida, and *Trypeta
femoralis* Thomson, California, are *Urelliae* (compare below).

The following species of Trypeta, described by earlier authors,
have not been identified by Mr. Loew; they are discussed in
Monogr. III, 325—338, and the descriptions are reproduced in
the Appendix to Vol. I, and Appendix II, to Vol. III. I reproduce
here the comments of Dr. Loew (as published, l. c.), with my
remarks on some of them, based on the examination of the
specimens in the Brit. Museum.
Acidusa Walker, List, etc. IV, 1014. — Florida [probably *Acrotoxa.* —
Lw.].
acutangula Thomson, Eugen Resa 583. — California [probably
Tephritis. — Lw.].

aurifera Thomson, Eugen. Resa, 585. — California [Subgenus *Ensina* — Lw.].

Avala Walker, List, etc IV, 1020 *(Urophora)*. — Jamaica. [Doubtful whether it belongs to Trypetidae or Ortalidae. — Lw.]. It is a small Ortalid.

Beauvoisii R. Desvoidy, Myod. 760 *(Prionella)*. — North America (?) [Same remark as the preceding species. — Lw.].

Dinia Walker, List, etc. IV, 1040 — Jamaica. [Perhaps allied to *Trypeta* (Hexachaeta eximia Wiedemann, or perhaps a bad description of a variety of this species. — Lw.].

femoralis Thomson, Eugen. Resa, 585. — California *[Urellia.* — Lw.].

genalis Thomson, Eugen. Resa, 585. — California. [Probably *Tephritis.* — Lw.].

marginepunctata Macquart, Hist. Nat. Dipt. II, 464. — Philadelphia. [Almost certainly a Trypetid; but it would be premature to identify it with *Carphotricha culta.* — Lw.].

Mevarna Walker, List, etc. IV, 1023. — Florida. *[Urellia.* — Lw.]. The specimen in the Brit. Mus. seems very like *T. solaris.*

Narytia Walker, List, etc. IV, 1020. — Florida; see my note[300].

obliqua Macquart, Hist. Nat. Dipt. II, 464, 14; Dipt. Exot. II, 3, 225, 6; Tab. XXX, f. 11 *(Tephritis)*. — Cuba. *[Acrotoxa.* — Lw.]. I saw the type in the Jardin des Plantes.

Ocresia Walker, List, etc. IV, 1016. — Jamaica. *[Acrotoxa.* — Lw.]. Yes!

scutellata Wiedemann, Auss. Zw. II, 494, 27. — Mexico. [A Trypetid of doubtful position. — Lw.].

villosa R. Desvoidy, Myod. 760, 2 *(Prionella)*. — United States. [Same remark as about *Avala.* — Lw.].

Macquart, Dipt. Exot. II, 3, 221 says that the european *Urophora quadrivittata* also occurs in Cuba. He can only mean *Urophora quadrifasciata* Meigen, and Schiner likewise understands it so, (compare his Dipt. Austriaca, Trypetidae, in the Verh. Zool. Bot. Ges. 1858, p. 657).

FAMILY LONCHAEIDAE. [309]

Palloptera.
Fallen, Ortalidae; 1820.

*Jucunda Loew, Centur. III, 55. — Sitka.
*superba Loew, Centur. I, 75. — Pennsylvania; Quebec, Canada.
*terminalis Loew, Centur. III, 54. — Sitka.

Lonchaea.
Fallen, Ortalidae; 1820.

caerulea Walker, List, etc. IV, 1004. — Georgia.

polita Say, J. Acad. Phil. VI, 183; Compl, Wr. II, 371. — Indiana, Massachusetts (Harr. Cat).

*rufitarsis Macquart, Dipt. Exot. 4e Suppl. 300, 3; Tab. XXVIII, f.
2. — North America. [The *L. tarsata* Fallen of Walker's List,
etc. IV, 1004, is probably this species.]

discrepans Walker, Trans. Ent. Soc. N. Ser. V, 322. — Mexico.
glaberrima Wiedemann, Auss. Zw. II, 475, 1. — West Indies.
nigra Wiedemann, Auss. Zw. II, 476, 3; Bigot, in R. de la Sagra etc.
827. — Brazil (Wied.); Cuba (Bigot).

FAMILY SAPROMYZIDAE.

Sapromyza. [310].

Fallen, Ortalidae; 1820.

Amida Walker, List, etc. IV, 988. — Georgia.
*bispina Loew, Centur. I, 79. — Nebraska.
*compedita Loew, Centur. I, 76. — Pennsylvania.
connexa Say, J. Acad. Phil. VI, 177, 1; Compl. Wr. II, 367. — Indiana.
*decora Loew, Centur. V, 96. — Lake George, New York; Quebec, Can.
*fraterna Loew, Centur. I, 77. — Pennsylvania
*lupulina Fabricius, Meigen, System. Beschr. V, 301 (*Lauxania*). —
Europe and North America (see Loew, Sillim. Journ. XXXVII, 318).
longipennis Meigen, System. Beschr. V, 300 (*Lauxania*). — Europe
and North America (according to v. d. Wulp, l. c.).
*macula Loew, Centur X, 82. — Texas.
notata Fallen; Loew, Dipt. Beitr. III, 40. — Europe and North
America (according to v. d. Wulp, l. c.).
*philadelphica Macquart, Dipt. Exot. II, 3, 191, 13. — Atlantic States.
*quadrilineata Loew, Centur. I, 78 — Pennsylvania.
resinosa Wiedemann, Auss. Zw. II, 456, 14. — Georgia.
*rotundicornis Loew, Centur. III, 56. — Sitka.
*stictica Loew, Centur. III, 58 — Distr. Columbia; Texas.
*tenuispina Loew, Centur. I, 80. — Nebraska.
*umbrosa Loew, Centur. III, 57. — Distr. Columbia.
*vulgaris Fitch, Reports, Vol. I, 800; Tab. I, f. 4 (*Chlorops*). —
Atlantic States.
Sapromyza plumata v. d. Wulp, Tijdschr. v. Ent. 2d Ser. 159. [311].

apta Walker, Trans. Ent. Soc. N. Ser. V, 321. — Mexico.
bipunctata Say, J. Acad. Phil. VI, 178, 2; Compl. Wr. II, 367. —
Mexico.
*cincta Loew, Centur. I, 81. — Cuba.
octopunctata Wiedemann, Auss. Zw. II, 454, 9. — West Indies.
sordida Wiedemann, Auss. Zw. II, 456, 12. — West Indies.

Pachycerina.

.Macquart, Hist. Nat. Dipt. II, 511; 1835.

*verticalis Loew, Centur. I, 82. — Florida.

Lauxania.

Latreille, Hist. Nat. des Crust. et des Ins. XIV, 390; 1804.

*cylindricornis Fabricius, Meigen, etc. — Europe and North America [Loew, Sillim. Journ. N. Ser. XXXVII, 318].

Elisae Meigen, System. Beschr. V, 297. — Europe and North America [Nova Scotia, Walker, List, etc. IV, 1003].

*encephala Loew, Centur. X, 83. — Texas.

*femoralis Loew, Centur. I, 89. — Georgia.

*frontalis Loew, Wien. Ent. Monatschr. II, 14. — Europe and North America (see Loew, Sillim. Journ, l. c. 318).

*flaviceps Loew, Centur. VII, 91. — Distr. Columbia.

*gracilipes Loew. Centur. I, 85. — Pennsylvania.

*manuleata Loew, Centur. I, 88. — Pennsylvania.

*opaca Loew, Centur. I, 84. — Florida.

*obscura Loew, Centur. I, 86. -- Atlantic States and Brit. America.

*trivittata Loew, Centur. I, 90. — Georgia.

nasalis Thomson, Eugen. Resa, 568. — California.

planiscuta Thomson, Eugen. Resa, 568. — California.

quatrisetosa Thomson, Eugen. Resa, 569. -- California.

*albovittata Loew, Centur. II, 79. — Cuba.

argyrostoma Wiedemann, Auss. Zw. II, 471, 3. — West Indies (South America, Schiner, Novara, 282).

*muscaria Loew, Centur. II, 87. — Cuba (South America, Schiner, Novara, 282).

*variegata Loew, Centur. II, 83. — Cuba (occurs als in South America, according to Schiner, Novara, 277, who places it in the genus *Physegenua*. Macq. (Dipt. Exot. 3e Suppl. 60), and has a long note on the subject.)

FAMILY PHYCODROMIDAE.

Coelopa.

Meigen, System. Beschr. VI, 194; 1830. [312].

*frigida Fallen, Hydrom. 6, 1. — Europe and North America (common on sea-beaches).

*nitidula Zetterstedt, Dipt. Scand. VI, 2173, 2; Stenhammar, Copromyz. 6. — Europe and North America.

.FAMILY HETERONEURIDAE.

Heteroneura.

Fallen, Agromyz.; 1823. [313].

*albimana Meigen, System. Beschr. VI, 128. — Europe and North America Loew, Sillim. J. XXXVII, 318).

* melanostoma Loew, Centur. V, 97. — White Mts., New Hampshire.
* latifrons Loew, Wien. Ent. Monatschr. IV, 82, 8; Centur. IV, 93. — Distr. Columbia.
* spectabilis Loew, Wien. Ent. Monatschr. IV, 82, 7; Centur. IV, 92. — Distr. Columbia.

Anthophilina.
Zetterstedt, Ins. Lapp. 785; 1840. ([314]).

* tenuis Loew, Centur. IV, 95. — Sitka.
* terminalis Loew, Centur. IV, 94. — White Mts., N. H. (erroneously „Carolina" in the Centuries).
* variegata Loew, Centur. IV, 96. — Distr. Columbia.

Ischnomyia.
Loew, Centur. IV, 97; 1863.
* vittata Loew, Centur. IV, 97. — Pennsylvania.

Trigonometopus.
Macquart, Hist. Nat. Dipt. II, 419; 1835.

* vittatus Loew, Centur. VIII, 98 (compare also Centur. Vol. II, 290 line 18 from the bottom, about the systematic location of this species). — Georgia.

FAMILY OPOMYZIDAE.
Balioptera.
Loew, Berl. Ent. Zeitschr. VIII, 347—356; 1864.
* lurida Loew, Centur. V, 98 *(Opomyza)*; Berl. Ent. Zeitschr. VIII, 356, where the species is referred to Balioptera. — Sitka.

Opomyza.
Fallen, Opomyzidae, 10; 1820. ([315]).
signicosta Walker, Trans. Ent. Soc. N. S. V, 320. — United States.

Scyphella.
R. Desvoidy, Myod. 650; 1850.

* flava Linné, Fallen, Dipt. Suec. Ortalid. 33. — Europe and North America (New York, on windows; see also Loew, Sillim. Journ. XXXVII, 318).

FAMILY SEPSIDAE. ([316]).
Sepsis.
Fallen, Ortalidae, 20; 1820.
referens Walker, List, etc. IV, 999. — North America.
similis Macquart, Dipt. Exot. 4e Suppl. 296, 4; Tab. XXVII, f. 11. — North America.

vicaria Walker, List, etc. IV, 998. — Florida.

discolor Bigot, in R. de la Sagra etc. 823. — Cuba.
*scabra Loew, Wien. Ent. Monatschr. V, 42. — Cuba.

ecalcarata Thomson, Eugen. Resa etc. 588. — California.

> Observation. For *Sepsis Guerinii* Bigot, see *Stenomacra Guerinii.*

Nemopoda.
Rob. Desvoidy, Myod. 743; 1830.

*cylindrica Fabricius; Meigen, System. Beschr. V, 290. — Europe and North America. [Harris's Catal. The species commonly found in New England seems to belong here.]
caeruleifrons Macquart, Dipt. Exot. 2e Suppl. 94. — Philadelphia.
minuta Wiedemann, Auss. Zw. II, 468, 4 *(Sepsis).* — New York. [Placed in *Nemopoda* by Loew *in litt.*]

FAMILY PIOPHILIDAE.
Mycetaulus.
Loew, Dipterol. Beitr. I, 37; 1845.

*longipennis Loew, Centur. IX, 100. — Huds. B. Terr.

Piophila.
Fallen, Heterom., 8; 1820.

*casei Linné, Meigen, System. Beschr. V, 395; Staeger, Groenl. Antl. 369. — Europe and North America (see Loew, in Sillim. Journ. XXXVII, 318).
nigriceps Meigen, System. Beschr. V, 397. — Europe and North America (see Loew, in Sillim. Journ. l. c.).
nigriceps Macquart, Dipt. Exot. 4e Suppl. 303; Tab. XXVIII, f. 6. — North America.
nitida v. d. Wulp, Tijdschr. v. Ent. 2d Ser. II, 160; Tab. V, f. 16—18. — Wisconsin.
petasionis L. Dufour, Ann. des Sc. Nat. 1844, 369. — Europe and North America (see Loew, in Sillim. Journ., l. c.).
pilosa Staeger, Groenl. Antl. 368, 52; Zetterstedt, Dipt. Scand. VI, 2514; Holmgren, Ins. Nordgroenl. 104. — Greenland.

concolor Thomson, Eugen. Resa, 596. — California.

Prochyliza.
Walker, List, etc. IV, 1045; 1849.

*xanthostoma Walker, List, etc. IV, 1045. — Huds. B. Terr. (Walk.); Distr. Columbia (O. S.).

Madiza.

Fallen, Oscinidae; 1820.

annulitarsis Zetterstedt, has been received from Wisconsin, according to Mr. v. d. Wulp, Tijdschr. N. S. IV, 80.

FAMILY DIOPSIDAE.

Sphyracephala.

Say, Amer. Entom. III, Tab. 52; 1828.

*brevicornis Say, J. Acad. Phil. I, 23; Compl. Wr. II, 3 *(Diopsis)*; Amer. Entom. III, Tab. 52; Compl. Wr. I, 116; Wiedemann, Auss. Zw. II, 563 *(Diopsis)*; id Achias etc. Tab. II, f. 3 *(id.)*; Gray, in Griffith's Anim. Kingd., Ins. etc. 774, Tab. 62, f. 2; Westwood, Trans. Linn. Soc Vol. XVII, 311, Tab. IX, f. 20 (copied from Say); Macquart, Hist. Nat. Dipt. II, 486 *(Diopsis)*; Loew, Zeitschr. f. Ges. Naturw. XLII, 101. — Atlantic States.

Sphyracephala subbifasciata Fitch, Reports, Vol, I, 70 [Loew l. c.].

FAMILY EPHYDRIDAE. [317]

SECTION I. NOTIPHILINA.

Dichaeta.

Meigen, System. Beschr. VI, 61; 1830.

*caudata Fallen, Meigen, System. Beschr. VI, 62; Loew, Monogr. I, 133. — Europe and North America [Massachusetts, White Mts., N. H.].

*brevicauda Loew, Neue Beitr. VII, 5; Monogr. I, 133. — Europe and North America (Middle States).

Notiphila.

Fallen, Hydromyz.; 1823.

*avia Loew, Zeitschr. f. d. Ges. Naturw. 1878 (March), 193. — Huds. B. Terr.

*bella Loew, Monogr. I, 135. — Middle States.

*carinata Loew, Monogr. I, 137. — Middle States.

*macrochaeta Loew, Zeitschr. f. d. Ges. Naturw. 1878 (March), 192. — Texas.

*pulchrifrons Loew, Centur. X, 84. — Texas.

*scalaris Loew, Monogr. I, 134. — Middle States.

*unicolor Loew, Monogr. I, 137. — Middle States.

*vittata Loew, Monogr. I, 134. — Middle States.

quadrisetosa Thomson, Eugen. Resa, etc. 594. — California.

* erythrocera Loew. Zeitschr. f. d. Ges. Naturw. 1878 (March), 194. — Cuba.

The following species were described as Notiphilae by Mr. Walker:
nitidula Fallen, Meigen; Walker, List, etc. IV, 1098. — Europe; Huds. Bay.
producta Walker, List, etc. IV, 1099. — Huds Bay.
repleta Walker, List, etc. 1099. — Huds. Bay.
solita Walker, Dipt Saund. 406. — United States.
transversa Walker, Dipt. Saund., 407. — United States.

Observation. For *Notiphila argentata* Walker see *Brachydeutera.*

Paralimna. (*)
Loew, Monogr. I, 138; 1862.

* appendiculata Loew, Monogr. I, 138. — Middle States.
* decipiens Loew, Zeitschr. f. d. Ges. Naturw. 1878 (March), 195. — Texas.

Discomyza.
Meigen, System. Beschr. VI, 76; 1830.

* balioptera Loew, Monogr. I, 140. — Cuba.

Psilopa.
Fallen, Hydromyz.; 1820.

* aeneo-nigra Loew, Zeitschr. f d. Ges. Naturw. 1878 (March), 196. — Texas.
* atra Loew, Monogr. I, 14?. — Middle States.
* atrimana Loew, Zeitschr. f. d. Ges. Naturw. 1878 (March), 197. — Distr. Columbia, Texas.
* nobilis Loew, Centur. II, 92. — Distr. Columbia.
* pulchripes Loew, Zeitschr. f. d. Ges. Naturw. 1878 (March), 197. — Texas.
* scoriacea Loew, Monogr. I, 142. — New York.

* aciculata Loew, Monogr. I, 142. — Cuba.
* caeruleiventris Loew, Monogr. I, 144. — Cuba.
* umbrosa Loew, Monogr. I, 143. — Cuba.

Discocerina.
Macquart Hist Nat. Dipt. II, 527; 1835.

* lacteipennis Loew, Monogr I, 145. — Distr. Columbia.
* leucoprocta Loew, Centur. I, 93; Monogr. I, 148. - Maryland.
* orbitalis Loew, Centur. I, 91; Monogr I. 147. — Distr. Columbia.
* parva Loew, Monogr. I, 146. — Distr. Columbia.
* simplex Loew, Centur. I, 92; Monogr. I, 147. — Maryland.

---

*) *Paralimna* appeared in the same year 1862 in the Ofvers. af K. Vet. Akad. Förh. p. 13, applied by Dr. Loew to three South African species. The genus, although introduced there for the first time, is not defined.

Athyroglossa.

Loew, Neue Beitr. VII, 12; 1860.

*glaphyropus Loew, Zeitschr. f. d. Ges. Naturw. 1878 (March), 198. — Texas.

SECTION II. HYDRELLINA.

Hydrellia.

R. Desvoidy, Myod. 790; 1830.

*conformis Loew, Centur. VIII, 73. — Newport, R I.
*formosa Loew, Centur. I, 94; I, 154. — Pennsylvania.
*hypoleuca Loew, Monogr. I, 151. — Middle States.
*ischiaca Loew, Monogr. I, 150. — Middle States.
*obscuriceps Loew, Monogr. I, 152. — Middle States.
*scapularis Loew, Monogr. I, 153. — Middle States.
*valida Loew, Monogr. I, 153. — Middle States.

Philygria.

Stenhammar, Ephydrin., 238; 1844.

*debilis Loew, Centur. I, 96; Monogr. I, 157. — Pennsylvania.
*fuscicornis Loew, Monogr. I, 155. — Middle States.
*opposita Loew, Centur. I, 95; Monogr. I, 156. — Distr. Columbia; Pennsylvania; Canada (Quebec).
vittipennis Zetterstedt, in Staeger's Groenl. Antl. 369. [Philygria. — Loew *in litt.*]

Hyadina.

Haliday, Ann. of Nat. Hist. III, 406; 1830.

*gravida Loew, Centur. IV, 98. — Sitka.

SECTION III. EPHYDRINA.

Pelina.

Haliday, Ann. Natur. Hist. III, 407; 1839.

*truncatula Loew, Zeitschr. f. d. Ges. Naturw. 1878 (March), 198. — Texas.

Ochthera.

Latreille, Hist. Nat. d. Crust. et d. Ins XIV; 1804.

*mantis Degeer, Loew, Monogr. I, 161. — Europe and United States.
*rapax Loew, Monogr. I, 162. — Carolina.
*tuberculata Loew, Monogr. I, 161. — Illinois.

*exsculpta Loew, Monogr. I, 160. — Cuba.

Observation. *Ochthera empiformis* Say, J. Acad. Phil. III, 99 is a Hemerodromia.

Brachydentera.
Loew, Monogr. I, 162; 1862.
*dimidiata Loew, Monogr. etc. I, 163. — Distr. Columbia; Cuba.
Notiphila argentata Walker, Dipt. Saund., 407 [Loew *in litt.*].

Parydra.
Stenhammar, Monogr Ephydr.; 1844.
*abbreviata Loew, Centur. I, 97; Monogr. I, 168. — Pennsylvania.
*appendiculata Loew, Zeitschr. f. d. Ges. Naturw. 1878 (March), 202. — Texas.
*bituberculata Loew, Monogr. I, 165. — Middle States.
*breviceps Loew, Monogr. I,.167. — Middle States.
*imitans Loew, Zeitschr. f. d. Ges. Naturw. 1878 (March), 201. — Massachusetts.
*limpidipennis Loew, Zeitschr. f. d. Ges. Naturw. 1878 (March), 201. — Distr. Columbia.
*paullula Loew, Monogr. I, 167. — United States.
*pinguis Walker, Dipt. Saund., 409 (*Ephydra*); Loew, Zeitschr. f. d. Ges. Naturw. 1878, March), 199. — Distr Columbia; Texas.
*quadrituberculata Loew, Monogr. I, 165. — Middle States.
*unituberculata Loew, Zeitschr. f. d. Ges. Naturw. 1878, (March) 200. — Distr. Columbia.
varia Loew, Centur. IV, 100. — Sitka.

Ephydra.
Fallen, Hydromyz.; 1820.
*atrovirens Loew, Monogr. I, 169. — Middle States.
brevis Walker, Trans. Ent. Soc. N. Ser. IV, 233. — United States.
halophila Packard, Proc. Essex Instit. VI, 46 (figure on page 48). — Illinois. ([318]).
lata Walker, Trans. Ent. Soc. N. S. IV, 233. — United States.
nana Walker, Trans. Ent. Soc. N. J. IV, 234. — United States.
*obscuripes Loew, Centur. VII, 92. — Massachusetts.
*subopaca Loew, Centur. V, 99. — Connecticut.

*crassimana Loew, Centur. VI, 88. — Mexico.
hians Say, J. Acad Phil. VI, 188; Compl. Wr. II, 371. — Mexico.
lutea Wiedemann, Auss. Zw. II, 593, 3 — West Indies.
Thomae Wiedemann, Auss. Zw. II, 593, 3. — St. Thomas.

Observation. *E. californica* and *gracilis* Packard, Am. J. Sc. and Art. 3d Ser. I, 103, from California, are described in the larva state only.

Scatella.
R. Desvoidy, Myod. 801; 1830.
*favillacea Loew, Monogr. I, 170. — Middle States.
*lugens Loew, Monogr. I, 171. — Middle States.
*mesogramma Loew, Centur. VIII, 74. — Newport, R. I.

*obsoleta Loew, Centur. I, 98; Monogr. etc, I, 172. — Distr. Columbia

*quadrata Fallen, Hydromyz. 5, 6; Schiner, Fauna Austr. II, 263. — Europe and North America (Loew, Sillim. Journ. etc.).

*sejuncta Loew, Centur. IV, 99. — Sitka.

*Stenhammari Zetterstedt, Dipt. Scand. V, 1842, 24. — Europe and North America (Loew, Sillim. Journ. XXXVII, 318).

stagnalis Meigen, in Staeger's Groenl. Antl. *(Ephydra);* Holmgren, Ins. Nordgroenl., 103. — Europe, Greenland.

> **Observation.** The following species seem also to belong to *Scatella.*
> Ephydra picea Walker, List, etc. IV, 1105. — Huds. B. Terr.
>   „   oscitans Walker, l. c. 1106. —   „
>   „   octonotata Walker, l. c. 1106. —   „
>   „   striata Walker, l. c. 1107. —   „
>   „   pentastigma Thomson, Eugen. Resa, etc. 591. — California.

Caenia.

R. Desvoidy, Myod. 800; 1830.

*spinosa Loew, Centur. V, 100. — Massachusetts, New York.

Ilythea.

Haliday, Ann. of Nat. Hist. III, 408; 1830.

*spilota Curtis, Brit Entom. 413; Schiner, Fauna Austr. II, 263. — Europe and North America (Loew, in Sillim. Journ. Vol. XXXVII, 318).

(?) *Ephydra oscitans* Walker, Trans. Ent. Soc. N. S. IV, 233. — United States. ([319])

FAMILY GEOMYZIDAE. ([320])

Diastata.

Meigen, System. Beschr. VI, 94; 1830.

*eluta Loew, Centur. III, 59. — Sitka.

*pulchra Loew, Centur. I, 100. — Pennsylvania.

tenuipes Walker, List, etc. IV, 1112. — Huds. B. Terr.

*vagans Loew *(in litt.).* — Europe and North America (N. Hampshire). I mention this name, because it occurs in Loew's typical collection and in my collection (now both in the Mus. Comp. Zool.); but I am not aware that the species has ever been described.

Diplocentra.

Loew, Centur. Vol. II, page 283; 1872; *Curtonotum* Macq., Dipt. Exot. II, 3, 193 (this name is preoccupied).

*helva Loew, Centur. II, 91. — British America.

FAMILY DROSOPHILIDAE.

Phortica.

Schiner, Wien. Ent. Monatschr. VI, 1862, December; *Amiota* Loew, Centur. II, 93; 1862, May; compare also Centur. Vol. II, page 288 ([321]).

*alboguttata Wahlberg, K. Vetensk. akad. handl. 1838, 22 (*Drosophila*).— Sweden and North America (Loew *in litt.*).
*humeralis Loew, Centur. II, 93 (*Amiota*). — Distr. Columbia.
*leucostoma Loew, Centur. II, 94 (*Amiota*). — Pennsylvania.

Stegana.

Meigen, System. Beschr. VI, 79; 1830.

*hypoleuca Meigen, System. Beschr. VI, 80. — Europe and North America (Loew, in Sillim. Journ. XXXVII, 318).
*nigra Meigen, System. Beschr. VI, 79; Tab. 58, f. 24, 25. — Europe and North America (Loew, l. c.).

Drosophila.

Fallen, Geomyz.; 1823.

*adusta Loew, Centur. II, 98. — Distr. Columbia.
albipes Walker, Dipt. Saund., 410. — United States.
*amoena Loew, Centur. II, 96. — Distr. Columbia.
*ampelophila Loew, Centur. II, 99 — Distr. Columbia; Cuba.
brevis Walker, Dipt. Saund, 411. — United States.
colorata Walker, List, etc. IV, 1010. — New York
decemguttata Walker, Dipt. Saund., 411. — United States.
*dimidiata Loew, Centur. II, 95. — Illinois.
fronto Walker, Dipt. Saund., 410. — United States.
funebris Meigen, quoted by Macquart, Dipt. Exot. 4e Suppl. 305, as occurring in Europe and North America.
*graminum Fallen, Geomyz. 8; Zetterstedt, Dipt. Scand. VI, 2560. — Europe and North America (Loew, Sillim. J. N. S. XXXVII, 313).
guttifera Walker, List, etc. IV, 1110. — Florida.
inversa Walker, Trans. Ent. Soc. N. Ser. V, 331. — United States.
linearis Walker, Dipt. Saund., 411. — United States.
minuta Walker, Dipt. Saund., 412. — United States.
*multipunctata Loew, Centur. VII, 93. — Distr. Columbia.
*obesa Loew, Centur. X, 85. — Texas.
quadrimaculata Walker, Dipt. Saund., 412. — United States.
*quinaria Loew, Centur. VI, 90. — New York.
*sigmoides Loew, Centur. X, 86. — Texas.
*terminalis Loew, Centur. III, 60. — Sitka.
*transversa Fallen, Geomyz. 6; Meigen, System. Beschr. VI, 84. — Europe and North America (Loew, in Sillim. J. N. S. XXXVII, 318).
*tripunctata Loew, Centur. II, 97. — Distr. Columbia.

valida Walker, Trans. Ent. Soc. N. Ser. IV, 232. — United. States.
* varia Walker, List, etc. IV, 1109. — Georgia.

* bimaculata Loew, Centur. VI, 91. — Cuba.
* flexa Loew, Centur. VI, 89. — Cuba.
 mexicana Macquart, Dipt. Exot. II, 3, 259, 4; Tab. XXXV, f. 1. —
  Mexico.
* obscuripennis Loew, Centur. VI, 92. — Cuba.
* punctulata Loew, Centur. II, 100. — Cuba.

 apicata Thomson, Eugen. Resa, etc. 597. — California.

  Observation. Walker, List, etc. 1107 has a *D. cellaris* Linné,
as common to Europe and North America. According to Schiner,
Dipt. Austr. II, 278, foot-note, Linné's *Musca cellaris* must be a
Phora, and Walker must have been in error both here and in
Ins. Brit. Dipt. II, 237, where he described a *Drosophila cellaris*
Linné.

FAMILY OSCINIDAE.

Crassiseta.

Von Roser, Verz. Württ. Dipt. Nachtrag; 1840; Loew, Dipterl.
Beitr. I, 48; 1845.

* costata Loew, Centur. III, 62. — Distr. Columbia.
* eunota Loew, Centur. X, 89. — Texas.
 formosa Loew, Centur. III, 61. — Distr. Columbia.
* longula Loew, Centur. III, 64. — Distr. Columbia.
* nigripes Loew, Centur. III, 63. — Distr. Columbia.
* nigricornis Loew, Centur. III, 65. — Distr. Columbia.

Gaurax.

Loew, Centur. III, 66; 1863.

* anchora Loew, Centur. VII, 94. — New York (inquilinous in cocoons
 of *Attacus cecropia*).
* festivus Loew, Centur. III, 66. — Pennsylvania.
* signatus Loew, Zeitschr. f. Ges. Naturw. 1876, 338. — Texas.

Hippelates.

Loew, Centur. III, 67; 1863.

* eulophus Loew, Centur. X, 88. — Texas.
* nobilis Loew, Centur. III, 67. — Illinois.
* plebejus Loew, Centur. III, 68. — Distr. Columbia.
* pusio Loew, Centur. X, 87. — Texas.

 genalis Thomson, Eugen Resa, etc. 608. — California.

*convexus Loew, Centur. VI, 94. — Cuba.
* dorsalis Loew, Centur. VIII, 75. — Cuba.
*flavipes Loew, Centur. VI, 95. — Cuba.
*pallidus Loew, Centur. VI, 93. — Cuba.

Oscinis.

Latreille, Nouveau Dict. d'Hist. Natur. XXIV, Tabl. Méthod 196; 1804. (³²²).

* atriceps Loew, Centur. III, 74. — Pennsylvania.
* carbonaria Loew, Centur. VIII, 76. — Distr. Columbia.
coxendix Fitch, Reports, Vol. I, 301. — New York.
*crassifemoris Fitch, Reports, Vol. I, 301. — New York. [Location
   doubtful; perhaps *Opetiophora?* — Lw.],
* decipiens Loew, Centur. III, 76. — Sitka.
*dorsalis Loew, Centur. III, 72. — Pennsylvania.
*dorsata Loew, Centur. Vol. II, page 291 in erratis.
   Oscinis dorsalis Loew, Centur. VIII, 77. — Newport, R. I.
* hirta Loew, Centur. III, 75. — Illinois.
*longipes Loew, Centur. III, 77. — Distr. Columbia
* nudiuscula Loew, Centur. III, 70. — Georgia.
soror Macquart, Dipt. Exot. 4e Suppl. 306, 5; Tab. XXVIII, f. 11
   (Chlorops). — North America.
* subvittata Loew, Centur. III, 78. — Distr. Columbia.
* trigramma Loew, Centur. III, 80. — Distr. Columbia.
*umbrosa Loew, Centur. III, 73. — Pennsylvania.
*variabilis Loew, Centur. III, 79. — Distr. Columbia.

*flaviceps Loew, Centur. III, 71. — Cuba.
* pallipes Loew, Centur. III, 69. — Cuba.

Meromyza.

Meigen, System. Beschr. V, 163; 1830.

* americana Fitch, Reports I, 299; Riley, First Report, Tab. II, f. 28.
— United States.

Ectecephala.

Macquart, Dipt. Exot. 4e Suppl. 280; 1850.

*albistylum Macquart, Dipt. Exot. 4e Suppl. 280, 1; Tab. XXV, f. 17.
— North America.

Opetiophora.

Loew, Centur. X, 90; 1872.

*straminea Loew, Centur. X, 90. — Texas.

Siphonella.

Macquart, Hist. Nat. Dipt. II, 584; 1835. (³²²).

*cinerea Loew, Centur. III, 81. — Florida.

*latifrons Loew, Centur. X, 91. — Texas.
obesa Fitch, Report I, 299. — New York.
plumbella Wiedemann, Auss. Zw. II, 574 (*Homalura*); placed among
the Siphonellae on the authority of Loew, Monogr. I, 46. —
West Indies.
*reticulata Loew, Centur. VIII, 78. — Cuba.

Chlorops.

Meigen, in Illig. Magaz. II, 278; 1803; the subgenera have been introduced and characterized by Mr. Loew in the Schles. Zeit. f. Entom.;
1866. ([322]).

Subgenus Centor.

*procera Loew, Centur. X, 92. — Connecticut.

Subgenus Haplegis.

*fossulata Loew, Centur. III, 82. — Cuba.

Subgenus Anthracophaga.

*eucera Loew, Centur. III, 85. — Distr. Columbia.
*maculosa Loew, Centur. X, 99. — Texas.
*sanguinolenta Loew, Centur. III, 84. — Carolina.

Subgenus Diplotoxa.

Compare about it: Loew, Centur. X, 98.
*alternata Loew, Centur. X, 97. — Texas.
*confluens Loew, Centur. X, 94. — Texas.
*microcera Loew, Centur. X, 95. — Texas.
*nigricans Loew, Centur. X, 98. — Texas.
*pulchripes Loew, Centur. X, 96. — Texas.
*versicolor Loew, Centur. III, 97. — United States and Canada.

*Gundlachi Loew, Centur. X, 93. — Cuba.

Subgenus Chlorops.

*crocota Loew, Centur. III, 89. — Pennsylvania.
*melanocera Loew, Centur. III, 91. — Distr. Columbia.
*mellea Loew, Centur. X, 100. — Texas.
*obscuricornis Loew, Centur. III, 90. — Distr. Columbia.
*producta Loew, Centur. III, 96. — Sitka.
*pubescens Loew, Centur. III, 88. — Florida.
*quinquepunctata Loew, Centur. III, 94. — Nebraska.
*Sahlbergi Loew, Centur. III, 95. — Sitka.
*sulphurea Loew, Centur. III, 83. — Brit. North America.
*unicolor Loew, Centur. lll, 93. - Mississippi.
*variceps Loew, Centur. III, 86. — Sitka.

Subgenus Chloropisca.

*grata Loew, Centur. III, 92. — Pennsylvania.

*trivialis Loew, Centur. III, 87. — Distr. Columbia.

Observation. About the species of Chlorops enumerated on page 85 of my first Catalogue, Mr. Loew communicates me the following remarks.

antennalis Fitch, Reports I, 300, see my note [311].

annulata Walker, List, etc. IV, 1119. — Huds. B Terr. [probably Chloropisca. — Loew].

assimilis Macquart, Dipt. Exot. 4e Suppl. 306, 3; Tab. XXVIII, f. 9. — North America (probably Diplotoxa. — Loew].

atra Macquart, Dipt. Exot. 4e Suppl. 307, 6; Tab. XXVIII, f. 12 [probably Eutropha; hardly Haplegis. — Loew].

bistriata Walker, List, etc. IV, 1120. — Huds. B. Terr. [apparently Chlorops, in the narrower sense — Loew].

perflava Walker, List, etc. IV, 1120 [perhaps Diplotoxa. — Loew].

proxima Say, J. Acad. Phil. VI, 187; Compl. Wr. II, 370. — Indiana.

soror Macquart, see *Oscinis soror*.

testacea Macquart, Dipt. Exot. 4e Suppl. 306, 4; Tab. XXVIII, f. 10. — North America [Chlorops, sensu strict. — Loew].

tibialis Fitch, Raports I, 300; Tab. I, f. 5. — New York.

vittata Wiedemann, Auss. Zw. II, 594, 1. — West Indies. [The plumose antennae render the position of this species in the family somewhat doubtful. As *Hippelates eulophus* alone, among all N. A. species, has such antennae, *C. vittata* may be a Hippelates. However South America possesses several Oscinidae with plumose antennae. — Loew.]

Elliponeura.
Loew, Centur VIII, 79; 1869.
*debilis Loew, Centur VIII, 79. — Distr. Columbia.

Gymnopa.
Fallen, Oscinid., 1820; *Mosillus* Latreille; 1804. (323).

nigroaenea Walker, Dipt. Saund., 413. — United States.

tarsalis Walker, l. c. — United States.

FAMILY AGROMYZIDAE.

Rhicnoëssa.
Loew, Wien. Ent. Monatschr. VI, 174. (324).
*albula Loew, Centur. VIII, 80. — Newport, R. I.
*coronata Loew, Centur VI, 98. — Georgia.
*parvula Loew, Centur. VIII, 81. — Newport, R. I.

Lobioptera.
Wahlberg, Oefvers. af K. Vetensk. Acad. Forh. 1847, 259.
*arcuata Loew, Zeitschr. f. Ges. Naturw. 1876, 339. — Long Island, N. Y.

*indecora Loew, Centur. VIII, 94. — Nebraska.

17

*lacteipennis Loew, Centur. VI, 97. — Cuba.
*leucogastra Loew, Centur. VIII, 95. — Cuba.
  Milichia leucogastra Loew, Wien. Ent. Monatschr. V, 43, 20.

Pholeomyia.
Bilimek, Verh. Zool. Bot. Ges. 1867, 903.
leucozona Bilimek, l. c. — Mexico.

Milichia.
Meigen, System. Beschr. VI, 131; 1830. ([321]).
*picta Loew, Centur. I, 99. — Georgia.

Cacoxenus.
Loew, Wien. Ent. Monatschr. 1858, 217. ([326]).
*semiluteus Loew, Centur. VIII, 97. — Cuba.

Aulacigaster.
Macquart, Hist. Nat. Dipt. II, 579; 1835. ([327]).
Amphycophora Wahlberg, Oefvers K. Svensk. Vet. Acad. Förh. 1847,
  p. 261—263; Tab. VII, f. 2.
Apotomella Leon Dufour, Ann Soc. Ent. de Fr. 1845. p. 455.
*rufitarsis Macquart, etc. For the description and full quotations see
  Schiner, Fauna Austr., Dipt. II, 270. — Europe and North America
  (Distr. Columbia; Texas. — Lw. in litt.].

Leucopis.
Meigen, System. Beschr. VI, 133; 1830.
*simplex Loew, Centur. VIII, 96. — New York.
*bella Loew, Centur. VI, 99. — Cuba.

Desmometopa.
Loew, Centur. VI, 96; 1865.
*latipes Meigen, etc. — Europe and North America (Distr. Columbia;
  Pennsylvania; Lw. in litt.).
*M nigrum Zetterstedt, Dipt. Scand. VII, 2743 (Agromyza). — Sweden;
  Malta, also Cuba (the latter Loew in litt.).
*tarsalis Loew, Centur. VI, 96. — Cuba.

Agromyza.
Fallen, Agromyz.; 1823.
*aeneiventris Fallen, etc. — Europe and North America [Loew in litt.].
*angulata Loew, Centur. VIII, 87. — Pennsylvania.
*coronata Loew, Centur. VIII, 89. — Pennsylvania.
invaria Walker, Trans. Ent. Soc. N. S. IV, 232 — United States.
jucunda v. d. Wulp, Tijdschr. v. Ent. 2d Ser. II, 161; Tab. V, t. 19,
  20. — Wisconsin.
*longipennis Loew, Centur. VIII, 90. — Distr. Columbia.
*magnicornis Loew, Centur. VIII, 86. — Pennsylvania.
*marginata Loew, Centur. VIII, 91. — Distr Columbia.

*melampyga Loew, Centur. VIII, 88. — Distr. Columb'a.
*neptis Loew, Centur. VIII, 93. — Nebraska.
*parvicornis Loew, Centur. VIII, 92. — Distr. Columbia.
*setosa Loew, Centur. ViI', 83. — Distr. Columbia.
*simplex Loew, Centur. VIII, 84. — Pennsylvania.
*tritici Fitch, Reports I, 303; Tab. II, f. 1. — New York.
*virens Loew, Centur. VIII, 85. — Pennsylvania.

pictella Thomson, Eugen. Resa, 609. — California.
platyptera Thomson, Eugen. Resa, 608. — California.

Odontocera.
Macquart, Hist. Nat. Dipt. II, 614; 1835.
*dorsalis Loew, Centur. III, 98. — Distr. Columbia.

Phyllomyza.
Fallen, Ochtidia; 1823.
*nitens Loew, Centur. VIII, 82. — Pennsylvania.

Ochthiphila.
Fallen, Ochtidia; 1823. ([328]).
lispina Thomson, Eugen. Resa, 599. — California.

Observation. *Ulidia metallica* Bigot, in R. de la Sagra etc. 825 belongs to the Agromyzidae, according to Loew, Monogr. III, 202; however in the same volume page 65, he says it may be a *Chrysomyza*, a genus allied to Ulidia.

FAMILY PHYTOMYZIDAE.

Phytomyza.
Fallen, Phytomyz.; 1823.
*clematidis Loew, Centur. III, 100. — Distr. Columbia.
diminuta Walker, Trans. Ent Soc. N. S. IV, 232. — United States.
*genualis Loew, Centur. VIII, 100. — Distr. Columbia.
*ilicicola Loew, Centur. Vol. II, 290. — Distr. Columbia.
Phytomyza ilicis Loew, Centur. III, 99 (change of name by Loew).
*nervosa Loew, Centur. VIII, 99. — Distr. Columbia.
solita Walker, Trans. Ent. Soc. N. Ser. V, 232. — United States.
obscurella Fallen, Phytomyz. 4, 8; Meigen, System. Beschr. VI, 191; Staeger, Groenl. Antl. 369, 55. — Europe and Greenland.

FAMILY ASTEIDAE.

Sigaloëssa.
Loew, Centur. VI, 100; 1865. ([329]).
*bicolor Loew, Centur. VI, 100. — Cuba.

Asteia.

Meigen, System. Beschr. V, 88, 1830; improved in *Astia* by Loew, Centur. VI, 100. ([330]).

tenuis Walker, Trans. Ent. Soc. Phil. V, 331. — United States.

FAMILY BORBORIDAE. ([331]).

Borborus.

Meigen, in Illiger's Magaz. II, 1803; *Copromyza* Fallen, Stenh.

annulus Walker, List, etc. IV, 1129. — Huds. B. Terr.

*equinus Fallen, Stenhammar, etc. — Europe and North America [Loew. Sillim. J. N. S. XXXVII, 318].

carolinensis R. Desvoidy, Myod. 811, 2 *(Scatophora)*. — Carolina.

*venalicius n. sp. see note ([332]). — Africa and Cuba |common, probably imported in slave-ships; about the specific identity, see Loew, Monogr. I, 47].

FAMILY PHORIDAE.

Trineura.

Meigen, Illiger's Magaz. II; 1803.

aterrima Fabricius, Meigen, etc.; Walker, List, etc. IV, 1138. — Europe; Huds. B. Terr. (Walker).

Gymnophora.

Macquart, Hist. Nat. Dipt. II, 631; 1835.

*arcuata Meigen, etc. — Europe and North America (Loew *in litt*).

Phora.

Latreille, Hist. Nat. des Crust. et des Ins. XIV; 1804.

*atra Fabricius, etc. — Europe and North America [Loew *in litt*.].
*clavata Loew, Centur. VII, 95. — Distr. Columbia.
fuscipes Macquart, Hist. Nat. Dipt. II, 627. — Europe and North America |Huds. B. Terr. Walker, List, etc. IV, 1136].
*incisuralis Loew, Centur. VII, 98. — Distr. Columbia.
*microcephala Loew, Centur. VII, 96. — Distr. Columbia.
*nigriceps Loew, Centur. VII, 99. — Distr. Columbia.
*pachyneura Loew, Centur. VII, 97. — Alaska.
*rufipes Meigen, System. Beschr. VI, 216. Europe and North America, Huds. B. Terr. [Walker, List, etc. IV, 1136; also Loew *in litt*.].

cornuta Bigot, R. de la Sagra etc. 827. — Cuba.
*scalaris Loew, Centur. VII, 100. — Cuba.

III. DIPTERA PUPIPARA.

FAMILY HIPPOBOSCIDAE. [333].

Olfersia.

Wiedemann, Auss. Zw. II, 605; 1830.

*americana Leach, Eprob. 11, 2, Tab. XXVII, f. 1—3 (*Feronia*); Wied.,
Auss. Zw. II, 606, 1; Macquart, Hist. Nat. Dipt. II, 641, 4. —
Georgia (Leach); Illinois, Massachusetts; Dallas, Texas (On *Bubo
virginianus, Buteo borealis.*)
Hippobosca bubonis Packard's Guide etc.. 417.
albipennis Say, J. Acad. Phil. III, 101; Compl. Wr. II, 87. (On *Ardea
Herodias.*)
*ardeae Macquart, Hist. Nat. Dipt. II, 640. — Europe and North
America [Loew, Sillim. J. XXXVII, 318].
brunnea Olivier, Encycl. Méthod. VIII, 544, 6 (*Ornithomyia*). —
Carolina.

mexicana Macquart, Dipt Exot. II, 3, 278, 5. — Mexico.
propinqua Walker, List, etc. 1141. — Jamaica.
sulcifrons Thomson, Eugen. Resa, etc. 611. — Panama.

Ornithomyia.

Latreille, Hist. Nat. des Crust. et des Ins. XIV, 402; 1804.

avicularia Linné, Leach, Meigen, etc. — Europe and North America
[the latter according to v. d. Wulp, Tijdschr. 2ᵈ Ser. IV, 80].
fusciventris Wiedemann, Auss. Zw. II, 611, 9. — Kentucky.
nebulosa Say, J. Acad. Phil. III, 102, 1; Compl. Wr. II, 87 (on *Strix
nebulosa*); Wiedemann, Auss. Zw. II, 610, 6. — North America.
*pallida Say, J. Acad. Phil. III, 103, 2; Compl. Wr. II, 87 (on *Sylvia
Sialis*); Wiedemann, Auss. Zw. II, 610, 7. — North America.

*erythrocephala Leach, Eprob. Ins. 13, 3; Tab. XXVII, f. 4—6;
Wiedemann, Auss Zw. II, 610, 5. — Brazil (Leach); Jamaica
(Walker, List, etc. IV, 1143); Cuba. (I received a specimen from
Quebec, Canada. — O. S.).
fulvifrons Walker, List, etc. IV, 1145. — Jamaica.
unicolor Walker, List, etc. IV, 1144. — Jamaica.
vicina Walker, l. c. 1144. — Jamaica.

Observation. *Ornithomyia laticornis* Macquart, Hist. Nat.
Dipt. II, 642, 3 etc., of my first Catalogue is omitted here, since
my attention was drawn to the *erratum* in the same volume,
where the locality: Cuba, is recognized as erroneous.

Novum genus? ([334]).

confluens Say, T. Acad. Phil. III, 103, 3; Compl. Wr. II, 87 *(Orni-thomyia confluenta)*; Wiedemann, Auss. Zw. II, 611, 8 (translation from Say). — Pennsylvania.

Lipoptena.

Nitsch, in Germ. Mag f. Ent. III, 310; 1818; *Leptotena* Macq.; *Haemobora* Curtis, etc.

depressa Say, J. Acad. Phil. III, 104; Compl. Wr. II, 88 *(Melophagus)*; Wiedemann, Auss. Zw. II, 614, 2. — Pennsylvania, on *Cervus virginianus*. [Referred to this genus by Loew *in litt.*]

Melophagus.

Latreille, Hist. Nat. des Crust. et des Ins. XIV, 402; 1804.

* ovinus Linné, Meigen, System. Beschr. VI, 236; Tab. 65, f. 16; Leach, Curtis, etc.; Fitch, Survey of Wash. Co. etc. 797. — Europe and North America. (See Loew, Sillim. J., l. c)

Hippobosca.

Linné, Fauna Suec.; 1761.

* equina Linné, etc.; Kirby, N. Am. Zool. Ins. 316. — Europe and North America. [See Loew, Sillim. Journ. N. S. XXXVII, 318.]

FAMILY NYCTERIBIDAE. ([335]).

Strebla.

Wiedemann, Analecta etc. 1824; Auss. Zw. II, 612; 1830.

* vespertilionis Fabricius, System. Antl. 339, 6 *(Hippobosca)*; Wiede-mann, Anal. Ent. 19, f. 7; Auss. Zw. II, 612, 1; Tab. X, f. 13; Macquart, Hist. Nat. Dipt. II, 637, 1; Tab. XXIV, f. 7. — South America (Fabr.); Jamaica (Walker, List etc IV, 1146); San Domingo, Cuba [Loew *in litt.*].

Strebla avium Macquart, Dipt. Exot. 5e Suppl. 127, 2. — San Domingo (on pigeons and parrots). [Loew *in litt.*]

Strebla Wiedemanni Kolenati, Horae Soc. Ent. Ross. II, 96; Tab. XV, f. 36 [Loew *in litt.*].

Megistopoda.

Macquart, Ann. de la Soc. Ent. de Fr. 1852, 331 - 333.

* Pilatei Macquart, Ann. Soc. Ent. Fr. 1852, 331; Tab. IV, Nr. 4. — Mexico (Macq.). — Cuba.

Megistopodia Pilatei Kolenati, Horae Soc. Ent. Ross. II, 89; Tab. XIV, f. 32.

Nycteribia.

Latreille, Hist. Nat. des Crust. et des Ins. XIV, 403; 1804. ([335]).

No N. A. species is as yet described. The M. C. Z. possesses a specimen from California.

NOTES.

1. **Cecidomyia.** On this family, the following papers may be consulted:

H. Loew. Dipterologische Beiträge IV, 1850

The same. Zur Kenntniss der Gallmucken, in the Linnaea Entomol. V, 1851.

J. Winnertz. Beitrag zu einer Monographie der Gallmücken, in Linnaea Entomol. VIII, 1854; with four beautiful plates.

The same. Heteropeza und Miastor, in the Verh. Zool. Bot. Gesellsch. 1869.

The same. Die Gruppe der Lestreminae, in the same volume.

Bergenstamm und Löw (Fr.), Synopsis Cecidomyiarum, in the Verh. Zool. Bot Gesellsch. 1876. A synopsis of all the literature on the subject; very accurate and complete.

C. R. Osten Sacken. On the North American Cecidomyidae. — In the Monographs of N. A. Diptera, Vol I (a survey of the previous publications concerning the classification, habits etc.).

In an inaugural Dissertation, entitled: Revision der Gallmücken, Münster 1877, Mr. F. A. Karsch changes the existing nomenclature of the Cecidomyidae, in virtue of the principle of priority. What we call now Cecidomyia, he calls Dasyneura Rondani; our Diplosis Loew, is his Cecidomyia Meigen; Clinorhyncha Loew is to be Ozirhyncus Rondani; Epidosis Loew is Porricondyla Rondani; Hormomyia Loew is Oligotrophus Latreille.

The general adoption of these changes does not seem at all desirable.

2. **Cecid. grossulariae Fitch** In the Monogr. I, p 7, Mr. Loew stated that this species is an *Asphondylia*, a statement which I repeated on faith, l. c. p. 189. Dr. Fitch's description renders it evident that his species is a true *Cecidomyia*. It is probable that, in making the above-quoted statement Mr. Loew had in his mind the european *Cecid. ribesii* Meigen, which, as appears from Meigen's description, must be an *Asphondylia*.

3. **Cecid. salicis batatas.** "This gall seems to agree in its structure with that of *Cecid. salicis* Schrank, on european willows." Bergenstamm & Löw, l. c. p. 71.

4. "The five kinds of leaf-accumulations and leaf-rosettes, which Mr. Walsh describes and which he attributes to his Cecidomyiae gnaphaloides, rhodoides, strobiloides, strobiliscus, coryloides, seem to be

the produce of the same species of Cecidomyia; the differences in the shape of the gall seem to be due, not to a specific difference among the insects, but to the specific difference of the willows on which they occur. The trifling differences between the flies, as described by Walsh, as well as the circumstance that each of those five forms of galls harbours only a single larva, strengthen this view. The european relative of this species, *Cecid. rosaria* Loew, likewise produces differently shaped galls on different species of willows." Bergenstamm and Löw, l. c p. 71.

5. **Cecid. salicis-strobiloides.** „This gall is the exact counterpart of the gall of *Cecid. rosaria* Loew, on the european Salix purpurea." Bergenstamm and Lów, l. c. p. 72.

6. Several of the galls which I described as occurring on hickories, as caryae, caryaecola, holotricha, persicoides, even tubicola, and other, undescribed forms, sometimes occur promiscuously, on the same leaf. It remains to ascertain, whether they are really produced by different species of Cecidomyia, or whether most of them are not merely modifications in shape and degree of pubescence, of the gall of *Diplosis caryae.*

7. **Cecidcmyia poculum** I am very much inclined now to believe that the larva of a Cecidomyia, which I found in the gall that I thus named, was a mere inquiline, and that the gall was the work of a Cynipid. The ground for my belief is, that there is an analogous gall in Europe, that of *Neuroterus lenticularis,* which frequently harbours inquilinous larvae of Cecidomyiae. As long as the gall is on the leaf, no larva of a Cynips can be found in it; it develops only when the gall falls to the ground. If my supposition is correct, this peculiarity of the gall of Neuroterus would explain why, in most cases, I did not find any larvae whatever in the gall *poculum.*

8. **Mycetophilidae.** For the definition of the genera see: Winnertz, Beitr. zu einer Monographie der Pilzmücken, in the Verh. Zool. Bot. Gesellsch. 1863, p. 637—964. Mr. Loew's species were all referred by him to the new genera formed by Winnertz The older species by Say, Wiedemann, etc., unless identified, I have left in the genera in which they were described.

9. **Empheria** is preoccupied by Hagen in the Psocidae, 1856. *Glaphyroptera* by Heer, fossil Buprestidae, 1852.

10. **Sciara** and **Trichosia.** Compare Winnertz, Beitrag zu einer Monographie der Sciarinen, in Verh. Zool. Bot. Gesellsch. 1867.

11. **Simulium** There is a monograph of this genus by Fries; compare also Zetterstedt, Meigen, Schiner.

12. **Bibio.** A monograph of the european species by Loew, in Linnaea Entomologica, I, p. 342. In quoting Geoffroy, here and elsewhere, I rely upon Schiner, because I possess only the *second* edition of Geoffroy.

The name *Bibio* was first introduced by Geoffroy in 1764; he included five species in it, three of which where Bibio's in the present meaning and two Psychodae

NOTES. **217**

The name *Hirtea* appeared first in Scopoli, Entomologia Carniolica 1763, where Hirtea longicornis Stratiomys strigata F.) is described. — For an unexplained reason, Fabricius, in the Supplement to his Entomologia Systematica, published in 1798, took up the name *Hirtea* (without any reference to Scopoli) and applied it to a number of species, the majority of which are Bibio's. At the same time, the majority of Fabricius's Bibio's are our Therevae, and Fabricius's Therevae are our Phasiae, Trichopodae etc.!

Meigen followed Fabricius's precedence about Hirtea in his earlier work: Klassification etc. (1804), and Fabricius quoted Meigen in his System. Antliatorum (1805). In his principal work, however, (1818) Meigen rejected the name *Hirtea*, and very properly adopted Geoffroy's earlier name *Bibio*. Later writers have followed Meigen's example, except Zetterstedt, who maintains the name Hirtea, for our Bibio.

It is very probable that Stratiomyia longicornis Scopoli (Syn. strigata Fabricius), which shows several peculiarities of structure, will, by and by, form a separate genus, and then *Hirtea* will be the proper name for that genus.

13. **Bibio articulatus** Say. According to Loew, Centur. V, 10, *Nota* this species belongs in the vicinity of *B. abbreviatus, fraternus, nigripilus,* but the descriptions, both of Say and of Wiedemann are not explicit enough for identification.

14. About **Plecia, Penthetria, Hesperinus,** etc. compare Loew, Berl. Entom Z. II, p. 101. Also by the same: Berichtigung der generischen Bestimmung einiger fossilen Dipteren, in the Zeitschr. f. Ges. Naturw. Vol. XXXII, p 80 (1868).

15. About **Scatopse,** see Loew, Linnaea Entom. I, p. 324, a monograph of the european species. Also another paper, by the same, in the Zeitschr. f. d. Ges. Naturw., Vol. XXXV. (1870).

16 The identity of **Arthria** Kirby with **Aspistes** and of *Arthria analis* with *Aspistes borealis* seems to me very probable, some discrepancies between the descriptions notwithstanding.

17. **Blepharoceridae.** Compare Loew's Monographic Essay La famiglia dei Blefaroceridi, in the Bollet. della Società Entom. Italiana, Vol. I, p. 85 (1869) — The same author's· Revision der Blepharoceridae (in the Schles. Zeitschr. f. Entomol. Neue Folge, Heft VI, Breslau 1877) is in the main a reproduction of the Italian paper, but being of later date contains several additions.

In the Monographs etc. IV, p. 3, I suggested the possibility of a relationship between the *Blepharoceridae* and the *Ptychopterina.* But since I know the Blepharoceridae better, I am less inclined to perceive that relationship In the structure of the eyes this family stands nearer to Simulium and Bibio.

18. **Asthenia americana** Walker, List, etc. I, p. 28, according to Loew, Monographs I, p 8, is not a Blepharocerid at all, and any one, who reads the description, will agree with this conclusion It seems furthermore that Mr. Walker's type is not to be found in its place at the British Museum; compare Mr. Haliday's note in the Bolletino della

Società Entomol. Italiana, Vol. I, p. 99. The fact that Mr. Walker had not the slightest idea of the true characters of this family, is further proved by his having described a true *Blepharocera* as an *Asyndulum*.

19. **Paltostoma** I will add to Dr. Schiner's description, that the palpi are distinct; the posterior tibiae bear *one* long, slender spur; ocelli large, distinct; eyes separated by a broad front, pubescent, facets of the same size on the whole surface; wings with a square anal angle, like that of the other species of the family and unlike their representation on the figure in the Novara-volume Altogether, the genus bears out the character of the family, as drawn by Loew (Revision der Blepharoceridae, p. 83).

The two mexican specimens, which I have seen, are much smaller than *P. superbiens* from South America, which I saw in Vienna, but there is a great deal of analogy in the coloring of the two, and they may possibly belong to the same species. In Turin I had no copy of the Novara work at hand, in order to compare the description with those specimens.

20. **Culex.** In the British Museum I found the following typical specimens of Mr. Walker's species: *excitans*, one specimen, *excrucians* two, *impatiens* four ($\male\female$), *impiger* two, *implacabilis* one, *provocans* two, *stimulans* one, *territans* two, *preturbans* one. Many of them are unrecognizable. **Culex cont:rrens** Walker, a fragment, is evidently *C. ciliatus; Culex sollicitans* is *C. taeniorhynchus*.

21. **Anopheles.** About the european species of this genus, compare Loew, Dipterol. Beiträge I.

22. **Chironomidae.** Mr. van der Wulp has made a particular study of this family and has introduced several new generic groups. Compare his articles in the Tijdschr. Entom. Nederl. Ver. 1859, T. 2, 1, p. 3—11; also l. c. in 1874; but especially the chapter on Chironomidae in his larger work: Diptera Neerlandica.

23. **Tanypus.** There is a Monographia Tanypodum Sueciae by Fries. 1823.

24. **Ceratopogon.** J. Winnertz, Beitrag zur Kenntniss der Gatt. *Ceratopogon*, in the Linnaea Entomol., Vol. VI (1852, contains a monograph of the european species, with remarkably fine plates. Unfortunately, Mr. Winnertz did not subdivide the genus in smaller genera, but left it, as it was, and still is, a congeries of heterogeneous forms. A beginning of such a subdivision may be found in Westwood's Synopsis, etc., p. 125; compare also Rondani, Prodr. I, p. 175, and v. d. Wulp, Diptera Neerlandica

25. On the **Tipulidae brevipalpi**, compare my Monograph, in the 4th Volume of the Monographs of North American Diptera, published by the Smithsonian Institution, in January 1869.

For many years, I have made a particular study of the Tipulidae, and of the *brevipalpi* especially. This study has enabled me to contribute something towards a better distribution of this group, but has, at the same time, thoroughly opened my eyes to the still remaining blanks in that classification. For from concealing there defects, I have carefully pointed them out in my volume. The Eriopterina especially, require a more thorough investigation, based on more abundant materials than I had at my disposal; the relations of *Goniomyia* to *Gonophomyia* must be more clearly defined; the genera *Cladura, Sigmatomera, Phyllolabis,* as they stand now, come within dangerous proximity of the *Limnophilina* and their true position is still a problem. In the Limnophilina, the numerous species of *Limnophila*, require a better grouping: I have shown, for instance, on p. 201 and 230, that the presence of four, or of five posterior cells, is an altogether secondary character and that some species with four cells, like *C. quadrata,* are very closely related to some other species, with five cells. Numerous hints of that kind will be found in my volume, hints which, at that time, it was not possible as yet to develop: but in order to be made use of, there hints must be sought in that volume, and not in the adaptations of my classification in other writers. Most of the entomologists who have adopted my classification, have become acquainted with it through Dr. Schiner's work. But that work was based on my earlier essay (1859), and does not contain the improvements, introduced in my later, and more voluminous, publication of 1869.

26. **Limnobia simulans.** I prefer to retain the name which I gave to this species: Mr. Walker's description is absolutely unrecognizable, as I have shown in Monogr. IV, p. 41.

27. **Trochobola argus.** This species hardly differs from the european *Trochobola annulata* Lin. (Syn. *imperialis* Loew). During my presence in London in July 1877 I had occasion again to see Linné's type of *Tipula annulata* in the Linn. Society and can only confirm the statement which I made after my previous visit to the same institution, twenty five years ago: that *Tipula annulata* Lin. is the same as *Limnobia imperialis* Loew. (See Stett. Ent. Zeitschr. 1857, p. 90.) The specimen is a fragment, but the supernumerary crossvein is distinctly visible on the wing. Thus much in answer to Prof. Zetterstedt's doubts in the Dipt. Scand. Vol. XIV, p. 6534. The fact that Prof. Zetterstedt, during his long dipterological career, never came across a swedish specimen of this insect, is curious. By and by it will be found there. In the mean time, Prof. Mik in Vienna showed me specimens which he caught in Upper Austria and in Gastein, Styria. The Imperial Museum in Vienna (Collect. Winthem), contains a specimen from Lyon, France. It seems to be a nothern and alpine species; and many alpine forms (for instance Parnassius Apollo), occur in the mountains of the Dauphiné not far from Lyon. Prof. Mik also found *Limnobia caesarea* O. S. near Gastein.

28. **Diotrepha** nov. gen. Related to *Orimarga* (compare the figure of the wing in Monographs, IV, Tab. I, f. 8), but the posterior branch of the fourth vein is not forked, so that there are only *three* posterior cells; the small crossvein is nearer to the apex of the wing; the great crossvein, on the contrary, is much nearer to the root of the wing, far anterior to the origin of the second vein. Being thus placed in a situation where the longitudinal veins come very close together, this

crossvein is short and may be easily overlooked. The wings are very narrow; the body delicate, the legs long and very slender; empodia distinct.

D. mirabilis n. sp. About 6 mm. long, brownish, very slender, with long, exceedingly delicate, white legs; the tips of the femora and of the tibiae, brown. — Georgia; Texas.

I am not able, at present, to give a better description of this species; still, its characters are to striking that it will be easily recognized. I first took it in Georgia, in 1858, and did not publish it, not knowing where to place it. Later, I sent it to Dr. Loew and did not have it before me at the time of the publication of Monographs, Vol. IV. During my visit to Dr. Loew in 1877, I saw the specimen again and took down a few notes about its characters, thinking that it was related to *Thaumastoptera* Mik. But I have seen the latter in Vienna since and have given up all idea of a relationship.

The type of *D. mirabilis* is now in the Mus. Comp. Zool. in Cambridge, Mass. I have seen a second specimen, apparently of the same species, taken by Mr. Boll in Texas. A specimen from Cuba in Mr. Loew's collection also seems to belong here.

The name *Diotrepha* means *fed by the Gods*.

29. **Rhypholophus fascipennis** Zett. According to Dr. Stein, who quotes Loew *in litt.*, this may be the same as the *R. phryganopterus* of Kolenati (Stein, in Stett. Ent. Zeitschr. 1873, p. 241).

30. **Erioptera** The characters of the subdivisions, established by me in this genus were explained in the Monogr. IV, 151—152. In their application to species from other parts of the world than North America, some of them will hold good, others will require to be remodelled. The subgenus *Erioptera* maintains all its characters in the european species *taenionota* M., *flavescens* F., *fuscipennis* M. (as I saw them named in Mr. Kowarz's collection). *Erioptera maculata* M. is a true *Acyphona*, agreeing in all generic characters with the american species of that subdivision. The definition which Dr. Loew gives of *Acyphona* (Beschr. Europ. Dipt. III, 50) is incomplete and therefore misleading; he evidently based it on my statements in Monogr. Vol. IV, p. 158 only, and overlooked the detailed character of the subgenus, as given on p. 151—152. His *Acyphonae* therefore, are not *Acyphonae* in my sense at all. *Molophilus* is a very well-defined form, existing in Europe and North America. The definition of *Mesocyphona* will require remodelling, as I have stated in the „Western Diptera", p. 199. I have not seen any european species, belonging in it. The structure of the forceps of the male, which untergoes very considerable modifications among the Eriopterae, in the surest guide towards the discovery of affinities; subdivisions, established without the use of that character, are worthless.

In the Monogr. Vol. IV, I have given my reasons for abandoning Dr. Schiner's arrangement of the *Eriopterina*. There is no reason for separating *Rhypholophus* from his *Dasyptera*; and, being united, the former name must be adopted as the earliest. *Trichosticha* Schiner is composed of the most heterogeneous elements: *T. maculata* is an

Acyphona; T. trivialis is a species which requires further study, and seems related to *Trimiera; T. icterica* has an altogether different organisation and has been placed by Loew in his genus *Lipsothrix* (Beschr. Europ. Dipt., Vol. III, p. 68); *T. imbuta* of which I had only a glimpse, seems to be an *Empeda;* the residue (*T. fuscipennis, flavescens, taenionota)* form the bulk of *Erioptera* Meigen, Division A, and should therefore retain that name, even in the ultimate subdivision of the genus: they are my *Eriopterae,* sensu strictiori.

These criticisms, will not, I hope, be considered disrespectful to those two writers, my seniors in Dipterology, and by far my superiors in the knowledge of most of its branches.

31. **Symplecta punctipennis.** Dr. Loew, in his Beschreibungen Europ. Dipteren III, p. 54, observes that Meigen, in his earlier work: *Klassification* etc. called the same species *hybrida,* a name which he afterwards changed, without explaining the reason, in *punctipennis.* Loew therefore recommends the reinstatement of that name, as the earliest. But why should we not, on the same ground, revive the generic name *Helobia* St. Fargeau, which is older than *Symplecta,* and call the species *Helobia hybrida?* And as *Symplecta punctipennis* has been used in all the works and catalogues of diptera in existence for more than half a century, we would never get rid of it, but would have to keep both names in our memory for ever. For this reason, I do not share the opinion of my esteemed friend and correspondent.

32. **Goniomyia.** I am aware of the existence of *Goniomya* Agassiz (Mollusca), but the derivation, at well as the termination of that name are different.

33. **Limnophila humeralis** Say. Journ. Acad. Phil. III, 22, 5; Compl. Wr. II, 47. Wiedemann unites this species with *L. tenuipes* Say, apparently deriving his opinion from the comparison of original specimens. Nevertheless, Say does not seem to have been of the same opinion. In a MSS. note in his handwriting, which I found in a copy of Wiedemann's Auss. Zw., which he had used, he refers *L. tenuipes* to *L. gracilis* Wied. The book is now in the library of the Academy of Natural Sciences in Philadelphia. (Compare also Monogr. etc IV, p. 41.) A specimen in the Winthem collection in Vienna, which I take to be the type of the description of *L. gracilis,* in labelled *tenuis* W.

34. **Anisomera.** About the european species, compare Loew in the Zeitschr. f. Ges. Naturw. Vol. XXVI (1865).

35. **Eriocera californica.** In describing this species in the Western Diptera, I mentioned that *Megistocera chilensis* Philippi, was, to all appearances, likewise an *Eriocera.* But I have seen it since in Mr. Bigot's collection; it is a *Megistocera,* that is a Tipulid and not a Limnobid.

36. **Ptychoptera.** The trophi of the larvae of this genus do *not* differ materially from those of the other Tipulidae; the characteristic dentate mentum is present. For this reason I am not inclined to follow Dr. Brauer in attaching to the fact, that the head of those larvae is not imbedded in the thoracic skin (as it is in other Tipulidae) such a

radical importance, as to justify the separation of the group as a distinct family. (Compare Verh. Zool. Bot. Ges. 1869, p. 844.)

37. **Ptychoptera metallica** Walk. The specimen in the Brit. Mus. is a mere fragment.

38. **Idioplasta.** In 1859 I had called this insect *Protoplasa;* in the Western Diptera, 1877, I adopted the more correct *Protoplasta.* But in the mean time, *Protoplasta* had been used in the Protozoa, so I prefer to give it up for *Idioplasta.*

I. Fitchii. I was quite recently that, for the first time, I saw a specimen of this insect again, after those two which I described twenty years ago. The specimen is in Mr. von Roeder's collection, in Hoym, Germany. It is a male, and has a forceps with very long branches. This proves that the specimens which I described and about the sex of which I was uncertain, were females. And it further proves that the female in this genus does not have the sabre-shaped, projecting ovipositor, which is usual among the Tipulidae. *Idioplasta,* in this respect, resembles *Bittacomorpha,* and differs from *Ptychoptera.*

The specimen in question was taken in Georgia, by Mr. Morrison, a collector who has the faculty of ferreting out the rarest insects, whatever country he undertakes to explore.

39. **Tipula.** Compare the important remarks on the structure of the genitals of *Tipula,* in Loew's Beschr. Europ. Diptern, Vol. III, p. 7—9.

40. **Tipula nodulicornis.** As to the synonymy of this species, I follow Mr. Schioedte's authority, although I expressed some doubts about it in the Proc. Bost. Soc. Nat. Hist. Dec. 6. 1876.

41: **Tipula casta** Loew, Syn. *cunctans* Say. There is some error at the end of Say's description, as the venation of a *Tipula* cannot well be like that of *Limnobia* (Geranomyia) *rostrata,* to which he apparently refers. This error prevented Dr. Loew from identifying Say's description.

42. **Mr. Walker's Tipulae.** After taking some notes from the types in the Brit. Mus. I hoped to establish the synonymy of some of Mr. Walker's species with Dr. Loew's. But upon comparing Mr. Walker's descriptions with the specimens, I found that they did not agree with what I thought I had seen. So I quote such synonymies with a query.

Tipula alterna Walk. I suspect the synonymy from a short note I made in London in 1859; Mr. Walker's description however renders it doubtful.

43. **Tipula fuliginosa.** Although this species is not rare, I have never seen the male yet.

44. **Ctenophora.** In the Proceedings Entom. Soc. Phil. May 1864, I published an article: Description of several new North America Ctenophorae; an unsatisfactory performance, because I attempted to work without sufficient material.

45. **Ptilogyna fuliginosa** Macquart (non Say) Dipt. Exot. I, 1, p. 46, 1; Tab. III, f. 2, is omitted, because it is an australian, and not a north american, species. I have seen the original type of Macquart s in Lille. It is a very well preserved female specimen, with pectinate

antennae, labelled North America. But I have also seen several specimens of the same species in Mr. Bigot's collection in Paris, all from Australia. Macquart taking the species for north american, had erroneously identified it with *Ctenophora fuliginosa* Say, which is a Tipula. Dr. Loew (Linn. Entom. V, p. 392) noticing this error, proposed to call this species *Ptilogyna Macquartii*. As it now appears that the species belongs to a different country, there is no reason for not calling it *Ptilogyna fuliginosa* Macquart, only striking out the quotation from Say. *Ptilogyna picta* Schiner, Novara, p. 38 from Sidney is the same species, as any one will perceive by comparing Dr. Schiner's description, with Macquart's figure.

46. **Bolbomyia.** The passage, quoted from Dr. Loew's „Bernstein u. Bernsteinfauna" reads as follows: „A second genus, more or less related to Ruppelia, may be placed among the Xylophagidae, its somewhat aberrant venation notwithstanding. I call it *Bolbomyia* and distinguish two species Characteristic is the shape of the antennae; the third joint consists of four or five divisions, the first of which is much larger and swollen." — The other passage, quoted from Silliman's Journal, only contains a remark about the difficulty of placing this species in any of the adopted families. A passage of the same import is that in the Monographs, Vol. I.

47. **Coenomyidae.** I restore this family, adopted by most of the previous authors, but suppressed in Loew's Monographs, Vol. I. It seems to me somewhat premature to unite it with the Xylophagidae.

47 a. The name **Sicus** was first used by Scopoli (1763), for a species of *Myopa*. — Fabricius, in the Supplement to his Entomologia Systematica (1798), arbitrarily misapplied it to *Coenomyia*, but the latter name having been published two years earlier by Latreille, was maintained.

Latreille (Hist. Nat. des Crust. et des Ins. 1804), used the same name *Sicus* in a third, altogether different, sense, for the genus now called *Tachydromia*. As such, it appears on Meigen's plate 23, in the third volume of his principal work. In the letterpress, Meigen rejects *Sicus* and maintains Tachydromia, introduced by himself in 1803. Latreille preserved the name *Sicus* (for Tachydromia) even in his last work, Familles Naturelles (1825)

Finally, Dr. Schiner revived *Sicus* for the species, for which it was originally intended by Scopoli.

48. **Arthropeas leptis** n. sp. Brownish-gray, wings unicolorous. slightly tinged with pale brownish-yellow. Length 6—7 mm.

Body brownish-gray, sparsely beset with minute yellowish, erect pile Thoracic dorsum brown, with two yellow lines, separating the three usual stirpes, the intermediate one of which is faintly geminate. Head dull grayish, but front and vertex brown, except a narrow gray margin along the orbit. Antennae blackish-brown. Legs brown, tibiae yellowish-brown; coxae grayish. Wings unicolorous, slightly tinged with pale-brownish; stigma brownish-yellow. Halteres yellow, with a brown knob.

Hab. White Mts., N. H. (E. P. Austin; his labels were marked: „woods" and „alpine"). Three females, only one of which is well preserved; the other is greazy; the third teneral, and for this reason of a uniformly reddish color.

This remarkable insect looks like a Leptid with the antennae of *Coenomyia.* I refer it to the genus *Arthropeas* Loew, Stett. Zeit. 1850, with which it seems to agree in the generic characters. It differs from the figures given by Dr. Loew, in having the anal cell open, the discal narrower, the posterior cells 2, 3, 4 longer. The second posterior cell is very narrow at base and the upper branch of the third vein is not bisinuate. I cannot at present compare this species to *A. americana,* and cannot therefore tell whether the structure of the face is the same in both. In *A. leptis* two deep, diverging furrows, run from the base of the antennae to the oral edge, and divide the face in three portions. Besides *A. sibirica, americana* and *leptis,* a species of the same genus, *A. nana,* occurs in amber. The doubts of Dr. Loew about the systematic position of *Arthropeas* are revealed in the fact, that he refers it to the *Coenomyidae* in the Stett. Zeit. and to the *Acanthomeridae* in the pamphlet: Der Bernstein und die Bernsteinfauna, although both papers appeared in the same year 1850.

The genus *Coenura* Bigot, from Chili (Ann. Soc. Entomol. de France, 1857) is most closely allied to *Arthropeas* and has even, in the coloring of the species described a certain family resemblance to *A. sibirica.* In fact it remains to be shown yet, in what the difference between the two genera consists.

49. **Beris.** Compare Loew, Stett. Entom. Z. 1846, p. 219 sqq.: Bemerkungen über die Gatt. *Beris.*

50. **Exaireta** Schiner. There exist the following, similar names: Exaerete, Hymenopt. 1848; Exaeretus, Hemipt. 1864; Exaereta, Coleoptera 1865. About the relation of *Exaireta* to *Diphysa* Macq. compare Nowicky, Beitrag zur Kenntniss der Dipterenfauna Neuzeelands, Krakau, 1875, p. 12.

51. About **Sargus** and the allied genera, see Loew's essay in Verh. Zool. Bot. Verein 1855. A great deal remains to be done as yet for the classification of the exotic species of *Sargina.* I did not attempt to refer the species which I have not seen to the newly-formed genera to which they may belong, but left them in the genus *Sargus* in the old acceptation.

52. As there is an earlier **Chrysomyia** R. Desvoidy, 1830, I revived the name of *Chloromyia* Duncan, in my Western Diptera, p. 212. Macquart himself acknowledged the priority of *Chrysomyia* Desvoidy in Ann. Soc. Ent. 1847, p. 75.

53. **Ptecticus.** In Mr. Loew's paper on *Sargus,* where this genus is introduced, it is always called Ptecticus; on the plate, it is called *Plectiscus,* and Gerstaecker (Entom Ber. 1855, p. 127) adopts the latter version. Mr. Loew told me that Ptecticus was the correct form.

54. **Oxycera** Compare on the european species a paper by Loew, in his Dipterol. Beiträge, I, p. 11 (1845).

Also by the same: die europ. Arten d. Gatt. *Oxycera,* in the Berl. Ent. Z. Vol. I, p. 21.

55. The paper by Gerstaecker referred to here is entitled: Beitrag zur Kenntniss exotischer Stratiomyiden, and is an important contribution to the classification of this family. The name **Euparyphus** can stay, although there is a much earlier genus *Euparypha* in the Mollusca, 1844.

56 Compare Loew, **Odontomyia,** in the Linnaea Entomologica, Vol. I, p. 467, a review of the european species.

57. **Odontomyia limbipennis.** The label in Macquart's handwriting in Mr. Bigot's collection bears *America,* with a query; the query is omitted in the Dipt Exot. I doubt that this is a north american species.

58. Compare **Stratiomys** by Loew, in Linn. Ent., Vol. I, p. 462. Review of the european species.

Also Gerstaecker, Linn. Ent. XI, p. 317, where some important remarks on exotic species will be found.

59. In Dr. Gerstaecker's article on exotic **Stratiomyidae** (Linn. Ent. Vol. XI, 1857) the genus *Cyphomyia* is treated monographically and with great completeness. He enumerates twenty four species.

A Synoptic List of the known *Cyphomyiae* is given by Bigot, Ann. Soc. Ent. 1875, p. 483.

60. **Clitellaria.** Compare Loew's remarks about this genus and *Ephippium,* in his Beschr. Europ. Diptern, Vol. III, p. 73.

61. There is a **Rondania** Bigot (Essai d'une Classific. 1853, Tipulida), and a still earlier *Rondania* R. Desvoidy 1850, Muscida.

62. A monograph of the european species of **Nemotelus** is given by Loew, in the Linn. Ent., Vol. I. See also Loew, Beschr. Europ. Dipt. II, p. 44, obs. 2.

63. Compare Loew: Revision d. Europ. **Pachygaster**-Arten, in the Zeitschr. f. Ges. Naturw. Vol XXXV; 1870.

64. Compare: Osten Sacken, Prodrome of a Monograph of the Tabanidae of the United States (in the Memoirs of the Boston Society of Natural History, Vol. II, 1876, p. 365—397 and p. 421—479; and a Supplement p. 555—560).

65. **Pangonia.** Compare: Notice sur le genre Pangonie, by Macquart, Ann. Soc. Ent. Fr. 1857, p. 429—438, Tab. XV; and Loew, Neue Dipt. Beitr. VI, p. 23; 1859 (european species).

Macquart, l. c. says that the genus Pangonia was established by Latreille, in the *Dict. d'Hist. Naturelle* of Déterville. I cannot now verify this quotation; at any rate the publication cannot have been earlier than 1802, because the dictionary bears the dates of 1802—1804.

66. **Silvius isabellinus** Wiedemann, the type of which I have seen in the Berlin Museum, is not a Silvius, but a *Pangonia.* It looks like a very pale-colored *Pangonia pigra* and may be that very species.

67. About the european species of *Chrysops,* compare: Loew, Verh. Zool. Bot. Ges. 1858, p. 613—634.

The knowledge of this genus and the porper method for the discrimination of the species date from this paper. Descriptions of earlier

18

writers, even those of the usually so accurate Wiedemann, are not to be relied on. I had an opportunity to convince myself of it, in Vienna. My examination of Wiedemann's types was confined to *Chrysops obsoletus,* Wied., as the type of *C. lugens* must be in Copenhagen, that of *plangens* in Berlin, and *C. flavidus* and *vittatus* cannot be doubtful; *C. fuliginosus,* which should be in Vienna, I did not find. *C. obsoletus* is represented in Winthem's collection by a single female, marked as a type. This specimen does not agree with Wiedemann's own description, because he compares the wings of *obsoletus* to those of *C. laetus* from Brazil, which species has *both* basal cells hyaline, while the typical specimen in question has the first basal cell brown and answers the description of my *C. morosus.* In Wiedemann's collection there are three specimens; one of them bears a label in Wiedemann's handwriting „*obsoletus m.*“; it agrees with the above-mentioned specimen in Winthem's collection; so does the second specimen; but the third (evidently the one to which Wiedemann alludes in his description as a variety, received from Pennsylvania) is a different species, I think that which I described as *univittatus* Macq In adjusting the nomenclature so as to bring it into agreement with these facts, we would only involve it into a hopeless confusion; and for this reason, it will be much preferable, I think, in this, as in other similar cases, to take the nomenclature of my Prodrome, however imperfect, as the basis for future work, and to let alone the older descriptions. This applies of course, *a fortiori,* to the descriptions of Macquart and Walker.

68. **Chrysops obsoletus.** Wiedemann's description, as I have shown in the preceding note, agrees with my *C. obsoletus,* but disagrees with the typical specimens in his own collection. Furthermore, one of these types (mentioned in the description as a variety), belongs to a different species. For the reason stated in that note, I do not change the nomenclature of my Prodrome.

69. **Chrysops quadrivittatus.** I did not possess this species, when I published my Prodrome. I found it since among the specimens from Dr. Heyden's collecting in Nebraska, which years ago, I had communicated to Dr. Loew.

70. On the european species of **Silvius,** see Loew, Wien. Ent Monatschr. 1858, p. 350; see also this genus in the same author's South African Fauna.

71. **Silvius gigantulus.** Mr. Loew mistook this species for a *Chrysops* and thus I overlooked it in preparing my Prodrome and described it again as *Silvius trifolium.* Mr. Loew's name has of course, the priority, although it is somewhat unbecoming, since the species would have been gigantic for a Chrysops, but is not for a Silvius.

72. **Tabanus carolinensis** Macq. I have seen the types in the Jardin des Plantes. I do not know the species.

73. **Tabanus flavocinctus** Bell. is *Tabanus zonalis;* it cannot well come from Mexico. The specimen had been received from the Museum in Paris, and an error of locality must have occurred.

74. **Tabanus nigropunctatus.** This is a regular Therioplectes, the eyes are pubescent, and not glabrous, as mentioned in the Saggio etc. Wiedemann notices the ocelligerous tubercle!

75. **Tabanus.** Compare Loew, in the Verh. Zool. Bot. Ges. 1858, p. 573—612; a paper on the european species.

I have taken great pains, in Paris and in Vienna, to verify my identifications of Macquart's and Wiedemann's descriptions of *Tabanus* and I have had the satisfaction of finding them justified in all instances, with the single exception of *T. sulcifrons* Macq. In examining Wiedemann's and Winthem's collections in Vienna, great care should be taken to discriminate the true types, from specimens that are not types, even when labelled in Wiedemann's own handwriting. I have explained in the Preface, some facts bearing on the distribution of the types in those collections. The types of Wiedemann's N. A. Tabani are now all in Winthem's collection. The Tabani in Wiedemann's collection are sometimes wrongly named. Thus *T. Reinwardtii* is represented by three specimens, which are not that species at all; Wiedemann described a female with spotted wings; those three specimens are males and have immaculate wings. The true type is in Winthem's collection. In the latter collection, there are likewise several wrongly named Tabani, of course, not types. *T. zonalis* is labelled *T. flavipes* Wied. with a query; the type of Wiedemann's description is in Copenhagen. *T. fuscopunctatus* Macq is labelled *variegatus* Fab. etc.

After having gone through the labor of examining so many types of earlier writers. I have become more than ever convinced of the necessity of basing our nomenclature *on recognizable descriptions* and not merely on typical specimens. And for this reason I have preferred to leave the nomenclature of my monograph, as much as possible, undisturbed, until another entomologist is in a position again to subject the whole genus to a thorough revision.

76. **Tabanus abdominalis** Fabr. is represented in the Museum of the Jardin des Plantes by two specimens, both of which have the first posterior cell closed, thus confirming the view I took of the synonymy in my *Prodrome.*

77. **Tabanus catenatus.** As I suspected in my Monograph, *T. catenatus* Walker is represented in the Brit. Mus. by specimens belonging to two different species; but it turns out upon examination of these specimens, that neither of them is my *T. catenatus.* One of them is the pale-colored variety of *T. turbidus* Wied., the other is *T. giganteus* (lineatus F.).

Thus *T. catenatus* Walker must be cancelled; *T. recedens* of the Brit. Mus. is my *catenatus;* but Walker's description *(cinereus* etc.) is not recognizable; my mention of it in Prodr. II, p. 434 was based upon a recollection, dating from my visit in the Museum in 1859. The species may remain as *catenatus,* O. S

78. **Tabanus hirtioculatus.** I have seen the original specimen in Mr. Bigot's collection and do not doubt the correctness of the synonymy.

Nevertheless, as Mr. Macquart's description is very unsatisfactory, I prefer to retain the name which I gave to this species.

79. **Tabanus cheliopterus** Rondani I have seen the original type of the description, preserved in the Royal Museum in Turin. It is a very much rubbed female specimen, which seems to belong to *T. fronto*. Of the white abdominal triangles, not a vestige is left, which explains their being omitted in the description.

80. **Tabanus imitans** Walker. (Syn. of *T. fuscopunctatus* Macq.). In order to understand Walker's description, it must be borne in mind that the *T. abdominalis*, to whom he compares it, is not that species at all, but the same *T. fuscopunctatus* Macq.

81. **Tabanus gracilis** Wied. Wiedemann's description was drawn from a single specimen, the hind legs of which were wanting. There are two specimens in the Vienna Museum (Winthem collection), one of which answers this description. It is of the size and shape of my *T. longus,* but more reddish, the wings more tinged with brownish etc. The abdominal pattern is very much faded. It seems to be a species which I do not know, but which is closely allied to my *longus*.

82. **Tabanus lineola** Macq. Dipt. Exot. I, 1, 146, 49 must be some other species than *lineola* Fab

83. **Tabanus sulcifrons**. The type, in Mr. Bigot's collection, is my *tectus*. As the description is sufficiently recognizable, I admit the priority. Macquart has *fulcifrons*, which, of course, is a misprint.

84. **Tabanus turbidus**. The type, now in Winthem's collection has very pale-colored wings.

85. **Tabanus unicolor**. The type in Mr. Bigot's collection is an unrecognizable specimen, perhaps *T. tener;* however there is an earlier *T. unicolor* Wied. from Brazil. Mr. Rondani (Archivio etc. Canestr. III, fasc. I, 1863) proposed to call the species *T. lateritius,* instead of *unicolor;* but the species, as a hopelessly doubtful one, be better cancelled.

86. **Tabanus variegatus** Fab. The type in Fabricius collection, from which Wiedemann's description was drawn, being probably destroyed, this will remain a doubtful species. The specimen in Winthem's collection *(not type)* is *T. fuscopunctatus* Macq It is very probable that my interpretation of Wiedemann's description is the correct one.

87. **Tabanus marginalis** Fab. Wiedemann says: „Die Art phrase habe ich nach einem sehr schön erhaltenen Exemplare des Wiener Museums verbessert etc." I looked for this specimen in the general collection, in Vienna, but could not find it. In the Winthem collection a specimen labelled *marginalis* Fab. var. and marked as type, is my *T. cerastes*. It cannot well be the specimen described by Wiedemann, because he would have noticed the peculiar structure of the antennae (at present, these are broken in the specimen). At any rate the *T. marginalis* of Fabricius is, and will remain a doubtful species, and be better dropped.

88. **Tabanus quinquevittatus**. In the Winthem collection (Vienna) there is a ♂ and a ♀ (both marked as types), from Savannah, and not

from Mexico. They look exceedingly like *costalis*. Of *T. costalis*, the types in Wiedemann's collection are very poor specimens, and for this reason, probably, his description is unrecognizable.

89. **Tabanus fulvescens** Walker. I have seen Walker's type in the Brit. Mus.; it is *T. bicolor* Wied. What I described as *T. fulvescens* is very probably only a variety of *T. bicolor*, with gray, instead of yellowish pleurae. A similar variety occurs in *T. fulvulus*.

90. **Tabanus Craverii.** May possibly be an *Atylotus*. The typical specimens, females, looked very much that way.

91. Mr. Loew *(in litt.)* proposes to divide in the *Leptidae* two sections:

    I. **Psammorycterina**, without facial swelling and with a strong spur on the front tibiae; genera: 1. *Pheneus*, as the typical genus, closely allied to: 2. *Psammorycter* (Syn. Vermileo); 3. *Triptotricha*.

    II. **Leptina**, with a facial swelling, but without spur on the front tibiae; all the other genera.

About *Leptidae* compare also *Frauenfeld*, Verh. Zool. Bot. Ges. 1867, p. 495.

92. **Leptis Servillei** Guérin. I suspect this is nothing but *Chr. ornata* Say. But the femora are said to be brown? The figure however does not show it.

93. **Atherix filia** Walker; is either *punctipennis* Say, or *plumbea* Say.

94. **Leptis cinerea** Bell according to the description, cannot well belong to *Leptis;* compare antennae, shape of anal cell etc. [Loew, *in litt.*]. The type in Mr. Bellardi's collection is, unfortunately, nearly destroyed only the thorax and wings are left.

95. **Spania edeta**; the specimen in the Brit. Mus. seems to be a real *Spania*, that is a Leptid with a stout, styliform arista.

96. **Glutops.** I am uncertain about the position of this extraordinary genus, but prefer this place to any other.

97. H. Loew's Monograph: Ueber die Europäischen Raubfliegen (Diptera Asilica), in the Linn. Ent Vol. II, III, IV: Suppl. in Vol. V, 1847—1851, laid the foundation to the systematic distribution of this family. This work was supplemented by him in numerous later publications, especially in the: Bemerkungen über die Familie der Asiliden, Berlin 1851, and Die Diptern-Fauna Südafrica's, Berlin 1860. About the exotic Asilidae, the following important papers by Dr. R. Schiner may be consulted:

    1. Die Wiedemannn'schen Asiliden (in the Verh. Zool. Bot. Ges. 1866, p. 649—722; Nachtrag, p. 845—848). The usefulness of this paper is somewhat impaired in consequence of the misapprehension under which it was written, about the distribution of Wiedemann types between the so-called Wiedemann's and the Winthem's collections, now both in the Vienna Museum. I have explained the whole matter in the preface to this volume. Some curious mistakes have arisen in consequence, as for instance, in the case of *Erax aestuans* (see my note 125). But Dr. Schiner's paper is nevertheless rendered invaluable by a survey of all the

genera of Asilidae (down to 1866) and the analytical tables for their determination, which it contains.

2. Neue oder wenig bekannte Asiliden des K. Zool. Hofcabinets in Wien (Verh. Zool. Bot. Ges. 1867, p. 355—412).

Mr. van der Wulp published a paper, about the Asilidae of the Eastern Archipelago. [Tijdschr. v. Ent. Vol. XV, 1872.]

98. **Gonypes nitidus.** Macquart quotes Tab. XII, f. 7; the comparison however of this figure with the descriptions of *G. nitidus* and *G. Audouinii* in the letterpress shows, that the figure refers to this latter species. The name *nitidus* must be dropped, having been used before; the name *G. gigas*, engraved on the plate instead of *G. Audouinii*, must likewise be erased. The passage in Loew, Linn. Entom. II, p. 395, proposing to adopt the name *gigas* for *nitidus*, was written before Macquart's mistake in the quotation of the figure had been discovered. Schiner did well in proposing a new name for the species. [Communicated by Loew *in litt.*] Mr. van der Wulp makes the same correction in Tijdschr. v. Entom. 1876, p. 172.

99. **Ceraturgus niger,** of which I saw the type in the Jardin des Plantes, looked like a *Taracticus* rather than a *Ceraturgus.* I have not examined it closely, but have had occasion to examine a similar, perhaps the same, species in the Berlin Museum, which is undoubtedly a *Taracticus.*

100. The Mus. Comp. Zool. possesses a number of specimens of a *Microstylum*, which is of the same size as *M. morosum*, but which Dr. Loew, to whom I communicated a specimen, considers a different species, and calls *M. pollens.* It is less intensely black than *morosum*, antennae and legs are often reddish-brown, the bristles on the sides of the thoracic dorsum are yellowish-white etc. As I had no opportunity to make a thorough comparative study of both species, I merely draw the attention of collectors to it.

M. pollens, like *M. morosum*, was taken at Dallas, Texas, by Mr. Boll.

101. **Stenopogon ochraceus** v. d. Wulp. The closed fourth posterior cell makes this species a Scleropogon But if I understand Mr. v. d. Wulp's letterpress, the front tibiae are armed with a spur. How can in this case the species be a Stenopogon?

102. There is an *Archilestes* Selys, Odonata 1862.

103. **Dizonias bicinctus** Loew. Loew describes the male. Specimens often occur without any trace of the white abdominal crossbands; they may however have disappeared since the death of the specimen. The type of *Dasypogon tristis* Walker, which I have seen in the Brit. Mus. is such a specimen.

The female of this species differs very considerably from the male and might easily be mistaken for a different species; I will therefore mention here that head, antennae, and thoracic dorsum are reddish-brown, and not black; the two abdominal crossbands yellow, and not white; legs brownish-red, more or less blackened· on the femora; wings brown; costal vein brownish-yellow. Both sexes were found flying together in the middle of May 1875 near Enterprise, Florida, by M. M. Hubbard and Schwartz.

Dr. Loew acknowledges that the description of *Dasypogon quadri-maculatus* Bellardi agrees with his *Dizonias bicinctus*. The only diffe-rence he finds, consists in the latter not having any white hairs on the front coxae, and having such hairs on the hypopygium. I have seen Mr Bellardi's type; is looks exactly like *D. bicinctus*. I have also seen specimens from the Southern States (in Mr. v. Roeder's collection), which were certainly *D. bicinctus*, although they had some white hairs on the fore-coxae. I doubt therefore the importance of this character, and believe that the synonymy of those two names can be safely assumed.

104. **Cyrtopogon.** To the description of *C. lyratus* n. sp., I add a more complete one of Walker's *C. Lutatius*, and also an analytical table for determining the five species hitherto known from New England.

1. { Scutellum flat, with very few, indistinct hairs *Lutatius* Walker
   Scutellum convex, with distinct, long, erect
   hairs (2)

2. { Third joint of the antennae red . . . . *marginalis* Loew
   Third joint of the antennae black (3) .

3. { Tibiae and tarsi altogether black . . . *lyratus* n. sp.
   Tibiae and tarsi more or less red or yellow (4)

4. { Tibiae red, the tip only black; the male
   with two large black spots on the wings *bimaculatus* Walk.
   Tibiae red at the base only; the male
   without large black spots on the wings *chrysopogon* Loew.

105. **Cyrtopogon Lutatius.**

Dasypogon Lutatius Walker, List, etc. II, p. 357.

Female. Legs black, bristles on the tibiae whitish; mystax white; abdominal segments, except the first, with interrupted crossbands of white pollen near the hind margin; wings hyaline. Length: 7,5 mm.

Front and face grayish pollinose, mystax white; antennae black. Thoracic dorsum clothed with a brown pollen, which forms the usual stripes; the humeral callosities and the sides of the dorsum are cover-ed with a more yellowish-gray pollen, which sometimes also extends more or less distinctly to the intervals between the dorsal stripes and the median line of the geminate stripe; a rather distinct, grayish-white spot on each side of the median geminate stripe, where the thoracic suture reaches it; scutellum rather flat, rugose, with but little hair; grayish-pollinose in the middle, black on the sides; pleurae grayish-pollinose; a shining black spot under the root of the wings; the fanlike fringe of hairs in front of the halteres seems to be mixed of whitish and black hairs. Halteres yellow. Abdomen of very nearly equal breadth (the seventh segment distinctly narrower), convex, black, moderately shining; with microscopic transverse rugosities; first segment with whitish-pollinose spots on the sides; segments 2—7 with crossbands of white pollen posteriorly; interrupted on segments 2—5, subinterrupted, nearly entire, on segments 6—7; they touch the hind margin of the segments on the sides, but diverge from it a little in the middle; the sides of

the abdomen, at the base, are clothed with white hairs; the surface
of the abdomen is clothed with short, microscopic pile, which, in a
certain light, appears golden-yellow. Legs black, tarsi more or less
dark chestnut-brown; femora with the usual white hairs, tibiae with
white bristles, the front pair with some black bristles on the underside.
Wings hyaline; a grayish tinge on the distal half is hardly perceptible;
venation normal.

Hab. Massachusetts; Cayuga lake, New York (Mr. Comstock); Nova
Scotia (Walk.). Two females.

Cyrtopogon lyratus n. sp. ♀. Legs, mystax and antennae altogether
black; thoracic dorsum with a very distinct pattern in whitish pollen.
Length: 13—14 mm.

Female. Head black, densely grayish-pollinose on the face, slightly
on the sides of the front; mystax altogether black; hairs on the occi-
put black above, white below; antennae black, third joint but little longer than the two preceding, taken together. The usual thoracic stripes
are dark brown, the white or yellowish pollen in their intervals forms
the following pattern: a median line, attenuated posteriorly; a figure in
the shape of a tuning-fork, having the end of the handle in front of
the scutellum, connected with the end of the median line; a broad stripe
on each side between the humeral and the antescutellar callosities, attenuated and abbreviated before reaching the latter; these lateral stripes
are twice connected by pollinose crossbands with the branches of the
tuning-fork, the second time, along the thoracic suture. Scutellum black,
with black pile; grayish pollinose anteriorly. Pleurae grayish-pollinose,
with a stripe of more dense silvery-gray pollen on the lower part; the
fanlike fringe of hairs in front of the halteres is black. Abdomen black,
shining, with a bluish reflection on the first five segments; each of these
has a large spot of white pollen on each side, against the posterior
margin; the sides of the abdomen are clothed with white hairs,
which become gradually shorter posteriorly and do not reach beyond
the fifth segment. Legs black; bristles on the tibiae black; femora with
long white hairs on the underside; the last pair also on the upper side.
near the base. Halteres reddish-yellow. Wings hyaline on their proximal half, including the discal cell; the distal half has a slight grayish
tinge; crossveins clouded with brown.

Hab. Catskill Mountain-House, N Y., July; White Mountains,
N. H Three females. The altogether black legs; the strong contrast
between the brown thoracic stripes and the whitish-pollinose intervals
between them; the altogether black beard etc. will help to distinguish
this species.

106. **Deromyia Philippi.** Verh. Zool. Bot. Ges. 1865, p. 705 is
erroneously referred by Gerstaecker, Entom. Ber 1867, p. 99, to *Plesiomma*
Macq. It has a spur on the front tibiae and must be very closely allied
to *Diogmites,* if not identical with it. Schiner (Die Wiedem. Asil., p. 653)
refers it to *Cyrtophrys* Loew.

107. **Dasypogon rufescens;** the synonymy rests on the assumption

(a very probable one), that Macquart overlooked the spurs on the frcnt tibiae.

108. **Diogmites umbrinus.** I am not quite sure whether the specimen of *Dasyp. basalis* Walker, in the Brit. Mus. belongs here or to *Diogmites discolor.*

109. **Diogmites annulatus** Bigot. This species does not belong to *Senobasis* Macq. from which it differs in the structure of the antennae and of the hypopygium. It may be placed provisionally in the genus *Diogmites*, however, as a separate section (Loew *in litt.*).

110. **D. brunneus.** Macquart's synonymy is not to be relied on, as he evidently mixed up several species of Diogmites.

111. **D. Duillius.** The description seems to betray a *Diogmites*, nevertheless certain statements render this interpretation doubtful; hence the isolated position given to this species. (Loew *in litt.*)

112. **Laphria lata.** I have seen the type in Lille and have taken a note, which enabled me to determine a specimen from Louisiana in the type-collection (now in the M. C. Z.).

113. **Laphria Alcanor** Walker, is the variety of *L. thoracica* which has the intermediate abdominal segments beset with yellow pile.

114. **Laphria affinis** Macq., the type of which I saw in Mr. Bigot's collection, looks very much like *L. thoracica* in the variety with altogether black abdominal pile. The description speaks of *white* hairs about the head, which do not exist in *L. thoracica*, but do not shake my belief in the synonymy.

11£ In the Banksian collection, preserved in the Brit. Mus. and containing the types of Fabricius, there is an *Asilus grossus*, with the reference: *Spec. Ins. Nr. 1.* The specimen bears a label *America*, and another label with the word *type*. This specimen is *Laphria tergissa* Say. In the *Species Insectorum* the locality is given simply as „*America*"; in the *Syst. Antl* we find „in America meridionali", evidently a later and probably erroneous addition. In both works however, the „Museum Dom. Banks" is quoted, as containing the type of the description.

116. **Laphria analis** Macq Synonymy hardly doubtful, although Macquart says: „les cinq premiers segments à poils jaunes".

117. **Laphria flavibarbis** Harris. The original type still exists in Dr. Harris's collection, in Boston. I do not think that it differs from *tergissa*. At any rate there is an earlier *L. flavibarbis*, by Macquart.

118. Schiner (l. c. p. 709) places **Laphria rubriventris** Macq., **L. formidolosa** Walk. and **xanthocnema** Wied. in the genus **Andrenosoma**. He is wrong about *rubriventris* which is a *Lampria*.

119. The genera of the Asilina are tabulated by Loew in the Linnaea Entom. III, p. 402 and IV, p. 148; also later in the Diptern-Fauna Südafrika's, p. 143. Compare also Schiner, Fauna Austriaca, Diptera, 1, p 142.

120. **Mallophora scopifer** Wied. It seems probable that Macquart's *M. scopifer* is not the same as Wiedemann's. Schiner, Verh. Zool Bot. Ges. 1866, p. 77, has a *M scopifer* Bell. non Wied. Cuba; which evidently means Macq. non Wied., as Bellardi has no *M. scopifer* at all

and never described any insects from Cuba. In the Diptera of the Novara Expedition, however, Schiner quotes Wiedemann's and Macquart's descriptions as synonymical. I follow Loew, *in litt.* and call the Cuban species *M. Macquartii.* Jaennicke has the same remark about the distinctness of the two species (Neue Ex. Dipt p. 54).

121. There is another **Trupanea** (Promachus) **apivora** Walk., Trans. Ent. Soc. N. S. V, p. 276, from Burmah, which has the same propensity for destroying bees. Mr. Walker's name having the priority, I have named Dr. Fitch's species P. F i t c h i i in the M. C Z. collection.

122. **Promachus quadratus.** Observe the misprint in Wiedemann's diagnosis: ♂ for ♀; correctly given in his Dipt. exot.

123. **Promachus fuscipennis.** The identity of Macquart's and Bellardi's species seems doubtful.

124. **Promachus quadratus** Bell. If this species does not turn out to be a synonym of some other, the name will have to be changed, on account of *P. quadratus* Wied.

125. **Erax aestuans.** I have seen Wiedemann's type in the Winthem collection; it is the *Erax aestuans* of the Mus. Comp. Zool. Schiner's statements (Verh. Zool. Bot Ges. 1866, p. 686) are based upon a misapprehension of the true type of Wiedemann, a misapprehension the source of which has been explained by me in the preface to this volume. But although the question of *Asilus aestuans* Wiedemann is thus settled, the identity of this species with *Asilus aestuans* of Linné and Fabricius may still be called in doubt, as the descriptions of both authors speak of *three* white segments on the abdomen of the male, while *A. aestuans* Wied. has only two. Harris's Ins. Inj. to Veget. 3d edit., Tab. I, f. 4, shows only *two* stripes. Compare also the note 128

126. **Erax ambiguus, interruptus, argyrogaster, maculatus.** Macquart's types of these species, which I have seen in the Museum in Lille and in Mr. Bigot's collection, look very much alike. However, I did not compare them with the descriptions; the latter, which I have read since, show that *argyrogaster* has a large male hypopygium, *ambiguus* a remarkably small one for an *Erax.* *E. maculatus,* judging from the figure, has likewise a large hypopygium. For the species which I have seen from Texas I preferred the name of *ambiguus,* as the most certain; the hypopygium of the male, in this species, is remarkably small for an *Erax.* I admit at the same time that the female of this species looks exactly like the figure of the female of *E. maculatus* in Macq. D. Exot. I, 2; Tab. IX, f 6. Schiner (Verh. Zool. Bot. Ges. 1867, p. 393) compares E. maculatus to its next relative, E striola, the specimens of both being from Brazil.

127. **Asilus apicalis.** Wiedemann's type, a female, was in his collection, but is no more in it. See Schiner, l. c. — Walker, List, etc. VII, p. 619, puts this species in the genus *Erax,* where indeed it may belong.

128. **Erax lascivus.** All that Schiner (Verh. Zool. Bot. Ges. 1866, p. 686, Nr. 63) says about this species, results from the misapprehension under which he was laboring. See my note 125.

129. **Eristicus** is preoccupied by Wesmael, in the Ichneumonidae, 1845.

130. **Proctacanthus fulviventris** Macquart. The length is said to be *four* lines, an evident misprint for *fourteen*, as appears from the comparison to *rufiventris* (Loew *in litt.*).

131. **Asilus agrion.** I have seen the original specimen in the Senckenberg Museum in Frankfort. It is nearly eaten up by *Anthrenus*, the abdomen being entirely gone, but it seems to be *Proctacanthus Milbertii;* compare however the description with the specimens of the latter.

132. **Asilus** is understood here in the wider sense, in order to include the species of former authors which I could not place anywhere else.

132. **Asilus apicalis** Bellardi. There is another *Asilus apicalis* Wied.; see *Erax.*

134. Both names, **Mochtherus** and **Itamus** are preoccupied by Schmidt-Goebel in the Carabida, in 1846. (See Marschall's Nomenclator.)

135. **Asilus gracilis** Wied. Very peculiar species, the type of which still exists in Vienna. Schiner (Verh. Zool. Bot. Ges. 1866, p. 686), is of opinion, that it may provisionally be placed in the genus *Mochtherus.*

135 a. **Ommatius.** Mr. Bigot has an article about this genus, with the list of all the described species, in the Annales Soc. Entom. 1875, p. 237—248.

136. **Ommatius marginellus.** Compare also Schiner, Verh. Zool. Bot. Ges. 1866, p. 682: „Very like *O. tibialis* but differs in the bristles of the mystax being black (and not snow-white as in *O. tibialis*) and those on the hind femora being of the same color (and not altogether or prevailingly yellow, as in *O. tibialis*) "

137. **Midaidae.** Compare the essay on this family by Gerstaecker in the Stett. Entom. Zeitschr 1868, p. 65--103 (with a plate): Systematische Uebersicht der bis jetzt bekannt gewordenen **Mydaiden.** Earlier monographs where given by Wiedemann and Westwood.

138. About **Mydas** and **Midas** see in Gerstaecker, l. c. With Wiedemann and others I prefer Midas.

139. **Midas audax.** O. Sacken, Bull. Buff. Soc. N. H. 1874, p. 186. ♂. — Black, second abdominal segment red *on the dorsal, as well as on the ventral side;* head, thorax and first abdominal segment with whitish hairs. *Length:* 23 mm. *Wing:* 18 mm.

Very like *M. clavatus* in its coloring, but easily distinguished by its smaller size, comparatively broader head, more cylindrical shape of the abdomen, by the red color of the second segment, which does *not* encroach anteriorly, on both sides, upon the first segment (as it does in *M. clavatus*), which exists on the ventral as well as on the dorsal side of the segment, and which is *not* interrupted on the dorsal side by a more or less distinct black spot; finally, by the whitish pubescence on the head, the thorax and the first abdominal segment. Head black, broader than the thorax, clothed with soft, white hairs, mixed with black ones; the white hair is especially apparent on the vertex and the sides of the front, also as a small tuft on each side under the antennae, near the orbit of the eye, and as a border round the clypeus.

Thorax black, opaque; the dorsum clothed with white hairs, forming four longitudinal bands, especially visible from a side view. First segment of the abdomen black, opaque, clothed with long, soft, erect white hair, which reaches down to the hind coxae; second segment shining, yellowish red, the remainder of the abdomen black, moderately shining. Venter black, except the second segment, which is yellowish red. Halteres and feet black, pulvilli brownish (of a darker color than in *M. clavatus*). Wings strongly tinged with brown, and with a slight purplish reflection. Venation like that of *M. clavatus*.

Belongs to Gerstaecker's first tribe, that is, it has spurs at the tip of the tibiae and a small cross-vein on the posterior border of the wing.

A single male discovered in the environs of Mammoth Cave in Kentucky, by Mr F. G. Sanborn, in June, 1874.

Midas carbonifer O. Sacken, l. c. ♂. — Altogether black, thorax opaque, abdomen shining, wings brown. *Length:* 22 mm. *Wing:* 18 mm

Black, front and epistoma shining, beset with black hair; antennae black, the expanded portion of the third joint brownish, and beset with a fine grayish pollen Thorax opaque above, showing two velvety black longitudinal lines. Abdomen black, shining, except the first joint, which is opaque. Feet black; ungues reddish, with black tips; hind tibiae beset with strong spines, except toward their base; terminal spur strong. Halteres black; wings dark brown, with a violet reflection; the brown somewhat fainter in the centre of several cells, and along the posterior margin. Small cross-vein on posterior margin present.

Habitat, Norton's Landing, Cayuga Lake, N Y. A single female taken in July by Mr J. H. Comstock. This species seems not unlike *M. crassipes* Westw. in coloring, but is much smaller, has much darker wings, an opaque (and not shining) thorax, etc. (I never saw Westwood's species.)

Midas chrysostomus O. Sacken, l. c. ♂. — Black, face with a tuft of golden hair, abdominal segments 2, 3, 4 with red margins posteriorly, legs black, wings tinged with brown. *Length:* 25—30 mm. *Wing:* 21 mm.

Black; the incrassated portion of third antennal joint dull reddish, except the tip, which is blackish. Face with a tuft of golden yellow hair. Thorax of a smoky black, opaque above. Abdomen black, shining, except the first segment, which is opaque; a narrow band on the posterior margins of the 2 d, 3 d and 4 th segments rufous, edged with yellow along the margin: on the 4 th segment this band is much narrower and somewhat indistinct in the middle. Feet black; hind tibiae with a strong spur; hind femora with two rows of short, but strong spines on the underside; ungues dull reddish, tipped with black. Halteres black. Wings strongly tinged with brown, although less so than in *M. clavatus*. Small cross-vein on posterior margin present.

Habitat, Dallas, Northern Texas. A single male collected by Mr. Boll. This species seems to have many characters in common with *M. fulvifrons* Illig. but it differs in the coloring of the abdomen.

140. Bibio illucens. Fabricius, in the System. Ent., perhaps in consequence of a *lapsus calami*, writes *illucens* for *filata* and *vice*

versa. In the Spec. Insect., as if becoming aware of his error, he correctly quotes System. Ent. 756, 1 (which in *B. illucens*) as a synonym of his *B. filatus.* Wiedemann, in *Monogr. Midar.,* and Westwood, *Arcana,* quote correctly *B. illucens,* System. Ent. 756, 1; Gerstaecker erroneously *B. filatus,* System. Ent. 757, 2 (which is *Hermetia illucens*).

141. **Nemestrinidae.** Dr. Loew (Dipternf. Südafr. p. 245) proposes to call this family *Hirmoneuridae;* Dr. Schiner (Novara, p. 105) opposes the change.

142. **Hirmoneura clausa.** Since describing this species, I have seen several specimens of a *Hirmoneura* brought by Mr. Morrison from Colorado. It has the second posterior cell open.

143. **Bombylidae.** In my *Western Diptera,* p. 225, I have given a synopsis of all the genera of this family hitherto found in the United States; and also, in the larger genera, a review of all the species, which may facilitate determination

144. **Anthrax californiae.** I could not find the original specimen in the Brit. Mus.

145. **Exoprosopa philadelphica.** This seems to be a small variety of *E. fascipennis;* I have met with such specimens several times.

146. **Exoprosopa rubiginosa.** Probably a denuded *E. fasciata;* anyhow a wretched description; the name be better dropped. (I have seen the type since writing this note and confirm my statement.)

147. **Exoprosopa eremita.** Is not this species only a variety of *E. pueblensis?*

148. **Exoprosopa ignifer.** Walker contradicts himself about this species; in the Dipt. Saund. p. 166 he places it among the species with two submarginal cells; later, he puts it in Wiedemann's Division I, the species of which have three such cells.

149. **Exoprosopa trimaculata** Walk. Same remark as in the preceding note.

150. **Anthrax.** A number of Macquart's species in this genus, especially of those with hyaline wings, will have to be cancelled, as the descriptions are absolutely unmeaning and evidently based on miserable, rubbed off specimens. Such are: *A. connexa, albipectus, gracilis.*

Of *Anthrax hypomelas* and *Bastardii* I have seen the types.

151. **Anthrax halcyon.** Macquart's specimen is from Carolina and may perhaps, belong to *A. Ceyx* Loew?

152. **Hemipenthes seminigra.** I suspect that this species is the same as *H. morioides* (Say). Compare O. Sacken, Western Dipt., p. 241.

153. **Argyramoeba georgica.** This synonymy is admissible on the supposition only that Macquart had a female before him, and not a male, as he states. The figure of the wing seems convincing. I do not quote *A. analis* (Say) Macquart, Dipt. Exot. II, 1, p. 67, 32, because I suspect that it is some other species.

154. **Argyramoeba fur** O. S. has the greatest resemblance to *A. binotata* Meigen, of Southern Europe (Fiume and Portugal).

155. **Stygia elongata** Say, *Lomatia elongata* Wied., is not a Lomatia as Wiedemann himself observes, but it is difficult to say,

what it is. It has the antennae of a Leptid, but, nevertheless, only *four* posterior cells. I saw the typical specimen in Vienna and it seemed to agree with Wiedemann's figure. It is singular that another specimen of this species has never turned up in the United States; it would have allowed a more thorough investigation than the fragile type in Vienna, which one is afraid to handle.

156. **Anisotamia eximia** Macq. I doubt very much whether this species is well placed in that genus of Macquart's own creation, but established originally for two African species. It has nothing to do with *Anthrax*, as the bifurcation of the second and third veins takes place long before the small crossvein. It belongs in the group of *Lomatina*, as characterized by me in the Western Dipt. p. 226, and may, at least temporarily, be placed in the genus *Oncodocera*.

157. **Bombylius.** About this and the related genera, see the elaborate paper by Loew, Neue Beiträge, III.

158. **Bombylius aequalis** Harris (nec Fabricius). I have omitted the species of Fabricius' in my list, because it is impossible to make anything of the short description, unless it means *B. fratellus*. Wiedemann's description refers to a different species, and Macquart's apparently again do a different one. The references are:

Fabricius, Mant. Ins. II, 365, 2; System. Antl. p. 128, 2.

Olivier, Encycl, méthod. I, 326, 2.

Wiedemann, Auss. Zw. I, 350, 32.

Macquart, Dipt. Exot. II, 1. 99, 34; Tab. VII, f. 3.

159. **Bombylius fulvita** is. The original type was from Mr. Bigot's collection. I saw two specimens there; the one is perhaps the same as *B. philadelphicus*; the other is *B. atriceps* Loew.

160. **Comastes.** *Bombylius basilaris* Wied. from Brazil and *B. ferrugineus* F. from S. Thomas belong to the genus *Comastes*. In establishing this genus, I was aware of the existence of *Comaster* Agassiz, Radiata, the derivation and termination of which are different.

161. **Bombylius brevirostris.** I saw Macquart's type in the Jardin des Plantes in Paris. *B. L'herminieri*, which is also there is, to all appearances, likewise *Sparnopolius fulvus*.

162. **Adelidea flava** Jaennicke, the type of which I have seen in Darmstadt, appeared to me like a small specimen of *Lordotus gibbus*. The description likewise, reads that way.

163. **Allocotus** Loew, 1872; *Allocotus* Mayr, Hemipt. 1864; *Allocota* Motchoulsky, Coleopt. 1854.

164. **Poecilognathus** Jaennicke, is simply *Phthiria*.

165. **Toxophora leucopyga.** I saw the type in Vienna; it has no longitudinal yellow stripe on the abdomen, thus resembling the figure of *fulva* Gray. Is the *Toxophora fulva*, described by me, which has such as stripe, a different species or a mere variety? I leave the question open.

166. **Epibates.** In establishing this genus, I overlooked the existence of *Eclimus* Loew, Stett. Ent. Z. 1844, which would have very nearly answered my purpose.

Eclimus, however, differs as follows:

1) the face and cheeks are much more projecting, the antennae are comparatively longer (compare the head of *Eclimus* as figured by Loew, Stett. Ent. Z. 1844; Tab. II, fig. 9, 10, with the *Epibates* by Burgess in Proc Boston Soc. N. H. 1878; Tab. IX, f. 1a);
2) the wings have no perceptible denticulations along the costa;
3) each abdominal segment is strongly coarctate at the base, the preceding segment having a corresponding swelling along the incisure; this is especially perceptible in *Eclimus perspicillaris* and *gracilis;* less so in *E. hirtus;*
4) the thorax in the male is not muricate.

I had an opportunity of comparing *Epibates muricatus* with the three species of *Eclimus* in Mr. v. Roeder's collection (in Hoym); probably the richest private collection of Diptera in Europe.

Thevenemyia Bigot has the shining thorax and the projecting face of *Eclimus*, and, at the same time, the muricate thoracic surface of *Epibates* (the latter is not mentioned in the description); it has a longer proboscis than either. These genera may, for the present, remain undisturbed, until a larger number of forms. belonging here, are discovered.

167. **Epibates niger.** The well-preserved male specimen in the Brit. Mus. shows the minute spines on the thoracic dorsum distinctly. I mention this to correct my statement in the Western Dipt., p. 274.

168. The latinized from **Thereva**, adopted universally, seems preferable to *Thereua* recommended by Mr. Loew. It is easier to pronounce like *Evangel* for *Euangel*, *Evander* for *Euander* etc. About the european species, compare Loew, Dipterol. Beiträge, II, 1847.

169. **Thereva candidata.** In Mr. Loew's diagnosis, read *clausa* for *aperta.*

170. There is an earlier **Thereva nervosa** Loew, 1847 (Loew *in litt.*).

171. About the european **Scenopinus**, compare Dr. Loew's article in the Verh. Zool. Bot. Ver. 1857; corrections and additions by the same, in Beschr. Europ. Dipt. III, p. 150—152. An earlier article by him, about the same genus, in the Stett. Ent Z. 1845, p. 312—315.

172. About the *Cyrtidae* there is a monograph by Erichson, in his *Entomographieen* (1840): Die Henopier.

Compare also Loew's: *Pithogaster,* eine neue Gattung der *Acroceriden* (Wien Ent. Mon. I, p. 33; 1857).

Westwood's: Descr. of some new exotic species of *Acroceridae* (in the Trans. Ent. Soc. V, p. 91—98; 1848). Another paper by the same in the same Transactions for 1876.

The name *Cyrtitae,* derived from the genus Cyrtus (κύρτος, humpbacked), I find was used by Newman, in his Grammar of Ent., 1841. *Cyrtidae* was adopted by Loew, in the Monogr Vol. I, instead of *Acroceridae* (Leach), *Henopidae* (Erichs), *Inflatae* (Meig.), *Vesiculosae* (Macq.). It certainly has more meaning than *Acroceridae*, derived from a character, the insertion of the antennae on the vertex, which is by no means universal in the family. *Henopidae* (*Henops*, one-eyed) was adopted by Erichson, in spite of the circumstance that the generic

name *Henops* had been given up; as this is contrary to the usual
practice in entomology, this family-name cannot well be maintained.

173. **Opsebius.** A more detailed definition of the genus is given
by Dr. Loew, in Beschr. Europ Dipt. II, p. 64. For the american
species, I have prepared the following analytical table:
- A. First posterior cell divided in two by a crossvein;
  - B. Anal cell closed; bases of the third and fourth posterior
    cells on the same line, or nearly so;
    - a. wings brownish . . . . . . *gagatinus* (Pennᵃ.);
    - aa. wings tinged with brownish, base and apex sub-
      hyaline . . . . *diligens* (Vancouv.)
  - BB. Anal cell open; third posterior cell shorter than the fourth
    - b. sixth vein prolonged to the margin of the
      wing . . . . *sulphuripes* (New York);
    - bb. sixth vein interrupted long before the margin of
      the wing. . . . . *paucus* (California).
- AA. First posterior cell not divided by a crossvein *inflatus* (Europe).

O. formosus Lw. (Provence), *O. pepo* Lw. (Spain), have the first
posterior cell divided by a crossvein; both, as well as *inflatus*, differ
from the american species in having the body *black* and *yellow* and not
uniformly black. (See Loew, l. c.).

O. perspicillaris Costa . . . . unknown to Loew.

174. **Hybos.** In the Brit. Mus. *H. duplex, triplex, purpureus,
subjectus* Walk. look very much like the same species. The two first,
as appears from the description, are certainly the same species. Observe
the careless wording of their diagnoses, where *pedibus* is used in two
different senses; once for *legs,* and afterwards for *tarsi!*

Hybos reversus is a different species and has the base of the
wings hyaline.

175. **Syneches and Syndyas.** The passage concerning these genera
in Loew, l. c., runs as follows: „The characteristic marks, which
distinguish *Syneches* from *Hybos,* consist in the shape of the head,
which is flattened in the region of the front; in the palpi being somewhat broader at the tip; in the shorter first longitudinal vein; in the
second vein taking its origin nearer the root of the wing, and ending
more steeply in its margin, than in the true species of *Hybos;* in the
somewhat shorter anal cell and in the usually spotted wings.“

„I take *Syneches* in this sense, and form alongside of it a new
genus, based on some species of Hybos from the Cape, in which the
fourth vein is almost indistinct before the discal cell and the origin of
the second vein is still more distant from the base of the wing, than in
those european species, which remain in the genus Hybos, so that the
origin of the third vein is very near that of the second. The name
Syndyas, which I give to this genus, is intended to allude to the
coalescence of the two cells, produced by the indistinctness of the first
section of the fourth vein.“

176. **Empina.** About the limits between this section and the
Hybotina, see in Loew, Fauna Südafrica's, p 258. Compare also his

papers on European Empidae, in the Berl. Entom, Zeitschr. Vol. XI, XII, XIII.

177. **Pachymeria.** See about it Loew's paper in the Wien. Ent. Mon. VIII, Novemb., where the two american species are also discussed.

178. Compare Loew, on **Microphorus** in the Schles. Z. f. Ent. 1863. On the relation between his genus and *Iteaphila*, see Loew, Beschr. Europ. Dipt. II, p. 250.

179. About the european species of **Drapetis** and **Stilpon** see Loew, Neue Beitr. VI, p. 33 The passage about *Stilpon* nov. gen. runs thus:

„Is separated from *Drapetis* on account of its front, which is of „equal breadth and not triangular; and of its arista, which is dorsal „and not apical."

180. Compare: Ueber die schlesischen Arten der Gatt. **Tachypeza** und **Microphorus** by H. Loew, in Schles. Z. f. Ent. 1863.

In this paper Dr. Loew protests against the substitution of *Platypalpus* Macquart for *Tachydromia* Div. B, Meigen.

The facts are these: Meigen. in his principal work, divides the genus Tachydromia in two sections, which he calls A and B; Macquart (Diptères du Nord etc. 1827), proposes to call the larger section B, *Platypalpus.*

Before being aware of this, Meigen, in his Vol. VI (1830), proposed to call the section A *Tachypeza,* leaving the name Tachydromia, to the larger section B. In his Vol. VII, p. 94 (1838), he maintains this arrangement against Macquart's, and points out that the name *Tachydromia* should, as a matter of right, remain to the larger section.

The question may be argued both ways. Zetterstedt and Loew (in the Schles. Zeitschr. 1863) take Meigen's view. Dr. Schiner takes the opposite ground, and adopts *Platypalpus* (Syn. Tachydromia, Div. B, Meigen) and *Tachydromia* (Syn. Div. A, Meigen and *Tachypeza,* Meigen). I follow Meigen's view, as a matter of expediency, waiving the doubtful question of right. Meigen's work being the foundation of Dipterology, it is better, I think, to preserve its nomenclature, as far as possible. *Platypalpus* moreover labours under the disadvantage of being a hybrid compound of a latin and a greek word.'

Sicus Latreille, cannot be maintained against the much earlier *Sicus,* Scopoli, which is a *Myopa* (compare note 47a).

181. On **Ardoptera,** see Loew, Wien. Entom. Monatschr. II, p. 7.

182. **Synamphotera.** In the Beschr. Europ. Dipt. II, 255, Mr. Loew characterizes this genus as follows:

Proboscis short, horny; palpi small, incumbent.

Antennae short, with an exceedingly short terminal style.

Legs slender, the anterior ones of the ordinary structure.

The third longitudinal vein of the wings has its anterior branch often connected by a crossvein with the second vein; discal cell elongated, emitting three veins towards the alar margin; the two posterior basal cells elongated; the posterior but very little shorter than the preceding; sixth longitudinal vein strong, reaching the alar margin.

19

183. On the european species of **Hemerodromia**, see Loew, Wien. Ent. Mon. 1864, p. 237.

184. An observation of the lamented B. D. Walsh may be worth recording here: „It may perhaps be worth while to add, that on the grape-vine where these *Erythroneurae* where swarming. I noticed a small and rather rare dipterous fly, the *Hemerodromia superstitiosa* of Say, very busily engaged. I caught him and put him in my collecting bottle, along with a number of leaf hoppers, and shortly afterwards saw him approach one slily, stick his beak into it, and suck it to death, without using previously his long raptorial front legs.“ (B. D. Walsh, Fire Blight, in the Prairie Farmer, Chicago Illin. 1862)

185. On the european species of **Clinocera**, see Loew, Wien. Ent. Mon. 1858, p. 238.

186. Compare H. L o e w: On the N. A. Dolichopodidae, in the Monographs of N. A. Diptera, Vol. II (1864), a monographic work on the north american genera and species of the family

The same author's earlier publication: *Die nordamericanischen Dolichopodiden* (in the Neue Beitrage, VIII, 1861) is superseded by the later one in English.

The classification of the family is chiefly due to Mr. Haliday (principally in Walker's Insecta Britannica, Diptera) and to Dr. Loew, in the Neue Beiträge, V, 1857 (Die Familie der Dolichopoden).

In a recent paper, Dipterologische Untersuchungen (Vienna 1878), Mr. Joseph Mik, describes twelve new genera, all european, and several new species of Dolichopodidae.

187. **Orthochile derempta** Walker, List, etc. III, p. 667, also in Monogr. II, p. 318, North America, is discussed by Mr. Loew, in Monogr. II, p. 115. It is certainly not an *Orthochile*, but from Mr. Walker's imperfect statements it is impossible to tell, where it belongs. The typical specimen, which I saw in London, looked very much like a *Chrysotus*.

188. About the definition of the genera **Hypophyllus**, **Hercostomus** and **Gymnopternus**, compare Loew, Beschr. Europ. Dipt. I, p. 278.

189. **Porphyrops signifer**, n sp. ♂. Tip of the arista expanded into a small lamel; body metallic green; feet yellow, except the hind tibiae and tarsi, which are black. Length, about 5 mm.

Bright metallic green; abdomen more golden green; the narrow face silvery; front green, with a white bloom; posterior orbits, below, with long white hair. Third antennal joint long and tapering, arista of nearly the same length as the joint, expanded at the tip into a small lamel. Feet yellowish, except the base of the coxae, which is blackish-gray; the end of the front tarsi brownish; upper part of the hind femora infuscated; hind tibiae and tarsi black. The front coxae, as well as the front and middle femora, are beset with long and delicate white hairs; there are remarkable small tufts of short hairs near the tip of each of the middle coxae. Halteres pale yellow; tegulae with yellowish cilia. Wings distinctly infuscated, more hyaline near the root.

Hab. Tarrytown, N. Y. July 1871; Manlius, in Western New York (J. H. Comstock).

This species resembles very much the european *Porphyrops antennatus* described and figured in the Ann. Soc. Entom. de France, 1835, p. 659; Tab. XX, c, as *Anglearia antennata.*

190. Mr. Kowarz has given important papers on the european species of **Chrysotus** in the Verh. Zool. Bot. Ges. 1874, and on **Medeterus** l. c. 1877.

191. **Chrysotus pallipes and obliquus.** According to Schiner, Novara, p. 221, these species have also been received from South America.

192. **Sympycnus.** There is a genus *Sympycna* Charp. 1840 (Neuropt.).

193. **Psilopus pallens.** This species, which is not uncommon along the Atlantic seaboard, and generally occurs about buildings, is the same as *P. albonotatus* Loew, from Rhodus. In Mr. Bergenstamm's collection in Vienna I saw a specimen from Barcelona, in Spain. Very probably, the species has been imported on ships to America.

194. **Microdon.** About the european species of this genus, see Loew, Verh. Zool. Bot. Ges. 1856.

195. **Ceratophyia fuscipennis Macq.** The genus *Ceratophyia* (Wiedemann, Anal. Ent. 1824; Auss. Zw. II, p. 79; Tab. IX, f. 5) is separated from Microdon on account of the absence of spines or even *tubercles* on the scutellum. This is not a sufficient reason for maintaining this genus, which in other respects, does not differ from a typical *Microdon.* The latter genus, as it is understood now, contains many species with much more important structural differences, and the existence of the genus *Ceratophyia*, until those other species are not likewise separated, is only misleading.

I believe I recognize *C. fuscipennis* Macq. in a specimen from Texas, in Dr. Loew's typical collection. It is recognizable by the length of its third antennal joint; in general appearance and coloring it looks very much like *Microdon globosus.* Macquart had it from „Philadelphia", but I do not quite trust his statements about localities and suspect that he sometimes labelled *Philadelphia* or *Baltimore* specimens which he had received from these cities, but which had a more southerly origin. (For instance *Lampria rubriventris* which is likewise frequently received from Texas, but which is labelled „Philadelphia" by Macquart.)

196. **Chymophila splendens Macq.**, Hist. Nat. Dipt. I, p. 486; Tab. XI, f. 3 (1834.; Dipt. Exot. II, 2, p. 10; Tab. I, f. 2. Philadelphia. Mr. Bigot, in whose possession the typical specimen of Macquart's description now is, makes the following statement about it (Annales Soc Entom. de France, 1858, p. 590): „The head of this specimen is glued on, and resembles that of Conops, while the body is that of an exotic *Microdon.*" We may with safety, therefore, strike out this genus and species from among the number of existing forms. (Osten Sacken, Bull. Buff. Soc. N. H., Nov. 1875.)

Since writing the above, I have seen the specimen and can only confirm the statement. The body seems to belong to *Microdon aurifix* Wied.

197. **Chrysotoxum.** About the european species, see Loew, Verh. Zool. Bot. Ver., 1856. Besides the enumerated species of *Chrysotoxum*, the following european species are quoted as occurring in North America: *bicinctum* Meigen, by Mr. Loew in Neue Dipterol. Beitr. IV, p. 18, together with *Helophilus pendulus, versicolor* and *floreus,* also european species. The statement about *Chr. bicinctum* is repeated by Loew, Verh. Zool. Bot. Ver. 1856, p. 614. None of these species has ever been found in N. A. since. and the statement seems to be based on an error of locality. The specimen of *Chr. bicinctum* on which the statement was based, is among the collection of Dr. Loew's North American types *C. fasciolatum* Deg., according to Walker, List, etc. III, p 541, was found in Huds. B. Terr. I would not trust this statement without comparing the specimens.

198. **Paragus aeneus.** „The name *aeneus* was given by Walker in 1849 when there existed an *aeneus* Meigen (1822), now considered a synonym of *tibialis* Fallèn". (Verall *in litt.*).

199. **Chrysogaster.** About the european species, compare Loew, Stett. Ent. Z. 1843, p. 204. sqq; also Wiener Entom. Monatsschr. I, p. 4. In the former article the author also gives his opinion on the nomenclature of the genera of *Syrphidae,* and on the confusion prevailing in it owing to the arbitrary changes, introduced by Fabricius, Fallèn and Zetterstedt (Eristalis Latr. = Syrphus Zett.; Syrphus Meig. = Scaeva Zett.; Eristalis Zett. = Chilosia Meig.).

200. **Chilosia.** On the european species, compare Loew, Verh. Zool. Bot. Ver. 1857.

201. **Syrphus Naso and Pacilus are Platychiri;** whether they differ from *P. peltatus* and *quadratus*, I am unable to tell, as I had no specimens for comparison when I saw the types in the Brit. Mus

202. **Leucozona.** There is a genus *Leucozonia* Mollusca, 1847, which however does not interfere with the other.

203. **Catabomba.** „The eyes of the male have an area of large facets in the upper and middle portion (a structure which I have not observed in any *Syrphus,* sensu stricto); the hypopygium of the male is . much smaller than in *Syrphus,* entirely concealed under the fifth segment; the front is remarkably convex in both sexes" (Osten Sacken, Western Diptera). The name is derived from καταβομβέω, „I am humming round." The european *Syrphus seleniticus* also belongs to *Catabomba; Syrphus melanostoma* Macq. Dipt. Exot. II, 2, p. 87, from Chile, likewise.

204. **Syrphus.** Compare my paper: On the N. A. species of the genus *Syrphus,* in the Proc. Bost. Soc. Nat. Hist., 1875.

205 **Syrphus Alcidice** Walker, List, etc. III, p. 579 (Huds. B. Terr.) is represented in the Brit. Mus. by three specimens, one of which belongs to the group of *S. lapponicus;* the two others have faint yellow spots on the second segment only, the remaining abdominal segments being dark metallic green, with an opaque black longitudinal line in the middle. It is either a species which I do not know, or a dark variety of some well-known one. The description refers to these latter speci-

mens, only the „four interrupted gray bands“, mentioned in it, were not seen by me.

206. **Syrphus geniculatus.** The type in the Jardin des Plantes is an unrecognizable fragment

207 **Syrphus lapponicus.** Whether this is a variable species or a group of closely allied species, I do not pretend to decide; see about it in my paper on *Syrphus*, but strike out whatever is said there about the synonymy with S. affinis Say The latter, as I recognized since, is *Catabomba pyrastri.*

208. **Syrphus arcvcinctus** Walker, List, etc. III, 580 (Huds. B. Terr.) is represented in the Brit. Mus. by two specimens, one of which is my *S. amalopis;* the other belongs to the group of *S. lapponicus.* The description is drawn from the latter specimen, the abdominal spots of *S. amalopis* being much more than „slightly curved“.

209. **Syrphus philadelphicus** Macq. and *Scaeva concava* Say are synonyms of either *S. ribesii* or *S. torvus.* The type of the former in the Jardin des Plantes is a very much soiled specimen. — The *S. concavus* in Wiedemann's collection in Vienna is *S. ribesii.* — I have no doubt now of the identity of *S. ribesii* with my *S. rectus.* Mr. Novicki (in his Beitr. z. Dipterenfauna Neuseelands, 1875), published another *Syrphus rectus,* in the very year of the publication of mine.

210. About **S. guttatus** Walker, Mr. Verrall writes me that it resembles *umbellatarum;* hence I place it as a doubtful synonym of the American *umbellatarum.*

211. **Syrphus adolescens** Walker, List, etc. III, p. 584 (Huds. B. Terr, Nova Scotia) is represented in the Brit Mus. by three specimens; one belongs to the group of *S. lapponicus;* the other (from N. Scotia) is *S. americanus;* the third is my *S. contumax.* The description was probably drawn from the latter, although it is very unmeaning.

212. **Didea fuscipes.** Differs from the European *D. fasciata* in the color of the legs only (Lw. Cent. IV, 82). *D. laxa* with its greenish color, is the representative of the European *D. alneti.*

Didea laxa O. Sacken ♂ ♀. Bull. Buff. Soc. l. c.).

The greenish or yellow cross-bands are attenuated on the sides and come in contact with the lateral margins of the abdomen. Length: 11—13 mm.

Female. — Face yellow, with a broad, brown stripe, front and vertex black; the former with gray dust on both sides. Antennae black. Thorax blackish-green, shining. Scutellum dull brownish-yellow, with a slight greenish or bluish metallic lustre pleurae with a whitish spot, beginning at the humerus and connecting almost at right angles with a similar spot in the middle of the pleura. Abdomen black, with two greenish-yellow or yellow spots and two cross-bands; the spots (on the second segment) are large, oval and in contact with the lateral margin; the cross-bands (on segments 3 and 4) have a triangular notch or excision on their hind margin (in some specimens they are altogether interrupted); on each side of the notch they are convex, so as to come in contact with the abdominal margin with less than their greatest

breadth; hind margin of the fourth segment margined with yellow.
Venter black, segments 2, 3, 4, each, with a broad yellow cross-band
at the base, coarctate in the middle. Legs yellow; proximal half of the
four anterior femora black; hind femora black, except at tip; hind tibiae
with a brown ring in the middle, sometimes expanding over the whole
tibia; tarsi more or less brown. Wings with a distinct grayish tinge,
stigma brownish; the third vein forms a distinct sinus, encroaching upon
the first posterior cell.

Male. — The white spots on the pleurae are less perceptible; the
cross-bands are sometimes interrupted in the middle, especially in the
smaller specimens. In one of the specimens the spots on the second
segment, as well as the interrupted cross-bands are separated from the
lateral margin by a distinct black interval.

Habitat, Lake Superior collect. A. Agassiz); Norway, Me. (S. I.
Smith); Mt. Washington, Alpine region (G. Dimmock). The largest lot
I received from Mr. H. K. Morrison, who collected it in the White
Mountains. Altogether I had fourteen males and an equal number of
females.

The cross-bands and spots on the abdomen usually are greenish,
like those of the European *D. alneti;* sometimes, however, they are
yellow.

D. laxa differs from *D. fuscipes* Loew in the shape of the abdominal cross-bands, which in the latter, become broader on each side, but
do not reach the margin; also in the color of the femora, etc.

213. **Sphaerophoria.** I restore this name, however incorrect its
termination may be, as *Melithreptus* was used long before 1840 for a
genus of birds.

214. **Allograpta.** „*Scaeva obliqua* Say, cannot well be placed in
any of the existing genera of this group. It does not possess the characteristic marks of *Mesograpta* (peculiar shape of the ocellar triangle
in the male, and peculiar coloring of the thorax); it has not the large
development of the hypopygium of the male of *Sphaerophoria;* it might
be placed among the species of *Syrphus* with a linear abdomen. But,
in the first place, these species will, sooner or later, have to be separated from the bulk of the genus; and, in the next place, *Scaeva obliqua*
possesses in the structure of the eyes of the male, and in the peculiar
markings of its abdomen, sufficient characters of its own. The eyes of
the male are divided in two parts by a well defined line, above which
the facets are larger than below; the line lies a little lower than the
antennae and thus divides the eye in two unequal parts, the upper one
of which is a little larger; its coloring, in life, is more red, the lower
half is more purplish. This character, very striking in life, is also
visible in dried specimens. I have not observed it in the species of
Syrphus, or of *Sphaerophoria,* or of *Mesograpta,* which I examined alive.
The name *Allograpta* is given in allusion to the peculiar coloration of
the typical species. *Scaeva emarginata* Say, which I do not possess, is
provisionally placed in the same genus. 1 suspect that more than one
Syrphus from Mexico and the West Indies belongs to the same group;

as for instance *S. delineatus* Macq., but, of course, it is impossible to judge from descriptions alone." (Reproduced from the Bull. Buff. Soc. N. H. 1876.) Since writing the above, I have discovered *Allograpta fracta*, n. sp. in California, which also shows the generic characters, as defined above. *Syrphus exoticus* Wied., Auss. Zw II, 136, is likewise an *Allograpta*.

215. **Xanthogramma felix** O. Sacken ♀. (Bull. Buff. Soc. l. c.)

Female. — Face and cheeks yellow (in all my specimens, except one, the face has the brownish-red tinge, which the faces of *Syrphi* sometimes assume); vertex dark metallic green, emitting a stripe of the same color, which reaches the base of the antennae, where it expands little; between this stripe and the eyes, the front is yellow. Antennae black, sometimes faintly reddish on the under side, near the suture of the second and third joints; third joint rather large, oval, blunt. Thoracic dorsum of a rather bright metallic green; on each side a yellow stripe runs from the humerus to the callosity near the scutel; the latter yellow, its extreme base and corners blackish or brown. Pleurae with a large, ill-defined yellow spot below the wings. First abdominal segment with a yellow spot each side (just under the halteres); the first cross-band (on the second segment) is either interrupted by a very narrow black line in the middle, or entire; the second band is coarctate in the middle, its hind margin being a shallow obtuse angle; the same may be said of the third band, except that the obtuse angle is deeper and often has a notch in the middle, which sometimes cuts the band in two; there is a narrow fourth band at the base of the fifth segment, encroaching upon the hind margin of the preceding segment; the fifth segment has a narrow yellow posterior margin. Legs yellow, hind legs black or brown, except the base of the femora and a space on both sides of the knees. Wings with a distinct brownish tinge on their distal half, anteriorly; stigma brownish; sometimes the whole wing has a brownish-yellow tinge. Length: $9^1/_2 - 10^1/_2$ mm

Habitat, Westpoint, N. Y., in Sept. 8—10, three females; Illinois; Pennsylvania. (The specimen from the latter locality is smaller, wings more hyaline, legs and antennae of a paler color.) The first and third band are as often interrupted as not; the second often shows a vestige of an interruption in the shape of an indistinct blackish line in the middle.

216. **Ocyptamus Amissas** Walker. In my List of N. A. Syrphidae, I took this for a synonym of *O. fuscipennis*. Since then I saw that Dr. Loew, in his N. A. collection, considered it a different species, and he may be right.

O. Radaca Walker, which I have seen in the Brit. Mus. is perhaps a synonym of *O. Amissas* or of *conformis* Loew; the posterior part of the wing is hyaline, traversed by a brown cross-band.

217. **Brachyopa vacua** O. Sacken ♀. (Bull. Buff. Soc. l. c.)

Brownish gray, thorax with three brown stripes; abdomen brown, its basal third whitish yellow, with a brown line in the middle; arista bare. Length: 8—9 mm.

Face. front and vertex densely clothed with a grayish pollen; lower part of the face very much projecting; a brownish stripe runs across the cheek, from the eye to the mouth; antennae brownish, grayish pollinose; arista bare, brown, reddish at base. Thoracic dorsum yellowish-gray, with three brown stripes; the intermediate one geminate and abbreviated posteriorly. Scutellum brownish-yellow. Abdomen brown, shining; first and second segments whitish yellow (as if translucent), the second brown posteriorly and with a longitudinal brown line in the middle. Legs grayish brown; hind femora slightly incrassate, on the under side with a brush of short spine-like bristles. Wings distinctly tinged with brownish, especially on the distal half, anteriorly; first posterior cell distinctly petiolate at the distal end, the petiole being equal in length to the small cross-vein.

Habitat, Quebec, Canada (Mr. F. X. Bélanger); a single male specimen. The interval between the distal ends of the first posterior and discal cells is a shallow sinus, and not a right angle, as in the following species.

Brachyopa notata O. Sacken, ♂ ♀. (Bull. Buff. Soc. l. c.)

Yellowish-ferruginous; abdomen with brown incisures and with a brown dorsal line; arista pubescent. Length: 5—6 mm.

Face and front pale yellowish, with a yellowish-silvery pollen; cheeks with a faint brownish stripe; antennae yellowish-ferruginous; arista yellowish-brown, pubescent; vertex yellowish-ferruginous. Thorax reddish above, clothed with a yellowish pollen, which leaves bare three reddish stripes; the intermediate one geminate. Scutellum reddish-yellow, nearly as long as it is broad; abdomen brownish-yellow, with the hind margins of the segments distinctly, but narrowly bordered with brown; lateral margins likewise brownish; in the middle of the back, a narrow, longitudinal brown stripe, sometimes interrupted at the incisures, in some specimens evanescent on the fourth segment. Halteres whitish. Legs brownish-yellow, hind tarsi brown. Wings somewhat tinged with brownish-yellow, more distinctly brownish on the apex and along the cross-veins at the distal ends of the first posterior and discal cells; first posterior cell short-petiolate at the distal end.

Habitat, White Mountains, N. H., beginning of July Two males and a female. In this species the interval between the distal ends of the first posterior and the discal cell is nearly a right angle.

I have a fourth specimen, a female, from Quebec (Mr. Bélanger), which is smaller, and very pale in coloring, without any brown stripe on the abdomen, the incisures but slightly infuscated, the wings almost hyaline, etc. I take it for a somewhat immature *B. notata*.

218. **Volucella Maximiliani.** When Brauer, in his Entom. Bericht für 1868, says that this species is a synonym of *Volucella americana* Wied., he probably means *V. mexicana* Macq., as a *V. americana* Wied. does not exist.

219. **Volucella fasciata and pusilla.** Until further evidence I do not unite these two species, Macquart's suggestion notwithstanding. The M. C. Z. has *pusilla* from Haulover, Florida, March 11 (M.M.

Hubbard and Schwarz); *fasciata* from Dallas, Texas, (Boll) and from Manitou, Colo, where I took it Aug 18.

220. **Temnocera.** Some of the species placed in the genus Volucella, may belong to *Temnocera*, as I do not quite understand the definition of this latter genus. Wiedemann (Auss. Zw. Preface to Volume II, p. X) was likewise doubtful about it.

221. **Eristalis albiceps** Macq. is a distinct species and looks like *E. seniculus* Loew, from Cuba. I have seen the type in Paris.

222. **Eristalis compactus** Walker has the whole leg red, while *E. atriceps* as described by Loew, has black femora. Nevertheless M. Walker's type, which I saw in London, struck me as being the same as *E. atriceps*. The question is therefore, whether the color of the legs is not variable, a question which I cannot solve here. (Heidelberg, Oct. 1877.)

223. **Eristalis Androclus** Walker, as I saw it in the Brit Mus. is a *Helophilus*. Nevertheless I retain the name as *E. Androclus* O. S. (non Walker), as I have referred to it in the Western Diptera and communicated it to many correspondents.

224. **Eristalis semimetallicus.** I have seen the type in Mr. Bigot's collection; it looked to me like *E. Bastardi*. It is possible however, that it is a closely allied, but different, species.

225. **Eristalis dimidiatus.** Macquart did not recognize *E. dimidiatus* Wied., and thus came to describe it, first as *niger* in the *Suites à Buffon*; then the male as *L'herminieri* and alongside of it, both sexes as *chalybeus* (Dipt. Exot. Vol. II); and then again the female as *incisuralis* (in the Supplem 4). That the eyes of the latter are described as *glabrous*, is erroneous, as all the known North American Eristalis have pubescent eyes, with the single exception of *E. aeneus*. I saw the types of *E. L'herminieri* and *chalybeus* in the Jardin des Plantes, and although I had no opportunity of comparing them with specimens or descriptions of *E. dimidiatus*, they did not shake the opinion I had previously formed of their synonymy. *E. incisuralis* I did not see.

226. **Eristalis flavipes,** Syn. Milesia barda Say ♀ (non ♂). The original type of Say's is still preserved in the Harris' collection in Boston. This synonymy explains the *brown spot* on the wings of the female, mentioned in Say's description, and which does not exist in the real female of *M. barda*.

227. **Syrphus oestriformis** Walker is a rather peculiar Eristalis, represented by a single specimen in the Brit. Mus.

228. **Eristalis tenax.** I took this species for the first time in Cambridge, Mass., in November 1875; also several specimens in Newport, R I., in October and November 1876. Since then, I have seen it from Georgia and Missouri (Collect. v. Roeder). It is strange that in my 20 years of North American collecting is had never occurred to me before.

229. **Eristalis philadelphicus.** The type, a single female, is in Mr. Bigot's collection; the yellow spots on the abdomen are somewhat

different from a typical *E. transversus*, but nevertheless I believe it to be the same species.

E. vittatus Macq. The description agrees with *E. transversus*, except that the eyes are said to be glabrous. But this statement is very probably erroneous, as, with the exception of *E. aeneus*, all the known N. A. *Eristalis* have pubescent eyes.

E. pumilus Macq., seems to be based on a very small specimen of *E. transversus*, in the variety with yellow anterior legs. I have not seen the type in Paris.

230. Eristal's Androclus, frater, chalepus Walker, which I have seen in the Britisch Museum, are *Helophili* of the group of *H. borealis*, *groenlandicus*, *glacialis*. As it was not possible for me to determine their synonymy, I have omitted them in the lists.

231. Plagiocera being preoccupied by Klug, (Hymenoptera 1834), Mr. Loew gave another name to this genus. It was, I suppose an oversight on his part, that he omitted to state that *Pteroptila* was merely a new name for an old genus. Schiner (Novara, 366) was right in suspecting it

232. Helophilus. Compare the paper on the European species of *Helophilus* by H. Loew, in the Stett. Ent. Zeitschr., Vol. VII; several North American species are described in it.

233. Helophilus stipatus and H. Anausis Walker. I saw both in the Brit. Museum. The former, I thought, was *Hel. lineatus* male. The latter, a greasy specimen, was undistinguishable, but the description shows it to be *H. lineatus*.

234. Helophilus obscurus. The patria as given by Mr. Loew in the Centuries (*Carolina*), was based upon an erroneous reading of the label.

235. Eumerus porcus Walker, which is a Helophilus, is a very peculiar species; it is represented in the Brit. Mus. by two ($\male$ and $\female$) wellpreserved specimens. I have never seen it elsewhere.

236. Helophilus susurrans Jaenn. The synonymy does not seem doubtful; only *Hinterrand* should be read instead of *Seitenrand* in the description; without this emendation the comparison with *H. pendulus* has no sense.

237. Teuchocnemis. Milesia Bacuntius Walker, and Pterallastes lituratus Loew, are closely allied and must be put in the same genus. Both have, in the male, curved hind tibiae, with a strong projecting spur *in the middle*, a character which is wanting in *Pterallastes thoracicus* Loew. The latter was described by Dr Loew in both sexes, and therefore must be considered as the type of the genus, while of *P. lituratus* Dr. Loew described only the female. Hence arose the necessity of establishing a new genus for the other two species.

238. Teuchocnemis Bacuntius. The specimens which I have from Texas do not quite agree with Mr. Walker s description of the thorax, nevertheless the identity is not doubtful.

239. Merodon Bautias Walker, is represented in the Brit. Mus. by a single male specimen; *M. bipartitus* by four specimens, two of which seem to be females of *M. Bautias;* the two others may be a different

species. The identification and synonymy of all the North American *Mallotae*, including even *posticata* and *Barda*, require a revision.

240. **Criorrhina armillata** O Sacken, Buff. Bulletin, 1 c. ♀.

Black, thorax bronze color, with fulvous pile; face, antennae, tip of femora, tibiae and three basal joints of tarsi, yellow; tibiae with a black ring in the middle. Length: 11—12 mm.

Face and front above the antennae honey-yellow; upper part of front and vertex blackish-bronze color, with fulvous pile; a black spot on the cheeks; antennae yellow-ferruginous, arista black. Thoracic dorsum and scutellum greenish-bronze color, clothed with erect fulvous pile; pleurae and pectus black. Abdomen black, shining, clothed with black pile; a tuft of yellow pile on each side at the base. Halteres yellow. Coxae and about two-thirds of the femora black; the end of the latter, the tibiae, except a black ring in the middle of each, and the three basal joints of the tarsi are of a saturate yellow; the two last tarsal joints black. The proximal two-thirds of the wings are tinged with yellowish, the remainder is gray; the latter coloring extends along the posterior margin as far as the axillary excision; within the yellow portion, there is a hyaline spot in the angle between the first and second veins (at the proximal end of the marginal cell); the veins near the root of the wings are all tinged with yellow.

Habitat. Quebec (Mr. Bélanger). A single female specimen.

241. **Crioprora, nov. gen.** In a note to his description of *Brachypalpus cyanogaster*, Mr. Loew observes, that this species holds the middle between *Brachypalpus* and *Criorrhina,* that it has a remarkably projecting face and would deserve the establishment of a new genus. Since the publication of my Western Diptera, I have seen Dr. Loew's type of *B. cyanogaster* and have perceived at once that it belongs to the same group with my *Pocota cyanella* and *P. alopex* from California, which I had doubtfully referred to St. Fargeau's genus *Pocota* (Western Diptera, p. 339). At the same time, I have also seen the european *Pocota apiformis*, the type of the genus, and have become aware that my two californian species, as well as *B. cyanogaster*, cannot be referred to *Pocota.* For this natural group of three species, I form therefore a new genus, and propose for it the name of *Crioprora* (κριοπρóßρος, with the face of a ram). The new group is characterized by the structure of its face, which forms a short snout, prolonged anteriorly, rather than downward, without tubercle in the middle and with an emargination at the tip; in the profile, the face is gently concave between the antennae and the oral edge.

Pocota is called by Schiner *Plocota* St. Fargeau; the latter author however calls the genus *Pocota*, probably from Πóκος sheepwool, and Ποκóω, to cover with wool. Since I made this correction in my Western Diptera, p. 339, Mr. Verrall has drawn my attention to the fact, that in Walker's Ins. Brit. Dipt. I, 238, as well as in the Index, in Vol. III, the genus is correctly called *Pocota*.

242. **Milesia Amithaon** Walker, which I saw in the Brit. Mus., looks very much like a *Brachypalpus.*

243. **Xylota.** Among the species, described by Mr. Walker, there are several, which I have never seen before, especially among those from the N. A. British possessions.

244. **Xylota Aepalius,** is not a Xylota; the specimen in the Brit. Mus. looks more like a *Brachypalpus.*

245. **Xylota vecors** O. Sacken, Bull. Buff. Soc. l. c. ♂ ♀.

Thorax brownish bronze-color, abdomen black; legs, including the coxae, ferruginous; end of hind femora, the hind tibiae and tarsi black. Length: 13—14 mm.

Face and cheeks black, with a greenish reflection and a delicate whitish down on the sides; antennae reddish-brown; front black, with some black, erect hairs. Thorax brownish bronze-color, with indistinct longitudinal greenish stripes; pubescence sparse, short, erect, brownish-yellow, mixed with black; a whitish-sericeous spot inside of the humeri; pleurae greenish-black, with blackish hairs; scutellum greenish bronze-color. Abdomen black, with a bluish or purplish reflection and scattered whitish and black hairs. Knob of halterses black, stem reddish. Legs ferruginous, including the coxae; the tip of the ungues brown; the distal third of the hind femora, the hind tibiae and hind tarsi black. Wings tinged with brownish, proximal half more hyaline; stigma dark brown.

Habitat, White Mountains, N. H. (E P. Austin and H. K. Morrison). Three males and two females.

In general appearance, this species is very like the European X. *femorata;* but it differs especially in the color of the coxae, which in the latter are black. Minor differences are that in X. *femorata* the wings are more uniformly colored, less tinged with brown on the distal half, the stigma paler, etc.

Xylota curvipes Loew? (Bull Buff. Soc. l. c.)

Among the specimens of *Xylota vecors* brought by Mr. Morrison from the White Mountains I found one, which is larger than the others (about 15 mm.), has altogether black coxae, the hind femora stronger and beset on the under side with yellowish hairs, longer and more conspicuous than similar hairs which exist in X. *vecors;* the hind tibiae, somewhat more strongly curved and ending in a short, stout spur; they are beset on the inner side with very conspicuous, long, erect black hairs; the halteres are altogether reddish; the antenlan arista dark brown, etc. Now all these characters, in which this specimen differs from X. *vecors,* belong to the European X. *curvipes* Loew, Neue Beitr. II, 19. As I have no specimen from the latter for comparison, I cannot settle the question of their identity, but I draw the attention of collectors to this undoubtedly distinct species. We have in this intance one of those curious cases of parallelism, as they so frequently occur between the two faunas. As X. *femorata* in Europe is supplemented by the closely resembling X. *curvipes,* the American representative of X. *femorata,* X. *vecors,* has alongside of it a species either identical with or closely resembling X. *curvipes.*

246. **Xylota euncida** Say. I am not sure whether I am right in

identifying this species with the one which is most common in New England, and agrees with Say's description, except that the antennae are more often dark than reddish; that the tarsi usually have the three last joints black, rarely two; the hind coxae in the male are armed with a spine. This last character prevents me from identifying this species with *X. quadrimaculata* Loew. I have not seen any original specimen of the latter. Loew seems to have identified *cjuncida*, as appears from the note in Centur. VI, 56. — Observe the genus *Microptoma* Westwood, Synopsis etc. p. 136, introduced for certain *Xylotae*.

247. On the European species of *Eumerus*, compare Loew, Stett. Ent. Z„ 1848, p. 108 and again Verh Zool. Bot, Ver., 1855.

248. **Novum genus?** I seems evident that *Xylota badia* Walker is no Xylota at all, and that *Milesia notata* Wiedemann must be placed in the same generic group with it. Not having the means of ascertaining whether this is a new genus, or not, I leave the question open. The synonymy of *Eristalis intersistens* Walker with *Xylota badia* Walker is doubtful, as the description of the face does not quite agree; it is principally based on my recollection of the type at the British Museum.

249. On **Chrysochlamis.** Compare Loew, Verh. Zool. Bot. Ver. 1857.

250. **Spilomyia.** Compare, Loew, Centur. V, 33, *Nota;* but insert the word *non* before *clausâ.*

251. **Temnostoma excentrica** Harris, and *T. aequalis* Lw. The latter, in all the numerous New England specimens which I have seen, has the femora black or brown, with the tips only more or less yellow. Harris describes the legs of his *Milesia excentrica* as „ochre-yellow, except the shanks and feet of the first pair, which are black". This agrees with some specimens from Illinois, which also have a more saturate-yellow abdomen and narrower black cross-bands than the New England specimens. The description of *M. excentrica,* which I prepared for the new edition of Harris' work was drawn from two western males of the above mentioned species. The female which I had before me at that time, was from Massachusetts, and I find now that I have a second female of the same kind from Lake superior; both differ from the western males (which I took for *T. excentrica),* as well as from *T. aequalis* in having two yellow dots on each side of the thoracic suture (like *T. alternans),* and not a yellow streak; the scutellum is darker, and its pubescence is black, not yellowish; the second abdominal segment has very little yellow, etc. — This may, after all, be the true *excentrica* Harris, although it is much rarer than *T. aequalis.* At all events I was wrong in uniting these females with those western males.

252. **Temnostoma Balyras.** The remark made by Mr. Jaennicke (Neue Exot. Dipt. p. 4) that the european *Temn. bombylans* occurs in North America, refers to this species. I adopt Mr. Walker's earlier name, under which I have distributed the insect to many collectors, the more so as the description is among the recognizable ones.

253. **Milesia limbipennis.** I have seen the type in Mr Bigot's

collection; it agrees with the specimen from Florida in the M. C. Z. Is it really a distinct species?

254. The history of this genus is as follows:

Sphecomyia. Latreille, Familles naturelles du *Règne* animal (1825), contains the name without any definition The definition appeared in the Dictionnaire classique d'histoire naturelle (by Rey and Gravier, publishers, in Paris), Vol. XV, p. 545 (1829), as follows:

Sphecomyia. Genre d'insectes de l'ordre des diptères, établi sur une seule espèce, rapporté de la Caroline par Bosc et très voisine de celui de *Chrysotoxe*, mais très distinct par un caractère unique dans cet ordre d'insectes, celui, d'avoir la soie des antennes insérée sur le *second* article; cet article, ainsi que le précedent est long, presque cylindrique; le troisième ou dernier, est beaucoup plus court. La soie est simple. Ce genre a été indiqué pour la première fois dans notre ouvrages sur les familles naturelles du règne animal, mais sans signalement. L'espèce qui lui a servi the type scra consacrée au célèbre naturaliste précité.

Latreille however never described this type of the genus, and it was Macquart who saw Bosc's and Latreille's original specimen in the Museum at Paris, and averred that is was the same as *Chrysotoxum vittatum* and *Psarus ornatus* of Wiedemann (Dipt. Exot. II, 2, p. 18, 1841).

Latreille's statement that the arista is inserted on the *second* antennal joint is, of course, erroneous. Macquart further mentions, l. c., that in the Berlin Museum this genus figures under the collection-name of *Epopter*. Gorski, in his Analecta ad Entomographiam, etc., 1852, proposes the generic name *Tyzenhausia* for the European species of the same genus. It occurs only in Eastern Europe (Sweden, Norway, Finland, Lithuania), and is very like the North American species. Wahlberg Ofvers. Vetensk. Acad. Forhandl., 1854, p. 155) gives a detailed description of it.

Mr. V. von Roeder, to whom I sent an american specimen of *S. vittata,* compared it to the eu opean *S. vespiformis.* He found only very slight differences, which would hardly justify a separation; his specimen of *vespiformis* (from Finland), has the yellow stripe on the pleura interrupted, which is not the case in the american *S. vittata;* the black cross-bands of the abdominal segments were broader in *vespiformis,* which, according to Mr. v. Roeder may be explained by the abdomen of his specimen being more drawn out. The figure, given by Gorski, certainly looks exactly like *S. vittata.* Still, Dr. Loew, if I recollect right, considered them as different species.

255. **Mixtemyia ephippium** O. Sacken, Bull. Buff. Soc. l. c. ♂.

Face yellow, with a brown stripe in the middle, which does not quite reach the antennae; the latter brown; second joint almost black; triangle of the vertex dark brown. Thorax dark brown; a brownish-yellow angular line runs from the scutellum, above the root of the wings, turning inside to follow the thoracic transverse suture and stopping before meeting the corresponding line on the other side; a less distinct angular line, on the anterior part of the thorax, begins on each side, at the yellow humeral tubercle, follows the anterior margin of the thorax and before reaching its middle, turns backwards; in the middle of the anterior margin, between the two angular lines, two delicate, short parallel yellow lines are perceptible. Scutellum brown in the middle, with yellow borders. Pleurae brown; a yellow spot above the root of the front coxae. Abdomen light brown; second segment with an arcuated yellow stripe, resting with its middle on the anterior, with its ends on the posterior margin, which is also yellow; the inside of the semi-circle thus formed, is dark brown, velvety; the third and fourth segments are clothed with a fine sericeous down; the third has a distinct tubercle in the middle and is margined with yellow posteriorly; the fourth is traversed by a yellow cross-band in the shape of an inverted V, the ends of which do not reach the lateral margins; hypopygium brown. Anterior half of the wings brown, the posterior hyaline; the anal cell, the second posterior, the discal and a part of the first posterior cell, as well as the whole posterior margin, including the alula, being hyaline (in *M. quadrifasciata* the second basal cell and the whole portion of the first basal, situated behind the spurious vein, are also hyaline). Legs; femora dark brown, the hind ones with a strong tooth on the underside; tibiae yellowish-brown, pale yellow at the base; front tarsi brown; middle and hind ones reddish-brown, two or three last joints brown.

Lenght: 12 mm. *Hab.,* Mexico.*)

256. Compare H. Loew's *Ceria* in his Neue Dipt. Beitr, I (1835).

257. See the papers by Loew:

1. Ueber die Ital Arten d. Gatt. *Conops*, in Dipterol. Beitr. III (1847).
2. *Conops*, in Neue Dipt. Beitr. I, p. 20 (1853); in the latter several N. A. species are described.

258. **Conops p'ctus** Fab. According to Loew, *in litt.* the *C. pictus* Wiedemann, Auss. Zw. II, 239, 7 is a different species from *pictus* Fab. In Macquart, the specimens, received from Serville, are *pictus* Fab.; the others *pictus* Wied.

259. **Stylomyia confusa** Westw. I have but little doubt about the identification of this species, Westwood's strictures on Fabricius's, Wiedemann's, and Macquart's descriptions notwithstanding. There is some confusion in Wiedemann's description, when he speaks of the *Hinterleibsgriffel* of the male. The Brazilian specimens may somewhat differ in coloring, or perhaps constitute a different species, in which case Say's name would have to be adopted for the North American species. (Since writing the above I found substantially the same statement by Loew, in Schaum's Jahresbericht 1851, p. 133.)

260. Dr. Schiner in the Verh. Zool. Bot. Ver. 1857 is in error when he states that the name **Stachynia** was introduced by Macquart in

---

*) **Observation.** The notes 196 — 200, 210, 212 — 214, 215, 217, 226, 232, 234, 236 -238, 240. 245—247, 249—251, 254, 255 are reprinted. with some emendations, from my List, of the North American Syrphidae, in the Bulletin of the Buffalo Society of Nat. History, Decemb. 1875.

the *Suites à Buffon;* an error however, which was due to Macquart himself, who did not allude to his previous publication.

261. There is a paper by F. Walker, Observations on the British species of **Pipunculidae** (Entom. Magaz., Vol. II, 1835, p. 262—270.) Also a survey of the swedish species by C. G. Thomson, in his *Opuscula entomologica,* Stockholm 1870, p. 109.

262. **Oestridae.** Compare Brauer, Monographie der Oestriden, Vienna 1863; with numerous plates of the imagos, larvae and pupae. The full synonymy of all the species enumerated will be found in this work, as well as the litterature.

263. **Tachinidae.** I have principally followed Schiner's distribution (in the Fauna Austriaca).

264. On **Ocyptera** see Loew, Stett. Ent. Z. 1844, p. 226, 266; also 1845, p. 170. Winnertz, Stett. Ent. Z. 1845, p. 33.

265. **Dejeania corpulenta.** I have seen Wiedemann's type in Vienna, which is my *D. vexatrix. D. rufipalpis* Macq., in Mr. Bellardi's collection, is the same species. I have been misled by Macquart's false identification of Wiedemann's species.

266. **Dejeania rutilioides.** I have seen Mr. Jaennicke's type in the Museum in Darmstadt and recognize in it the Tachinid which I mentioned in the Western Diptera, p 3-4, line 8 from the end.

267. **Tachina vivida.** Mr. Harris described this species in 1841; there existed at that time a *Tachina vivida* Wiedemann, Auss. Zw. II, p. 312 (1830). Wiedemann's *Tach. abrupta* would thus have the priority, if its identity with *Tach. vivida* Harris was ascertained.

268. For **Micropalpus flavitarsis** Macq. and *ornatus* Macq., as well as for a considerable number of other south american species, Dr. Schiner (l. c.) introduces the genus *Saundersia,* as these species have nothing in common with *Micropalpus,* but the rudimentary palpi.

269. I take **Nemoraea** in the sense of Schiner as embracing *Erigone* and other genera of R. Desvoidy.

270. **Exorista** in the sense of Schiner, involves the genera *Lydella, Zenillia, Carcellia* and in part *Winthemia* of Rob. Desvoidy. I have also included in it all the species which Mr. Walker described under the head of *Aplomyia* R. D. Myod. p. 184, for the reason that Rob. Desvoidy calls this genus intermediate between his *Winthemia* and *Carcellia* and that, in his later work (Dipt. des envir. de Paris, I, p. 459) he adopts for the type of the genus *Tachina confinis* Fallen, Zetterstedt, which is an *Exorista.*

271. **Belvoisia bifasciata.** The larva, according to Macquart, was bred by Boisduval from the chrysalis of *Cerocampa regalis;* Mr. Riley obtained it from *Dryocampa rubicunda* Fabr.

272. **Metopia.** I take this genus in the sense of Schiner as embracing *Araba* and *Ophelia* of Rob. Desvoidy.

272 a. A detailed definition of the genus **Eurygaster** and of its relationship to other genera of Tachinidae, is given by Nowicky, Beitrag z. Kenntniss d. Dipterenfauna Neuseelands, Krakau 1875, p. 23.

273. Compare: Monograph upon the British species of **Sarcophaga,**

or flesh-fly, by R. H. Meade in the Entomologist's Monthly Magazine, Vol. XII, p. 216. (February — May 1876); also Rondani, *Sarcophagae italicae.*

Mr. Meade had the kindness to examine a collection of Sarcophagae from North America, (belonging to the Museum of Comparative Zoölogy) for the purpose of comparing them to the european species. He arranged the collection according to the plan, adopted in his monograph and made out 24 distinct species of the restricted genus *Sarcophaga* (with black palpi) and four species belonging to the genera *Peckia* Desv. (Phrissopoda Macq.), *Cynomyia* Desv. and *Theria* Desv. He adds: „I am doubtful whether any of the species is absolutely identical with a european species, unless it be with *Sarcophaga similis,* which closely resembles *S. carnaria.* There is no specimen in your collection, however, exactly like the true *S. carnaria,* so common in Europe. — There are some striking points of difference between the Sarcophagae of America and Europe generally, the chief of which is that in the former, the species with one or both anal segments red or yellow, largely predominate, while among the latter, those with the anal segments black or gray, are much more numerous than those with the red."

The specimens alluded to as resembling *S. similis* Meade, were collected in the Rocky Mountains, Colorado and on the northern shore of Lake Superior.

273 a. Idia. Compare Loew, Die europäischen Arten der Gattung Idia (Stett. Entom. Z. 1844, p. 15—25).

274. **Calliphora mortisequa.** Kirby says: „this seems to be the american representative of *Musca vomitoria"* and states the differences. However, the cheeks being described as red, he must mean either *M. erythrocephala* or its representative.

275. **Calliphora obscoena.** Eschscholz says: „exceedingly like *Musca carnivora."* *M. carnivora* Fabr. = *Calliphora vomitoria.*

276. **Calliphora stygia** Schiner, Novara, p. 309, observes and probably with good reason, that Fabricius meant *New-Zealand* and not Newfoundland. Schiner had a number of specimens from Sydney, agreeing exactly with Fabricius's and Wiedemann's descriptions.

277. On the distrubution of Anthomyiidae in genera, compare:

R o n d a n i, Dipterologiae Italicae Prodromus, Vol. VI, Parma 1877.

R. H. M e a d e, On the arrangement of the British Anthomyidae (Entomologists Monthly Magazine, February, March 1875), where a useful analytical table of the genera is given.

L o e w, Die deutschen Arten d. Gatt. *Azelia* R. Desv. (Entomologische Miscellen, herausgegeben vom Schles. Entom. Ver. 1874. 41 pages)

Compare also Haliday's note, in Westwood's Synopsis, p. 143.

R. H. Meade Esq. in Bradford, Yorkshire England, has had the kindness to examine a collection of North American Anthomyiae, sent to him by me. The result of this examination is embodied in an article:

20

Notes on the Anthomyiidae of North America. (Entomologists Monthly Magazine, April 1878, p. 250—252.)

He sums up his comparison as follows:

„On looking over the collection, it struck me, in the first place, „that the number of species was small in proportion to the number of „specimens; and next, that the number of smaller and feebler species „was greater in proportion to that of the larger and more highly „developed forms, than occurs in Europe. I only determined 121 species „in the collection. There where few, if any, peculiar forms among them; „they could all be arranged in the same g·nera as the european species; „they had the same sombre colours and ordinary forms, which are so „familiar to us; and many of the common european kinds where so „closely represented, that it was difficult to say, in some instances, „whether they were exactly the same, or closely analogous species."

278. Schiner, Fauna Austr., Dipt. I, p. 644, quotes **Anthomyia brassicae** Bouché as a synonym of *A. ruficeps* Meig, but with a doubt.

279. Schiner, l. c. p. 643, quotes **A. ceparum** as a synonym of *A. antiqua* Meig

280. M. Walsh describes in the same place the larva-stages of two other **Homalomyiae, H. Leydii** and **H. Wilsonii,** the imago of which is not known.

281. **Dialyta.** About this genus, see Loew, Wien. Entom. Mon. II, p. 152.

282. **Lispe.** On this genus comp. Loew, Stett. Zeitung, 1847, p. 23—32.

283. About the systematic location of **Schoenomyza,** compare Loew, Centur. X, 73, nota.

284. **Cordylura.** Compare Haliday's note in Westwood's Synopsis, p. 143—144; see also *Scatophaga* ibid. There is a paper by Prof. C. Rondani, Scatophaginae Italicae.

285. Schiödte (Berl. Ent. Zeit. 1859, p. 153) seems to be in doubt about the interpretation of the **Musca stercoraria** of O. Fabricius, as well as of the two following species, *M. scybalaria* and *cloacaris* (Fn. Groenl. 161—163).

286. Compare the monographic essay by Loew: **Ueber d. Europ. Helomyzidae,** in the Schl. Zeitschr. f. Entom. 1859.

287. **Blepharoptera defessa.** The detestable figure appended to my description of this species, was published without my knowledge and consent.

288. See the paper: On the North American **Sciomyzidae,** by H. Loew, in the Monogr. of N. A. Diptera, I, p. 103.

289. „**Tetanocera Boscii** is characterized so insufficiently, that there is no possibility to identify it. *T. canadensis* is also unknown to me. *T. guttularis* Wied. is mentioned by Macquart as a native of N. Am., but I must consider this statement as a mistake, since the characters he gives do not agree with the description of *T. guttularis* Wied.; but

what species he has mistaken for *T. guttularis* I have not as yet made out." Loew, Monogr. I, p. 108.

290. **Loxocera.** On the european species, see Loew, Schles. Ent. Zeit. 1857.

291. **Calobata lasciva** Fab, Wied. = *albimana* Macq. I assume the synonymy on the authority of Schiner, who had the advantage of comparing Wiedemann's types. I do not pretend to decide, whether Macquart is right in referring to the same species the specimens from Cuba, Philadelphia, Java and Port Jackson.

As to *Taeniaptera trivittata*, Macquart, Dipt. Exot. II, 3, p. 240, says: „The genus *Taeniaptera*, which I established in the *Suites à Buffon*, has for type a species allied to some exotic *Calobatae*. I suppress it." The reason is not given, but the probable cause may have been the loss of the original specimen, which would explain why Macquart, in giving up the genus, never mentions the species again. I look upon the synonymy of *C. albimana* Macq. (which is a *Taeniaptera* in Macquart's sense), with T. trivittata Macq. as certain. Compare also Loew, Beschr. Eur. Dipt. III, p. 254.

292. About the european, as well as the exotic **Micropezae**, compare Loew, Berl. Ent. Zeit. XII, 1868, p. 161—167, also pag. 393.

293. The third volume of the Monographs of the N. A. Diptera (1873) contains a monograph of the N. A. **Ortalidae** by Dr. Loew, with an introduction, concerning the classification of the Ortalidae in general, and a review of the work of previous authors on the same subject; however, no notice is taken of the new genera published by Dr. Schiner (Novara etc.); nor of Prof. Rondani's *Ortalidinae italicae*. The article by Dr. Loew: *Die N. A. Ulidina*, in the Berl. Ent. Zeitschr. 1867, p. 283, was the precursor of his larger publication, but also contains South-American species.

294. **Oxycephala fenestrata** and **O. fuscipennis.** I have seen the types of both in the Museum of the Jardin des Plantes. *O. fenestrata* seems to be a different species.

295. **Pyrgota valida.** When Mr. Loew set aside this name, as a mere catalogue-name, he overlooked its publication by Mr. Harris in the Ins. Inj. to Vegetation.

296. **Ortalis Ortoeda** The specimens in the Brit. Museum bearing this name are *Chaetopsis aenea*.

297. **Herina splendens.** I owe this synonymy to Mr. v. Roeder.

298. **Urophora nigriventris** Macquart. Dr Loew, in the Monogr. etc. Vol. III, p. 337, says about this species that it is a Trypetid of doubtful systematic position; but not an *Urophora*. Macquart's description made me suspect that this was simply *Camptoneura picta*. As I had overlooked this species, while examining Mr. Bigot's collection in Paris, I wrote to him about it, and he kindly informed me, that „after a careful comparison of the types in his collection, labelled in Macquart's own handwriting, he finds no difference between *U. nigriventris* Macq. and *Camptoneura picta* Macq."

299. **Tephroncta humilis.** In the Monographs, III, p. 125; Mr. Loew rejects the earlier name given to this species by Mr v. d. Wulp, on the ground that „it has been preoccupied by Fabricius". This cannot be sustained, as neither of the two generic names, *Herina* or *Tephronota* existed at the time of Fabricius.

300. **Trypeta Narytia** Walker. There are four specimens in the Brit. Mus.; two of them are *Chaetopsis aenea*, and one of these bears Walker's label „Narytia", the two others, marked „Florida, Doubleday", seem to be *Tephronota humilis*.

301. **Euxesta annonae**; Schiner, Novara etc., p. 283, places this species in the genus *Amethysa* Macquart (Hist Nat. Dipt. II, p. 440) together with *Urophora aenea* Macq. (l. c, p. 458), from Columbia, S. America.

302. **Idiotypa** Foerster, Proctotrypidae 1856, has the priority.

303. See the papers of Mr. Loew: „On the North American Trypetidae" in the Monogr. of the N. A. Dipt, Vol. I, and „Review of the N. A. Trypetina", in the Monogr. etc., Vol III. On the european Trypetae, see the large work of Mr. Loew: Die Europäischen Bohr-fliegen, Wien 1862; in folio, with 26 plates of magnified photographs. The literature about the Trypetidae will be found in Schiners: *Diptera Austriaca, Die Oesterr. .Trypetiden;* Wien, 1858.

304. Schiner (Novara etc., p. 263) draws attention to the probable identity of *Leptoxys* with *Anastrepha*. But this identity seems certain, owing to the fact that Macquart himself, in the Dipt. Exot. II, 3, p. 216, mentions the *Dacus serpentinus* Wied. as belonging to *Leptoxys*. Macquart, l. c. improves *Leptoxyda* in the more correct *Leptoxys*. (I find in Agassiz „Index universalis" *Leptoxys* Rafinesque, 18 . ., Mollusca.)

305. **Eurosta**, Loew, 1873; *Eurostus*, Dallas, Hemipt. 1851.

    Peronyma, Loew, 1873; *Peronymus*, Peters, Volitantia, 1868.

    Euxesta, Loew, 1867; *Euxestus*, Wollaston, Erotyl. 1858.

    Euolena, Loew, 1873; *Evolenes*, Le Conte, Carab. 1853.

    Pterocalla, Rondani, 1848; *Pterocallis*, Passerini, Hemipt. 1863.

All these names do not interfere with each other, according to my opinion, and can remain. Should a change be thought necessary, add the syllable *Neo*.

306. **Icaria** Saussure, Vespidae 1858, has the priority.

307. **Aspilota** Foerster, Braconida 1862.

308. **Trypeta alba.** Mr. Riley told me that he bred it from seeds of *Vernonia*. I found it abundantly on the flowers of that plant.

309. About the systematic position of the **Lonchaeidae**, and especially of the genera *Palloptera* and *Lonchaea*, compare Loew, in Monogr. etc. III, p. 8—10. — About the european species of *Palloptera,* compare Loew, Schles Entom. Zeitschr. 1857. Do not overlook Haliday's note about these genera in Westwood's Synopsis of the genera of British Insects, p. 150, at the end of Vol. II. of his Introduction.

310. Compare Loew: **die Europ. Arten der Gatt. Sapromyza** in his

Dipterol. Beiträge, III, p. 25 (1847). Also some further remarks in Schles. Entom. Zeitschr. 1857; also *Drepanophora*, n. gen. of Sapromyzidae, in Berl. Ent. Zeitschr. XIII, p. 96. See also Haliday's note, quoted above, in Nr. 309.

311. **Sapromyza vulgaris** Fitch *(Chlorops)*. It is easy to recognize this species in the description of Dr. Fitch and in the figure. The description of *Chl. antennalis* Fitch evidently contains some clerical error, as it describes the antennae as plumose and alludes to those of *Chl. vulgaris* as *not* plumose, while the latter are represented as plumose in the figure. Mr. Loew followed the letterpress and not the figure, and hence called *antennalis* the species in which I recognize *vulgaris*. (See Loew, Zeitschr. f. Ges. Naturw. XXXVII, p. 117.)

312. About **Coelopa**, compare *Stenhammar*, Copromyzinae Scandinaviae, 1853.

313. About the species of **Heteroneura** occurring in Europe, compare Loew, Wien. Ent. Monatsschr., Vol. I, 1857, p. 51, and Berl. Ent. Zeitschr. VIII, p. 334—346.

314. Loew, Centur. Vol. II, p. 289, proposes to revive, instead of **Anthophilina**, the older name of this genus *Anthomyza* Fallen, Specim. Entomol. 1810. The same argument is adduced by him in the Jahrb. d. k. k. Gel. Ges. in Krakau, Vol. XLI. But it seems to me that *Anthomyza* is too much like *Anthomyia* and that there is a serious objection against using names, so nearly alike, in the same order of insects. Furthermore, as the name Anthomyza has been used by Zetterstedt *in the sense of Anthomyia*, its reinstatement, in a different acceptation, would be misleading. We have therefore the choice between *Leptomyza* Macq. (1835) and *Anthophilina* Zetterstedt (1838). Dr. Schiner adopted the former, which, I suppose is the right course; but until the question is decided, I retain the three north american species under the name of *Anthophilina*, under which they where originally published by Dr. Loew.

315. On the european **Opomyzae**, see Loew, Berl Ent. Zeitschr. IX, 1865, p. 26—33. On *Baalioptera*, l c. VIII, 1864, p. 347—356. The subgenus *Tethina* Haliday, in Westwood's Synopsis, p. 152, seems to have been overlooked.

316. **Sepsidae.** The following papers may be consulted:

1. Walker, F. Observations on the British *Sepsidae* (Ent. Magaz. 1833, p. 244—256.
2. Loew, H. Ueber die Gatt. *Saltella* überhaupt etc. (Stett. Ent. Z. 1841, p. 182—193). Contains useful systematic and historic data about *Sepsidae* in general.
3. Staeger, C. Systematisk Fremstelling af den danske faunas Arter at Antliatslaegten *Sepsis* (Kröyer's Tidskr. 1845, p. 22—36).
4. Van der Wulp. Jets over de in Nederland waargenomen *Sepsinen*. (Tijdschr. v. Ent. Ser. 1, Vol. VII, p. 129—144, with a plate.

317. **Ephydridae**, as preferable to *Ephydrinidae* is adopted by Loew, in Centur. Vol. II.

On this family, consult the following papers:

Haliday, Remarks on the generic distribution of the british Hydromyzidae (Annals of Nat. Hist. 1839, Vol. III).

Stenhammar, Forsok till Gruppering och Revision af de Svenska Ephydrinae, in the Kongl. Vet. Ac. Handl 1844.

H. Loew, On the North American Ephydrinidae, in the Monogr. etc. I, p. 129 (1862), where a definition of the genera will be found.

H. Loew, Die Europäischen Ephydrinidae, Neue Dipt. Beitr. VII, 1860. This paper, together with the preceding are very important.

H. Loew, Die Gattung Canace, in the Berl. Ent. Z 1874, where some further suggestions about the classification will be found.

318. **Ephydra halophila** Packard. The name cannot stand, as there is *Caenia halophila* v. Heyden, which is an *Ephydra*.

319. **Ephydra oscitans** Walker. Whether the synonymy that I suggest is adopted or not, the name must be dropped, as there is another and earlier *E. oscitans*, also by Walker in List etc. IV, p. 1106 (see under *Scatella*).

320. On the european **Geomyzidae**, compare Loew, Berl. Ent. Z. IX, 1865, p. 14—25; on *Diastata*, ibid. VIII, p. 357—368.

321. **Phortica** Schiner is not interfered with by *Phorticus* Stål, Reduvida 1860. *Amiota* Loew was published in the same year with *Phortica*, a few months earlier, but has never been characterized. Ten years after its publication, a few words of explanation appeared in the Centuries, Vol. II, p. 288, to establish its identity with **Phortica**.

322. **Chlorops, Oscinis, Siphonella.** About the relation of these genera to each other and their respective limits, compare Loew, Wien. Ent. Monatschr. Vol. II, the article: Zwanzig neue Dipteren, in' the note to No. 11, *Oscinis gilvipes*.

For the subdivisions of *Chlorops*, in the sense of Macquart, see Loew, Ueber die bisher in Schlesien aufgefundenen Arten der Gattung *Chlorops*, in the Schles. Zeitschr. f. Ent. 1866. Contains much more than its title implies, and is an elaborate monograph of the genus.

323. In the Jahrbuch der K. K. gelehrten Ges. in Krakau (1870), p. 15, Mr. Loew says that *Gymnopa*, on account of its venation, should be placed among the *Ephydridae*. But as he does not state to what group in that family it should be referred, and as, in the list of Diptera, appended to that same article, *Gymnopa* is left in its old place among the *Oscinidae*, I will follow his example here. In the same place Mr. Loew, explains why the older name of the genus, *Mosillus*, should be rejected. Whether his grounds are sufficient, I do not pretend to decide; but that *Mosillus* has not been entirely overlooked between its publication in 1804 and its reinstatement by Schiner, is proved by a curious passage in the Preface of Wiedemann's Auss. Zw., I, p. XI (1828), in which he speaks of *Mosillus* as something wellknown to him, and refers to it (erroneously?) the *Sargus aeneus* of Fabricius.

An earlier article by Mr. Loew on *Gymnopa* (Stett. Ent. Z. 1848) discusses the european species, and not the systematic position of the genus.

324. About **Rhicnoëssa** and its european species, see Loew, Berl. Ent. Z. 1865, p. 34.

325. **Milichia.** Compare Loew, Stett. Ent. Zeitung 1843, p. 310, 322.

326. **Cacoxenus.** About this genus and the related *Milichia*, *Lobioptera* etc., compare Loew, Wiener Ent. Mon. 1858, p. 213.

327. **Aulacigaster.** I place it among the *Agromyzidae*, on the authority of Loew *in litt.*

328. **Ochthiphila,** Compare Schiner, Verh. Zool. Bot. Ges. 1867, p. 325; also Loew, Wien. Ent. Mon., 1858, p. 219, in the article about *Cacoxenus.*

329. **Sigaloëssa,** compare Schiner, Novara etc., p. 238, where some further remarks about the genus will be found.

330. About **Asteia** or **Astia,** compare Loew, Berl. Ent. Zeitschr. II, p. 114, where a new genus *Periscelis* is introduced.

331. Compare Stenhammar, Copromyzinae Scandinaviae, Stockholm 1855; (originally in Vetensk. Akad. Förhandl. 1853, p. 257—442): A monograph of the family, including the genera *Coelopa, Copromyza, Limosina, Sphaerocera, Orygma.*

An earlier paper by Haliday: British species of the dipterous tribe *Sphaeroceridae;* in the Entom. Magaz. 1836.

332. **Borborus venalicius,** n. sp. Head brownish-red, vertex darker brown; several whitish-pollinose dots on the front, near the eyes, and on the vertex; antennae brownish-red. Thorax brown, with longitudinal rows of dots of gray pollen; a pair of similar spots at the tip of the scutellum. Abdomen blackish, hind margins of the segments whitish. Wings faintly tinged with yellowish; a transverse brownish spot at the base of the submarginal cell and another at the tip of the third vein. Legs yellowish; femora darker; front tibiae with one, middle and hind tibiae with two dark brown rings. Length: 2—3 mm.

Hab. Cuba. Dr. Loew *(in litt.)* informs me that this is an african species; and as I found it abundantly in Cuba, it seems probable that it was brought over in slave-ships.

333. **Hippoboscidae.** Compare:

1. W. E. Leach, On the genera and species of Eproboscideous Insects. (In the Mem. Werncrian Society, Edinb. 1818, p. 547—566, with three plates; the memoir was presented in 1810).

2. Rondani, Hippoboscita Italiana. (In the Bolletino Soc. Entom. Ital. 1878; at my writing the paper is announced as being in the press.)

334. **Ornithomyia confluenta** Say will, I suppose, form a new genus, on account of its peculiar venation. An apparently different species of the same group was found by Mr. Wm. Holden on *Accipiter fuscus,* near San José, Cal. (M. C. Z.).

335. Compare:

1. Westwood, Nycteribia, a genus of wingless insects, in the Trans. Zool. Soc., Vol. I, p. 275 (1834).

2. **Kolenati, Beiträge z.** Kenntniss der Phthirio-Myiarien; Versuch einer Monographie der Aphanipteren, Nycteribien und Strebliden (in the Horae Entom Rossicae, Vol. II, 1863, p 11—109, with XV plates), a very superficial performance according to Gerstaecker's opinion (Entom. Bericht für 1864—65, p. 126). The combination of *Aphaniptera* and *Nycteribiae* into a common sub-division is certainly an absurdity.

3. **Gerstaecker, Sitzungsb. d. Ges. d. Naturforsch. Freunde in Berlin,** 18. Februar 1862, on the existence of halteres on Nycteribiae (extracted in Gerstaecker, Entom. Ber. 1862, p. 215).

ADDITIONS AND CORRECTIONS.

I. To the list of **Authorities** add :—

COSTA, Achille.—In Annuario del Museo Zool. Univ. di Napoli, II, p. 151, 1864.

N. sp. *Systropus Sallei* and *S. funereus*, both without indication of locality, but both evidently Mexican; the first, a species very frequently met with in collections (also in the M. C. Z.); the second, a synonym of *S. foenoides*, Westm.

I discovered these descriptions accidentally, in looking over Mr. Bigot's library. The diagnosis of *S. Sallei*, which I reproduce, will be fully sufficient for its recognition.

Systropus Sallei.—Niger, antennis, peristomate, thoracis vitta utrinque antice T-formi maculisque duabus ad scutelli angulos baseos, pedibus anterioribus basi excepta, et posticorum apice femorum et tibiarum tarsorumque articulo primo flavis; metatuorace flavo, maculis quatuor rectangulis nigris; abdominis segmentis 1—4 infra pallidis; alis cinereohyalinis, venis fuscis.—Long. mill. 22.

II. **Dates of the first publication of genera.**—In preparing this Catalogue for the press I did not have Latreille's works at hand, I had to rely on Schiner, but have discovered the following errors since :—

Phora was published in Latreille, Précis, etc., 1796.

Simulium, Beris, Pipunculus, Scenopinus, Ochthera, Ornithomyia, Melophagus, Nycteribia appeared in Latreille, Hist. N. des Crust. et des Ins. Vol. III, 1802 (and not Vol. XIV, 1804, as Dr. Schiner has it).

Asyndulum, Rhyphus, Hermetia, Psazus, Paragus, Milesia, Eristalis, Ploas, Ocyptera, Phasia, Oscinis, Sepedon, Tephritis, Lauxania appeared in the Dictionn. d'Hist. Natur., Déterville, Vol. XXIV, 1804, and also in Hist. Nat. des Crust. et des Ins., Vol. XIV, in the same years 1804. The publication in the Dictionnaire is generally quoted as the earlier one; it would be better, perhaps, to quote both.

(265)

In all these cases Agassiz's Nomenclator gives the correct dates. *Echinomyia*, Duméril, was published in 1801; in giving the date 1798, I was led into error by the obituary notice of Duméril, in the Annales de la Soc. Entom. de France, 1860, p. 653, where that date is given.

The name *Tetanocère* appears for the first time in the same publication of Duméril's (1801), but is translated *Tetanocerus* in his Zool. Analyt., 1806. Latreille adopted it as *Tetanocera* in his Hist. Natur. des Crust. et des Ins., Vol. III (1802). Schiner is again in error here.

On page 223, in the note 47*a*, sixth line, for Latreille, H. N., etc., 1804, read Latreille, Précis, etc., 1796.

III, p. 17. Family **Blepharoceridæ.**

Since my arrival in Europe I have had opportunities of a closer study of the Blepharoceridæ, and have come to the conclusion, that *Bleph. yosemite* should rather be considered a *Liponeura*, its broad front being in this case a character of higher order than the differentiation of the facets of the eyes in two portions (with larger and smaller facets). I published this fact in an article entitled, *Bemerkungen über Blepharoceriden* (Deutsche Entomol. Monatschr., 1878, p. 405–416), in which many other remarks, supplementary to Loew's *Revision*, etc., are incorporated.

In looking over Mr. Bigot's collection in Paris, I observed in it an undescribed Blepharocerid (a female), likewise from California, and very remarkable for having the venation exactly like *Liponeura yosemite*, although its contiguous eyes make it a *Blepharocera*. A deep groove divides the eyes in two portions, but there is no strip without facets, as in the two species of Blepharocera hitherto described. The identity of the venation of this species, which I call *Bl. ancilla*, with that of *L. yosemite*, would seem to prove that it is the venation, which in this case is a character of higher order than the structure of the front. Many such discoveries would tend to obliterate the limit between the genera *Blepharocera* and *Liponeura*.

Blepharocera ancilla, n. sp.; female; Gray; thoracic dorsum brownish, with paler longitudinal lines; abdomen brownish, incisures yellowish; antennæ brownish-yellow, brownish towards the tip; legs brownish-yellow; tips of femora brownish; tarsi brown; knob of halteres infuscated; wings subhyaline; veins brownish-yellow; venation similar to that of *Lipon. yosemite*. Length, 7 mm.

Hab. California (collection of Mr. Bigot, in Paris).

The *antennæ* have nothing unusual in their structure; they are a little longer than the head, 14-jointed; first joint short, nearly of the same length with the second, but a little stouter; first joint of the flagellum a little longer than the two following joints taken together; the other joints short-cylindrical, becoming gradually shorter towards the tip; the last inverted-turbinate; pubescence of antennæ short, without any longer hairs.

Eyes contiguous in front of the ocelli, slightly diverging lower down, and leaving room for a narrow, triangular front between them. A deep groove divides each eye in two halves; the upper portion, having the larger facets, is a little longer than the lower portion. A strip destitute of facets is not perceptible in that groove.

Legs rather strong, especially the hind femora; front coxæ at a considerable distance from each other; those of the middle pair are more approximate, those of the hind pair are contiguous; hind tibiæ with a pair of distinct spurs, the inner one by far the longest; spurs on middle tibiæ very minute; I do not see any on the front pair; hind tarsi equal in length to $\frac{2}{3}$ or $\frac{3}{4}$ of the hind tibiæ.

Wings like those of the other Blepharoceridæ as to shape, secondary venation, and chitinous incrassation in the axillary excision. Submarginal cell short, provided with a petiole about equal in length to the abbreviated vein of the posterior margin. A crossvein connects the second vein with the fourth; another crossvein connects the fourth with the base of the large fork of the fifth vein. (In other words, the venation is like that of *Liponeura yosemite;* also like that figured in Loew, Revision, etc., fig. 5, with the exception, as to the latter, of the structure of the submarginal cell, as stated above.)

The *ovipositor* consists of two short, rather obtuse lamels.

IV, p. 92, **Comastes.**—The genus *Heterostylum*, Macq., 3d Suppl., p. 35, is the same as *Comastes*. The principal character, assigned to it by Macquart, pubescence of the third antennal joint, has no existence in reality; Macquart mistook dust for a pubescence! I saw the original type in Mr. Bigot's collection. I do not think that under such circumstances the older name has any claim to priority, especially in this case, where that name is derived from the very character whose existence is disproved.

V, p. 134, below **Helophilus polygrammus,** Loew, is a synonym of *H. mexicanus*, Macq. I saw many Mexican specimens in Mr. Bigot's collection.

VI, p. 181, line 16 from bottom. Strike out the (?) before *Oxycephala maculipennis;* I saw Macquart's type in Mr. Bigot's collection.

INDEX.

Ablautatus, 67
Acanthina, 50
Acanthomera, 51
Acanthomeridae, 51
Acidia, 189
Acidogona, 192
Aciura, 191
Acnemia, 11
Acrocera, 98
Acrosticta, 185
Acrotaenia, 191
Acrotoxa, 189
Actora, 178
Acyphona, 28
Aëdes, 19
Agromyzidae, 209
Agromyza, 210
Allophyla, 175
Allodia, 11
Allograpta, 126
Amalopis, 34
Amphicnephes, 181
Anacampta, 184
Anastoechus, 92
Andrenosoma, 77
Anisomera, 33
Anisopogon, 68
Anopheles, 19
Anorostoma, 175
Anthomyia, 168
Anthomyidae, 164
Anthophilina, 198
Antocha, 27
Anthracophaga, 208
Anthrax, 87
Apelleia, 98
Aphoebantus, 91

Apiocera, 85
Archilestris, 68
Arctophila, 130
Ardoptera, 106
Argyra, 112
Argyramoeba, 89
Aricia, 164
Arthropeas, 43
Ascia, 126
Asilidae, 65
Asilus, 81
A'sphondylia, 5
Aspistes, 17
Asteia, 212
Asteidae, 211
Asyndetus, 113
Asyndulum, 9
Atarba, 27
Atherix, 64
Athyroglossa, 202
Atomogaster, 170
Atomosia, 74
Atylotus, 62
Aulacigaster, 210
Azelia, 170

Baccha, 127
Balioptera, 198
Baumhaueria, 153
Belvoisia, 153
Beris, 44
Bibio, 14
Bibiocephala, 17
Bibionidae, 14
Bittacomorpha, 36
Blacodes, 71
Blepharocera, 17

Blepharoceridae, 17
Blepharoneura, 191
Blepharopeza, 154
Blepharoptera, 175
Bolbomyia, 42
Boletina, 10
Bolitophila, 8
Bombylidae, 85
Bombylius, 91
Borboridae, 212
Borborus, 212
Brachydeutera, 203
Brachyopa, 128
Brachypalpus, 136
Brachystoma, 100

Cacoxenus 210
Caenia, 204
Callinicus, 68
Calliphora, 159
Callomyia, 142
Callopistria, 184
Calobata, 179
Campsicnemus, 114
Camptoneura, 183
Campylomyza, 8
Carphotricha, 192
Caricea, 171
Catabomba, 122
Cecidomyia, 3, 6
Cecidomyidae, 3
Centor, 208
Cephenomyia, 144
Ceratopogon, 22
Ceraturgus, 66
Ceria, 139
Ceroplatus, 9
Cerotainia, 74
Ceroxys, 184
Chaetopsis, 186
Chasmatonotus, 22
Chauna, 51
Chilosia, 121
Chionea, 29
Chironomidae, 20
Chironomus, 20
Chloromyia, 45
Chloropisca, 208
Chlorops, 208
Chordonota, 50
Chortophila, 169
Chrysochlamys, 138
Chrysochlora, 45

Chrysogaster, 121
Chrysomyia, 162
Chrysonotus, 45
Chrysopila, 63
Chrysops, 52
Chrysotimus, 116
Chrysotoxum, 120
Chrysotus, 113
Chyliza, 179
Cistogaster, 146
Cladura, 31
Clavator, 71
Clinocera, 106
Clitellaria, 50
Clytia, 154
Coelometopia, 187
Coelopa, 197
Coenomyia, 43
Coenomyidae, 43
Coenosia, 171
Comastes, 92
Coniceps, 187
Conopidae, 140
Conops, 140
Copestylum, 130
Cordylura, 172
Cordyluridae, 172
Corethra, 20
Crassiseta, 206
Crioprora, 136
Criorrhina, 136
Cryptolabis, 30
Ctenophora, 41
Culex, 18
Culicidae, 18
Cuterebra, 144
Cylindrotoma, 35
Cynomyia, 158
Cyphocera, 149
Cyphomyia, 49
Cyrtidae, 98
Cyrtoma, 104
Cyrtoneura, 163
Cyrtopogon, 69

Dalmania, 141
Dasyllis, 74
Daulopogon, 70
Degeeria, 154
Dejeania, 147
Dermatobia, 145
Desmometopa, 210
Dexia, 155

Dexidae, 155
Diachlorus, 55
Diacrita, 183
Dialysis, 43
Dialyta, 171
Diamesa, 20
Diaphorus, 113
Diastata, 204
Dichaeta, 200
Dichelacera, 55
Dicolonus, 68
Dicranomyia, 24
Dicranota, 35
Dicranoptycha, 27
Didea, 124
Dilophus, 15
Dioctria, 66
Diogmites, 72
Diomonus, 9
Diostracus, 112
Diotrepha, 27
Dipalta, 87
Diplocentra, 204
Diplosis, 4
Diplotoxa, 208.
Discocerina, 201
Discomyza, 201
Ditomyia, 8
Dixa, 41
Dixidae, 41
Dizonias, 68
Docosia, 11
Dolichopeza, 40
Dolichopodidae, 107
Dolichopus, 107
Doros, 126
Drapetis, 105
Drosophila, 205
Drosophilidae, 205
Drymeia, 166
Dryomyza, 178

Eccritosia, 81
Echinomyia, 149
Ectecephala, 207
Echthodopa, 66
Elephantomyia, 26
Elliponeura, 209
Elliptera, 27
Empeda, 30
Emphysomera, 83
Empidae, 99
Empis, 100

Ensina, 193
Ephydra, 203
Ephydridae, 201
Epibates, 95
Epicypta, 12
Epiphragma, 31
Epiplatea, 187
Epitriptus, 82
Epochra, 189
Erax, 79
Eriocera, 34
Erioptera, 28
Eriphia, 167
Eristalis, 131
Ervia, 147
Euaresta, 194
Eudicrana, 10
Eulonchus, 99
Eumerus, 137
Eumetopia, 187
Euparyphus, 46
Eupeodes, 122
Eurosta, 192
Eurygaster, 154
Euryneura, 50
Euthera, 154
Eureta, 191
Euxesta, 185
Exoprosopa, 85
Exorista, 151,

Fucellia, 174

Gastrophilus, 142
Gaurax, 206
Geomyzidae, 204
Geranomyia, 25
Geron, 94
Gloma, 104
Glutops, 65
Gnophomyia, 30
Gnoriste, 10
Gonia, 150
Goniomyia, 30
Graphomyia, 160
Gymnochaeta, 149
Gymnopa, 209
Gymnophora, 212
Gymnopternus, 110
Gymnosoma, 146

Haematopota, 55
Haplegis, 208
Helomyza, 174

Helomyzidae, 174
Helophilus, 133
Hemerodromia, 106
Hemipenthes, 89
Hercostomus, 112
Hermetia, 46
Hesperinus, 16
Heteromyia, 23
Heteromyza, 176
Heteroneura, 197
Heteroneuridae, 193
Hexachaeta, 188
Hilara, 103
Himantostoma, 146
Himeroessa, 181
Hippelates, 206
Hippobosca, 214
Hippoboscidae, 213
Hirmoneura, 85
Holcocephala, 70
Holopogon, 70
Holorusia, 37
Homalomyia, 170
Hoplolabis, 29
Hormopeza, 104
Hyadina, 202
Hyalomyia, 145
Hybos, 99
Hydrellia, 202
Hydromyza, 173
Hydrophoria, 165
Hydrophorus, 115
Hydrotaea, 165
Hygroceleuthus, 107
Hylemyia, 167
Hypoderma, 143
Hystricia, 148
Hystrisyphona, 148

Icterica, 193
Idana, 183
Idia, 159
Idioplasta, 36
Illigeria, 156
Ilythea, 204
Ischnomyia, 198
Iteaphila, 101

Jurinia, 148

Lampria, 76
Laphria, 75
Laphystia, 77

Lasia, 99
Lasioptera, 5
Lasiops, 166
Lasiosoma, 10
Lastaurus, 73
Lauxania, 197
Leja, 11
Lepidomyia, 138
Lepidophora, 94
Lepidoselaga, 55
Leptidae, 62
Leptis, 64
Leptochilus, 91
Leptogaster, 65
Leptomidas, 83
Leptopeza, 104
Leucopis, 210
Leucostola, 113
Leucozona, 122
Liancalus, 115
Limnobia, 25
Limnophila, 31
Limnophora, 166
Lipoptena, 214
Lispe, 171
Lissa, 180
Lobioptera, 209
Lomatia, 90
Lonchaea, 195
Lonchaeidae, 195
Lonchoptera, 118
Lonchopteridae, 118
Longurio, 37
Lophonotus, 82
Lophosia, 147
Lordotus, 93
Loxocera, 178
Lucilia, 160
Lyroneurus, 113

Machimus, 82
Macrocera, 8
Macroceromys, 43
Madiza, 200
Mallophora, 77
Mallota, 135
Masicera, 152
Medeterus, 116
Megapoda, 73
Megaprosopus, 156
Megarrhina, 18
Megistopoda, 214
Melanophora, 156

Melanostoma, 121
Melophagus, 214
Merodon, 135
Meromyza, 207
Mesembrina, 159
Mesocyphona, 29
Mesograpta, 125
Metopia, 153
Metoponia, 43
Microchrysa, 45
Microdon, 119
Micropalpus, 149
Micropeza, 180
Micropezidae, 179
Microphorus, 102
Microphthalma, 156
Microstylum, 67
Midaidae, 83
Midas, 83
Milesia, 139
Milichia, 210
Miltogramma, 153
Mixogaster, 119
Mixtemyia, 139
Molophilus, 29
Musca, 163
Muscidae, 159
Mycetaulus, 199
Mycetobia, 8
Mycetophila, 12
Mycetophilidae, 8
Mycothera, 12
Myennis, 184
Myiolepta, 128
Myopa, 141
Myospila, 164
Myrmecomyia, 182

Nemestrinidae, 85
Nemopoda, 199
Nemoraea, 150
Nemotelus, 50
Neaspilota, 192
Neoempheria, 9
Neoeristicus, 81
Neoexaireta, 44
Neoglaphyroptera, 10
Neoidiotypa, 187
Neoitamus, 82
Neomochtherus, 82
Neorondania, 50
Nicocles, 71
Nothomyia, 45

Notiphila, 200
Notogramma, 185
Nycteribia, 214
Nycteribidae, 214

Ochthera, 202
Ochthiphila, 211
Ocnaea, 98
Ocydromia, 100
Ocyptamus, 127
Ocyptera, 146
Odontocera, 211
Odontomyia, 47
Oecacta, 23
Oecothea, 176
Oedaspis, 191
Oedemagena, 143
Oedicarena, 190
Oedopa, 185
Oestridae, 142
Oestrus, 143
Olfersia, 213
Ommatius, 83
Oncodes, 99
Oncodocera, 90
Oncomyia, 141
Opetiophora, 207
Ophyra, 166
Opomyza, 198
Opomyzidae, 199
Opsebius, 98
Ornithomyia, 213
Orphnephila, 23
Orphnephilidae, 23
Ormia, 163
Ortalidae, 181
Orthoneura, 121
Oscinidae, 206
Oscinis, 207
Ospriocerus, 67
Oxycera, 46

Pachycerina, 196
Pachygaster, 51
Pachymeria, 101
Pachyrrhina, 39
Palloptera, 195
Paltostoma, 17
Pangonia, 52
Pantarbes, 92
Paraclius, 111
Paracosmus, 93
Paragus, 120

Paralimna, 201
Parydra, 203
Pedicia, 34
Pelastoneurus, 111
Pelina, 202
Penthoptera, 34
Peronyma, 190
Phasia, 145
Phalacrocera, 36
Pheneus, 63
Philonicus, 82
Philopota, 99
Philygria, 202
Pholeomyia, 210.
Phoneutisca, 105
Phora, 212
Phoridae, 212
Phorocera, 152
Phortica, 205
Phrissopoda, 158
Phthinia, 10
Phthiria, 93
Phycodromidae, 197
Phyllolabis, 33
Phyllomyza, 211
Phytomyza, 211
Phytomyzidae, 211
Pialoidea, 98
Piophila, 199
Piophilidae, 200
Pipiza, 120
Pipunculidae, 142
Pipunculus, 142
Plagioneurus, 114
Plagiotoma, 190
Platychirus, 122
Platycnema, 142
Platypeza, 142
Platypezidae, 142
Platyura, 8
Plecia, 16
Plectromyia, 35
Plesiastina, 8
Plesiomma, 66
Ploas, 93
Pogonosoma, 75
Pollenia, 160
Polydonta, 135
Polylepta, 9
Polymedon, 111
Porphyrops, 112
Prochyliza, 199
Proctacanthus, 81

Promachus, 78
Prosena, 155
Pseudatrichia, 93
Pseudorus, 76
Psila, 179
Psilidae, 178
Psilocephala, 95
Psilocurus, 70
Psilopa, 201
Psilopus, 116
Psilota, 120
Psychodidae, 23
Psychoda, 23
Ptecticus, 45
Pterallastes, 135
Pterocalla, 184
Pterodontia, 93
Pteroptila, 133
Ptilocera, 154
Ptiolina, 64
Ptychoptera, 36
Pycnopogon, 69
Pyrellia, 162
Pyrgota, 181
Pyrophaena, 122

Rhachicerus, 42
Rhagoletis, 191
Rhamphidia, 26
Rhamphomyia, 102
Rhaphidolabis, 35
Rhaphiomidas, 85
Rhaphium, 112
Rhicnoëssa, 209
Rhingia, 123
Rhipidia, 25
Rhymosia, 11
Rhyphidae, 41
Rhypholophus, 23
Rhyphus, 41
Rivellia, 182

Sapromyza, 196
Sapromyzidae, 196
Sarcophaga, 157
Sarcophagidae, 157
Sargus, 44
Saropogon, 73
Saucropus 116
Scatella, 203
Scatina, 174
Scatophaga, 173
Scatopse, 16

Scellus, 115
Scenopinidae, 97
Scenopinus, 97
Schoenomyza, 171
Sciara, 12
Sciomyza, 176
Sciomyzidae, 176
Sciophila, 9
Scleropogon, 68
Scopolia, 154
Scoliocentra, 175
Scyphella, 198
Senotainia, 153
Seoptera, 185
Sepedon, 178
Sepsidae, 199
Sepsis, 198
Sericocera, 156
Sericomyia, 130
Sigaloéssa, 211
Sigmatomera, 31
Silvius, 55
Simulidae, 14
Simulium, 14
Siphonella, 207
Somomyia, 162
Somula, 138
Spania, 65
Sparnopolius, 93
Sphaerophoria, 125
Sphageus, 68
Sphecomyia, 139
Sphegina, 126
Sphyracephala, 200
Spilogaster, 165
Spilographa, 190
Spilomyia, 138
Stegana, 205
Steneretma, 187
Stenomacra 187
Stenomyia, 187
Stenopa, 189
Stenopogon, 67
Stenopterina, 182
Stichopogon, 70
Stictocephala, 184
Stilpnogaster, 83
Stilpon, 104
Stomoxys 159
Straussia, 189
Stratiomyia, 48
Stratiomyidae, 43
Strebla, 214

Stygeropis, 40
Stylogaster, 140
Subula, 42
Symplecta, 30
Sympycnus, 114
Synamphotera, 106
Synarthrus, 112
Syndias, 100
Syneches, 100
Syntemna, 10
Syritta, 137
Syrphidae, 119
Syrphus, 123
Systoechus, 92
Systropus, 94

Tabanidae, 52
Tabanus, 57
Tabuda, 97
Tachina, 151
Tachinidae, 145
Tachydromia, 105
Tachypeza, 105
Tachytrechus, 111
Tanypus, 21
Taracticus, 72
Temnocera, 130
Temnostoma, 138
Tephritis, 193
Tephrochlamys, 176
Tephronota, 183
Tetanocera, 177
Tetanops, 183
Tetragoneura, 10
Tetropismenus, 183
Teuchocnemis, 135
Teucholabis, 27
Theresia, 156
Thereva, 96
Therevidae, 95
Therioplectes, 56
Thevenemyia, 95
Tipula, 37
Tipulidae, 24
Tolmerus, 83
Toxophora, 95
Toxorrhina, 26
Toxotrypana, 181
Trichocera, 33
Trichonta, 11
Trichopoda, 145
Trichosia, 3
Triglyphus, 120

Trigonometopus, 108
Trimicra, 29
Trineura, 212
Triodites, 90
Triogma, 35
Triptotricha, 62
Tritoxa, 182
Tritozyga, 7
Trochobola, 26
Tropidia, 136
Trypeta, 188, 190
Trypetidae, 188

Ula, 35
Ulidia, 185
Ulomorpha, 33

Urellia, 194
Volucella. 128
Wahlbergia, 147

Xanthochlorus, 116
Xanthogramma, 126
Xestomyza, 97
Xylophagidae, 42
Xylophagus, 42
Xylota, 136
Xysta, 146

Zodion, 141
Zonosema, 190
Zygomyia, 12
Zygoneura, 13

SMITHSONIAN MISCELLANEOUS COLLECTIONS.

——— 321 ———

THE TONER LECTURES

INSTITUTED TO ENCOURAGE THE DISCOVERY OF NEW TRUTHS
FOR THE ADVANCEMENT OF MEDICINE.

LECTURE VII.

THE NATURE OF REPARATORY INFLAMMATION IN ARTERIES AFTER LIGATURE, ACUPRESSURE, AND TORSION.

BY

EDWARD O. SHAKESPEARE, A.M., M.D.,
OF PHILADELPHIA.

DELIVERED JUNE 27, 1878.

WASHINGTON:
SMITHSONIAN INSTITUTION.
MARCH, 1879.

ADVERTISEMENT.

———

THE "Toner Lectures" have been instituted at Washington, D. C., by Joseph M. Toner, M.D., who has placed in charge of a Board of Trustees, consisting of the Secretary of the Smithsonian Institution, the Surgeon-General of the United States Navy, and the President of the Medical Society of the District of Columbia, a fund, "the interest of which is to be applied for at least two annual memoirs or essays relative to some branch of medical science, and containing some new truth fully established by experiment or observation."

As these lectures are intended to increase and diffuse knowledge, they have been accepted for publication by the Smithsonian Institution in its "Miscellaneous Collections."

SPENCER F. BAIRD,
Secretary Smithsonian Institution.

SMITHSONIAN INSTITUTION,
Washington, April, 1879.

LECTURE VII.

Delivered June 27, 1878.

THE NATURE OF REPARATORY INFLAMMATION IN ARTERIES AFTER LIGATURE, ACUPRESSURE, AND TORSION.

By Edward O. Shakespeare, A.M., M.D., of Philadelphia.

Gentlemen : Hemorrhage has formed a favorite theme for study from time immemorial. Its nature and the most efficient means for its arrest have commanded the earnest attention of the most distinguished physicians in all ages and in all lands. Yet in despite of the labors of centuries, in despite of the triumphant march of modern surgery, and the countless revelations of the microscope, it must even to-day be admitted with humility that the hand of man is again and again raised in vain to stay the puissance of this hydra-headed foe. The arrest of hemorrhage, therefore, still remains a subject of the most vital importance. But the time at our disposal does not admit of a discussion of the general question; it does not even permit of a very thorough treatment of a single one of its phases.

HISTORY.

Let us preface our own investigations with a few words concerning the work of our predecessors. Jean Louis Petit, so far as I can learn, was the first who made any systematic attempt to determine the cause of the frequent secondary hemorrhages after wounds and amputations, and to discover a more efficient method of applying the ligature (Mémoires de l'Académie Royale des Sciences, 1731–1732). Since no experiments are related, it may be inferred that his observations were such as

1   (1)

opportunity permitted him to make upon the human body at interrupted intervals.

M. Petit thought as follows:—

After ligation or compression of an artery a clot is generally formed above the place or on the cardiac side of the ligation, or the point of compression.

The constitution and density of the coagulum are varied in different portions by reason of the massing together in places of the several elements composing it—the position of the corpuscles and the fibrin being determined by their specific gravity relative to that of the liquor sanguinis.

It is more advantageous that the clot should be formed of the white part (lymph) only, than that it should consist of a mixture of the lymph and the red globules.

The clot in a short time becomes as firmly united to the sides of the artery as the granulation-tissue which forms cicatrices is to the lips of wounds. This intimate union once formed, not only is secondary hemorrhage prevented, but the clot in this state remains and disappears only as cicatrices diminish, in proportion to their condensation.

M. Morand (Sur les changements qui arrivent aux artères coupées ; où l'on fait voir qu'ils contribuent essentiellement à la cessation de l'Hémorrhagie. Mémoires de l'Académie Royale des Sciences, tome liii. Année 1736) communicated some observations and conclusions upon the subject matter of the foregoing essays of M. Petit, le Chirurgien.

The paper concludes with the following sentence, which is in reality a formulation of his opinion concerning nature's mode of stopping blood, viz.:—

"The changes which take place in the arteries (retraction and contraction of the walls) contribute, then, to the cessation of hemorrhage conjointly with the clot, generally in every case ; and if it is possible that the artery alone or the clot alone can do so, the cases which may be cited in proof thereof will be extremely rare."

Mr. Sharp, a few years later (Operations of Surgery, 1739), entertained and taught principles governing the checking of hemorrhage very similar to those advanced by the last-named investigator.

M. Pouteau is the next person we find publishing the results of investigations relating to the healing of bloodvessels (Mélanges de Chirurgie, 1760).

He concluded, " That when an artery is divided, a coagulum does not always form; that the retraction of the artery has not yet been demonstrated; that the retraction of the walls is not more effectual for the arrest of hemorrhage than is the presence of a clot; that the presence of a clot is only a very weak and subsidiary means toward that end; that the infiltration and swelling of the cellular membrane at the circumference of the cut extremity of the artery offer the chief obstruction to the bleeding; that by exciting and aiding in a more rapid and general induration of that membrane, the use of the ligature is valuable for the arrest of hemorrhage."

The name of Kirkland appears next upon the list of those who have endeavored by a series of observations to penetrate the ways which nature adopts for the cure of a wounded vessel (Essay on the Methods of Suppressing Hemorrhages from Divided Arteries, 1763).

His opinions may be formulated as follows:—

The hemorrhage from a very considerable artery is easily and effectually suppressed by merely making a perpendicular pressure upon the end of the vessel for a few minutes.

The bleeding is not suppressed by congealed blood, but by the vessel being quite closely contracted for near an inch or more from its extremity.

Interruption of the passage of the blood for a while is all that is required from art.

Gooch (Chirurgical Works, 1766) turned his attention only a year or two later to the subject before us.

Mr. White (Cases in Surgery) agreed with Pouteau, Kirk. land, and Gooch in rejecting Petit's theory of a coagulum as not at all probable. He concluded that the formation of a coagulum is only incidental, and is of no use whatever except under particular circumstances.

Hunter believed in the adhesive inflammation of all the tissues of the vessels. He considered that the clot adheres to the walls, and undergoes organization.

John Bell (Principles of Surgery, 1801) also ranged himself on the side of those who opposed the views of Petit and Morand. He thought that hemorrhage is always permanently prevented by the changes which take place in the surrounding cellular tissue, and by adhesive inflammation of the arterial walls themselves.

J. Thomson, of Edinburgh, made some observations upon the effect of ligation.

The next to be mentioned in chronological order is the classic work of J. F. D. Jones, M.D. (A treatise on the process employed by nature in suppressing the hemorrhage from divided and punctured arteries, and on the use of the ligature, etc., 1805). The completeness of this man's experiments, and the apparent soundness of his judgment upon the principles to be deduced from his results, succeeded in settling, at least for a lengthened period, the much-vexed question which he set himself to solve. Indeed, such have been the closeness and accuracy of his investigations that, even to-day, his excellent monograph remains admittedly the authority upon the means which nature adopts for the suppression of hemorrhage. The occasion is taken here to acknowledge our indebtedness to his paper for much of this history.

With respect to spontaneous arrest of hemorrhage from divided vessels, Jones states that for the reason that the formation of the internal blood-clot is uncertain, or that when formed it rarely fills the canal of the artery, or if it fills the canal does not adhere to its internal coat, it is not to be ranked among the means which nature employs for the suppression of hemorrhage, for in ordinary accidents it contributes nothing to those means.

The permanent changes which take place in an artery and in the circulation through the limb, in consequence of the application of the ligature, are precisely similar to those after the division of an artery. Some of the effects of tying an artery appear to be the following : to excite inflammation in the middle and internal coats by having cut them through, and, consequently, to give rise to the effusion of lymph (colorless clot), by which the wounded surfaces are united and the canal is rendered impervious; to produce an inflammation on the corresponding external surface of the artery, and at the same time, by the exposure and inevitable wounding of the surrounding parts, to occasion inflammation in the latter and an effusion of lymph which covers the artery and forms the surface of the wound.

According to Jones, it is a fact that in most cases only a slender clot is formed at first, which gradually becomes larger by successive coagulations of the blood, and it is for this reason that the clot is always at first of a tapering form, with its base at the extremity of the artery. But the formation of this coagulum is of little consequence, for soon after the application of the ligature the extremity of the artery begins to inflame. The wounded internal surfaces of its canal being kept in close contact by the ligature adhere, when this portion of the artery is transformed into an impervious and, at first, slightly conical sac. It seems to be entirely owing to the effusion of lymph that this adhesion is effected.

Hodgson (Diseases of the Arteries and Veins, 1815) contended that the veins are liable to all those morbid changes which are common to the soft parts in general, but the membranous lining of those vessels is peculiarly susceptible of inflammation.

Bouillaud (Archives Générales, 1824, série vi., tome 5) maintained the organization of the thrombus and its adhesion to the walls of the vessel, as also did Ribes (Revue Médicale Française et Étrangère, 1825, tome 3), as well as Roche and Sanson (Nouveaux Éléments de Pathologie. Medico-Chirurgicale, Paris, 1826).

Scarpa (Memoria sulla ligatura della principali arteori, edizione 1825) has occasionally observed, two or three days after the application of the ligature, the adhesion of the walls without the intervention of a clot.

Gendrin (Histoire Anatomique des Inflammations, 1826, tome ii.) perhaps deserves mention here, since a theory respecting the mode of formation of lymph in inflamed vessels had derived much of its support from an often-cited experiment which he reported. He claimed that the inner coat of veins affords a concrete layer of lymph which obliterates the vascular canal.

Ebel (De natura medicatrice sicubi arteriæ vulneratæ et ligatæ fuerunt, Guersa, 1826) denied that the internal coagulum takes any part in the organizing process, and affirmed his belief in its disorganization and disappearance.

Cruveilhier (Anatomie Pathologique, 1829) spoke of the disappearance of the thrombus by absorption

The next great communication on the subject of hemorrhage came from M. Amusat (On a new method of arresting hemorrhage from large vessels without the aid of the ligature. Académie Royale de Médecine, 1829). The conception of this new method was first suggested to his mind by the long-recognized fact that torn wounds do not bleed. The development of this suggestion was worked out by experiments upon animals. The perfected plan was applied upon the human subject, and the practice of torsion was introduced and published to the world.

Blandin (Journal Hebdomadaire, mai 1830) accepted the organization of the thrombus, and its thorough adhesion to the vessel walls, and Mannec (Traité Théorique et Pratique de la Ligature des Artères, 1832) was in accord with this view; on the other hand, Walther (Système de Chirurgie, 1833) and M. Lobstein (Pathologische Anatomie, 1834) were of the opposite opinion.

W. B. Costello, a former pupil of M. Amusat, followed the communication of his master by a paper read before the Westminster Medical Society on "Torsion of Arteries for the purpose of Arresting Hemorrhage" (London Lancet, March 8, 1834), in which experiments upon dogs were detailed.

Stilling (Ueber Bildung und Metamorphose des Thrombus in verletzten Blutgefäsen, Eisenach, 1834) repeated and corroborated the researches of Petit. He saw the adhesion of the clot to the wall of the vessel, its pyramidal shape, considered eighteen hours as about the length of time requisite for its formation, and admitted, with Moran, the action of the ligature on the two inner tunics of the vessel.

Pirogoff (Ueber die Durchshneidung der Achillessehne. Dorpat, 1840) defended the general proposition that fibrin possesses a power of self-organization.

Zwicky (Metamorphose des Thrombus, 1845) recognized fibrin as a formative element in the process of organization. For him fibrin forms a plastic exudation upon the inner wall of the vessel, and effects the growth of the latter to the thrombus. The fibrin found in the thrombus as one of its elements likewise soon organizes itself there. He observed the formation of vessels in the thrombus.

Both Castelnau and Notta (De la cicatrisation des artères. Gazette des Hôpitaux, 1851, No 13, 14) confessed to the same opinion, and claimed further that the thrombus is subject to purulent degeneration.

Thierfelder (De regeneratione tendinum, 1852) is to be ranged with Pirogoff and the others who admit the formative power of fibrin.

Henry Lee (On the deposition of fibrin in the lining membrane of veins. Med.-Chir. Transactions, 1852) did not think that in Gendrin's experiments sufficient care had been taken to exclude the possible presence of a small blood-clot. He devised a method by which this dilemma could be avoided, and he aspired to put the question of the rôle of the vessel wall at rest forever by performing a solitary experiment. This author concluded that the blood coagulum is necessary to the presence of inflammation, and that it acts as a foreign body, the inflammation excited by it being a natural process for its elimination. This inflammation begins in the outer, and thence extends to the inner coat; extends *to* the lining membrane of the vein, and not *from* it.

Boner (Die Regeneration der Sehnen. Virchow's Arch., 1854) acquiesced in the independent formative power of the fibrin wherever found.

Rokitansky (Pathologische Anatomie, 1856) regarded the walls of the vessel as the origin of the material which finally fills the lumen and becomes organized.

Meckel (Microgeologie, Herausgegeben von Billroth, 1856) was among the first in this connection who began to perceive in the white blood-corpuscle an element which might possess capabilities that should not be entirely overlooked in the examination of these processes. However, he neither ascribed to the leucocyte any great rôle, nor yet denied to it a power of organization.

Virchow (Canstatt Jahresbericht, Bd. 1, I. 31) advanced the opinion that the white blood-corpuscle as a formative element carries the fibrin.

Simpson (Acupressure as a new hæmostatic process. Royal Society of Edinburgh, 1859) claimed that the acupressed vessel is closed by adhesive inflammation of its inner walls.

Bogdonowsky (Medicinische Zeitschrift (Russia), 1862) summarized the results of experiments upon varicose veins by the declaration that the thrombus formed by injecting into the veins liquor ferri sesquichloridi acts as a foreign body, and can only degenerate; that the vessel is obliterated at the expense of its walls.

Ardreef (Ueber das Blutkörperchen in histologischer Beziehung, St. Petersburg, 1862) affirmed that he had observed the transition of the red blood-corpuscles into the white, and subsequently the formation of connective tissue from these.

Koslowsky (Untersuchung ueber die Strabotomie. St. Petersburg, 1863) and Rindfleisch (Apoplexia cerebri. Arch. d. Heilkd. von Wagner, 1863) confirmed this observation.

Billroth (Allgem. chirurg. Pathologie und Therapie, 1863), studying the thrombus with the microscope, made observations which he considered to be a demonstration of the truth of Virchow's suspicion of the formative activity of the white blood-corpuscle.

Schmidt (Ueber den Faserstoff und die Ursache seiner Gerinnung. (Russische) Militär-Mediçin Journal, 1863) contended that the length of time requisite for the formation of a coagulum after ligation depends upon the various conditions affecting the coagulability of the blood. He conceived that a fibrino-plastic substance exists in the vessel walls, and, that after its destruction, by any injury to the walls for instance, it is endued with the capability so to act upon the fibrinogenous substance of the blood as to condition, from the latter, coagulation.

Janowitsch Tschiansky (Dissertatio, St. Petersburg, 1864) repeated and confirmed the experiment and conclusions of Bogdonowsky.

O. Weber (Ueber die Vascularisation des Thrombus. Berliner

Klin. Wochenschrift, 1864) made a series of experiments upon dogs and rabbits, directed to the determination of the mode of vascularization of the thrombus. He concluded from his numerous observations:—

1st. That the red corpuscles and the fibrin degenerate and disappear.

2d. That the white cells by means of their peculiar movements during the first hours undergo a change into bodies of a peculiar shape, and very soon become transformed into spindle-form cells.

3d. That in the first four days the extremities of the prolongations are seen uniting and forming a network, taking position in lines having every semblance of vessels.

4th. That the younger vessels are generally formed in the periphery of the thrombus.

5th. That by the end of the third or fourth week, the vessels of the thrombus have formed a union with those of the adventitia. At the place of ligature where the intima and media are lacerated, the vessels of the adventitia pass directly into the thrombus; farther away from the ligature, they reach the thrombus by penetrating the intima.

6th. That by the fiftieth to the sixtieth day the whole thrombus, especially its periphery, is full of bloodvessels. A single large one is often seen in the centre.

7th. That these vessels subsequently close up.

Forster (Handbuch der Pathólog. Anatomie, 1865) denied the organization of the thrombus, and believed that the healing and final obliteration of the veins are due to a growth of the walls.

Stricker (Ueber das Leben der farblosen Blutkörperchen. Sitzungs-berichte der Akademie der Wissenschaften, 1867) admitted the viability of the white corpuscle in the thrombus, but did not affirm its formative power.

Obolensky (Ueber die Organisation des Blutes. Protokoll

des Vereins russicher Aerzte, 1867) attempted to test the various theories concerning the organizing power of the white blood-cells. His observations were made on a clot of blood and a large number of white blood-corpuscles, which were taken from one frog and placed under the skin of another. He found by this experiment that the whole mass—the red and the white blood-corpuscles, as well as the fibrin—underwent degeneration by fatty metamorphosis. The red blood disks first decolorized, and then degenerated. On the fourth day, there remained of the clot only pigment and fatty particles.

Bubnoff (Ueber die Organisation des Thrombus. Central Blatt, No. 48, 1867), under the direction of Von Recklinghausen, performed three series of experiments, aiming at the tracing of the movements of the white blood-corpuscles in the organization of venous thrombi.

1st Series. Ligation of jugular vein. Rubbed vermilion on exterior wall of the vessel. Result—the colorless corpuscles penetrate the wall of the vessel, absorb the vermilion, reach the thrombus, and then organize themselves, the color not disappearing.

2d Series. Ligature of one jugular. .Twelve hours afterward injection of vermilion into the other jugular. Result—vermilion did not reach the thrombus.

3d Series. Two ligatures on one vein. An injection of vermilion into the thrombus. Result—the white corpuscles did not absorb the vermilion.

Conclusions: 1st. The thrombus organizes only by means of the white blood-corpuscles, which penetrate the vessel-wall.

d. The white blood-corpuscles do not reach the thrombus directly by way of the blood-current.

3d. The cells of the vessel-wall are probably concerned in the organization.

Waldeyer (Zur pathologischen Anatomie der Hautkrankheiten, Virchow's Archiv, Band xl., 1867) has affirmed his belief that the tunica intima takes an active if not the sole part in the organization of the thrombus.

Thiersch (Chir. von Pitha und Billroth) would be named among those who, *a priori*, incline to the view that the epithelial cells and the nuclei of the different lamellæ of the intima should be considered as formative elements, active in the organization of thrombi.

Henry Lee and Lionel S. Beale (On the repair of arteries and veins after injury. Med. Chir. Trans., vol. i., 1867) studied the phenomena following a puncturing wound of an artery, and found that a colorless fibrin-like material fills the wound. It consists mainly of colorless blood-corpuscles derived from the blood in the lumen of the wounded vessel. This forms layer after layer, a temporary tissue. The subsequent changes which take place in this fibrin-like material, and effect the permanent closure of the wound, they did not investigate ; but they were convinced, *a priori*, that the formation of a new permanent fibrous tissue results from the masses of germinal matter (colorless corpuscles) of the temporary adventitious tissue above mentioned, and not from the masses of "germinal matter" of the arterial tunics, or of the vasa vasorum.

Hewson (Pennsylvania Hospital Reports, vol. i., 1868) made a careful study of several specimens from human arteries after acupressure had been performed. Longitudinal section of the acupressed vessel showed the opposite surfaces of internal coat glued together by lymph. No clot beyond the point of pressure, and no laceration of the internal coat. What struck most forcibly was the extent of the exudation which had taken place upon the internal coat, and even outside of it. The thickening extended nearly a half inch above, gradually diminishing up to the first branch.

Tschausoff (Ueber den Thrombus bei der Ligatur, in dem "Verein russicher Naturforscher" vorgelesen, Protokoll des

Vereins russicher Aerzte, 1868, and Archiv für Klinische Chirurgie, Band II., 1869) published an exhaustive paper—which we have extensively used in this history—on thrombi after ligature. The author states that—the thrombus never organizes; the muscle-fibres of the media are never concerned in the organization, and the same may be said of the epithelium; changes are soon observed in the wall of the vessel, and in all its tissues; growths from the walls encroach upon the lumen of the vessel; this newly formed tissue, both of the wall and of the lumen, is rich in vessels; from the arterial wall the vessels go direct into the lumen, which either altogether or in part is closed up; the development of the vessels progresses at the same time with that of the newly formed tissue; the circulation of the blood comes from the wall of the vessel itself.

The author repeated the experiments of Bubnoff, and failed to obtain the same results. He states that in five experiments, colored form-elements were perceivable in the thrombus, but they were small, in large numbers, and without definite form. They somewhat resembled altered red blood-corpuscles.

Bryant (On the torsion of arteries as a means of arresting hemorrhage. Experiments. Med. Chir. Trans., vol. ii., 1868, and On torsion of arteries, a description of some models made to illustrate the effects of torsion, Guy's Hosp. Rep., Series III., vol. xv.) believed that in torsion the twist of the cellular coat of an artery, the division and subsequent retraction, incurvation, and adhesion of the middlle coat, and the coagulation of the blood in the vessel as far as the first branch, are the three points upon which temporary as well as permanent safety depends. In his opinion the permanent safety of acupressure rests upon the last point alone, and the temporary effects upon the pressure produced by the needle.

Kocher (Ueber die feineren Vorgänge bei der Blutstillung durch Acupressur, Ligatur, und Torsion. Archiv für Kli-

nische Chirurgie, No. 11, 1869) made a number of experiments
and microscopic studies upon the mode of permanent arrest of
hemorrhage, by the use of the ligature, acupressure, and acutorsion. He found that by the employment of each of these
means the presence of a clot in the vessel was usually secured;
that these clots were sufficient to arrest hemorrhage temporarily from a small vessel; that in acupressure or in acutorsion they are gradually formed and increase slowly in size, not
usually being sufficiently large by the end of forty-eight hours
to check the bleeding from a large vessel.

Cornil and Ranvier (Manuel d'Histologie Pathologique,
1869) have carefully examined the method of healing in an
artery after ligature.

With respect to the *double ligature* of *veins*, they think that
what Bubnoff claims to have observed is incontestable. But
never in a single ligature of arteries and veins, where the
bottom of the wound had been smeared with vermilion, did
they see the latter penetrate through the walls of the vessel.
En résumé, they declare that the definitive obliteration of
arteries after ligature is effected by a neoplasm, the point of
departure of which is the arteritis consecutive to the traumatic
lesion. As to the clot, it disappears by a series of retrogres-
sive alterations similar to those which the blood goes through
when it escapes from the vessels into the tissues.

Durante (Entzündung der Gefässwände. Med. Jahrbüch.,
Band III., 1871, and Recherches expérimentales sur l'organisation du caillot dans les vaisseaux, Arch. de Physiologie Nor-
male et Pathologique, tome iv., 1872) has conducted a most
careful and thorough examination of the still unsettled question as to what are the organizing elements active in the processes inaugurated by the ligature of an artery.

He admits the formation of a temporary and a permanent
clot. The former is of blood, and is not homogeneous; it
gradually disappears. The latter is a colorless clot formed

mainly of epithelioid cells; it effects the permanent closure of
the vessel. He thinks that his preparations clearly demonstrate
that in the case of the single ligature the organizing elements
of the permanent clot are derived from the tunica intima. In
the double ligature, on the contrary, the internal membrane in
the portion limited by the two threads becomes modified. The
substitution of the temporary by the permanent clot is accomplished more slowly, since the coagulated blood must produce
mortification of the internal membrane, and later become the
irritating agent of the middle and external membrane.

Following the experiments of Bubnoff, Durante declares that
never in the single ligature, when the coloring matter has
been simply placed upon the vessel, has he been able to find, in
the clot or in the wall of the vessels, cells containing granules of vermilion. The same is true of the double ligature
when the inflammation has not yet destroyed the limit between the walls of the vessel and the surrounding tissue. If
the vermilion is gently applied to the walls of the vessel, the
coloring matter remains for many days at the periphery of
the artery, and in transverse sections it appears as a line distinct and continuous, at the surface of the adventitia. But
when the walls of the vessel become confounded with the
neighboring tissues by the progress of inflammation, the vermilion may be recognized here and there in the midst of the
tunics. In the single ligature, if the greatest possible care is
taken, it is easy at the end of the twelfth day still to perceive
the vermilion limited to the perivascular connective tissue;
but after prolonged and somewhat rough friction, he has been
able, at the end of a few hours, to demonstrate on the jugular
vein of rabbits, exposed and included between two ligatures,
that there exist in the middle of the clot granules of vermilion *in a free state*. The walls of the vein were infiltrated
with similar granules. The same manœuvre practised upon
the arteries causes the vermilion to reach only as far as the

muscular tunic   "It is then by mechanical penetration, and thanks to the thinness of the venous walls, that the particles of vermilion have travelled as far as the clot." Durante thinks the origin of the formative cells in the single ligature is to be found in the endothelium, or the ramified cells of the internal membrane.

Baumgarten (Centralblatt für die Medicinischen Wissenschaften, No. 34, 1876) publishes the very latest investigation upon the healing of arteries. He sums up the substance of his researches as follows:—

1st. The so-called organization of red thrombi is due to two distinct processes: first, a proliferation of the arterial endothelium; and, second, an invasion from without of connective tissue elements from which the new bloodvessels are solely formed.

2d. The part played by the clot in the organization is *nil ;* an occasional fragment of encapsuled pigment is the only remnant it leaves.

A. Pitres (Recherches expérimentales sur le mode de formation et sur la structure des caillots qui déterminent l'hémostasie. Arch. de Phys. Nov. et Path., 1876) affirms that hæmostasis is usually spontaneously secured by a clot composed of three distinct portions: an external part, which is a simple blood coagulum, and which is only an accessory; a middle part, which is lodged in the wound, or lumen of the vessel, which is the most constant and active agent of spontaneous hæmostasis, and which consists almost exclusively of white blood-corpuscles; and an interior clot, which has merely an accidental office, and which has a complicated constitution.

SUMMARY OF PREVALENT OPINIONS.

We have seen that notwithstanding the apparent confidence and sometimes even dogmatism with which the leading pathologists have published opinions and advocated theories concerning the organization of the blood, there has at no time been a unanimity of opinion among the investigators whose labors have furnished the most important observations bearing upon the process of healing after wounds of bloodvessels, and that no less than four of the latest publications which have been furnished by the pens of most distinguished pathologists directly contradict the assumption of Billroth and Rindfleisch concerning the activity of the wandering cells in the organization of thrombi.

In commencing the relation of our own personal observations, perhaps it may be proper to state at the outset that the conclusions which we believe to be legitimate deductions from the facts which shall be reported are, in many important points, at variance with some opinions generally admitted by the scientific world to be well established.   If these deductions shall stand irrefuted it will become necessary to modify greatly the present prevalent opinions concerning the nature of inflammation.   It is not our intention, however, upon this occasion, to discuss the nature of inflammation in general.   The question of inflammation will be raised only by indirection, and will be limited to the inflammatory processes as they are seen in wounded arteries.

It is by the light of pathological histology alone that we propose to examine to-night " The Nature of Reparatory Inflammation in Arteries after Ligature, Acupressure, and Torsion."

Just here let it be premised that if our conclusions are not in accord with views considered as established, it cannot be charged by the defenders of the latter that our investigation has been undertaken or conducted with an unfavorable bias. Until the completion of our experimental study of ligation, no authorities upon the pathological histology of the subject had been examined by us other than Rindfleisch and Billroth. Their opinions upon this subject had, up to the time of the

2

inauguration of this investigation, been in our mind unquestioned.

Before entering into details of our own studies, it seems advisable to summarize the opinions concerning the intimate nature of the healing process in arteries after ligation which are at present supported by the weight of authority, and are consequently accepted by the medical world as beyond dispute.

Billroth, in his celebrated work on surgical pathology, says that after ligation the plugging of an artery by a blood-clot is only a provisional attempt on the part of nature to arrest hemorrhage. The thrombus does not remain in the same condition for all future time, but it becomes transformed into cicatricial tissue, shrinks and atrophies, when the artery at the point of division has become solid by the complete fusion of this cicatricial mass with the walls of the vessel. For the completion of this process months, and even years, are required. In what these changes of the blood-clot actually consist, the microscope gives valuable evidence. The clot is homogeneous throughout, that is to say, there is no stratification or grouping of the blood-disks, either of the white or red; but, on the contrary, they are scattered evenly through the entire coagulum.

It is the further development of the colorless cells of this clot which secures the definite termination of the whole process. Since the blood-clot, consisting of cells and coagulated fibrin, is at first a non-vascular cellular tissue, which can only at first maintain its existence in thin layers, it is apparent (and observation confirms this) that large blood-clots are not organized at all or only in their peripheral layers, while they disintegrate in the centre.

What are those cells which organize the thrombus? and whence do they come?

On a previous page of this distinguished author's most excellent work a sentence appears which places Billroth upon even more radical ground with respect to the Cohnheim theory than the celebrated author thereof himself takes.

These are the exact words: "*All young cells* which in inflammation are found abnormally in the tissues *are wandering white corpuscles.*" Himself replying to the questions above propounded, the Vienna Professor makes use of the following unequivocal language: "After having abandoned the idea of proliferation of stable tissue cells in inflammation, we can no longer talk of the *proliferation of the intima* in the old sense." And again: "I have no doubt that they originate from the white blood-cells, which have been partly inclosed in the thrombus, and partly may have wandered into it, according to the observations of Von Recklinghausen and Bubnoff." As to the ultimate origin of the wandering cells Billroth conceives their factories to be the lymph glands, and remotely the stable cells of the connective tissue.

The great German pathologist Rindfleisch in the main accords with Billroth respecting the formation and organization of the thrombus in a ligated vessel.

They agree that it is formed suddenly, and that it is unstratified, there being no accumulation of numbers of white cells in places; it is homogeneous throughout.

They also agree that the blood-clot is organized; that the white corpuscles (wandering cells) are the organizing elements; that the red disks slowly degenerate and disappear.

Rindfleisch believes that the thrombus is largest immediately after coagulation, which takes place almost immediately after the blood is placed out of circulation by the constriction of the ligature; and in this also accords with Billroth.

Both authors think that the clot becomes gradually converted into ordinary cicatricial tissue, and that through the cavernous metamorphosis of this the clot and the vessel-wall surrounding it are at length converted into a mere cord or thin band of dense fibrous tissue, the only remains of the previous blood channel.

To summarize briefly, the standard authorities publish in very positive terms the opinions which they believe have been indisputably demonstrated, viz.: that immediately after the ligature of an artery a blood-clot is formed which plugs the lumen of the vessel, generally up to the level of the first collateral branch; that this clot is homogeneous, unstratified, is formed at one time, and is larger during the first hours than at any other period; that it offers a temporary barrier to the flow of blood; that soon the blood-clot thus formed, itself becomes organized and supplied with its own vessels, which form a communication first with the lumen above the clot, next with the vasa vasorum mainly at the bottom of the clot; that it is by this organization and vascularization of the temporary blood-clot, and the intimate union of this newly formed tissue with the vessel-walls, that the lumen of the wounded artery becomes permanently closed against the blood current; and that the organizing elements are solely and exclusively the white blood-corpuscles and their descendants—either those which are caught in the clot at the time of its formation, or those which may have wandered into it afterward, or more probably those derived from both sources. It is thus that their ideas concerning the nature of the inflammatory process in wounded arteries are made to coincide entirely with and to give some further support to Cohnheim's theory of inflammation in general.

PERSONAL OBSERVATIONS.

More than three years ago, at the request of my friend Prof. Agnew, of the University of Pennsylvania, and for his benefit, I (with the surgical assistance of my friend Dr. Wm. Mastin, of Mobile, who was at that time an Interne of University Hospital) traversed experimentally some of the ground which O. Weber had gone over while making his researches relative to the march of the vascularization of a thrombus in a ligated vessel. To my surprise, I was unable in any single instance to find under the microscope appearances in my sections of thrombi which could to my mind be fairly considered as confirmatory of the foregoing statements of Billroth and Rindfleisch, either as to the formation, the constitution, the organization, the vascularization, or the obliteration of the blood-clot.

After having given to Dr. Agnew the results of the examination which I had undertaken for him, I inaugurated for myself a more thorough and systematic experimental study of the whole question of the manner and the means by which a ligated artery is healed. This second investigation was conducted at odd times, and had extended in this way over a year and a half, when, at the commencement of January of 1877, I determined to embody the results I had attained in an essay, which secured for me the award of the Warren Triennial Prize for that year. Since then, as occasion has offered, I have from time to time added to the number of my experiments and observations, and have pushed them into the question of the healing of an artery after acupressure and torsion as well. The conclusions based upon the entire series of observations are more comprehensive than those derived from my first study, and, in a few points, are slightly different.

Upon the healing of arteries four regular series of preparations have been secured, the experiments principally being performed upon the femoral arteries of young, vigorous dogs.

In obtaining the *first series* the following order has been observed, viz.: the artery was exposed and tied in continuity with the ordinary silk ligature, the thread being allowed to remain on the vessel until the animal was killed, or until it came away without assistance. The subjects were then killed in rotation at such times as to afford preparations of their arteries, 24, 36, 48, and 94 hours, and 5, 8, 10, 15, and 21 days after ligation. Each number of this series was duplicated, at least once, and several of them, such as 24, 48, 94 hours, and 10 and 15 days, were repeated two or three times and oftener.

A *second series* was begun, but owing to pressure of other engagements was not entirely completed. This series was intended to supply as full a number of preparations as the first. The procedure followed in the preparation of this series consisted in a slight modification of the ordinary manner of performing ligation: the ligature was applied in the continuity of the vessel in the usual manner; immediately afterward the vessel was compressed an eighth of an inch above the point of ligature, by means of an ordinary pair of dressing forceps, so as to moderately rub together opposite points of the inner surface of the internal membrane of the vessel, and thus produce at these points a sufficient irritation, at the same time avoiding if possible any rupture of the inner tunic. A number of preparations from this were obtained, varying from three to ten days.

The *third series* consisted of a limited number of preparations to show the method of healing after *limited torsion.*

A *fourth series* was obtained, the number of preparations also being much smaller than the first. They were intended to supply a full series for the satisfactory study of the process of healing after *acupressure.* In performing acupressure, the procedures known as the third and fourth methods were adopted — the third being done in the continuity, and the fourth after the division of the vessel. The needle was allowed to remain in the tissue until the specimens were hardened for examination. I may say here that a *fifth series* was also commenced, wherein specimens were to be obtained to show the results following a mere occlusion of an artery in continuity, by moderate pressure produced by the inclusion of the artery for a few hours between the arms of a small serre-fine. The experiments for this last series were made upon the femoral and carotid arteries of good-sized rats.

The various operations upon dogs were performed during anæsthesia; those upon rats after they had been bridled and tied down to a board. The vessels containing thrombi five days old and upward were generally injected with Beale's Prussian blue fluid. In all cases the vessel operated upon was removed immediately after the death of the animal, extreme care being taken to avoid pressing upon or stretching that portion of the vessel which contained the thrombus. Immediately after removal from the animal the specimens were usually placed in dilute alcohol, which was subsequently gradually strengthened from day to day by the addition of small quantities of strongest alcohol. Occasionally a specimen was hardened in chromic acid or Müller's fluid.

The specimens were allowed to remain undisturbed in the hardening agent until they had become thoroughly firm and hard.

After that they were placed for a day in absolute alcohol. They were subsequently removed from this and saturated with oil of cloves, and were then imbedded in a mixture of about one part of benzine to twelve or sixteen parts of paraffine. Thin sections, both longitudinal and transverse, were then made from each specimen. Generally all such sections were subsequently stained with carmine, and temporarily prepared for microscopic examination by being mounted whole in oil of cloves, or by being torn apart by needles for examination of their isolated elements. A few gold and silver preparations were also made. It may be stated at this point, that the original drawings which illustrate my own part of the labors chronicled in these pages are not mere diagrams, but are actual copies of objects in the field of the microscope, traced by myself, as accurately as possible, by the aid of a good camera.

FIRST SERIES. — Microscopic examination of sections from the *first series* demonstrated the fact that the apparent sequential order of the various phenomena exhibited throughout the series, presented a marked uniformity.

Nearly every preparation twenty-four hours old showed, under a low power and in longitudinal sections, a blood-clot, not unusually extending as far as the first collateral branch. This clot was usually egg-shape, and it did not fill the entire calibre of the vessel. Ordinarily it was adherent to the vessel-wall only at one side, while it was slightly separated from the opposite side. It did not extend quite down to the point of ligature, for the bottom of the little cup formed by the constricting action of the thread upon the arterial walls was generally covered over several layers deep with colorless cells, and it was upon this cushion of colorless cells that the butt-end of the blood-clot rested. The outer surface of this cup-shape cushion of colorless cells was everywhere closely adherent to the inner membrane of the vessel-walls. At the sides this cup-shape cushion extended along the inner surface of the vessel-wall for a considerable distance from the ligature—occasionally up as far or even farther than the apex of the blood-clot. The bottom of the blood-clot was adherent to the bottom of this cup-shape cushion of colorless cells, and it was also adherent to one of its sides. To avoid confusion, I shall hereafter refer to the blood coagulum as the *blood* or *fibrinous clot*, and in distinction shall speak of the cup-shape cushion of colorless cells as the *cellular* or *plastic clot*. The number of the colorless cells of the plastic cup or clot, or, in other words, the thickness of its walls, rapidly decreased in proportion to the remoteness from the point of ligation. Concerning the constitution of the fibrinous or blood clot, the declaration is emphatically made that, when viewed in longitudinal section, in not one solitary instance in any of these series was it observed to be homogeneous in structure; but that, on the contrary, when so viewed *every blood-clot* presented the most *unmistakable* appearance of *lamination* or stratification. This feature was uniformly present throughout the whole of the first series, as well as throughout all of the others. It may be stated, however, that some of the transverse sections did not present this appearance of lamination. Fig. 2, although drawn under a low power from a preparation of forty-eight hours, fairly represents the stratified appearance of all these blood-clots when seen upon section in profile.

In order as much as possible to avoid repetition, the further discussion of the constitution of the blood or fibrinous clot will be deferred until we consider the structure of that of forty-eight hours.

Recurring once more to the plastic or cellular clot of twenty-four hours, running through the accumulation of colorless cells at the bottom of the vessel are to be found narrow, highly refractory bands, evidently portions of the elastic layer of the split and lacerated tunica intima. Dissociation of the plastic clot with needles shows the great majority of the cells constituting it to be flat, swollen, granular, and generally oval, with ordinarily one moderately large and round or slightly oval granular nucleus. Sometimes these cells contain a large nucleus with a constriction in the middle; sometimes two or more smaller nuclei; occasionally the body of the cell itself shows a tendency to the same constriction. They often possess a transverse diameter twice as large as that of the white blood-corpuscle, and a longitudinal diameter sometimes three and occasionally even four times as great as the latter. The general arrangement of these cells seems to have special relation to the plane of the elastic layer of the tunica intima, whether this layer occupy its accustomed position relative to the media, or whether it be found scattered through the cellular accumulation in the shape of the previously mentioned bands. While the disposition of these cells evidently is to

flatten themselves upon the elastic layers parallel with the surface of the latter, still through the whole of the accumulation
cells can be seen occupying every conceivable position, and,
consequently, presenting widely varying profiles. In consequence of being viewed in profile, many of the cells appear to
be spindle-shape. Interspersed among these epithelioid cells are
to be found also many round granular cells, precisely similar
in size and general features to the white blood-corpuscles. In
still greater number are to be seen round or polygonal granular
cells twice and even three times the size of the latter. Besides
these three general types of cell elements, a few red blood-
corpuscles can be distinguished here and there. Examining
this cellular accumulation throughout its whole extent, it was
observed that, in proportion as the distance from the ligature
increased, the endothelial cells along the sides of the vessel
indicated a smaller degree of activity or irritation. The tunica
media nowhere, except at and immediately above the situation
of the thread, showed decided signs of increased activity. It
might be judicious to remark, however, that in the portion of
the media immediately beneath the elastic layer of the intima
and in the neighborhood of the ligature, possibly the cells may
have exhibited slight traces of irritation. At this date, then,
the plastic or cellular clot mainly consisted of an accumulation
of epithelioid or, more correctly speaking, endothelioid cells
and their progeny. In the tunica adventitia, especially near
the ligature, and in the surrounding connective tissue a considerable cellular increase had commenced.

Fig. 1 represents the femoral artery of a dog twenty-four
hours after ligature. A transverse section just above the level
of the bottom of the blood-clot, which has fallen out while
handling, and which has not been drawn. *a.* Adventitia, not
much cellular increase at this level. *c.* Surrounding cellular
or connective tissue, showing greater increase of cell elements.
m. Media not perceptibly altered. *e.* Elastic folds of the inti-

ma; highly refractive, very distinct, also apparently unaltered. *p.* Thick layer of colorless cells closely adhering to each other and to the elastic layer of the intima, entirely filling up the crypts made by the folds of the latter; dissociation demonstrated these cells to be of the same general character as those described above. The section has passed through the sides of the plastic cup or clot.

Preparations from this same series thirty-six hours old, in the main presented similar characteristics. It is only necessary to remark that the thickness of the plastic clot at the bottom and sides of the arterial stump had considerably increased, and that a comparatively greater number of the cells had assumed the oval or spindle outline. The cellular infiltration of the adventitia and surrounding connective tissue had become much more decided. Now, also, one could speak a little more positively concerning a slight irritation of the protoplasm immediately outside of the elastic layer of the intima. The elastic layer itself still showed no change; neither did the muscular elements of the media.

Of the same series, the preparations next in order of date are those containing thrombi forty-eight hours old. Careful study of these demonstrated the following. The cells of the plastic clot presented changes which were a progression of those already noted in the two younger clots. The size of the plastic clot was found to be considerably increased. Some of the cells constituting it were spindle-form, and numbers of them now possessed one, sometimes two or even more, slender and somewhat lengthened processes. Occasionally two or more cells were united together by a long process, and then a tendency to the formation of a cellular network could be made out. The nucleus of many was oval or oblong, and frequently there were two or more round nuclei in the cell. The elastic bands of the tunica intima were still to be seen near the bottom. These and the elastic layer of the intima in its proper position now for the first time appeared to have undergone some change. Their index of refraction had slightly lessened, and their substances had begun to imbibe the carmine—previously they had remained entirely unstained. Through growth of the cellular covering of the intima, the walls of the plastic cup or clot now generally extended some distance above the position of the blood or fibrinous clot, sometimes as far as the first collateral branch. The cellular infiltration of some of the tissues in the neighborhood of the ligature was very decided. The adventitia and the adjacent connective tissue, as also to a slight extent the media, here presented points approaching to a purulent infiltration—an obvious preparation for the separation of the ligature. This infiltration extended some distance above and below the ligature; but in proportion as the distance from the latter increased, the infiltration became more and more limited to the internal portion of the adventitia and to the external layer of the media. No tendency of the capillaries or other vessels of the vaso vasorum, which were as yet entirely confined to the outer coat and the external layers of the media, to send projecting loops toward the lumen of the artery could be observed. A transverse section, extending through the vessel at such a level that its plane passed immediately below the bottom of the blood-clot, showed a considerable cellular increase in those inner layers of the media in apposition with the elastic layer of the intima. This cell increase could still be discovered even in cross-sections at the level of the apex of the blood-clot, but there it was not well marked.

Fig. 3. Preparation forty-eight hours old. Transverse section extending through plastic clot. High power. c. Cellular tissue, showing cell increase. a. Adventitia, also showing increase of cell elements, but not so markedly. m. Media, in its inner layer showing considerable cell proliferation. e. Folds of elastic layer of intima still very distinct and highly refractive, yet showing a tinge of carmine which cannot be so distinctly seen in younger preparations. *e'*. Elastic bands from the lacerated intima, not so highly refractive or so free from carmine-staining as the preceding. *P*. The cellular elements of plastic clot, which when separated by needles correspond in outline and character with their description previously detailed.

Now we come to the consideration of the blood or fibrinous clots.

It has already been stated that Fig. 2, although drawn from a preparation forty-eight hours old, fairly represents the stratified appearance of all of these blood-clots. It can be seen by a glance at the thrombus represented in Fig. 2, in longitudinal section, that the clot is stratified, and that the strata are so placed that, if judged from their position alone, one would naturally conclude that the strata have been deposited at four or more different epochs. It is not to be expected that the blood caught by the ligature in the end of the stump of the artery should, against experience, form at one time four separate coagula, distinct and superimposed. The burden of proof must rest upon him who will attempt to support the assumption that the deposition of four distinct portions of the blood-clot has been simultaneous. Moreover, the different portions of this blood-clot, when studied closely and with a high magnifying power, bear internal evidence of a diversity of age. They present ocular proof that they are of different density and firmness; in other words, that the fibrin in the lower has contracted more than it has in the higher portions. The condition of the protoplasmic elements which the different portions contain also adds a confirmation to the inference that the contents of the lower have been longer placed aside from the circulation than have those of the higher. Considering all these indications then, it would appear that there is reason for the belief that the four portions of the blood coagulum under discussion have been set aside from the circulation at four different periods, and that there has been a succession of depositions from below upward, so that the bottom portion has been first and the top last formed. While speaking of the differences shown by the several portions of this fibrinous or blood coagulum, it may be well to mention that there is still a further want of homogeneity besides that for which a mere difference in age will account.

The three lower portions of this blood-clot, aside from changes due to differences of age, have a similar structure; their elements are similarly arranged. But the fourth portion, constituting the apex of the clot, is, respecting the arrangement of its elements, of very different constitution; but more of the peculiarity of this portion anon.

What now follows has reference only to the three lower portions of the blood-clot. As has been already remarked, each of the lower portions appears to have been similarly constituted. Their similarity in constitution appears to indicate that they have been formed in a similar manner. A detailed description of one of them will suffice for all. Each of the three lower portions itself appeared, at first glance, to be formed of from two to four or more strata, successively and interruptedly superimposed. But a more careful examination under a higher power proved that the edge of a stratum could be traced in an uninterrupted serpentine course from the bottom to the top of the portion. Still closer inspection demonstrated the existence of another unexpected phenomenon, viz., the middle portion or line of such a serpentine stratum was composed almost entirely of red blood-corpuscles, a very few white ones being intermingled, while the borders of the stratum were mainly composed of a network of bands of fibrin whose prevalent direction was parallel with that of the middle line of the stratum. In the meshes of this fibrinous reticulum were numberless white blood-corpuscles and a few red ones. The serpentine course of the stratum was such that between the lateral bends the border

of the stratum was in contact with that of the coil next above
or below—adjacent coils being bound together by intervening
bands of fibrin. The meshes formed by these cross-bands also
were filled with numbers of white blood-cells, scarcely any red
ones.

What is the significance of this interesting serpentine lamellation of each of those three lower portions of the blood-clot?

Before proceeding to the solution of this question, let it be
again distinctly understood that in the examination of this clot
of 48 hours we are not directing our attention to an exceptional
formation, but, so far as my observation goes, to a typical
blood-coagulum, such as usually forms when conditions are
favorable to healing in arteries after ligature. The only exceptions as yet found have been limited to cases where it was
impossible to discover the slightest sign of an attempt at healing, or where the first collateral branch happened to be given
off immediately above the ligature, in which case there generally was no blood-clot at all.

Let us recur now to the serpentine lamellation of these portions of the fibrinous clot.

Possibly the following observations made upon the large vessels of the mesentery and tongue of the living frog may contribute something toward an explanation.

The abdomen of a curarized frog was opened at a convenient
point, and a loop of intestine was withdrawn. The latter was
so placed as to bring to view in the field of the microscope one
of the mesenteric arteries. By carefully stretching the exposed
loop the velocity of the circulation was easily reduced to a convenient slowness. The most important fact obtained by this
experiment may be best stated by detailing that portion of the
observation which relates to it. By stretching the intestinal
loop not only could the blood-current be slowed, but, by the
employment of a little more force, it would be arrested entirely,
and by continuing the strain a few moments it could be even

reversed. During such a reversal of the current, the fork of one of the large arteries was brought into the field. Instead of the backward-flowing blood columns intermingling with each other at the fork where the smaller branches joined the larger trunk, and then travelling toward the heart in one solid round and homogeneous cylinder, it was observed that as far as the field of view extended the blood on the proximal side of the fork continued to flow backward in two distinct streams. Sometimes indeed the two currents travelled with different velocities. These two separate currents appeared to preserve their individuality, and as nearly as possible the shape which characterized them while within the smaller branches. They were, in fact, two separate and distinct cylinders of flowing blood contained within the lumen of the larger arterial trunk, still preserving by their inherent tendencies, or by the viscosity of their elements, the relative positions in which these elements had previously travelled. So far as the corpuscles of a column of blood moving in a vessel are concerned, we know their relative position; the mass of red corpuscles generally occupies the centre, while the greatest number of white blood-cells are near the periphery of the column. Thus the capability of arterial blood, when flowing sluggishly, of receiving and for some time retaining forms impressed by a narrow mould, received ocular demonstration.

The tongue of a frog was next operated upon. It was drawn out and fixed conveniently for observation. One of the medium-size arteries of the organ, at a point where the vessel gave off a branch about half the size of the main trunk, was arranged for study by placing it in the field of the microscope. By means of a delicate *serre-fine* the main trunk of the selected vessel was compressed at a position a little below the branch in such a manner that the point of compression, the collateral branch, and the intervening portion of the main trunk were all in the field and well seen at the same time. Almost immediately after compression of the main trunk the collateral branch commenced to dilate. Confining the attention to what was taking place in the main trunk between the branch and the point of compression, it was noticed that for a short time the calibre of that portion of the artery remained unaltered, and that during this time the blood within it, suddenly arrested and placed out of circulation by the compression, underwent no visible change in the position of its elements relative to themselves or to the walls of the vessel. The only movement which could be perceived at that time was that which was due to the regular impulse of the heart. Soon, however, this portion of the vessel began to dilate, reaching finally to nearly twice its original diameter. The concurrent change in the included blood-column was curious and highly instructive. As the calibre of the vessel increased, the blood-column did not correspondingly fill out the widening space by attempting to increase its diameter while shortening from above downward. No doubt this shortening and spreading out to some extent took place. But if it did so, it was to a greatly insufficient degree, for the column began to assume a curve. As the lateral resistance of the vessel-wall was removed and the heart continued to impel the column from above, this curve gradually shortened and bent more and more until the bands became finally flattened against each other, and the column was coiled in the widened lumen similarly to the successive coils of a rope or of a condensing pipe. Subsequently this clot was examined under a higher power, when the serpentine strata, of which it was composed, and the relation of their elements were found to present the same characteristics, except for age, as have already been stated for the lower portion of the blood-clot of 48 hours. Before dismissing these observations it may not be amiss to remark that no accumulations of white corpuscles sticking in masses to the walls either at the side or bottom of the vessel were seen.

3

The above observations were several times repeated, usually with the same result. They left in my mind but little doubt that the three lower portions of the clot of 48 hours, and the same appearance of the other blood coagula found in my preparations, were produced in a similar manner.

The different portions of the clot of 48 hours were bound together rather firmly by intercrossing bands of fibrin in the same manner, although not so tightly, as the previously mentioned juxtaposed bands of the serpentine lamella were united. The coagulum was found to be more adherent to one side of the vessel than to the other. This union also was effected by bands of fibrin, similar to the preceding.

We now come to another remarkable feature in the construction of this particular thrombus, which, so far as my observation has gone, is only to be seen occasionally. The fourth portion—that which formed the apex of the thrombus—had a constitution different from that of the preceding. It appeared to be composed of three distinct layers, separately superimposed. Furthermore, each layer corresponded in homogeneity of structure to the description which Billroth and Rindfleisch have given of the whole of the recently formed thrombus. They were, so to speak, homogeneous throughout—no massing of red or white corpuscles anywhere, not the slightest appearance of stratification. Moreover, there were to be remarked throughout the separate coagula constituting this portion of the thrombus a small number of flat ovoid cells with clear contents, the nucleus slightly oval, and the quantity of protoplasm large in proportion to the size of the nucleus. The long diameter of these cells was often three or four times that of the neighboring white blood-corpuscles. These flat cells were more numerous in the superior layer, and more scarce in the lower stratum. Besides this difference in the strata composing this upper portion, it was also to be noted that the lowest was the largest, while the highest was the smallest stratum. Not the slightest sign of a tendency to organization was recognizable here.

Fig. 2 represents a longitudinal section of a 48-hour thrombus in the femoral artery of a dog, low power.

a. Adventitia. *m.* Media. *p.* Plastic clot. *e.* Intima. *d.* Blood-clot, three lower laminated portions. *f.* Apex of blood-clot—different in structure from the three lower portions. *g.* Bands of fibrin uniting the blood-clot to the vessel-walls rather tightly on one side, loosely on the other. *b.* Small collateral branch.

Fig. 4. Apex of the thrombus represented in Fig. 2, magnified 200 diameters. *a.* Top of third laminated portion of thrombus. *f.* Lower stratum of the homogeneous clot constituting the apex. *f'.* Middle stratum. *f''.* Upper stratum. The white corpuscles are seen at regular intervals, and a few epithelial cells are present.

The 94-hour thrombus supplied the preparations for the succeeding examination. It was found that the blood-clot now extended a little higher. Its constitution was similar to that of 48 hours, except that it was not capped with an apex of homogeneous formation. The plastic clot had much increased in thickness, both at the bottom and sides of the vessel. The thickening of the cellular layer of the intima extended high up the walls of the artery. The cells constituting the plastic clot were somewhat larger and more spindle-shape, with larger and longer processes than before. Some tendency to form a foundation for the development or vessels might be inferred from a rather uncertain arrangement of some of the spindle-form cells in rows. Sections were made from three thrombi of this age. In those from one of them the plastic clot was observed to send shoots a short distance into the divisions between the laminated portions of the blood-clot. The latter presented no other signs of organization. In those from the other two preparations this relation between the plastic and fibrinous

clot was not to be seen, and no trace of any tendency to organization of the blood-clot could be made out. In the plastic
clot only slight traces of the previously mentioned elastic bands
from the intima could be observed. Yet the elastic layer
of the intima where its relation to the media had been undisturbed by violence was sharply defined and not much changed.
Neither had the protoplasm in the media immediately beneath
suffered much visible increase. The cellular infiltration of the
adventitia and media near the ligature had materially advanced
—a still further preparation for the separation of the thread.

Fig. 5. A faithful representation of a highly magnified view
of a transverse section of a thrombosed femoral artery of a
dog, ninety-four hours after ligature. The section passed
through the middle of the plastic clot. An attempt to loosen
the thrombus from its attachment to the arterial wall had been
successfully made, thus performing without the aid of needles
a dissociation of the cells which were next the intima. a. Adventitia. m. Media. e. Elastic folds of intima perfectly defined,
and showing as yet not much if any tendency toward breaking
down. p. Oval- and lozenge-shape cells of the plastic portion
of the thrombus, their outlines, processes, and nuclei being well
seen.

The next stage of the healing process was made out from
the examination of four preparations, viz.: two arteries at
eight days, and two at ten days after ligature. The general
result may be stated thus : In some cases, granulations springing from the plastic clot have penetrated nearly every crack
and crevice of the blood-clot. In these cases the blood coagulum has formed early and firm attachments to the vessel-wall.
It consequently occupies a height above the point of ligation
nearly identical with that which it occupied at its first formation, the increase of the plastic formation finding vent in the
honeycombing of the blood-clot rather than by uplifting the
latter. In other cases the growth of the plastic clot finds the

additional room it requires by slowly uplifting and pushing before it the blood-clot, which had formed only loose lateral attachments. Under the latter circumstance, I have never found in any part of the blood coagulum the slightest tendency to organization. In all the preparations of this date, the plastic clot was found to be nearly double the size of the average clot last described. The cells were nearly all spindle-form, many of them possessing long processes. A number of large stellate cells were also observed. A considerable number of blood capillaries and vascular channels could now be discerned. These were in connection above with the open lumen of the artery; but in no place could an anastomosis with the vaso vasorum be made out. In longitudinal sections, the elastic layer of the intima could be distinctly traced without the slightest breach or interruption from the top of the section down to within an extremely short distance of the point of ligation, and it appeared in its whole extent to be still tough and resistant. Neither was the media vascularized; the vessels from the adventitia could not be traced inward beyond the exterior lamellæ of the muscular coat.

The preparations which exhibited the above-described invasion of the cracks and crevices of the blood-clot by granulations springing from the plastic clot, demonstrated the fact that these granulations also were composed of tissue identical in structure with that of the formation from which they sprung. They were not, however, vascularized. In cross-section of the granulations it was impossible to distinguish any appearance which could indicate the occupation of their axis by a capillary.

Fig. 6. Transverse section of the femoral artery of a dog, eight days after ligature, highly magnified. *a.* Adventitia. *m.* Media. *e.* Elastic layer of the intima, still sharply defined. *p.* Granulations springing from the mass of cells developed from the cellular elements of the intima; they consist of spindle-cells, the direction of whose long axis in the main ob-

serves a parallelism to the axis of the granulation. The surface of the granulation is covered with one or two layers of
epithelioid cells; not the slightest sign of a capillary loop
occupying the axis of the granulation, nor the least trace of a
vessel to be seen anywhere in the inner layers of the media,
preparing to send a vascular loop through the elastic layer of
the intima.

It could not be found that the clot possessed any vascular
communication with the vasa vasorum at this stage. The
blood which permeated the plastic clot travelled by way of the
previously mentioned capillaries and blood-channels, and was
supplied from the open artery above the thrombus.

Preparations from thrombi fifteen days old exhibited only a
more complete development of the conditions shown to be present in the last-discussed stage of organization. I will merely
add that the blood coagulum, when lifted up from its proper bed
by the growth of the plastic clot, still remained, at this date,
as at first formed. No changes other than those of the inevitable consequences of contraction of the fibrin were to be remarked. The clots were attached to the top of the organized
or plastic clot only by their base. When, on the other hand,
the blood-clot had remained in its primitive position, firmly
attached to the walls of the artery, the previously mentioned
granulations had so increased in number and size as to cause,
probably by pressure, a progressive degeneration of the red
blood disks, and their slow disappearance by granular disintegration and absorption. Preparations for the establishment
of an anastomosis between the vessels of the clot and those
of the walls were now for the first time definitely observed.
The capillaries at the bottom of the plastic clot had by cavernous dilatation become enlarged almost into sinuses. Opposite
to these enlarged capillaries, beyond them, and on the other
side of the intima and media, similar varices had been formed
from the vasa vasorum. A loop from one of these varices

would occasionally be seen extending toward the intima, but would not be observed to reach the latter.

At this time, however, there was no tendency of any vessel to pass into the now thoroughly vascularized clot from the media, by penetrating, at the sides of the clot, the well-defined elastic layer of the internal lining of the arterial walls.

The last study of this series was made upon preparations from the femoral arteries of dogs, twenty and twenty-five days after ligature.

All that need be said of the thrombi twenty days old is that the two previously mentioned modes of growth of the plastic clot had reached a still further development. A complete anastomosis between the vessels of the clot and those of the walls had now been established at the bottom of the clot, by the before-mentioned varices sending toward each other capillary loops, which passed through ruptures in the intima, and which united together forming a network. Even now there was no visible advance toward the establishment of a vascular anastomosis between the vessels of the walls and those of the clot directly through the sides of the artery. At the sides of the vessel the elastic layer of the intima still appeared to be intact, or but little softened. The end of the artery had already begun to shrink by reason of the transition of the spindle-cells of the organized clot into cicatricial tissue. As this contraction continues the stump of the artery assumes a conical shape, and the organized clot slowly disappears by cavernous transformation.

In those blood-clots twenty and twenty-five days old which are found attached to the top of the plastic clot, no decided metamorphoses are yet observable. The red disks often have not even become decolorized or shrunken. Those blood coagula which become occupied by trabeculæ of the plastic clot generally at this date have disappeared, the only remains of them being small masses of colored granules occupying some of the intertrabecular spaces. Frequently, however, considerable masses of decolorized red disks can be seen filling out the spaces, while the trabeculæ are stained and infiltrated with numerous colored granules.

Fig. 7. Vascularized tissue obliterating the lumen of a femoral artery of a dog twenty-one days after ligature, injected with Beale's blue, low power. *a.* Adventitia. *m.* Media. *p.* Vascularized granulation-tissue, the dark lines in which represent bloodvessels which are seen to be in communication above with the open lumen (*L*) of the artery. *v.* Varix in the cellular new formation below the point of ligature, the same being developed from the vasa vasorum. *p, v.* Similar varix in the bottom of the plastic clot. The two varices communicate by means of small capillaries passing between breaks in the elastic layer (*e*) of the intima. *i.* Thickened intima. This thickening extends up to the first collateral branch.

In other and a little older preparations, the communication between the varices was accomplished by one or two tolerably large trunks.

Fig. 8. Longitudinal section of the femoral artery of a dog twenty-five days after ligature, injected. Low power. *a.* Adventitia. *m.* Media. *m'.* Media at end of artery where ligature was applied. *c.* Cellular tissue. *e.* Elastic layer of intima at side of artery, where it appears unbroken and unchanged. *I.* Thickened cellular portion of intima, on a level with blood-clot. *V.* Varices in the cellular tissue at end of the artery. *T.* Large trunk which establishes the anastomosis of external vessels with those of the clot. A few smaller vessels pass directly from the varices to the capillaries at the sides of the plastic formation obstructing the lumen. *p.* Thoroughly vascularized plastic clot, now showing commencing cavernous transformation. Up the centre of this is seen to pass a large vascular stem.

It is observable, both in this figure and in the one immediately preceding, that there is a rich capillary plexus extending from the bottom to the top of the plastic clot. *d, d'*, is a blood-clot showing the serpentine lamellation and exhibiting no sign of approaching organization or degeneration. It has been uplifted from its original position by the growth of the plastic clot. *g.* Fibrous filaments which probably served the function of bands of union between the clot and the arterial walls when the former was first deposited. At present the blood-clot has no attachment except at its base, where, with considerable firmness, it is united to a cellular mass (*h*) which itself is an outgrowth from the intima and from the top of the vascularized clot. This cellular mass (*h*) is permeated by large channels through which blood can freely pass. *L.* Open lumen of the artery.

SECOND SERIES. — The *second series* of experiments was instituted with the object of learning, if possible, what proportional part those wandering cells which may have reached the interior of the ligated artery, through the ruptures in the intima caused by the ligature, may have borne in the healing process as above described.

The sections from all of these preparations presented very uniform pictures. Each one showed the presence of two distinct blood-clots; the one above the point of compression by the forceps, the other between that point and the position of the ligature. At the same time they demonstrated the fact that these double blood-coagula were similar in constitution to those stratified clots found after the usual application of the ligature. They further showed that up to ten days there was no disposition in them to organize. The preparation five days old exhibited below the bottom of the lower blood-clot a very slight accumulation of plastic material. The cells of which the latter consisted were in the main similar to leucocytes, which had probably wandered in through the laceration in the coats produced by the ligature. Besides these, and confined mostly to the neighborhood of the elastic layer of the intima, were a number of cells similar to those previously described as generally present after the ordinary application of the ligature, but neither the leucocytes nor the epithelioid cells seemed to be possessed of any great degree of activity. The media and adventitia in this neighborhood were the seat of a very lively cellular infiltration. The lower blood-clot appeared very completely to fill out that portion of the calibre of the artery in which it was located. The most striking phenomena were observed at the level of the point of compression by the forceps. At this point there were very decided indications of a lively state of activity in the intima and innermost layers of the media. At the point of compression, and in a decreasing degree a little above and below it, an accumulation upon the intima of the same kind of cells which constituted the previously described plastic clot of 24 hours, was very noticeable. In longitudinal section, this accumulation, having its greatest depth at the point of compression, formed a considerable promontory which projected from each side into the lumen. The elastic layer of the intima at this point was more deeply stained with carmine than were the portions more remote. It was also to be noticed that the elastic layer in this situation was slightly bulged inward by a tumefaction and a cellular infiltration of the inner layers of the media. This cellular infiltration of the inner layers of the media was limited to the inner lamellæ, and was not even here decided. There was no decided increase of protoplasmic elements, either in the external layers of the media, or in the adventitia. Nor was there any other appearance leading to an inference that there had been any wandering of white blood-corpuscles from the vasa vasorum.

Sections from the preparations eight and ten days old showed only a further advance of the same process. The two opposite promontories projecting into the lumen where pressure had been applied, had met and formed an extensive union. They consisted almost entirely of spindle and stellate cells with long and anastomosing processes. They were observed to be permeated by a capillary and canalicular vascular network. At this early date the vessels from this plastic formation had extended as far as the inner layers of the media to a depth corresponding to the extent of the cellular infiltration above alluded to, but had not gone further outward. There was not the least trace of an anastomosis having yet been established with the vasa vasorum. The vasa vasorum of the adjacent adventitia did not yet exhibit any tendency to send vascular loops into the media. The blood coagula above and below this point of activity showed the usual serpentine lamellation, and presented no appearance of progressive organization. At the point of ligature the vessel-walls and the connective tissue were in a state of purulent infiltration, the ligature having nearly ulcerated through.

Fig. 9. Thrombus ten days old. A typical view, in longitudinal section, of the condition invariably found to be present after ligation in this manner. Low power.

A. Position of ligature. B. Level of application of forceps. a. Adventitia. m. Media. c. Cellular tissue. p. Cellular formation at the bottom of clot, non-organized and apparently not larger than such an accumulation usually is at five days; it consists mainly of cells similar to white blood-corpuscles; only a few epithelioid cells are scattered through it, and applied along the elastic layer of the tunica intima; no granulations springing from it penetrate the crevices of the laminated clot (d) immediately above.

The blood-clot (d) is seen to be formed of two separate portions of coagulum, exhibiting the previously named serpentine lamellation. This blood-clot is firmly adherent at the bottom, but possesses only slight bands of union with the lateral walls of the vessel. L. Lumen of the vessel. While manipulating this section, a blood-clot similar to d fell out from the position, L. This clot was adherent, though not very strongly, to the top of p'', and it had no lateral attachment whatever.

On a level with B the enormously thickened intima, p', and the growing inner layers of the media are more or less blended. Large granulations arise from this tissue and project inward, entirely obliterating the lumen. They often meet and unite, forming a trabecular network with very small narrow interstices through which flows the blood; p'' consists of such a trabeculated mass. The structure of the granulations themselves is cellular, in fact identical with those granulations which form the plastic clot after the ordinary ligature. $v.$ Capillary vessels and small blood canals in the inner layer of the media and the thickened intima: they are in communication with the intertrabecular spaces; the latter open into the lumen of the artery, and receive and return their blood thence. $e.$ Position of elastic limiting layer between the intima and media. Only traces of this elastic layer, however, can be discovered here; immediately above and below the point of compression it is well defined. The cellular portion of the tunica intima is very much thickened.

THIRD SERIES.—The *third series*, consisting of a few preparations where *limited torsion* had been performed upon the femoral arteries of dogs, showed a process of healing similar in very many respects to that described for the *second series.*

At the point where the artery was seized and compressed by the limiting forceps, was to be seen the same growth of the plastic clot springing mainly from the irritated intima, as was described and represented in Fig. 9. The principal difference between the preparations from the two series in question was located in the lower end of the arterial stump, and was due to the mechanical difference between the operation for ligature as performed in the modified way, and that usually followed while performing *limited torsion.* In the latter, if the operation is properly done, by means of the twisting forceps, the external tunic of the vessel is formed into a kind of knot, so to speak, while the middle and inner coats are separated from the adventitia for a slight distance, and are curved inward, thus forming a more or less perfect valve a small distance below the point of seizure by the limiting forceps. By pressure of the limiting forceps the internal tunic of the artery is rubbed together a little distance above the end of the arterial stump, as in the operation for the second series. This is the point where the healing process is again most active, where the granulations spring from the proliferating intima, and where, by the union of the latter and the subsequent changes which have already been mentioned, the lumen of the vessel is first permanently closed.

In the space below the point of compression by the limiting forceps (that part of the lumen of the artery included between the point of compression above and the incurved walls below) there was the same fibrinous clot having a serpentine lamellation and showing no signs of organization, and immediately below it the same accumulation of colorless cells represented at p just above the ligature in Fig. 9.

The incurved media was early infiltrated with a great number of cells, and the twisted adventitia still more abundantly showed this infiltration.

The healing in these cases seemed to progress with about the same rapidity as in cases forming the second series.

FOURTH SERIES.—The *fourth series* of experiments was directed toward the determination of the sequence of phenomena after the flow of blood in an artery has been arrested by the temporary use of the needle. As was previously stated, the third and fourth methods of applying acupressure were followed. The number of preparations constituting this series was also somewhat limited. An examination of the few made has led to the conviction that the process of healing after acupressure is very similar to that which secures the obliteration of the ligated artery. The sections examined show that the blood coagula in all have been fashioned in accordance with the same general law previously enunciated for ligature, and that there has been a similar although very much less marked increase in the cells of the intima at the point of greatest irritation, which in the third method is at the locus of the needle.

Fig. 10 represents a thrombus after *acupressure* (third method), 36 hours old. Low power.

a. Adventitia. *m.* Media. *n.* Position of needle. *p.* Plastic clot at the bottom. *d.* Stratified clot above. *l.* Portion of lumen now free; when the section was made this was occupied by a recent unstratified or homogeneous blood-clot which fell out during handling.

In the preparations obtained by acutorsion (or the fourth method), the chief difference from the preceding was that the processes were more active. In all the preparations of this series the plastic clot seemed to be the sole organizing agent, the blood coagula to be inert or passive.

FIFTH SERIES.—The *fifth and last series* consisted of a few preparations obtained by compressing, between the arms of serre-fines, the femoral and common carotid arteries of good strong rats—the pressure being continued from two to four hours. In some of these preparations there was evidence that the channel of the artery had been restored soon after removal of the pressure. In some, however, the lumen of the vessel remained permanently occluded. In the latter the surfaces of the intima brought into contact by the serre-fines remained adherent, and a blood and plastic coagulum similar to those seen after acupressure by the third method were observed. The plastic clot here also played the same rôle as in the former series, but the inflammatory process, as might have been expected, was even less advanced than in the case of acupressure.

In concluding the discussion of the five series of experiments above related, let us again call attention to the almost unvarying uniformity throughout all of them, of apparently *one method of healing; i. e.,* by means of the *organization and vascularization of the plastic clot alone.*

Concerning the collateral circulation there is, as far as I know, no dispute. Since the time of Porter it has been well established that there are two species of collateral circulation, a direct and an indirect, which, however, may both be present in the same instance.

Respecting the length of time required for the perfect establishment of the collateral circulation, the following observation may have some significance.

A loop of intestine of a curarized frog was withdrawn from the abdominal cavity and placed under the microscope, so that the artery running along the inner curve of the gut was in the field. Numerous small capillaries were observed to come off from it and run around the intestine immediately beneath the serous covering. These capillaries gave off numerous branches which united with each other. The blood was now interrupted in its wonted course through the artery by pressing the point of a needle upon the latter, about half way between the places of departure of two adjacent capillaries. Immediately the portion of the artery on the distal side of the compressing needle became empty and contracted for a little distance; at the same time the proximal end commenced to dilate. Isochronous with this, the nearest capillary on the proximal side began rapidly to dilate; in the space of a few seconds the blood in it went by jerks, showing the arterial impulse. A few seconds later the lateral anastomosing branches also began to dilate rapidly. Later still, the first capillary branching from the artery on the distal side of the point of compression began to return its blood into the artery, at first slowly, then more rapidly, and finally with an arterial impulse.

By this time (certainly not more than twenty seconds after the first interruption of the arterial current), the anastomosing capillaries which had established the collateral circuit were nearly as wide as the artery itself, and were beating quite as violently. The arterial flow beyond this point did not now seem to be at all affected. The establishment of the collateral circulation in the frog's tongue, after the experiment related some distance above, did not take place so rapidly, since at least fifty seconds were necessary for its free establishment.

What is the origin of the cells which constitute the organizable plastic clot?

After the study of our preparations we have no doubt that the great masses of them are derived immediately from the endothelial and other cell elements of the tunica intima, by a process of proliferation excited partly by the irritation caused by the ligature, the needle, or forceps, and stimulated by the unwonted supply of nutrient material constantly retained within their reach, in consequence of the sluggish movements of the fluids of the blood.

Whence come those colorless elements which have been brought from some distance by the blood current—those both of the plastic and of the fibrinous clots of a thrombus such as we have been considering?

Let us consider first the migrated leucocytes, whose presence in the plastic clot in considerable numbers I have previously mentioned, and to whose agency Billroth and Rindfleisch ascribe, in a great degree, that organization of the fibrinous clot which they believe in.

Do they come directly through the walls of the vessel, or do they come principally, by way of the arterial current, from above the thrombus? Bubnoff declares that many of the white blood-corpuscles found in a blood coagulum after ligature of a vein, have travelled directly through the vessel-wall. Billroth repeated the experiments of Bubnoff, and extended them to thrombosed arteries. He admits, in a general way, the conclusions of the latter, but while stating that he has found the vermilion granules in the midst of the blood-clot in the carotid artery of a rabbit, he says that *they are free*, i. e., not contained within the body of leucocytes. Rindfleisch also accepts Bubnoff's conclusions. Tschausoff has repeated the experiments of Bubnoff, and has declared that he has been unable to confirm the observations of the latter. Durante, after an elaborate series of experiments, contests the conclusions of Bubnoff (*vide* pp. 15, 16), as also do Cornil and Ranvier.

Thus we have seen that not only have the observations of Bubnoff concerning the source of the organizing elements of the thrombus failed to receive exact confirmation by the experiments and observations of any one of the previously named investigators, but that, on the contrary, no less than five most excellent observers, after carefully repeating and somewhat extending his experiments, have flatly contradicted him in many important particulars.

It therefore seems to me that, in the face of these negative results and positive assaults, neither the observations and conclusions of Bubnoff, respecting the migration and functions of the white blood-corpuscles found in the lumen of the ligated vessels, nor the theories of others based thereon, should stand for one moment.

We are, then, forced to the conclusion that if any leucocytes at all have wandered into the clot, they could only have come from the blood in the lumen of the artery above the thrombus.

As to the function of the leucocytes found in the blood or fibrinous clot, it is so nearly nil, as we have already seen, that whatever it may be it cannot save that clot from inevitable destruction. As to whether or not those leucocytes found in the plastic clot have any mission to perform, I have no facts to offer, and therefore refrain from advancing assumptions.

We next inquire into the genesis of those colorless cells

4

which may, by way of the blood current, have travelled to the thrombus from some distance.

The question of the genesis of the colorless cells of the blood has for years called forth the most indefatigable efforts of the most eminent microscopists, and has taxed the genius of the greatest physiologists of the age. Yet we are far from possessing an entirely satisfactory solution of the problem

It is, however, generally admitted that in the spleen, in the liver, in the lymphatic glands, and, according to some, in the red marrow of bones, the rate of increase of these cells is more rapid than elsewhere. It has, consequently, been claimed that each of those organs has something special to do with their generation. It has also been demonstrated by Striker and by others that the stable cells of the connective tissue may physiologically give origin to cells which enter the lymphatic circulation, and which cannot, by any means at present known, be distinguished from lymph corpuscles.

The lymph corpuscles themselves have been observed to increase during their own proper circulation, and it is generally admitted that whenever the circulation is sufficiently slowed and oxygen is present in sufficient quantity, their self-propagation is by no means infrequent.

The following observation constrains me to recognize an additional source of supply, especially very considerable during the existence of inflammation.

Fig. 11 represents a capillary of the mesentery of a frog, nine hours inflamed and magnified three hundred diameters, afterward amplified. e. Capillary walls. l. Leucocytes or wandering cells, external to the walls. g. Cells of adventitia swollen and granular. f. Capillary endothelia granular and swollen, their prominent bellies encroaching considerably upon the lumen of the vessel. The arrow indicates the direction of the blood current. a, d, i. Colorless corpuscles adherent to the walls. d. Is rather firmly bound to the wall by means of a bud penetrating the latter. *i*. A corpuscle adherent at the point of union of two adjacent endothelial cells. *k*. An unattached white corpuscle. *a*. A white corpuscle, adhering tightly to the upper end of an endothelial cell (*b*). At the commencement of the observation, this cell (*b*) was flatly applied to the capillary wall as the other endothelial cells now are, but its upper extremity showed the slightest possible separation from the lower point of the next endothelial cell above. The upper point of this cell (*b*) appeared a little thicker than that of its higher neighbor. The blood current was sluggish, and at intervals interrupted. Occasionally for a few moments the current would move on with considerable energy. At the point of observation, besides the obstruction to the circulation by the swollen endothelia, the current was impeded by the adherent white corpuscles. The relative position of the corpuscles was such that, at the time when the current was forced forward with some impetus, the points of the red blood disks went with considerable momentum against the chink, and were violently jammed into the angle formed by the upper surface of the adherent white corpuscle (*a*) and the surface of the endothelial cell above it. In the attempt to pass on, these red blood disks must perforce bend around the white corpuscle (*a*). The tendency of these forces was evidently to loosen and to pry out from its bed the upper end of the endothelial cell (*b*).

During an energetic increase in the velocity of the current this was actually observed to take place. After that, the next violent movement of the blood current sufficed to detach the whole cell and to carry it off in advance of the other elements. When the movement slowed again, it was observed that the place of former attachment of this cell (*b*) was void of its endothelial covering. I have observed the above-described phenomena on one other occasion.

Now any one who carefully examines the course of capillaries in the inflamed mesentery of a frog will meet, at not very infrequent points, with just such appearances of the interior of the vessel as the detachment of an epithelial cell will go far to explain the significance of. I confess that I am inclined to believe that this appearance would occur much more frequently but that soon a white corpuscle possessing unusual viscosity, fastening itself there, spreads out and fills the void. It will be remembered that, *à propos* of the apex of the forty-eight hour blood-clot of a ligated artery, a number of epithelioid cells present in the clot were both described and figured.

As a possible explanation of their presence, we may suppose that they may have been detached from the irritated intima at or above the level of their location in the clot.

Once admitting, in inflammations affecting the inner lining of bloodvessels, this detachment of swollen and irritated epithelium, it may be claimed as a necessary consequence that those cells must appear in appreciably increased numbers in the blood. Now precisely this is found to be true respecting the blood in the inflamed stump of a ligated artery; on the other hand, it has not yet been observed of the blood in more general inflammations. It may be affirmed respecting the latter cases that because the expected increase is not apparent the theory has at once been placed *hors de combat.*

But does it necessarily follow that herein is an insuperable objection? These swollen granular epithelial cells which are displaced from their position on the internal lining membrane of a vessel are in a state of irritation. What should happen to their shape after being set free in the blood current? Under this condition undoubtedly their tendency would be to assume a spherical outline, and, if they should remain suspended sufficiently long in the flowing blood, it is probable that every trace of their original form would be obliterated.

Concerning the changes which an endothelial cell may pass through under somewhat similar conditions, is an observation

of Cornil and Ranvier (Manuel d'Histologie Pathologique) on
the behavior of the endothelial covering of the trabeculæ of
the omentum of adult animals.

They found that after inflammation has been artificially excited, the peritoneal fluid becomes cloudy and contains many
cellular elements somewhat similar to pus-corpuscles; others
more voluminous, having one or more oval nuclei; and intermediate cells between these two.   In cells which are applied to
the trabeculæ are observed all the phenomena of multiplication.
The multiplication is such that the hypertrophied cells form
projections on the trabeculæ; or they are adherent to it, at one
time by a large surface, at another by a single point; they
become detached, and may continue to live and vegetate isolated in the peritoneal fluid.   Their protoplasm, which is
soft and granular, is susceptible of taking the most varied
forms and of giving birth to amœboid prolongations and to
new cells.   After five or six days the majority of the detached voluminous and turgid cells reapply themselves to the
trabeculæ, while presenting projecting bellies.   They shrink,
flatten themselves against the trabeculæ, present a protoplasm
more or less similar to that of their primitive type, and may
assume later the appearance of endothelium.

I conceive it possible for the endothelium of the vascular
tract to undergo similar metamorphoses.

Applying the foregoing to the subject before us, it seems
probable that in addition to the ordinary white corpuscles of
the blood and their immediate descendants, there may be present, both in the plastic clot and in the blood of a ligated
artery, other somewhat similar, often larger, corpuscles, which
are the metamorphosed endothelial cells of the lining membrane or their descendants.

Furthermore, it is highly probable that among the epithelioid
cells which constitute the mass of an organizing clot, and
which spring, in the main, from the endothelial cells of the ad-

joining tunica intima, there may be no very inconsiderable number of endothelial cells detached from the arterial wall above, and but little changed.

CONCLUSIONS.

The foregoing study has led me to the following conclu-sions:—

1st. After the ligation of an artery, if the first collateral branch above is sufficiently distant, a blood coagulum generally forms at the bottom of the arterial stump, but not always.

2d. The formation of this blood coagulum, when conditions are favorable to healing, is not sudden. Frequently the struc-ture of this fibrinous clot proves it to have been deposited at interrupted intervals. The blood-clot is, therefore, often larger some hours or days after its first formation, than when it is first deposited. See Fig. 2.

3d. The portions of the fibrinous clot which have been deposited at interrupted intervals have usually a stratified aspect. The blood-clot is not homogeneous in structure.

4th. The blood or fibrinous clot does not undergo a genuine organization or vascularization. It acts only as a temporary barrier to the course of the blood, and as a foreign body, whose tendency is first to produce a certain amount of irrita-tion in the adjacent internal coat of the vessel, and to finally disappear after slow disintegration.

5th. The healing of an artery ligated after the ordinary method is effected by the organization, vascularization, and subsequent cicatricial metamorphosis of a plastic formation which grows between the blood-clot and the ligature, and which is mainly composed of colorless endothelioid cells.

6th. The origin of the cells of the plastic clot is to be referred chiefly to a proliferation of the endothelium and sub-jacent cellular elements of the tunica intima, between the point of ligation and the first collateral branch above. See Figs. 1, 5, and 6.

7th. The rapidity of the healing process is usually proportionate to the growth of the plastic clot.

8th. The growth of the plastic clot is at first somewhat stimulated by the presence of a fibrinous or blood clot. The presence of the latter is not essential, for the formation and organization of the plastic clot occasionally take place without it.

9th. The plastic clot begins to present signs of commencing vascularization as early as the sixth day.

10th. The organizable or plastic clot is vascularized at first independently of the vasa vasorum. Some days before any trace of a vascular communication between the plastic clot and the vasa vasorum can be discovered, the former is thoroughly permeated by a rich capillary network which is in communication with the open lumen above the thrombus, by means of blood channels or sinuses of considerable size located mostly in the superior portion of the plastic clot. The first vascular formation generally appears in the peripheral portions of this clot

11th. Usually between the fifteenth and the thirtieth day after ligation an anastomosis is established between the vessels of the clot and those of the walls of the artery. The communication is established at the bottom of the arterial stump where the intima and media have been cut through by the ligature. At this date the elastic layer of the intima, from the top nearly to the bottom of the clot, is sharply defined, presents little evidence of softening, and offers no perforation for the establishment of a lateral anastomosis between the vasa vasorum and the vessels of the clot directly through the sides of the artery. See Fig. 7.

12th. The plastic clot, by a gradual metamorphosis into cicatricial tissue, and by a subsequent cavernous transformation of the latter, finally disappears—the only remains of the vessel and of the clot being a tough fibrous cord.

13th. If the blood-clot, during the first days of its formation, become firmly adherent to the vessel-wall, the increase in the size of the plastic clot causes granulations springing from it to grow into the crevices and channels of the blood coagulum. Through the continued invasion of the blood-clot by these granulations, and their increase in thickness, the blood-clot disintegrates in consequence of the gradually increasing pres. sure, and is finally absorbed. See Fig. 6.

14th. But if the blood coagulum form only slight connections with the walls, the plastic clot, while increasing in size during the process of organization, gradually uplifts the blood-clot and pushes the latter before it. In these cases, as late as the twentieth day, when organization and vascularization have been nearly completed in the plastic clot, not the slightest in. dication of change, except that naturally due to the contraction of fibrin, is to be seen in the uplifted blood-clot. See Fig. 8.

15th. If, in addition to the usual method of applying a ligature, compression be produced upon the walls of the artery a short distance above the point of ligation, in such a manner as to slightly rub together opposite points of the surface of the internal limiting membrane without rupturing the latter, and to excite at that place an irritation, the plastic clot mainly forms at that point instead of at the level of the ligature, and the obliteration of the lumen of the vessel and the permanent arrest of hemorrhage are more rapidly and more certainly secured. A practical application of the same procedure to the usual methods of performing acupressure may, à priori, be expected to secure similarly good results. See Fig. 9.

16th. The process of healing in an artery after limited torsion has been performed is, in its essentials, identical with that mentioned in the preceding paragraph.

17th. The process of healing in an artery after acupressure does not essentially differ (except in its slowness) from that which is usually seen after simple ligation has been done. In consequence of the slowness of the healing process present in acupressure, either limited torsion, simple ligation, or the modified ligation above described, would seem in general more reliable. See Fig. 10.

18th. By the compression of an artery for a few hours be-tween the arms of a pair of forceps or of a serre-fine applied directly to the vessel-walls, an inflammation may be excited through the agency of which the lumen of an artery may be permanently obliterated. This inflammatory process does not differ materially from that which is present after simple acu-pressure. This procedure should, à priori, be peculiarly use-ful when the vessel-walls are diseased, as in atheroma or in aneurism.

19th. The organizing elements which are active in the heal-ing of arteries are neither the white blood-corpuscles which are a part of the fibrinous clot at the time of its formation, nor those which may wander into it afterward; nor are they principally the so-called white corpuscles of the blood and their progeny, which may have wandered into the plastic clot.

20th. The so-called wandering cells, which may be found in any part either of the plastic or of the blood clot, seldom, if ever, reach their destination by escaping from the vasa vasorum and passing directly through the vessel-walls.

21st. The endothelium which lines the inner surface of the arteries and capillaries may be considered the source of origin of some of the increased number of colorless elements of the blood in local inflammation. From this conclusion naturally issues the corollary, that the endothelia in general may be con-sidered as some of the possible physiological progenitors of the colorless elements of the blood. See Fig. 11.

22d. In the inflammatory processes through the agency of which an artery is healed after ligature, acupressure, or torsion, the stable cells of the tunica intima play a very important—probably the most important rôle.

EXPLANATION OF PLATE I.

Fig. 1.

A transverse section of the femoral artery of a dog, 24 hours after ligature. High power. See pp. 26–27. (First Series.)

Fig. 2.

A longitudinal section of a 48-hour thrombus in the femoral artery of a dog. Low power. See pp. 25, 35. (First Series.)

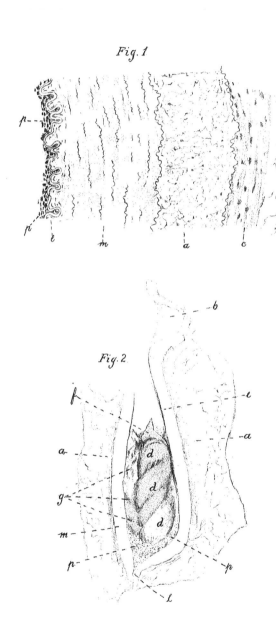

Fig. 1

Fig. 2

EXPLANATION OF PLATE II.

Fig. 3.

Transverse section passing through the plastic portion of a clot in the femoral artery of a dog. Preparation 48 hours old. High power. See pp. 28–29. (First Series.)

Fig. 4.

Apex of the thrombus represented in Fig. 2, magnified 200 diameters. See p. 35. (First Series.)

Fig. 3.

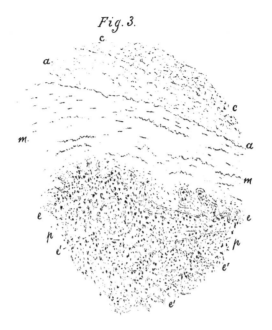

Fig. 4.

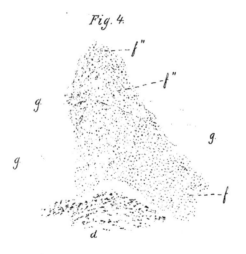

EXPLANATION OF PLATE III.

Fig. 5.

A faithful representation of a highly magnified view of a transverse section of a thrombosed femoral artery of a dog, 94 hours after
ligature. See p. 36. (First Series.)

Fig. 5.

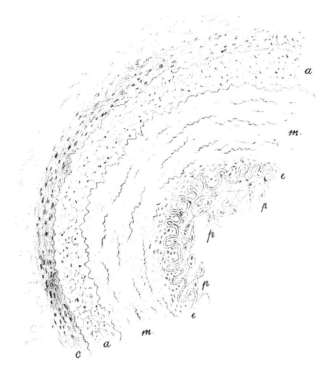

EXPLANATION OF PLATE IV.

FIG. 6.

A transverse section of the femoral artery of a dog, 8 days after ligature. High power. See pp. 37–38. (FIRST SERIES.)

FIG. 7.

Vascularized tissue obliterating the lumen of a femoral artery of a dog, 21 days after ligature. Injected. Low power. See p. 40. (FIRST SERIES.)

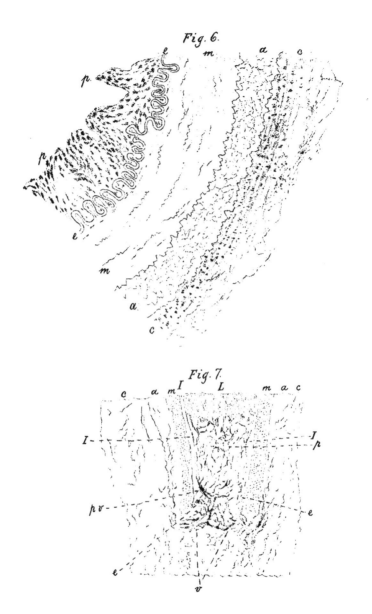

Fig. 6.

Fig. 7.

EXPLANATION OF PLATE V.

FIG. 8.

A longitudinal section of a femoral artery of a dog, 25 days after ligature. The blood or fibrinous clot has been uplifted from its primitive position. Injected. Low power. See pp. 40–41. (FIRST SERIES.)

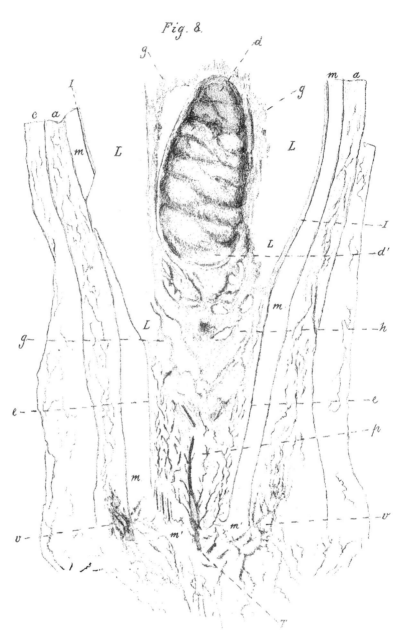

Fig. 8.

EXPLANATION OF PLATE VI.

FIG. 9.

A thrombus 10 days old.   Longitudinal section.   Modified ligature.   Low power.   See pp. 43–44.   (SECOND SERIES.)

(68)

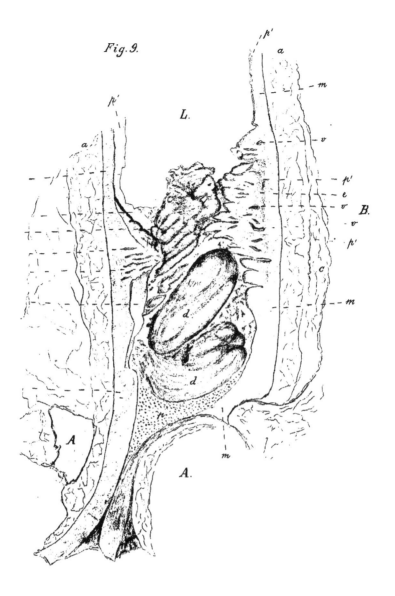

Fig. 9.

EXPLANATION OF PLATE VII.

FIG. 10.

A thrombus after acupressure (third method), 36 hours old.
Low power. See p. 46. (FOURTH SERIES.)

FIG. 11.

A capillary of the mesentery of a frog, 9 hours inflamed. High
power. See pp. 50–51.

(70)

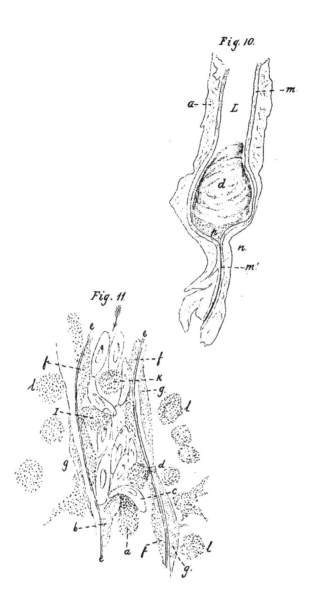

Fig. 10.

Fig. 11.

CIRCULAR RELATIVE TO SCIENTIFIC AND LITERARY EXCHANGES.

The Smithsonian Institution, among its operations for the increase and diffusion of knowledge among men, organized, many years ago, a system of exchanges, for the purpose of more readily distributing its own publications and of receiving returns for its library; at the same time offering its services to other establishments requiring similar facilities. It has enlarged this system, continuously, until it has become of such magnitude as to require for its maintenance one-fourth of the entire income from the Smithsonian fund. The greater part of this increase is on account of the transmissions of the departments and bureaus of the United States Government. It is no longer possible for the Institution to meet these expenses, and a small charge will, hereafter, be made on all matter sent by and received for these public offices. For the present no such payment will be required of learned societies or individuals, unless their transmissions are of unusual magnitude, although the right to make a charge is reserved.

To facilitate the business connected with the system of the Smithsonian exchanges the following rules have been adopted:

1. Transmissions through the Smithsonian Institution for foreign countries to be confined exclusively to books, pamphlets, charts, and other printed matter, sent as DONATIONS or EXCHANGES, and not to include those procured by purchase. The Institution and its agents will not receive for any address apparatus and instruments, philosophical, medical, etc., (including microscopes,) whether purchased or presented; nor specimens of natural history, except where special permission from the Institution has been obtained.

2. The Departments or Bureaus of the United States Government to pay to the Smithsonian Institution five cents per pound on their packages, which includes all expense of boxing, shipping, and transportation.

3. A list of the addresses and a statement of contents of each

sending·to be mailed to the Smithsonian Institution at or before the time of transmission.

4. Packages to be legibly addressed and to be endorsed with the name of the sender and their contents.

5. Packages to be enveloped in stout paper, and securely pasted or tied with strong twine—never sealed with wax.

6. No package to a single address to exceed one-half of one cubic foot in bulk.

7. To have no enclosures of letters.

8. To be delivered to the Smithsonian Institution or its agents free of expense.

9. To contain a blank acknowledgment, to be signed and returned by the party addressed.

10. Should *returns* be desired, the fact is to be explicitly stated on or in the package.

Unless these conditions are complied with, the parcels cannot be forwarded by the Institution.

SPENCER F. BAIRD,
Secretary Smithsonian Institution.

SMITHSONIAN INSTITUTION,
WASHINGTON, *January* 1, 1879.

BUSINESS ARRANGEMENTS

OF THE

SMITHSONIAN INSTITUTION.

Washington, January 1, 1879.

The annual meeting of the BOARD OF REGENTS is held on the third Wednesday in January.

The annual meeting of the "ESTABLISHMENT" is held on the first Tuesday in May.

The meetings of the EXECUTIVE COMMITTEE are held on the second Monday of January, April, July, and October.

The general business of the Institution, under direction of the Secretary, is in charge of the CHIEF CLERK, and applications are to be made to the latter for publications, supplies, service of laborers, leave of absence, keys, &c.

The exhibition halls are open to the public from 9 a. m. to 4.30 p. m. every day in the year, except Sundays.

The business offices are open from 9 a. m. to 4 p. m.

The work-rooms and shops are open from 7.30 a. m. to 4.30 p. m.

No smoking allowed in the public halls.

Employees entrusted with keys will be held responsible for them, and no one will be allowed to procure a duplicate key without permission.

Gas is not to be left burning in unoccupied rooms.

Receipts are to be given for any public property received by employees.

The printing of the publications of the Institution, the blank forms, circulars, labels, etc., is in charge of the Chief Clerk, who will keep a record of each article, showing its title, author, commission of reference, name of printer, number and character of illustrations, number of copies printed, reception of proofs, &c., &c.

A record is to be kept of each wood-cut, plate, or illustration, and the latter are to be properly numbered and arranged in cases, and these, with stereotype plates, are to be in charge of the Chief Clerk.

The Corresponding Clerk is to prepare letters or answers to communications as directed by the Secretary or Chief Clerk; is to make references as required; to have charge of the current letter-copy books; to superintend copying letters; to make the proper enclosures, and direct and seal the envelopes.

He is also to direct the filing of letters and documents attended to, and the indexing and binding of letters received and written.

He is to prepare orders on the Document Clerk for publications promised.

The correspondence attended to is to be filed daily in alphabetical boxes, and bound in volumes as may be necessary.

Applications for volumes of the "Smithsonian Contributions to Knowledge" and the "Miscellaneous Collections" are to be made to the Secretary; for parts of series and for annual reports to the Chief Clerk.

The Document Clerk is to fill orders for publications, and forward them by mail, messenger; or otherwise, as directed.

The Document Clerk is to have charge of the stock of all printed matter belonging to the Institution; to take account of the same in the month of July annually; to report when the supply of any work or blank is nearly exhausted; to keep a sample book of circulars, blanks, labels, &c., &c.

Transmissions through the Smithsonian Institution for foreign countries are to be confined exclusively to books, pamphlets, charts, and other printed matter, sent as DONATIONS or EXCHANGES, and not to include those procured by purchase. The Institution and its agents will not receive apparatus and instruments, philosophical, medical, etc., (including microscopes,) whether purchased or presented; nor specimens of natural history, except where special permission from the Institution has been obtained.

The Departments or Bureaus of the United States Government to pay to the Smithsonian Institution five cents per pound on their packages, which includes all expense of boxing, shipping, and transportation.

A list of the addresses and a statement of contents of each sending to be mailed to the Smithsonian Institution at or before the time of transmission.

Packages to be legibly addressed and to be endorsed with the name of the sender and their contents.

Packages to be enveloped in stout paper, and securely pasted or tied with strong twine—never sealed with wax.

No package to a single address to exceed one-half of one cubic foot in bulk.

To have no enclosures of letters.

To be delivered to the Smithsonian Institution or its agents free of expense.

To contain a blank acknowledgment, to be signed and returned by the party addressed.

Should *returns* be desired, the fact is to be explicitly stated on or in the package.

Unless these conditions are complied with, the parcels cannot be forwarded by the Institution.

Packages for distribution in the United States, and for all other parts of the world, are to be received, recorded, acknowledged, and forwarded by the Exchange Clerk.

Boxes for England, France, and Germany are to be forwarded every two months, and for other countries as often as the accumulations render it necessary.

Books, pamphlets, maps, periodicals, etc., intended for the Smithsonian library are to be delivered to and recorded by the Librarian.

Such books as are designated by the Secretary are to be sent to the Library of Congress, but all articles received must remain in the office of the Librarian at the Institution at least one week for examination.

No book or other article belonging to the library shall be taken until entered in the register by the Librarian.

Books from the Library of Congress are to be obtained by written application, on the proper forms, to the Librarian, and approved by the Secretary.

A record of books lent shall be kept by the Librarian, who is to see that they are duly returned.

The Transportation Clerk is to take charge of all boxes, barrels, and packages delivered at the Institution; record their size, weight, number, nature, address, from whom received, cost of freight, &c., and to collect charges on packages for individuals. He is to enter, stamp, and send by express, railroad, steamer, &c., all packages except those to foreign countries; to have charge of empty boxes and packing material, and assist the Librarian and Exchange Clerk whenever required.

No checks are to be drawn except for the payment of accounts which have been examined and approved by the Secretary.

The Institution will not be responsible for the payment of any bills contracted without a written order from the Secretary or Chief Clerk.

Orders are to be returned by the party furnishing the article required, with the cost stated, and the receipt of the person to whom it was delivered.

Bills presented are to be examined by the Accountant, to see that the calculations are correct and the voucher in proper form, prices reasonable, and the articles delivered.

Receipts signed by clerks or agents are inadmissible, unless accompanied by a power of attorney, showing the legal authority of the party signing to receipt for the money.

The payment of bills and salaries is to be made on the twenty-fifth day of each month.

Cash from sales of old material, publications, postage stamps, &c., to be deposited with the Accountant.

The amount required to pay bills in foreign countries is to be ascertained by the Accountant, and the statement of the bank as to cost to be kept with the voucher.

A statement is to be made to the Secretary monthly of the receipts and expenditures.

A quarterly examination of all vouchers, books, checks, &c., will be made by the Executive Committee, who certify to the condition of the accounts and make an annual report to the Board of Regents.

The Superintendent of the Building is to have general direction of laborers, and keep account of their time and the nature of the work performed; to have charge of the repairs of the building, the roofs, gutters, grounds, water and gas-pipes, plugs, stop-cocks, hose, water-closets, wash-stands, sinks, stoves, flues, building materials, tools, hardware, trucks, wheelbarrows, ladders, furniture, clocks, storage of boxes; to make an inventory annually, on 1st July, of property; to make frequent examination of the fire-plugs, hose, and buckets, and see that they are kept in good order; to drill all employees in the use of fire-alarm signals, hose, and the protection of the building in case of fire, and to see that the keys are kept in their proper places; to have charge of waste paper and dispose of it from time to time, turning over the proceeds to the Accountant; to have charge of the watch-stations and daily reports of the night-watchman, and to see that a watchman is always on duty in the building to answer the front-door bell at any hour, day or night.

The Janitor is to open the building at 9 a. m. and close it at 4.30 p. m., ringing a bell five minutes before the time for closing; to direct visitors to different parts of the building, and to point out objects of special interest; to prevent the entrance of improper or disorderly persons, to secure order in the public halls, and to guard the property of the Institution; to see that all doors and windows are fastened at the time the building is closed and on the approach of a storm.

The Messenger is to bring the mail at 9 a. m. and 2 p. m., and take it at 1.30 and 4.30 p. m., daily, except Sunday; to carry messages and packages, as required by the Secretary or Chief Clerk; to assort the mail and place the letters in the lock-boxes; to take

charge of letters, &c., for persons temporarily connected with the
Institution; to make press copies of letters; to stamp all mail sent
out with the name of the Institution, and affix the necessary postage stamps; to have charge of postage stamps and envelopes, and
make returns of sales to the Accountant; to give proper directions
to visitors.

EXPLANATION OF SYMBOLS USED IN BUSINESS OF THE INSTITUTION.

ACTION TO BE TAKEN.

S. For files of the Smithsonian Institution.

F. For files of the United States Fish Commission.

a. Prepare answer.

r. To be read, and contents noted.

c. Personal conference desired by the Secretary.

t. To be translated.

r. As a second letter—to be returned as soon as possible.

f. To be filed in general correspondence, or under special head designated.

The combination of two letters shows that double action is to be taken: as—*a. r.* Answer and return; *t. r.* Translate and return; *r. r.* Read and return; *S. f.* File in Smithsonian correspondence.

ABBREVIATIONS FOR REFERENCE OF LETTERS, &c.

BAIRD, S. F. _____B.

BEAN, T. H. _____ ____Bn.

BOEHMER, G. H._____ Br.

BESSELS, E. _____ ____Bs.

BROWN, S. C._____S. C. B.

BROWN, S. G. _____S. G. B.

CUSHING, F. H. _____F. H. C.

DAINGERFIELD, Miss_____Dd.

DIEBITSCH, H. _____D.

DALL, W. H._____Dl.

ENDLICH, F. M. _____E.

FOREMAN, E. _____Fn.

GASS, HENRY_____G.

GOODE, G. B. _____G. G.

GILL, T. N. _____T. G.

GILL, HERBERT A. _____H. G.

GRIFFIN, Miss M. E._____M. G.

HORAN, HENRY ___._____H.

LEECH, D._____L.

MASON, O. T. _____Mn.

MILNER, J. W._____M.

RHEES, W. J. _____R.

RAU, C._____C. R.

RIDGWAY, ROB'T_____R. R.

STOERZER, Mrs. L. _____L. S.

SHINDLER, A. Z._____A. Z. S.

SMILLIE, T. W. _____T. S.

TAYLOR, W. B. _____T.

TURNER, Miss J. A. _____J. T.

YOUNG, C. B._____Y.

SMITHSONIAN MISCELLANEOUS COLLECTIONS.

334

LIST

OF

DESCRIBED SPECIES

OF

HUMMING BIRDS.

BY

DANIEL GIRAUD ELLIOT.

REPRINTED FROM A SYNOPSIS OF THE TROCHILIDÆ IN THE SMITHSONIAN
CONTRIBUTIONS TO KNOWLEDGE.

WASHINGTON:
SMITISONIAN INSTITUTION.
1879.

ADVERTISEMENT.

The present List of Described Species of Humming Birds, has been reprinted with some changes from the Classification and Synopsis of the Trochilidæ by D. G. Elliot, published in the Smithsonian Contributions to Knowledge. Its object is to facilitate the labelling of the specimens of humming birds in the Museum of the Institution, as also to serve the purposes of a check-list of the species.

SPENCER F. BAIRD,
Secretary Smithsonian Institution.

WASHINGTON, May, 1879.

(iii)

LIST

OF DESCRIBED

SPECIES OF HUMMING-BIRDS.

The page references refer to Elliot's Synopsis of the *Trochilidæ*.

	PAGE
1. Eutoxeres, *Reichenbach*	2
1. Eutoxeres aquila (*Bourc.*)	3
2. Eutoxeres heterura, *Gould*	3
3. Eutoxeres condamini (*Bourc.*)	3
2. Rhamphodon, *Lesson*	4
4. Rhamphodon nævius (*Dumont*)	4
3. Androdon, *Gould*	5
5. Androdon æquatorialis, *Gould*	5
4. Glaucis, *Boie*	5
6. Glaucis hirsuta (*Gmel.*)	6
7. *Glaucis dorhni (*Bourc.*)	7
8. Glaucis antoniæ (*Bourc.*)	7
9. Glaucis leucurus (*Linn.*)	7
10. Glaucis cervinicauda (*Gould*)	8
11. Glaucis ruckeri (*Bourc.*)	8
5. Doleromya, *Bonaparte*	8
12. Doleromya fallax (*Bourc.*)	9
6. Phæoptila, *Gould*	9
13. Phæoptila sbrdida, *Gould*	10

PAGE

7. Phæthornis, *Swainson* 10

    14. Phæthornis bourcieri (*Less.*) 12
    15. Phæthornis philippi (*Bourc.*) 12
    16. Phæthornis yaruqui (*Bourc.*) 13
    17. Phæthornis guyi (*Less.*) 13
    18. Phæthornis emiliæ (*Bourc.*) 13
    19. Phæthornis augusti (*Bourc.*) 14
    20. Phæthornis pretrii (*Less.*) 14
    21. Phæthornis superciliosus (*Linn.*) 14
    22. Phæthornis longirostris (*Less.*) 15
    23. Phæthornis hispidus (*Gould*) 16
    24. Phæthornis syrmatophorus, *Gould* 16
    25. Phæthornis anthophilus (*Bourc.*) 16
    26. Phæthornis eurynome (*Less.*) 17
    27. Phæthornis squalidus (*Temm.*) 17
    28. Phæthornis longuemareus (*Less.*) 18
    29. Phæthornis adolphi, *Gould* 18
    30. Phæthornis griseigularis, *Gould* 18
    31. Phæthornis striigularis, *Gould* 19
    32. Phæthornis idaliæ (*Bourc. and Muls.*) . . . 19
    33. Phæthornis pygmæus (*Spix.*) 20
    34. Phæthornis episcopus, *Gould* 20
    35. Phæthornis nigricinctus, *Lawr* 20

8. Eupetomena, *Gould* 21

    36. Eupetomena macroura (*Gmel.*) 21
    37. Eupetomena hirundo, *Gould* 22

9. Sphenoproctus, *Cabanis and Heine* 22

    38. Sphenoproctus pampa (*Less.*) 23
    39. Sphenoproctus curvipennis (*Licht*) 23

10. Campylopterus, *Swainson* 23

    40. Campylopterus largipennis (*Bodd.*) 24
    41. Campylopterus obscurus, *Gould* 25
    42. Campylopterus rufus, *Less.* 25
    43. *Campylopterus hyperythrus, *Caban.* 25
    44. Campylopterus lazulus (*Bonnatt.*) 26
    45. Campylopterus hemileucurus (*Licht.*) 26
    46. Campylopterus ensipennis (*Swains.*) 26
    47. Campylopterus villavicencio (*Bourc.*) 27
    48. Campylopterus phainopeplus, *Salvin*, Ibis (1879), p. 202.
    49. Campylopterus cuvieri (*Delattr.*) 27
    50. Campylopterus roberti (*Salvin*) 28

PAGE

11. Aphantochroa, *Gould* 28
   51. *Aphantochroa gularis, Gould* 28
   52. Aphantochroa cirrochloris (*Vieill.*) 29
   53. Aphantochroa hyposticta, *Gould* 29

12. Cæligena, *Lesson* 29
   54. Cæligena clemenciæ, *Less.* 30
   55. Cæligena henrici (*Less.*) 30
   56. Cæligena viridipallens (*Bourc. and Muls.*) . . . 31
   57. Cæligena hemileuca (*Salvin*) 31

13. Lamprolæma, *Reichenbach* 32
   58. Lamprolæma rhami (*Less.*) 32

14. Oreopyra, *Gould* 33
   59. Oreopyra calolæma, *Salvin* 33
   60. Oreopyra leucaspis, *Gould* 33
   61. Oreopyra cinereicauda, *Lawr.* 34

15. Oreotrochilus, *Gould* 34
   62. Oreotrochilus pichincha (*Bourc. and Muls.*) . . 35
   63. Oreotrochilus chimborazo (*Delattr.*) 35
   64. Oreotrochilus estellæ (*D'Orb. and Lafr.*) . . . 36
   65. Oreotrochilus leucopleurus, *Gould* 36
   66. Oreotrochilus melanogaster, *Gould* 36
   67. Oreotrochilus adelæ (*D'Orb. and Lafr.*) . . . 37

16. Lampornis, *Swainson* 37
   68. Lampornis violicauda (*Bodd.*) . . . .) . 38
   69. Lampornis mango (*Linn*) 39
   70. Lampornis prevosti (*Less.*) 39
   71. Lampornis viridis (*Aud. and Vieill.*) 40
   72. Lampornis veraguensis. *Gould* 40
   73. Lampornis gramineus (*Gmel.*) 40
   74. Lampornis calosoma, *Elliot* 41
   75. Lampornis dominicus (*Linn.*) 41

17. Eulampis, *Boie* 42
   76. Eulampis holosericeus (*Linn.*) 42
   77. Eulampis jugularis (*Linn.*) 43

PAGE

18. Lafresnaya, *Bonaparte* 43
    78. Lafresnaya flavicaudata (*Fraser*) 44
    79. Lafresnaya gayi (*Bourc. and Muls.*) 44

19. Chalybura, *Reichenbach* 45
    80. Chalybura buffoni (*Less.*) 45
    81. *Chalybura urochrysea, *Gould* 46
    82. Chalybura isauræ (*Gould*) 46
    83. Chalybura melanorrhoa, *Salvin* 47
    84. Chalybura cæruleiventris (*Reichenb.*) 47

20. Florisuga, *Bonaparte* 47
    85. Florisuga mellivora (*Linn.*) 48
    86. Florisuga fusca (*Vieill.*) 48

21. Petasophora, *Gray* 49
    87. Petasophora anais (*Less.*) 50
    88. Petasophora thalassina (*Swains.*) . . . 51
    89. Petasophora cyanotis (*Bourc.*) 51
    90. Petasophora corruscans, *Gould* 51
    91. Petasophora rubrigularis, *Elliot* 51
    92. Petasophora serrirostris (*Vieill.*) 52
    93. Petasophora delphinæ (*Less.*) 52

22. Panoplites, *Gould* 53
    94. Panoplites jardini (*Bourc.*) 53
    95. Panoplites flavescens (*Lodd.*) 54
    96. Panoplites mathewsi (*Bourc.*) 54

23. Phæolæma, *Reichenbach* 55
    97. Phæolæma rubinoides (*Bourc.*) 55
    98. Phæolæma æquatorialis, *Gould* 55

24. Clytolæma, *Gould* 56
    99. Clytolæma rubinea (*Gmel.*) 56
   100. Clytolæma aurescens, *Gould* 57

25. Iolæma, *Reichenbach* 57
   101. *Iolæma luminosa 58
   102. Iolæma schreibersi (*Bourc.*) 58
   103. Iolæma frontalis, *Lawr.* 59
   104. *Iolæma whitelyana, *Gould* 59

PAGE

26. **Sternoclyta,** *Gould* 59
   105. Sternoclyta cyaneipectus, *Gould* 60

27. **Eugenes,** *Gould* 60
   106. Eugenes fulgens (*Swains.*) 60
   107. Eugenes spectabilis (*Lawr.*) 61

28. **Urochroa,** *Gould* 61
   108. Urochroa bougueri (*Bourc.*) 62

29. **Eugenia,** *Gould* 62
   109. Eugenia imperatrix, *Gould* 62

30. **Lampraster,** *Taczanowski* 63
   110. *Lampraster branicki,Taczan.* 63

31. **Heliodoxa,** *Gould* 63
   111. Heliodoxa jacula, *Gould* 64
   112. Heliodoxa jamesoni (*Bourc.*) 65
   113. Heliodoxa leadbeateri (*Bourc.*) 65

32. **Pterophanes,** *Gould* 66
   114. Pterophanes temminckii (*Boiss.*) 66

33. **Patagona,** *Gray* 67
   115. Patagona gigas (*Vieill.*) 67

34. **Docimastes,** *Gould* 68
   116. Docimastes ensiferus (*Boiss.*) 68

35. **Diphlogæna,** *Gould* 69
   117. Diphlogæna iris (*Gould*) 69
   ·118. Diphlogæna hesperus, *Gould* 70

36. **Helianthea,** *Gould* 70
   119. Helianthea isaacsoni (*Parzud.*) 71
   120. Helianthea typica (*Less.*) 71
   121. Helianthea bonapartii (*Boiss.*) 72
   122. Helianthea eos, *Gould* 72
   123. Helianthea lutitiæ (*Delattr. and Bourc.*) . . . 73
   124. Helianthea violifera (*Gould*) 73

PAGE

125. Helianthea osculans, *Gould* 73
126. Helianthca dichroura, *Taczan.* 74

37. Bourcieria, *Bonaparte* 74

127. Bourcieria inca, *Gould* 75
128. Bourcieria conradi (*Bourc.*) 76
129. Bourcieria insectivora (*Tschudi.*) 76
130. Bourcieria fulgidigula, *Gould* 76
131. Bourcieria torquata (*Boiss.*) 77
132. Bourcieria traviesi (*Muls. and Verr.*). . . . 77
133. Bourcieria wilsoni (*Delattr. and Bourc.*) . . . 77
134. *Bourcieria purpurea (*Gould*) 78
135. Bourcieria assimilis (*Elliot*) 78
136. Bourcieria prunelli (*Bourc. and Muls.*) . . . 78
137. Bourcieria cæligena (*Less.*) 79
138. Bourcieria columbiana (*Elliot*) 79
139. Bourcieria boliviana (*Gould*) 79

38. Hemistephania, *Reichenbach* 80

140. Hemistephania johannæ (*Bourc.*) 80
141. Hemistephania ludoviciæ (*Bourc. and Muls.*) . . 81
142. Hemistephania rectirostris (*Gould*) . . . 81
143. Hemistephania euphrosinæ (*Muls. and Verr.*) . . 81
144. Hemistephania veraguensis (*Salv.*) 82

39. Floricola, *Elliot* 82

145. Floricola longirostris (*Vieill.*) 83
146. Floricola albicrissa (*Gould*) 83
147. Floricola constanti (*Delattr.*) 84
148. Floricola leocadiæ (*Bourc.*) 84

40. Lepidolarynx, *Reichenbach* 84

149. Lepidolarynx mesoleucus (*Temm.*) 85

41. Heliomaster, *Bonaparte* 86

150. Heliomaster furcifer (*Shaw*) 86

42. Heliotrypha, *Gould* 86

151. Heliotrypha viola, *Gould* 87
152. Heliotrypha exortis (*Fras.*) 87
153. Heliotrypha micrastur (*Gould*) 88
154. Heliotrypha barrali, *Muls. and Verr.* . . . 88

43. Heliangelus, *Gould* . . . . PAGE 89
    155. Heliangelus clarissæ (*DeLong*) 89
    156. Heliangelus strophianus (*Gould*) 90
    157. Heliangelus spencei (*Bourc.*) 90
    158. Heliangelus amethysticollis (*D'Orb. and Lafr.*) . 90
    159. Heliangelus mavors, *Gould* 91

44. Urosticte, *Gould* 91
    160. Urosticte ruficrissa, *Lawr.* 91
    161. Urosticte benjamini (*Bourc.*) 92

45. Eustephanus, *Reichenbach* 92
    162. Eustephanus galeritus (*Mol.*) 93
    163. Eustephanus fernandensis (*King*) 93
    164. Eustephanus leyboldi, *Gould* 94

46. Topaza, *Gray* 94
    165. Topaza pella (*Linn.*) 95
    166. Topaza pyra, *Gould* 95

47. Aithurus, *Cabanis and Heine* 96
    167. Aithurus polytmus (*Linn.*) 96

48. Hylonympha, *Gould* 97
    168. Hylonympha macrocerca, *Gould* 97

49. Thalurania, *Gould* 98
    169. Thalurania glaucopis (*Gmel.*) 99
    170. Thalurania columbica (*Bourc. and Muls.*) . . 99
    171. Thalurania furcata (*Gmel.*) 99
    172. Thalurania furcatoides, *Gould* 100
    173. Thalurania nigrofasciata (*Gould*) 100
    174. *Thalurania jelskii, *Taczan.* 101
    175. *Thalurania watertoni (*Bourc.*) 101
    176. Thalurania refulgens, *Gould* 101
    177. Thalurania eriphile (*Less.*) 101
    178. Thalurania hypochlora, *Gould* 102
    179. Thalurania bicolor (*Gmel.*) 102

50. Mellisuga, *Brisson* 103
    180. Mellisuga minima (*Linn.*) 103

PAGE

51. Microchera, *Gould* 104
    181. Microchera albocoronata (*Lawr.*) 104
    182. Microchera parvirostris, *Lawr.* 104

52. Trochilus, *Linnæus* 105
    183. Trochilus colubris, *Linn.* 105
    184. Trochilus alexandri, *Bourc. and Muls.* . . . 106

53. Calypte, *Gould* 106
    185. Calypte costæ (*Bourc.*) 107
    186. Calypte annæ (*Less.*) 107
    187. Calypte helenæ (*Lemb.*) 108

54. Selasphorus, *Swainson* 108
    188. *Selasphorus floresii, *Gould* 109
    189. Selasphorus platycercus (*Swains.*) 109
    190. Selasphorus ardens, *Salvin* 110
    191. Selasphorus flammula, *Salvin* 110
    192. Selasphorus rufus (*Gmel.*) 110
    193. Selasphorus scintilla, *Gould* 111
    194. Selasphorus henshawi, *Elliot* 111
    195. Selasphorus torridus, *Salvin* 112

55. Catharma, *Elliot* 112
    196. Catharma orthura (*Less.*) 112

56. Atthis, *Reichenbach* 113
    197. Atthis heloisæ (*Less.*) 113
    198. Atthis ellioti, *Ridgw.* 114

57. Stellula, *Gould* 114
    199. Stellula calliope, *Gould* 115

58. Rhodopis, *Reichenbach* 115
    200. Rhodopis vesper (*Less.*) 115
    201. *Rhodopis atacamensis, *Leyb.* 116

59. Heliactin, *Boie* 116
    202. Heliactin cornuta (*Max.*) 116

60. Calothorax, *Gray* 117
    203. Calothorax pulchra, *Gould* 118
    204. Calothorax lucifer, *Swains.* 118

PAGE

61. Acestrura, *Gould* 119
   205. Acestrura mulsanti (*Bourc.*) 119
   206. *Acestrura decorata (*Gould*) 119
   207. Acestrura heliodori (*Bourc.*) 120
   208. *Acestrura micrura (*Gould*) 120

62. Chætocercus, *Gray* 120
   209. Chætocercus jourdani (*Bourc.*) 121
   210. Chætocercus rosæ (*Bourc. and Muls.*) . . . 121
   211. Chætocercus bombus, *Gould* 122

63. Thaumastura, *Bonaparte* 122
   212. Thaumastura cora (*Less. and Garn.*) . . 123

64. Doricha, *Reichenbach* 123
   213. Doricha enicura (*Vieill.*) 124
   214. Doricha elizæ (*Less. and Delattr.*) 125
   215. Doricha bryantæ, *Lawr.* 125
   216. Doricha evelynæ (*Bourc.*) 125
   217. Doricha lyrura, *Gould* 126

65. Myrtis, *Reichenbach* 126
   218. Myrtis fanny (*Less.*) 127
   219. Myrtis yarrelli (*Bourc.*) 127

66. Tilmatura, *Reichenbach* 128
   220. Tilmatura duponti (*Less.*) 128

67. Smaragdochrysis, *Gould* 129
   221. Smaragdochrysis iridescens, *Gould* 129

68. Ptochoptera, *Elliot* 129
   222. *Ptochoptera iolæma (*Pelz.*) 130

69. Calliphlox, *Boie* 130
   223. Calliphlox amethystina (*Gmel.*) 130
   224. Calliphlox mitchelli (*Bourc.*) 131

70. Lophornis, *Lesson* 131
   225. Lophornis stictolophus, *Salv. and Elliot* . . . 133
   226. Lophornis delattrii, *Less.* 133
   227. Lophornis regulus, *Gould* 133

PAGE

228. Lophornis ornatus (*Bodd.*) 134
229. Lophornis gouldi (*Less.*) 134
230. Lophornis magnificus (*Vieill.*) 135
231. Lophornis helenæ (*Delattr.*) 135
232. Lophornis adorabilis, *Salvin* 135
233. Lophornis verreauxi (*Bourc.*) 136
234. Lophornis chalybea (*Vieill.*) 136

71. Gouldia, *Bonaparte* 137
235. Gouldia popelairii (*DuBus.*) 138
236. Gouldia langsdorffi (*Bourc. and Vieill.*) . . . 138
237. Gouldia conversi (*Bourc. and Muls.*) 139
238. Gouldia lætitiæ (*Bourc.*) 139

72. Discura, *Reichenbach* 140
239. Discura longicauda (*Gmel.*) 140

73. Steganura, *Reichenbach* 141
240. Steganura underwoodi (*Less.*) 142
241. Steganura melananthera, *Jard.* 142
242. Steganura solstitialis, *Gould* 142
243. Steganura peruana, *Gould* 143
244. Steganura addæ (*Bourc.*) 143
245. *Steganura cissiura (*Gould*) 144

74. Loddigesia, *Gould* 144
246. *Loddigesia mirabilis (*Bourc.*) 145

75. Lesbia, *Lesson* 145
247. Lesbia gouldi (*Lodd.*) 146
248. Lesbia nuna (*Less.*) 147
249. Lesbia eucharis (*Bourc.*) 147
250. Lesbia amaryllis (*Bourc.*) 148

76. Zodalia, *Mulsant* 149
251. *Zodalia ortoni (*Lawr.*) 149
252. *Zodalia glyceria (*Bonap.*) 150

77. Cynanthus, *Swainson* 150
253. Cynanthus forficatus (*Linn.*) 151
254. Cynanthus mocoa (*Delattr. and Bourc.*) . . . 152

PAGE

73. Sappho, *Reichenbach* 153
   255. Sappho sparganura (*Shaw*) 154
   256. Sappho phaon (*Gould*) 154
   257. *Sappho caroli (*Bourc.*) 155

79. Oxypogon, *Gould* 155
   258. Oxypogon lindeni (*Parzud.*) 156
   259. Oxypogon guerini (*Boiss.*) 156

80. Oreonympha, *Gould* 157
   260. Oreonympha nobilis, *Gould* 157

81. Rhamphomicron, *Bonaparte* 158
   261. Rhamphomicron olivaceus, *Lawr.* 158
   262. Rhamphomicron heteropogon (*Boiss.*) . . . 158
   263. Rhamphomicron herrani (*Delattr. and Bourc.*) . . 159
   264. Rhamphomicron stanleyi (*Bourc. and Muls.*) . . 159
   265. Rhamphomicron ruficeps (*Gould*) 160
   266. Rhamphomicron microrhynchum (*Boiss.*) . . . 160

82. Avocettinus, *Bonaparte* 161
   267. Avocettinus eurypterus (*Lodd.*) 161

83. Avocettula, *Reichenbach* 162
   268. Avocettula recurvirostris (*Swains.*) 162

84. Metallura, *Gould* 163
   269. Metallura opaca (*Licht.*) 163
   270. Metallura jelski, *Caban.* 164
   271. *Metallura chloropogon (*Caban. and Hein.*) . . 164
   272. Metallura eupogon, *Caban.* 164
   273. Metallura æneicauda (*Gould*) 165
   274. Metallura primolina, *Bourc.* 165
   275. Metallura williami (*Bourc. and Delattr.*) . . . 165
   276. Metallura tyrianthina (*Lodd.*) 166
   277. Metallura smaragdinicollis (*D'Orb. and Lafr.*) . . 166

85. Chrysuronia, *Bonaparte* 167
   278. Chrysuronia humboldti (*Bourc. and Muls.*) . . 168
   279. Chrysuronia œnone (*Less.*) 168
   280. Chrysuronia josephinæ (*Bourc. and Muls.*) . . 169
   281. Chrysuronia eliciæ (*Bourc. and Muls.*) . . . 169
   282. Chrysuronia chrysura (*Less.*) 169

PAGE

86. Augastes, *Gould* 170
    283. Augastes lumachellus (*Less.*) 170
    284. Augastes superbus (*Vieill.*) 171

87. Phlogophilus, *Gould* 171
    285. Phlogophilus hemileucurus, *Gould* 172

88. Schistes, *Gould* 172
    286. Schistes personatus, *Gould* 173
    287. Schistes geoffroyi (*Bourc. and Muls.*) . . 173

89. Heliothrix, *Boie* 174
    288. Heliothrix auritus (*Gmel.*) 174
    289. Heliothrix auriculatus (*Licht.*) 175
    290. Heliothrix barroti (*Bourc.*) 175

90. Chrysolampis, *Boie* 176
    291. Chrysolampis moschitus (*Linn.*) 176

91. Bellona, *Mulsant and Verreaux* 178
    292. Bellona cristata (*Linn.*) 178
    293. Bellona exilis (*Gmel.*) 179

92. Cephalolepis, *Loddiges* 179
    294. Cephalolepis delalandi (*Vieill.*) 180
    295. Cephalolepis loddigesi (*Gould*) 180

93. Adelomyia, *Bonaparte* 181
    296. Adelomyia cervina, *Gould* 181
    297. Adelomyia inornata, *Gould* 181
    298. Adelomyia chlorospila, *Gould* 182
    299. Adelomyia melanogenys (*Fraser*) . . . 182

94. Anthocephala, *Cabanis and Heine* 183
    300. *Anthocephala floriceps (*Gould*) 183

95. Abeillia, *Bonaparte* 183
    301. Abeillia typica (*Bonap.*) 184

96. Klais, *Reichenbach* 184
    302. Klais guimeti (*Bourc. and Muls.*) . . . 184

PAGE

97. Aglæactis, *Gould* 185
   303. Aglæactis cupripennis (*Bourc. and Muls.*) . . 186
   304. Aglæactis caumatonota, *Gould* 186
   305. Aglæactis castelnaudi (*Bourc. and Muls.*) . . . 187
   306. Aglæactis pamela (*D'Orb. and Lafr.*) . . . 187

98. Eriocnemis, *Reichenbach* 188
   307. Eriocnemis derbiana (*Delattr. and Bourc.*) . . 189
   308. Eriocnemis assimilis, *Elliot* 189
   309. Eriocnemis aureliæ (*Bourc. and Muls.*) . . . 190
   310. Eriocnemis squamata, *Gould* 190
   311. Eriocnemis lugens (*Gould*). 190
   312. Eriocnemis alinæ (*Bourc.*) 191
   313. Eriocnemis mosquera (*Bourc. and Delattr.*) . . 191
   314. *Eriocnemis glaucopoides (*D'Orb. and Lafr.*) . . 191
   315. Eriocnemis luciani (*Bourc.*) 192
   316. Eriocnemis cupreiventris (*Fraser*) 192
   317. *Eriocnemis sapphiropygia, *Taczan.* 193
   318. Eriocnemis chrysorama, *Elliot* 193
   319. Eriocnemis godini (*Bourc.*) 193
   320. Eriocnemis vestita (*Longuem.*) 193
   321. Eriocnemis smaragdinipectus, *Gould* . . . 194
   322. Eriocnemis nigrivestis (*Bourc.*) 194
   323. Eriocnemis dyselius, *Elliot*. 194

99. Pantèrpe, *Cabanis and Heine* 195
   324. Panterpe insignis, *Cab. and Hein.* 195

100. Uranomitra, *Reichenbach* 195
   325. Uranomitra quadricolor (*Vieill.*) 196
   326. Uranomitra violiceps (*Gould*) 196
   327. Uranomitra viridifrons, *Elliot* 197
   328. Uranomitra cyanocephala (*Less.*) 197
   329. Uranomitra microrhyncha, *Elliot* 197
   330. Uranomitra franciæ (*Bourc. and Muls.*) . . . 197
   331. *Uranomitra cyanicollis (*Gould*) 198

101. Leucippus, *Bonaparte* 198
   332. Leucippus chionogaster (*Tschudi*) 199
   333. Leucippus chlorocercus, *Gould* 199

102. Leucochloris, *Reichenbach* 200
   334. Leucochloris albicollis (*Vieill.*) 200

PAGE

103. Agyrtria, *Reichenbach* 201

    335. Agyrtria niveipectus *Caban. and Hein.* . . . 202
    336. Agyrtria leucogaster (*Gmel.*) 202
    337. Agyrtria viridiceps (*Gould*) 203
    338. Agyrtria milleri (*Bourc.*) 203
    339. Agyrtria candida (*Bourc. and Muls.*) . . . 203
    340. *Agyrtria norrisii (*Bourc.*) 204
    341. Agyrtria brevirostris (*Less.*) 204
    342. *Agyrtria compsa, *Heine* 204
    343. *Agyrtria neglecta, *Elliot* 205
    344. Agyrtria bartletti (*Gould*) 205
    345. *Agyrtria nitidifrons (*Gould*) 205
    346. *Agyrtria cœruleiceps (*Gould*) 206
    347. Agyrtria tephrocephala (*Vieill.*) 206
    348. Agyrtria tobaci (*Gmel.*) 206
    349. Agyrtria fluviatilis (*Gould*) 207
    350. Agyrtria apicalis (*Gould*) 207
    351. *Agyrtria maculicauda (*Gould*) 207
    352. *Agyrtria luciæ (*Lawr.*) 208
    353. Agyrtria nigricauda, *Elliot* 208
    354. Agyrtria nitidicauda, *Elliot* 208

104. Arinia, *Mulsant* 209

    355. *Arinia boucardi, *Muls.* 209

105. Elvira, *Mulsant and Verreaux* 210

    356. Elvira cupreiceps (*Lawr.*) 210
    357. Elvira chionura (*Gould*) 210

106. Callipharus, *Elliot* 211

    358. Callipharus nigriventris (*Lawr.*) 211

107. Eupherusa, *Gould* 212

    359. Eupherusa poliocerca, *Elliot* 212
    360. Eupherusa eximia (*Delattr.*) 212
    361. Eupherusa egregia, *Scl. and Salv.* 213

108. Polytmus, *Brisson* 213

    362. Polytmus thaumantias (*Linn.*) 214
    363. Polytmus viridissimus (*Vieill.*) 214
    364. Polytmus leucorrhous, *Scl. and Salv.* . . . 215

15

109. Amazilia, *Lesson* 216

365. Amazilia pristina, *Gould* 217
366. Amazilia leucophæa, *Reichenb.* 218
367. Amazilia alticola, *Gould* 218
368. Amazilia dumerili (*Less.*) 218
369. Amazilia cinnamomea (*Less.*) 219
370. *Amazilia graysoni, *Lawr.* 219
371. Amazilia yucatanensis (*Cabot*) 219
372. Amazilia fuscicaudata (*Fraser*) 220
373. Amazilia viridiventris (*Reichenb.*) 220
374. *Amazilia ocai, *Gould* 221
375. Amazilia beryllina (*Licht.*) 221
376. Amazilia edwardi (*Delattr.*) 221
377. Amazilia niveiventris (*Gould*) 222
378. Amazilia mariæ (*Bourc.*) 222
379. *Amazilia cyanura, *Gould* 223
380. *Amazilia iodura (*Saucer.*) 223
381. Amazilia lucida, *Elliot* 223
382. Amazilia erythronota (*Less.*) 224
383. Amazilia feliciæ (*Less.*) 224
384. Amazilia sophiæ (*Bourc. and Muls.*) . . . 224
385. Amazilia warszewiczi (*Caban. and Hein.*) . . 225
386. Amazilia saucerottii (*Bourc. and Delattr.*) . . 225
387. Amazilia cyanifrons (*Bourc.*) 225
388. *Amazilia elegans (*Gould*) 226

110. Basilinna, *Boie* 226

389. Basilinna leucotis (*Vieill.*) 227
390. Basilinna xantusi (*Lawr.*) 227

111. Eucephala, *Reichenbach* 227

391. Eucephala grayi (*Delattr. and Bourc.*) . . 228
392. *Eucephala smaragdo-cærulea, *Gould* . . 229
393. *Eucephala cæruleo-lavata, *Gould* . . . 229
394. *Eucephala scapulata, *Gould* 229
395. *Eucephala hypocyanea, *Gould* 230
396. Eucephala subcærulea, *Elliot* 230
397. Eucephala cærulea (*Vieill.*) 230
398. *Eucephala chlorocephala (*Bourc.*) . . . 231
399. *Eucephala cyanogenys 231

112. Timolia, *Mulsant* 231

400. Timolia lerchi (*Muls. and Verr.*) 232

PAGE

113. Juliamyia, *Bonaparte* 232
 401. Juliamyia typica, *Bonap.* 233
 402. Juliamyia feliciana (*Less.*) 233

114. Damophila, *Reichenbach* 233
 403. Damophila amabilis (*Gould*) 234

115. Iache, *Elliot* 234
 404. Iache latirostris (*Swains.*) 235
 405. Iache magica (*Muls. and Verr.*) 235
 406. Iache doubledayi (*Bourc.*) 235

116. Hylocharis, *Boie* 236
 407. Hylocharis lactea (*Less.*) 236
 408. Hylocharis sapphirina (*Gmel.*) 236
 409. Hylocharis cyanea (*Vieill.*) 237

117. Cyanophaia, *Reichenbach* 237
 410. Cyanophaia cæruleigularis (*Gould*) 238
 411. Cyanophaia goudoti (*Bourc.*) 239
 412. *Cyanophaia luminosa (*Lawr.*) 239

118. Sporadinus, *Bonaparte* 240
 413. *Sporadinus bracei, *Lawr.* 240
 414. Sporadinus elegans (*Vieill.*) 241
 415. Sporadinus riccordi (*Gerv.*) 241
 416. Sporadinus maugæi (*Vieill.*) 242

119. Chlorostilbon, *Gould* 242
 417. Chlorostilbon auriceps, *Gould* 243
 418. Chlorostilbon caniveti (*Less.*) 243
 419. Chlorostilbon pucherani (*Bourc. and Muls.*) . . 244
 420. Chlorostilbon splendidus (*Vieill.*) 244
 421. Chlorostilbon haberlini (*Reichenb.*) 245
 422. Chlorostilbon angustipennis (*Fraser*) . . . 245
 423. Chlorostilbon atala (*Less.*) 246
 424. Chlorostilbon prasinus (*Less.*) 246

120. Panychlora, *Cabanis and Heine* 247
 425. Panychlora poortmani (*Bourc.*) 247
 426. Panychlora aliciæ (*Bourc. and Muls.*) . . . 248
 427. *Panychlora stenura, *Caban. and Hein.* . . 248

INDEX.

Abeil ia, 12
Acestrura, 9
Adelomyia, 12
Aglæactis, 13
Agyrtria, 14
Aithurus, 7
Amazilia, 15
Androdon, 1
Anthocephala, 12
Aphantochroa, 3
Arinia, 14
Atthis, 8
Augastes, 12
Avocettinus, 11
Avocettula, 11

Basilinna, 15
Bellona, 12
Bourcieria, 6

Callipharus, 14
Calliphlox, 9
Calothorax, 8
Calypte, 8
Campylopterus, 2
Catharma, 8
Cæligena, 3
Cephalolepis, 12
Chætocercus, 9
Chalybura, 4
Chlorostilbon, 16
Chrysolampis, 12
Chrysuronia, 11
Clytolæma, 4
Cyanophaia, 16
Cynanthus, 10

Damophila, 16
Diphlogæna, 5
Discura, 10
Docimastes, 5
Doleromya, 1
Doricha, 9

Elvira, 14
Eriocnemis, 13
Eucephala, 15

Eugenes, 5
Eugenia, 5
Eulampis 3
Eupetomena, 2
Eupherusa, 14
Eustephanus, 7
Eutoxeres, 1

Floricola, 6
Florisuga, 4

Glaucis, 1
Gouldia, 10

Heliactin, 8
Heliangelus, 7
Helianthea, 5
Heliodoxa, 5
Heliomaster, 6
Heliothrix, 12
Heliotrypha, 6
Hemistephania, 6
Hylocharis, 16
Hylonympha, 7

Iache, 16
Iolæma, 4

Juliamyia, 16

Klais, 12

Lafresnaya, 4
Lampornis, 3
Lampraster, 5
Lamprolæma, 3
Lepidolarynx, 6
Lesbia, 10
Leucippus, 13
Leucochloris, 13
Loddigesia, 10
Lophornis, 9

Mellisuga, 7
Metallura, 11
Microchera, 8
Myrtis, 9

Oreonympha, 11
Oreopyra, 3
Oreotrochilus, 3
Oxypogon, 11

Panoplites, 4
Panterpe, 13
Panychlora, 16
Patagona, 5
Petasophora, 4
Phæolæma, 4
Phæoptila, 1
Phæthornis, 2
Phlogophilus, 12
Polytmus, 14
Pterophanes, 5
Ptochoptera, 9

Rhamphodon, 1
Rhamphomicron, 11
Rhodopis, 8

Sappho, 11
Schistes, 12
Selasphorus, 8
Smaragdochrysis, 9
Sphenoproctus, 2
Sporadinus, 16
Steganura, 10
Stellula, 8
Sternoclyta, 5

Thalurania, 7
Thaumastura, 9
Tilmatura, 9
Timolia, 15
Trochilus, 8
Topaza, 7

Uranomitra, 13
Urochroa, 5
Urosticte, 7

Zodalia, 10

2

MAY, 1879.

ocieties publishing Transactions; L. Libraries exclusively, though the others also have librar
ies; M. Museums; O. Observatories; T. Technical Schools; U. Universities.
This arrangement is alphabetical by towns and not by States.

Town	State	No.	Institution	Type
any	N. Y.	1	Albany Institute	S.
"	"	2	N. Y. State Agricultural Society	S.
"	"	3	N. Y. State Library	L.
"	"	4	State Museum of Natural Sciences	M.
egheny	Pa.	5	Observatory	O.
napolis	Md.	6	U. S. Naval Academy	T.
n Arbor	Mich.	7	University of Michigan	U.M.
"	"	8	Observatory	O.
anta	Ga.	9	City Library	L.
ltimore	Md.	10	Johns Hopkins University	U.
"	"	11	Maryland Academy of Sciences	S.
"	"	12	Peabody Institute	L.
hlehem	Pa.	13	Packer University	T. U.
omington	Ind.	14	The " Owen Cabinet "	M.
ston	Mass.	15	Amer. Academy of Arts and Sciences	S.
"	"	16	Amer. Statistical Association	S.
"	"	17	Athenæum	L.
"	"	18	Mass. Institute of Technology	T.
"	"	19	Boston Natural History Society	S. M.
"	"	20	Public Library of the City	L.
"	"	21	State Library	L.
ooklyn	N. Y.	22	Brooklyn Library	L.
unswick	Me.	23	Bowdoin College	U.
ffalo	N. Y.	24	Buffalo Society of Natural Sciences	M. S.
rlington	Vt.	25	University of Vermont	U.
mbridge	Mass.	26	Harvard University	U.
"	"	27	Lawrence Scientific School	T.
"	"	28	Museum of Comparative Zoölogy	M. S.
"	"	29	Observatory	O.
"	"	30	Peabody Museum	M.

It was designed to limit this list to one hundred of the principal institutions of the Unite
.es; but this number is slightly exceeded.

Charleston	S. C.	31	College of Charleston
"	"	32	Library Society
"	"	33	Medical School of South Carolina
Charlottesville	Va.	34	University of Virginia
Chicago	Ill.	35	Chicago Academy of Sciences
"	"	36	Observatory
"	"	37	Public Library
Cincinnati	Ohio.	38	Observatory
"	"	39	Public Library
"	"	40	Zoölogical Society
Cleveland	"	41	Kirtland Society of Natural Sciences
Clinton	N. Y.	42	Observatory
Columbia	S. C.	43	University of South Carolina
Columbus	Ohio.	44	State Library
Davenport	Iowa.	45	Academy of Natural Sciences
Des Moines	"	46	State Library
Dubuque	"	47	Iowa Institute of Science and Arts
Easton	Pa.	48	Pardee Scientific School
Glasgow	Mo.	49	Observatory
Hanover	N. H.	50	Dartmouth College
"	"	51	Observatory
Harrisburg	Pa.	52	State Library
Hartford	Ct.	53	Amer. Philological Association
"	"	54	Watkinson Reference Library
Hoboken	N. J.	55	Stevens Institute of Technology
Indianapolis	Ind.	56	Public Library
Iowa City	Iowa.	57	State University
Ithaca	N. Y.	58	Cornell University
Lansing	Mich.	59	State Library
Lawrence	Kans.	60	Academy of Science
Lexington	Va.	61	School of Civil and Mining Engin'g
Louisville	Ky.	62	Public Library
Madison	Wis.	63	State Historical Society
"	"	64	Wis. Acad. of Sciences, Arts & Letters
Middletown	Ct.	65	Wesleyan University
Minneapolis	Minn.	66	University of Minnesota
Nashville	Tenn.	67	State Library
Newark	N. J.	68	New Jersey Historical Society
New Brunswick	"	69	Rutgers Scientific School
New Haven	Ct.	70	Academy of Arts and Sciences
"	"	71	American Oriental Society
"	"	72	Observatory

		ıɔ	ca ʊ ɔ ˙ ɯɔeuɪ _____	._ ._
	"	74	Sheffield Scientific School _____	T.
	"	75	Yale College _____	U.
ns _____		76	Academy of Sciences _____	S.
	"	77	State Library _____	L.
City___	N. Y.	78	American Geographical Society _____	S.
___	"	79	American Institute ._._____	S.
___	"	80	American Institute of Architects ___	S.
___	"	81	Amer. Museum of Natural History __	M.
___	"	82	American Society of Civil Engineers	S.
___	"	83	Astor Library_____	L.
___	"	84	College of Physicians and Surgeons _	M.
___	"	85	Columbia College_____	U.
___	"	86	Cooper Union_____	T.
___	"	87	Lenox Library _____	L.
___	"	88	Mercantile Library_____	L.
___	"	89	Metropolitan Museum of Art_____	M.
___	"	90	New York Academy of Sciences_____	S.
___	"	91	New York Historical Society _____	S. M.
___	"	92	New York Society Library_____	L.
___	"	93	New York School of Mines _____	T.
_____	Ill.	94	Illinois Museum of Natural History	M.
_____	Cal.	95	University of California _____	U.
hia _____	Pa.	96	Academy of Natural Sciences _____	S. M.
_____	"	97	American Philosophical Society ____	S.
_____	"	98	Franklin Institute_____	S.
_____	"	99	Library Company of Philadelphia__	L.
_____	"	100	Mercantile Library_____	L.
_____	"	101	Pennsylvania Historical Society_____	S.
_____	"	102	University of Pennsylvania _____	U.
_____	"	103	Zoölogical Society of Philadelphia __	M.
h _____	"	104	Mercantile Library _____	L.
_____	Me.	105	Portland Society of Natural History	S. M.
_____	N. J.	106	College of New Jersey_____	U.M.
_____	"	107	Green School of Science_____	T.
_____	"	108	Observatory_____	O.
e _____	R. I.	109	Athenæum_____	L.
__ ____	"	110	Brown University_____	M.U.
_____	N. C.	111	State Library _____	L.
_____	Va.	112	State Library_ _____	L.
_____	Mo.	113	Mo. School of Mines and Metallurgy	T.
ıto _____	Cal.	114	State Library _____	L.

4

Louis_____	Mo.	115	Academy of Science _____	S.
" _____	"	116	Mercantile Library_____	L.
ɔm _____	Mass.	117	Am. Assoc. for Advancement of Sci.	S.
' _____	"	118	Essex Institute _____	S. M
' _____	"	119	Peabody Academy of Sciences_____	S.
Francisco ____	Cal.	120	California Academy of Sciences_____	S.
" ____	"	121	Woodward's Zoölogical Institute ___	M.
ingfield _____	Ill.	122	State Library _____	L.
y _____	N. Y.	123	Rensselaer Polytechnic Institute_____	T.
hington _____	D. C.	124	American Medical Association_____	S.
" _____	"	125	Army Medical Museum _____	M.
ᛣ _____	"	126	Corcoran Gallery of Art _____	M.
_____	"	127	Library of Congress _____	L.
.. _____	"	128	National Academy of Sciences_____	S.
.. _____	"	129	Philosophical Society of Washington	S.
" _____	"	130	Smithsonian Institution _____	S. M.
t Point_____	N. Y.	131	Military Academy _____	T.
·cester_____	Mass.	132	American Antiquarian Society _____	S. M.

NCIPAL GOVERNMENT DEPARTMENTS AND BUREAUS IN
CITY OF WASHINGTON.

133 Agriculture, Department of
134 Census Office.
135 Coast Survey.
136 Education, Bureau of
137 Engineer Bureau, War Department.
138 Entomological Commission.
139 Fish Commission.
140 Geological Surveys.
141 Hydrographic Office.
142 Interior Department.
143 Land Office.
144 Light House Board.
145 Marine Hospital Service.
146 Medical Department, U. S. A.
147 National Museum.
148 Nautical Almanac Office.
149 Naval Observatory.
150 Navigation, Bureau of
151 Navy Department.
152 Ordnance Office.
153 Patent Office.
154 Post Office Department.
155 Quartermaster General's Office.

CLASSIFIED INDEX.

LIBRARIES:

. **3,** 9, 12, 17, **20,** 21, 22, 32, **37, 39,** 44, 46, 52, 54, 56, 59,
62, 67, 77, **83, 87,** 88, 92, **99,** 100, 104, 109, 111, 112,
114, 116, 122, **127,** 133, 135, 136, 137, 142, 146, 149, 151,
153, 157, 159, 160.

MUSEUMS:

. **4,** 7, 14, **19, 24, 28, 30,** 31, 33, **35,** 40, 41, 47, **50,** 60,
63, 64, **65, 73, 81,** 84, **89,** 91, 94, **96,** 103, 105, **106,**
110, 118, 121, 125, 126, 130, 132, 133, 135.

OBSERVATORIES:

. **5, 8, 29,** 36, 38, **42,** 49, 51, 72, 108, **149.**

SOCIETIES OR INSTITUTIONS PUBLISHING TRANSACTIONS:

. 1, 2, 11, **15,** 16, **19,** 24, 28, **35,** 41, **45,** 47, 53, 60, 63, 64,
68, **70,** 71, **76, 78,** 79, 80, 82, **90,** 91, **96, 97,** 98, 101,
105, 115, 117, 118, **119, 120,** 124, **128,** 129, **130, 132,**
133, 134, 135, 136, 137, 138, 139, 140, 141, 142, 143, 144,
145, 146, 147, 148, 149, 150, 151, 152, 153, 154, 155, 156,
157, 158, 159, 160.

TECHNICAL SCHOOLS:

.. **6,** 13, **18,** 29, 34, 48, **55,** 58, 61, 69, **74,** 86, 93, 107, 113,
123, **131.**

UNIVERSITIES:

.. 7, **10,** 13, 23, 25, **26,** 31, 34, 43, 50, 57, 58, 65, 66, **75, 85,**
95, 102, 106, 110.

SMITHSONIAN MISCELLANEOUS COLLECTIONS.

——————— 344 ———————

LIST

OF

PUBLICATIONS

OF THE

SMITHSONIAN INSTITUTION,

JULY, 1879.

WASHINGTON, D. C.:

JULY, 1879.

LIST

OF

PUBLICATIONS OF THE SMITHSONIAN INSTITUTION,

To July, 1879.

Where no price is affixed the work cannot be furnished, it being out of print or not yet published.

Publications marked * do not appear in the Contributions, Collections, or Reports.

No.	Author.	Title.		Pages.	Date.	Price.
A	....	Journal of Regents,	8vo.*	32	1846	
B	....	Report of Organization Committee	8vo.*	32	1847	
C	....	Digest of Act of Congress,	8vo.*	8	1847	
D	Dallas, G. M.	Address at Laying Corner Stone,	8vo *	8	1847	
E	Henry, Jos.	Exposition of Bequest,	8vo.*	8	1847	
F	....	First Report of Secretary,	8vo.*	48	1848	
G	....	Report of the Institution,	8vo.*	38	1847	
H	....	Second Report of Institution,	8vo.*	208	1848	
I	....	Third Report of Institution,	8vo.*	64	1849	
J	....	Programme of Organization,	4to.*	4	1847	
K	....	Correspondence, Squier & Davis,	8vo.*	8	1848	
L	....	First Report of Organization Committee,	8vo.*	8	1846	
M	....	Reports of Institution up to Jan. 1849,	8vo.*	72	1849	
N	....	Officers, Regents, Act, &c.,	8vo.*	14	1846	
O	....	Act to establish Smithsonian Institution,	8vo.*	8	1846	
P	Owen, R. D.	Hints on Public Architecture,	4to.*	140	1849	
Q	....	Check List of Periodicals,	4to.*	28	1853	
1	Squier & Davis,	Ancient Monuments of Mississippi Valley,	S. C. I,	346	1847	
2	....	Smithsonian Contributions to Knowledge,	S. C. I,	346	1848	

No.	Author.	Title.		Pages.	Date.	Price.
3	Walker, S. C.	Researches, Planet Neptune	S. C. ii,	60	1850	
4	Walker, S. C.	Ephemeris of Neptune for 1848,	S. C. ii,	8	1849	
5	Walker, S. C.	Ephemeris of Neptune for 1849,	S. C. ii,	32	1849	
6	Walker, S. C.	Ephemeris of Neptune for 1850,	S. C. ii,	10	1850	
7	Walker, S. C.	Ephemeris of Neptune for 1851,	S C. ii,	10	1850	
8	Downes, John	Occultations in 1848,	4to.*	12	1848	
9	Downes, John	Occultations in 1849,	4to.*	24	1848	
10	Downes, John	Occultations in 1850,	4to.*	26	1849	
11	Downes, John	Occultations in 1851,	S. C. ii,	26	1850	
12	Lieber, Francis	Vocal Sounds of L. Bridgeman,	S. C. ii,	32	1850	
13	Ellet, Charles	Physical Geography of U. S.	S. C. ii,	64	1850	
14	Gibbes, R. W.	Memoir on Mosasaurus,	S. C. ii,	14	1850	
15	Squier, E. G.	Aboriginal Monuments of N. Y.	S. C. ii,	188	1850	
16	Agassiz, Louis	Classification of Insects,	S. C. ii,	28	1850	
17	Hare, Robert	Explosiveness of Nitre,	S. C. ii,	20	1850	
18	Gould, Jr., B. A.	Discovery of Neptune,	8vo.*	56	1850	
19	Guyot, A.	Directions for Meteorological Observations,	8vo.*	40	1850	
20	Bailey, J. W.	Microscopic Examination of Soundings,	S. C. ii,	16	1851	
21	....	Annual Report of Smithsonian Institution for 1849	8vo.	272	1850	
22	Gray, Asa	Plantæ Wrightianæ,	S. C. iii,	146	1852	
23	Bailey, J. W.	Microscopic Observations in S. Carolina, Georgia, and Florida,	S. C. ii,	48	1851	
24	Walker, S. C	Ephemeris of Neptune, 1852. Appendix I,	S. C. iii,	10	1853	
25	Jewett, Chas. C.	Public Libraries of United States,	8vo.*	210	1851	.5
26	....	Smithsonian Contributions to Knowledge,	S. C. ii,	464	1851	
27	Booth, J. C. and Morfit, C.	Improvements in Chemical Arts,	M. C. ii,	216	1852	.5
28	....	Annual Report of Smithsonian Institution for 1850,	8vo.	326	1851	
29	Downes, John	Occultations in 1852,	S. C. iii,	34	1851	

No.	Author.	Title.	Page	Date	Price
30	Girard, Charles	Fresh-Water Fishes of N. America	S. C. III,	80 1851	
31	Guyot, A.	Meteorological Tables,	M. C. I,	212 1852	·
32	Harvey, Wm. H.	Marine Algæ of North America. Part I,	S. C. III,	152 1852	
33	Davis, Chas. H.	Law of Deposit of Flood Tide,	S. C. III,	14 1852	.75
34	....	Directions for Collecting Specimens,	M. C. II,	40 1859	free
35	Locke, John	Observations on Terrestrial Magnetism,	S. C. III,	30 1852	
36	Secchi, A.	Researches on Electrical Rheometry,	S. C. III,	60 1852	
37	Whittlesey, Ch.	Ancient Works in Ohio,	S. C. III,	20 1851	
38	....	Smithsonian Contributions to Knowledge,	S. C. III,	564 1852	
39	....	Smithsonian Contributions to Knowledge,	S. C. IV,	426 1852	
40	Riggs, S. R.	Dakota Grammar and Dictionary,	S. C. IV,	414 1852	
41	Leidy, Joseph	Extinct American Ox,	S. C. V,	20 1852	
42	Gray, Asa	Plantæ Wrightianæ. Part II,	S. C. V,	120 1853	
43	Harvey, Wm. H.	Marine Algæ of North America. Part II,	S. C. V,	262 1853	10.00
44	Leidy, Joseph	Flora and Fauna within Living Animals,	S. C. V,	68 1853	
45	Wyman, Jeffries	Anatomy of Rana Pipiens,	S. C. V,	52 1853	
46	Torrey, John	Plantæ Fremontianæ,	S. C. VI,	24 1853	
47	Jewett, Chas. C.	Construction of Catalogues of Libraries,	8vo.*	108 1853	.50
48	Girard, Charles	Bibliotheca Americana Historico Naturalis,	8vo.*	68 1852	
49	Baird, S. F. and Girard C.	Catalogue of Serpents,	M. C. II,	188 1853	1.00
50	Stimpson, Wm.	Marine Invertebrata of Gr. Manan	S. C. VI,	68 1853	1.50
51	....	Annual Report of Smithsonian Institution for 1851,	8vo.	104 1852	
52	Coffin, Jas. H.	Winds of the Northern Hemisphere,	S. C. VI,	200 1853	
53	Stanley, J. M.	Portraits of N. American Indians,	M. C. II,	76 1852	
54	Downes, John	Occultations in 1853,	S. C. VI,	36 1853	

No.	Author.	Title.		Pages.	Date.	Price.
55	....	Smithsonian Contributions to Knowledge,	S. C. v,	538	1853	
56	....	Smithsonian Contributions to Knowledge,	S. C. vi,	476	1854	
57	....	Annual Report of Smithsonian Institution for 1852,	8vo.	96	1853	
58	Leidy, Joseph	Ancient Fauna of Nebraska,	S. C. vi,	126	1853	
59	Chappelsmith, J.	Tornado in Indiana,	S. C. vii,	12	1855	.25
60	Torrey, John	Batis Maritima,	S. C. vi,	8	1853	
61	Torrey, John	Darlingtonia Californica,	S. C. vi,	8	1853	
62	Melsheimer, F.E.	Catalogue of Coleoptera,	8vo.*	190	1853	2.00
63	Bailey, J. W.	New Species of Microscopic Organisms,	S. C. vii,	16	1854	.50
64	....	List of Foreign Correspondents of Smithsonian Institution,	M. C.	16	1856	
65	....	Registry of Period. Phenomena,	folio,*	4	1854	
66	....	Annular Eclipse, May 26, 1854	M. C.	14	1854	
67	....	Annual Report of Smithsonian Institution for 1853,	8vo.	310	1854	
68	Mitchell, B. R. & Turner, W. W.	Vocabulary of Jargon of Oregon,	8vo.*	22	1853	
69	....	List of American Correspondents of Smithsonian Institution,	8vo.*	16	1853	
70	Lapham, I. A.	Antiquities of Wisconsin,	S. C. vii,	108	1855	
71	Haven, S. F.	Archæology of the United States,	S. C. viii,	172	1856	
72	Leidy, Joseph	Extinct Sloth Tribe of N. America,	S. C. vii,	70	1855	
73	....	Publications of Societies in Smithsonian Library,	S. C. vii,	40	1855	.25
74	....	Catalogue of Smithsonian Publications,	M. C. v,	52	1862	
75	....	Annual Report of Smithsonian Institution for 1854,	8vo.	464	1855	
76	....	Smithsonian Contributions to Knowledge,	S. C. vii,	252	1855	
77	....	Annual Report of Smithsonian Institution for 1855,	8vo.	440	1856	
78	....	Smithsonian Contributions to Knowledge,	S. C. viii,	556	1856	

No.	Author.	Title.		Pages.	Date.	Price.
79	Runkle, John D.	Tables for Planetary Motion,	S. C. ix,	64	1856	1.00
80	Alvord, Benj.	Tangencies of Circles and Spheres,	S. C. viii,	16	1856	1.00
81	Olmsted, D.	Secular Period of Aurora Borealis	S. C. viii,	52	1856	
82	Jones, Joseph	Investigation on A. Vertebrata,	S. C. viii,	150	1856	1.50
83	Meech, L. W.	Relative Intensity of Heat and Light of the Sun,	S. C. ix,	58	1856	1.25
84	Force, Peter	Auroral Phenomena in North Latitudes,	S. C. viii,	122	1856	1.25
85	....	Publications of Societies in Smithonian Library. Part II,	S. C. viii,	38	1856	.25
86	Mayer, Brantz	Mexican History and Archæology	S. C. ix,	36	1856	
87	Coffin, Jas. H.	Psychrometrical Tables,	M. C. i,	20	1856	.25
88	Gibbs, W. and Genth, F. A.	Ammonia Cobalt Bases,	S. C. ix,	72	1856	1.00
89	Brewer, Th. M.	North American Oology. Part I,	S. C. xi,	140	1857	5.00
90	Hitchcock, E.	Illustrations of Surface Geology,	S. C. ix,	164	1857	4.00
91	....	Annual Report of Smithsonian Institution for 1856,	8vo.	468	1857	
92	....	Smithsonian Contributions to Knowledge,	S. C. ix,	482	1857	
93	....	Meteorological Observations for 1855,	8vo.*	118	1857	
94	Runkle, John D.	Asteroid Supplement to New Tables for $b\frac{(h)}{s}$,	S. C. ix,	72	1857	1.00
95	Harvey, Wm. H.	Marine Algæ of North America. Part III,	S. C. x,	142	1858	6.00
96	Harvey, Wm. H.	Marine Algæ of North America. 3 parts complete,	4to.	568	1858	20.00
97	Kane, E. K.	Magnetic Observations in the Arctic Seas,	S. C. x,	72	1859	1.00
98	Bowen, T. J.	Yoruba Grammar and Dictionary,	S. C. x,	232	1858	4.00
99	....	Smithsonian Contributions to Knowledge,	S. C. x,	462	1858	12.00
100	Gillis, J. M.	Eclipse of the Sun, Sept. 7, 1858,	S. C. xi,	22	1859	.50
101	Hill Thos.	Map of Solar Eclipse, Mar. 15, '58,	8vo.*	8	1858	.15
102	Osten Sacken, R.	Catalogue of Diptera of North America,	M. C. iii,	112	1858.	

No.	Author.	Title.		Pages.	Date.	Price.
103	Caswell, A.	Meteorological Observations, Providence, R. I.,	S. C. xii,	188	1860	2.00
104	Kane, E. K.	Meteorological Observations in Arctic Seas,	S. C. xi,	120	1859	
105	Baird, S. F.	Catalogue of North American Mammals,	4to.*	22	1857	.50
106	Baird, S. F.	Catalogue of North American Birds,	4to.*	42	1858	.50
107	....	Annual Report of Smithsonian Institution for 1857,	8vo.	438	1858	
108	Baird, S. F.	Catalogue of N. American Birds,	M. C. ii,	24	1859	.25
109	....	Annual Report of Smithsonian Institution for 1858,	8vo.	448	1859	
110	....	Annual Report of Smithsonian Institution for 1859,	8vo.	450	1860	
111	....	Smithsonian Contributions to Knowledge,	S. C. xi,	506	1859	12.00
112	....	Smithsonian Contributions to Knowledge,	S. C. xii,	540	1860	12.00
113	Bache, A. D.	Magnetic and Meteorological Observations at Girard Coll. Pt. I,	S. C. xii,	22	1859	.25
114	Sonntag, A.	Terrestrial Magnetism in Mexico,	S. C. xi,	92	1859	1.25
115	....	Report on Invention of Electro-Magnetic Telegraph,	M. C. ii,	40	1861	free
116	Rhees, Wm. J.	List of Public Libraries, &c.	8vo.*	84	1859	
117	....	Catalogue of Publications, &c., in Smithsonian Library,	M. C. iii,	264	1859	2.00
118	Morris, John G.	Catalogue of Lepidoptera of North America,	M. C. iii,	76	1860	1.00
119	Whittlesey, Ch.	Fluctuations of Level in N. A. Lakes,	S. C. xii,	28	1860	1.00
120	Hildreth, S. P. and Wood, J.	Meteorological Observations at Marietta, O.,	S. C. xvi,	52	1867	1.00
121	Bache, A. D.	Magnetic and Meteorological Observations at Girard Coll. Pt. II,	S. C. xiii,	28	1862	.25
122	....	Smithsonian Miscellaneous Collections,	M. C. i,	738	1862	
123	....	Smithsonian Miscellaneous Collections,	M. C. ii,	715	1862	
124	....	Smithsonian Miscellaneous Collections,	M. C. iii,	772	1862	

No.	Author.	Title.		Pages.	Date.	Price.
125	....	Smithsonian Miscellaneous Collections,	M. C. iv,	760	1862	
126	Le Conte, John L.	Coleoptera of Kansas and New Mexico,	S. C. xi,	64	1859	1.25
127	Loomis, E.	Storms in Europe and America, Dec. 1836,	S. C. xi,	28	1860	1.25
128	Lea, Carpenter, &c.	Check List of Shells in N. America	M. C. ii,	52	1860	.25
129	Kane, E. K.	Astronomical Observations in the Arctic Seas,	S. C. xii,	56	1860	1 00
130	Kane, E. K.	Tidal Observations in the Arctic Seas,	S. C. xiii,	90	1860	1.50
131	Smith, N. D.	Meteorological Observations in Arkansas from 1840 to 1859,	S. C. xii,	96	1860	1.25
132	Bache, A. D.	Magnetic and Meteorological Observations at Girard Coll. Pt.III	S. C. xiii,	16	1862	.25
133	Morris, John G.	Synopsis of Lepidoptera of North America. Part I,	M. C. iv,	386	1862	2.00
134	Hagen, H.	Synopsis of Neuroptera of North America,	M. C. iv,	368	1861	
135	Mitchell, S. W.	Venom of the Rattlesnake,	S. C. xii,	156	1860	
136	Le Conte, John L.	Classification of Coleoptera of North America,	M. C. iii,	312	1862	1.50
137	....	Circular to Officers of Hudson's Bay Co.,	M.C. viii,	6	1860	
138	Morgan, L. H.	Circular as to Degrees of Relationship,	M. C. ii,	34	1860	
139	....	Collecting Nests and Eggs of North American Birds,	M. C. ii,	34	1861	free
140	Le Conte, John L.	List of Coleoptera of North America. Part I,	M. C. vi,	82	1866	.7
141	Loew, H. and Osten Sacken	Monographs of Diptera. Part I,	M. C. vi,	246	1862	1.5
142	Binney, W. G.	Bibliography of North American Conchology. Part I,	M. C. v,	658	1863	3.0
143	Binney, W. G.	Land and Fresh-Water Shells of North America. Part II,	M. C. vii,	172	1865	1.2
144	Binney, W. G.	Land and Fresh-Water Shells of North America. Part III,	M. C. vii,	128	1865	1 0
145	Prime, Temple	Monograph of American Corbiculadæ,	M. C. vii,	92	1865	.7

44

No.	Author.	Title.		Pages.	Date.	Price.
146	M'Clintock, Sir F. L.	Meteorological Observations in the Arctic Seas,	S. C. xiii,	164	1862	1.50
147	....	Annual Report of Smithsonian Institution for 1860,	8vo.	448	1861	
148	....	Directions for Meteorological Observations,	M. C. i,	72	1860	
149	....	Annual Report of Smithsonian Institution for 1861,	8vo.	464	1862	
150	....	Annual Report of Smithsonian Institution for 1862,	8vo.	446	1863	
151	....	Smithsonian Contributions to Knowledge,	S. C. xiii,	558	1863	12.00
152	Carpenter, P. P.	Lectures on Mollusca,	8vo.*	140	1861	
153	Guyot, A.	Tables, Meteorological and Physical,	M. C. i,	638	1859	3.00
154	....	List of Foreign Correspondents of Smithsonian Institution,	M. C. v,	56	1862	
155	Whittlesey, Ch.	Ancient Mining on Lake Superior	S. C. xiii,	34	1863	
156	Egleston, T.	Catalogue of Minerals,	M. C. vii,	56	1863	.50
157	....	Results of Meteorological Observations from 1854 to 1859,	4to.*	1270	1861	
158	....	Smithsonian Miscellaneous Collections,	M. C. v,	774	1864	
159	Mitchell, S. W. & Morehouse, G. R.	Anatomy and Physiology of Respiration in Chelonia,	S. C. xiii,	50	1863	1.00
160	Gibbs, G.	Instructions for Ethnology and Philology,	M. C. vii,	56	1863	.25
161	Gibbs, G.	Dictionary of the Chinook Jargon	M. C. vii,	60	1863	.50
162	Bache, A. D.	Magnetic and Meteorological Obs at Girard Coll. Pt. IV, V,& VI,	S. C. xiii,	78	1862	1.00
163	....	Circular on History of Grasshoppers,	M. C. ii,	4	1860	
164	....	Smithsonian Museum Miscellanea	M.C. viii,	88	1862	.50
165	Allen, H.	Monograph of the Bats of North America,	M. C. vii,	110	1864	
166	Bache, A. D.	Magnetic Survey of Pennsylvania	S. C. xiii,	88	1863	1.00
167	Le Conte, Jno. L.	New Species of North America Coleoptera,	M. C. vi,	180	1866	1.00
168	....	Circular Relative to Birds from Middle and South America,	M.C. viii,	2	1863	free

No.	Author.	Title.		Pages.	Date.	Price.
169	....	Smithsonian Miscellaneous Collections,	M. C. vi,	888	1864	
170	....	Comparative Vocabulary,	4to.*	20	1863	free
171	Loew, H.	Monograph of the Diptera of North America. Part II,	M. C. vi,	372	1864	
172	Meek, F. B. and Hayden, F. V.	Palæontology of the Upper Missouri. Part I,	S. C. xiv,	158	1865	
173	Dean, John	Gray Substance of the Medulla Oblongata,	S. C. xvi,	80	1864	
174	Binney, W. G.	Bibliography of North American Conchology. Part II,	M. C. ix,	302	1864	2.00
175	Bache, A. D.	Mag. and Met. Observ. at Girard Coll. Parts VII, VIII, & IX,	S. C. xiv,	72	1864	1.00
176	....	Circular, Collecting North American Shells,	M. C. ii,	4	1860	free
177	Meek, F. B.	Check List of Invertebrate Fossils of North America,	M. C. vii,	42	1864	.25
178	....	Circular to Entomologists,	M.C. viii,	2	1860	
179	....	Catalogue of Publications of Societies,	M. C. ix,	596	1866	
180	Draper, H.	Construction of a Silvered Glass Telescope,	S. C. xiv,	60	1864	
181	Baird, S. F.	Review of American Birds in Smithsonian Museum,	M. C. xii,	454	1866	2 00
182	....	Results of Meteorological Observations from 1854–1859. Vol. II,	4to.*	546	1864	2 50
183	Meek, F. B.	Check List of Invertebrate Fossils of North America,	M. C. vii,	34	1864	
184	....	Smithsonian Contributions to Knowledge,	S. C. xiv,	490	1865	12.00
185	....	List of Birds in Mexico, &c.,	8vo *	8	1863	
186	Bache, A. D.	Mag. and Met. Observ. at Girard College. Parts X, XI, & XII,	S. C. xiv,	42	1865	.50
187	....	Annual Report of Smithsonian Institution for 1863,	8vo.	420	1864	1.00
188	....	Annual Report of Smithsonian Institution for 1864,	8vo.	450	1865	1.00
189	Scudder, S. H.	Catalogue of Orthoptera of North America,	M C. viii,	110	1868	1 0
190	....	Queries Relative to Tornadoes,	M. C. x,	4	1865	free

No.	AUTHOR.	TITLE.		PAGES.	DATE.	PRICE.
191	....	Smithsonian Miscellaneous Collections,	M. C. VII,	878	1865	
192	Leidy, Joseph	Cretaceous Reptiles of the U. S.,	S. C. XIV,	142	1865	
193	....	Duplicate Shells from Expedition of Capt. Wilkes,	8vo.*	4	1865	
194	Binney, W. G & Bland, T.	Land and Fresh-Water Shells of North America. Part I,	M.C. VIII,	328	1869	
195	....					
196	Hayes, I. I.	Physical Observations in the Arctic Seas,	S. C. XV,	286	1867	
197	Whittlesey, Ch.	Glacial Drift of Northwestern States,	S. C. XV,	38	1866	
198	Kane, E. K.	Physical Observations in the Arctic Seas. Complete,	4to.*	340	1860	
199	Newcomb, S.	Orbit of Neptune,	S. C. XV,	116	1866	
200	Conrad, T. A.	Check List of the Invertebrate Fossils of North America,	M. C. VII,	46	1866	.25
201	Stimpson, Wm.	Hydrobiinæ and Allied Forms,	M. C. VII,	64	1865	.50
202	Pumpelly, R.	Geological Researches in China, Mongolia, &c.	S. C. XV,	173	1866	3.50
203	....	List of Works published by Smithsonian Institution,	M. C. VII.	12	1866	
204	Cleaveland, P.	Meteorological Observations, Brunswick, Me., 1807–1859,	S. C. XVI,	60	1867	1.00
205	....	Circular for Archæology and Ethnology,	M.C. VIII,	2	1867	free
206	....	Smithsonian Contributions to Knowledge,	S. C. XV,	620	1867	12.00
207	....	Relative to Scientific Investigations in Russian America,	M.C. VIII,	10	1867	
208	Pickering, Chas.	Gliddon Mummy Case in Smithsonian Institution,	S. C. XVI,	6	1862	.50
209	....	Annual Report of the Smithsonian Institution for 1865,	8vo.	496	1866	
210	....	Arrangement of Families of Birds in Smithsonian Institution,	M.C. VIII,	8	1866	
211	....	Smithsonian Contributions to Knowledge,	S. C. XVI,	498	1870	12.00

No.	AUTHOR.	TITLE.		PAGES.	DATE.	PRICE.
212	....	Smithsonian Miscellaneous Collections,	M.C. VIII,	921	1869	5.00
213	....	Smithsonian Miscellaneous Collections,	M. C. IX,	898	1869	5.00
214	....	Annual Report of Smithsonian Institution for 1866,	8vo.	470	1867	
215	....	Annual Report of Smithsonian Institution for 1867,	8vo.	506	1868	
216	....	Photograph Portraits of North American Indians,	M. C. XIV,	42	1867	.25
217	Hoek, M.	Meteoric Shower, 1867, Nov. 13,	8vo.*	4	1867	
218	Morgan, L. H.	Systems of Consanguinity and Affinity,	S. C. XVII,	616	1869	
219	Osten Sacken, R.	Monograph of Diptera of North America. Part IV,	M.C. VIII,	358	1869	2.00
220	Swan, Jas. G.	Indians of Cape Flattery,	S. C. XVI,	118	1869	
221	Coffin, James H.	Orbit, &c., of Meteoric Fire Ball, July 20, 1860,	S. C. XVI,	56	1869	1.00
222	Schott, Chas. A.	Tables of Rain and Snow in United States,	S.C.XVIII,	175	1872	3.00
223	Gould, B. A.	On the Transatlantic Longitude,	S. C. XVI,	110	1869	1.00
224	....	Annual Report of Smithsonian Institution for 1868,	8vo.*	473	1869	
225	....	List of Foreign Correspondents of Smithsonian Institution,	8vo.*	53	1869	
226	....	List of Publications of Smithsonian Institution,	8vo.	34	1869	
227	Gill, Theod.	Families of Mollusks,	M. C. X,	49	1871	.25
228	....	Annual Report of Smithsonian Institution for 1869,	8vo.	430	1871	1.00
229	....	Smithsonian Contributions to Knowledge,	S. C. XVII,	616	1871	
230	Gill, Theod.	List of Families of Mammals,	M. C. XI,	104	1872	.2
231						
232	Stockwell, J. N.	Secular Variations of Orbits of Planets,	S. C. XVIII	220	1872	2.0
233	Ferrel, Wm.	Converging Series, Ratio of Diameter, and Circum. of Circles,	S.O.XVIII,	6	1871	.5
234	Baird, S. F.	Circular Relative to Food Fishes.	M. C. X,	14	1871	free

No.	Author.	Title.		Pages.	Date.	Price.
235	....	Circular Relative to Thunder-storms,	M. C. x,	2	1871	free
236	....	Circular Relative to Altitudes,	M. C. x,	2	1871	free
237	....	Circular Relative to Lightning-rods,	M. C. x,	4	1871	free
238	Rhees, Wm. J.	List of American Libraries, and Public Institutions,	M. C. x,	256	1872	1.00
239	Harkness, Wm.	Magnetic Observations on the Monadnock,	S. C. xviii,	226	1872	2.00
240	Barnard, J. G.	Problems of Rotary Motion,	S. C. xix,	74	1872	
241	Wood, H. C.	Fresh-Water Algæ of N. America,	S. C. xix,	272	1872	7.50
242	Clark, H. J.	Lucernariæ. 11 plates,	S. C. xxii,	138	1878	7.50
243	....	List of Foreign Correspondents of Smithsonian Institution,	M. C. x,	63	1872	
244	....	Annual Report of Smithsonian Institution for 1870,	8vo.	494	1871	1.00
245	....	Check List of Smithsonian Publications to July, 1872,	M. C. x,	21	1872	free
246	....	Smithsonian Contributions to Knowledge,	S. C. xviii,	643	1872	12 00
247	Gill, Theod.	List of Families of Fishes,	M. C. xi,	96	1872	
248	Hilgard, E. W.	Geology of Lower Louisiana,	S. C. xix,	38	1872	
249	....	Annual Report of Smithsonian Institution for 1871,	8vo.	473	1872	
250	....	Smithsonian Miscellaneous Collections,	M. C. x,	913	1873	5.00
251	Carpenter, P. P.	Monograph of Chitonidæ,		...	...	...
252	Carpenter, P. P.	American Mollusca,	M. C. x,	446	1873	
253	Tryon, G. W.	Monograph of Strepomatidæ,	M. C. xiii,	490	1873	
254	De Saussure, H.	Monograph of Hymenoptera,	M. C.	430	1875	2.00
255	Clarke, F. W.	Specific Gravity Tables,	M. C. xii,	272	1873	1.00
256	Loew, H.	Monograph Diptera. Part III,	M. C. xi,	381	1873	2.00
257	....			...	...	...
258	Watson, S.	Botanical Index,	M. C. xv,	484	1878	2.00
259	Jones, Jos.	Antiquities of Tennessee,	S. C.	181	1876	3.00
260				...	...	...

No.	Author.	Title.		Pages.	Date.	Price.
261	Packard, A. S.	Directions for Collecting and Preserving Insects,	M. C. xi,	60	1873	free
262	Newcomb, S.	Orbit of Uranus,	S. C. xix,	296	1873	3.0
263	....	Astronomical Telegram Circular,	M. C. xii,	4	1873	...
264	LeConte, J. L.	New Species Coleoptera. Part II,	M. C. xi,	74	1873	
265	LeConte, J. L.	Classification Coleoptera. Part II,	M. C. xi,	72	1873	
266	Woodward, J. J.	Toner Lecture I. Cancerous Tumors,	M. C. xv,	44	1873	.2
267	Swan, J. G.	Haidah Indians,	S. C. xxi,	20	1874	1.0
268	Coffin, J. H.	Winds of the Globe,	S. C. xx,	781	1876	
269	Habel, Simeon,	Sculptures of Santa Lucia Cosumalwhuapa in Guatemala. 8 plates,	S. C.	94	1878	2.0
270	Osten Sacken, C. R.	Catalogue of Diptera of North Am.,	M. C.	324	1878	2.0
271	....	Annual Report of Smithsonian Institution for 1872,	8vo.	456	1874	...
272	....	Smithsonian Contributions,	S. C. xix,	660	1874	12 0
273	....	Miscellaneous Collections,	M. C. xi,	789	1874	5.0
274	....	Miscellaneous Collections,	M. C. xii,	767	1874	5.0
275	....	Annual Report of Smithsonian Institution for 1873,	8vo.	452	1874	1.0
276	Clarke, F. W.	Specific Heat Tables. Part II,	M. C. xiv,	58	1876	.5
277	Schott, C. A.	Temperature Tables,	S. C. xxi,	360	1876	3.0
278	....	Check List of Smithsonian Publications,	M. C.	24	1874	...
279	DaCosta, J. M.	Toner Lecture III. The Heart,	M. C. xv,	32	1874	.2
280	Alexander, S.	Harmonies of Solar System,	S. C. xxi,	104	1875	1.0
281	Newcomb, S.	Planetary Motion,	S. C. xxi,	40	1874	.7
282	Wood, H. C.	Toner Lecture IV. Study of Fever,	M. C. xv,	50	1875	.2
283	Gill, Theod.	Catalogue of Fishes,	M. C. xiv,	56	1875	
284	....	Smithsonian Contributions,	S. C. xx,	794	1876	12.0
285	....	Smithsonian Contributions,	S. C. xxi,	543	1876	12 0
286	....	Annual Report of Smithsonian Institution for 1874,	8vo.	416	1875	1.0
287	Rau, Charles,	Archæological Collection, Nat. Museum,	S. C.	118	1876	2.0

No.	Author.	Title.	Pag	Dati	Pric
288	Clarke, F. W.	Specific Gravity Tables. Supp. I, M. C. xiv,	62	1876	.5
289	Clarke, F. W.	Tables, Expansion by Heat M. C. xiv,	58	1876	.5
290	....	List of Smithsonian Publications, 8vo. xiii,	12	1876	...
291	Brown-Séquard,	Toner Lecture II. The Brain, M. C. xv,	26	1877	.2
292	Cope, E. D.	Batrachia. Bulletin National Museum, No. 1, M. C. xiii,	106	1875	.5
293	Kidder, J. L., Coues, E.	Birds Kerguelen Island. Bulletin National Museum, No. 2, M. C. xiii,	61	1875	.5
294	Kidder, J. L., and others,	Nat. Hist. Kerguelen Island. Bulletin Nat. Museum, No. 3, M. C. xiii,	122	1876	.5
295	Lawrence, G. N.	Birds of Mexico. Bulletin Nat. Museum, No. 4, M. C. xiii,	56	1876	.5
296	Goode, G. B.	Fishes of Bermuda. Bull. Nat. Museum, No. 5, M. C. xiii,	82	1876	.5
297	Goode, G. B.	Classification of Animal Resources, etc. Bulletin Nat. Museum, No. 6, M. C. xiii,	139	1876	.5
298	....	Annual Report of Smithsonian Institution for 1875, ...	422	1876	1.0
299	....	Annual Report of Smithsonian Institution for 1876, 8vo.	...	...	...
300	Keen, W. W.	Toner Lecture V. Continued Fevers, M. C. xv,	72	1877	.2
301	....	Check List of Smithsonian Publications to July, 1877, M. C. xiv,	72	1877	free
302	Adams,	Toner Lecture VI. Subcutaneous Surgery, M. C. xv.	20	1877	.2
303	Streets, Thos. H.	Natural History of Hawaiian and Fanning Islands and Lower California. Bulletin National Museum, No. 7, M. C. xiii,	172	1877	.5
304	Dall, Wm. H.	Index to names applied to Brachiopoda. Bulletin Nat. Mus., No. 8, M. C. xiii,	88	1877	.2
305	Jordan, David S.	North Amer. Ichthyology, No. 1. Review of Rafinesque's N. Am. Fishes. Bulletin of Nat. Mus., No. 9, M. C. xiii,	53	1877	.2
306	Jordan, David S.	North Amer. Ichthyology, No. 2. Notes on Cottidae, &c. Bull. Nat. Mus., No. 10. 45 plates, M. C. xiii,	120	1877	.5
307	Gill, Theodore,	Fishes of Western North America. Bulletin of Nat. Mus., No. 11, M. C.	...	...	...

No.	Author.	Title.		Pages.	Date.	Price.
308	Jordan, D. S., and Brayton, A. W.	North Amer. Ichthyology, No. 3. Distribution of Fishes of S. C., Ga., and Tenn. Bulletin Nat Museum, No. 12,	M. C.	237	1878	.50
309	....	List of Foreign Correspondents of the Smithsonian Institution to Jan. 1878,	M. C. xv,	120	1878	.25
310	Barnard, J. G.	Internal Structure of the Earth,	S. C.	19	1877	.25
311	Holden, Edw. S.	Index Catalogue of Books Relating to Nebulæ,	M. C. xiv,	126	1877	.50
312	....	Smithsonian Miscellaneous Coll. (Bull. N. M., 1–10.) 45 plates,	M. C. xiii,	982	1878	5.00
313	Eggers, Baron,	Flora of St. Croix and Virgin Islands. Bull. Nat. Mus., No. 13,	M. C.	...	...	...
314	....	Smithsonian Miscellaneous Collections.	M. C. xiv,	911	1878	5.00
315	....	Smithsonian Miscellaneous Collections.	M. C. xv,	880	1878	5.00
316	....	Circular in Reference to American Archæology,	M. C. xv,	15	1878	free
317	Elliot, D. G.	Classification and Synopsis of Trochilidæ,	S. C.	289	1879	5.00
318	Dall, Wm. H.	Remains of Man from Caves in Aleutian Islands. 10 plates,	S. C.	44	1878	2.00
319	....	Circular. Inquiries Relative to Crawfish and Crustacea,	M. C. xv,	8	1878	free
320	....	Circular Relating to Collections of Living Reptiles,	M. C. xv,	2	1878	free
321	Shakespeare, E. O.	Nature of Reparatory Inflammation in Arteries after Ligature, etc. 7 plates,	M. C.	74	1879	.25
322	....	Smithsonian Miscellaneous Collections.	M. C. xvi,	...	...	...
323	....	Annual Report of the Smithsonian Institution for 1877,	8vo.	500	1878	...
324	....	Circular relative to Scientific and Literary Exchanges,	M. C.	2	1879	free
325	....	Circular. Business Arrangements of the Smithsonian Institution,	M. C.	7	1879	free
326	Goode, G. Brown,	Catalogue of Collection of Animal Resources and Fisheries of U. S. Bulletin Nat. Museum, No. 14,	M. C.	367	1879	.50

No.	AUTHOR.	TITLE.		PAGES.	DATE.	
327	Smithson, James,	Scientific writings of,	M. C.	...	1879	...
328	Rhees, Wm. J.	Smithsonian Institution. Documents Relative to its Origin and History,	M. C.	1027	1879	3.(
329	....	Smithsonian Institution. Journal of Board of Regents and Reports of Committees,	M. C.	...	1879	...
330	Rhees, Wm. J.	Smithson and his Bequest,	M. C.	...	1879	...
331	Rau, Chas.	The Palenque Tablet,	S C.	...	1879	...
332	....	Proceedings of the Nat. Museum for 1878,	M. C.	...	...	...
333	....	Proceedings of the Nat. Museum for 1879,	M. C.	...	...	...
334	Elliot, D. G.	List of Described Species of Humming-Birds,	M. C.	22	1879	.2
335	....	List of Principal American Libraries, Museums, Societies, etc.,	M. C.	6	1879	fre
336	....	Smithsonian Miscellaneous Collections.	M. C. XVII,	...	...	...
337	....	Smithsonian Miscellaneous Collections.	M.C. XVIII,	...	...	...
338	....	Smithsonian Miscellaneous Collections.	M. C. XIX,	...	...	...
339	....	Smithsonian Miscellaneous Collections.	M. C. XX,	...	...	...
340	....	Smithsonian Contributions to Knowledge,	S. C. XXII,	...	...	...
341	....	Annual Report of Smithsonian Institution for 1878,	8vo.	...	...	...
342	Kumlien, L.	Contributions to Natural History of Arctic America. Bulletin of the National Museum, No. 15,	M. C.	...	...	...
343	Jordan, D. S.	Synopsis of the Fishes of the U. S. Bull. Nat. Mus., No. 16,	M. C.	...	...	...
344	....	Check List of Smithsonian Publications,	M. C.	16	1879	fre
345	....	Annual Report of Smithsonian Institution for 1879,	8vo.	...	...	...